THE OXFORD HANDBOOK OF

NON-SYNOPTIC WIND STORMS

THE OXFORD HANDBOOK OF

NON-SYNOPTIC WIND STORMS

Edited by
HORIA HANGAN
and
AHSAN KAREEM

OXFORD
UNIVERSITY PRESS

OXFORD
UNIVERSITY PRESS

Oxford University Press is a department of the University of Oxford. It furthers the University's objective of excellence in research, scholarship, and education by publishing worldwide. Oxford is a registered trade mark of Oxford University Press in the UK and certain other countries.

Published in the United States of America by Oxford University Press
198 Madison Avenue, New York, NY 10016, United States of America.

Library of Congress Cataloging-in-Publication Data
Names: Hangan, Horia, editor. | Kareem, Ahsan, editor.
Title: The Oxford handbook of non-synoptic wind storms /
edited by Horia Hangan and Ahsan Kareem.
Description: New York : Oxford University Press, 2021. |
Includes bibliographical references.
Identifiers: LCCN 2021003190 (print) | LCCN 2021003191 (ebook) |
ISBN 9780190670252 (hardback) | ISBN 9780190670276 (epub) |
ISBN 9780190670269 (ebook) | ISBN 9780190096465 (ebook other)
Subjects: LCSH: Windstorms—Handbooks, manuals, etc.
Classification: LCC QC942 .O94 2021 (print) | LCC QC942 (ebook) |
DDC 551.55—dc23
LC record available at https://lccn.loc.gov/2021003190
LC ebook record available at https://lccn.loc.gov/2021003191

1 3 5 7 9 8 6 4 2

Printed by Sheridan Books, Inc., United States of America

This handbook is dedicated to the memory of Professor Giovanni Solari, who was a champion of recognizing the need to address non-synoptic winds and led a major European Research Council project "THUNDERR" to address these issues. His several contributions to this volume are greatly appreciated.

—Horia Hangan and Ahsan Kareem

Contents

SECTION 2 IN SITU DATA ACQUISITION FOR NSWS

SECTION 3 MODELING METHODS FOR NSWS

SECTION 4 STRUCTURAL AND OTHER IMPACTS OF NSWS

SECTION 5 RISK MODELS RELATED TO NSWS

SECTION 6 REDUCTION OF EXPOSURE TO NSWS

SECTION 7 CONCLUSION

Contributors

Howard B. Bluestein School of Meteorology, University of Oklahoma

Massimiliano Burlando Department of Civil, Chemical and Environmental Engineering, University of Genoa

Jubayer Chowdhury WindEEE Research Institute, Western University, Canada

Branko Grisogono Department of Geophysics, Faculty of Science, University of Zagreb, Croatia

Horia Hangan WindEEE Research Institute, Western University, Canada

Patrick Hawbecker Project Scientist, National Center for Atmospheric Research

Han Ping Hong Department of Civil and Environmental Engineering, Western University, Canada

Qian Huang Department of Civil and Environmental Engineering, Western University, Canada

Julian Hunt University College London

Ahsan Kareem NatHaz Modeling Laboratory, University of Notre Dame

Hrvoje Kozmar Faculty of Mechanical Engineering and Naval Architecture, University of Zagreb, Croatia

Dae Kun Kwon NatHaz Modeling Laboratory and Center for Research Computing (CRC), University of Notre Dame

Franklin T. Lombardo Department of Civil and Environmental Engineering, University of Illinois at Urbana-Champaign

Junji Maeda Division of Architecture and Urban Design, Kyushu University, Japan

Timothy P. Marshall Haag Engineering Co.

Kishor C. Mehta Department of Civil, Environmental, & Construction Engineering, Texas Tech University

Edmund P. Meredith Institut für Meteorologie, Freie Universität Berlin

Takashi Nomura Department of Civil Engineering, College of Science and Technology, Nihon University, Tokyo, Japan

Kirsten D. Orwig Natural Catastrophe Specialist, Property & Specialty, Swiss Re Corporate Solutions

Maryam Refan Technical Director, Rowan Williams Davies & Irwin Inc. (RWDI)

Djordje Romanic WindEEE Research Institute, Western University, Canada; Department of Civil, Chemical and Environmental Engineering, University of Genoa, Italy; Department of Atmospheric and Oceanic Sciences, McGill University, Montreal, Quebec, Canada

Partha P. Sarkar Department of Aerospace Engineering, Iowa State University

John L. Schroeder Department of Geosciences, Texas Tech University

Kevin M. Simmons Austin College

Giovanni Solari Department of Civil, Chemical and Environmental Engineering Polytechnic School University of Genoa, Italy

Takashi Takeuchi Department of Architecture, Kobe University, Japan

Yukio Tamura School of Civil Engineering, Chongquing University

Eriko Tomokiyo Faculty of Advanced Science and Technology, Kumamoto University, Japan

Robert J. Trapp Department of Atmospheric Sciences, University of Illinois at Urbana-Champaign

Uwe Ulbrich Institut für Meteorologie, Freie Universität Berlin

J. Arn Womble Insurance Institute for Business & Home Safety

Teng Wu Department of Civil, Structural and Environmental Engineering, University at Buffalo

Alexander S. Zickar Department of Civil and Environmental Engineering, University of Illinois at Urbana-Champaign

Delong Zuo Department of Civil, Environmental, and Construction Engineering, Texas Tech University

Foreword

JULIAN HUNT

Historically, before the nineteenth century, the winds that were studied by meteorologists tended to be focused on a range of isolated characteristic events such as clouds, gusts, and vortices and also their relations between each other, such as sea breezes and variations of wind directions with height. Patterns of wind storms over oceanic distances between countries were already being studied at the time of Aristotle and recorded on the Greek "tower of the winds" in the center of Athens. Later, in the early nineteenth century, detailed classifications of types of clouds were first proposed by Lamark and Howard in Europe, while the turbulent eddying motions over land and the sea surface were famously tabulated in Admiral's Beaufort's "wind scales."

But it was not until the middle of the nineteenth century before the great development in meteorology took place, when the link was made between the horizontal variation of surface-level barometric pressure and the observations of continental-scale "synoptic" wind patterns over significant distances (the Buys-Ballot law). This was the basis for the different kinds of practically important observations of wind speeds, ocean waves, and turbulence and systematic forecasts for different kinds of storms and their consequential damage impacts on sailing vessels, steam-driven ships, and, occasionally, on land-based structures, notably railway bridges, such as the Tay Bridge (1879).

The next stage in the forecasting of storm damage began with more comprehensive records for wind speed profiles and turbulence measurements in major storms and other meteorological dangers, like icebergs in foggy atmospheric conditions, which led in 1912 to the sinking of the passenger ship *RMS Titanic* when it collided with a large iceberg and sank. The following meteorological and oceanographical expedition into the Newfoundland area of the disaster that followed involved detailed measurements taken from a research yacht, which led to a new understanding about atmospheric wind and temperature profiles. G. I. Taylor, one of the leading scientists on board, used this opportunity to begin describing the atmospheric boundary layer (ABL) structure and also developing turbulence theory, both for "synoptic" winds measured over wide areas and also for complex non-synoptic winds, such as frontal winds with significant variations of wind speed and temperature gradients.

Subsequent studies of ABL structure made in field experiments between 1920s and 1980s over flat terrain in Europe, Russia, and the United States, led to universally accepted statistical data for slowly varying synoptic conditions. Many of the basic ideas and fundamental formulae for these conditions resulted from the statistical and interscale analysis of fluid dynamical concepts developed by L. Prandtl, von Karman, L. F. Richardson, A. N. Kolmogorov, A. M. Obukhov, and others. Initially these general results established the structure of the lower part of the boundary layer, in particular the universal self-similar turbulence profiles (including spectra and higher orders). With the development of higher measurement platforms and remote sensing instruments in the 1960s, it became possible to extend the wind height measurements to increasing heights above the top of the boundary layer (over 100–1,000 m).

The results also showed typical transitions from the lower, turbulent, stratified, and convective turbulent boundary layer to upwardly propagating internal waves, which can be in the form of slanting wave-like stable fronts. In other situations, clouds and significant non-synoptic convective motions are sometimes generated above the boundary layer. Some types of these upper level motions can penetrate downward (forming downbursts) into the surface boundary layer, causing intense vertical and horizontal gusts, with horizontal gust scales ranging from the depth of the boundary layer thickness (i.e., 1 km) to much larger, stronger, curving synoptic-scale vortical storm structures—typically located along the northern hemisphere jet stream—that are typical of weather forecasts at temperate latitudes.

The highest intensity local flows over large scales are non-synoptic vortical storms—mostly, but not exclusively, over land/coastal areas. On ABL scales these "tornado" vortices driven by disturbed shear layers and also by local convection have diameters ranging from a few meters to hundreds of meters, and strengths that, according to Fujita's classification, exhibit peak winds that can exceed 200 m/s. There are generally random distributions between how tornadic vortices occur in groups. But their large-scale synoptic distribution tends to be systematically concentrated in certain areas of countries, such as the Midwest and Southeast United States and the mid latitudes in Russia and China, away from mountainous areas.

The most energetic and damaging vortical wind storms are tropical cyclones (TCs) that form over warm oceans (south and north respectively of 10 degrees latitude) but can move over a limited distance onto adjacent land surfaces. They tend to have a strong characteristic wind and thermodynamic structure that forms with a central quiet zone with weak waves but that is surrounded by intense turbulent circulating wind jets that reach maximum at a radius of about 30 kilometers. The strongest winds, up to 100–200 m/s, typically extend vertically up to about 10 kilometer and extend radially over about 1,000 kilometers. Near the equator, the TCs only form over latitudes for about 300 kilometers north and south of the equator, which is where the Earth's rotation (or Coriolis force) is most dynamically significant. A more complex feature of TCs occurs where there are interactions, as defined by the spectra, between the atmospheric flow in the atmosphere and the dynamics and thermodynamics of the surface waves and their production of spray droplets. A recent high-amplitude TC (Hayan in 2013, in the Philippines) occurred in which the TC impacted strongly enough onto shallow coastal waters that the shallow seas were dispersed into the atmosphere, and the winds were strong enough to transport bottom sediment and rocks onto the land, causing great destruction to local communities.

Other types of intense NSWS are different and more focused than types described earlier. The differences in wind structures in such NSWS can occur in specific areas defined by different types of orography (including surface roughness and urban areas). In particular, characteristic synoptic scale winds result from flows over isolated mountains and also over large-scale mountain ranges. Other features of NSWS are associated with different types of wind shear over the terrain.

For future planning of the effects of storms and other environmental hazards, it is necessary to consider how climate change may affect global and synoptic changes in NSWS. Some indications from current research have shown how, at both temperate and polar latitudes, the effect of climate change is strengthening and focusing non-synoptic storms along the jet stream. An equally important example is the effects of global sea level rise on non-synoptic scale impacts, one example being the worsening of coastal waves and flooding, with massive subsequent adverse impacts on urban communities.

Experimental Simulations and Modeling Methods for Synoptic and Non-Synoptic Wind Storms

Laboratory studies for the broad range of ABL characteristic of synoptic winds are generally conducted in special types of elongated wind tunnels in which the entering flow is uniform and smooth in the space upwind of the "working section." The entry flow is usually mechanically driven by wind propellers, as in short-length aeronautical wind tunnels (AWT). ABL wind tunnels differ significantly from AWT in that, following a defined development distance (DD) along the wind tunnel, the flow is artificially distorted by a distribution of vortical and fence-like obstacles fixed to the floor or sometimes by upward or cross-flow vortical jets. The objective is to simulate in the working section downwind of DD the complex eddy structure characteristic of the ABL and to ensure that this structure develops slowly downwind of DD. Many wind tunnel studies (such as those pioneered by J. E. Cermak and A. G. Davenport in North America in 1960s and intercomparisons reported at the Euromech conference in 1975) have shown that in these types of well-designed and calibrated wind tunnels, the wind shear and turbulence correspond compared to the basic types of ABT data that naturally evolved in typical synoptic conditions.

However, these wind tunnels constructed for ABL can also be modified to simulate certain non-synoptic features of the boundary layer flow, such as inhomogeneous weak vertical motions and inhomogeneous flows over mountains or urban areas.

For inhomogeneous, non-synoptic flows with characteristic strong vortical and also vertical motions (e.g., tornadoes, TCS, downbursts, weather fronts), new laboratory modeling approaches have been developed that combine the characteristic non-synoptic and synoptic storm winds. In the past laboratories had to choose between unidirectional wind tunnels, pure vortices, and gravity current models of downbursts depending on the type of flow to be studied. But now the WindEEE laboratory (unique to Western University in London, Ontario, Canada) has been developed that combines gradually unidirectional turbulent ABL (non-synoptic) shear flows with a combination of vortical and turbulent flows.

Complementary to experimental simulations have been recent developments of conceptual and computational modeling of non-synoptic, high-amplitude atmospheric flows that have enabled wind engineering researchers to describe and deliver data for design and experimental projects. Examples using analytical, descriptive, and statistical approaches have identified particular types of eddy mode depending on the type of overall flow structures (i.e., synoptic and non-synoptic) and the connected correlation with concepts of flow around structures and vibrations and damage to structures.

Damage Caused by Impacts of Synoptic and Non-Synoptic Winds

Previous studies of complex atmospheric winds can be linked to the dynamics and deformation of solid and deforming materials if we are to understand and predict the movements and deformation of structures of different forms. These forms range from bluff, quasi-rigid obstacles to highly elongated flexible shapes (e.g., wind turbines and electric cables). Understanding and modeling wind impacts on solid and flexible static and moving objects are of particular concern both in urban areas and in natural environments, as when winds impact on people and on moving vehicles. The consequences can be dysfunction and even disintegration.

To model and simulate the major types of wind damage on large-scale structures, it is necessary to consider the detailed aspects of both mean and turbulent wind flows and also detailed pressure and surface stresses. Although the unsteady aspect of the flow affects the inertial "added mass" forces, their contribution to dynamics of structures is generally weaker than was generally assumed in the earlier wind engineering research of the 1960s. Of course the vibrations (and their modes) of a typical engineering structure itself are critical and have to be related to the eddy structure (following A. Kareem).

In extreme conditions, where buildings break up, dislodged parts of structures may take the form of dangerous "missiles," as when tropical cyclones impact on city centers. In such situations, the drag, lift, inertial forces, and added mass forces are all significant, following the early explanation by G. I. Taylor and W. Tollmien's analyses of airships moving in accelerating flows in the 1910–1920s; later research has also shown the significant contribution of lift forces on moving objects.

The modeling of forces and stresses on solid and flexible structures of most types and scales can be classified on a system map of normalized eddy length/time scales (e.g., Hunt and Savill 2005). This is defined by the ratio L~ of the turbulent spatial integral scale Lx to the size of the obstacle D (i.e., L~ = Lx/D). This parameter can be considered a function of the normalized time scale (i.e., T~ = T/TL), where T = D/U is the travel time, and TL = Lx/u′ is the typical eddy time. Thus, where typical significant eddy sizes L~ and T~ are both of order 1 or less in a fluid–structure system, the turbulence in the flow around the structure is in quasi-equilibrium (e.g., for typical wind-sensitive structures). Understanding this locally synoptic equilibrium contributes to the accuracy of modeling and laboratory experiments.

However, for a very large building structures (such as open or covered stadia) in built-up urban areas (i.e., where typically the length scales of pressure fluctuations have a length scale Lp larger than that of the local turbulence L and a longer time scale), small eddies move and evolve over the structure. In this flow the effect of a large building structure can reinforce the synoptic length scales of the turbulent flow.

By contrast, as vortical tornadoes develop on local diameter scales Lv (~30 m), they also interact on a longer but limited (non-synoptic) distance. Such motions interact weakly with the surrounding ABL out to a significantly greater distance than the length scale Lx at the height of the vortex (Lx~ 100 m).

This structure of interactions between tornadoes and boundary layer scale turbulence differs from interactions with synoptic larger scale interactions, which have a reduced intensity of interactions.

Societal Impacts

Studies of the technical and geographical aspects of non-synoptic storms and related non-standard environmental extremes also need to be integrated with studies of the socioeconomic impact of hazards and their impact on society. Note that there are also significant feedback connections from socioeconomic influence to technical studies, for example through relating increased financial investment to improved atmospheric monitoring of extreme hazards. Another aspect concerns a more thorough understanding of societal responses to such hazards and the critical socioeconomic investments in input for forecasting, planning, and operational and long-term investments.

Studies of these socioeconomic evaluations and technical and scientific responses vary significantly between regions and countries, and between economic and social interests. Increasingly, in the future, these aspects of hazards, including wind storms, must be better understood better by all sections of society, including business if programs of actions are to become progressively more effective economically and socially.

SECTION 1

INTRODUCTION: NON-SYNOPTIC WIND STORMS (NSWS)

CHAPTER 1

INTRODUCTION

HORIA HANGAN AND AHSAN KAREEM

THE opportunity offered by Oxford University Press to produce an Oxford Handbook on non-synoptic wind systems and their implications to wind hazard came at the right moment. This may be defined as the timely intersection between the importance of the subject matter, the accumulation of a critical mass of knowledge, and the opportunity for critical debate before the subject matter completely matures.

In a book chapter dedicated to hurricanes in the United States, Pielke and Pielke (1997) defined the problem of hazard reduction as minimizing vulnerability. In turn, vulnerability can be expressed as a function of (1) the incidence of the storm itself, which includes its intensity, frequency, and geographical occurrence; and (2) the exposure to the storm event, which includes the population concentration, the property value, and the overall degree of preparedness to withstand such events. By reducing the vulnerability to wind storms, we attempt "to prevent natural hazards from becoming disasters" (National Science Foundation, 2015) by adopting a risk mitigation approach. The framework of the present handbook for non-synoptic winds follows the same approach.

Traditionally, wind engineering deals with storm events and their impact on buildings and structures. The developments in wind engineering research can then be translated in new building code implementations that have a direct impact on our degree of preparedness to face these events. Alan Davenport's wind loading chain takes into account "the combined effects of the local wind climate, which must be described in statistical terms; the local wind exposure, which is determined by the terrain roughness and topography; the aerodynamic characteristics of the building shape; and the potential for load increases due to possible wind-induced resonant vibrations" in order to determine the wind load and thereafter the wind-induced responses on buildings and structures (Davenport, 2002; Isyumov, 2012). A recent expansion of this has been introduced as a generalized loading chain, which encompasses non-synoptic winds.

During the past two decades, it has come to the attention of the wind engineering community that in certain parts of the world much of the damage due to wind can be attributed to non-synoptic, local wind storm events with main manifestations like tornadoes and downbursts. In North America, the damage due to these winds reaches more than 65 percent of the total wind damage, and there are no guidelines or code implementations to deal with these catastrophic events. Current research is in its first phase of addressing these type of events, including their characterization, simulation, loading and collapse modes effects on buildings and structures, as well as socioeconomic implications.

We have then jumped back to rediscover and sometimes reinvent what has been done before in this area of science and engineering. Pioneering work carried out in the 1970s and 1980s in relation to downburst (Fujita, 1985, 1990), but even more so tornadoes (Fujita, 1981) has been brought back to life and analyzed, while some of this work has been refurbished in more practical ways to be reclaimed (e.g., Letchford, Mans, & Chay, 2002). Within that initial opportunistic occasion, many fundamentals have been forgotten and replaced by rather new empiricism than constructive further scientific investigation. It took some time to clarify again the fundamental science and organize ourselves to make substantial progress afterward. The work of Theodore Fujita again became seminal to our own collective work (McDonald, 2001).

The term *non-synoptic winds* has been used to encompass an entire range of wind systems that drastically depart from synoptic, large-scale wind systems. The main differences relate to the spatial and temporal features, including three-dimensionality, stationarity, as well as the associated statistics (e.g., Hangan, Romanic, & Jubayer, 2019; Kareem & Wu, 2013). In this handbook, we refer, but not limit ourselves, to the following type of winds as non-synoptic in nature: downbursts, tornadoes, gust fronts, sea or lake breezes, squalls, gravity waves, shammals, foehns, and katabatic and gap winds. As this handbook relates to the implication of these types of winds to wind hazard, more emphasis is placed on higher wind speed non-synoptic wind events.

Often, these non-synoptic winds are very different from the synoptic winds in terms of climatology and surface flow structure as well as in terms of the resulting aerodynamic loading and structural responses induced on buildings, structures, and the natural habitat. Luckily, at the same time, there was an innovation in investigation tools spanning from meteorology to wind engineering and risk analysis (Hangan et al., 2017). Therefore, the understanding and investigation of this class of wind phenomena and their effects on the human and natural environment is currently underway. Next, we briefly introduce the differences between synoptic and non-synoptic winds.

1. Climatology

Synoptic, or extratropical cyclonic and anticyclonic winds, are large-scale phenomena dominated by the geostrophic balance between pressure gradient and Coriolis forces spanning meso-alpha to macro-beta scales and therefore extending over hundreds to thousands of kilometers. The large scales also imply that the curvature of the isobars when considering synoptic wind systems became secondary, and these systems are most of the time simulated as straight winds.

In contrast, non-synoptic winds occupy the micro-alpha to meso-beta scales corresponding to scales that extend between hundreds of meters and hundreds of kilometers, and they are dominated by small curvatures and relatively fast translations. These characteristics translate in a pronounced three-dimensional (3D) flow structure dominated by intense vorticity and therefore the inability to simulate these as straight winds in wind tunnels.

2. Surface Layer

The surface layer of synoptic wind systems has the typical characteristics of the planetary (or atmospheric) boundary layer (ABL), with a monotonically increasing mean velocity profile characterized by either log laws or power laws. The ABL-associated turbulence field is stationary and can be described by fairly constant levels of turbulence intensity (with the exception of the immediate ground proximity) by Gaussian probability distribution functions (PDFs) as well as by power spectra with a well-defined inertial layer.

The non-synoptic winds have a much more complex and 3D structure, which involves the presence of concentrated vertical (tornado) or annular (downburst) vorticity. This pronounced vortex structure produces in most cases vertical velocity flows. From the turbulence point of view, nonstationarity is another factor that dominates some of these winds, which means that an overall mean value has no particular meaning, and that the analyses need to be adjusted accordingly. Extending toward higher order statistics, the PDFs characterizing the non-synoptic winds are in many situations non-Gaussian, and their power spectra sometimes show a departure from the well-known inertial range -5/3 slope.

3. Aerodynamics

As a result of their different natures, the resulting aerodynamic loading due to synoptic and non-synoptic winds will be different. Bluff body aerodynamics for ABL wind flows clearly shows that the largest negative peak pressures occur near the edges or corners of the bodies representative of buildings and are associated with very localized vortices.

It is becoming clearer that in tornadoes and downburst flows the aerodynamic loading is governed by the larger scale vertical structures present in these flows and the resulting uplift or downdraft velocity components. As a result, the aerodynamic loading zones on buildings change significantly between synoptic and non-synoptic winds, with the potential for more damage and therefore a need for revaluation of design loads in codes and standards.

4. Structural Responses

Design wind loads based on conventional approaches in codes and standards would not accurately describe loads introduced by fundamentally different gust front wind events that depart from typical boundary layer winds in both their kinematics and dynamics. In an effort to begin working toward establishing a new procedure for wind load effects in thunderstorm winds, one relies on using the time domain time history analysis. Alternatively, the frequency domain utilizing recently established the generalized loading chain, which builds on the Davenport chain by including provisions to capture nonstationary features in thunderstorm winds (Kareem, Hu, Guo, & Kwon, 2019), can be utilized. Earlier efforts in this context have offered means to capture unique features of thunderstorms using model-based and data-driven perspectives with a promise of a resilience-based wind design practice to enhance disaster resilience of built environment (Kwon & Kareem, 2019; Solari, De Gaetano, & Repetto, 2015).

One such approach is the introduction of the gust front factor, which builds on the conventional gust effect factor in codes and standards and corrects for the features in thunderstorms that depart from those in extratropical or synoptic winds. These include changes in the kinematics and dynamics of the velocity field, effects of sudden application of wind gust, nonstationarity, and possible modifications in aerodynamics in thunderstorm-related wind events. The gust front factor reduces to unity in case wind conditions change to extratropical or synoptic winds (Kwon & Kareem, 2009, 2013, 2019).

From the fundamental science perspective, this area is in a nascent state. There is a clear need to better understand and characterize the full-scale non-synoptic winds; to properly simulate these events either numerically or physically; to assess the difference in loading between these events and the straight-line synoptic winds that have been part of the wind engineering practice for more than six decades now; to determine their statistics and associated risks; and finally, to apply all these to design via guidelines, codes, risk, and socioeconomic impact models.

The handbook presented here is intended to bring these issues to light and stimulate thought processes, research, and steps needed to develop a response to non-synoptic wind-related hazards proposing a strategic vulnerability frame, providing clear definitions of the problems to be tackled, and identifying the best suited tools and methodologies to address these problems in an effort to maximize collaborative planning between the technical and associated socioeconomic aspects. Based on this framework, the handbook has been structured in the seven main sections discussed next.

Section 1 defines the various non-synoptic wind storms (NSWSs) mostly from a meteorological point of view. It encompasses not only downbursts and tornadoes as the most important, but also shammals, foehns, and katabatic and gap winds. It then reviews the current status of knowledge and research for each of these NSWS types of winds, pointing out the knowledge gaps and therefore research opportunities. Finally, the NSWS future evolution and trends are discussed in relation to climate change.

Section 2 presents the most important full-scale data acquisition measurement techniques used for NSWSs. It analyzes the completeness of each of these techniques and talks about future possible developments. It also addresses the current NSWS databases and future possible developments.

Section 3 reviews the main modeling methods for NSWSs, pointing out their relative advantages and disadvantages. The review presents the main analytical models, mesoscale and microscale numerical models, as well as conventional and novel physical simulations. It also addresses the relation between modeled and full-scale NSWSs.

Section 4 analyses the impacts of NSWSs on buildings and structures, aiming at the adaptation of the "Alan Davenport wind engineering chain" to non-synoptic winds. It addresses aerodynamic loading under NSWS events and the loading comparison between synoptic and non-synoptic events. It then progresses toward new structural analysis methods framed for NSWSs and follows toward damage and loss prediction. Other problems, such as flying debris and NSWS–tree interactions, are also addressed.

Section 5 describes the framework and the state of knowledge in risk models related to NSWSs and the development of capacity factors for non-synoptic winds as well as the need for database access.

Section 6 addresses the exposure to NSWSs through prediction models, engineering guidelines, and codes. It also reveals socioeconomic aspects as well as those related to preparedness and decision-making. Finally, it also introduces how non-synoptic winds influence the insurance aspects.

Section 7 concludes by analyzing the present research and knowledge outcomes as well as gaps in relation to meteorological characterization, modeling, structural interactions, and socioeconomics aspects of NSWSs.

We hope that this Oxford handbook on non-synoptic wind systems and their implications to wind hazard will provide an overall state of the art of the research in this area and a reference point for further exploration.

We would like to thank all contributors for their time and sharing of their knowledge in this book. We thank also the advisory group for their insights and review of the material; special thanks to Ms. Priscilla De Luca and Ms. Elisa Yaquian, whose perseverance and dedication were crucial in making all this happen.

References

Davenport, A. G. (2002). Past, present and future of wind engineering. *Journal of Wind Engineering and Industrial Aerodynamics, 90*, 1371–1380.

Fujita, T. (1981, August). Tornadoes and downbursts in the context of planetary scales. *Journal of the Atmospheric Sciences, 38*(8), 1511–1534.

Fujita, T. (1985, February). *The downburst: Microburst and macroburst* (Research Paper No. 2010). Chicago, IL: Satellite and Mesometeorology Research Project. Retrieved from National Technical Reports Library at https://ntrl.ntis.gov/NTRL/dashboard/searchResults/titleDetail/PB85148880.xhtml

Fujita, T. (1990). Downburst: Meteorological features and wind field characteristics. *Journal of Wind Engineering and Industrial Aerodynamics, 36*, 75–86.

Hangan, H., Refan, M., Jubayer, C., Romanic, D., Parvu, D., LoTufo, J., & Costache, A. (2017). Novel techniques in wind engineering. *Journal of Wind Engineering and Industrial Aerodynamics, 171*, 12–33.

Hangan, H., Romanic, D., & Jubayer, C. (2019). Three-dimensional, non-stationary and non-Gaussian (3D-NS-NG) wind fields and their implications to wind-structure interaction problems. *Journal of Fluids and Structures, 91*, 102583.

Isyumov, N. (2012). Alan G. Davenport's mark on wind engineering. *Journal of Wind Engineering and Industrial Aerodynamics, 104–106*, 12–24.

Kareem, A., Hu, L., Guo, Y., & Kwon, D. K., (2019). Generalized wind loading chain: Time-frequency modeling framework for nonstationary wind effects on structures. *Journal of Structural Engineering, 145*(10), Article 04019092.

Kareem, A., & Wu, T. (2013). Wind-induced effects on bluff bodies in turbulent flows: Nonstationary, non-Gaussian and nonlinear features. *Journal of Wind Engineering and Industrial Aerodynamics, 122*, 21–37.

Kwon, D. K., & Kareem, A. (2009). Gust-front factor: A new framework for wind load effects on structures. *Journal of Structural Engineering, 135*(6), 717–732. doi:10.1061/(ASCE)0733-9445(2009)135:6(717)

Kwon, D. K., & Kareem, A. (2013, March). Generalized gust-front factor: A computational framework for wind load effects. *Engineering Structures, 48*, 635–644. doi:10.1016/j.engstruct.2012.12.024

Kwon, D. K., & Kareem, A. (2019, November). Towards codification of thunderstorm/downburst using gust front factor: Model-based and data-driven perspectives. *Engineering Structures, 199*, Article 109608.

Letchford, C. W., Mans, C., & Chay, M. T. (2002). Thunderstorms—Their importance in wind engineering (a case for the next generation wind tunnel). *Journal of Wind Engineering and Industrial Aerodynamics, 90*, 1415–1433.

McDonald, J. (2001). T. Theodore Fujita: His contribution to tornado knowledge through damage documentation and the Fujita scale. *Bulletin of the American Meteorological Society, 82*(1), 63–72.

National Science Foundation. (2015). *Engineering for natural hazards program*. http://www.nsf.gov/funding/pgm_summ.jsp?pims_id=505177

Pielke, R. A., Jr., & Pielke, R. A., Sr. (1997). Vulnerability to hurricanes along the US Atlantic and Gulf coasts: Considerations of the use of long-term forecasts. In D. H. F. Diaz & D. R. S. Pulwarty (Eds.), *Hurricanes* (pp. 147–184). Berlin, Germany: Springer.

Solari, G., De Gaetano, P., & Repetto, M. P. (2015, August). Thunderstorm response spectrum: Fundamentals and case study. *Journal of Wind Engineering and Industrial Aerodynamics, 143*, 62–77. doi:10.1016/j.jweia.2015.04.009

CHAPTER 2

THE TYPES OF NON-SYNOPTIC WIND SYSTEMS

HOWARD B. BLUESTEIN

1. INTRODUCTION

SYNOPTIC-SCALE disturbances in midlatitudes (we do not consider the tropics or polar latitudes here) are characterized by horizontal scales (L) of thousands of kilometers and time scales (τ) of days. The advective time scale (the time it takes an air parcel to traverse the distance spanned by the disturbance) of synoptic-scale disturbances is on the order of L/U, which is around 10^6 m/10 m s^{-1} = 10^5 s, which is on the order of 1 day, where the horizontal scale is about 1,000 km, and the typical wind speed is about 10 m s^{-1}. Synoptic-scale weather systems such as cyclones and anticyclones and troughs and ridges are responsible for our everyday weather (Bluestein, 1992). By non-synoptic wind systems (NSWSs), we refer here to those having space and time scales that are subsynoptic, that is, are less than thousands of kilometers and less than days. The "systems" in wind systems refer loosely to weather phenomena that produce wind speeds near the ground strong enough to inflict damage; we consider only what happens near the ground (<100 m above ground level [AGL]). We generally consider winds in excess of about 27.5 m s^{-1} as strong enough to inflict some damage, recognizing that some structures can withstand even higher wind speeds, while some structures are weaker, experiencing damage from wind speeds as low as about 15 m s^{-1}.

Nonsynoptic-scale wind storms (NSWSs) associated with convective storms have horizontal scales of about 10 km and wind speeds of about 10 m s^{-1}. Thus, the advective time scale of convective storms is about 10^4 m/10 m s^{-1} = 1,000 s, which is on the order of tens of minutes. Subconvective storm features are even smaller and have even shorter time scales. Tornadoes, for example, have advective (orbit) time scales of around (π) × 300 m/50 m s^{-1}, which is 20 s or less, where the diameter of the core of the tornado is about 300 m.

On the other hand, subsynoptic-scale wind storms not directly associated with convective storms may be produced by gravity waves or the local downward transport of air having high momentum or by jets. In some instances, the gravity waves or jets may be related to flow over mountains or flow downward over gently sloping surfaces. These NSWSs have structure on the mesoscale, having horizontal scales of approximately 100 km and time scales of hours. Often, they are mesoscale features embedded within synoptic-scale systems.

Sometimes, however, the distinction between subsynoptic-scale wind storms associated with convective storms and those not associated with convective storms cannot be clearly made. For example, a feature may be a byproduct of a convective storm, but outlast the parent convective storm. We therefore consider separately three types of wind storms:

1. Those found in convective storms;
2. Those found in the absence of convective storms; and
3. Those found in the absence of convective storms, but from phenomena produced earlier by convective storms.

The first two are the main focus of this chapter. The last is noted briefly and contained within the second.

2. Convective Storm-Related Phenomena

Wind-storm–related phenomena produced by convective storms include tornadoes and waterspouts, mesocyclones, gust fronts, and downbursts/microbursts. Dust devils will also be included, though in this case convection is a major mechanism, precipitation and convective clouds are not present, and the wind speeds are usually not strong enough to produce any significant damage. There is usually the distinction made between strong straight-line winds and the strong, swirling winds in tornadoes. Climatology in the United States for the former is given and discussed in Doswell, Brooks, and Kay (2005); climatology of the latter is found in Brooks, Doswell, and Kay (2003).

2.1. Tornadoes

Tornadoes, rapidly rotating columns of air pendant from cumuliform clouds (Figure 2.1), are the most intense of all NSWS events (Bluestein, 2013); they have horizontal, ground-relative wind speeds, measured by Doppler radars near the ground, ranging from about 30 m s^{-1} up to about 135 m s^{-1}.

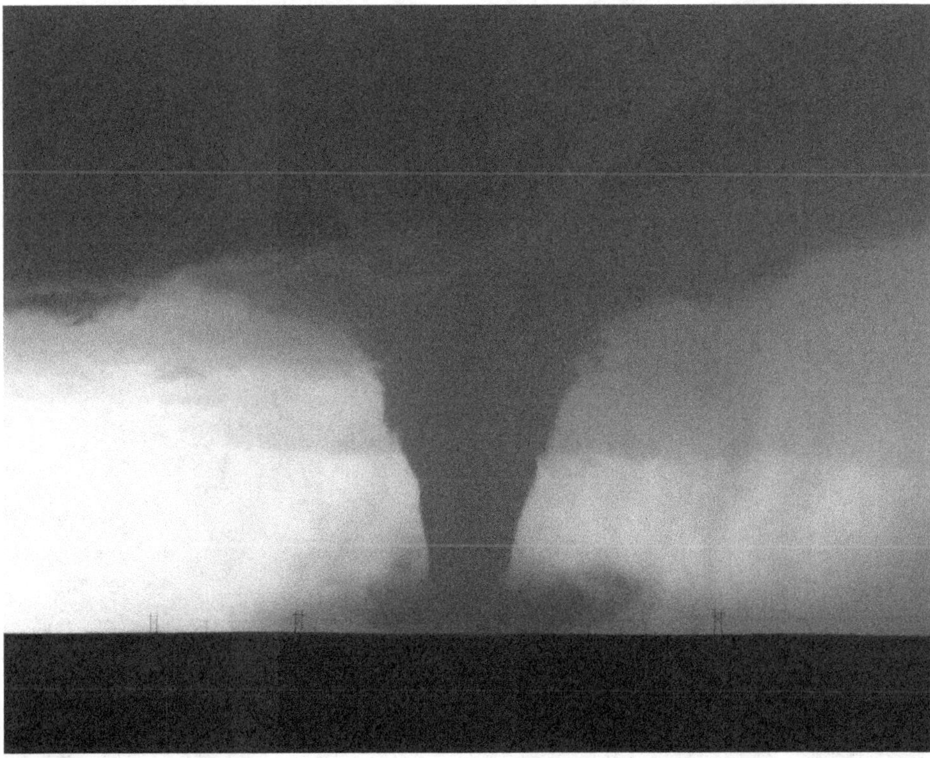

FIGURE 2.1 Tornado near Dodge City, Kansas, on May 24, 2016. (Photograph provided by the author.)

In numerical simulations and according to theoretical considerations and some photogrammetric analyses, the vertical velocity in tornadoes can be as strong as the horizontal wind speeds. In the absence of Doppler radar or in situ measurements, wind speeds are usually estimated based on damage indicators (e.g., the types of objects damaged) and the extent of damage inflicted and are rated on the Enhanced Fujita Scale, from EF0 to EF5 (Table 2.1) (Edwards et al., 2013).

These estimates require knowledge of the structural integrity of the objects damaged; in addition, if there are no structures present (e.g., over an open field), then it is difficult to produce an estimate. Doppler radars can detect vortices such as tornadoes as a region of strong lateral shear in the Doppler velocity field at a constant range from the radar (Figure 2.2) (Donaldson, 1970) in the signal from radiation backscattered by rain, hail, and debris. Because tornadoes are relatively narrow compared to the beamwidth of a typical Doppler radar, the "tornadic vortex signature" (TVS) is often defined by strong gate-to-gate shear along a constant range from the radar unless the tornado is relatively broad or close to the radar (Brown, Lemon, & Burgess, 1978). It is only on very rare occasions that Doppler radar measurements are made within 10 m of the ground.

The most intense tornadoes are spawned by supercells, convective storms characterized by a large, rotating updraft; they have a component of propagation normal to the

Table 2.1 Comparison of the original Fujita (F) Scale for estimating wind speeds in tornadoes based on damage and the initial "derived" Enhanced Fujita (EF) Scale and the operational EF Scale. (From Edwards et al., 2013. © 1989 AMS: https://www.ametsoc.org/ams/index.cfm/publications/ethical-guidelines-and-ams-policies/ams-copyright-policy/.)

Fujita Scale			Derived EF Scale		Operational EF Scale	
F Level	Fastest 1/4 Mile in mph (m s^{-1})	3-s Gust in mph (m s^{-1})	EF Level	3-s Gust in mph (m s^{-1})	EF Level	3-s Gust in mph (m s^{-1})
0	40–72 (18–32)	45–78 (20–35)	0	65–85 (29–38)	0	65–85 (29–38)
1	73–112 (33–50)	79–117 (35–52)	1	86–109 (38–49)	1	86–110 (38–49)
2	113–157 (51–70)	118–161 (53–72)	2	110–137 (49–61)	2	111–135 (50–60)
3	158–206 (71–92)	162–209 (72–93)	3	138–167 (62–75)	3	136–165 (61–74)
4	207–260 (93–116)	210–261 (94–117)	4	168–199 (75–89)	4	166–200 (74–89)
5	261–318 (117–142)	262–317 (117–142)	5	200–234 (89–105)	5	>200 (>89)

mean vertical shear in the troposphere (Rotunno & Klemp, 1982).[1] As a result of the rotation, a horizontal cross section of which is called a *mesocyclone* (a mesocyclone in general is a column of rotating air about 2–5 km in diameter) (Fujita, 1981), the updraft, which sustains the storm, propagates away from where precipitation falls, allowing supercells to live much longer than the time it takes an air parcel to enter the cloud base, travel to the tropopause (near the top of the storm), and then exit the storm; this time is generally more than 50 min, which is the typical life span of an "ordinary" cell. In an ordinary cell, the downdraft produced when precipitation falls into unsaturated air destroys the updraft.

Supercells tend to form in an environment of high vertical wind shear in the lower half of the troposphere (Weisman & Klemp, 1984) (> about 18 m s^{-1} over the lowest 6 km; Thompson, Mead, & Edwards, 2007). The rotation of the updraft is caused by the interaction between a buoyant updraft and the relatively deep environmental vertical shear in the lower half of the troposphere (Rotunno & Klemp, 1982). While many supercells form in an environment of relatively high potential buoyant energy (measured as convective available potential energy [CAPE], which is the vertically integrated buoyancy), often more than 1,500 to 5,000 j kg^{-1}, they can also occur when the shear is high, but CAPE is low (< about 500 j kg^{-1}), especially in the southeastern part of the United States, at night, during the cool season (Davis & Parker, 2014; Sherburn & Parker, 2014), and in

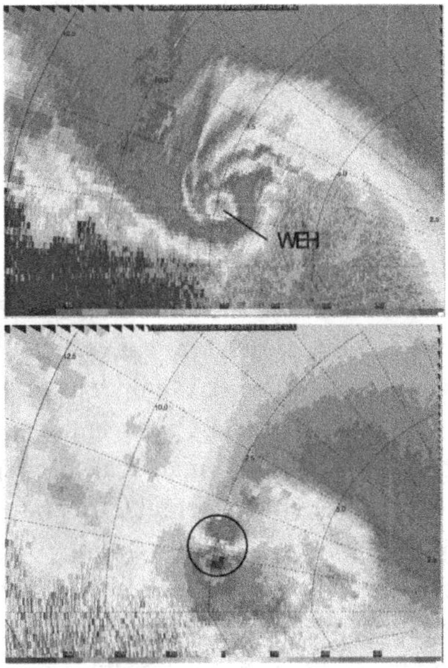

FIGURE 2.2 Doppler radar vortex signature (black circle) in a tornado near Dodge City, Kansas, on May 24, 2016. From the RaXPol (rapid-scanning, X-band [3-cm wavelength], polarimetric) mobile Doppler radar at 3° elevation angle. (a) Radar reflectivity in decibels of Z (dBZ) (color scale at bottom); a weak-echo hole (WEH) is located at the center of the tornado in the "hook." (b) Doppler velocity (m s^{-1}) (color scale at bottom); e.g, red/light gray indicates flow away from the radar; purple/dark gray indicates flow toward the radar. Range rings are shown in kilometers. (Image provided by the author.)

landfalling tropical cyclones (Baker, Parker, & Eastin, 2009[2]). There is some modeling evidence that supercell convection in landfalling tropical cyclones and in some midlatitude, synoptic-scale disturbances (Kennedy, Wescott, & Scott, 1993) can occur when the CAPE is focused at low levels (McCaul & Weisman, 2001). "Low-topped" supercells or "mini-supercells" may occur under these circumstances.

Not all supercells produce tornadoes; in fact, in one study it was reported that less than 25 percent of them do (Trapp, Stumpf, & Manross, 2005; Trapp, Tessendorf, Godfrey, & Brooks, 2005), though it may be even less than that. It is not completely understood why some supercells produce tornadoes, but most do not. Based on observations and highly idealized numerical simulations, it appears as if strong low-level vertical shear (Markowski & Richardson, 2014) and relatively high boundary-layer humidity (producing a relatively low lifting condensation level [LCL] [Rasmussen & Blanchard, 1998], as manifested by a relatively low cloud base) are necessary for the production of strong low-level mesocyclones or tornadoes in supercells.

When the humidity is relatively high, the potential for strong downdrafts is small. It has been found from numerical simulation experiments that the main source of vorticity for the low-level mesocyclones that spawn the tornadoes in supercells is horizontal vorticity produced baroclinically, along the forward flank of the supercell at low levels, where unsaturated air is cooled as precipitation falls into it and evaporates (producing a "cold pool"), and contrasts with the warmer air outside the storm (Rotunno & Klemp, 1985).[3] When the cold pool produced near the ground in a supercell is too strong, the low-level mesocyclone is advected too far away from the main updraft in the parent supercell; when the cold pool is too weak, there is not enough horizontal vorticity produced baroclinically to form a robust enough low-level mesocyclone (Markowski & Richardson, 2014). On the other hand, the source of vorticity for a mesocyclone produced aloft is large-scale environmental shear (horizontal vorticity) in the lower half of the troposphere (Rotunno & Klemp, 1985). Since vertical shear above the boundary layer is usually associated with a horizontal temperature gradient, ample deep-layer shear is accompanied by at least a modest horizontal temperature gradient. Thus, synoptic-scale baroclinicity, in addition to small-scale baroclinicity near the ground, is important. Frontal zones, along which horizontal temperature gradients are enhanced, can thus be a favorable location for supercell formation.

Tornadoes are also spawned by non-supercell (ordinary-cell) convective storms (Houston & Wilhelmson, 2007; Wakimoto & Wilson, 1989) when the vertical shear is too weak for supercells, but the CAPE is relatively high. In some of these storms, the tornado forms while the parent convective storm is growing and may not even have much, if any, precipitation. They tend to appear as rotating columns of dust underneath a rotating cloud base or as narrow condensation funnels. The parent convective storms form along a preexisting band of low-level convergence and vertical vorticity (Figure 2.3); they often appear as a series of vortices strung out along a line.

The source of rotation in these tornadoes, sometimes colloquially called "landspouts," is the preexisting vorticity, which is usually manifest as a zone of strong lateral shear. This shear (vorticity) can be found along fronts and is produced by the action of convergence underneath the rising branch of a vertical circulation forced by the tightening up of the temperature gradient along the front (Bluestein, 1993). Zones of shear can also be

FIGURE 2.3 Illustration of small convective clouds growing (left panel) along a convergent boundary between opposing airstreams, along which there is cyclonic shear and periodic cyclonic vortices (A, B, and C). The clouds grow (middle panel), and an updraft from one of the buoyant, growing convective clouds becomes superimposed on a cyclonic vortex below (right panel); convergence underneath the updraft increases the vorticity of the cyclonic vortex, perhaps becoming strong enough to become a tornado. (From Wakimoto & Wilson, 1989. © 1989 AMS: https://www.ametsoc.org/ams/index.cfm/publications/ethical-guidelines-and-ams-policies/ams-copyright-policy/.)

found along mesoscale features created in large part as a result of the slope of the ground or changes in topography, such as the "Denver cyclone" (Wilczak & Glendening, 1988). Strong lateral shear is also found along the edge of convective storm outflow and may become unstable owing to lobe and cleft instabilities (Lee & Wilhelmson, 1997a, 1997b). Supercells, on the other hand, can form in an environment with no preexisting vorticity but instead produce vorticity themselves.

Tornadoes are also produced in quasilinear (mesoscale) convective systems (QLCSs or MCSs, respectively), which are also called "squall lines," particularly along the leading edge, where there might be a series of them (Figure 2.4) (Mahale, Brotzge, and Bluestein, 2012).

They also have been observed within the dominant, cyclonic member of "bookend" vortices, the "mesovortex," or embedded near the mesovortex. (See further discussion on bow echoes in Section 2.4.) The bookend vortices are produced as horizontal vorticity associated with low-level environmental shear is tilted onto the horizontal along the ends of the convective system (especially early on before an evaporatively cooled air mass has been produced) or horizontal vorticity baroclinically generated along the leading edge of the cold pool (especially later on in the life of the MCS) (Figure 2.5).

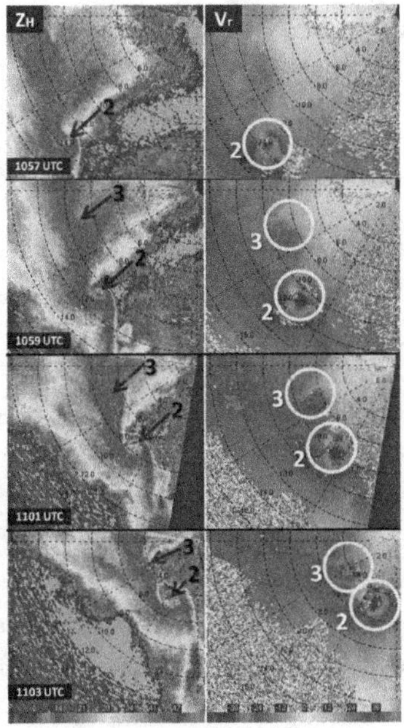

FIGURE 2.4 Evolution of tornadoes in a QLCS in central Oklahoma on April 2, 2010, as depicted by a low-powered, X-band (3-cm wavelength) Doppler radar at Rush Springs, Oklahoma, at 1° elevation angle. Left panels show radar reflectivity (note the swirl-shaped echoes; in dBZ); right panels show Doppler velocity (m s^{-1}); yellow/light gray circles enclose vortex shear signatures. The vortices are numbered. UTC, coordinated universal time. (Adapted from Mahale et al., 2012. © 2012 AMS: https://www.ametsoc.org/ams/index.cfm/publications/ethical-guidelines-and-ams-policies/ams-copyright-policy/.)

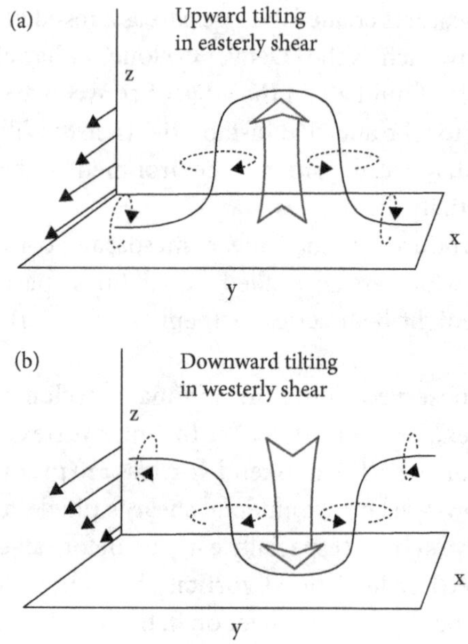

FIGURE 2.5 Schematic showing how horizontal vorticity associated with (a) easterly vertical shear (wind profile depicted at the left sides of each panel) can be tilted by an updraft (open arrow) to produce a cyclonic (north)—anticyclonic (south) couplet. (b) How a downdraft acting on westerly vertical shear can also produce the same cyclonic–anticyclonic couplet. Panel (a) represents what happens in an MCS later in its life when the baroclinic generation of horizontal vorticity along a north–south oriented outflow boundary dominates, with the cold air lying to the west. Panel (b) represents what happens before a cold pool has built up in intensity enough to dominate the generation of horizontal vorticity. (From Weisman & Davis, 1998. © 1998 AMS: https://www.ametsoc.org/ams/index.cfm/publications/ethical-guidelines-and-ams-policies /ams-copyright-policy/.)

Because QLCSs tend to occur later in the diurnal cycle of convection, after the cold pools from many convective storms congeal, QLCS tornadoes often form during the evening or early morning. In one numerical experiment, it was found that a "tornado-like vortex" formed as frictionally generated horizontal vorticity near the ground in the vicinity of strong outflow produced a rotor (see further discussion in Section 3.1), while the updraft portion of the rotor was the source of stretching of the vorticity in a mesovortex into the tornado-like vortex (Schenkman, Xue, & Shapiro, 2012).

Tornadoes that occur over water are called waterspouts. It is really not known how many waterspouts are actually supercell tornadoes and how many are non-supercell tornadoes because we do not have many radar observations over the open ocean, away from the coast. We do know, however, that many waterspouts that occur near the coastline of Florida are non-supercell tornadoes and are relatively weak (Golden, 1971); non-supercell waterspouts have been seen over the Great Lakes (van den Broeke & van den Broeke, 2015); Lake Pochartrain (see YouTube videos); the Great Salt Lake (Simpson et al., 1991); and elsewhere. Of special interest are waterspouts that make landfall, becoming tornadoes and inflicting damage. There are also instances of tornadoes moving over water becoming waterspouts.

The vertical structure of tornadoes is now considered. In the "surface friction layer," where there is an approximate balance between the frictional drag force at the surface and the pressure gradient force (Wilson & Rotunno, 1986), much of the airflow near the ground, in the lowest 10 m or perhaps even less (Kosiba & Wurman, 2013), is directed radially inward. It is difficult for radars to observe what is happening right near the ground, owing to ground clutter (Kosiba & Wurman, 2013; Wurman, Kosiba, & Robinson, 2013); evidence for this radially inward flow comes from photographs of the downed trees in relation to the tracks of tornadoes (Karstens, Gallus, Lee, & Finley, 2013; Snyder & Bluestein, 2014). Swirling motions are confined to regions above the surface friction layer. Radially inward flowing air turns upward near the axis of rotation as a jet of rapidly rising air. Such a vortex is known as an *endwall vortex* (Figure 2.6, "low-swirl" panel) (Rotunno, 2013).

There are compensating downward motions outside the tornado, and such a tornado vortex is known as a "one-celled" vortex. When the vortex becomes intense near the ground and decreases in intensity aloft, a downward-directed pressure gradient force

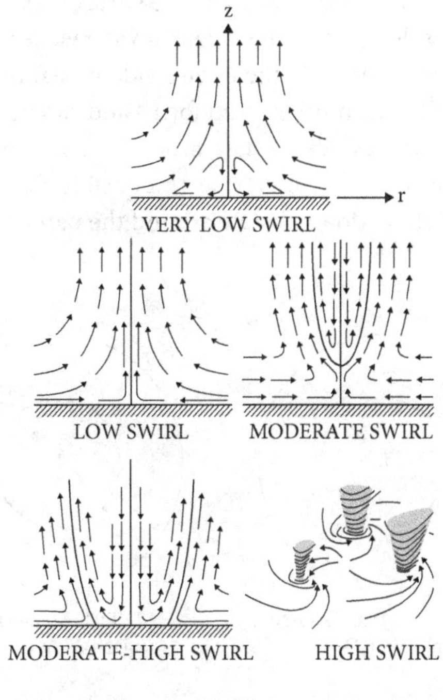

FIGURE 2.6 Idealized depictions of the wind fields in cross sections cutting through a tornado, beginning with very low swirl, in successive panels up to a high-swirl vortex with secondary vortices. Note how, beginning with the moderate-swirl tornado, a downdraft is produced aloft at the center, which reaches the ground in the moderate–high swirl case. (Adapted from Davies-Jones, 1986, by permission of the University of Oklahoma Press.)

develops, which forces a downdraft in the middle of the tornado. The rising jet of air can be very rapid.

For the next 100 m or so above the surface friction layer, in the "inertial layer," the dominant forces are the pressure gradient force and centrifugal force, but the flow is not balanced, and there are radial and vertical transports of momentum (Wilson & Rotunno, 1986). The tornado's boundary layer is composed of the surface friction layer and the inertial layer. Above the boundary layer, the tornado is in approximate "cyclostrophic balance"; that is, there is an approximate balance between the pressure gradient force and the centrifugal force (Figure 2.7).

Centrifugal waves form when there is an imbalance between the inward-directed pressure gradient force and the outward-directed centrifugal force. When the rising jet coming out of the surface friction layer is very strong, wind speeds may exceed the vertical propagation speed of centrifugal waves. Such a flow is called *supercritical*. When the supercritical flow slows down aloft, owing to the downward-directed pressure gradient force, the flow may become "subcritical," that is, slower than vertically propagating centrifugal waves. When and where this happens, there is "vortex breakdown," which is analogous to the hydraulic jump phenomenon found in fluids when flow transitions from supercritical to subcritical (Benjamin, 1962), in this case faster than or slower than shallow-water gravity waves, respectively. During vortex breakdown, the flow makes a rapid transition from laminar to turbulent, and the vortex widens.

A vortex in which the level of vortex breakdown reaches the ground is called a *drowned* vortex (Maxworthy, 1973). As the vortex widens near the ground, an annulus of very high lateral shear of the azimuthal wind forms and can become unstable, forming small-scale "secondary vortices" that rotate around a common axis (Rotunno, 1984; Ward, 1972). These multiple vortices can be responsible for very high wind speeds locally. Most multiple vortices do not rotate around the parent vortex more than once.

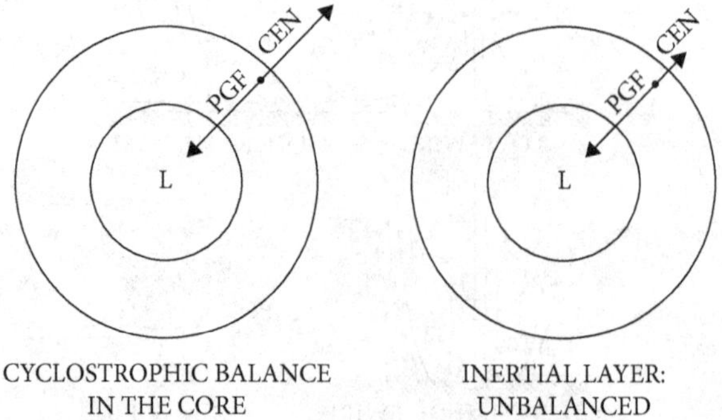

FIGURE 2.7 Schematic representation of cyclostrophic balance [left panel; inward directed pressure gradient force (PGF) is balanced by the centrifugal force (CEN)]. In the inertial layer, where the azimuthal velocity is slowed down as an indirect result of frictional drag at the surface and its effect through vertical mixing, the CEN is less than it is in the core, while the PGF stays the same (right panel). In this case, air has a component of acceleration inward toward the axis of rotation. (Image provided by the author.)

In a multiple-vortex tornado, there is rising motion near the radius of maximum wind (RMW) but sinking motion at the center and far outward from the annulus of rising motion. Such a vortex is known as a two-celled vortex. Some tornadoes make transitions from single-celled tornadoes to two-celled, multiple-vortex tornadoes and back. A high *swirl ratio* (an approximate measure of the ratio of the amount of swirling, azimuthal flow to radial flow) favors a multiple-vortex tornado, while a low swirl ratio favors a single-vortex tornado (Figure 2.6).

The azimuthal winds inside the "core" of a low-swirl tornado increase with distance from the center until the core radius, the RMW. Outside the core radius, the azimuthal winds drop off approximately as the inverse of the distance from the center (Tanamachi, Bluestein, Lee, Bell, & Pazmany, 2007). Such a wind profile, which has been observed from both photogrammetric analyses of debris motion and mobile Doppler radar observations, is called a *combined Rankine vortex*. An idealized combined Rankine vortex has a perfect linear variation of azimuthal wind speed with radius out to the RMW and a drop-off in wind speed as the inverse of the radius beyond the RMW. Thus, a combined Rankine vortex has solid-body rotation and constant vorticity within the RMW and potential flow (no vorticity) outside the RMW, where cyclonic curvature vorticity cancels out anticyclonic shear vorticity. In nature, the transition at the RMW from solid-body rotation to potential flow is smoother (not discontinuous), probably owing to horizontal diffusion. Such a wind profile is referred to as a *Burgers–Rott vortex* (Figure 2.8).

The maximum wind speed in a tornado depends in large part on the intensity of the updraft that drives it, which comes from buoyant energy. The maximum wind speed V_{max} in a tornado (having a combined Rankine vortex profile) based on thermodynamic considerations alone is called the *thermodynamic speed limit* (Fiedler & Rotunno, 1986):

$$V_{max} = (CAPE)^{1/2} \tag{1}$$

The speed limit may be exceeded by as much as a factor of two, especially near the ground, in the boundary layer, where the radially inward-directed pressure gradient force overwhelms the outward-directed centrifugal force owing to the frictional drag of air near the ground. In this case, a ring of air having a fixed amount of angular momentum from its environment may be brought inward closer to the center of rotation (through "inertial overshoot") than possible if the flow were in cyclostrophic balance; if angular momentum is conserved, the azimuthal wind speed must increase. Thus, for a large CAPE of 5,000 J kg^{-1}, the maximum wind speed of approximately twice the thermodynamic speed limit (actually 5/3 according to Rotunno, 2013) is 140 m s^{-1}, which is close to the maximum observed by mobile Doppler radars of about 135 m s^{-1}. The highest wind speeds in a tornado occur near the level of vortex breakdown when it is near the ground (Fiedler & Rotunno, 1986).

The actual ground-relative wind speed can be affected significantly not only by the structure of the tornado, but also by the speed of the translation of the tornado (Mahale, Bortzge, & Bluestein, 2012). When a tornado is carried along by a rapidly moving parent storm or by a parent mesocyclone, the ground-relative winds on the approaching-velocity side of the tornado are enhanced, while the ground-relative winds on the receding-velocity side are

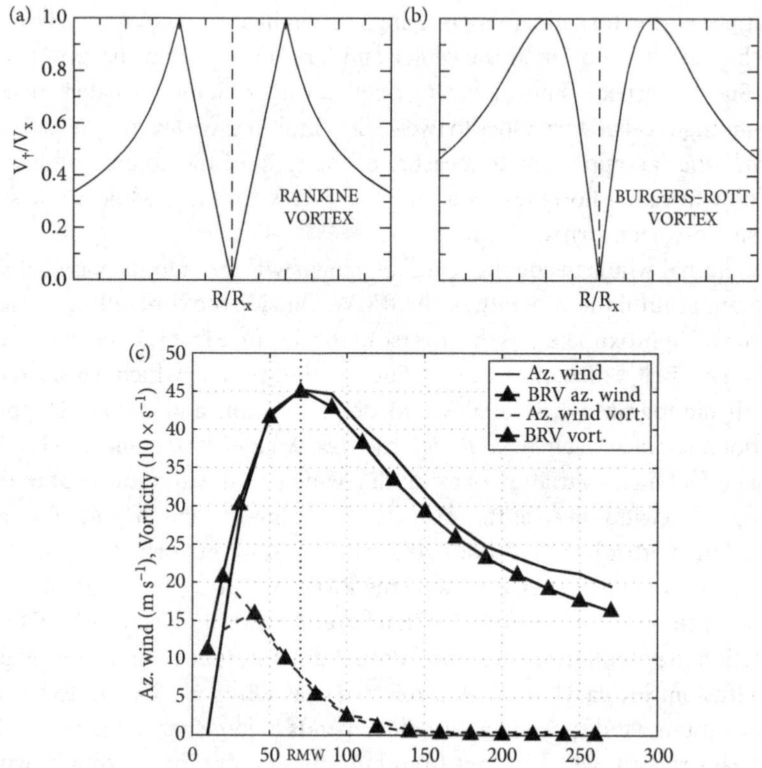

FIGURE 2.8 Radial profiles of the azimuthal wind (V_t) for a one-celled Rankine combined vortex (a) and Burgers–Rott vortex (BRV) (b). Azimuthal winds are normalized by the peak azimuthal wind velocity (V_x); R is the distance from the axis of rotation; R_x is the core radius (from Brown & Wood, 2012). (c) Time-averaged (over about 3.5 min) azimuthally averaged azimuthal wind (solid line) as a function of distance (m) from the axis of rotation at low altitude in a tornado near Stockton, Kansas, on May 15, 1999, based on data from a W-band (3-mm wavelength), mobile Doppler radar at about 1° elevation angle (~100 m AGL). Also plotted in (c) is the profile of a Burgers–Rott vortex fit to the maximum wind speed and radius of maximum wind (line with arrows). Vorticity of the azimuthal wind is plotted by a solid line with triangles for a Burgers–Rott vortex and by a dash–dot line for the actual azimuthally averaged azimuthal wind. (From Tanamachi et al. 2007. © 2007 AMS: https://www.ametsoc.org/ams/index.cfm/publications/ethical-guidelines-and-ams-policies/ams-copyright-policy/.)

suppressed. For this reason, the highest ground-relative wind speeds and great damage tend to be found on the right side of the tornado path (in the Northern Hemisphere). Furthermore, ground-relative wind speeds may be augmented in secondary vortices on the approaching side of the vortices. So, even if two tornadoes are identical in structure and "intensity," the ground-relative winds may be relatively enhanced in the one that is translating along the ground more rapidly. The speed of the tornado is determined in part by the wind in the environment of the storm.

Most tornadoes rotate cyclonically (in the same sense as the rotation of Earth) in a counterclockwise manner in the Northern Hemisphere. However, occasionally anticyclonically rotating tornadoes form in supercells. Most of these are paired with cyclonically rotating tornadoes or with mesocyclones in right-moving supercells (those that move to the right of the mean tropospheric wind), but are weaker than their cyclonically rotating partners (Figure 2.9) (Bluestein, French, Snyder, & Houser, 2016).

On very rare occasions, anticyclonically rotating tornadoes occur in left-moving supercells with meso-anticyclones (Bluestein, 2013). The meso-anticyclones are extremely rare because left-moving, anticyclonic supercells require an environmental hodograph that curves in counterclockwise manner with height; it is much more common, owing to the effects of friction and warm advection, for a supercell environment to be characterized by a hodograph that curves in a clockwise manner with height, especially at low altitudes.

Sometimes, mesocyclones, though relatively broad in scale, may produce winds strong enough to inflict damage (Wakimoto, Cai, & Murphey, 2004). Such mesocyclones are sometimes identified as tornadoes because they are rapidly rotating columns of air causing damage, but they may not be characterized by an endwall vortex like smaller scale tornadoes or may not be visible because they are embedded in precipitation. Furthermore, a rapidly translating mesocyclone may contain very strong winds on the

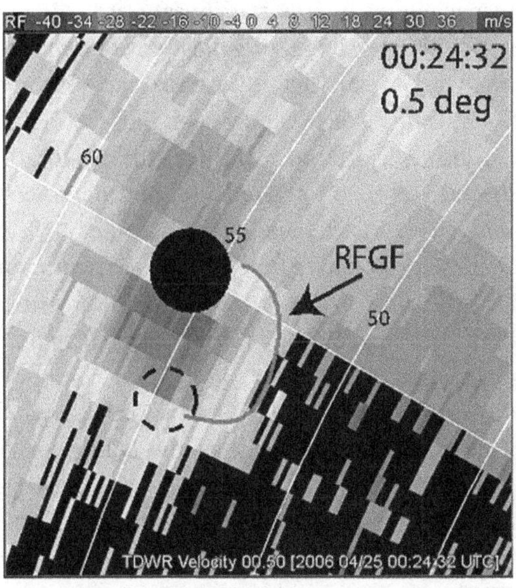

FIGURE 2.9 The dissipation of a cyclonic tornado and formation of an anticyclonic tornado in a supercell near El Reno, Oklahoma, on April 24, 2006, as depicted by the Doppler velocity field from a scan by an FAA (Federal Aviation Administration) C-band (5-cm wavelength) Doppler radar south of Oklahoma City, Oklahoma. The cyclonic vorticity shear signature associated with the tornado is indicated by the solid circle; the anticyclonic shear signature associated with the developing anticyclonic tornado is indicated by the dashed circle. The leading edge of the rear-flank gust front (RFGF) is indicated by the curved, solid red/light gray line joining the cyclonic and anticyclonic shear signatures. Range rings are plotted in kilometers. (Adapted from Bluestein et al. 2016. © 2016 AMS: https://www.ametsoc.org/ams/index.cfm/publications/ethical-guidelines-and-ams-policies/ams-copyright-policy/.)

approaching side capable of doing damage, even though the storm-relative azimuthal wind speeds are modest. For example, a mesocyclone with subtornadic, maximum, storm-relative azimuthal wind speeds of 25 m s^{-1} may create ground-relative wind speeds as high as 40 m s^{-1} if the parent storm is moving at 15 m s^{-1}. Such a translating mesocyclone could inflict EF1 damage on the approaching side of the mesocyclone.

Tornadoes overall have been reported in every state of the United States, but most occur in a swath over the central United States. Typically, they appear early in the year (winter and early spring) farther south and later in the season (spring and summer) farther to the north, as the core of strongest westerly winds aloft migrate northward with the season. Other regions of frequent tornadoes include the Gulf states of the southeast and Florida and eastern Colorado. The types of tornadoes vary with geographical area. For example, the plains of the central United States harbor more supercell tornadoes, while Florida has more non-supercell tornadoes because the relatively stronger shear necessary for supercell formation is found more often at higher latitudes.

2.2. Gust Fronts

When precipitation, especially rain, falls into unsaturated air owing to its weight, it evaporates and the air cools, further increasing the negative buoyancy. Snow falling aloft, where it is much colder, melts as it falls below the freezing level and also cools. Hail melts also as it falls below the freezing level and may cool. But, the main mechanism for producing a pool of cool air near the ground typically is the evaporation of rain. The negatively buoyant downdraft then hits the ground and, in order to conserve mass, spreads out laterally (Wakimoto, 1982). As the air becomes cooler than its environment, the pressure at the surface increases relative to that in the environment because cold air is denser than warm air at the same pressure. A horizontal pressure gradient force then develops that is directed from the cold side to the warm side and forces the cold air to expand into the warm air (Figure 2.10).

The straight-line wind at the leading edge of a two-dimensional cold pool becomes relatively strong and gusty; the leading edge of the cold pool is called a gust front (Charba, 1974). The gust front is an atmospheric manifestation of a more general phenomenon in fluid dynamics known as a *density current* or *gravity current*. The

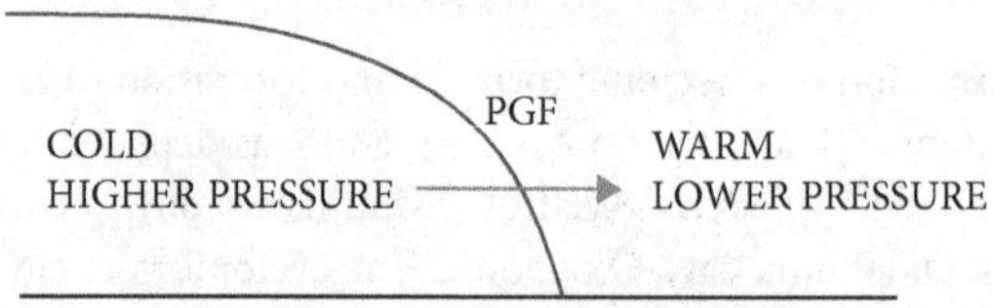

FIGURE 2.10 Idealized representation of a vertical cross section through a density current/gust front. PGF is the horizontal pressure gradient force, directed from the cold air to the ambient, warmer air. The green arrow represents the direction of the movement of the density current/gust front. (From Bluestein, 2013. Republished by permission of Springer Nature: Springer/Praxis, *Severe Convective Storms and Tornadoes: Observations and Dynamics* by Howard B. Bluestein, © 2013.)

stronger the difference between the cold air and the warm air in the environment and the greater the depth of the cold pool, the higher the wind speed will be. The speed of a gust front in still air (a "resting atmosphere"), in which rotation is negligible, U, is given approximately by the following (Benjamin, 1968)[4]:

$$U = 1.4\ [g\ h\ (\theta_0 - \theta_1 / \theta_0]^{1/2} \tag{2}$$

where g is the acceleration of gravity, h is the depth of the cold pool, and θ_0 and θ_1 are the potential temperature inside the cold pool and outside the cold pool in the environment, respectively, and the potential temperature is given as

$$\theta = T\left(p_0 / p\right)^{C_p/R} \tag{3}$$

where T is temperature; p_0 is the pressure at a reference level, usually 1,000 hPa; p is the pressure; C_p is the specific heat at constant pressure (1,004 J kg^{-1} K^{-1}); and R is the gas constant for dry air (287 J kg^{-1} K^{-1}). It is assumed that the gust front moves as a material surface, bounded by the leading edge of cold air. In this formula, turbulent mixing between the cold, dense air mass and the warmer, less dense air mass of the environment is not considered, and the stratification is discontinuous (i.e., there are two different fluids of differing density, unlike the real atmosphere, in which density decreases continuously with height).

Damaging winds may therefore occur at low levels when the cold pool is relatively intense and deep. Strong gust fronts are favored by situations in which precipitation falls into a deep, dry layer. If the gust front forms in an environment in which there is a component of wind in the same direction as that in which the gust front is moving, then the surface wind speeds can be even higher. In many instances, the passage of the gust front is followed by heavy precipitation. However, in some instances, when the convective storm has dissipated, there is only light precipitation or no precipitation at all.

In the reference frame of the leading edge of the gust front in a resting atmosphere, air approaches the gust front from the warm side and comes to a halt as it hits the leading edge of the gust front. In order for this to happen, the nonhydrostatic, dynamic part of the pressure must be relatively high, so that air encounters a horizontal pressure gradient force in the direction opposite to that in which it is moving. Air then rises over top of the cold pool, where vertical shear is enhanced. The vertical shear may become large enough that there is turbulence as some mixing occurs between the cold air below and warmer air aloft. Thus, there can be very gusty winds at the top of the gust front. Small horizontal vortices known as Kelvin–Helmholtz waves may also form (Droegemeier & Wilhelmson, 1987).

2.3. Downbursts

Downbursts are intense, localized downdrafts produced in convective storms; these downbursts range from almost 1 km to about 10 km in diameter (Fujita, 1981).

Downbursts that are less than 4 km in diameter are called *microbursts*. Their peak winds persist for about 2–5 min. They may occur alone, in conjunction with gust fronts, or to the rear of gust fronts, as separate pulses of downward-flowing air. When a downdraft reaches the ground, the dense air spreads out and diverges, producing strong, potentially damaging straight-line winds (Figure 2.11).

This field of diverging air may be detected as a strong gradient in the Doppler wind component along the radar beam (*not* at constant range, as for a TVS or mesocyclone signature). Microbursts are three dimensional, unlike gust fronts, which are usually considered two dimensional.

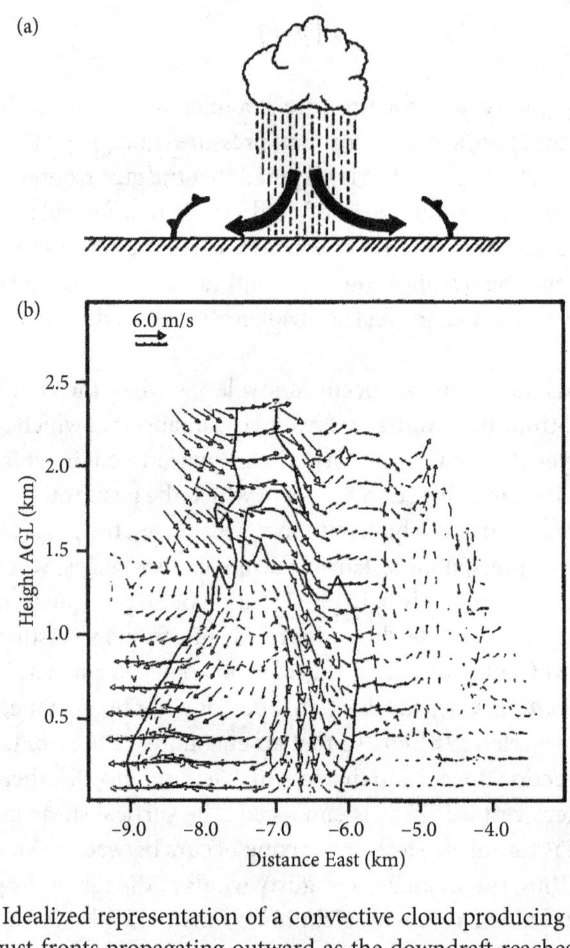

FIGURE 2.11 (a) Idealized representation of a convective cloud producing precipitation and a downdraft, with gust fronts propagating outward as the downdraft reaches the ground (from Rotunno et al., 1988). (b) Vertical cross section of the perturbation wind field (deviation from the mean wind, so motion of the parent storm has been removed) in a microburst reaching the ground in northeastern Colorado on May 31, 1984. Based on dual-Doppler analyses from a fixed network of X-band (3-cm wavelength) Doppler radars. (From Parsons & Kropfli, 1990. © 1990 AMS: https://www.ametsoc.org/ams/index.cfm/publications/ethical-guidelines-and-ams-policies /ams-copyright-policy/.)

There are two types of microbursts, "dry" and "wet." The dry microbursts are driven mainly by the negative buoyancy created by evaporative cooling of raindrops and the cooling from the melting frozen precipitation (Wakimoto, 1985). Dry microbursts are favored by environments that are relatively dry, with a deep dry-adiabatic layer underneath the cloud base. They occur often in the arid, mountainous regions of the western United States during the warm season, but also occur in the plains during the summer. One can often see virga under the cloud base of the parent convective storm.

Wet microbursts are driven by negative buoyancy associated with heavy precipitation loading with contributions from the entrainment of dry environmental air, sublimation of ice crystals, melting of ice crystals, and evaporation; they are therefore favored by very moist environments, such as are found during the summer around the rim of the Gulf Coast (Atkins & Wakimoto, 1991).

Downdrafts strong enough to produce microbursts could also be driven by downward-directed pressure gradient forces above surface vortices, which are associated with locally low pressure, as in "occlusion" downdrafts in supercells (Trapp & Weisman, 2003; Wakimoto, Liu, & Cai, 1998). Microbursts can occur in both ordinary cell convective storms and supercells. There is sometimes a relationship between them and tornadoes. In some instances, the impinging of air from a microburst onto a gust front can lead to tornadogenesis (Bluestein, Weiss, & Pazmany, 2003) or just mesocyclogenesis while a previous mesocyclone/tornado occludes (Adlerman, Droegemeier, & Davies-Jones, 1999). The latter process is referred to as *cyclic mesocyclogenesis* and is responsible for some supercells producing a series of tornadoes and a resulting long track of damage.

At the rear edge of MCSs, within the trailing anvil and stratiform precipitation area, precipitation may descend from a relatively high altitude and fall into unsaturated air, where it cools to the wet bulb temperature until all the rain has evaporated. If the sinking air is cooler than its environment, it continues to sink, owing to negative buoyancy, while it warms at the dry-adiabatic lapse rate (at 10 K km^{-1}). If the air near the surface is relative cool and stable, then there may be enough kinetic energy of the negatively buoyant air to pass through the low-level stable layer as it loses its negative buoyancy and reaches the ground. If it does, the surface temperature may increase considerably. Such a phenomenon is known as a *heat burst*, which can be associated with strong and potentially marginally damaging surface winds, similar to those from a microburst (Bernstein & Johnson, 1994). Heat bursts are very localized, and when they occur at night can effect substantial warming.

2.4. Bow Echoes

In QLCSs, a precipitation-induced cold pool forms, along with a gust front. When the environment has some vertical shear, especially at low levels, the horizontal vorticity associated with the shear may be advected toward the gust front from the environment (in the reference frame of the gust front). If it is exactly counteracted by the generation of baroclinic horizontal vorticity of the opposite sense, at the leading edge of the gust front, then this configuration favors an erect updraft at the leading edge of the gust front (Figure 2.12) (Rotunno, Klemp, & Weisman, 1988).

This configuration, which is common during the early stages of an MCS, is conducive for the continued triggering of new convective cells at the leading edge. These cells move rearward with respect to the leading edge of the gust front and evaporate precipitation into unsaturated air, thus reinforcing and enhancing the cold pool. In this idealized two-dimensional geometry, there is a flow of air that ascends from the front to the rear of the QLCS (Houze, Rutledge, Biggerstaff, & Smull, 1989).

At the rear of the storm (as opposed to the front), on the other hand, there is cloud material condensing and warming aloft above the cold pool below, especially as buoyant air is advected rearward aloft. Horizontal vorticity is therefore generated baroclinically at the back edge aloft that is opposite in direction to the horizontal vorticity that is generated below, owing to heating due to condensation aloft and the cooling due to evaporation and melting below. The result is the superposition of two horizontal vortices, which leads to a flow from the rear to the front of the QLCS at midlevels, separating the layer of the clouds aloft and the cold pool below (Figure 2.13) (Weisman, 1992).

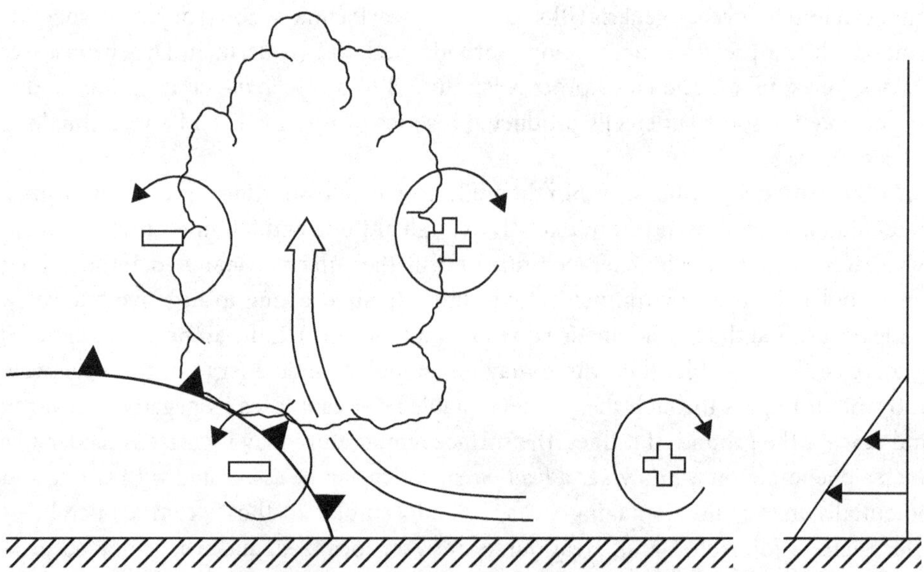

FIGURE 2.12 Demonstration of how low-level shear (insert to right) can make air go straight up, even though the baroclinic generation of horizontal vorticity at the leading edge of a gust front makes the air go up over the leading edge of the gust front at an angle. Vorticity at low levels directed into the figure ("+," about a clockwise direction) from the environment is advected toward the gust front, while baroclinically generated horizontal vorticity directed out from the figure ("–," about a counterclockwise direction) counteracts the advection, so that air streams more directly upward, leading to an increased likelihood of new convective cell formation there. (From Rotunno et al., 1988. © 1988 AMS: https://www.ametsoc.org/ams/index.cfm/publications/ethical-guidelines-and-ams-policies/ams-copyright-policy/.)

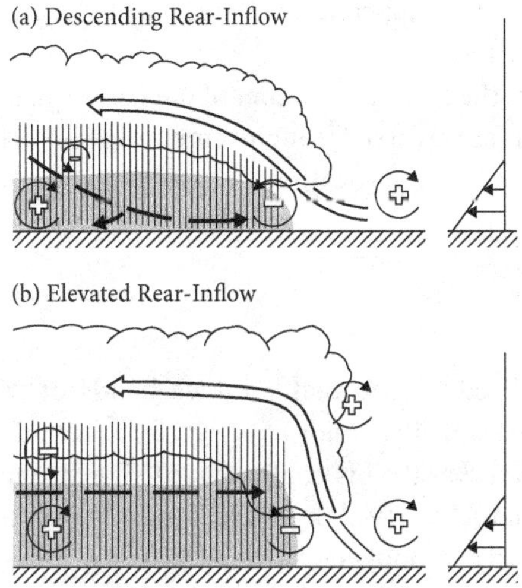

FIGURE 2.13 Idealized illustration of how a descending rear-inflow jet develops when (a) the generation of horizontal vorticity aloft ("–," out from the figure, in a counterclockwise direction) is greater than the generation of horizontal vorticity below ("+," into the figure about a clockwise direction) at the rear edge of the QLCS and consequently a descending rear inflow develops. (b) An elevated rear-inflow jet develops when the rate of generation of horizontal vorticity aloft is approximately the same as that below at the rear edge. (From Weisman, 1992. © 1992 AMS: https://www.ametsoc.org/ams/index.cfm/publications/ethical-guidelines-and-ams-policies/ams-copyright-policy/.)

This rear-to-front jet of air is called the *rear-inflow jet*[5] (Smull & Houze, 1987). When the magnitude of the rate at which horizontal vorticity produced baroclinically below associated with the rear edge of the cold pool is much greater than the magnitude of the rate at which horizontal vorticity is produced baroclinically aloft associated with the rear edge of the cloud, the rear-inflow jet descends and may reach the ground (Weisman, 1992), resulting in strong surface winds at or to the rear of the leading gust front in the QLCS (Trapp & Weisman, 2003). This situation is more likely during the later stages of a QLCS. Earlier, before the cold pool has built up to an even higher intensity, horizontal vorticity associated with the vertical shear under the rear-inflow jet (generated baroclinically by the QLCS itself, not advected from the environment) is advected toward the front of the QLCS, without descending, where it counteracts the baroclinically generated horizontal vorticity at the leading edge to produce a more erect updraft. Elevated rear-inflow jets are favored by high shear and high CAPE, while descending rear-inflow jets are favored by weak-to-moderate shear and weak–moderate CAPE.

When considered in three dimensions, not just two dimensions, it is seen that bookend (or "line-end") vortices form at the opposite ends of a QLCS oriented north to south owing to the tilting of baroclinically generated horizontal vorticity at the leading edge of the gust front, at the northern and southern edges (Weisman, 1993). Therefore, in a descending rear-inflow jet, when bookend vortices form close together (Figure 2.14), the QLCS assumes a bowed out appearance. The associated radar echo structure is called a *bow echo* (Przybylinski, 1995).

Late in the QLCS's life, the cyclonic member of the vortex pair dominates. Bow echoes and strong surface winds are favored by environments in which there are both relatively high CAPE and strong low-level vertical shear.

2.5. Derechos

Rapidly moving, long-lived QLCSs that produce bands of widespread, strong, and potentially damaging straight-line winds at the surface are called by the Spanish name (for "straight ahead") *derechos* (Johns & Hirt, 1987). The strong surface winds may be associated with strong baroclinicity along the leading edge of the leading gust front, the descending rear-inflow jet and bookend vortices, and downbursts. Wind speeds as high as 50 m s^{-1} may occur. Tornadoes and microbursts may also be embedded within derechos. *Derechos are basically QLCSs/squall lines, whose strong straight-line winds may be associated with the phenomena described in* Sections 2.2–2.4 (Corfidi, Coniglio, Cohen, & Mead, 2016).

2.6. Dust Devils

Dust devils are rapidly rotating columns of air in clear air, like tornadoes, but not pendant from cumuliform clouds. There does not appear to be a preferred sense of rotation (Sinclair, 1969). They are forced by dry thermals in the surface boundary layer in which the air at the ground is heated rapidly and intensely enough to form a superadiabatic layer. The background vorticity that is stretched to become dust devils may be from frictionally induced lateral shear associated with local changes in surface roughness or may be associated with vertical shear (horizontal vorticity) associated with Rayleigh–Bénard convective cells that is tilted onto the vertical (Kanak, Lilly, & Snow, 2000). Dust devils are common over the desert areas of the western United States, especially during the

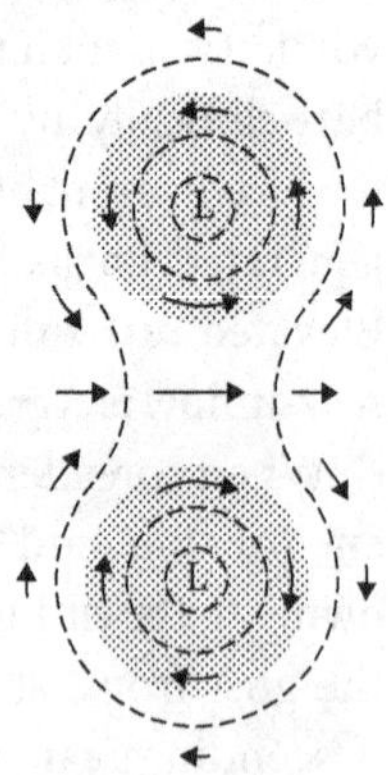

FIGURE 2.14 Idealized representation of how a jet of enhanced rear-to-front flow can be a result of the superposition of two nearby "bookend," counterrotating vortices. (From Weisman, 1993. © 1993 AMS: https://www.ametsoc.org/ams/index.cfm/publications/ethical-guidelines-and-ams-policies/ams-copyright-policy/.)

warm season. Generally, they are too weak to produce damage (Ryan & Carroll 1970; Sinclair, 1973), although on rare occasion minor damage has been reported. On Doppler radar imagery, dust devils look like small tornadoes (Bluestein, Weiss, & Pazmany, 2004). Most are short lived; the longest lived ones last on the order of minutes.

3. Nonconvective Storm-Related Phenomena

Many NSWSs not associated with convective storms are associated with high-amplitude gravity waves. Gravity waves, in this case, are undulations in the height of the top of a denser, more stable layer of air below them; the vertical restoring force the oscillations is buoyancy. Since gravity waves are dispersive (i.e., since their phase speed depends on the wavelength), then if a wide spectrum of waves is excited, they tend not to travel very far from their source as a coherent disturbance. In some circumstances, especially when a layer aloft does not permit gravity wave propagation, upward-propagating waves may be reflected back downward and resonate with the upward-propagating waves, keeping waves together for longer distances. Wind storms associated with gravity waves are not only often associated with mountains, but also may be triggered in or by convective storms.

Other nonconvective wind storms are also associated with negative buoyancy (drainage winds), but not as waves. Yet other con-convective wind storms are caused by the downward mixing of high-momentum air from above and through inertial oscillations. A readily available climatology of all types of damaging, nonconvective winds in the United States is currently unavailable, to the best of my knowledge.

3.1. Downslope Wind Storms

Vertically propagating gravity waves may be triggered when air is forced to flow up and over mountains (Durran, 1990). Gravity wave oscillations make the air bounce up and down locally, but mass is not transported horizontally by the wave, which, as visualized by the height of the top of the stable layer, only appears to propagate laterally. (Gust fronts/density currents/gravity currents, on the other hand, actually transport mass horizontally, not just up and down.) Under some circumstances, a wave with a very high amplitude is produced on the lee side (Klemp & Lilly, 1975). When this happens, especially when there is transition from subcritical (wind speeds less than that of shallow-water gravity waves) to supercritical (wind speeds greater than that of shallow-water gravity waves) flow right at the top of the mountain ridge, there is a jump downstream from the mountains, and very strong wind speeds flowing in the downslope direction are observed just upstream from the jump (also called a "hydraulic jump") (Figure 2.15).

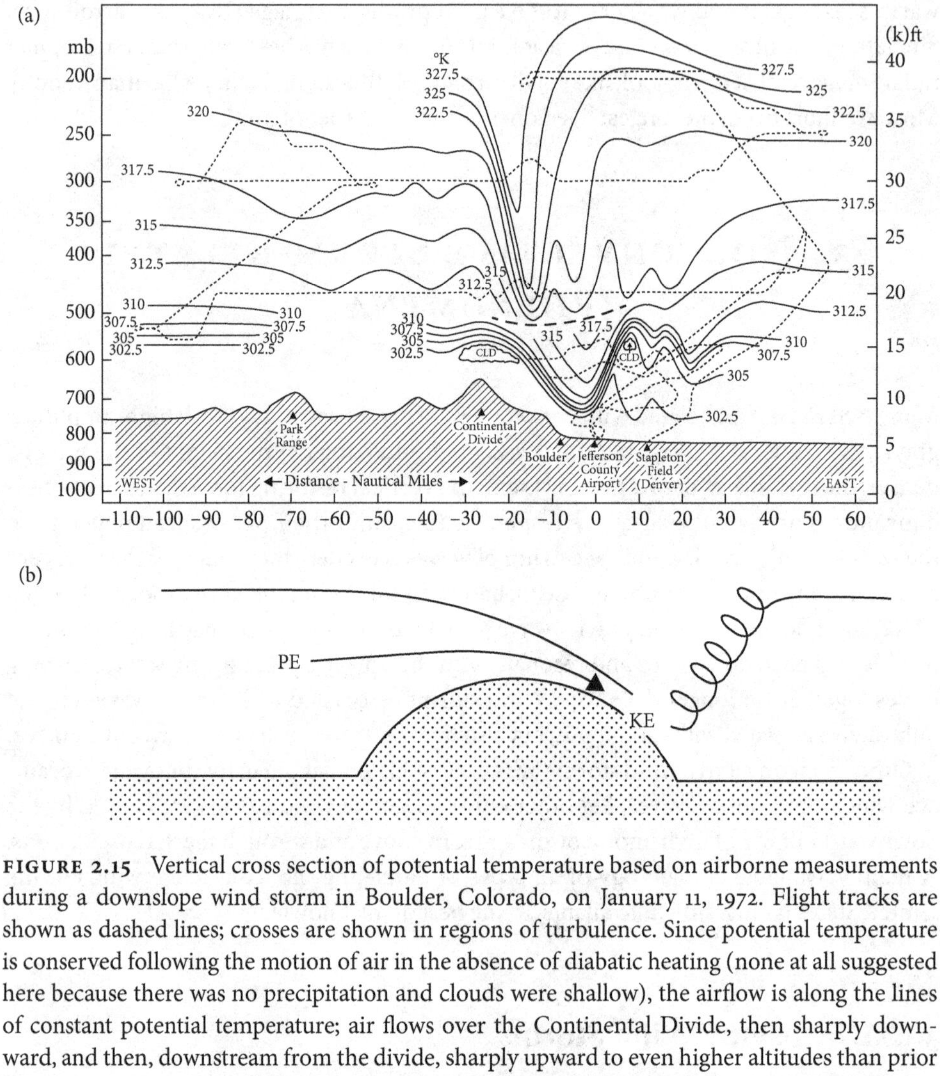

FIGURE 2.15 Vertical cross section of potential temperature based on airborne measurements during a downslope wind storm in Boulder, Colorado, on January 11, 1972. Flight tracks are shown as dashed lines; crosses are shown in regions of turbulence. Since potential temperature is conserved following the motion of air in the absence of diabatic heating (none at all suggested here because there was no precipitation and clouds were shallow), the airflow is along the lines of constant potential temperature; air flows over the Continental Divide, then sharply downward, and then, downstream from the divide, sharply upward to even higher altitudes than prior to being lifted over the divide, to the west (from Klemp & Lilly, 1975). (b) In the hydraulic jump, two-fluid model, air gains kinetic energy (KE) at the expense of potential energy (PE) even as it rises over a mountainous obstacle and continues to increase as it descends on the lee side until turbulence ensues and the air jumps back upward. In this model, the flow is subcritical (slower than shallow-water gravity waves) up to the top of the mountain, but switches to supercritical (faster than shallow-water gravity waves) at the top of the mountain. (From Durran, 1990. © 1990 AMS: https://www.ametsoc.org/ams/index.cfm/publications/ethical-guidelines-and-ams-policies/ams-copyright-policy/.)

However, in the hydraulic jump theory there are just two fluids of differing density (e.g., water and air), which do not mix; in the real atmosphere, density decreases continuously with height, as was noted in Section 2.2. An alternative theory emphasizes the importance of an inversion near mountain-top level in the upstream environment and if the gravity waves triggered reverse phase in the troposphere (Klemp & Lilly, 1975). In

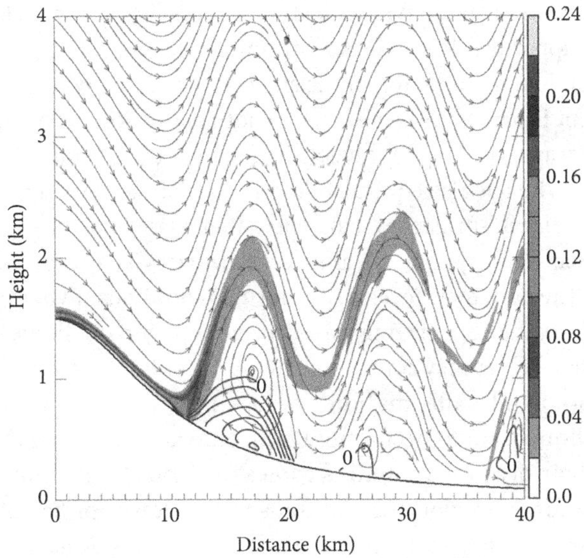

FIGURE 2.16 Rotors downstream from a mountain, modeled numerically in two dimensions (for trapped waves, not for vertically propagating waves, in this case). Strong horizontal vorticity (s^{-1}) (into the figure) is shaded; solid lines are isotachs for wind velocities with components from right to left (easterly) (in m s^{-1}). (From Doyle & Durran, 2002, reproduced in Doyle & Durran, 2004. © 2002 AMS: https://www.ametsoc.org/ams/index.cfm/publications/ethical-guidelines-and-ams-policies/ams-copyright-policy/.)

some favored geographical areas, such is along the Front Range of northeastern Colorado, winds may often gust during the winter season to 35 m s^{-1} or more, and sometimes as high as 60 m s^{-1} during these wind storms.

A downslope wind storm that occurs that is relatively warm is called a *chinook* or *foehn*. In this case, relatively warm air, typically found along and downstream from an upper-level ridge of high pressure where there is subsidence, descends dry adiabatically (at the rate of 10 K km^{-1}), and preexisting snow cover may rapidly melt and sublimate. When relatively cool air, for example, to the rear of the trough of low pressure aloft, descends, the downslope wind storm is called a *bora*.

Strong winds and strong vertical wind shear may also be produced by horizontal vortices called *rotors* and *subrotors*. These rotors at the ground are regions of reversed flow underneath gravity waves (Figure 2.16) (Doyle & Durran, 2004). Visually, they often appear to be lines of ragged-looking clouds parallel to the mountains.

3.2. Bores

Bores are gravity waves produced when a gravity current/density associated with a sharp synoptic-scale cold front or convective storm gust front, sea breeze, or land breeze lifts a layer of stable air just ahead of it, producing a local rise in the height of the top of the stable layer *ahead* of the gravity/density current (Figure 2.17) (Haghi, Parsons, & Shapiro, 2017; Knupp, 2006; Rottman & Simpson, 1989).

The local rise in height of the top of the stable layer occurs when the flow relative to the density current is partially blocked (when not enough kinetic energy is available to lift the air all the way to the top of the density current; this is equivalent to the Froude number being < 1, which is also equivalent to the flow being subcritical). Bore passage is characterized by a rise in surface pressure and either a rise or no change in temperature (density current passages, by contrast, are marked by rises in surface pressure, but a *drop* in temperature). The bore may develop, as a result of nonlinear dynamics (of the wind), into an undular bore (i.e., one having waves) or even larger amplitude *solitary waves* or *solitons*. The solitary waves may contain the highest amplitude wave at the leading edge, with decreasing amplitudes after the bore has passed by; such a group of solitary waves is called a soliton. Bores tend to occur more frequently at night or in the early morning hours, when the air near the ground is coolest and most stable. In some instances, high-momentum air from aloft may be transported down to the surface, producing potentially damaging winds (Bluestein & Snyder, 2015). Bores have been modeled by three nonmixing fluids of different density, while in nature, as for gust fronts, there is a continuously decreasing density stratification with height.

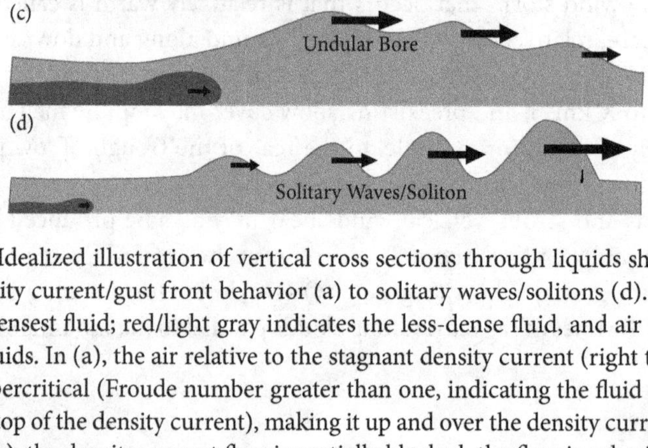

FIGURE 2.17 Idealized illustration of vertical cross sections through liquids showing a transition from density current/gust front behavior (a) to solitary waves/solitons (d). Blue/dark gray indicates the densest fluid; red/light gray indicates the less-dense fluid, and air is on top of the two denser liquids. In (a), the air relative to the stagnant density current (right to left) is totally unblocked, supercritical (Froude number greater than one, indicating the fluid can be lifted all the way to the top of the density current), making it up and over the density current and flowing to its rear. In (b), the density current flow is partially blocked, the flow is subcritical, and some fluid is raised ahead of the density current as a non-undular bore. In (c), waves develop in the bore and propagate ahead of the density current as an undular bore; in (d), smaller scale solitary waves/solitons develop. (From Haghi et al., 2017. © 2017 AMS: https://www.ametsoc.org/ams/index.cfm/publications/ethical-guidelines-and-ams-policies/ams-copyright-policy/.)

3.3. Haboobs

Haboobs are wind storms associated with gust fronts that persist after a convective storm has dissipated. In the desert southwest, where the atmosphere is relatively dry, strong gust fronts may be produced in convective storms (Idso, Ingram, & Pritchard, 1972) owing to the high cloud bases associated with a deep, well-mixed boundary layer. The leading edge of the gust front may loft dust, producing dust storms with very reduced visibility, driven by very strong surface winds; wind speeds as high as 36 m s^{-1} have been observed. Haboobs are in a sense nonconvective phenomena but produced as a result of earlier convection. Haboobs could also trigger bores at night.

3.4. Katabatic Winds

As a result of the diurnal variation in heating over a sloping surface such as along a mountain or even along a meteor crater or the edge of a glacier, there is a diurnal oscillation in the wind. The wind is driven baroclinically by diabatic processes such as heating during the day and radiational cooling at night. Air flows in the upslope direction during the day and "drains" in the downslope direction at night (Figure 2.18). The drainage wind at night is called a *katabatic* wind (the upslope flow during the day is an *anabatic* wind). The katabatic wind may be strong and potentially damaging when the slope of the ground is steep or the temperature change (cooling) is relatively large.

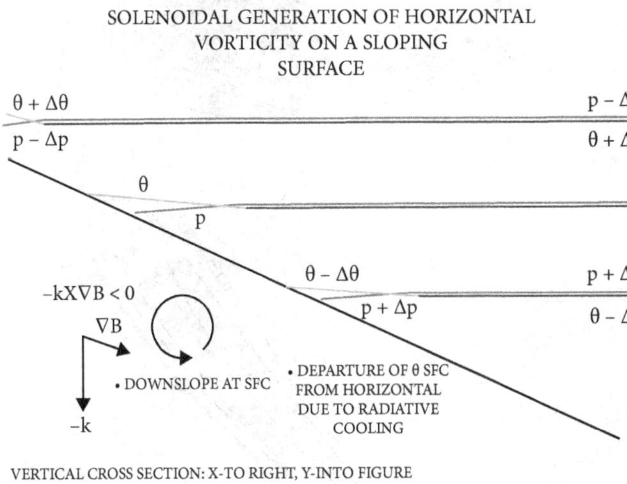

FIGURE 2.18 Dynamics of a katabatic wind driven by nocturnal radiational cooling on a sloping surface. Pressure surfaces are mainly horizontal, but surfaces of constant potential temperature bulge upward to the west as a result of the cooling. The gradient of buoyancy is therefore directed to the right and slightly downward, near the ground. The rate of generation of horizontal vorticity baroclinically is $-\mathbf{k} \times \nabla B$, which is out from the figure (about a counterclockwise direction). **k** is a unit vector pointing upward, B is the buoyancy, and ∇ is the gradient operator. (Image provided by the author.)

When radiation is the main mechanism for producing the negatively buoyant air on the sloping surface, the downslope drainage wind speeds are only about 1 m s^{-1} (Horst & Doran, 1986). However, the katabatic winds may be enhanced when evaporative cooling from precipitation falls into unsaturated air at higher elevation and produces much more negative buoyancy. Also, when a katabatic drainage wind channels into a canyon, the flow is enhanced. The last two mechanisms might be responsible for potentially marginal damaging winds.

3.5. Sting Jets

Intense, potentially damaging winds are sometimes observed within synoptic-scale cyclones in midlatitudes equatorward of the surface low center, particularly over the ocean. Some are referred to as *sting jets* and are found in the region of the "bent-back

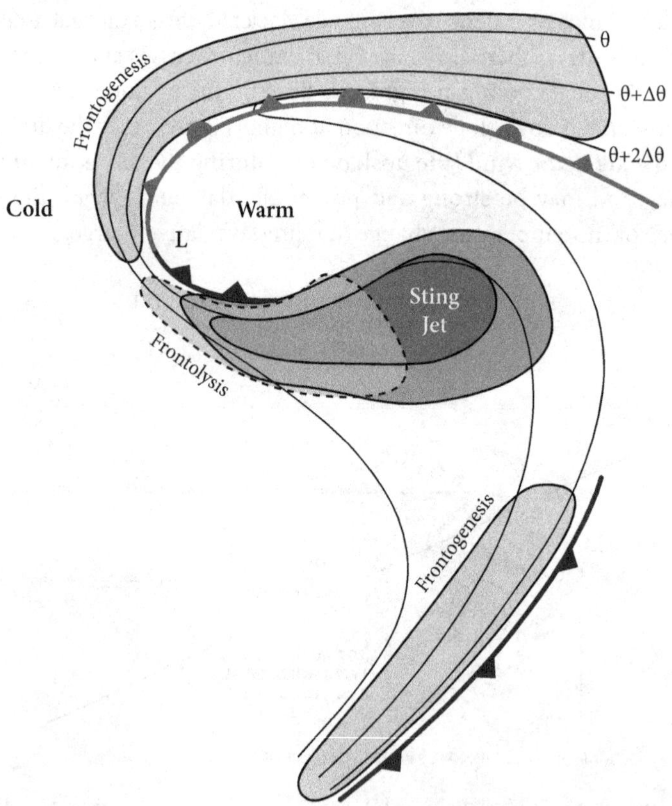

FIGURE 2.19 Conceptual model of a sting jet (orange/light gray shading) with respect to center of a cyclone (low-pressure area L), cold front (line with teeth), occluded front (line with alternating teeth and half circles), areas of frontogenesis and frontolysis (shaded in blue/dark gray), and potential temperature isotherms in the lower troposphere, about 925 hPa. In the area of frontolysis, the warmer air is sinking, and the cooler air is rising. (From Schultz & Sienkiewicz, 2013. © 2013 AMS: https://www.ametsoc.org/ams/index.cfm/publications/ethical-guidelines-and-ams-policies/ams-copyright-policy/.)

occluded front" (Figure 2.19) (Schultz & Sienkiewicz, 2013). Sting jets may last for hours and produce wind speeds as high as about 50 m s^{-1} over a mesoscale area.

It is thought that sting jets are produced when higher momentum air from aloft (associated with an upper-level jet) is transported downward by a frontolytic vertical circulation on the warm side of a front; it is thought that the transition from frontogenesis along the occluded front to frontolysis along the bent-back occlusion is of vital importance. Downward mixing of higher momentum air aloft is facilitated by heating from the ocean surface and cold advection over the relatively warm ocean, which leads to a nearly dry-adiabatic lapse rate.

3.6. Dryline Bulges

Drylines are boundaries between relatively warm, dry air of continental origin and relatively cool, moist air of maritime origin, over sloping terrain. The air mass to the

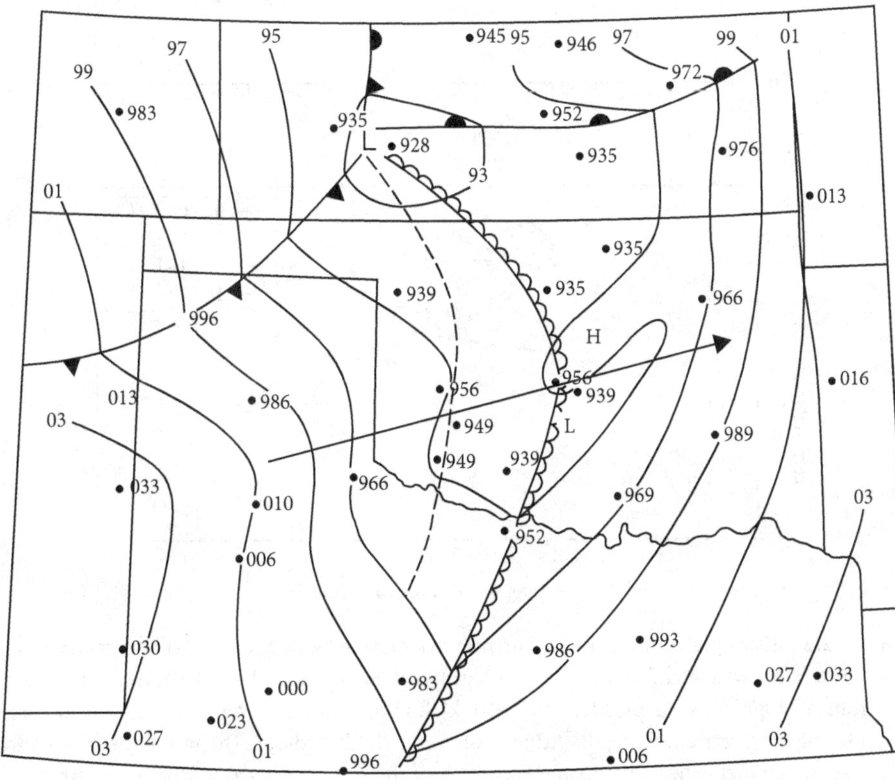

FIGURE 2.20 Example of a dryline bulge in Oklahoma, 15:00 Central Daylight Time, June 8, 1974; dryline (scalloped line); cold front (line with teeth); warm front (line with half circles); arrow indicates approximate location of a jet at 700 hPa and the dryline bulge. Isobars of altimeter settings (solid lines, 9XX or 10XX hPa, where XX indicates the numbers shown in the figure). (Adapted from McCarthy & Koch, 1982. © 1982 AMS: https://www.ametsoc.org/ams/index.cfm/publications/ethical-guidelines-and-ams-policies/ams-copyright-policy/.)

west of the dryline is characterized by a deep, dry-adiabatic layer. Dryline bulges occur when a jet aloft is present west of the dryline, so that a zone of higher momentum air from aloft is mixed down to the surface (Figure 2.20) (McCarthy & Koch, 1982). Along a dryline bulge, where the dryline is advected along the direction of the jet most rapidly, surface wind speeds may be high enough to create a dust storm and inflict damage.

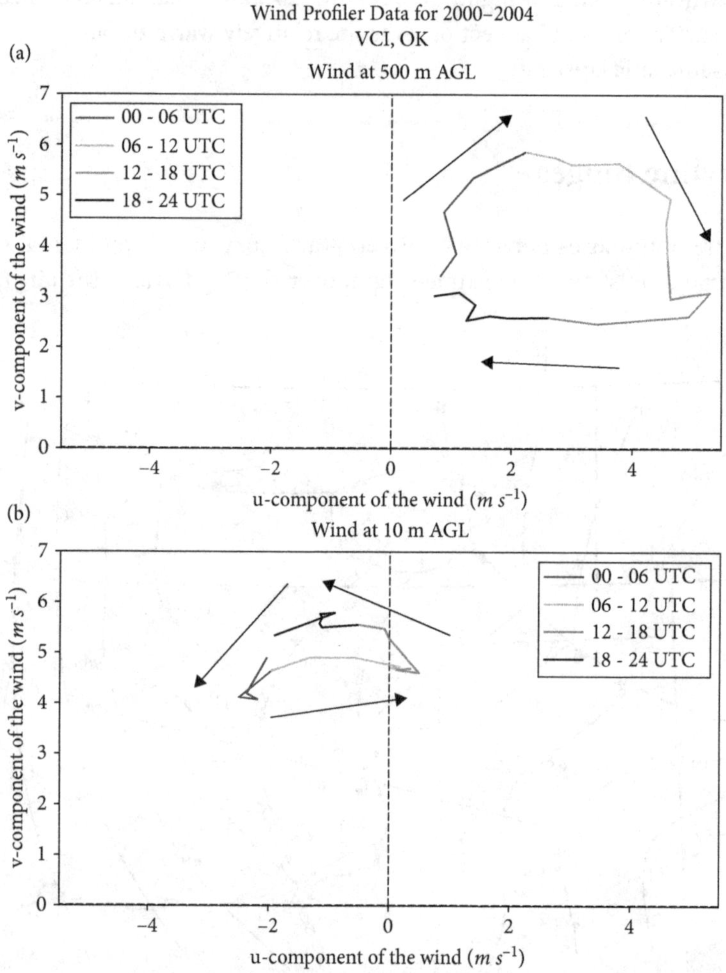

FIGURE 2.21 Example of a clockwise-turning hodograph with time at (a) 500 m AGL and (b) a counterclockwise turning with time hodograph at anemometer level, 10 m AGL; from wind data from a Doppler wind profiler at Vici, Oklahoma, averaged over the years 2000–2004. Owing to the long-term average, disturbances have been filtered out. The tail of the mean wind vector for each panel is near the geometric center of the hodographs as a function of time. The segments of the hodographs are color coded by time of day. The red/light gray lines, for example, are from early evening until about midnight. The arrows indicate the sense of rotation of the perturbation wind with time. (From Bluestein et al., 2018. © 2018 AMS: https://www.ametsoc.org/ams/index.cfm/publications/ethical-guidelines-and-ams-policies/ams-copyright-policy/.)

3.7. The Nocturnal Low-Level Jet

The nocturnal low-level jet (NLLJ) is a relative maximum in wind in the boundary layer, typically found during the warm season late at night or early in the morning, in the plains of the United States. It is usually a southerly or southwesterly jet centered on about 500 m AGL. It may be as strong as 25–30 m s^{-1} or more.

It is the result of both the diurnal variation of heating along the gently sloping surface of the plains (which slopes upward to the west) and the diurnal variation of vertical mixing of momentum as the ground is heated by the sun during the day (when vertical mixing is allowed) and cooled at night owing to outgoing radiation (when vertical mixing is inhibited) (Shapiro, Fedorovich, & Rahimi, 2016). When the synoptic-scale wind is from the south, the perturbation wind (the deviation of the wind from its diurnal average) turns in a clockwise manner with time above 50–100 m (Figure 2.21), but turns in a counterclockwise manner with time at 10 m AGL (anemometer level) (Bluestein, Romine, Rotunno, Reif, & Weiss, 2018). The clockwise turning of the wind with time above anemometer level leads to a wind speed maximum there late at night and early in the morning.

When the NLLJ is strongest, vertical mixing is least effective, so the strongest winds remain elevated. Potentially damaging winds associated with the NLLJ are therefore most likely for high-rise buildings over several hundred meters in height.

Notes

1. For a representation of this, see Figure 15a in Weisman (1993) (available at https://journals.ametsoc.org/doi/pdf/10.1175/1520-0469(1993)050%3c0645:TGOSLL%3e2.0.CO;2). The three-dimensional (3D) schematic shows how midlevel, counterrotating vortices (cyclonic, "+" to the south; anticyclonic, "–" to the north) are produced by tilting of horizontal vorticity in the environment (solid black lines, about which curved ribbon arrows are pointing; vortex lines point to the north) by an updraft (solid, white streamline) below midlevels. A profile of unidirectional wind in the environment is shown at the far upper left; easterly flow, weakening with height, then becomes westerly and increases in speed up to midlevels. The westerly (unidirectional) vertical wind shear is associated with the horizontal vorticity, which points toward the north. The cyclonic vortex labeled + is the mesocyclone. Updrafts are driven underneath both cyclonic and anticyclonic (–) vortices by dynamically driven pressure-gradient forces, and a precipitation-loaded downdraft forms in between the vortices, resulting in the propagation of the updrafts normal to the vertical shear. When the vertical shear is not unidirectional, one of the vortices is favored over the other owing to effects not explained by this illustration.
2. In this case, the CAPE may have been higher owing to the intrusion of dry air aloft.
3. For a representation, see Figure 2.2 in Shabbott and Markowski (2006) (available at https://journals.ametsoc.org/doi/pdf/10.1175/MWR3131.1). The schematic shows how a low-level mesocyclone is produced from baroclinically produced horizontal vorticity near the warm–cool boundary produced at the interface between the evaporatively cooled air underneath the forward-flank downdraft (FFD) and the ambient, warmer air to the south

and southwest. In this example, environmental low-level vorticity is associated with westerly shear. Horizontal vorticity is produced pointing toward the west at the interface between the cool and warm air. Air having this property flows toward the updraft, where it is tilted upward to produce a mesocyclone. The rear-flank gust front is indicated southwest of the low-level mesocyclone.

4. In the original article, the formula for *U* was given in terms of density, not potential temperature.
5. Actually it was called just "rear-inflow" originally.

References

Adlerman, E. J., Droegemeier, K. K., & Davies-Jones, R. (1999). A numerical simulation of cyclic mesocyclogenesis. *Journal of Atmospheric Sciences, 56*, 2045–2069.

Atkins, N. T., & Wakimoto, R. M. (1991). Wet microburst activity over the southeastern United States: Implications for forecasting. *Weather and Forecasting, 6*, 470–482.

Baker, A. K., Parker, M. D., & Eastin, M. D. (2009). Environmental ingredients for supercells and tornadoes within Hurricane Ivan. *Weather and Forecasting, 24*, 223–244.

Benjamin, T. B. (1962). Theory of vortex breakdown phenomena. *Journal of Fluid Mechanics, 14*, 593–629.

Benjamin, T. B. (1968). Gravity currents and related phenomena. *Journal of Fluid Mechanics, 31*, 209–248.

Bernstein, B. C., & Johnson, R. H. (1994). A dual-Doppler radar study of an OK PRE-STORM heat burst event. *Monthly Weather Review, 122*, 259–273.

Bluestein, H. B. (1992). *Synoptic-dynamic meteorology in midlatitudes. Vol. I: Principles of kinematics and dynamics.* New York, NY: Oxford University Press.

Bluestein, H. B. (1993). *Synoptic-dynamic meteorology in midlatitudes. Vol. 2: Observations and theory of weather systems.* New York, NY: Oxford University Press.

Bluestein, H. B. (2013). *Severe convective storms and tornadoes: Observations and dynamics.* Cham, Switzerland: Springer/Praxis.

Bluestein, H. B., French, M. M., Snyder J. C., & Houser, J. B. (2016). Doppler radar observations of anticyclonic tornadoes in cyclonically rotating, right-moving supercells. *Monthly Weather Review, 144*, 1591–1616.

Bluestein, H. B., Romine, G. S., Rotunno, R., Reif, D. W., & Weiss, C. C. (2018). On the anomalous counterclockwise turning of the surface wind with time in the plains of the United States. *Monthly Weather Review, 146*, 467–484.

Bluestein, H. B., & Snyder, J. C. (2015). An observational study of the effects of dry air produced in dissipating convective storms on the predictability of severe weather. *Weather and Forecasting, 30*, 79–113.

Bluestein, H. B., Weiss, C. C., & Pazmany, A. L. (2003). Mobile Doppler radar observations of a tornado in a supercell near Bassett, Nebraska, on 5 June 1999. Part I: Tornadogenesis. *Monthly Weather Review, 131*, 2954–2967.

Bluestein, H. B., Weiss C. C., & Pazmany, A. L. (2004). Doppler radar observations of dust devils in Texas. *Monthly Weather Review, 132*, 209–224.

Brooks, H. E., Doswell, C. A., III, & Kay, M. P. (2003). Climatological estimates of local daily tornado probability for the United States. *Weather and Forecasting, 18*, 626–640.

Brown, R. A., Lemon, L. R., & Burgess, D. W. (1978). Tornado detection by pulsed Doppler radar. *Monthly Weather Review, 1*, 29–38.

Brown, R. A., & Wood, V. T. (2012). The tornadic vortex signature: An update. *Weather and Forecasting, 27*, 525–530.

Charba, J. (1974). Application of gravity current model to analysis of a squall-line gust front. *Monthly Weather Review, 102*, 140–156.

Corfidi, S. F., Coniglio, M. C., Cohen, A. E., & Mead, C. M. (2016). A proposed revision to the definition of "derecho." *Bulletin of the American Meteorological Society, 97*, 935–949.

Davies-Jones, R. P. (1986). Tornado dynamics. In E. Kessler (Ed.), *Thunderstorm morphology and dynamics* (pp. 197–236). Norman, OK: University of Oklahoma Press.

Davis, J. M., & Parker, M. D. (2014). Radar climatology of tornadic and nontornadic vortices in high-shear, low-CAPE environments of the mid-Atlantic and southeastern United States. *Weather and Forecasting, 29*, 828–853.

Donaldson, R. J., Jr. (1970). Vortex signature recognition by a Doppler radar. *Journal of Applied Meteorology, 9*, 661–670.

Doswell, C. A., III, Brooks, H. E., & Kay, M. P. (2005). Climatological estimates of daily local nontornadic severe thunderstorm probability for the United States. *Weather and Forecasting, 20*, 577–595.

Doyle, J. D., & Durran, D. R. (2002). The dynamics of mountain-wave-induced rotors. *Journal of the Atmospheric Sciences, 59*, 186-201.

Doyle, J. D., & Durran, D. R. (2004). Recent developments in the theory of atmospheric rotors. *Bulletin of the American Meteorological Society, 85*, 337–342.

Droegemeier, K. K., & Wilhelmson, R. B. (1987). Numerical simulation of thunderstorm outflow dynamics. Part I: Outflow sensitivity experiments and turbulence dynamics. *Journal of Atmospheric Sciences, 44*, 1180–1210.

Durran, D. R. (1990). Mountain waves and downslope winds. In W. Blumen (Ed.), *Atmospheric processes over complex terrain* (Meteorological Monographs, 23, pp. 59–83). Boston, MA: American Meteorological Society.

Edwards, R., LaDue, J. G., Ferree, J. T., Scharfenberg, K., Maier, C., & Coulbourne, W. L. (2013). Tornado intensity estimation. *Bulletin of the American Meteorological Society, 94*, 641–653.

Fiedler, B. H., & Rotunno, R. (1986). A theory for the maximum windspeeds in tornado-like vortices. *Journal of the Atmospheric Sciences, 43*, 2328–2340.

Fujita, T. T. (1981). Tornadoes and downbursts in the context of generalized planetary scales. *Journal of Atmospheric Sciences, 38*, 1511–1534.

Golden, J. H. (1971). Waterspouts and tornadoes over South Florida. *Monthly Weather Review, 99*, 146–154.

Haghi, K. R., Parsons, D. B., & Shapiro, A. (2017). Bores observed during IHOP_2002: The relationship of bores to the nocturnal environment. *Monthly Weather Review, 145*, 3929–3946.

Horst, T. W., & Doran, J. C. (1986). Nocturnal drainage flows on simple slopes. *Boundary-Layer Meteorology, 34*, 263–286.

Houston, A. L., & Wilhelmson, R. B. (2007). Observational analysis of the 27 May 1997 central Texas tornadic event. Part II: Tornadoes. *Monthly Weather Review, 135*, 727–735.

Houze, R. A., Jr., Rutledge, S. A., Biggerstaff, M. I., & Smull, B. F. (1989). Interpretation of Doppler weather radar displays of midlatitude mesoscale convective systems. *Bulletin of the American Meteorological Society, 70*, 608–619.

Idso, S. B., Ingram, R. S., & Pritchard, J. M. (1972). An American haboob. *Bulletin of the American Meteorological Society, 53*, 930–935.

Johns, R. H., & Hirt, W. D. (1987). Derechos: Widespread convectively induced windstorms. *Weather and Forecasting, 2*, 32–49.

Kanak, K. M., Lilly, D. K., & Snow, J. T. (2000). The formation of vertical vortices in the convective boundary layer. *Quarterly Journal of the Royal Meteorological Society, 126*, 2789–2810.

Karstens, C. D., Gallus, W. A., Jr., Lee, B. D., & Finley, C. A. (2013). Analysis of tornado-induced tree fall using aerial photography form the Joplin, Missouri, and Tuscaloosa–Birmingham, Alabama, tornadoes of 2011. *Journal of Applied Meteorology and Climatology, 52*, 1049–1068.

Kennedy, P. C., Wescott, N. E., & Scott, R. W. (1993). Single-Doppler radar observations of a mini-supercell tornadic thunderstorm. *Monthly Weather Review, 121*, 1860–1870.

Klemp, J. B., & Lilly, D. K. (1975). The dynamics of wave-induced downslope winds. *Journal of Atmospheric Sciences, 32*, 320–339.

Knupp, K. R. (2006). Observational analysis of a gust front to bore to solitary wave transition within an evolving nocturnal boundary layer. *Journal of Atmospheric Sciences, 63*, 2016–2035.

Kosiba, K. A., & Wurman, J. (2013). The three-dimensional structure and evolution of a tornado boundary layer. *Weather and Forecasting, 28*, 1552–1561.

Lee, B. D., & Wilhelmson, R. B. (1997a). The numerical simulation of non-supercell tornadogenesis. Part I: Initiation and evolution of pretornadic mesocyclone circulations along a dry outflow boundary. *Journal of Atmospheric Sciences, 54*, 32–60.

Lee, B. D., & Wilhelmson, R. B. (1997b). The numerical simulation of nonsupercell tornadogenesis. Part II: Evolution of a family of tornadoes along a weak outflow boundary. *Journal of Atmospheric Sciences, 54*, 2387–2415.

Mahale, V. N., Bortzge, J. A., & Bluestein, H. B. (2012). An analysis of vortices embedded within a quasi-linear convective system using X-band polarimetric radar. *Weather and Forecasting, 27*, 1520–1537.

Markowski, P. M., & Richardson, Y. P. (2014). The influence of environmental low-level shear and cold pools on tornadogenesis: Insights from idealized simulations. *Journal of Atmospheric Sciences, 71*, 243–275.

Maxworthy, T. (1973). A vorticity source for large-scale dust devils and other comments on naturally occurring columnar vortices. *Journal of Atmospheric Sciences, 30*, 1717–1722.

McCarthy, J., & Koch, S. E. (1982). The evolution of an Oklahoma dryline. Part I: A meso- and subsynoptic-scale analysis. *Journal of Atmospheric Sciences, 39*, 225–236.

McCaul, E. W., & Weisman, M. L. (2001). The sensitivity of simulated supercell structure and intensity to variations in the shapes of environmental buoyancy and shear profiles. *Monthly Weather Review, 129*, 664–687.

Parsons, D. B., & Kropfli, R. A. (1990). Dynamics and fine structure of a microburst. *Journal of Atmospheric Sciences, 47*, 1674–1692.

Przybylinski, R. W. (1995). The bow echo: Observations, numerical simulations, and severe weather detection methods. *Weather and Forecasting, 10*, 203–218.

Rasmussen, E. N., & Blanchard, D. O. (1998). A baseline climatology of sounding-derived supercell and tornado forecast parameters. *Weather and Forecasting, 13*, 1148–1164.

Rottman, J. W., & Simpson, J. E. (1989). The formation of internal bores in the atmosphere: A laboratory model. *Quarterly Journal of the Royal Meteorological Society, 115*, 941–963.

Rotunno, R. (1984). An investigation of a three-dimensional axisymmetric vortex. *Journal of Atmospheric Sciences, 41*, 283–298.

Rotunno, R. (2013). The fluid dynamics of tornadoes. *Annual Review of Fluid Mechanics, 45*, 59–84.

Rotunno, R., & Klemp, J. B. (1982). The influence of the shear-induced pressure gradient on thunderstorm motion. *Monthly Weather Review, 110*, 136–151.

Rotunno, R., & Klemp, J. B. (1985). On the rotation and propagation of simulated supercell thunderstorms. *Journal of Atmospheric Sciences, 42*, 271–292.

Rotunno, R., Klemp, J. B., & Weisman, M. L. (1988). A theory for strong, long-lived squall lines. *Journal of Atmospheric Sciences, 45*, 463–485.

Ryan, J. A., & Carroll, J. J. (1970). Dust devil wind velocities: Mature state. *Journal of Geophysical Research, 75*, 531–541.

Schenkman, A. D., Xue, M., & Shapiro, A. (2012). Tornadogenesis in a simulated mesovortex within a mesoscale convective system. *Journal of Atmospheric Sciences, 69*, 3372–3390.

Schultz, D. M., & Sienkiewicz, J. M. (2013). Using frontogenesis to identify sting jets in extratropical cyclones. *Weather and Forecasting, 28*, 603–613.

Shabbott, C. J., & Markowski, P. M. (2006). Surface in situ observations within the outflow of forward-flank downdrafts of supercell thunderstorms. *Monthly Weather Review, 134*, 1422–1441.

Shapiro, A., Fedorovich, E., & Rahimi, S. (2016). A unified theory for the Great Plains nocturnal low-level jet. *Journal of Atmospheric Sciences, 73*, 3037–3057.

Sherburn, K. D., & Parker, M. D. (2014). Climatology and ingredients of significant severe convection in high-shear, low-CAPE environments. *Weather and Forecasting, 29*, 854–876.

Simpson, J., Roff, G., Morton, B. R., Labas, K., Dietachmayer, G., McCumber, M., & Penc, R. (1991). A Great Salt Lake waterspout. *Monthly Weather Review, 119*, 2741–2770.

Sinclair, P. C. (1969). General characteristics of dust devils. *Journal of Applied Meteorology, 8*, 32–45.

Sinclair, P. C. (1973). The lower structure of dust devils. *Journal of Atmospheric Sciences, 30*, 1599–1619.

Smull, B. F., & Houze, R. A., Jr. (1987). Rear inflow in squall lines with trailing stratiform regions. *Monthly Weather Review, 115*, 2869–2889.

Snyder, J. C., & Bluestein, H. B. (2014). Some considerations for the use of high-resolution mobile radar data in tornado intensity determination. *Weather and Forecasting, 29*, 799–827.

Tanamachi, R. L., Bluestein, H. B., Lee, W.-C., Bell, M., & Pazmany, A. (2007). Ground-based velocity track display (GBVTD) analysis of W-band Doppler radar data in a tornado near Stockton, Kansas, on 15 May 1999. *Monthly Weather Review, 135*, 783–800.

Thompson, R. L., Mead, C. M., & Edwards, R. (2007). Effective storm-relative helicity and bulk shear in supercell thunderstorm environments. *Weather and Forecasting, 22*, 102–115.

Trapp, R. J., Stumpf, G. J., & Manross, K. L. (2005). A reassessment of the percentage of tornadic mesocyclones. *Weather and Forecasting, 29*, 680–687.

Trapp, R. J., Tessendorf, S. A., Godfrey, E. S., & Brooks, H. E. (2005). Tornadoes from squall lines and bow echoes. Part I: Climatological distribution. *Weather and Forecasting, 20*, 23–34.

Trapp, R. J., & Weisman, M. L. (2003). Low-level mesovortices within squall lines and bow echoes. Part II: Their genesis and implications. *Monthly Weather Review, 131*, 2804–2823.

Van den Broeke, M. S., & Van den Broeke, C. A. (2015). Polarimetric radar observations from a waterspout-producing thunderstorm. *Weather and Forecasting, 30*, 329–348.

Wakimoto, R. M. (1982). The life cycle of thunderstorm gust fronts as viewed with Doppler radar and rawinsonde data. *Monthly Weather Review, 110*, 1060–1082.

Wakimoto, R. M. (1985). Forecasting dry microburst activity over the High Plains. *Monthly Weather Review, 113*, 1131–1143.

Wakimoto, R. M., Cai, H., & Murphey, H. V. (2004). The Superior, Nebraska, supercell during BAMEX. *Bulletin of the American Meteorological Society, 85*, 1095–1106.
Wakimoto, R. M., Liu, C., & Cai, H. (1998). The Garden City, Kansas, storm during VORTEX95. Part I: Overview of the storm's life cycle and mesocyclogenesis. *Monthly Weather Review, 126*, 372–392.
Wakimoto, R. M., & Wilson, J. W. (1989). Non-supercell tornadoes. *Monthly Weather Review, 117*, 1113–1140.
Ward, N. B. (1972). The exploration of certain features of tornado dynamics using a laboratory model. *Journal of Atmospheric Sciences*, 29, 1194–1204.
Weisman, M. L. (1992). The role of convectively generated rear-inflow jets in the evolution of long-lived mesoconvective systems. *Journal of Atmospheric Sciences*, 49, 1826–1847.
Weisman, M. L. (1993). The genesis of severe, long-lived bow echoes. *Journal of Atmospheric Sciences, 50*, 645–670.
Weisman, M. L., & Davis, C. A. (1998). Mechanisms for the generation of mesoscale vortices within quasi-linear convective systems. *Journal of Atmospheric Sciences*, 55, 2603–2622.
Weisman, M. L., & Klemp, J. B. (1984). The structure and classification of numerically simulated convective storms in directionally varying wind shears. *Monthly Weather Review, 112*, 2479–2498.
Wilczak, J. M., & Glendening, J. W. (1988). Observation and mixed-layer modeling of a terrain-induced mesoscale gyre: The Denver cyclone. *Monthly Weather Review, 114*, 270–284.
Wilson, T., & Rotunno, R. (1986). Numerical simulation of a laminar end-wall vortex and boundary layer. *Physics of Fluids*, 29, 3993–4005.
Wurman, J., Kosiba, K., & Robinson, P. (2013, June). In situ, Doppler radar, and video observations of the interior structure of a tornado and the wind–damage relationship. *Bulletin of the American Meteorological Society*, 835–846.

CHAPTER 3

ENGINEERING PERSPECTIVE OF TORNADOES

KISHOR C. MEHTA

1. Introduction

A tornado is a meteorological phenomenon that is randomly developed and has a short life. It is a rotating column of air which can achieve devastating wind speed inside its core. It forms on land and is part of some thunderstorms. The main concerns about this phenomenon are making accurate prediction about its generation, potential track, intensity in terms of damaging wind speeds, interaction with buildings and structures, and considerations of structural designs that can withstand tornado-force winds.

The first organized documentation of tornadoes started with Dr. Theodore Fujita, Professor of Meteorology at the University of Chicago, in 1970. He developed the Fujita F-scale to rate tornado intensity (Fujita, 1971). Subsequently, he developed the FPP scale to record the intensity, path length, and path width of each storm (Fujita 1973). Annually there are more than 1,000 tornado storms recorded in the contiguous United States, with numbers generally varying between 800 and 1,400.

The first official documentation of tornado damage to buildings and structures was conducted by the engineering faculty at Texas Tech University in the aftermath of the tornado of May 11, 1970, which impacted the city of Lubbock, Texas (Mehta et al. 1971). In that tornado 26 people lost their lives, more than 500 people were injured, and the storm destroyed or severely damaged roughly 250 businesses, 600 apartments, and 8,800 residences. The tornado winds also damaged a 20-story office building in the city (McDonald 1970).

Since the organized meteorological and engineering studies of 1970s, research and development in the understanding of tornadoes and their impact have progressed over the past five decades. One of the early comprehensive reports on the engineering perspective of tornadoes was published by the National Severe Storm Laboratory in 1977 (Minor et al. 1977). This 200-page report presents an in-depth understanding of the tornado phenomena and its interactions with the built environment from an engineering perspective. It considers the experiences of researchers undertaking damage investigations in the field in 36 events including some of the tornadoes that occurred in the super-outbreak of April 3–4, 1974 (Mehta et al. 1975). Future investigators pursuing research on the impact of tornadoes on buildings and structures will find these early reports beneficial.

This introductory chapter on tornadoes presents a brief discussion on the prediction of tornadoes, numerical and experimental research on tornadoes, assessment of tornado intensity in terms of wind speeds, tornado–structure interaction, and building design to withstand tornadoes. Several chapters in this handbook cover some of these topics in depth, and readers are referred to these chapters.

2. Prediction and Response

Tornadoes are formed from thunderstorms. However, the question remains why some thunderstorms form tornadoes and others do not. The chapter in this volume by Howard Bluestein has an in-depth discussion on tornadoes and their parent convective storms (thunderstorms). This scientific discussion is designed for researchers involved in pursuing atmospheric science research in tornado genesis.

The Storm Prediction Center, part of the National Oceanic and Atmospheric Administration (NOAA), issues forecasts for tornado watches based on meso-scale meteorological parameters. The main ingredients necessary for a tornado to form is a confluence of fronts of moist and warm air, dry air, and cool air in the atmosphere. Prediction of these meso-scale atmospheric parameters has reached a stage where a tornado watch over an area can be predicted several hours in advance. The issuance of *tornado watch* does not mean that there will be a tornado in the area, and it does not indicate where the tornado will be formed. A watch is issued to encompass a relatively large area of several kilometers in width and length. According to NOAA, "A Tornado Watch means conditions are favorable for tornadoes and severe thunderstorms in and close to the watch area. Persons in these areas should be on the lookout for threatening weather conditions and listen for later statements and possible warnings" (www.spc.noaa.gov/products/watch/2020/ww0137.html).

The local National Weather Service (NWS) office issues a *tornado warning*. Issuance of a warning means that a tornado has been sighted by observers or that a tornado is imminent in the signature of Doppler radar. Action recommended by NWS advisory is to (1) seek shelter in a pre-designated safe zone immediately, (2) (if driving) park the car and go outside, (3) keep a radio or TV tuned to a local station for updates, and (4) stay alert for sirens, but do not leave your safe zone or shelter (https://www.spc.noaa.gov/products/wwa/). One of the major questions that persists is to determine how people react to tornado warnings issued by the NWS.

On May 22, 2011, one of the most devastating tornadoes in US history killed 159 people and injured thousands of others in Joplin, Missouri. According to the National Weather Service (NWS), this storm was a warned disaster in that the residents of Joplin received advance notice and critical information about the tornado prior to its arrival. The city sirens alerted the residents of approaching tornado. Unfortunately, some residents for various reasons decided against protective action. In some instances this decision may have proved fatal (Kuligowski et al 2014).

This subject of population response to tornado warnings requires further research. It relates to the type and wording of the warnings and the psychology of those who do or do not take action. The subject is a complicated one and deserves interdisciplinary research in the future.

3. Tornado Hazard

It is important to understand tornado hazard in terms of probability of occurrence of the storm at a given location or in a given geographical area. Tornadoes cover a relatively small area because they are small and short-lived. From an engineering point of view, we want to know what wind speeds are likely to be experienced at a location. Basically, *tornado hazard* is the probability of wind speed occurrence at a certain location.

It is very difficult to measure wind speeds inside a tornado because of its random occurrence and short life. Attempts have been made to measure tornado wind speeds by placing hardened instruments in the path of tornadoes (Winn et al. 1999; Karsten et al. 2010; Wurman et al. 2013). Also, Doppler radar and photogrammetric analysis are used to assess wind speeds inside a tornado (Wakimoto et al. 2011). These attempts are isolated and do not provide wind speeds for all tornadoes. Dr. Ted Fujita developed F-scale to estimate wind speeds based on damage (Fujita 1971), and this scale is applicable to all tornadoes. Later, he developed FPP scale to estimate wind speed and area of coverage by a tornado based on visual inspection of damage. Using area of coverage and wind speeds in all tornadoes that occur, it is possible to generate tornado hazard models (ANSI/ANS-2.3–2011 2016; DOE-STD-1020 2016).

Details on the development of the Fujita F-scale are given in the chapter by Marshall and Womble in this handbook, so they are not repeated here. However, it was found during damage investigations of the super outbreak of tornadoes on April 3–4, 1974 and during subsequent damage investigations of other storms that the F-scale did not provide sufficient resolution of damages correlated with each F-scale rating and that wind speeds in the F3, F4, and F5 categories were too high (Mehta et al. 1976; Phan & Simiu 1998). Recognizing the need to address the limitations of the original F-scale,

Texas Tech University (TTU) Wind Science and Engineering (WISE) Research Center personnel proposed a project to examine the scale's limitations, revise and enhance the Fujita Scale, and attempt to gain a consensus from the meteorological and engineering communities. Reaching a consensus on the revision of the F-scale was important because the US NWS archives data on tornado events.

A forum of 23 persons representing meteorologists and engineers provided direction to develop the Enhanced Fujita EF-scale (McDonald & Mehta 2001). The EF-scale has 28 building and structure categories called *damage indicators* (DIs). Within each DI there are *degree of damage* (DOD) and estimated wind speeds (WISE 2006). The developed EF-scale is correlated with the original F-scale to preserve the data archived by the USNWS.

There is a paradigm shift from the F-scale rating process to assess the intensity of a tornado to the EF-scale rating. In the F-scale, the wind speed ranges were developed based on a line connecting wind speeds associated with Beaufort force 12 and Mach number 1. The line was divided into 12 segments on a logical but arbitrary basis (see the chapter by Marshall and Arn in this Handbook). The damage scenarios were fit into each F-scale category without any serious consideration of the engineering implications to buildings. Hence, the F-scale rating is assigned based on observed damage first, and then wind speed range is judged. In the EF-scale, wind speed is estimated based on the DOD for a specific DI. Thus, the EF-scale assigns estimated wind speed first and then determines the EF-scale rating. The second improvement in the EF-scale process is that there are 28 DIs identified. For each DI, there are 6–10 DODs. Thus, there are more than 200 levels of damages (DODs) with their associated estimated wind speeds, and so the accuracy of assigning the EF-scale is improved significantly. The EF-scale process has the potential of adding more DIs and DODs as more information becomes available. The third improvement in the EF-scale rating system is the lowering of wind speeds in EF4- and EF5-rated tornadoes. As early as the mid-1970s, it was judged that the wind speed ranges associated with F4 and F5 tornadoes may be too high (Mehta et al. 1976; Minor et al. 1977). There was little evidence in the tornado damage investigations that the recorded damage required wind speeds associated with the given F4 and F5 wind speed ranges. With this consideration of wind speed ranges for EF4 and EF5, the risk analysis for tornadic winds is likely to be more accurate. A caveat is necessary with regard to the EF scale. The wind speeds related to the EF scale are wind speeds estimated from damage; they are not measured. Continued research is needed to validate the wind speed values associated with each DI and DOD, and some of this necessary research is underway, as indicated in the references. If there is no DI in the path of a tornado, it is not possible to estimate the wind speed. Also, if the DI is totally destroyed, there is no way to estimate the maximum wind speed in the tornado at that location. Direct measurements using probes placed in the path of tornadoes and/or remote sensing using Doppler radar technology can provide definitive wind speed values. With this thought in mind, the upper boundary for the wind speed range for EF5 is not specified.

4. Tornado–Structure Interaction

The effects of tornadic winds on buildings and structures can be studied through physical experiments in the laboratory or through numerical simulation. It has not been possible to put an experimental building with sensors in the path of a tornado as yet, and this may not be possible for any foreseeable future.

Most of the wind–structure interaction data that are used for current design procedures are obtained in a boundary layer wind tunnel using synoptic wind. In recent years, there have been developments in experimental research in wind tunnels that generate non-synoptic winds. A new generation of physical simulators has emerged in the past decade. These new simulators have evolved from the research-oriented *tornado vortex chambers* (TVCs). These chambers use a multifan assembly to generate a tornado-like vortex. Details on these facilities and their capabilities are given in the chapter on novel physical simulators by Hangan, Refan, Sarkar, and Zuo in this handbook. A brief summary of facilities is given here.

Currently, there are three experimental facilities for tornado–building interaction in North America. These three are at Iowa State University under the direction of P. Sarkar (Haan et al. 2008), at Texas Tech University under the direction of D. Zuo and D. James (Tang et al. 2018), and the WindEEE Dome at Western University under the direction of H. Hangan (Refan et al. 2014). Each of these facilities is able to produce a tornado-like vortex that can be translated to represent a moving tornado. Vortices in these facilities are generated using a combination of fans at the top to create updraft and vanes and/or fans to produce rotating winds at the base (near the ground). The core diameters of the vortex range from 1 to 2 m. The building (or structure) model is subjected to winds from these vortices. Modeling parameters control the size of the model in each facility.

Numerical simulations of the tornado vortex have been done by several researchers in the meteorological community. The section "Simulations of Tornadoes and Their Parent Storms" in the Oxford Handbooks Online article "Tornadoes and Their Parent Convective Storms" by H. Bluestein has description of research in the numerical simulation of tornado. However, research on the numerical simulation of tornado–structure interaction is very limited.

The effects of tornadoes on buildings were examined at Lawrence Livermore Laboratory using computers to determine the wind forces on structures (Wilson 1977). A two-dimensional fluid dynamics computer code and a vortex model were used to determine external pressures on a building frame and its components. Following this preliminary work, several attempts have been made to develop tornado vortex–structure interactions to assess frame and component forces (McDonald & Selvam 1986; Selvam 2002; Selvam & Millett 2005). In these efforts, Selvam used large-eddy simulation for a tornado vortex and finite element simulation for a building or structure. Additional recent work on computational fluid dynamics for tornado–structure interaction is being pursued by Selvam and his colleagues (Gorecki & Selvam 2015; Strasser et al. 2016; Yousef et al. 2018).

Bitsuamlak at Western University has also pursued research in the fluid dynamics modeling of tornado vortex–structure interaction. His group reviewed the state of art and wrote a paper that helps us to understand the progress being made in tornado vortex–structure interaction (Dagnew & Bitsuamlak 2013). Progress has been made but there is a lot of room for investigators pursuing research in this field (Nasir et al. 2014; Nasir & Bitsuamlak 2016).

5. Design for Tornadoes

Structural design for tornadoes is a challenging subject because of our lack of a complete understanding of the tornado wind field (including turbulence) and tornado–structure interactions. The effects of tornado on a building or structure are multifaceted and include wind speed and direction varying with time, rapid atmospheric pressure change that occurs with the passage of a tornado, and flying debris that impacts the structure. Even with these multifaceted effects and lack of complete understanding of tornado effects, criteria have been developed for the design of nuclear facilities and people shelters.

For nuclear power plant design, the Nuclear Regulatory Commission (NRC) staff has determined that the design-basis tornado wind speeds for new reactors should correspond to the exceedance frequency of 10^{-7} per year (calculated as a best estimate) (NRC Regulatory Guide 1.76 2007). Based on this criteria and using tornado data records from between 1950 and 2003, the design basis wind speeds vary from 103 m/s (230 mph) and 72 m/s (160 mph) in three zones of the country (Ramsdell 2007). The NRC Regulatory Guide 1.76 provides the following parameters for tornadoes: (1) maximum wind speed, (2) translational speed, (3) maximum rotational speed, (4) radius of maximum rotational speed, (5) pressure drop, and (6) rate of pressure drop for design purposes. These design parameters are based on statistics from recorded tornadoes and engineering judgment. The NRC Guide 1.76 also provide a Design-Basis Tornado Missile Spectrum and Maximum Horizontal Speeds for each of the three regions. These design parameters suggest that there is room for research in developing design criteria for nuclear power plants that can resist tornadoes.

Guidelines for building occupant protection from tornadoes were published by the Defense Civil Preparedness Agency, US Department of Defense in 1975 (Mehta et al. 1975). These were interim guidelines for the design of interior areas of buildings such as hallways. The first edition of Federal Emergency Management Agency (FEMA) P-320, *Taking Shelter from the Storm: Building a Safe Room for Your Home or Small Business*, was issued in 1998 (FEMA 2014). More than 1 million hardcopies of this publication were distributed free of charge by FEMA. Design wind speeds varied from

130 mph to 250 mph in four zones of the country. These design wind speeds were developed from the recorded statistics of tornadoes and engineering judgment.

Since 1998, FEMA has upgraded FEMA 320 using the design criteria for occupant protection developed in a consensus standard of the International Code Council (ICC), with the support of the National Storm Shelter Association (NSSA) (ICC 500 2014). The latest version of FEMA 320 was published in 2014 (FEMA 2014). In addition, FEMA developed and published FEMA P-361, *Safe Rooms for Tornadoes and Hurricanes: Guidance for Community and Residential Safe Rooms,* in 2015 (FEMA 2015). These publications by FEMA are intended for use by homeowners, builders, contractors, and design professionals. Recognizing the uncertainties in understanding of tornado wind fields and tornado–structure interactions, the design criteria are by nature conservative. As the understanding of tornadoes improves, design for occupant protection will change with new criteria and with new materials.

6. Conclusion

This chapter provides brief description of prediction and response of tornado, tornado hazard, tornado-structure interaction and design for tornado. The chapter points out where additional research is needed. The reader is directed to other chapters in this handbook and the references cited for more details.

References

ANSI/ANS-2.3–2011. (reaffirmed in 2016). Estimating tornado, hurricane, and extreme straight-line wind characteristics at nuclear facility sites. La Grange Park, IL: American Nuclear Society.

Dagnew, A., & Bitsuamlak, G. T. (2013). Computational evaluation of wind loads on buildings: A review. *Wind and Structures, 16*(6), 629–660.

DOE Standard 1020. (2016). *Natural Phenomena Hazards Analysis and Design Criteria for DOE Facilities*. DOE-STD-1020-2016. Washington, DC: US Department of Energy.

FEMA P-320. (2014). Taking shelter from the storm: Building a safe room for your home or small business. https://www.fema.gov/fema-p-320-taking-shelter-storm-building-safe-room-your-home-or-small-business

FEMA P-361. (2015). Safe rooms for tornadoes and hurricanes: Guidance for community and residential safe rooms. https://www.fema.gov/media-library-data/1467990808182-0272256

Fujita, T. T. (1971). *Proposed characterization of tornadoes and hurricanes by area and intensity* (SMRP Research Rep. 91). Chicago, IL: University of Chicago Press.

Fujita T. T. (1973). Experimental classification of tornadoes in FPP scale (SMRP Rep. 98). Chicago, IL: University of Chicago Press.

Gorecki, P. M., & Selvam, R. P. (2015). Rankin combined vortex interaction with rectangular prism. *International Journal of Computational Fluid Dynamics, 29*, 120–132.

Haan Jr, F. L., Sarkar, P. P., & Gallus, W. A. (2008). Design, construction and performance of a large tornado simulator for wind engineering applications. *Engineering Structures, 30*, 1146–1159.

ICC 500/NSSA. (2014). Standard for the design and construction of storm shelters. https://codes.iccsafe.org/content/ICC5002014

Karstens, C. D., Samaras, T. M., Lee, B. D., Gallus, W. A., Jr., & Finley, C. A. (2010). Nearground pressure and wind measurements in tornadoes. *Monthly Weather Review, 138*, 2570–2588.

Kuligowski, E., Lombardo, F. T., Phan, L. T., Levitan, M. L., & Jorgensen, M. P. 2014. Final report, National Institute of Standards and Technology (NIST) technical investigation of the May 22, 2011, Tornado in Joplin, Missouri. http://nvlpubs.nist.gov/nistpubs/NCSTAR/NIST.NCSTAR.3.pdf

McDonald, J. R. (1970). *Structural response of a twenty-story building to the Lubbock Tornado.* Storm Research Report 01. Lubbock, TX: Texas Tech University.

McDonald, J. R., & Mehta, K. C. (2001). *Summary report of the Fujita Scale forum*. Lubbock, TX: Wind Science and Engineering, Texas Tech University Press.

McDonald, J. R., & Selvam, R. P. (1986). Application of boundary element method to tornado–structure interaction problem. In F. R. Payne et al. (Eds.), *Integral methods in science and engineering* (pp. 359–367). New York: Hemisphere Publishing Corporation.

Mehta, K. C., Minor, J. E., McDonald, J. R., & Sanger, A. (1971). *Response of structural systems to the Lubbock storm,* SRR 03. Lubbock, TX: Texas Tech University. http://lubbocktornado1970.com/pdf/ResponseStructuralSystems_TTUResearch.

Mehta, K. C., Minor, J. E., & McDonald, J. R. (1976). Wind speed analysis of April 3–4, 1974 tornadoes. *Journal of the Structural Division, ASCE, 102*(ST9), 1709.

Mehta, K. C., Minor, J. E., & McDonald, J. R. (1975). *Interim guidelines for building occupant protection from tornadoes and extreme winds.* DCPA. Department of Defense, TR-83A (reissued by FEMA). Washington, DC: FEMA.

Mehta, K. C., Minor, J. E., McDonald, J. R., Manning, B. R., Abernethy, J. J., & Koehler, J. F. (1975). *Engineering aspects of the tornadoes of April 3–4, 1974*. Report to the National Academy of Engineering, National Academy of Sciences, NAE. Washington, DC: National Academy of Sciences.

Minor, J. E., McDonald, J. R., & Mehta, K. C. (1977). *The tornado: An engineered-oriented perspective* (NOAA Tech. Memo. ERL-NSSL-82). Reprinted February 1993 Texas Scientific Services Division, Southern Region. Fort Worth, TX: US Department of Commerce.

Nasir, Z., & Bitsuamlak, G. T. (2016). *Tornado like vortex induced load on a bluff-body with and without openings.* Presented at the 8th International Colloquium on Bluff-Body Aerodynamics and Applications, Boston, USA.

Nasir, Z., Bitsuamlak, G., & Hangan, H. (2014). *Computational modeling of tornadic load on a bluff-body.* Presented at the 6th International Symposium on Computational Wind Engineering, Hamburg, Germany.

NRC Regulatory Guide 1.76. (2007). Design-basis tornado and tornado missiles for nuclear power plants. https://www.nrc.gov/docs/ML0703/ML070360253.pdf

Phan, L. T., & Simiu, E. (1998). *The Fujita tornado intensity scale: A critique based on observations of the Jarrell tornado of May 27, 1997* (NIST Tech. Note 1426). Gaithersburg, MD: US Department of Commerce.

Ramsdell, J. V., Jr. (2007). *Tornado climatology of the contiguous United States* (PNNL-15112 [rev.2], NUREG/CR-4461). Washington, DC: US Nuclear Regulatory Commission.

Refan, M., Hangan, H., & Wurman, J. (2014). Reproducing tornadoes in laboratory using proper scaling. *Journal of Wind Engineering and Industrial Aerodynamics, 135*, 136–148.

Selvam, R. P. (2002). Computer modeling of tornado forces on buildings (invited paper). In C. K. Choi et al. (Eds.), *Proceedings of the Second International Symposium on Advances in Wind and Structures* (pp. 105–114). Seoul: Techno-Press.

Selvam, R. P., & Millett, P. C. (2005). Large eddy simulation of the tornado–structure interaction to determine structural loadings. *Wind and Structures, 8*, 49–60.

Strasser, M. N., Yousef, M. A. A., & Selvam, R. P. (2016). Defining the vortex loading period and application to assess dynamic amplification of tornado-like wind loading. *Journal of Fluids and Structures, 63*, 188–209.

Tang, Z., Feng, C., Wu, L., Zuo, D., & James, D. L. (2018). Characteristics of tornado-like vortices simulated in a large-scale ward-type simulator. *Boundary-Layer Meteorology, 166*, 327–350.

Wakimoto, R. M., Atkins, N. T., & Wurman, J. (2011). The LaGrange tornado during VORTEX2. Part I: Photogrammetric analysis of the tornado combined with single-Doppler radar data. *Monthly Weather Review, 139*, 2233–2258.

Wilson, T. (1977). *Tornado structure interaction: A numerical simulation* (Report: UCRL-52207, Distribution Category: UC-11, 80). Lawrence Livermore Laboratory. Berkeley: University of California Press. https://inis.iaea.org/collection/NCLCollectionStore/_Public/09/352/9352625.pdf

Winn, W. P., Hunyady, S. J., & Aulich, G. D. (1999). Pressure at the ground within and near a large tornado. *Journal of Geophysical Research, 104*(D18), 22,067–22,082.

WISE. (2006). A recommendation for an enhanced Fujita scale (EFscale). Texas Tech University, National Wind Institute. https://www.spc.noaa.gov/faq/tornado/ef-ttu.pdf

Wurman, J., Kosiba, K., & Robinson, P. (2013). In situ, Doppler radar, and video observations of the interior structure of a tornado and the wind–damage relationship. *Bulletin of the American Meteorological Society, 94*, 835–846.

Yousef, M. A. A., Selvam, R. P., & Prakash, J. (2018). A comparison of the forces on dome and prism for straight and tornadic wind using CFD model. *Wind and Structures, 26*, 369–382.

CHAPTER 4

GROUNDBREAKING CONTRIBUTIONS TO DOWNBURST MONITORING, MODELING, AND DETECTION

MASSIMILIANO BURLANDO
AND DJORDJE ROMANIC

1. Introduction

THUNDERSTORMS are associated with a variety of wind systems and circulations such as downbursts, tornadoes, and mesocyclones, to name a few. The majority of thunderstorm winds can reach high wind speeds and pose a threat to human safety, aviation, and structures on the earth's surface (Solari, 2019). As demonstrated in this chapter, the research on thunderstorm winds and downbursts, in particular, was motivated by their hazardous nature to aircraft during the takeoff and landing stages of the flight.

As the title suggests, this chapter focuses on the review of the studies that in our opinion provided the initial and groundbreaking contributions to the discovery, measurements, modeling, and detection of thunderstorm downbursts. The studies presented hereafter are relevant today and are used as the benchmark for more recent works.

2. The Downburst Discovery

While, according to Gedzelman (1990), one of the first evidence of recognition and possible discovery of the downburst outflow dates back to the work of Leonardo da Vinci at the beginning of the sixteenth century in his drawing *A Deluge*, the first systematic and scientific research on thunderstorms and related wind fields took place immediately after World War II when four US government agencies funded the Thunderstorm Project, which was conducted in Florida during the summer of 1946 and in Ohio during the summer of 1947. Braham (1996, 1997) stated that the interest in thunderstorm research that led to the Thunderstorm Project arose mainly from an accident that occurred in 1943 to an American Airlines DC-3 aircraft, which crashed in a thunderstorm. The crash investigation carried out by the Civil Aeronautics Board in 1944 concluded that the accident was due to the loss of the aircraft's altitude in a downdraft, which was followed by strong tailwinds in storm conditions. This description of the flow field is surprisingly similar to the phenomenon later named "downburst" by Theodore Fujita.

During the Thunderstorm Project, meteorological measurements of thunderstorm winds and cloud dynamics were carried out using five airplanes, which made measurements inside the thunderstorms; a surface network of weather automated stations; and radiosonde and radar stations located along the perimeter of thunderstorm outflows to probe the three-dimensional (3D) structure of thunderstorms and track their movement. From a large number of detected and probed thunderstorms, 179 storms were selected for detailed study, which led to the first physical model of the so-called air mass thunderstorm. The life cycle of this type of thunderstorm consists of three different stages (Byers & Braham, 1949): (1) The first stage, dominated by an updraft, with the maximum measured vertical speed of 26 m s^{-1}; (2) the mature stage, characterized by the development of a downdraft nearby the updraft, with the maximum measured descending speed of 17 m s^{-1}; and (3) the dissipating stage, during which only a weak downdraft remains before the storm dissipates.

Despite Byers and Braham (1949) being mostly cited in the literature for this three-stage model of thunderstorm evolution in time, they also reported the first scientific description of a downburst outflow (Figure 4.1). In Section 3.2 of Byers and Braham (1949), the authors described the outflow field produced by the downdraft by saying that, "When the cold downdraft of a cell reaches the surface layers of the atmosphere, it spreads out in a fashion similar to that of a fluid jet striking a flat plate," (p. 53) also associating for the first time the thunderstorm outflow to the impinging jet model that would be widely used as the basic state to describe analytically such phenomenon, in both the meteorological and wind engineering field. They also anticipated the concept of stationary and traveling downbursts, reintroduced later by Fujita, by stating that, "In relatively slow-moving storms, the outflow is almost radial"; they continued, "In most instances, however, the outflow field is asymmetrical,

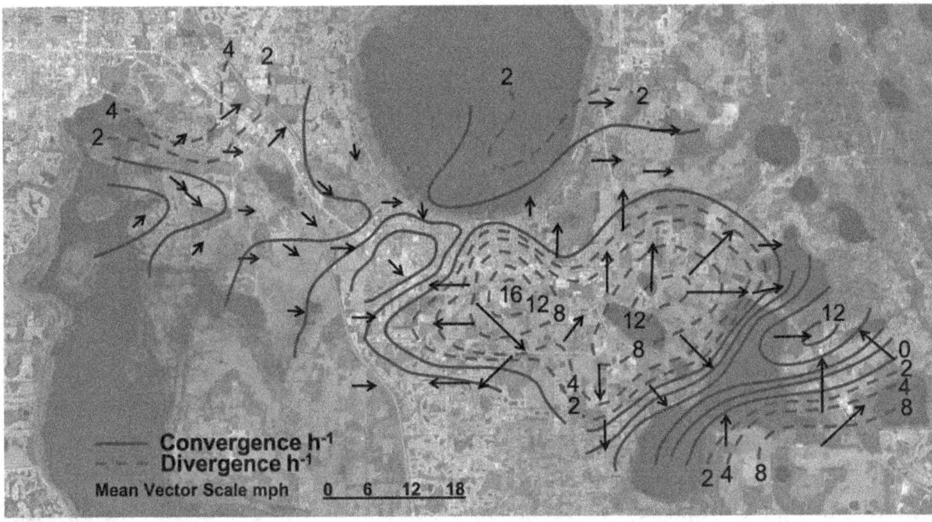

FIGURE 4.1 Illustration of the radial outflow produced from the divergence center at the surface due to a downdraft. This is probably the first example of downburst outflow described in the literature (reconstruction from Byers & Braham, 1949).

with wind speeds on the downstream side substantially higher than those on the upstream side." (p. 53)

Therefore, it is probably correct to claim that Byers and Braham (1949), analyzing the measurements collected during the Thunderstorm Project, created the conditions to understand properly the real nature of downbursts and gave important clues concerning downburst detection from radar and ground-based stations that would consequently become the next generation of low-level wind shear alert systems (LLWASs) (Goff, 1980; Wolfson, DiStefano, & Forman, 1987; Wolfson, DiStefano, & Fujita, 1985) and modern Doppler radars (Eilts, 1987; Roberts & Wilson, 1989) for operational nowcasting of aviation safety conditions.

In addition, in 1951 Horace R. Byers wrote a letter to Fujita, who was at that time at the Kyushu Institute of Technology in Japan, to involve him in a collaboration to study thunderstorms by making use of the US Air Force radar photographs. Byers had noticed the research carried out by Fujita (1951) and considered his contribution valuable and original, stating that "[Fujita] deserves credit for noting the importance of the thunderstorm downdraft and outflowing cold air" (Fujita, 1992). Therefore, Byers himself was the person who "fished out [Fujita] of postwar Japan to the University of Chicago," and he was called by Fujita his "fatherly mentor professor."

In the aftermath of the Thunderstorm Project, a new series of accidents involving commercial aircraft during takeoff and landing occurred in the United States and worldwide engendered some specific studies on thunderstorm-induced downburst outflows in the 1970s. Similar to the Thunderstorm Project in respect to employed measurement techniques and field campaigns, three new comprehensive and specific programs were carried out in the subsequent years: (1) the Northern Illinois Meteorological Research on Downburst (NIMROD) Project (Fujita, 1978, 1985); (2) the Joint Airport Weather Studies (JAWS) Project (Fujita, 1985; McCarthy, Wilson, & Fujita, 1982); and (3) the Microburst and Severe Thunderstorm (MIST) Project (Atkins & Wakimoto, 1991; Dodge, Arnold, Wilson, Evans, & Fujita, 1986), operated within the COHMEX (Cooperative Huntsville Meteorological Experiment; Dodge et al., 1986) together with the Satellite Precipitation and Cloud Experiment (SPACE) Project and the FAA-Lincoln Laboratories Operational Weather Studies (FLOWS) Project (Wolfson et al., 1985). The Cooperative Convective Precipitation Experiment (CCOPE) Project (Knight, 1982), as well as the Convective Initiation and Downburst Experiment (CINDE) Project (Wilson et al., 1988), are also worth mentioning for their contribution to this research field. Most of the observations and intuitions that were later used for the development of various downburst models have been based on the data gathered in these research projects.

Many occurrences of this phenomenon that was visually observed and measured by remote sensing techniques and mesometeorological stations documented the physical structure of downdrafts and their evolution into downbursts by producing divergent outflows. The analyses of the geometrical dimensions and kinematics of downdrafts and downbursts led to a classification of these phenomena into macrobursts and microbursts according to their damaging wind extension. The role of the superposition of the environmental flow onto a radial outburst was disclosed as an explanation for asymmetric damage patterns. Precipitation loading was discovered to be one of the triggering mechanisms of downdraft formation, while the drier environmental air entrainment and melting/evaporation of precipitation were considered more relevant for downdraft acceleration in wet and dry downbursts, respectively.

3. Physical Comprehension of a New Phenomenon

As reported in Fujita (1985), his first intuition for the structure of downbursts arose in 1974 after an aerial survey of the damage due to tornadoes in West Virginia in the United States. At that time, he was already considered as an expert on tornado damage detection and analysis, and his Fujita-Pearson wind scale was widely used for assessing tornado strength based on the amount of damage it caused (Abbey, 1976; Fujita, 1971, Fujita & Pearson 1973). In West Virginia, however, he noticed an unusual pattern of tree damage that, unlike the swirling pattern commonly observed in the wake of tornadoes, resembled a starburst (Figure 4.2). One year later, in 1975, he supported the idea that an intense starburst-like flow might have caused the Eastern 66 accident at John F. Kennedy (JFK) International Airport in New York City and coined the term *downburst* for such a phenomenon. The given term indicates a strong downdraft that induces an outburst of damaging winds on or near the ground.

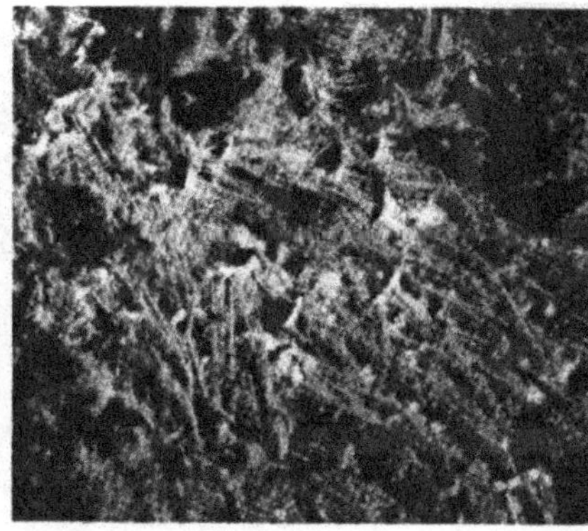

FIGURE 4.2 An example of the different footprints left by the passage of a tornado on a forest (left) (courtesy of Maximilian Conrad) and a starburst of uprooted trees (right) visible from aerial views (from Fujita, 1990).

It is impressive that this intuition was initially based only on the visual analysis of photographs of damage patterns without the support of direct measurements. The very first scientific methodology to observe downbursts was mostly qualitative, as described by Fujita (1978) in the *Manual of Downburst Identification for Project NIMROD*. Several downburst events related to aircraft accidents that happened in the period 1975–1977 and one in 1956 (Caracena, 1978; Fujita, 1976; Fujita & Byers, 1977; Fujita & Caracena, 1977), as well as various damage to ground structures and forests that occurred in 1970–1977, were documented and described using the photogrammetric surveys and aerial photographs. Afterward, the analyses performed by Fujita not only were qualitative but also quantitative. Interestingly, Fujita (1979) reported that the mesonetwork (Fujita, 1963) of Project NIMROD was placed to the west of Chicago, Illinois, because this area was cultivated with large fields of corn crops that were grown annually, and the corn stalks were used as low-cost indicators of downburst winds. Therefore, the project's operation period from May to June 1978 was accordingly chosen to be concurrent with the growing season of corn.

3.1. The NIMROD Project Outcomes

During the NIMROD Project, Fujita was a pioneer in using remote sensing techniques—that is, radar Doppler and satellite images—to study and measure downbursts in an attempt to relate such small-scale phenomena to the parent storm. According to the analysis of the accident at JFK Airport in New York City reported by Fujita (1976) and Fujita and Byers (1977), he related the formation of downbursts to "bow echoes" (named after Fujita 1978; see also Weisman, 2001) and, in particular, to the evolution of this reflectivity pattern into a spearhead-like shape where the bulge of the echo was regarded as a high-wind indicator. Later, Fujita (1979) gave somewhat of an explanation for this

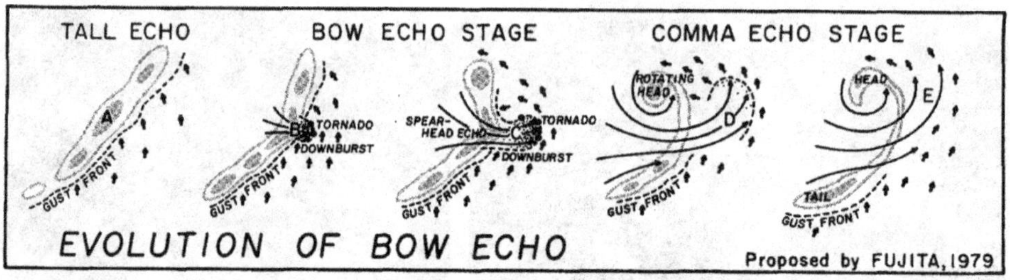

FIGURE 4.3 Evolution of a straight-line echo into a bow echo and then into a comma-shaped echo with a cyclonically rotating head. This evolution is induced by the downflow produced by a thunderstorm downburst (from Fujita, 1981b; initially proposed by Fujita, 1979).

phenomenon. Namely, he argued that the downburst itself is the cause of the bulge and not vice versa, as it was thought before (Hamilton, 1970).

The model to explain such a pattern had already been conceptually described by Fujita (1978) based on the northern Wisconsin downburst of July 4, 1977. In this event, the warming of the downburst wake, as observed in a series of infrared satellite images, was explained through the three-stage evolution of the wake flow behind the downburst (Figure 4.3). In the first stage—the so-called speed-up stage—the wake flow intensifies by sucking air from behind and from the anvil top. This descent of air from higher elevations results in its adiabatic warming. The second stage is the "peak speed" stage, during which the bow echo takes the shape of a spearhead. Finally, the "slow-down" stage follows when the wake flow decelerates, bringing the bulge upward of the anvil top. This process results in the consequent cooling of the downburst wake as seen from satellite images and the deformation of the spearhead echo into a comma echo.

Moreover, Fujita (1978) speculated that the formation of tornadoes is often associated with nearby downbursts. He argued that the presence of a downburst can influence and modify tornadoes' trajectory, as well as their flow structure, and sometimes cause their formation at the external edge of the outflow where the air may swirl cyclonically or anticyclonically. Fujita (1979) proposed a model of a supercell mesocyclone storm with alternating stages of tornado and downburst formation in which the hook echo at a low level and the high-level echo evolve according to a sequence of updrafts and downdrafts that trigger tornadoes and downbursts cyclically. The relationship between cool downdraft outflows and the formation of low-level vertical vorticity, which may in principle generate a tornado, was later investigated and better clarified by Rotunno and Klemp (1985), who showed that in supercell-like convection the latter derives from the baroclinic generation of horizontally oriented vorticity along the cold air front (see also Parker & Dahl, 2015).

Fujita (1978) also made a first attempt to classify downbursts according to their spatial scale and intensity. He achieved this through the definition of a subclass of downbursts called microbursts, which are defined as downbursts with path lengths less than 5.1 km (3.16 miles). This initial classification was refined later by Fujita (1981b) with five spatial scales—maso, meso, miso, moso, and musoscale (arranged in the order of the vowels)—were proposed to cover the range of airflow motions in the atmosphere. Each scale cascaded down two orders of magnitude from the upper limit represented by Earth's equator length (4×10^7 m), which was assumed to be the maximum dimension of the masoscale. On the other hand, the lowest scale of turbulence was proposed to be on the order of millimeters. According to this classification, downbursts were mainly associated with the mesoscale, whereas microbursts were categorized as misoscale phenomena. Therefore, the larger size of microbursts that was proposed earlier (i.e., 5.1 km) by Fujita (1978) was revised and reduced to 4 km.

This aspect was better clarified by Fujita and Wakimoto (1981), with an attempt to relate downburst-induced damage patterns to the physical phenomenon that caused it. In this article, a microburst at the misoscale is thought to be a structure embedded into a mesoscale downburst, and it often produces the smaller swaths of diverging winds, called "burst swaths." Further, downbursts were supposed to be embedded into downburst clusters, which in turn were thought to be a part of a family of clusters produced by a larger storm. This cascading concept, represented in Figure 4.4, is similar to a *matryoshka* spanning four orders of magnitude in space (see also the multiscale analysis of the weather conditions during the Pan American World Airways Flight 759 accident on July 9, 1982, reported by Caracena, Maddox, Purdom, Weaver, & Greene, 1983).

Due to their smaller scale and inherently more difficult detection as well as their often greater intensity than downbursts, Fujita considered microbursts to be more dangerous than the larger scale downbursts, particularly for aviation safety. For these reasons, Fujita focused his attention on this scale of downburst-like outflows in the following series of documents and papers. Fujita (1980, 1981a) presented some results from several microbursts detected during the NIMROD experiment and measured using a Doppler radar. In particular, Fujita (1981a) proposed a conceptual model of a microburst descending toward the ground (Figure 4.5), and, for the first time, he distinguished between a stationary and a traveling microburst, as well as between a warm- and a cold-sector microburst. Fujita (1981a) described how the microburst's inclination and the buoyancy effect at the ground might influence the outflow intensity and the height above the terrain at which the maximum horizontal (outflow) wind speed occurred (Figure 4.5).

3.2. The JAWS Project Outcomes

Wilson and Wakimoto (2001) described that the second fundamental project on downbursts, that is, the JAWS Project, was planned during the 19th Conference on Radar Meteorology in Miami, Florida, in 1980. The project was conducted in May–August 1982 in the Denver, Colorado, area slightly to the north of Stapleton International Airport. As described by McCarthy et al. (1982), the principal focus of the JAWS Project was convective microburst events, which require instability conditions of the lower atmosphere to allow the development of updrafts and downdrafts. During the NIMROD field

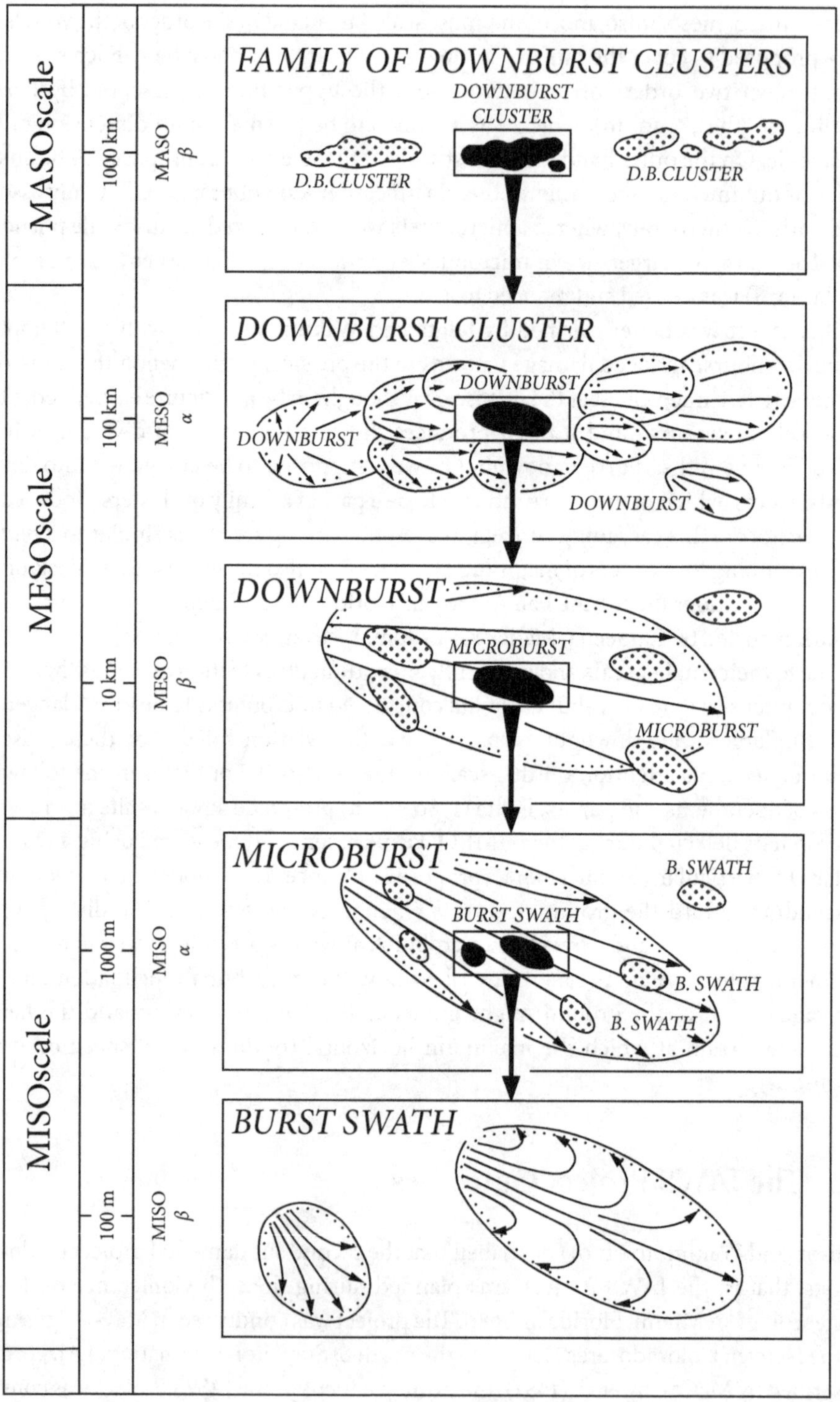

FIGURE 4.4 Matryoshka-like scheme showing the different scales of motion involved in the downburst formation, which highlights the structured nature and generation mechanism of the smaller scale damage patterns (i.e., microburst and burst swath) (from Fujita & Wakimoto, 1981). B. = burst; D.B. = downburst.

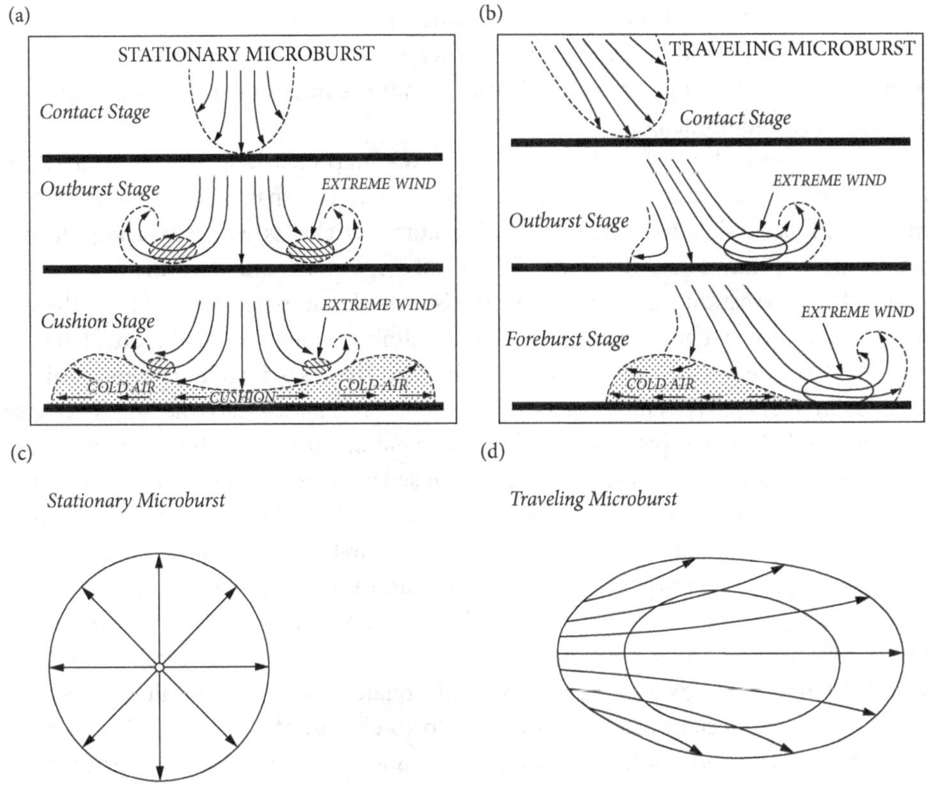

FIGURE 4.5 Scheme of stationary (a, c) and traveling (b, d) downburst models elaborated by Fujita (1981a, 1985) depicted through vertical and planar projections. The cylindrical symmetry of the cold air spreading out from the stationary downburst is lost when the downburst vertical axis gets inclined.

campaign, Doppler radars were spaced 60 km apart from each other in order to maximize the probability to detect downbursts. Due to the ground clutter and the curvature of Earth's surface, low-level winds over 30 km could not be retrieved by radars, which prevented the concurrent measurements of the wind fields close to the ground by more than one instrument. In JAWS, however, the focus was on capturing the entire 3D structure of a microburst during its entire life cycle. Therefore, the radars were installed with the spacing less than 30 km between two adjacent radars. Also, the project was planned to serve as a test bed for the evaluation of the performances of several real-time wind shear detection and warning systems for aviation traffic safety.

The first preliminary results of JAWS were presented by Fujita and Wakimoto (1983) at the 21st Conference on Radar Meteorology in Edmonton, Alberta, Canada. In that paper, downbursts were classified into macroburst (with the temporal resolution between 5 and 20 min and with the outflow larger than 4 km) and microburst (lasting 2–5 min and with the outflow below 4 km). Further, microbursts were categorized into wet and dry depending on the 0.01-inch (0.254-mm) threshold of surface precipitation between the onset and the end of the high winds. Moreover, the microbursts were also divided into surface and midair microbursts depending on their maximum descending height. This classification describes whether or not the microburst will be detected by ground-based anemometers.

The observational proof of dry microbursts was particularly important in aviation because these events are less detectable by Doppler radars than wet microbursts. Dry microbursts occur below apparently benign cumulus clouds which typically do not evolve into tall thunderstorm clouds. Streaks of virga shafts are often the only visible mark of these descending airflows. While the NIMROD campaign reported that the wet microbursts only spawned from deep cumulonimbus clouds, the subsequent results from JAWS showed that cumulus congestus showing streaks of virga can produce microbursts as well. These types of microbursts were observed underneath the long streaks of precipitation below the spreading anvil of a dissipating cumulonimbus. The dominant mechanism for their formation is the cooling caused by raindrop evaporation in the subsaturated environment. Figure 4.6 shows the radiosounding of the atmospheric thermodynamic conditions measured during the dry downburst that occurred in Elko, Nevada, with a deep layer of very dry air in the lower levels and a very high cloud base.

Importantly, the comparison between NIMROD and JAWS microbursts revealed two different mechanisms of downburst formation. The first mechanism, which seems to be typical for the US Midwest, is predominantly related to synoptic-scale forcing and linked to the presence of precursor thunderstorm clouds, often organized into larger scale structures like bow or hook echoes. The second mechanism of downburst formation that is mostly observed over the US High Plains is related to the local effects that play a major role in triggering convective motions, with diabatic solar heating being the most important contributing factor. Besides, the observations did not find a relationship between the microburst maximum wind speed and the amount of rain. This result

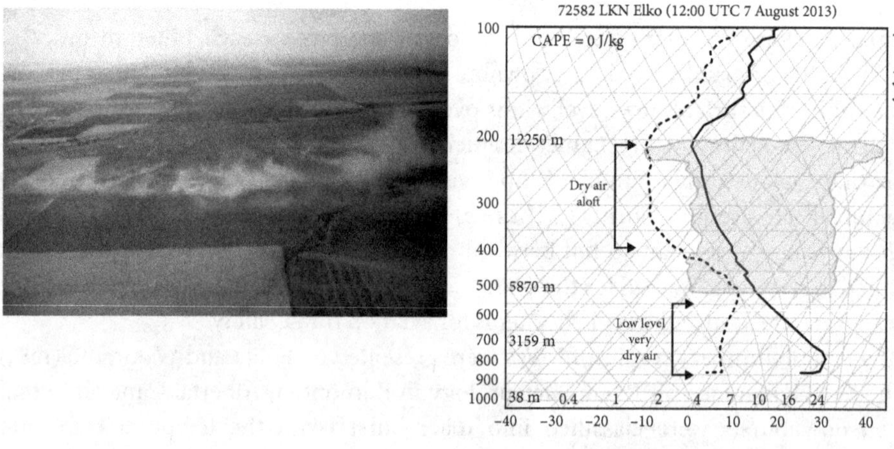

FIGURE 4.6 Photo of a ring of dust due to a dry downburst taken during the JAWS Project (left) and the typical atmospheric conditions during which a dry downburst occurs (right). Photograph from Fujita (1985).

suggests that wet and dry microbursts are equally dangerous for air traffic, and that precipitation drag cannot be considered one of the main factors for determining downdraft fall speeds. Taking all these findings into consideration, Fujita and Wakimoto (1983) concluded that probably a third project executed in an area of different climatic conditions concerning NIMROD and JAWS would be required to verify whether the results obtained to that point could be generalized to different geographical contexts.

3.3. Results from the MIST Project: The Thermodynamic Mechanisms Behind Wet Downburst Formation

After JAWS, the important advances to the body of knowledge of the downburst life cycle and evolution came from the MIST Project. As suggested by Fujita and Wakimoto (1983), this project was planned with a goal to measure downbursts in a humid environment that was different from those of NIMROD and JAWS. The selected location of the field campaign was an area nearby Huntsville, Alabama, during the summer of 1986. Although that summer was not as humid as expected, several very peculiar downbursts were captured, such as the so-called Monrovia microburst, which was described and studied in great detail by Fujita (1992). The microphysical characteristics of the parent cloud and the thunderstorm-induced microburst wind field were measured by a few Doppler radars. These measurements provided the time evolution of the vertical velocities within the cloud and enabled the observation of the updraft throughout the cloud lifecycle, that is, from the initial developing stage of the storm to the formation of the downdraft, its subsequent touchdown, and finally the microburst-to-macroburst evolution at the ground. The strong midlevel entrainment that occurred during the downdraft descent was reported by superposing radar reflectivity measurements (precipitation core region at 60 dBZ [decibels relative to Z]), the vertical velocities within the cloud, and the aerial photographs of the cloud evolution. This result for the first time confirmed that the large-scale entrainment of dry air into the towering cumulus is the main mechanism of downdraft enhancement in warm-based air mass thunderstorms. The entrainment and formation of downdraft profoundly altered the thermodynamics of cumulonimbus clouds.

Entrainment of dry air aloft was recognized as a typical radio-sounding pattern for wet downburst formation (Caracena & Maier, 1987) above a deep and almost saturated layer with a nearly pseudoadiabatic lapse rate (Figure 4.7). When the cooler and drier air mixes with the warmer saturated updraft, the mixed parcel starts subsiding because of negative buoyancy and generates a strong downdraft, whose intensity can be estimated utilizing thermodynamic diagrams. Kingsmill and Wakimoto (1991) studied this aspect further and determined that the downdraft appeared to be initiated by precipitation loading and intensified by the negative thermal buoyancy due to entrainment. In contrast to the main characteristics of microbursts over the High Plains in Colorado during JAWS (Wakimoto, 1985), the moist-adiabatic subcloud layer became an important

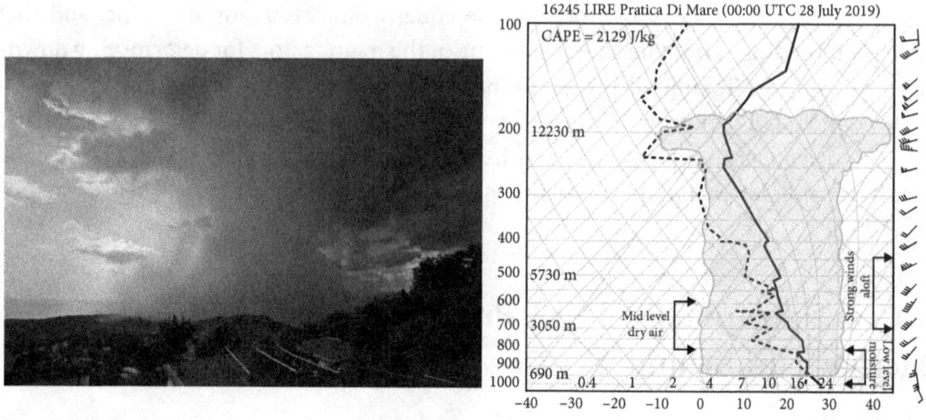

FIGURE 4.7 Photo of a downdraft and wet downburst originated from a tall cumulonimbus cloud over Genoa, Italy (left), and the typical atmospheric conditions during which a wet downburst occurs (right) (photograph by William Demasi and Jacopo Zannoni; courtesy of ZenaStormChaser and Jacopo Zannoni).

precursor for microburst formation in Alabama (MIST Project) because the increased moisture in this layer may result in a larger virtual temperature difference between downdraft and environment and consequently increase the negative buoyancy and accelerate the downdraft descent (Srivastava, 1987).

3.4. Beyond NIMROD, JAWS, and MIST Projects

Some of the most important outcomes of approximately 10 years of Fujita's intensive research on downbursts were summarized in his research paper, *The Downburst: Microbursts and Macroburst* (Fujita, 1985). In addition to detailed documentation and description of the most relevant downburst events, Fujita (1985) reported a statistical analysis of important downburst parameters measured during the NIMROD, JAWS and MIST projects (e.g., downburst duration, direction and intensity of peak winds, and variation in temperature, humidity, and pressure at the surface associated with downburst passage), as well as the description of the types of microbursts and characteristics of their parent clouds.

However, the 3D and time-dependent analysis of the wind fields in microburst outflows throughout their life cycle—which was one of the objectives of the JAWS observations—was not directly accomplished by Fujita. Instead, Wilson, Roberts, Kessinger, and McCarthy (1984) made use of multiple Doppler radars from the JAWS Project and calculated several geometric and kinematic parameters that described the horizontal and vertical structure of microbursts (Figure 4.8). For instance, they concluded that an average downdraft was about 1 km wide, and it is spread out below 1 km from the surface. The outflow maximum differential wind speed of 24 m s^{-1} was achieved 6.4 min after touchdown over a distance of about 3 km at the height of 75 m above ground. The

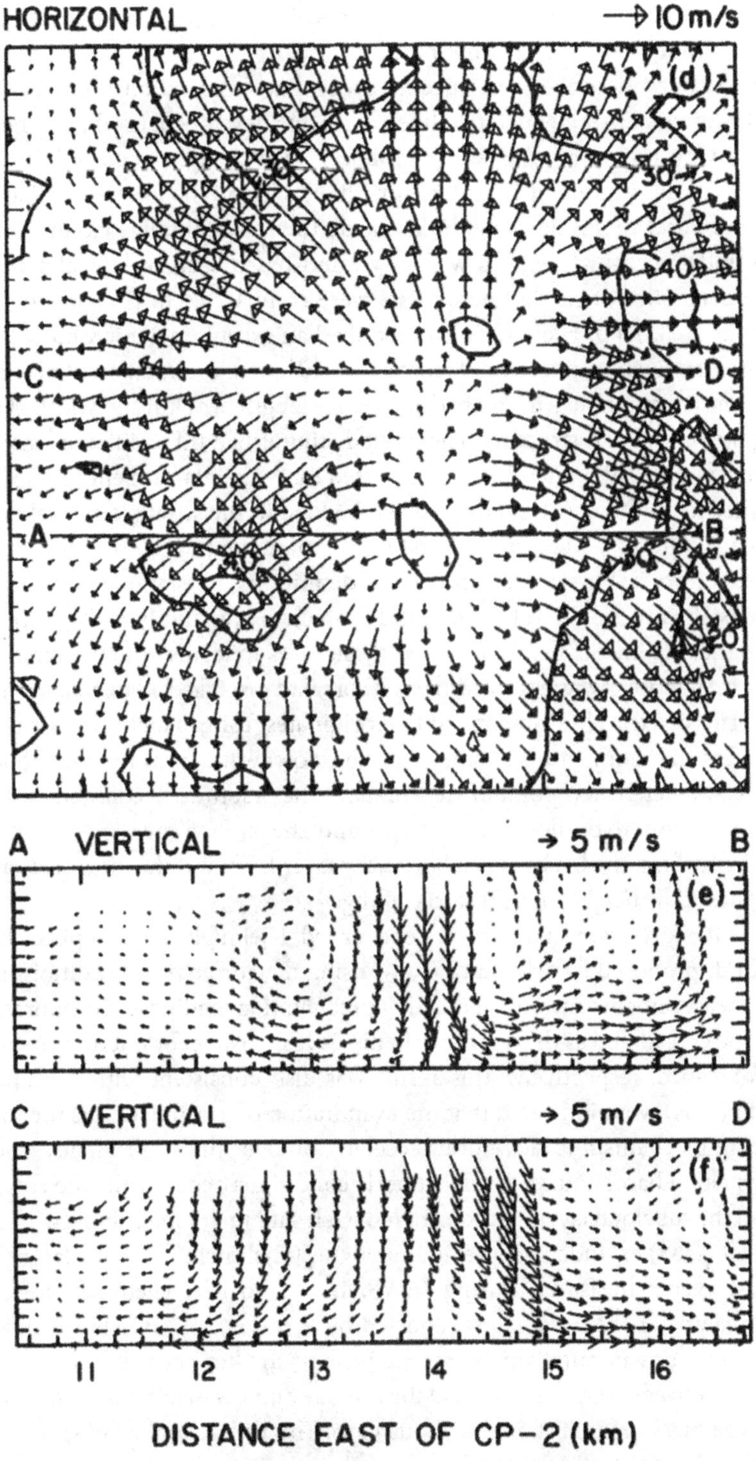

FIGURE 4.8 Doppler radar measurements of wind field and reflectivity inside the microburst measured during JAWS on July 14, 1982: The upper panel shows the horizontal divergence (mean flow subtracted) produced by the thunderstorm outflow at the surface (50–100 m above ground); the lower panels are vertical sections along lines A–B and C–D (from Wilson et al., 1984).

ring vortex was generally not symmetric in terms of intensity, which indicates that most of the events were translating downbursts.

Hjelmfelt (1987) described one specific event that occurred on June 22, 1982, near Denver, Colorado, during which two lines of storms developed, producing heavy rain, hail, and giving rise to several wet microbursts. The analysis reported by Hjelmfelt focused on the description of the microburst life cycle and the forcing mechanism that induced the onset of strong downdrafts inside towering cumulus clouds, and it was based on radar measurements as well as ground-based meteorological stations. The main characteristics of single microbursts were comparable to the ones reported by Wilson et al. (1984), but Hjelmfelt also described an exceptional microburst line made up of at least 20 separate microbursts that achieved a maximum differential wind speed of 27 m s^{-1}, lasted for about 2 h, and had an average length of 21 km and width of 6 km. By studying the microburst structure, Hjelmfelt distinguished three different categories of microbursts: (1) isolated microbursts immersed in weak environment flows with radially symmetric outflows; (2) microbursts embedded in the strong low-level flows with asymmetric outflows; and (3) microburst lines. Apart from the third category, the first two types of microbursts are similar to definitions of stationary and traveling downbursts already proposed by Byers and Braham (1949) and Fujita (1981a), with the exception that Hjelmfelt (1987) described these cases in terms of downburst and low-level flow interaction, whereas the other authors had proposed an explanation of these flows in terms of a traveling downburst that originates from an inclined-axis downdraft that transfers the momentum of storm motion downward from the midtroposphere. This was a completely new point of view that, for the first time, accounted for three different contributions to the flow field at the ground: the outflow after downdraft impingement at the surface, the background boundary layer flow, and the momentum that the air parcel has at the height where it starts falling.

As far as the forcing mechanisms are concerned, Hjelmfelt (1987) applied the model of downdraft proposed by Srivastava (1985) using meteorological conditions coherent with real measurements and concluded that precipitation loading and evaporative cooling could be considered as the main mechanisms responsible for microburst formation and maintenance, respectively. This result was also consistent with the findings of Wakimoto (1985), who indicated that the evaporation of precipitation is the most relevant forcing mechanism to downdraft acceleration over the High Plains, where relatively high cloud bases and drier atmospheric conditions occur, while the entrainment of air from the subcloud layer was considered irrelevant in dry microbursts.

Hjelmfelt (1988) extended his previous analyses (Hjelmfelt, 1987) to a larger set of 27 microburst events measured during JAWS. In this article, Hjelmfelt presented an updated analysis of the parameters reported by Wilson et al. (1984). The new results showed that the downdraft diameter ranged from 1.5 to 3 km, the maximum descending speeds were between 6 and 22 m s^{-1}, and the average outflow maximum differential wind speed was 24 m s^{-1} over a distance of about 3 km. This differential wind speed was typically observed 7.5 min after the touchdown and at a height between 50 and 100 m above ground. The outflow symmetry analysis once again confirmed that the radially advancing ring vortices were not symmetric in terms of intensity, and the minimum differential wind speed was half to one third the maximum velocity differential.

The downbursts analyzed by Hjelmfelt (1988) were both wet and dry microbursts, which allowed for a broader investigation of the variability of parameters that are directly related to the water content in downdrafts. The analyzed events included the microbursts developed underneath isolated and moderate-to-intense cumulus clouds, virga shafts, and lines of convective clouds. The reflectivity in the precipitation core region ranged from 15 to 65 dBZ (with an average value of 30 dBZ), the cloud base was on average at 3.3 km above ground in the low-reflectivity cases and 2.3 km above ground in higher reflectivity cases. Hjelmfelt (1988) suggested that the vertical outflow profiles should be approaching a wall jet flow (Bradshaw & Love, 1959) at a distance from the touchdown of 1.5–2.0 times the downdraft radius and found a strong similarity between microburst outflow profiles and wall jet profiles. Finally, he discussed the issue of symmetry in microbursts embedded in strong directional flow, pointing out that the measured downburst wind velocity is a composition of the storm motion aloft, which determines the microburst translation, and the low-level flow in the subcloud environmental shear. He found that, after removal of the low-level environmental flow, microbursts recover their original symmetry so that they can be considered as the superposition of an isolated microburst and the environmental flow.

Figure 4.9, which was originally reported in Hjelmfelt (1988), can still be considered as one of the reference downburst models commonly cited and used even in the literature to represent the structure, geometrical, and kinematic characteristics of downburst outflows. This model is coherent with the previous vertical wind profile measurements

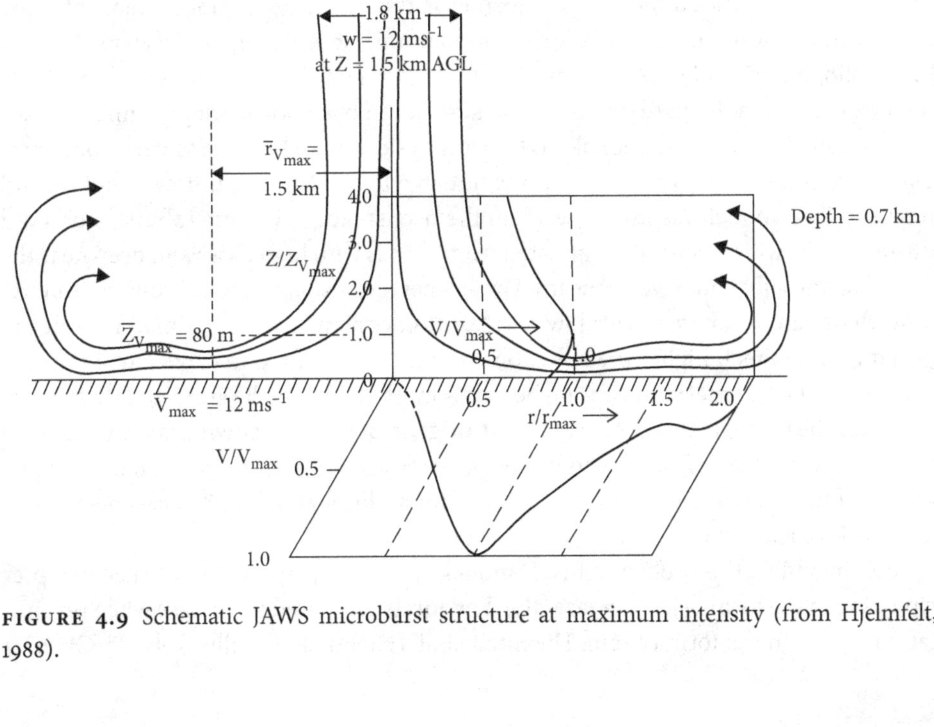

FIGURE 4.9 Schematic JAWS microburst structure at maximum intensity (from Hjelmfelt, 1988).

taken by Goff (1976)—who was also the first scholar to describe the nose-like shape of velocity profiles in thunderstorm outflows—and Sherman (1987), as well as with the thunderstorm gust front analyses carried out by Charba (1974) and Wakimoto (1982).

4. Modeling

After the NIMROD and JAWS projects provided the scientific community with direct observations of downbursts, the investigation of this phenomenon commenced by the means of numerical models. Two numerical approaches proceeded in parallel. On the one hand, meteorologists developed models to replicate the dynamics and thermodynamics of real events to better understand the physical complexity of downbursts and their parent clouds; on the other hand, aviation meteorologists and engineers developed simpler kinematic models for use in real-time piloted simulation studies for aircraft operations in wind shear conditions. The next sections examine the main outcomes derived from both these aforementioned approaches. Also, the gravity current models are discussed because they have largely been used to study the gust front part of downburst outflows.

4.1. Cloud and Subcloud Models

The numerical simulation of real downburst events was carried out through the use of full-cloud and subcloud models. The former is time-dependent models that solve the prognostic equations of conservation of momentum, mass, energy, and water substance in its solid, liquid, and gaseous forms in the atmosphere. Srivastava's (1985) model was the first model that focused on numerical simulation of cloud-developing microbursts. He developed a one-dimensional (1D) microphysical model that considered only precipitation in the form of rain, and it was not able to resolve the outflow structure and intensity. This model was used to explain the occurrence of dry microbursts observed during JAWS, which typically originated from clouds with high bases and deep near the dry adiabatic subcloud layer beneath. Under these atmospheric conditions, an intense downdraft can occur even with low precipitation contents and no rainfall rates at the ground, while much higher precipitation contents are required to drive intense downdraft when the thermal stratification becomes more stable. As the atmospheric stability increases, heavier precipitation is needed to generate strong downdrafts that change their nature gradually from dry to wet downbursts. A possible mechanism for increasing the amount of precipitation is to account for additional microphysical effects, such as melting of ice particles.

After including this process in his 1D model, Srivastava (1987) demonstrated that precipitation in the form of ice increases the downdraft compared to the case when precipitation is only in the form of rain. Hjelmfelt et al. (Hjelmfelt, Orville, Roberts, Chen, &

Kopp, 1989; Helmfelt, Roberts, & Orville, 1986) confirmed the importance of the combined melting of graupel and rain evaporation in initiating and intensifying downbursts, respectively, employing a two-dimensional (2D) cloud model that the authors used to simulate the microburst line that developed on July 13, 1982, during JAWS (Hjelmfelt, 1988). Very similar conclusions were drawn by Orville, Farley, Chi, and Kopp (1989), who used a 2D cloud model to simulate eight microburst events that took place during JAWS and other experimental projects, namely, the COHMEX (Dodge et al., 1986), the CCOPE, the Texas High Plains Experiment (HIPLEX) (Smith et al., 1984), and the Alberta Hail Project (Renick, 1975). Afterward, these results were categorically confirmed through radar observations in Atlas, Ulbrich, and Williams (2004). Also, they emphasized the importance of small-size hail particles in large concentrations for increasing microburst intensity.

One of the first examples of 3D cloud modeling can be traced to the work carried out at Colorado State University, where the dynamics and thermodynamics of downburst-producing clouds were investigated using the numerical cloud model developed by Tripoli and Cotton (1982). Knupp and Cotton (1985), based on the comprehensive survey of available literature on convective cloud downdraft structure, indicated that four different types of downdrafts (Figure 4.10, left) can be identified within or nearby cumulonimbus clouds (Cb), which typically develop under weak-to-moderate environmental wind shear conditions (i.e., non-supercell convection). They pointed out that only the low-level downdraft in the precipitation core (PR) systematically reaches the ground, while the other types of downdrafts usually exist at higher levels. According to Squires (1958), penetrative downbursts (P) are often initiated by the strong entrainment (E) of subsaturated environmental air that typically occurs in the downshear side of the Cb. Next, downdrafts also exist at the edges of the cloud or updrafts as a consequence of convective motions (L), and last, a larger scale downdraft region (R) that exists downshear from the cloud is induced by cloud-scale circulations (Fritsch, 1975).

Following these results, Knupp (1987, 1988) focused on the low-level downdraft kinematic structure and dynamic/thermodynamic processes that govern these types of downdrafts. Based on both observations and numerical results, he developed a generalized low-level downdraft conceptual model in which he distinguished between an up-down branch that generates within or near the updraft inflow sector at a low level in the downshear flank of Cb (indicated as a_1 in Figure 4.10, right) and certain midlevel branches (indicated as b and c in Figure 4.10, right) that are generated within the atmospheric layer between the top of the boundary layer (~2 km) and the height where the PR downdraft originates (~4 km). This height is approximately at the height of the melting level.

In addition, Knupp (1989) used the same cloud model to investigate the initiation of low-level downdrafts concerning environmental conditions that are typical for dry and wet microbursts. The former originate from weakly precipitating cumulus congestus with a high cloud base up to 3 km above a deep and dry boundary layer, whereas the latter occur below tall cumulonimbus clouds with a low cloud base and a shallow moist environment beneath. Both conditions were reproduced numerically using the same

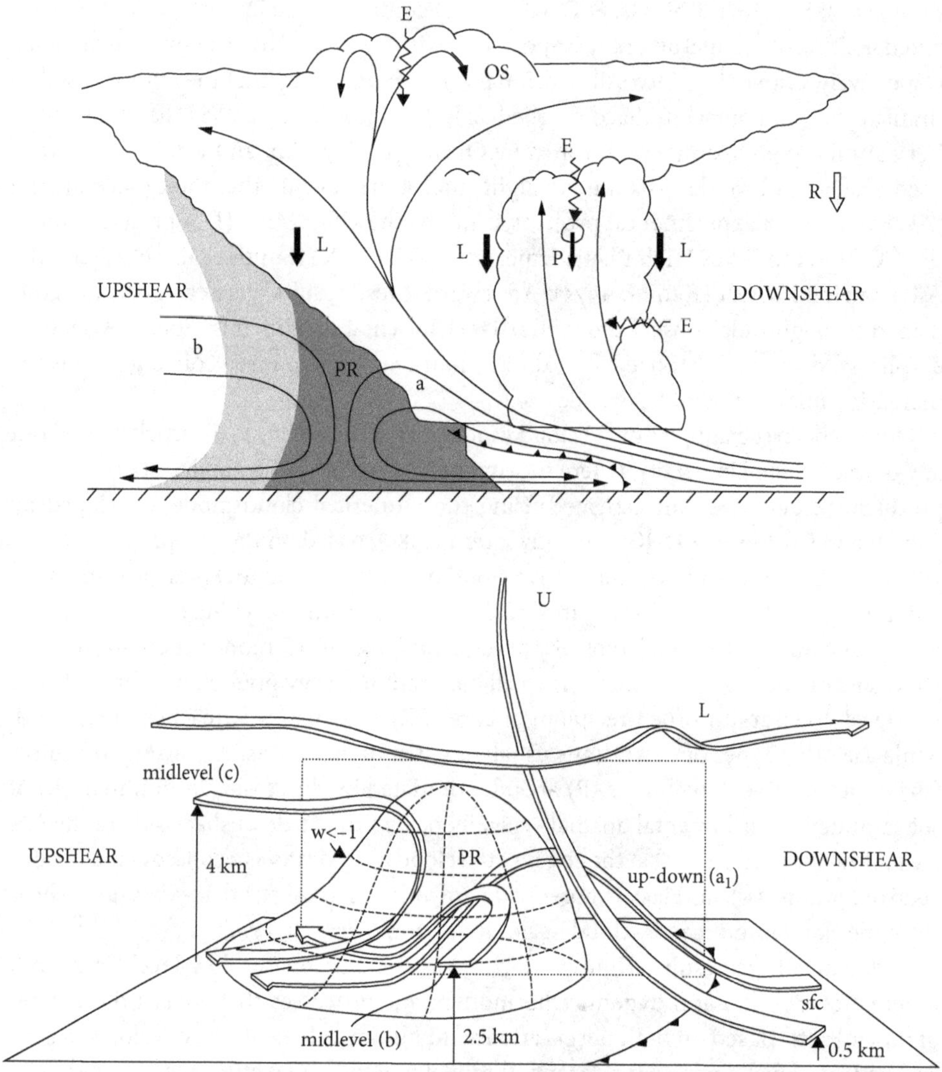

FIGURE 4.10 Left: Schematic of updraft, downdraft, and entrainment (E) flows inside a typical cumulonimbus cloud (Cb), within a weak-to-moderate wind shear environment. Downdraft occurs in the precipitation core (PR), at cloud/updraft edges (L), in the region (R) downshear from the cloud induced by cloud-scale convection. The position of penetrative downdrafts (P) is also shown. OS is overshooting at the top of the cloud. All flows are storm relative (from Knupp & Cotton, 1985). Right: Schematic of the low-level PR downdraft kinematic model corresponding approximately to the layer between the ground and 5,000 m in the figure to the left, upshear from the updraft (U). L is the ambient flow outside the Cb (from Knupp, 1987).

convection initiation mechanism (i.e., a circular convergence within the boundary layer) and different soundings of temperature and mixing ratio profiles. Negative buoyancy produced by the overall effect of precipitation loading, melting, and evaporation were shown to be the main downward-forcing mechanism for the low-level downdrafts (see also Straka & Anderson, 1993) in both dry and wet microbursts. For this reason, the arrival of precipitation below the melting level was identified as a fundamental aspect of the downdraft initiation process. Downdraft occurrence is therefore closely related to precipitation initiation, turbulent transport, and microphysical processes within clouds.

More or less concurrently with the research activities on cloud modeling carried out at Colorado State University, Proctor (1987a, 1987b) at the NASA (National Aeronautics and Space Administration) Langley Research Center, formulated an axisymmetric version of the Terminal Area Simulation System (TASS) derived from the former 2D model of Proctor (1985). The model was used to replicate two isolated and stationary microbursts that occurred on June 30, 1982, during JAWS and on August 2, 1985, nearby the Dallas–Fort Worth Airport (Fujita, 1986). The model properly reproduced the microburst primary ring vortex that expanded radially outward (Proctor, 1988). The simulations also showed the formation of a secondary vortex with opposite vorticity to the primary ring vortex. The formation of the secondary vortex was due to the frictional effect at the ground that acts to retard the air pushed outward by the outburst front (Figure 11a). This feature was not revealed, for instance, in the previous 2D microburst simulation with a free-slip boundary condition at the surface (Krueger & Wakimoto, 1985). Moreover, a temperature decrease and a high-pressure zone—called a pressure nose—underneath the downdraft axis were created by the expansion of the primary ring vortex (see also Straka & Anderson, 1993). The simulation produced the typical nose-like vertical velocity profile (Figure 11b) that was in agreement with the laboratory wall jet profile (Bakke, 1957), as well as measurements from the NIMROD campaign (Fujita, 1981b).

The use of cloud models represented a significant advance in the research field on downbursts because it allowed the investigation of detailed dynamics and thermodynamics that influence the characteristics of these phenomena. The TASS, for instance,

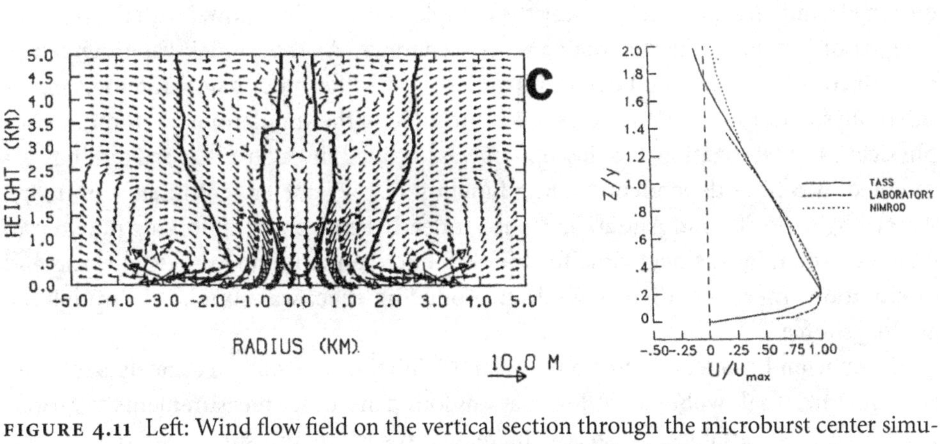

FIGURE 4.11 Left: Wind flow field on the vertical section through the microburst center simulated by a cloud model 7 min after downdraft touchdown. Bold solid lines correspond to 10- and 60-dBZ radar reflectivity contours. The dashed line encloses the area with temperature departures from ambient greater than 1 K (from Proctor, 1988). Right: Nondimensional vertical profiles of microburst outflow taken at the radius of peak outflow. Velocity is nondimensionalized by the peak speed U_{max}, and height is nondimensionalized by the half-velocity height *y*, corresponding to the height where speed is 0.5 U_{max} (from Proctor, 1988).

was used to carry out a sensitivity analysis on downburst intensity, spatial extension, and time evolution as a function of vertical ambient lapse rate and humidity distribution, downdraft diameter, precipitation loading, as well as the type and duration of precipitation (Proctor, 1989). It was demonstrated that the existence of a deep and stable layer, on the order of 1,000 m, may prevent the downburst from reaching the surface, consequently producing the midair microburst that was previously described by Fujita (1981b) and Fujita and Wakimoto (1983). The study also demonstrated the quintessential role of ambient temperature and humidity in the amplification or weakening of downburst intensity.

Yet another aspect that was addressed in Proctor (1989) was the role of downdraft dimension in the dynamics of their outflows. The study reported that an optimal diameter for penetrative downdrafts, common to nonprecipitating convective clouds that generate dry microbursts (see also Paluch, 1979, and Emanuel, 1981) is approximately 1 km. At this diameter, the microburst's strength is maximized due to the minimization of entrainment and the effects of pressure-gradient forces acting against buoyancy forces. In conclusion, Proctor (1989) asserted that the most favorable scenario for an intense downdraft is a small cell composed of a deep layer of graupel or snow above the melting level in dry microbursts and a precipitation shaft of several kilometers in diameter heavily loaded with hail and rain in wet microbursts. Finally, according to the outcomes described, he also proposed an index for the wet microburst potential that can be considered as the first attempt of microburst prediction. Proctor's microburst index was later improved by Wolfson (1990) and afterward by McCann (1994). Today, this index is known as the WINDEX (abbreviated from "wind index").

The pioneering research by Proctor (1988, 1989) showed that high-resolution numerical models (~50 m or higher horizontal resolution) were capable to reproduce the detailed structure of downburst outflows. Anderson, Orf, and Straka. (1992) designed a numerical model to study the complex near-surface flows, with the particular goal of investigating the smaller scale dynamics. As the model was intended to reproduce only the subcloud dynamics under specific thermodynamic forcing, neither microphysics nor radiation processes was included in the simulations. Without microphysical parameterization, the downburst was initiated through a thermal forcing represented by a time-dependent cooling function (Droegemeier & Wilhelmson, 1987). According to Straka and Anderson (1993), indeed, the mechanical forcing due to precipitation loading was negligible for the thermodynamic forcing due to melting and evaporation. Therefore, their downdraft-producing effect was simulated through the cooling source.

The subcloud model by Droegemeier and Wilhelmson (1987) accurately described the signature of a downburst outflow that was found in surface measurements of various meteorological variables. A steady increase in surface pressure due to the nonhydrostatic effects and increase of wind speed precede the temperature drop that occurs after the passage of the radially advancing front. As the front passes, the surface pressure starts to rise, and the horizontal wind speed increases significantly. This model of wind speed increase is per Hjelmfelt (1988) (Figure 4.9). However, the wind speed and pressure fluctuations also intensify after the front passage due to the 3D Kelvin–Helmholtz instability in the shear layer overhead (i.e., between the radially advancing outflow and the relatively calm ambient air). This signature of a downburst outflow was previously reported by Mueller and Carbone (1987) in the data coming from three different field campaigns (Phoenix II; Maypole in 1984; and Convection Initiation, 1984). Moreover, Sherman (1987) found a similar pattern in the measurements associated with the passage of a weak downburst over an instrumented tower at Bald Hills, near Brisbane, Australia.

4.2. Outflow Kinematic Models

The simplest model of a downburst outflow is a steady flow against a flat plate with or without friction and without buoyancy (Bakke, 1957; Batchelor, 1970; Bradshaw & Love, 1959; Glauert, 1956). This type of simulation is commonly referred to as an impinging jet model (Zhu & Etkin, 1985). This model predicts a high-pressure area at the touchdown position that was first reported in the Thunderstorm Project (Byers & Braham, 1949) during the mature stage of thunderstorm cells. In NIMROD and JAWS, closer spacing among ground stations allowed measurements with higher precision. These field campaigns have shown that the high-pressure region typically increases the surface pressure for 2–4 mb (millibar) over the areas of a few kilometers and for 2 to 5 min.

Fujita (1985) reported that the surface pressure at the center of the Andrews Air Force Base (AFB) microburst was 3 mb higher than in the ambient air. The application of the Bernoulli equation, however, showed that the reported order of magnitude of pressure gradients in the impinging jet model is not enough to generate wind velocities as high as 60 m s^{-1}, such as the one observed in the Andrews AFB microburst. Moreover, in some cases, low pressures have been observed during microburst events. These peculiar cases suggest that a more complex model is required to properly explain the dynamics of microbursts. The analysis of the pressure field in the Andrews AFB microburst revealed that an increase of the pressure gradient by a factor of approximately 5 was associated with the low-pressure ring that surrounded the central high pressure.

A similar finding was also reported by Lin and Hughes (1987) based on the analysis of dual-Doppler radar data during a microburst that occurred during the JAWS campaign. The low-pressure ring is related to the existence of the primary vortex ring, which was often measured in microbursts (Bedard & Caplan, 1987; Caracena, 1982). The primary vortex ring was also well documented in photographs from the NIMROD and JAWS campaigns; in laboratory experiments (Didden & Ho, 1985; Fujita, 1986; Yao & Lundgren, 1996); and numerical simulations (Proctor, 1988). The expanding vortex ring at the base of a downdraft increases its vorticity as it expands (stretches) radially due to the conservation of angular momentum. This relationship between vortex ring expansion and vorticity increase explains the observed strengthening of the radial outflow as the vertical jet expands at the surface. The mechanism that acts against ring expansion is the surface friction.

The very first model proposed to depict a downdraft for flight simulation was the ring vortex downburst model of Ivan (1986). This stationary model simulates the mature stage of downbursts according to the circulatory pattern observed in the JAWS microburst of August 5, 1982. The same model was further improved by Schultz (1990), who superimposed two vortex rings to reproduce the microburst that caused the crash of Delta Flight 191 on August 2, 1985, at the Dallas–Forth Worth International Airport, causing 133 fatalities (Caracena, Ortiz, & Augustine, 1986; Fujita, 1986). Based on the event reconstruction by Wingrove and Bach (1987), this microburst consisted of a larger vortex ring containing smaller vortex rings. The multiple vortex ring model proposed by Schultz was able to capture with significant accuracy the flow variations experienced by the aircraft AA539 that flew across the microburst after the Delta 191 accident (Figure 4.12). The vortex rings produced in the wind shear that occurs at the external edge of downdrafts and expands at the ground according to a primary image ring pair model (Lamb, 1945) act as a dissipation mechanism of the downdraft itself. A single vortex ring is enough to dissipate the downdraft energy when the latter is small, whereas multiple vortices occur for long-lasting downdrafts as a result of 3D Kelvin–Helmholtz instabilities producing sequences of toroidal vortices, which appear as pulsations in the microburst evolution (Anabor, Rizza, Nascimento, & Degrazia, 2011; Droegemeier & Wilhelmson, 1986, 1987; Simpson, 1969). Importantly, though, this phenomenon should not be confused with the cases of multiple microbursts embedded into larger downdrafts or mesoscale storm systems that produce single primary vortex rings mutually interacting at the ground (Anderson et al., 1992; Hjelmfelt, 1988; Hjelmfelt et al., 1989; Vermeire, Orf, & Savory, 2011).

Another model development motivated by aircraft accidents in downbursts was initiated by Oseguera and Bowles (1988). Their model represents an axisymmetric stagnation point flow that satisfies the mass continuity equation, which was solved based on the velocity profiles obtained from the TASS model (Proctor 1987a, 1987b). Although the model evaluation in their work was promising, Vicroy (1991) demonstrated that some of the assumptions (i.e., the profile function is taken from the TASS model) were sufficiently different from the observations, and the model often provided inaccurate estimates of downdrafts in real applications. Therefore, Vicroy (1991) improved the Oseguera and Bowles (1988) model and compared the new model—the so-called OBV model (from Oseguera, Bowels, and Vicroy)—against the NIMROD and JAWS data. Since neither of the two models accounted for the translation velocity of microbursts, Vicroy (1991) subtracted the translation speed of 15 m s^{-1} from the NIMROD microburst before comparing it against the results from the analytical model. Nevertheless, the improved model still underestimated the near-surface velocities in the region below the profile nose when compared against the mean downburst velocity profile from the JAWS Project. All these early contributions to analytical modeling of downbursts were focused on the mean flow and turbulent fluctuations were not considered.

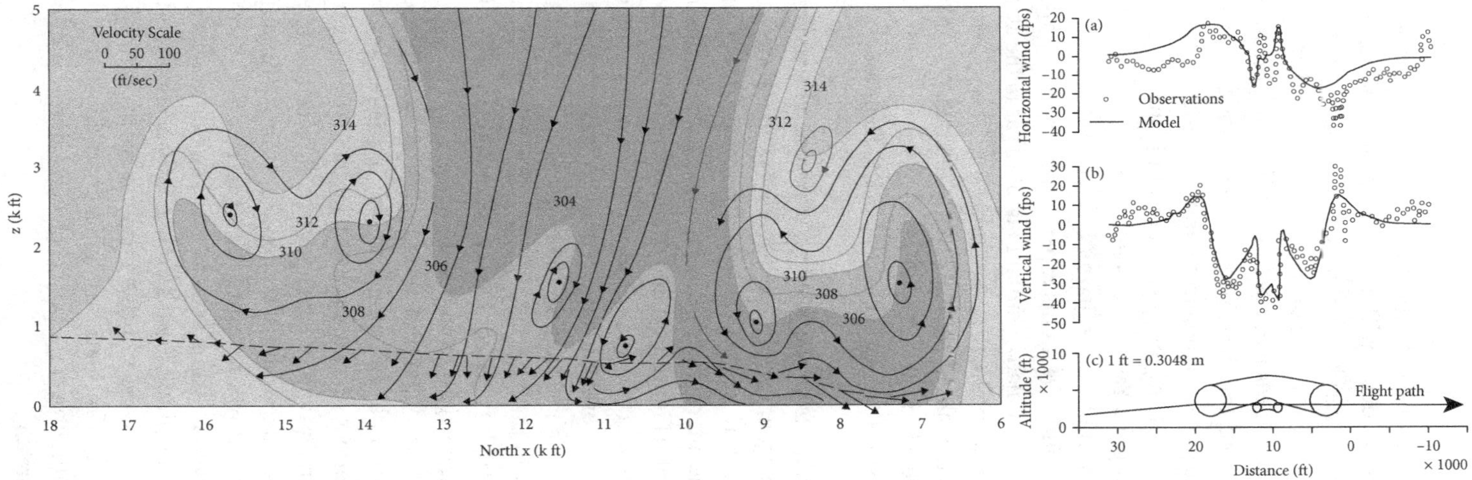

FIGURE 4.12 Left: Vertical section of airflows in the Dallas–Fort Worth microburst based on the analysis by Bach and Wingrove (1985), showing streamlines, velocity vectors along the Delta 191 flight path (dashed line), and potential temperature (K) in contours with numbers (from Caracena et al., 1986). Right: Multiple vortex ring model (c) and wind components along the flight path of AA539 (a, b) (from Schultz, 1990).

4.3. Gravity Currents as Gust Front Models

A gravity current is a high-density fluid that flows along the horizontal plane and displaces an ambient fluid of lesser density. The dominant force that drives the gravity current is the pressure gradient across the interface that separates the two fluids. In this situation, the hydrostatic pressure inside the current is greater than that in the ambient fluid. This pressure gradient force is balanced by the dynamic pressure induced at the frontal edge of the outflow and the shear stress at the boundaries (Charba, 1974). The boundary that separates the warmer ambient air and the colder outflow is called a thunderstorm gust front. An example of the gravity current is the release of a mass of salt water into fresh water (Lundgren, Yao, & Mansour, 1992).

Simpson (1969) demonstrated the similarity between the frontal region of gravity currents and thunderstorm outflows. In his experiments, the Richardson number (i.e., the buoyancy to wind shear ratio) had a similar magnitude as in the atmosphere. Under the assumption that the gradient Richardson number at the upper boundary of the thunderstorm gravity current is about 0.2, Hall, Neff, and Frazier (1976) proposed a heuristic model that linked the wind shear in the thunderstorm gravity current and temperature drop associated with the passage of the front. The model depicts the linear relationship between the two.

By synthesizing the findings from the previous experimental studies on gravity currents (Benjamin, 1968; Keulegan, 1958; Middleton, 1966; von Kármán, 1940) and applying the results to the meteorological phenomena of a gust front, Charba (1974) also noticed the similarity in the internal density structure, surface geometry, and flow pattern between gravity currents (Figure 4.13) and the gust front of the May 31, 1969, thunderstorm in central Oklahoma. However, some differences were also noticed. First, the head height of the leading edge of the advancing outflow was significantly higher in the gravity current (59 percent higher) than in the real event. Similarly, the overhanging nose was 88 percent higher in the case of the gravity current than in the observed gust front. Previously, Simpson (1972) noticed that the overhanging nose is the common feature of both gravity currents and gust front outflows.

Charba (1974) further reported that the entrainment and mixing process in the upper region of the gravity current inflow is similar to that in the real downburst outflows. When it comes to the dynamics of cold outflow, Charba (1974) demonstrated that the theoretically and experimentally derived expression for displacement speed of the gravity current nicely also describes real downburst outflow. For example, the displacement speed of the May 31, 1969, gust front was measured to be 20 m s^{-1}, while the displacement equation theoretically derived by von Kármán (1940) and others gave results between 17.5 and 22.7 m s^{-1}. Yet another similarity between gravity currents and real downburst outflows is the existence of the lobe and cleft instability at the leading edge of the two currents (Simpson, 1969, 1972). This static instability arises due to the tilting of the overhanging nose of the cold outflow over the pocket of warm air that is situated in a shallow layer near the surface (Figure 4.13).

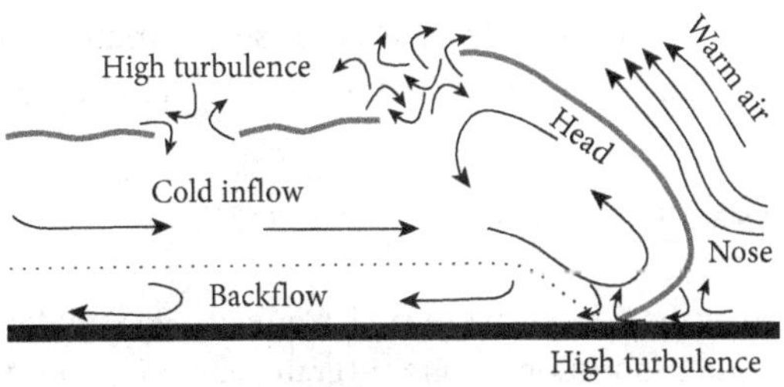

FIGURE 4.13 Schematic structure of a gravity current based on the results from Simpson (1969, 1972). The thick curvy line represents the gravity current boundary, while the black dotted line indicates the depth of a layer in which backflow develops. The decrease of the friction boundary layer depth close to the gravity current edge occurs due to the increased pressure gradient in the flow (i.e., pressure decreases in the direction of the motion) (Goff, 1976).

Lundgren et al. (1992) proposed a method for scaling laboratory-produced gravity currents to full-scale downbursts using the equivalent spherical radius of the two outflows to define the proper length scale. The timescale was calculated as the function of length scale and density differences between the two flows. The density ratios that are typically observed in the atmosphere (1.97×10^{-2}; Charba, 1974) provide very low wind speeds in the laboratory replica due to the Froude number mismatch between the two cases. Therefore, the gravity current in the work of Lundgren et al. (1992) geometrically and temporally scaled 1:7000 and 1:23, respectively, against the DL 191 microburst that occurred at Dallas-Fort Worth Airport on August 2, 1985 (Fujita, 1986).

5. Concluding Remarks

This chapter reviewed the selected contributions to downburst discovery, modeling, and detection in the early stages of downburst research. The original contributions from the works of Byers and Braham, Fujita, Wilson, Hjelmfelt, Charba, Simpson, and others from the period between approximately the second half of the 1940s to the mid-1990s were chronologically reviewed and their main findings presented. One of the major focuses was on the research findings from the field campaigns NIMROD Project (Fujita, 1978, 1985) and the JAWS Project (Fujita, 1985; McCarthy et al., 1982). In addition to providing the first picture of downburst outflows in the real atmosphere, the results from these and a few other field campaigns from that time were also used for calibration and validation of various cloud and subcloud models, analytical expressions, and gravity current experiments. Without a doubt, the published results from these two projects are still among the most used and cited sources of data in today's literature on downburst outflows.

We hope that this study will extend (or trigger) the reader's interest in thunderstorm downbursts and associated phenomena. In that case, the reader is highly encouraged to review the studies mentioned on their own because the material presented therein is vastly richer than the brief overview that is presented in this chapter.

Acknowledgments

This research was funded by the European Research Council (ERC) under the European Union's Horizon 2020 research and innovation program (grant agreement no. 741273) for the project THUNDERR—detection, simulation, modeling, and loading of thunderstorm outflows to design wind-safer and cost-efficient structures—supported by an advanced grant (AdG) 2016.

References

Abbey, R. F. (1976). Risk probabilities associated with tornado windspeeds. In *Proceedings of the Symposium on Tornadoes: Assessment of Knowledge and Implications for Man* (pp. 177–236), Lubbock, TX: Texas Tech University, June 22–24.

Anabor, V., Rizza, U., Nascimento, E. L., & Degrazia, G. A. (2011). Large-eddy simulation of a microburst. *Atmospheric Chemistry and Physics, 11*, 9323–9331.

Anderson, J. R., Orf, L. G., & Straka, J. M. (1992). A 3-D model system for simulating thunderstorm microburst outflows. *Meteorology and Atmospheric Physics, 49*, 125–131.

Atkins, N. T., & Wakimoto, R. M. (1991). Wet microburst activity over the southeastern United States. *Weather Forecasting, 6*, 470–482.

Atlas, D., Ulbrich, C. W., & Williams, C. R. (2004). Physical origin of a wet microburst: Observations and theory. *Journal of Atmospheric Sciences, 61*, 1186–1196.

Bach, R. E., & Wingrove, R. C. (1985). Estimation of the winds along the flight path along the Delta L1011 accident at the Dallas-Fort Worth Airport on August 2, 1985. Special report to NTSB, Moffett Field, CA: NASA Ames Research Center.

Bakke, P. (1957). An experimental investigation of a wall jet. *Journal of Fluid Mechanics, 2*, 467–472.

Batchelor, G. K. (1970). *An introduction to fluid dynamics*. London, UK: Cambridge University Press.

Bedard, A. J., & Caplan, S. J. (1987, March). Microburst vorticity AIAA-87-0440. Paper presented at the 25th AIAA Aerospace Sciences Meeting, Reno, NV.

Benjamin, T. B. (1968). Gravity currents and related phenomena. *Journal of Fluid Mechanics, 31*, 209–248.

Bradshaw, P., & Love, E. M. (1959). *The normal impingement of a circular air jet on a flat surface* (R&M 3205). Aeronautical Research Council. London, UK: Her Majesty's Stationery Office.

Braham, R. R. (1996). The thunderstorm project: 18th conference on severe local storms luncheon speech. *Bulletin of the American Meteorological Society, 77*, 1835–1845.

Braham, R. R. (1997). Thunderstorms and the thunderstorm project. *National Weather Digest, 21*, 24–31.

Byers, H. R., & Braham, R. R. (1949). *The thunderstorm*. Washington, DC: US Department of Commerce.

Caracena F. (1978). *A comparison of two downbursts of different mesoscales*. Preprints of the Conference on Weather Forecasting and Analysis and Aviation Meteorology, Silver Spring, MD, American Meteorological Society, 293–300.

Caracena, F. (1982). Is the microburst a large vortex ring imbedded in a thunderstorm downdraft? *EOS Transactions of the American Geophysical Union*, *63*, 89.

Caracena, F., Maddox, R. A., Purdom, J. F. W., Weaver, J. F., & Greene, R. N. (1983). *Multiscale analysis of meteorological conditions affecting Pan American World Airways flight 759* (Technical Memo ERL ESG-2). Boulder, CO: NOAA.

Caracena, F., & Maier, M. W. (1987). Analysis of a microburst in the FACE meteorological mesonetwork in South Florida. *Monthly Weather Review*, *115*, 969–985.

Caracena, F., Ortiz, R., & Augustine, J. A. (1986). *The crash of Delta flight 191 at Dallas-Fort Worth International Airport on 2 August 1985: Multiscale analysis of weather conditions* (Technical Report ERL 430-ESG 2). Boulder, CO: NOAA.

Charba, J. (1974). Application of gravity current model to analysis of squall-line gust front. *Monthly Weather Review*, *102*, 140–156.

Didden, N., & Ho, C.-M. (1985). Unsteady separation in a boundary layer produced by an impinging jet. *Journal of Fluid Mechanics*, *160*, 235–256.

Dodge, J., Arnold, J., Wilson, G., Evans, J., & Fujita, T. T. (1986). The Cooperative Huntsville Meteorological Experiment (COHMEX). *Bulletin of the American Meteorological Society*, *67*, 417–419.

Droegemeier, K. K., & Wilhelmson, R. B. (1986). Kelvin-Helmholtz instability in a numerically simulated thunderstorm outflow. *Bulletin of the American Meteorological* Society, *67*, 416–417.

Droegemeier, K. K., & Wilhelmson, R. B. (1987). Numerical simulation of thunderstorm outflow dynamics. Part I: Outflow sensitivity experiments and turbulence dynamics. *Journal of Atmospheric Sciences*, *44*, 1180–1210.

Eilts, M. D. (1987). Low altitude wind shear detection with Doppler radar. *Journal of Applied Meteorology and Climatology*, *26*, 96–106.

Emanuel, K. A. (1981). A similarity theory for unsaturated downdrafts within clouds. *Journal of Atmospheric Sciences*, *38*, 1541–1557.

Fritsch, J. M. (1975). Cumulus dynamics: Local compensating subsidence and its implications for cumulus parameterization. *Pure and Applied Geophysics*, *113*, 851–867.

Fujita, T. T. (1951). Micro-analytical study of thunder-nose. *Geophysical Magazine* (Japan), *22*, 78–88.

Fujita, T. T. (1963). Analytical mesometeorology: A review. *Meteorological Monographs*, *5*, 77–125.

Fujita, T. T. (1971). *Proposed characterization of tornadoes and hurricanes by area and intensity* (SMRP Research Paper 91). Chicago, IL: University of Chicago.

Fujita, T. T. (1976). *Spearhead echo and downburst near the approach end of a John F. Kennedy Airport runway, New York City* (SMRP Research Paper 137). Chicago, IL: University of Chicago.

Fujita, T. T. (1978). *Manual of downburst identification for project NIMROD* (SMRP Research Paper 156). Chicago, IL: University of Chicago.

Fujita, T. T. (1979). Objectives, operation, and results of Project NIMROD. In *Preprints of 11th Conference on Severe Local Storms* (pp. 259–266). Kansas City, MO: American Meteorological Society.

Fujita, T. T. (1980). *Miso-scale structure of downbursts depicted by Doppler radar* (SMRP Research Paper 181). Chicago, IL: University of Chicago.

Fujita, T. T. (1981a, January). Microbursts as an aviation wind shear hazard (AIAA-81-0386). In *Preprints AIAA 19th Aerospace Sciences Meeting*, St. Louis, MO.

Fujita, T. T. (1981b). Tornadoes and downbursts in the context of generalized planetary scales. *Journal of Atmospheric Sciences, 38*, 1511–1534.
Fujita, T. T. (1985). *The downburst* (SMRP Research Paper no. 210). Chicago, IL: University of Chicago.
Fujita, T. T. (1986). *DFW microburst on August 2, 1985* (SMRP Research Paper no. 217). Chicago, IL: University of Chicago.
Fujita, T. T. (1990). Downbursts: Meteorological features and wind field characteristics. *Journal of Wind Engineering and Industrial Aerodynamics, 36*, 75–86.
Fujita, T. T. (1992). *The mystery of severe storms* (WRL Research Paper no. 239). Chicago, IL: University of Chicago.
Fujita, T. T., & Byers, H. R. (1977). Spearhead echo and downburst in the crash of an airliner. *Monthly Weather Review, 105*, 129–146.
Fujita, T. T., & Caracena, F. (1977). An analysis of three weather-related aircraft accidents. *Bulletin of the American Meteorological Society, 58*, 1164–1181.
Fujita, T. T., & Pearson, A. D. (1973). *Results of FPP classification of 1971 and 1972 tornadoes.* Preprints, 8th Conf on Severe Local Storms, Denver, CO, American Meteorological Society, 142–145.
Fujita, T. T., & Wakimoto, R. M. (1981). Five scales of airflow associated with a series of downbursts on 16 July 1980. *Monthly Weather Review, 109*, 1511–1534.
Fujita, T. T., & Wakimoto, R. M. (1983, September). Microbursts in JAWS depicted by Doppler radars, PAM, and aerial photographs. In *Preprints 21st Conference on Radar Meteorology* (pp. 638–645), September 19–23, Edmonton, Canada, American Meteorological Society.
Gedzelman, S. D. (1990). Leonardo da Vinci and the Downburst. *Bulletin of the American Meteorological Society, 71*, 649–655.
Glauert, M. B. (1956). The wall jet. *Journal of Fluid Mechanics, 1*, 625–643.
Goff, R. C. (1976). Vertical structure of thunderstorm outflows. *Monthly Weather Review, 104*, 1429–1440.
Goff, R. C. (1980). *Low-level wind shear alert system (LLWSAS)* (FAA-RD-80-45). Washington, DC: Federal Aviation Administration.
Hall, F. F., Neff, W. D., & Frazier, T. V. (1976). Wind shear observations in thunderstorm density currents. *Nature, 264*(5585), 408–411.
Hamilton, R. E., (1970). *Use of detailed intensity radar data in mesoscale surface analysis of the 4 July 1969 storm in Ohio*. Preprints, 14th Conf. on Radar Meteorology, Tucson, AZ, American Meteorological Society, 339–342.
Hjelmfelt, M. R. (1987). The microburst of 22 June 1982 in JAWS. *Journal of Atmospheric Sciences, 44*, 1646–1665.
Hjelmfelt, M. R. (1988). Structure and life cycle of microburst outflows observed in Colorado. *Journal of Applied Meteorology, 27*, 900–927.
Hjelmfelt, M. R., Orville, H. D., Roberts, R. D., Chen, J. P., & Kopp, F. J. (1989). Observational and numerical study of microburst line-producing storm. *Journal of Atmospheric Sciences, 46*, 2731–2743.
Hjelmfelt, M. R., Roberts, R. D., & Orville, H. D. (1986, September). *Observational and numerical study of microburst line-producing storm.* Paper presented at Joint Sessions of 23rd Conference on Radar Meteorology and Conference on Cloud Physics, Snowmass, CO, American Meteorological Society, J77–J80.
Ivan, M. (1986). A ring-vortex downburst model for flight simulations. *Journal of Aircraft, 23*, 232–236.

Keulegan, G. H. (1958). *Twelfth progress report on model laws for density currents: The motion of saline fronts in still water.* Washington, DC: US Department of Commerce, National Bureau of Standards.

Kingsmill, D. E., & Wakimoto, R. M. (1991). Kinematic, dynamic, and thermodynamic analysis of a weakly sheared severe thunderstorm over Northern Alabama. *Monthly Weather Review, 119*, 262–297.

Knight, C. A. (1982). The Cooperative Convective Precipitation Experiment (CCOPE), 18 May–7 August 1981. *Bulletin of the American Meteorological Society, 63*, 386–398.

Knupp, K. R. (1987). Downdrafts within High Plains cumulonimbi. Part I: General kinematic structure. *Journal of Atmospheric Sciences, 44*, 987–1008.

Knupp, K. R. (1988). Downdrafts within High Plains cumulonimbi. Part II: Dynamics and thermodynamics. *Journal of Atmospheric Sciences, 45*, 3965–3982.

Knupp, K. R. (1989). Numerical simulation of low-level downdraft initiation within precipitating cumulonimbi: Some preliminary results. *Monthly Weather Review, 117*(7), 1517–1529.

Knupp, K. R., & Cotton, W. R. (1985). Precipitating convective cloud downdraft structure: An interpretive survey. *Review of Geophysics, 23*, 183–215.

Krueger, S. K., & Wakimoto, R. M. (1985, October–November). Numerical simulation of dry microbursts. In *Preprints of the 14th Conference on Severe Local Storms* (pp. 163–166). Indianapolis, IN: American Meteorological Society.

Lamb, H. (1945). *Hydrodynamics* (6th ed.). New York, NY: Dover.

Lin, Y.-J., & Hughes, R. G. (1987). Structural features of a microburst-producing storm in Colorado Revealed by JAWS Dual-Doppler Radars. *Journal Atmospheric Sciences, 44*, 3640–3655.

Lundgren, T. S., Yao, J., & Mansour, N. N. (1992). Microburst modelling and scaling. *Journal of Fluid Mechanics, 239*, 461–488.

McCann, D. W. (1994). WINDEX—A new index for forecasting microburst potential. *Weather Forecasting, 9*, 532–541.

McCarthy, J., Wilson, J. W., & Fujita, T. T. (1982). The Joint Airport Weather Studies Project. *Bulletin of the American Meteorological Society, 63*, 15–22.

Middleton, G. V. (1966). Experiments on density and turbidity currents: I. Motion of the head. *Canadian Journal of Earth Sciences*, 3(4), 523–546.

Mueller, C. K., & Carbone, R. E. (1987). Dynamics of a thunderstorm outflow. *Journal of the Atmospheric Sciences, 44*(15), 1879–1898.

Orville, H. D., Farley, R. D., Chi, Y.-C., & Kopp, F. J. (1989). The primary cloud physics mechanisms of microburst formation. *Atmospheric Research, 24*, 343–357.

Oseguera, R. M., & Bowles, R. L. (1988). *A simple, analytic 3-dimensional downburst model based on boundary layer stagnation flow* (NASA Technical Memorandum 100632). Hampton, VA: National Aeronautics and Space Administration, Langley Research Center.

Paluch, I. R. (1979). The entrainment mechanism in Colorado Cumuli. *Journal of Atmospheric Sciences, 36*, 2467–2478.

Parker, M. D., & Dahl, J. M. L. (2015). Production of near-surface vertical vorticity by idealized downdrafts. *Monthly Weather Review, 143*, 2795–2816.

Proctor, F. H. (1985, June). *Numerical simulation of precipitation induced downbursts.* Presented at the Second International Conference on the Aviation Weather System (pp. 291–298), Montreal, Canada, American Meteorological Society.

Proctor, F. H. (1987a). *The Terminal Area Simulation System. Volume I: Theoretical formulation* (NASA Contractor Rep. 4046). Washington, DC: NASA.

Proctor, F. H. (1987b). *The Terminal Area Simulation System. Volume II: Verification experiments* (NASA Contractor Rep. 4047). Washington, DC: NASA.

Proctor, F. H. (1988). Numerical simulations of an isolated microburst. Part I: Dynamics and structure. *Journal of Atmospheric Sciences, 45*, 3137–3160.

Proctor, F. H. (1989). Numerical simulations of an isolated microburst. Part II: Sensitivity experiments. *Journal of Atmospheric Sciences, 46*, 2143–2165.

Renick, J. H. (1975). The Alberta Hail Project. *Journal of Weather Modification, 7*, 1–6.

Roberts, R. D., & Wilson, J. W. (1989). A proposed microburst nowcasting procedure using single-Doppler radar. *Journal of Applied Meteorology, 28*, 285–303.

Rotunno, R., & Klemp, J. (1985). On the rotation and propagation of simulated supercell thunderstorms. *Journal of Atmospheric Sciences, 42*, 271–292.

Schultz, T. A. (1990). Multiple vortex ring model of the DFW microburst. *Journal of Aircraft, 27*, 163–168.

Sherman, D. J. (1987). The passage of a weak thunderstorm downburst over an instrumented tower. *Monthly Weather Review, 115*, 1193–1205.

Simpson, J. E. (1969). A comparison between laboratory and atmospheric density currents. *Quarterly Journal of the Royal Meteorological Society, 95*, 758–765.

Simpson, J. E. (1972). Effects of the lower boundary on the head of a gravity current. *Journal of Fluid Mechanics, 53*(4), 759–768.

Smith, P. L., Dennis, A. S., Silverman, B. A., Super, A. B., Holroyd, E. W., III, Cooper, W. A., Mielke, P. W. Jr., Berry, K. J., Orville, H. D., & Miller, J. R., Jr. (1984). HIPLEX-1: Experimental design and response variables. *Journal of Climate and Applied Meteorology, 23*, 497–512.

Solari, G. (2019). *Wind science and engineering: Origins, developments, fundamentals and advancements*. New York, NY: Springer.

Squires, P. (1958). Penetrative downdraft in cumuli. *Tellus, 10*, 381–389.

Srivastava, R. C. (1985). A simple model of evaporatively driven downdraft: Application to microburst downdraft. *Journal of Atmospheric Sciences, 42*, 1004–1023.

Srivastava, R. C. (1987). A model of intense downdrafts driven by the melting and evaporation of precipitation. *Journal of Atmospheric Sciences, 44*, 1752–1773.

Straka, J. M., & Anderson, J. R. (1993). Numerical simulations of microburst-producing storms: Some results from storms observed during COHMEX. *Journal of Atmospheric Sciences, 50*, 1329–1348.

Tripoli, G. J., & Cotton, W. R. (1982). The Colorado State University three-dimensional cloud/mesoscale model—Part I: General theoretical framework and sensitivity experiments. *Journal de Recherches Atmospheriques, 16*, 185–220.

Vermeire, B. C., Orf, L. G., & Savory, E. (2011). A parametric study of downburst line near-surface outflows. *Journal of Wind Engineering and Industrial Aerodynamics, 99*, 226–238.

Vicroy, D. D. (1991). *A simple, analytical, axisymmetric microburst model for downdraft estimation* (NASA Technical Memorandum 104053 DOT/FAA/RD-91/10). Hampton, VA: National Aeronautics and Space Administration, Langley Research Center.

von Kármán, T. (1940). The engineer grapples with nonlinear problems. *Bulletin of the American Mathematical Society, 46*(8), 615–683.

Wakimoto, R. M. (1982). The life cycle of thunderstorm gust fronts as viewed by Doppler radar and rawinsonde data. *Monthly Weather Review, 110*, 1060–1082.

Wakimoto, R. M. (1985). Forecasting dry microburst activity over the High Plains. *Monthly Weather Review, 113*, 1131–1143.

Weisman, M. L. (2001). Bow echoes: A tribute to T. T. Fujita. *Bulletin of the American Meteorological Society, 82*, 97–116.

Wilson, J. W., Moore, J. A., Foote, G. B., Martner, B., Rodi, A. R., Uttal, T., & Wilczak, J. M. (1988). Convection Initiation and Downburst Experiment (CINDE). *Bulletin of the American Meteorological Society, 69*, 1328–1348.

Wilson, J. W., Roberts, R. D., Kessinger, C., & McCarthy, J. (1984). Microburst wind structure and evaluation of Doppler radar for airport wind shear detection. *Journal of Climate and Applied Meteorology, 23*, 989–915.

Wilson, J. W., & Wakimoto, R. M. (2001). The discovery of the downburst: T. T. Fujita's contribution. *Bulletin of the American Meteorological Society, 82*, 49–62.

Wingrove, R. C., & Bach, R. E. (1987). *Severe winds in the DFW microburst measured from two aircraft* (AIAA Paper 87-2340-CP), AIAA Guidance, Navigation and Control Conference, Monterey, California, USA, American Institute of Aeronautics and Astronautics (AIAA), 477–482.

Wolfson, M. M. (1990). *Understanding and predicting microbursts* (Unpublished PhD thesis). Massachusetts Institute of Technology, Cambridge, MA.

Wolfson, M. M., DiStefano, J. T., & Forman, B. E. (1987). *The FLOWS (FAA-Lincoln Laboratory Operational weather Studies) automatic weather station network in operation* (Project Report ATC-134). Lexington, MA: Lincoln Laboratory, Massachusetts Institute of Technology.

Wolfson, M. M., DiStefano, J. T., & Fujita, T. T. (1985, October–November). Low-altitude wind shear characteristics in the Memphis, TN area based on Mesonet and LLWAS data. In *Preprints, 14th Conference on Severe Local Storms* (pp. 322–327). Indianapolis, IN: American Meteorological Society.

Yao, J., & Lundgren, T. S. (1996). Experimental investigation of microbursts. *Experiments in Fluids, 21*, 17–25.

Zhu, S., & Etkin, B. (1985). Model of the wind field in a microburst. *Journal of Aircraft, 22*, 595–601.

CHAPTER 5

CHARACTERISTICS OF DOWNSLOPE WIND STORMS IN THE VIEW OF THE TYPICAL ATMOSPHERIC BOUNDARY LAYER

HRVOJE KOZMAR AND BRANKO GRISOGONO

1. Introduction

Civil engineering infrastructure has been commonly designed according to relevant standards and codes—this is obligatory for the contractors and the owners. An important portion of those standards and codes focuses on wind-induced damage and structural collapse, whereas the wind characteristics are, for the most part, represented as the atmospheric boundary layer (ABL). At the same time, only in North America it has been observed that 65 percent of the wind-induced damage is due to non-synoptic, local wind storm events that may be very different from the ABL.

At this moment, there are no clear guidelines or code implementations on the characteristics of local, mountainous wind storms. It is therefore of great interest to learn more about those winds and provide their characteristics in a form usable for engineers. The wind storms in mountainous regions have been previously thoroughly assessed in the meteorology and geophysics communities; however, their focus is quite different from that in the wind engineering field, and the available data on characteristics of mountainous wind storms are of marginal relevance for engineering applications.

Some people say, consider Earth greatly downscaled to a tennis ball, it will be smoother than a billiard ball. But there is a crux: The relevant atmosphere here is shallow, its density scale is only about 8.5 km deep compared, for example, with Earth's radius; thus, it is extremely sensitive to vertical motions, at least for two major reasons. One is high amounts of water vapor, so that any vertical motion might lead to air saturation, eventual condensation, and deep convection, including organized, non-synoptically driven airflows such as multicells, supercells, squall lines, mesoscale meteorological systems, and more (these may generate, e.g., tornadoes, flash, flooding, etc.). Another reason—largely the subject of this chapter—is that generally strong static stratification of the atmosphere, which also keeps the air close to Earth, generates buoyancy waves and resists vertical motions (e.g., Smith 1979, 1985). Hence, the air must often undertake large horizontal excursions, and, when inevitable, it yields strong-to-severe downslope winds, which may easily exceed hurricane speeds (>33 m/s) and generate severe turbulence near the surface and aloft. Meanwhile, the world's population is largely affected by mountainous conditions because mountains cover 25 percent of Earth's surface, they are home to 26 percent of the world's total population, and they generate 32 percent of the surface runoff. Moreover, plateaus and hills cover another 21 percent of Earth's surface, contain an additional 20 percent of the population, and yield 19 percent of the runoff (e.g., Chow, de Wekker, & Snyder, 2013).

Different, yet significant, regionally decisive airflows are generated by mountains; these are usually segregated into thermally driven (e.g., katabatic and anabatic winds with wind speed commonly smaller than 10 m/s, except for Antarctica) and dynamically (mechanically) driven circulations (e.g., Bora, Chinook, Foehn, Helm, Mistral, Santa Ana, Zonda, etc.). For a recent review of the former type, see Zardi and Whiteman (2013); for the latter type, see Jackson, Mayr, and Vosper (2013) and a special issue of *Wind and Structures* (Kozmar, 2017). These winds, especially severe downslope wind storms, together with the related gap winds and cold-air pools, determine the ABL local properties, loads on engineering structures, dispersion of pollutants, road safety and overall traffic, crop damage, and more (e.g., Baklanov et al., 2011). To reword, strong downslope wind storms related to large-amplitude mountain waves (e.g., severe Foehn) can easily cause significant damage that is typically not considered by engineering design, standards, and codes.

Simultaneously, these mountain-wave breakings in the atmosphere are most often related to wild hydraulic-like jumps (e.g., Dörnbrack, 1998; Enger & Grisogono, 1998; Klemp & Durran, 1987; Smith, 1987) or rotors, even with dangerous subrotors with excessive turbulence, in the lee of mountains (Doyle & Durran, 2004, 2007; Doyle et al., 2009), or roll vortices with a horizontal wavelength of about 1 km, comparable to the corresponding ABL depth, which generate a great variability of wind speeds over short distances, comparable to street sizes or less. Rotors (intense low-level vortices with horizontal rotation axes that are observed parallel and downstream of a mountain related to large-amplitude mountain wave) immediately correspond to the adverse pressure gradient in the mountain lee, due to large-amplitude waves, yielding to the ABL separation and more (e.g., Grubišić & Billings, 2008; Sachsperger, Serafin, & Grubišić, 2015; Strauss, Serafin, & Grubišić, 2016).

In fact, one of the most important recent megaprojects related to mountainous severe weather is T-REX, summarized by Grubišić et al. (2008). In T-REX (Terrain-Induced Rotor Experiment), the most advanced observational and modeling tools to assess airflows from Earth's surface to the lower stratosphere over the Sierra Nevada in the United States, were used to unveil wild rotors in the air (see, e.g., https://www.eol.ucar.edu/node/666/publications). It is appealing that such rotors, induced by strong-to-severe downslope winds, were already studied more than 130 years ago by the geophysical community (Mohorovičić, 1889) and as a follow-up (Grubišić & Orlić, 2007). However, so far these winds have not been accordingly considered by engineering standards and codes.

For a review of stratified airflow over orography see Smith (2002). A fresh and timely set of papers related to mountainous weather was compiled by Teixeira, Kirshbaum, Ólafsson, Sheridan, and Stiperski (2016). This work indicated that field campaigns like MAP, T-REX, Materhorn, COLDPEX, and i-Box, famous field experiments over the Alps, Sierra Nevada, broader region of Utah and the Alps again (all accessible via web and the papers cited), containing a wealth of mountain meteorology measurements, are only starting to be explored in depth. That should lead to important advances in understanding weather in mountainous regions, as well as mixing and pollutant dispersion over complex terrain. The latter is sometimes still mistreated by the geophysical community by using a first-order K theory for parameterizing turbulent fluxes, which is not appropriate for that purpose (e.g., Pope, 2000; Wyngaard, 2010).

The scope of the present chapter is to address characteristics of some major mountainous winds in comparison with the standard ABL, that is, Santa Ana (West Coast of North America), Foehn (European Alps), downslope winds of Liguria (Europe), Bora (Adriatic region in Europe), downslope winds of Japan (Asia), and Zonda (Andes in Latin America). The information presented is expected to enhance creating a platform for implementing the characteristics of local mountainous winds in the international standards and codes.

Structurally, we opted for a scientific review study instead of a pure or applied corresponding timely paper, for which there are ample results in terms of field campaigns, laboratory experiments, and advanced computational simulations. The option chosen is because the community yearns urgently for a better leveling off and a better mutual correspondence between the engineering and geophysical, say, meteorological and climatological communities due to rapid recent climate changes, for an optimized collaboration, for better mutual shares in education, just to mention a few.

2. The Essence of Mountainous Wind Storms

Although interactions between thermally forced and mechanically (dynamically) driven airflows are possible, and in fact such interactions are ubiquitous during the course of the day (e.g., Poulos, Bossert, McKee, & Pielke, 2000, 2007), it appears that the latter flow type is generally stronger (e.g., Chow et al., 2013; Smith, 1979). Hence, we focus on that.

To put it most boldly and simply, those most severe downslope winds appear wherever a resonance between the incoming stratified flow (wind speed U and buoyancy frequency N) and the underlying orography (height H), such as an elongated mountain, occurs (leave aside various, often secondary yet important, details, e.g., the width and length of the mountain, eventually multilayered atmosphere, etc.). The vertical Froude number, $Fr = U/(NH)$, or its inverse, which is dimensionless mountain height, $\hat{H} = HN/U$, describes the resonance. Typically, for $0.15 \leq Fr \leq 1.2$, large mountain waves break due to the resonance (e.g., Hunt, Feng, Linden, Greenslade, & Mobbs, 1997; Klemp & Durran, 1987; Ólafsson & Bougeault, 1996; Smith, 1985, 1987) and produce severe winds below the wave-breaking region (Figure 5.1).

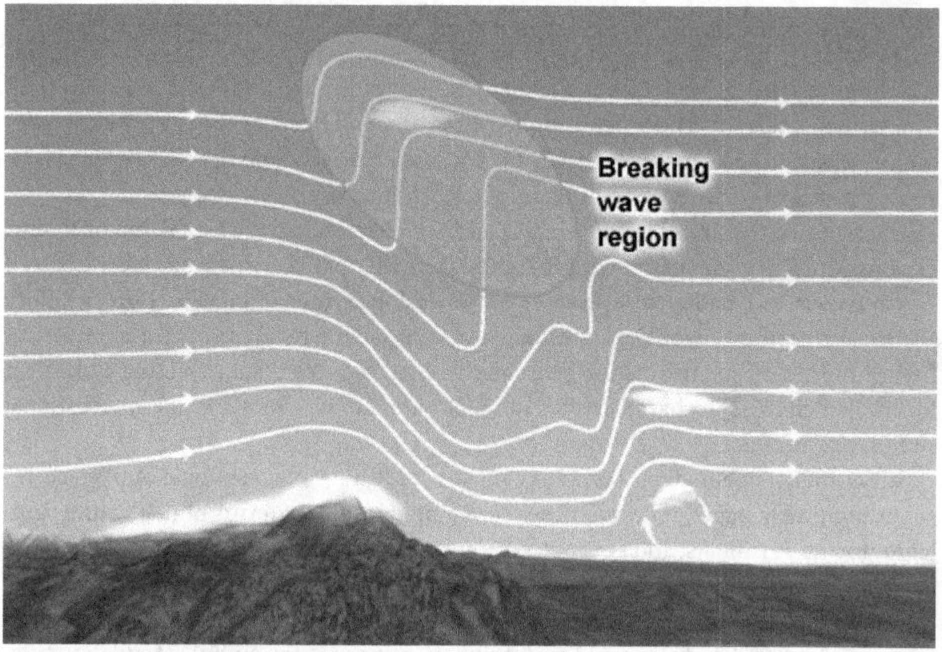

FIGURE 5.1 Sketch of a large mountain wave that breaks. Downslope wind storm occurs in the lee of the mountain, where the streamlines are the closest; there and below is a shooting flow. The corresponding supercritical flow usually ends with a hydraulic-like jump or an undular bore, both sometimes followed by a rotor (also drawn). The sketch originally provided for Boulder-Denver, Colorado, in the United States, wind storm also applies to other similar winds, such as Bora, Foehn, Santa Ana, and Sierra Nevada wind storms; wind storms of Japan; and so on wherever noticeable flow blocking occurs. (Source: COMET® website at http://meted.ucar.edu/ of the University Corporation for Atmospheric Research [UCAR], sponsored in part through cooperative agreement[s] with the National Oceanic and Atmospheric Administration [NOAA], US Department of Commerce [DOC]. © 1997–2017 University Corporation for Atmospheric Research. All rights reserved.)

At the lower range of those critical Froude numbers, wave breaking occurs only at the flanks of mountains and is more localized. Somewhat above that critical range of *Fr*, waves only steepen but do not overturn and break. For $Fr \geq 2$, the mountain waves become increasingly linear and smooth, without patches of turbulence. Almost needless to say, the linear wave theory cannot explain mountain wave breaking (e.g., Smith, 1979); the theory is practically invalid for $Fr \leq 1$, theoretically even before the quantitative errors may go beyond a factor of two and more; yet, some qualitative points based on linear wave theory are still valid (e.g., Hunt et al., 1997; Smith, 2002). It is conceivable, understandable, and theoretically proven that for mountain shapes prone to yield severe downslope wind storms, their height on the windward site should gradually increase, so that a strong standing wave can build up (for a ridge $H = 1$ km, 20 km wide, typical $U = 10$ m/s and $N = 0.01$/s, it takes about 3 hr to make a strong mountain wave, based on a rough hydrostatic estimate of the vertical component of the wave group velocity); on the contrary, the lee side ought to be rather steep (Lilly & Klemp, 1979; Smith, 1977, 1979). All that is closely related to significant airflow blockings by orographical obstacles.

Mountain pressure drag (MPD) is one of the best measures of mountain effects, including nonlinearities and gross turbulence, on incoming stratified airflows (Smith, 1978; Teixeira, 2014; Teixeira, Miranda, & Cardoso, 2008). It is an integral of the horizontal pressure difference over a mountain in the main wind direction perpendicular to the mountain. The point is that for severe downslope wind storm values of the MPD, the wave drag is a large contributor to the mountain pressure drag, by a factor of three and more beyond their linearly corresponding values (Durran, 1986, 1990). Unfortunately, the MPD is not readily available from regular numerical weather prediction (NWP) models; however, it can and it should be included for simple, obtainable, and reliable preliminary diagnostics of mountainous weather effects for various engineering applications. Furthermore, the MPD behavior in future climates could be estimated from regional climate models that use horizontal resolution below 10 km. One of the recommendations of this study is to use the MPD diagnostics and to apply it to certain areas of interest, that is, to the areas prone to severe wind storms to reliably estimate wind loads on engineering infrastructure.

Going back to the processes tackled, Figure 5.2 shows one typical setup with $Fr \sim 0.5$ for severe turbulence in the lee of a mountain. The largest near-surface wind speeds occur in the immediate lee (already discussed), while gusty, unsteady turbulence appears several kilometers further downwind and gradually dies out (not shown) a few tens of kilometers downwind of the obstacle (e.g., Klemp & Lilly, 1978; Lilly, 1978).

In the immediate lee of a mountain where a severe wind storm occurs, there is another plausible, yet frequent, process that generates additional turbulence—quasiperiodic oscillations in the mean wind speed (i.e., pulsations). These often come from Kelvin-Helmholtz instabilities, that is, secondary atmospheric waves' instabilities, although other mechanisms are possible as well (Scinocca & Peltier, 1993, 1994). The streamline on the right-hand side in the lee of the mountain develops upward similar to a sine function due to the observed complex physical phenomena that include rotors and subrotors. As an example, Figure 5.3 shows such pulsations in the Bora wind at the Adriatic

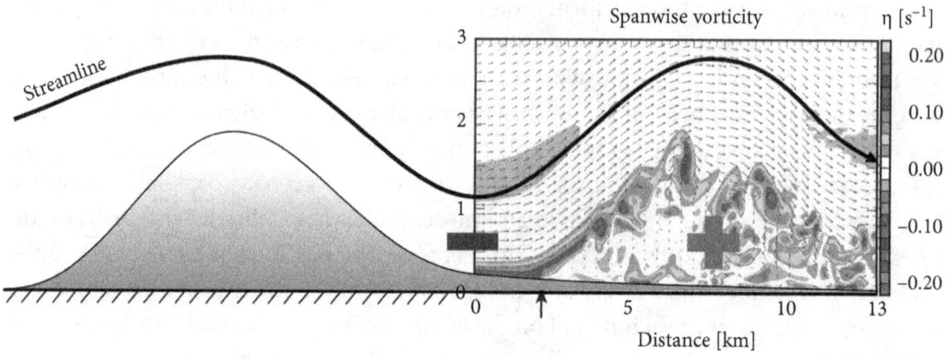

FIGURE 5.2 Severe downslope wind storm near the ground (minus sign corresponding to $U \sim$ 25 m/s) ending with a rotor and subrotors around (plus sign). The airflow is from left to right, and a large-amplitude mountain wave is generated due to a resonance between the incoming airflow and mountain. Intensive horizontal vorticity snapshot is marked as η having both positive and negative huge values ~ 0.2/s; these are a thousand times larger than a typical vertical (i.e., dominant) value for an atmospheric weather system. Wilderness and severity of this unsteady vorticity is largely due to 3D subrotors and their turbulence; there, vertical accelerations may go up to $\pm 5g$ (g is acceleration due to gravity) (adapted from Doyle and Durran, 2007, and with courtesy of Johannes Sachsperger).

Coast. The most exquisite thing about Bora is not its mean wind speed but its frequent gustiness, which occurs at several distinct frequency ranges (e.g., Belušić, Žagar, & Grisogono, 2007) while reaching hurricane speeds. One way to conceive and comprehend such a shooting flow (e.g., Klemp & Durran, 1987) is a variable and three-dimensional (3D), localized, low-level jet prone to its own instabilities. It is important to note that Šoljan et al. (2018) showed that Bora turbulence properties do not depend on the synoptic setup.

Other strong winds can be generated by partial flow separation in the lee of mountains (buoyancy forces negligible), generation of trapped lee waves, upstream blocking of near-surface cold air, and interaction between mountain ridge/range for a stable ABL (Chow et al., 2013; Smith, 1978, 1979); the last two include gap winds and barrier jets. The main feature of gap winds is the asymmetry between the region upstream of the gap entrance and downstream of the gap. The upstream layer flowing through the gap is thick and slow (in hydraulic terminology, it is subcritical). Approaching the gap, it gets thinner, accelerates, and becomes supercritical downstream of the gap. In that sense, especially with maximum $U \sim 25$ m/s, it appears almost as an atmospheric wave-breaking phenomenon, yielding the severe wind in the lee; however, gap winds do not require a complex vertical airflow structure, including stratification effects. We mention barrier jets also, one of at least four different kinds of tropospheric low-level jets (coastal, nocturnal, etc.) (see, e.g., Rogers et al., 1998, and Jackson et al., 2013).

In general, low-level jets are important because they transport air properties over long distances, sometimes over thousands of kilometers; moreover, due to their intrinsic

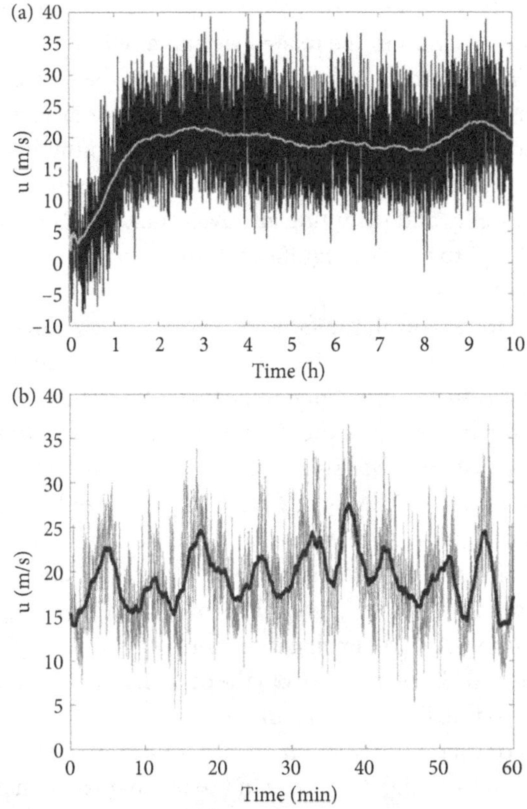

FIGURE 5.3 Pulsations of the Bora wind speed, the northeastern Adriatic coast, Croatia. The along-Bora wind component in the town of Senj is shown: (a) 1-s time series of a part of a Bora episode (black) with 1-h mean superimposed (gray); (b) 1-h expanded view of the sixth hour of the episode of the 1-s series (gray) with 2-min mean superimposed (black) showing pulsations (adapted from Grisogono and Belušić, 2009).

instabilities, they may affect ABL properties significantly, for example through top-down turbulence (e.g., Wyngaard, 2010). A barrier jet is an elevated wind maximum on the windward side of a mountain obstacle that blows parallel to the barrier around the ABL top. Barrier jets appear where stably stratified flow approaches an extratropical mountain obstacle and the airflow is blocked by the barrier for hours or longer. Those types of winds are typically but not necessarily more localized winds, at least in terms of two coordinates, compared to severe downslope wind storms (e.g., Jackson et al., 2013). Where the mountain range is sufficiently long and perpendicular to the incoming airflow, say, at least 150 km and more, then the rotation of Earth becomes important, namely, the Coriolis force (Hunt, Olafsson, & Bougeault, 2001; Smith, 1979).

On the northern hemisphere, when the mean westerly airflow approaches such a long barrier and it inevitably senses a lower pressure on its left-hand side (i.e., toward north [due to a geostrophic balance between the pressure gradient force and Coriolis force]), it naturally turns there. This causes a stronger airflow at the northern flanks of the barrier, including anticyclonic rotation in the related lee and a relatively stronger low-level jet, all compared to the flow at and after the southern flank of the barrier (cyclonic rotation, a weaker jet, and perhaps more humidity). That pertains, for example, to the Rockies in the United States, to some extent to the Scandinavian mountains and the Alps, and more; however, it is difficult to observe that effect due to numerous other influences, such as detailed orography and incoming airflow structures. The opposite sense of the airflow turning while approaching a significant mountain occurs at the southern hemisphere (e.g., Smith, 1979).

It is inconceivable to review all the famous types of severe downslope wind storms and strong gap winds in this chapter. For instance, severe Chinook and Foehn appear with partial or almost complete blocking of the lower troposphere airflow due to 3-km high mountains or even higher; thus, the air from the higher troposphere, or sometimes even from the tropopause or lower stratosphere, must subside to compensate for the blocking (e.g., Hunt et al., 1997; Lilly, 1978; Lilly & Klemp, 1979; Mayr et al., 2018; Richner & Hächler, 2013; Smith, 1979). Of course, some of the air flows around the terrain slopes and over its flanks. These winds are usually warm since the air from high aloft has a larger potential temperature than the air in the ABL. A potential release of precipitation may add to that warming. A list of explanations and definitions of Foehn-related terms was given by Richner and Hächler (2013). On the other hand, if the mountains are lower, say $H \sim 1$ km, only partial airflow blocking may occur.

Typical examples of that phenomenon are severe Bora types of flow. Dynamically speaking, there is no substantial difference in the mechanisms; that is, all three kinds of downslope wind storms are very similar, although their details vary. Bora-type winds are usually cold. First, because the compensating air from aloft is relatively closer to the ABL, when compared to Chinook and Foehn, so that there is no substantial downslope warming in Bora. Furthermore, Bora (from the old Greek language for northerly flows) usually comes from northerly directions, which brings generally cooler air in the northern hemisphere. To add a point, the Adriatic Bora, which blows more frequently in winter, is usually induced by a high-pressure system over central Europe or Siberia, which brings cooler air to the Mediterranean, including the Adriatic Sea. There are two other major synoptic setups for Bora, but they also end up with cold winds over the relatively warmer coastal sea in winters (e.g., Jurčec, 1981). It is unfortunate that some websites, including Wikipedia, classify Bora wrongly as a falling (e.g., katabatic) wind instead of a downslope wind storm.

It is interesting that the term *Bora* is also applied to similar winds in other parts of the world. That includes, but it is not limited to, Novaya Zemlya in the Arctic, Novorossiysk at the northern coast of the Black Sea, the Gulf of Tehuantepec on the Pacific coast of Mexico, Salt fiord and Bals fiord in northern Norway, and more (e.g., Yoshino, 1976). Another curious point about Bora is that it was classified as a downslope wind storm only 34 years ago (Klemp & Durran, 1987; Smith, 1987). Then, the whole concept for this wind changed (from a falling, i.e., mostly thermally driven flow) into dynamically forced airflow. The mountain wave overturning that occurs beneath the temperature inversion appears as the most important factor in producing the strong-to-severe response. This overturning generates a shooting flow over the lee slope and resembles the hydraulic-like flow, undular bore, and more (e.g., Chow et al., 2013; Sachsperger et al., 2015). The next section outlines the most important currently known points on the discrepancies in characteristics of mountain wind storms and the ABL that still remain to be entirely accounted for in the major international engineering standards and codes.

3. Downslope Wind Storms Versus Typical Atmospheric Boundary Layer

3.1. Thermal Stratification

Thermal stratification in the ABL is a consequence of the balance between the thermally and mechanically induced mixing of airflows in the lower atmosphere. It may substantially influence wind characteristics at small wind velocities (i.e., smaller than 5 m/s) and therefore change patterns in air pollution dispersion and dilution, urban micrometeorology, and so on. The ABL thermal stratification can be (very) stable, neutral, or unstable depending on the heat gradient between the ground, water, or ice surface and the airflows above it (Figure 5.4; see, e.g., Serafin et al., 2018).

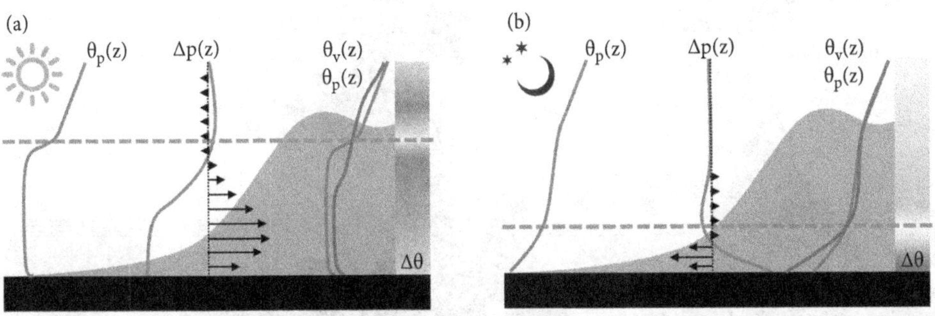

FIGURE 5.4 Stable, neutral, and unstable types of ABL thermal stratification (after Serafin et al. 2018); thermally driven winds in a valley with horizontal ground surface. (a) During the day, the atmosphere in the valley (ϑ_v profile) is warmer than that over the plain (ϑ_p profile). The largest imbalances ($\Delta\vartheta = \vartheta_v - \vartheta_p$) are often found near the mountaintop level. Horizontal pressure differences ($\Delta p = p_v - p_p$) result quasi-hydrostatically from the vertically integrated temperature imbalances and are typically largest at the valley ground surface. The wind responds to the pressure gradient and is decelerated by friction near the ground. Deeper convective mixing in the valley causes an elevated cool anomaly, which may drive an upper level return flow. (b) The opposite situation occurs at night (image used in accordance with Creative Commons Attribution 4.0 International License).

Rotach and Zardi (2007) described numerous meteorological and even hydrological issues about the ABL over complex terrain, especially over the European Alps during the MAP project. Their work was put in a broader perspective by Serafin et al. (2018). One of the key issues is that there is frequent inadequacy of the Monin-Obukhov similarity theory (e.g., Grisogono & Oerlemans, 2001), which has significant consequences not only for data processing, but also for the NWP and climate models.

Numerous further issues about weather forecasting and regional climate modeling over complex terrain, which definitely is an unsettled problem, were tackled, for example, by Zhong and Whiteman (2008) and Chow et al. (2013). Various issues arise, ranging from numerical stability aspects, to choice of non-orthogonal coordinate frame, and to parameterizations of subgrid-scale processes (radiation, clouds, fog, precipitation, [sub]surface processes and turbulence). Figures 5.5 and 5.6 add on ABL complexity by sketching mountainous low-level airflow conditions; needless to say, these situations are not easy to predict by operational NWP models (e.g., Sun et al., 2015). While Figure 5.5 displays the frequent multilayer structure of the ABL over complex terrain, including multiple temperature inversions, Figure 5.6 sketches the corresponding stably stratified nocturnal ABL. The latter includes, yet it is not limited to, katabatic (thermally driven downslope) wind, low-level jet, fine-scale turbulence, and air-pollution issues. All that may interact with a downslope wind storm (e.g., Poulos et al., 2000; Rotach & Zardi, 2007) or even be flushed away (e.g., Jackson et al., 2013).

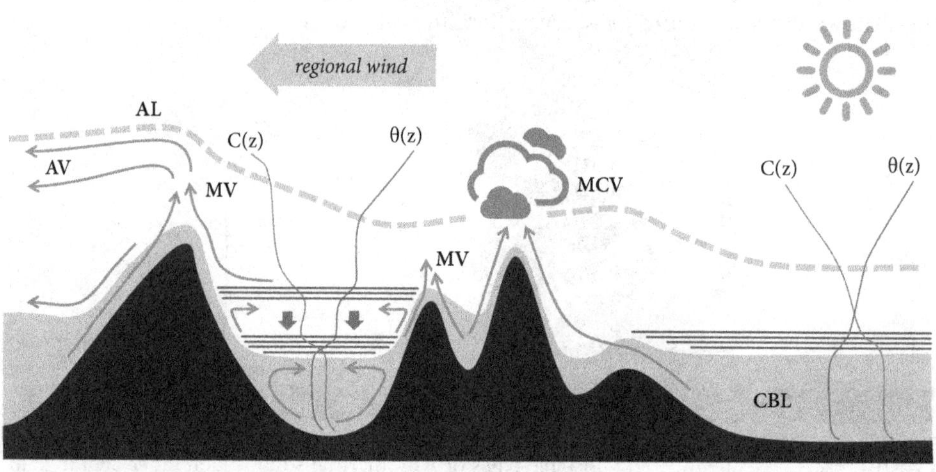

FIGURE 5.5 Similar to Figure 5.4 but for the ABL over mountainous terrain yielding to numerous flow complexities (adopted from Serafin et al., 2018). Exchange processes in the daytime boundary layer over mountainous terrain. Gray shading indicates the ground-based mixed layer (convective boundary layer, CBL). MV, AV, and MCV denote, respectively, mountain venting, advective venting, and mountain cloud venting. Arrows indicate airflow, while $C(z)$ and $\vartheta(z)$ indicate vertical profiles of pollutant concentration and potential temperature, respectively. Horizontal lines represent layers with enhanced static stability, which favor the separation of upslope flows from the ground. Down-pointing arrows represent valley–core subsidence. The dashed line indicates the top of the regional aerosol layer (AL) (image used in accordance with Creative Commons Attribution 4.0 International License).

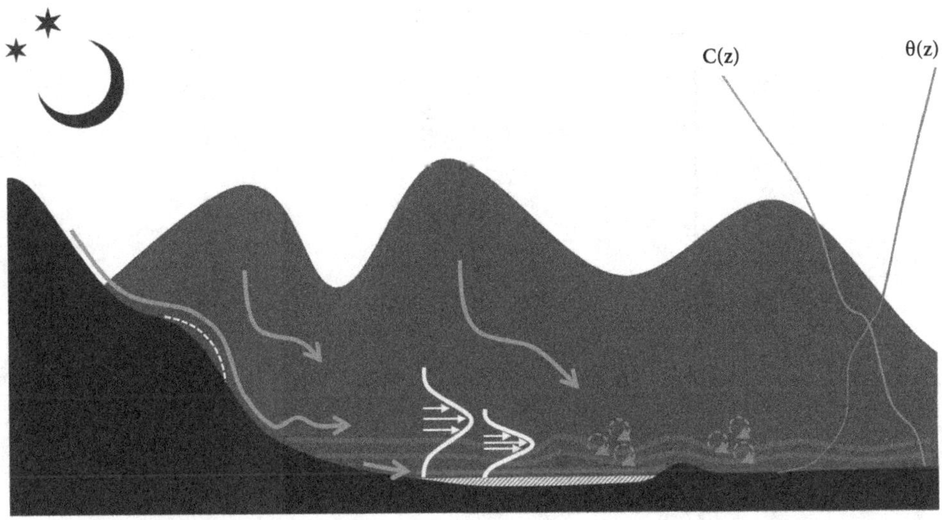

FIGURE 5.6 Similar to Figures 5.4 and 5.5 but for the nighttime (stable) ABL over mountainous terrain. A review of interactions between the stable ABL and short buoyancy waves, as indicated here, can be found in the work of Sun et al. (2015) (adopted from Serafin et al., 2018; image used in accordance with Creative Commons Attribution 4.0 International License).

On the other hand, there are strong indications that the downslope wind storms are thermally neutrally stratified due to intense mechanical mixing of the airflows as they develop over complex mountainous terrains, whereas the thermal effects are negligible (e.g., Lepri, Kozmar, Večenaj, & Grisogono, 2014; Lepri, Večenaj, Kozmar, & Grisogono, 2017) (Figure 5.7). In Figure 5.7, the thermal stratification is described using the dimensionless parameter ζ (also called a stability parameter, e.g., Stull, 1988). Positive ζ implies a statically stable stratification, negative ζ implies a statically unstable stratification, and ζ equal to zero implies statically neutral stratification. For Bora-like flows (and very likely for other downslope wind storms as well), ζ is nearly zero, statistically speaking, thus indicating the near-neutral thermal stratification of the atmosphere in the 30-min temporary mean surface layer close to ground surface.

Nevertheless, for some other downslope winds (e.g., Santa Ana), various profiles of the potential temperature were observed (e.g., Figure 5.8 courtesy of Fovell & Cao, 2017), where the surface-layer static stratification may significantly depart from near neutrality for a while.

3.2. Wind Velocity Time History

In wind engineering applications that are commonly performed using experimental, theoretical, and analytical approaches, the wind velocity in the ABL is usually taken as quasi-steady (e.g., Figure 5.9 courtesy of Dyrbye & Hansen, 1997), as long wave meandering plays a small role in wind effects on structures. In urban environments close to

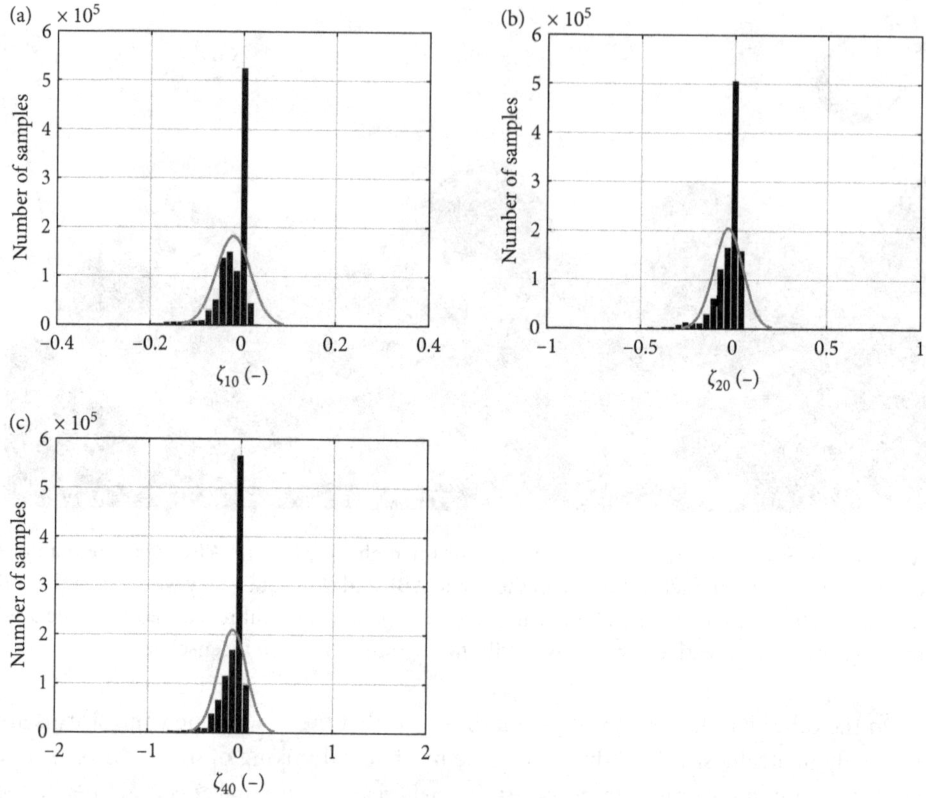

FIGURE 5.7 Typical example of statistically neutral static stratification of the atmospheric surface layer in a downslope wind storm; Bora neutral stratification of the atmosphere (Lepri et al., 2014) (copyright © 2014 Techno Press).

the ground, where the atmospheric turbulence is typically, but not necessarily, at its maximum, the standard deviation of the ABL velocity fluctuations to the local mean wind velocity (commonly known as the turbulence intensity) is around 40 percent, while for downslope wind storms, the wind velocity during the wind gust may be five times larger than the mean wind velocity (e.g., Durran, 1986).

The fluctuation periods of the flow velocities in the ABL, which are of the order from less than a minute to over 11 min and more (e.g., Belušić, Pasarić, Pasarić, Orlić, & Grisogono, 2006), may be substantially shorter for mountainous winds and therefore relevant for wind engineering applications. For example, downslope wind storm gusts typically exhibit a quasi-periodic behavior, as Bora pulsations emerge in periods usually between 3 and 11 minutes (Belušić, Pasarić, & Orlić, 2004). It is expected that those characteristic periods may be quite different for other mountainous winds; some examples are provided for Santa Ana (Fovell & Cao, 2017) in Figure 5.10; European Alps (Cantelli, Monti, Leuzzi, Valerio, & Pilotti, 2017) in Figure 5.11; Ligurian downslope wind (Burlando, Tizzi, & Solari, 2017) in Figure 5.12; downslope wind in Japan (Kusaka & Fudeyasu, 2017) in Figure 5.13; and Zonda (Loredo-Souza, Wittwer, Castro, & Vallis, 2017) in Figure 5.14.

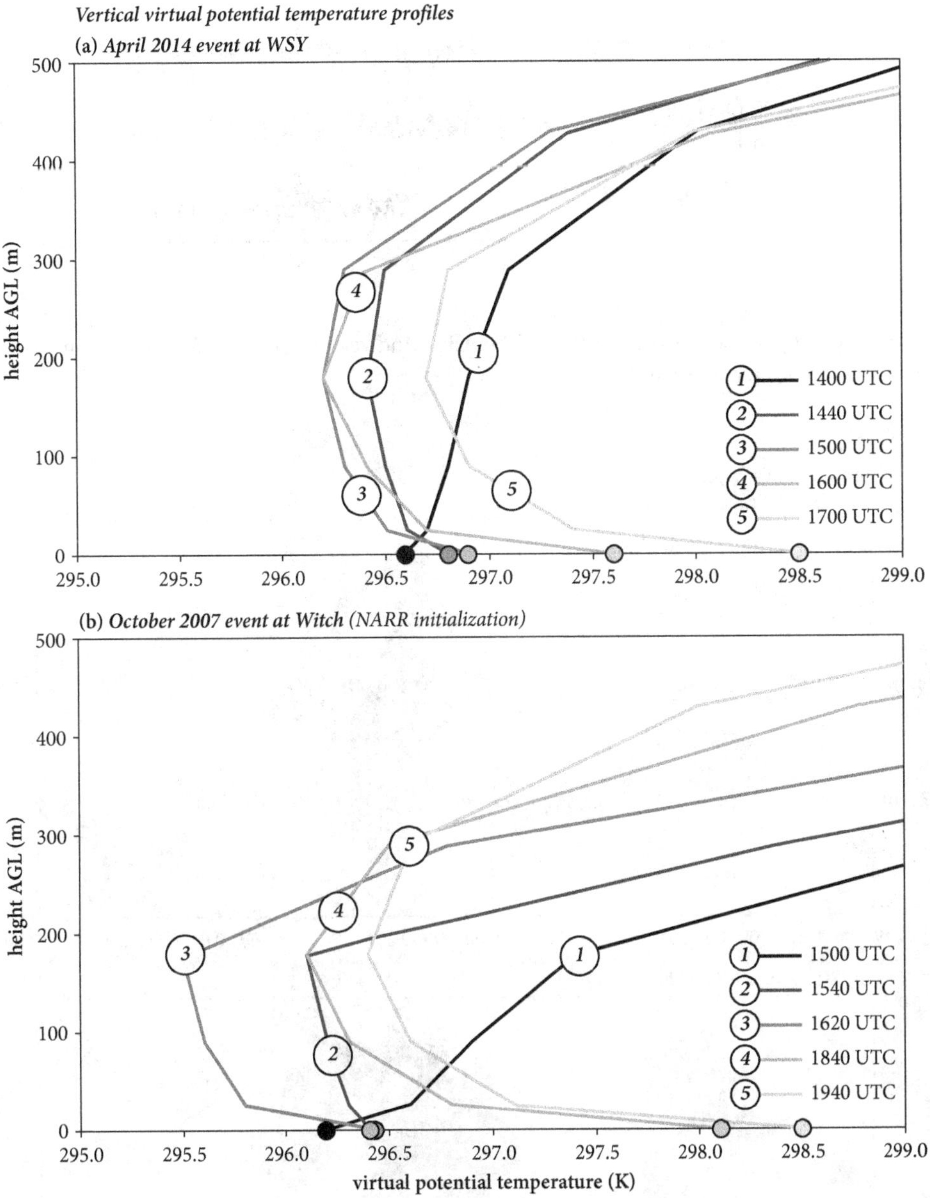

FIGURE 5.8 Thermal stratification of a Santa Ana event (Fovell & Cao, 2017). Also note that this type of wind promotes wildfires, similar to Bora, Chinook, and more (copyright © 2017 Techno Press).

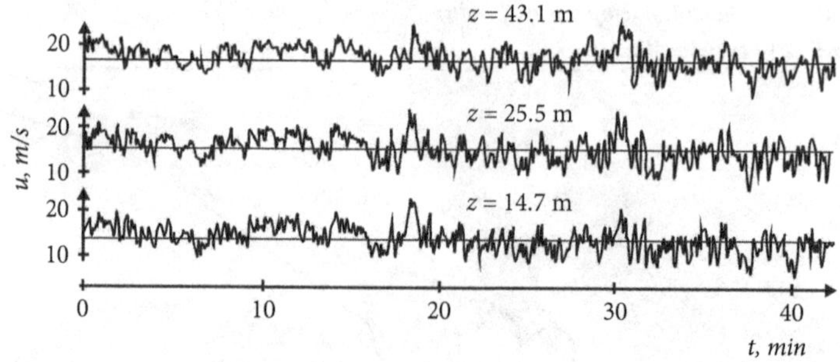

FIGURE 5.9 Quasi-steady time history of the ABL wind velocity (Dyrbye & Hansen, 1997; after Sigbjörnsson, 1974; copyright © 1997 Wiley).

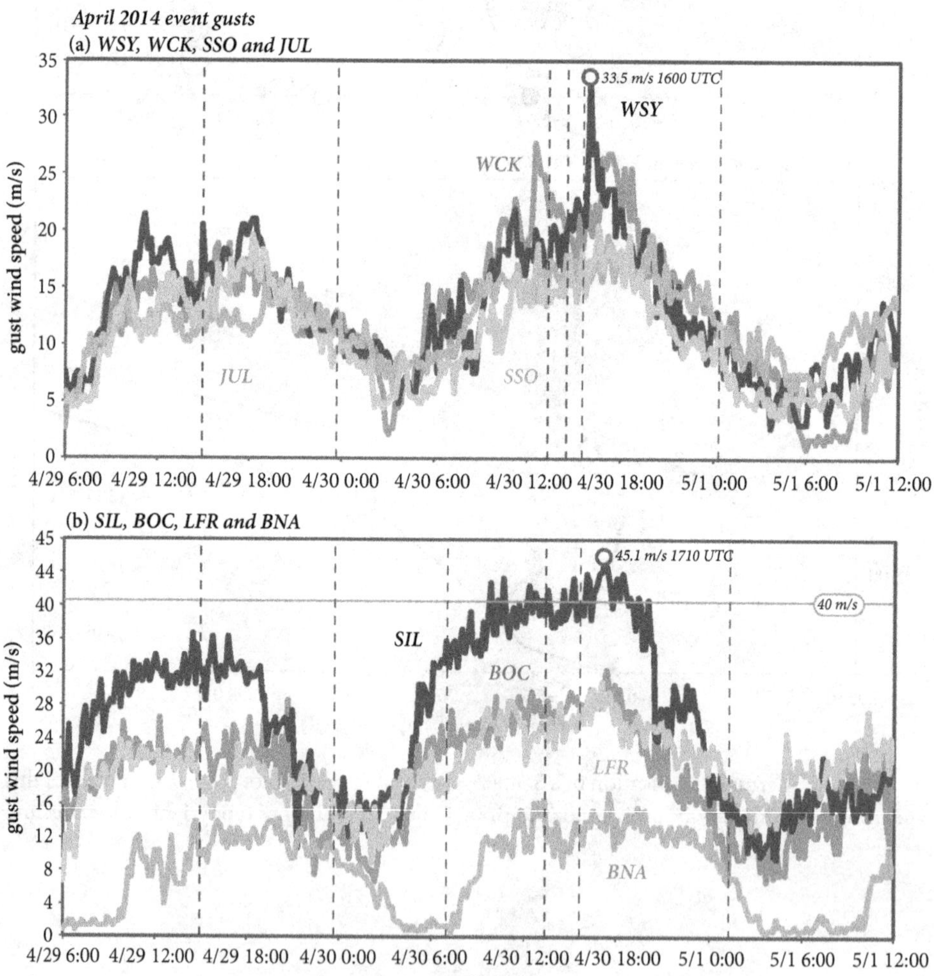

FIGURE 5.10 Velocity time history for Santa Ana (Fovell & Cao, 2017; © Techno Press 2017).

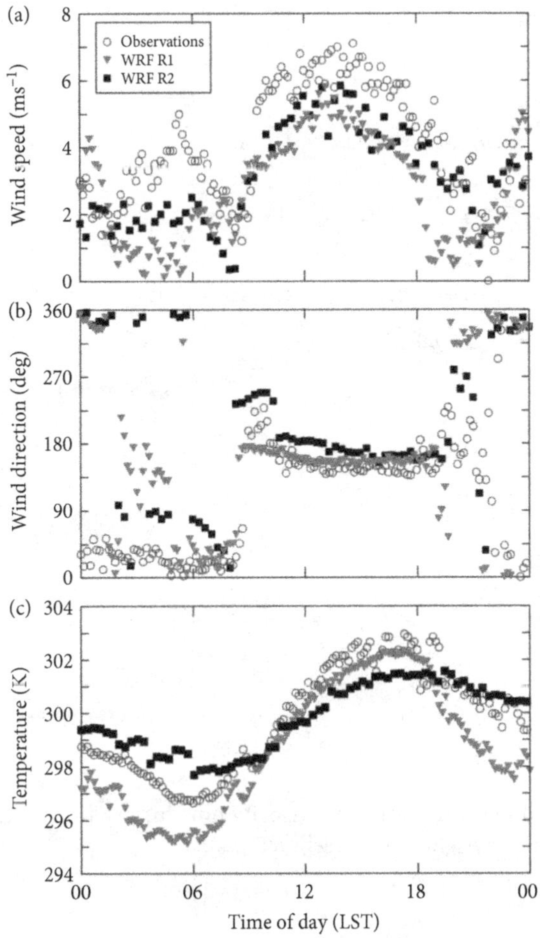

FIGURE 5.11 First panel from the top: Velocity time history for mountain winds in the Alps (Cantelli et al., 2017): comparisons between simulated (solid symbols) and measured (open circles): (a) wind speed, (b) wind direction, and (c) air temperature on July 14, 2010 (all data refer to 2 m above the lake surface; copyright © 2017 Techno Press). WRF: Weather Research Forecasting model.

3.3. Wind Velocity Profiles

The engineers have been commonly using the logarithmic and the power laws to represent the vertical profile of the mean wind velocity in the ABL (e.g., Simiu & Scanlan, 1996; Dyrbye & Hansen, 1997; Holmes, 2015; and Figure 5.15 adapted from Kozmar (2011a). The power law (Hellman, 1916) proved to represent well the mean velocity profile throughout the entire ABL, while the logarithmic law (Thuillier & Lappe, 1964) is valid within the surface layer (lowest 10–15 percent of the ABL). However, this might not be the case for mountainous winds, where a horizontal homogeneity assumption is typically not fulfilled and where the surface layer can be less than 1 m deep (e.g., Baklanov

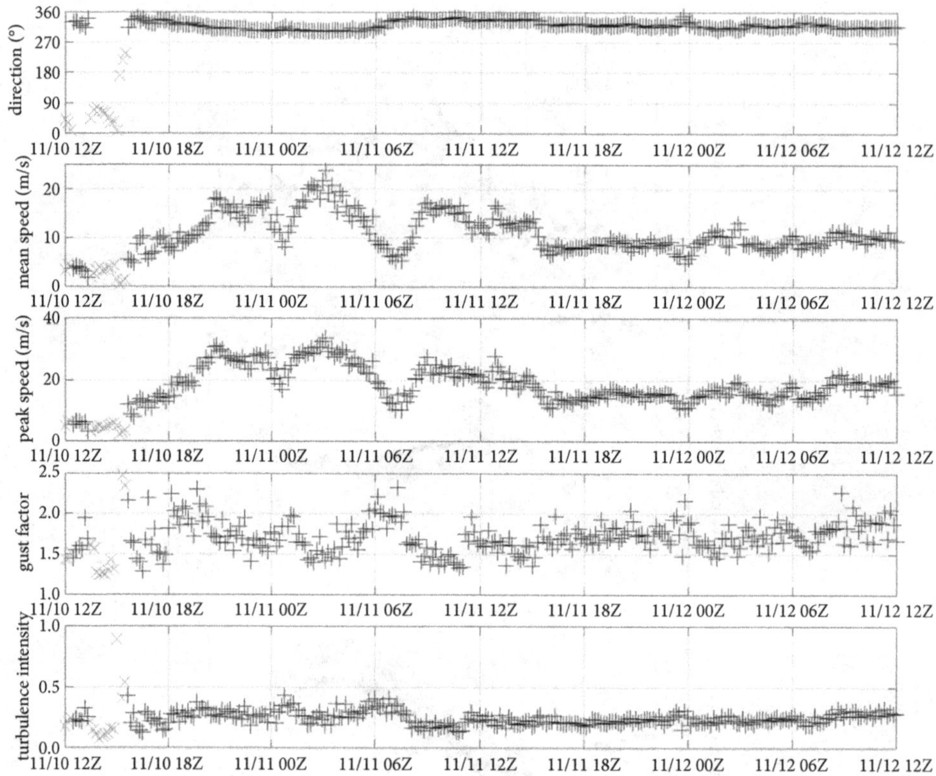

FIGURE 5.12 Second panel from the top: velocity time history for Ligurian downslope wind (Burlando et al., 2017; copyright © 2017 Techno Press).

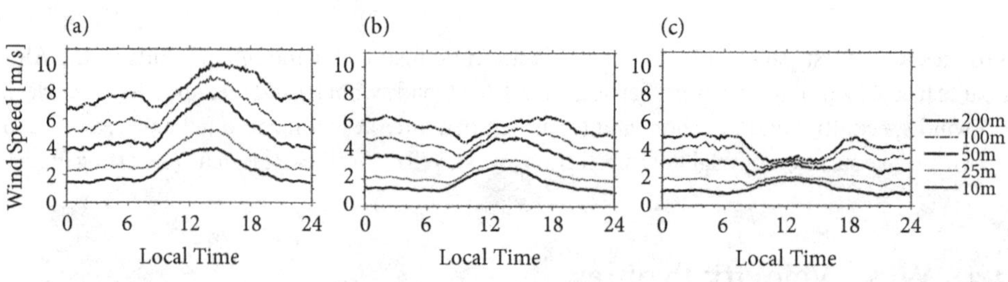

FIGURE 5.13 Velocity time history for downslope wind in Japan (Kusaka & Fudeyasu, 2017; copyright © 2017 Techno Press).

et al., 2011; Grisogono & Oerlemans, 2001). There, the mean wind may have a component from gravity as well.

While the vertical velocity profile close to the ground may agree well with the logarithmic and the power laws (e.g., Lepri et al., 2017), there are substantial discrepancies from those laws further away from the ground (e.g., Burlando et al., 2017; Kusaka & Fudeyasu, 2017; Loredo-Souza et al., 2017—all not shown; but as shown by Belušić & Klaić, 2004; Fovell & Cao, 2017, in Figures 5.16 and 5.17, respectively).

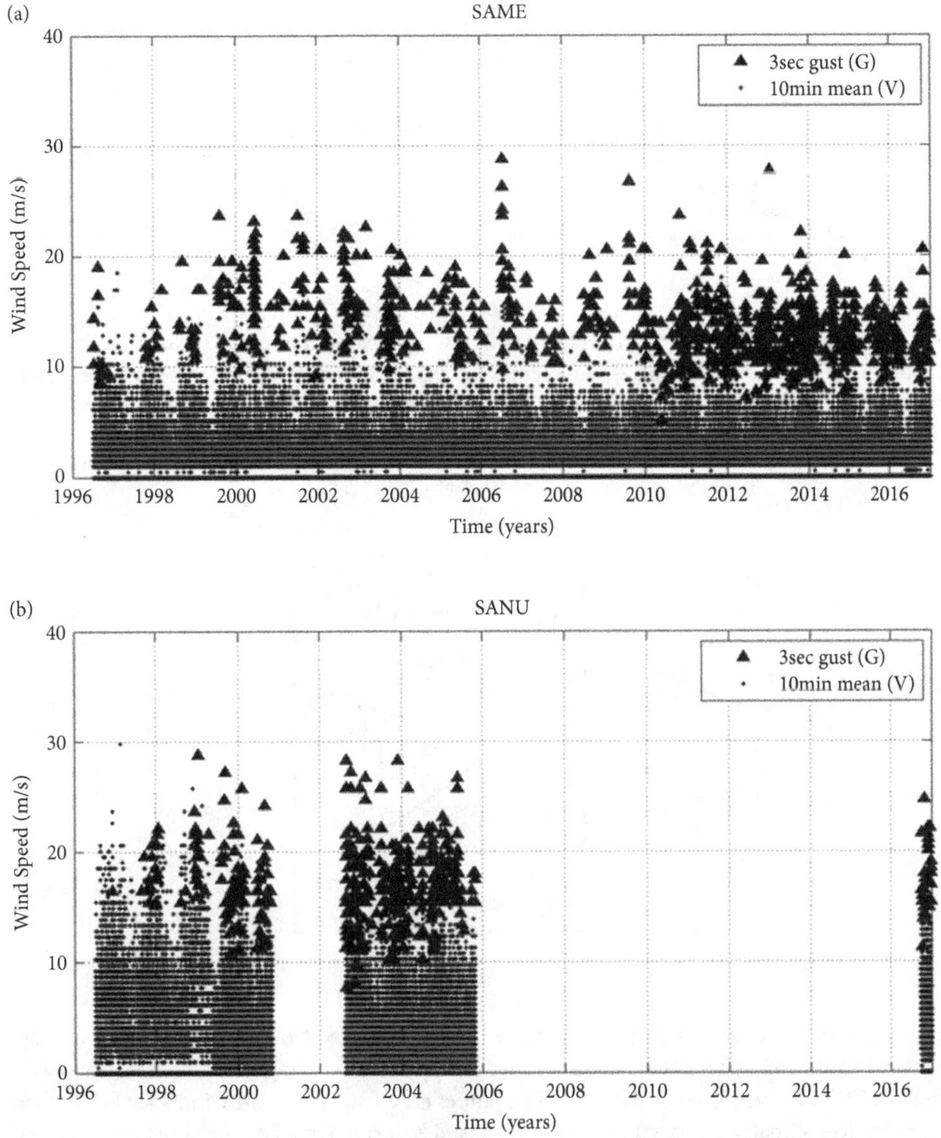

FIGURE 5.14 Velocity time history for Zonda wind in the Andes (Loredo-Souza et al., 2017; copyright © 2017 Techno Press).

3.4. Integral Turbulence Parameters

From the engineering perspective, integral turbulence parameters are important for the dynamic performance of complex engineering infrastructures. For example, for wind loading of structures, turbulence intensity and length scales have been commonly assessed, while Reynolds shear stress is important as well when addressing air pollution transport, dispersion, and dilution. It is therefore necessary to account for discrepancies

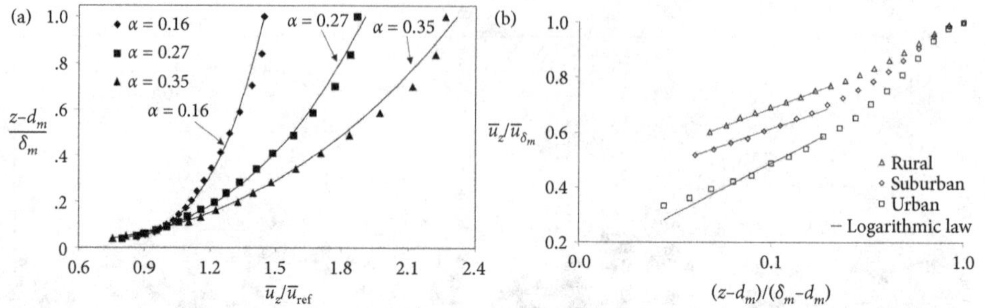

FIGURE 5.15 (a) Power law and (b) logarithmic law approximations of the mean velocity profile (solid lines) after Kozmar (2011a); dots represent experimental results obtained in a boundary layer wind tunnel (copyright © 2011 Springer-Verlag).

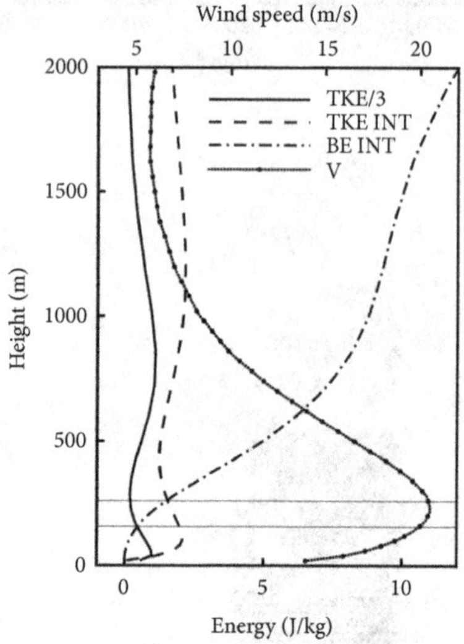

FIGURE 5.16 Vertical velocity profile of the Bora wind adapted from Belušić and Klaić (2004). Vertical profiles of 1/3 TKE (turbulence kinetic energy, corresponding to the local vertical TKE), mean TKE in a given layer (TKE INT), buoyant energy (BE INT), and wind speed (V) at the Island of Krk, Croatia, averaged over all Bora episodes during January 2002. The two horizontal lines toward the bottom of the figure mark the upper limits of the lower bound and gust estimate.

between the integral turbulence parameters in the ABL and downslope wind storms. For the ABL, engineers have been commonly using recommendations from international standards and codes, such as the Engineering Science Data Unit [ESDU], Eurocode, American Society of Civil Engineers (ASCE), where the turbulence intensity is given for various terrain types and commonly follows the trend that the longitudinal to lateral to vertical turbulence intensity is 1:0.75:0.50 (e.g., Dyrbye & Hansen, 1997) or similar. The turbulence intensity has a distinct peak immediately above the ground and decreases monotonically as the height increases (Figure 5.18), while turbulence length scales increase as the height increases above the ground surface (Figure 5.19 after Kozmar, 2011b).

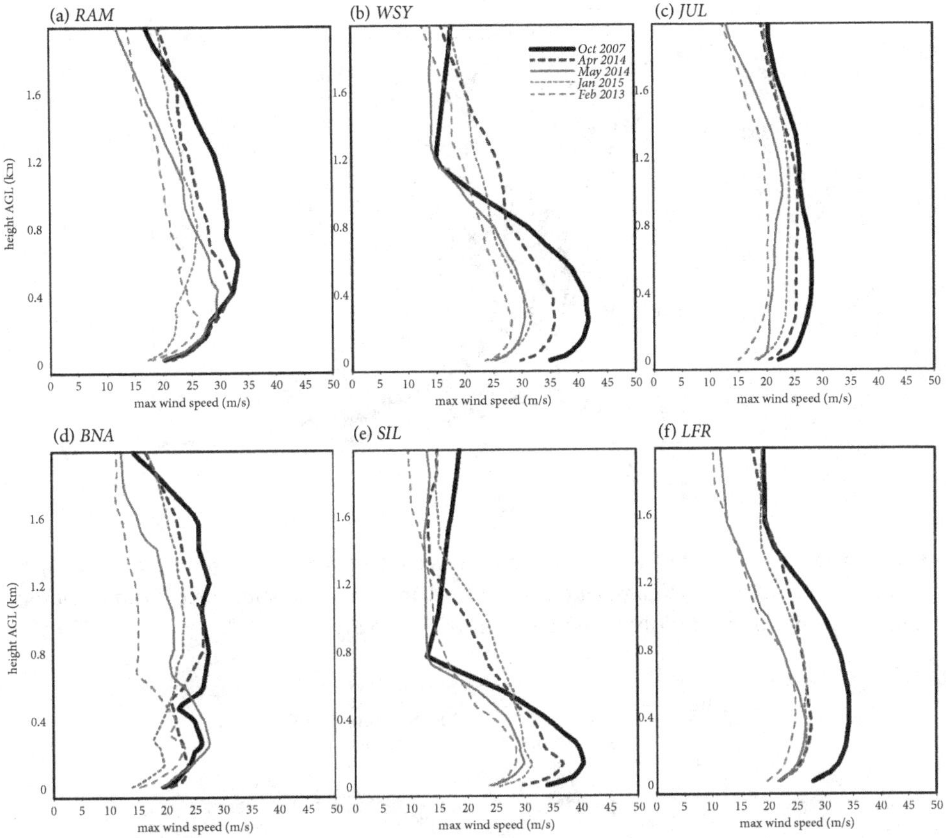

FIGURE 5.17 Vertical velocity profile of the Santa Ana wind from Fovell and Cao (2017) (copyright © 2017 Techno Press).

The Reynolds shear stress remains at some constant value with increasing height throughout the surface layer and then decreases with further increases in height. This range of the constant shear stress (Prandtl constant-flux layer) may vary with changing ground surface roughness (i.e., it remains constant higher above rougher surfaces; however, not higher than the surface-layer top [usually 10–15 percent of the ABL thickness, e.g., Garratt, 1992; Stull, 1988]). The question is how these characteristic ABL profiles behave in downslope wind storms since already weaker (e.g., katabatic winds; e.g., Grisogono & Oerlemans, 2001) showed a typical surface layer was nonexistent (or it was less than 1 m deep).

In downslope wind storms, not to mention low-level jets, there are some important differences from the ABL with respect to the integral turbulence parameters that might change and broaden the concept of the engineering perspective on the ABL. While the turbulence intensity profiles appear to agree with international standards closer to the ground surface (Figure 5.20), they exhibit behavior uncommon for the ABL higher above the ground surface, as the turbulence intensity profiles remain fairly flat (i.e., turbulence intensity does not decrease with increasing the height; Figure 5.21).

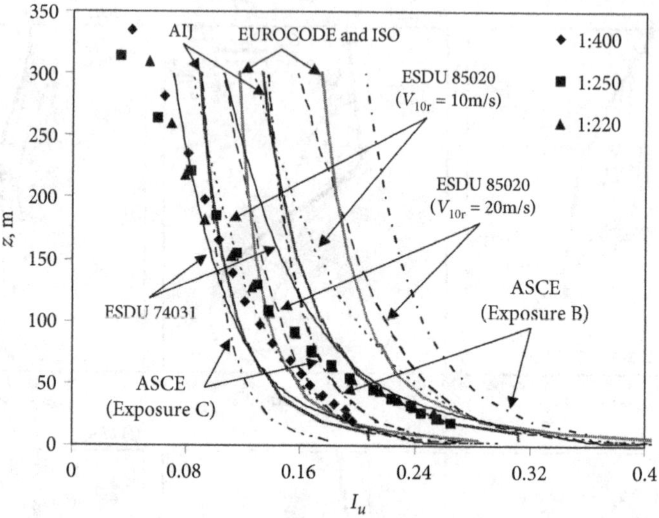

FIGURE 5.18 Profiles of the longitudinal turbulence intensity I_u in the 1:400, 1:250, and 1:220 suburban ABL wind tunnel simulations compared to values recommended in international standards, from Kozmar (2011b) (copyright © 2011 Techno Press).

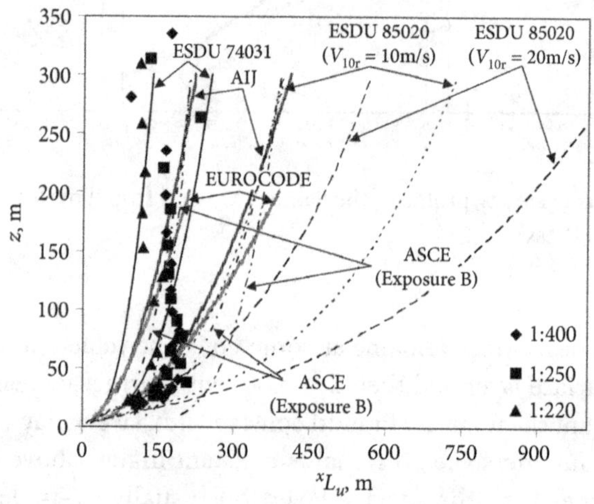

FIGURE 5.19 Profiles of the integral length scales of turbulence xL_u in the 1:400, 1:250, and 1:220 suburban ABL wind tunnel simulations compared to values recommended in international standards, from Kozmar (2011b) (Copyright © 2011 Techno Press).

The constant-flux layer is not exhibited as is the case for the ABL (e.g., Figure 5.22; Lepri, Večenaj, Kozmar, & Grisogono, 2015).

The longitudinal turbulence length scales are generally in good agreement with the international standards ESDU 85020 (ESDU, 1985); EN 1991 Eurocode (European Union, 2005); and ASCE 7-05 (ASCE, 2006). The lateral and vertical turbulence length scales however are considerably larger compared to the standard ABL values (e.g., Figure 5.23).

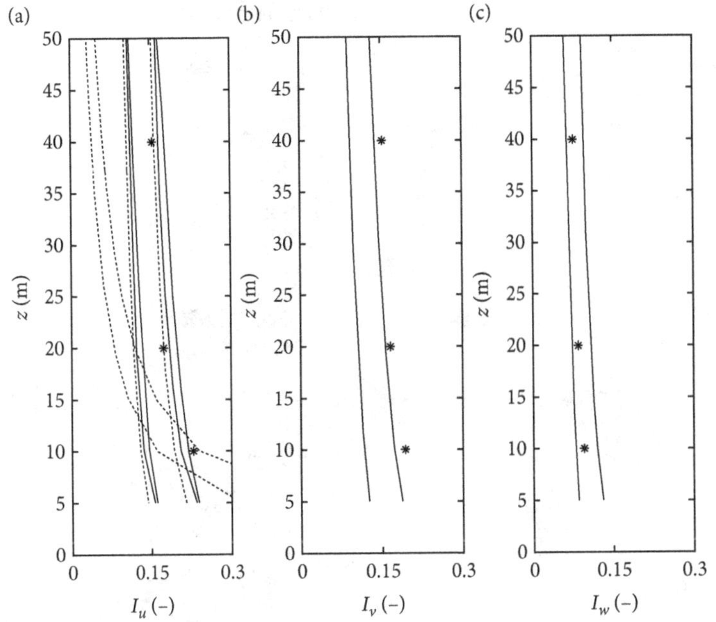

FIGURE 5.20 Vertical profile of turbulence intensity for Bora after Lepri et al. (2015). Vertical profiles of observed turbulence intensity for the averaging period of 17 min in the (a) x, (b) y, and (c) z direction calculated through the entire time record and compared with the values in international standards with the tolerance range of ± 20 percent. Legend: Star is average turbulence intensity; gray solid line is ESDU 85020 (1985) for z_0 = 0.03 m, black solid line is ISO 4354 (International Organization for Standardization [ISO], 1997) and EN 1991 Eurocode 1 (2005); black dotted line is ASCE 7-05 (2006); gray dotted line is Architectural Institute of Japan (AIJ) (2006) (copyright © 2015 Elsevier Ltd; all rights reserved).

Even larger discrepancies in integral turbulence parameters in comparison with the typical ABL may be expected higher up in the atmosphere, similarly as for the mean wind velocity profiles, which are in better agreement with the ABL closer to the ground in the surface layer than higher in the atmosphere (e.g., Smith, 2002). These discrepancies may be particularly relevant for tall structures (e.g., tall buildings and wind turbines), which are designed according to the ABL characteristics because in mountainous regions those structures may experience quite different wind loads when subjected to downslope wind storms.

3.5. Wind Velocity Power Spectra

Engineering structures have commonly been designed assuming some standard models for the velocity power spectra (e.g., Davenport, 1967; Harris, 1970; von Kármán, 1948; e.g., Figure 5.24 [adapted from Kozmar, 2012]), which are commonly also a part of the international standards and codes. A properly selected wind velocity power spectrum is

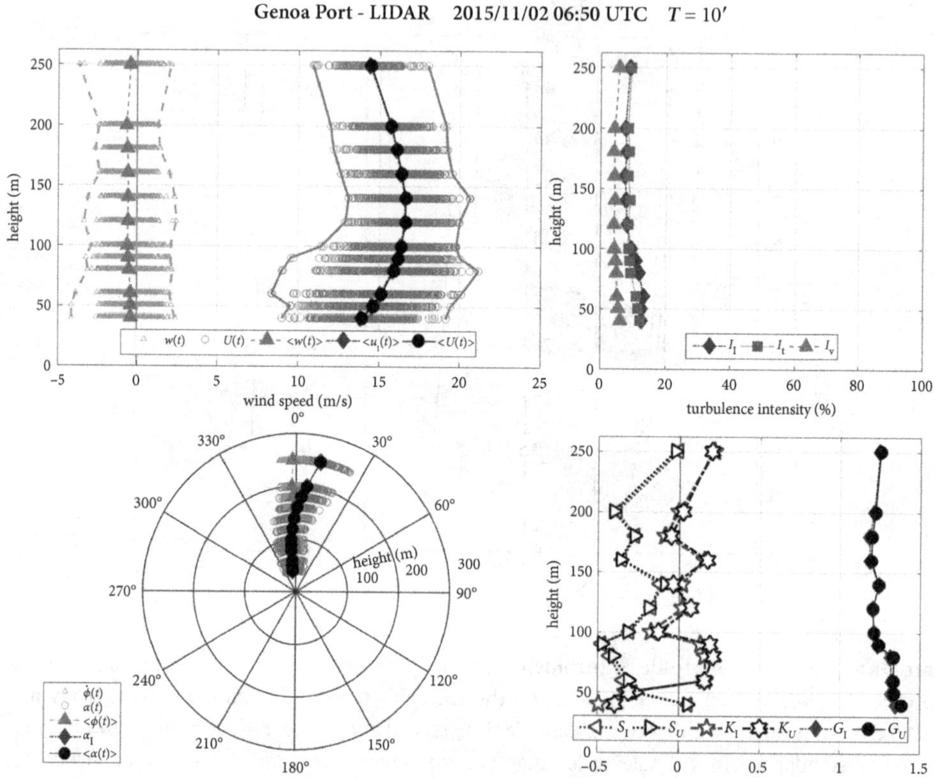

FIGURE 5.21 Top right panel: Vertical profile of turbulence intensity for Ligurian downslope wind (Burlando et al., 2017). Wind velocity field measured on November 2, 2015, at 06:50–07:00 UTC by the LiDAR in the port of Genoa: profiles of 10-min mean horizontal (<U>, circles), longitudinal (<u_l>, diamonds), and vertical (<w>, triangles) wind components and their variability up to 250 m above ground (top left); longitudinal I_l, transversal I_t, and vertical I_v turbulence intensities (top right); directions of the 10-min mean horizontal (<α>, circles), longitudinal (α_l, diamonds), and vertical (Φ, triangles) wind components and their variability as a function of height (bottom left); skewness S, excess kurtosis K, and gust factors G of the longitudinal (subscript l) and mean (subscript U) components (bottom right) (copyright © 2017 Techno Press).

very important as input information when designing structural dynamics to extend the lifetime and enhance the safety of those structures. Of particular interest is the frequency range shorter than a minute.

In areas prone to downslope wind storms, not to mention low-level jets and tropical cyclones, some other types of wind velocity spectra may be expected (e.g., Li et al., 2014). Likewise, a recent study by Babić, Večenaj, and de Wekker (2017) showed that the standard atmospheric spectra models can be, and often are, modified above the complex terrain by some local effects, mostly induced by the orography. This includes local mesoscale phenomena such as rotors and mountain waves. It is therefore necessary to improve the standards and codes by implementing those other types of wind spectra to be used in areas prone to downslope wind storms.

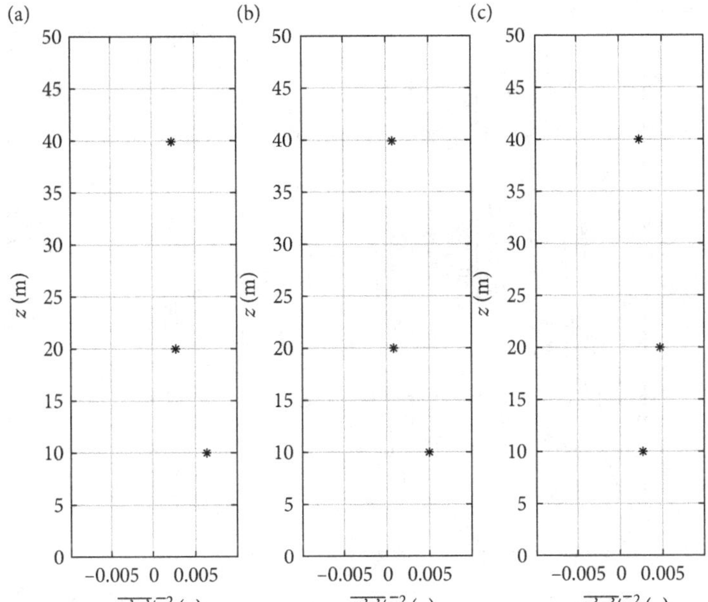

FIGURE 5.22 Vertical profile of the Reynolds shear stress for Bora after Lepri et al. (2015). Vertical profiles of the (a) $-\overline{u'w'}$, (b) $-\overline{u'v'}$, (c) $-\overline{v'w'}$. Reynolds shear stress components for the averaging period of 17 min calculated through the entire time record and normalized using the 17-min average wind velocity in the main (x) wind direction at the corresponding height for each height level (copyright © 2015 Elsevier Ltd.; all rights reserved).

4. Concluding Remarks and Perspectives

Characteristics of downslope wind storms in comparison with the typical ABL were critically reviewed. It was shown that the current wind engineering standards do not entirely keep up with the atmospheric physics of downslope wind storms. In particular, the atmosphere near the ground surface is neutrally thermally stratified, and the wind velocity fluctuations are much stronger in downslope wind storms than in the standard ABL. The vertical profile of wind velocity in the main wind direction close to the ground in the surface layer agrees well with the logarithmic and the power laws in accordance with international standards and codes; however, there are substantial discrepancies from those laws higher from the ground. The vertical profiles of the turbulence intensity agree with international standards closer to the ground surface, while they exhibit behavior uncommon for the ABL higher above the ground surface, as the turbulence intensity profiles remain fairly flat (i.e., they do not decrease with increasing height). Also, the constant-flux layer is not as exhibited in downslope wind storms as it is the case in the ABL. While the longitudinal turbulence length scales are generally in good

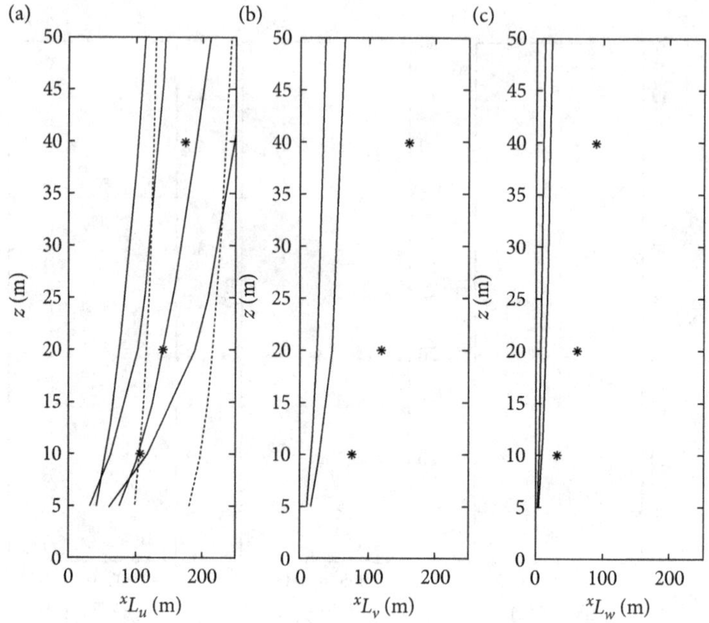

FIGURE 5.23 Vertical profiles of turbulence length scales for Bora after Lepri et al. (2015). Vertical profiles of turbulence length scales (a) ^xL_u, (b) ^xL_v, and (c) ^xL_w component for the averaging time 17 min calculated through the entire time record compared with the values in international standards with tolerance range of ± 30 percent. Legend: Star is observed turbulence length scale; gray solid line is ESDU 85020 (1985) with z_o = 0.03 m; black solid line is EN 1991 Eurocode 1 (2005); black dotted line is ASCE 7-05 (2006) (copyright © 2015 Elsevier Ltd.; all rights reserved.

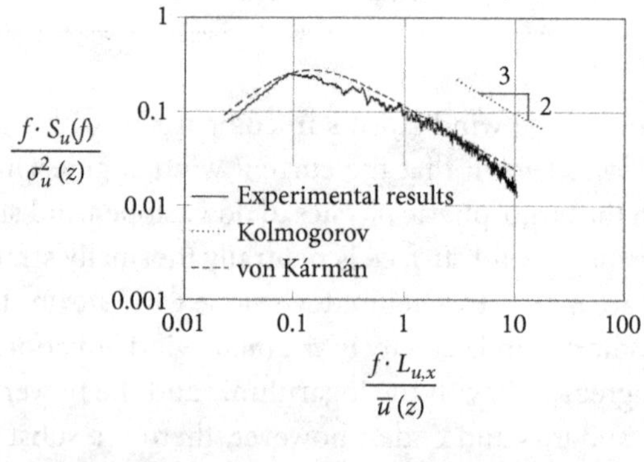

FIGURE 5.24 Power spectral density of wind velocity fluctuations in the typical ABL (Kozmar, 2012; copyright © 2011 Springer Science Business Media BV).

agreement with the international standards, the lateral and vertical turbulence length scales are considerably larger compared to the standard ABL values. There is a need for semiempirical models of velocity power spectra in downslope wind storms comparable to the existing ABL models. It needs to be mentioned, though, that more work is required to further support those findings.

This new and necessary knowledge may possibly be acquired using field measurements, small-scale laboratory experiments, computational simulations, and analytical modeling. Field measurements would need to be performed in locations typical for occurrence of downslope wind storms simultaneously at a large number of meteorological towers in various heights at large resolution frequency to fully account for the complex 3D high frequent-flow phenomena of downslope wind storms.

Small-scale laboratory measurements have been commonly performed in boundary-layer wind tunnels that are not entirely suitable to model downslope wind storms; some common issues are Reynolds number similarity, test section blockage, and the large inertia of one big fan commonly used in wind tunnels of that type. Improvements in laboratory small-scale simulation of downslope wind storms can be achieved by using boundary-layer wind tunnels substantially larger than those commonly used, for example, at Politecnico di Milano (Italy) and Southwest Jiaotong University (China); transient flow field simulators, such as at Miyazaki University (Japan), the University of Notre Dame (United States), and Florida International University (United States); 3D testing chambers, such as at Western University (Canada). Characteristics of downslope wind storms may be well complemented by using the results obtained in tornado simulators, such as at Iowa State University (United States), Tongji University (China), or the University of Birmingham (United Kingdom).

While steady Reynolds-averaged Navier–Stokes (RANS) approach is not suitable for computational modeling of downslope wind storms, possibilities are available in unsteady RANS and large eddy simulations (LES), whereas direct numerical simulations (DNS) would be the optimal solution; however, possibilities with this approach are still not that feasible due to high demand for computer power that is currently not available. Meteorological state-of-the-art models almost invariably use the unsteady RANS approach for operational purposes; hence, it is still inconceivable that the downslope wind storms will be predicted accurately in the near future, not to mention their climate perspectives. The latter means the place, time of the onset, duration, and cessation as well as the intensity of severe winds and the corresponding state of the ABL. In short, the meteorological community is still struggling in making more regular detailed forecasts of local, intensive winds by using more advanced (e.g., LES) methods.

Acknowledgments

Dr. Željko Večenaj, Faculty of Science, University of Zagreb, is thanked for his constructive criticism and a comment regarding Figure 5.24. The Croatian Science Foundation HRZZ-IP-2016-06-2017 (WESLO) support is gratefully acknowledged. This work was partly financed within the Croatian–Swiss Research Program of the Croatian Science Foundation and the Swiss National Science Foundation with funds obtained from the Swiss–Croatian Cooperation program, project SWALDRIC (IZHRZo-180587). A final form of this work was completed when B.G. visited E. Schrödinger Institute, Vienna, Austria, in 2020.

References

American Society of Civil Engineers (ASCE). (2006). *Minimum design loads for buildings and other structures ASCE/SEI 7-05*. Reston, VA: ASCE.

Architectural Institute of Japan (AIJ). (2006). *AIJ recommendations for loads on buildings*. Tokyo, Japan: AIJ.

Babić, N., Večenaj, Ž., & de Wekker, S. F. J. (2017). Spectral gap characteristics in a daytime valley boundary layer. *Quarterly Journal of the Royal Meteorological Society, 143*, 2509–2523.

Baklanov, A., Grisogono, B., Bornstein, R., Mahrt, L., Zilitinkevich, S., Taylor, P., . . . Fernando, H. J. S. (2011). The nature, theory and modeling of atmospheric planetary boundary layers. *Bulletin of the American Meteorological Society, 92*, 123–128. http://journals.ametsoc.org/toc/bams/92/2

Belušić, D., & Klaić, Z. B. (2004). Estimation of bora wind gusts using a limited area model. *Tellus, Series A: Dynamic Meteorology and Oceanography, 56*(4), 296–307.

Belušić, D., Pasarić, M., & Orlić, M. (2004). Quasi-periodic bora gusts related to the structure of the troposphere. *Quarterly Journal of the Royal Meteorological Society, 130*, 1103–1121.

Belušić, D., Pasarić, M., Pasarić, Z., Orlić, M., & Grisogono, B. (2006). On local and non-local properties of turbulence in the bora flow. *Meteorologische Zeitschrift*, 15, 301–306.

Belušić, D., Žagar, M., & Grisogono, B. (2007). Numerical simulation of pulsations in the bora wind. *Quarterly Journal of the Royal Meteorological Society, 133*, 1371–1388. https://www.doi.org/10.1002/qj.129

Burlando, M., Tizzi, M., & Solari, G. (2017). Characteristics of downslope winds in the Liguria region. *Wind and Structures, 24*(6), 613–635.

Cantelli, A., Monti, P., Leuzzi, G., Valerio, G., & Pilotti, M. (2017). Numerical simulations of mountain winds in an alpine valley. *Wind and Structures, 24*(6), 565–578.

Chow, F. K., de Wekker, S. F. J., & Snyder, B. J. (Eds.). (2013). *Mountain weather research and forecasting*. Dordrecht, Netherlands: Springer.

Davenport, A. G. (1967). Gust loading factors. *Journal of Structural Division: ASCE 93*, 11–34.

Dörnbrack, A. (1998). Turbulent mixing by breaking gravity waves. *Journal of Fluid Mechanics, 375*, 113–141.

Doyle, J. D., & Durran, D. R. (2004). The MAP room: Recent developments in the theory of atmospheric rotors. *Bulletin of the American Meteorological Society, 58*, 337–342.

Doyle, J. D., & Durran, D. R. (2007). Rotor and subrotor dynamics in the lee of three dimensional terrain. *Journal of the Atmospheric Sciences, 64*, 4202–4221.

Doyle, J., Grubišić, V., Brown, W. O. J., de Wekker, S. F. J., Dörnbrack, A., Jiang, Q., . . . Weissmann, M. (2009). Observations and numerical simulations of subrotor vortices during T-REX. *Journal of the Atmospheric Sciences, 66*, 1229–1249.

Durran, D. R. (1986). Another look at downslope windstorms. Part I: The development of analogs to supercritical flow in an infinitely deep, continuously stratified fluid. *Journal of Atmospheric the Sciences, 43*, 2527–2543.

Durran, D. R. (1990). Mountain waves and downslope winds. In W. Blumen (Ed.), *Atmospheric processes over complex terrain* (pp. 59–81). Boston, MA: American Meteorological Society.

Dyrbye, C., & Hansen, S. (1997). *Wind loads on structures*. New York, NY: Wiley.

Enger, L., & Grisogono, B. (1998). The response of bora-type flow to sea surface temperature. *Quarterly Journal of the Royal Meteorological Society, 124*, 1227–1244.

Engineering Science Data Unit (ESDU). (1985). *Characteristics of atmospheric turbulence near the ground. Part II: Single point data for strong winds (neutral atmosphere)* (Data Item No. 85020). London: ESDU.

European Union. (2005). *Eurocode 1: Actions on structures—General actions: Part 1-4: Wind actions.* Retrieved from https://www.phd.eng.br/wp-content/uploads/2015/12/en.1991.1.4.2005.pdf

Fovell, R. G., & Cao, Y. (2017). The Santa Ana winds of Southern California: Winds, gusts, and the 2007 Witch fire. *Wind and Structures, 24*(6), 529–564.

Garratt, J. R. (1992). *The atmospheric boundary layer.* Cambridge, UK: Cambridge University Press.

Grisogono, B., & Belušić, D. (2009). A review of recent advances in understanding the meso- and micro-scale properties of the severe Bora wind. *Tellus, 61A*, 1–16.

Grisogono, B., & Oerlemans, J. (2001). A theory for the estimation of surface fluxes in simple katabatic flows. *Quarterly Journal of the Royal Meteorological Society, 127*, 2725–2739.

Grubišić, V., & Billings, B. J. (2008). Summary of the Sierra Rotors Project wave and rotor events. *Atmospheric Science Letters, 9*(4), 176–181. https://doi.org/10.1002/asl.200

Grubišić, V., Doyle, J. D., Kuettner, J., Moggs, S., Smith, R. B., Whiteman, C. D., . . . Chow, F. K. (2008). The terrain-induced rotor experiment. *Bulletin of the American Meteorological Society, 89*, 1513–1533.

Grubišić, V., & Orlić, M. (2007). Early observations of rotor clouds by Andrija Mohorovičić. *Bulletin of the American Meteorological Society, 88*, 693–700.

Harris, R. I. (1970). *The nature of the wind.* Seminar on Modern Design of Wind-Sensitive Structures. Construction Industry Research & Information, CIRIA, London, pp. 29–35.

Hellman, G. (1916). Über die Bewegung der Luft in den untersten Schichten der Atmosphäre. *Meteorologische Zeitschrift, 34*, 273–285.

Holmes, J. D. (2015). *Wind loading of structures.* Boca Raton, FL: CRC Press.

Hunt, J. C. R., Feng, Y., Linden, P. F., Greenslade, M. D., & Mobbs, S. D. (1997) Low-Froude-number stable flows past mountains. *Il Nuovo Cimento, 20*, 261–272.

Hunt, J. C. R., Olafsson, H., & Bougeault, P. (2001). Coriolis effects on orographic and meso-scale flows. *Quarterly Journal of the Royal Meteorological Society, 127*, 601–633.

International Organization for Standardization (ISO). (1997). *Wind actions on structures* (ISO 4354). Geneva, Switzerland.

Jackson, P. L., Mayr, J., & Vosper, S. (2013). Dynamically driven winds. In: F. K. Chow, S. F. J. de Wekker, & B. J. Snyder (Eds.), *Mountain weather research and forecasting* (pp. 121–218). Dordrecht, Netherlands: Springer.

Jurčec, V. (1981). On mesoscale characteristics of bora conditions in Yugoslavia. *Pure and Applied Geophysics, 119*(3), 640–657.

Klemp, J. B., & Durran, D. R. (1987). Numerical modelling of Bora winds. *Meteorology and Atmospheric Physics, 36*, 215–227.

Klemp, J. B., & Lilly, D. K. (1978). Numerical simulation of hydrostatic mountain waves. *Journal of the Atmospheric Sciences, 35*, 78–107.

Kozmar, H. (2011a). Characteristics of natural wind simulations in the TUM boundary layer wind tunnel. *Theoretical and Applied Climatology, 106*, 95–104.

Kozmar, H. (2011b). Wind-tunnel simulations of the suburban ABL and comparison with international standards. *Wind and Structures, 14*(1), 15–34.

Kozmar, H. (2012). Physical modeling of complex airflows developing above rural terrains. *Environmental Fluid Mechanics, 12*, 209–225.

Kozmar, H. (2017). Characteristics of downslope windstorms for engineering applications [Special issue]. *Wind and Structures, 24*(6).

Kusaka, H., & Fudeyasu, H. (2017). Review of downslope windstorms in Japan. *Wind and Structures, 24*(6), 637–656.

Lepri, P., Kozmar, H., Večenaj, Ž., & Grisogono, B. (2014). A summertime near-ground velocity profile of the Bora wind. *Wind and Structures, 19*(5), 505–522.

Lepri, P., Večenaj, Ž., Kozmar, H., & Grisogono, B. (2015). Near-ground turbulence of the Bora wind in summertime. *Journal of Wind Engineering and Industrial Aerodynamics, 147*, 345–357.

Lepri, P., Večenaj, Z., Kozmar, H., & Grisogono, B. (2017). Bora wind characteristics for engineering applications. *Wind and Structures, 24*(6), 579–611.

Li, L., Kareem, A., Hunt, J., Xiao, Y., Zhou, C., & Song, L. (2014). Turbulence spectra for boundary-layer winds in tropical cyclones: A conceptual framework and field measurements at coastlines. *Boundary-Layer Meteorology, 154*(2), 243–263.

Lilly, D. K. (1978). A severe downslope windstorm and aircraft turbulence event induced by a mountain wave. *Journal of the Atmospheric Sciences, 35*, 59–77.

Lilly, D. K., & Klemp, J. B. (1979). The effects of terrain shape on nonlinear hydrostatic mountain waves. *Journal of Fluid Mechanics, 95*, 241–261.

Loredo-Souza, A. M., Wittwer, A. R., Castro, H. G., & Vallis, M. B. (2017). Characteristics of Zonda wind in South American Andes. *Wind and Structures, 24*(6), 657–677.

Mayr, G. J., Plavcan, D., Armi, L., Elvidge, A., Grisogono, B., Horvath, K., . . . Zängl, G. (2018). The community Foehn classification experiment. *Bulletin of the American Meteorological Society* [online]. https://doi.org/10.1175/BAMS-D-17-0200.1

Mohorovičić, A. (1889). Interessante Wolkenbildung űber der Bucht von Buccari (with a comment from the editor J. Hann). *Meteorologische Zeitschrift, 24*, 56–58.

Ólafsson, H., & Bougeault, P. (1996). Nonlinear flow past an elliptic mountain ridge. *Journal of the Atmospheric Sciences, 53*, 2465–2489.

Pope, S. B. (2000). *Turbulent flows*. Cambridge, UK: Cambridge University Press.

Poulos, G. S., Bossert, J. E., McKee, T. B., & Pielke, R. A. (2000). The interaction of katabatic flow and mountain waves. Part I: Observations and idealized simulations. *Journal of the Atmospheric Sciences, 57*, 1919–1936.

Poulos, G. S., Bossert, J. E., McKee, T. B., & Pielke, R. A. (2007). The interaction of katabatic flow and mountain waves. Part II: Case study analysis and conceptual model. *Journal of the Atmospheric Sciences, 64*, 1857–1879.

Richner, H., & Hächler, P. (2013). Understanding and forecasting alpine Foehn. In: F. K. Chow, S. F. J. de Wekker, & B. J. Snyder (Eds.), *Mountain weather research and forecasting* (pp. 219–260). Dordrecht, Netherlands: Springer.

Rogers, D. P., Dorman, E. C., Edvards, K. A., Brooks, I. M., Melville, W. K., Burk, S. D., . . . Schanot, A. J. (1998). Highlights of coastal waves 1996. *Bulletin of the American Meteorological Society, 79*, 1307–1326.

Rotach, M., & Zardi, D. (2007). On the boundary-layer structure over highly complex terrain: Key findings from MAP. *Quarterly Journal of the Royal Meteorological Society, 133*, 937–948.

Sachsperger, J., Serafin, S., & Grubišić, V. (2015). Lee waves on the boundary-layer inversion and their dependence on free-atmospheric stability. *Frontiers in Earth Science, 3*, 14–24. https://doi.org/10.3389/feart.2015.00070

Scinocca, J. F., & Peltier, W. R. (1993). The instability of Long's stationary solution and the evolution toward severe downslope windstorm flow. Part I: Nested grid numerical simulations. *Journal of the Atmospheric Sciences, 50*, 2245–2263.

Scinocca, J. F., & Peltier, W. R. (1994). The instability of Long's stationary solution and the evolution toward severe downslope windstorm flow. Part II: The application of finite-amplitude local wave-activity flow diagnostics. *Journal of the Atmospheric Sciences, 51*, 623–653.

Serafin, S., Adler, B., Cuxart, J., De Wekker, S. F. J., Gohm, A., Grisogono, B.,... Zardi, D. (2018). Exchange processes in the atmospheric boundary layer over mountainous terrain. *Atmosphere, 9*(3), 102.

Sigbjörnsson, R. (1974). *On the theory of structural vibrations due to natural wind* (Report R 59). Kongens Lyngby, Denmark: Department of Structural Engineering, Technical University of Denmark.

Simiu, E., & Scanlan, R. H. (1996). *Wind effects on structures: Fundamentals and applications to design*. New York, NY: Wiley.

Smith, R. B. (1977). The steepening of hydrostatic mountain waves. *Journal of the Atmospheric Sciences, 34*, 1634–1654.

Smith, R. B. (1978). A measurement of mountain drag. *Journal of the Atmospheric Sciences, 35*, 1644–1654.

Smith, R. B. (1979). The influence of mountains on the atmosphere. *Advances in Geophysics, 21*, 87–230. https://doi.org/10.1016/S0065-2687(08)60262-9

Smith, R. B. (1985). On severe downslope winds. *Journal of the Atmospheric Sciences, 42*, 2597–2603.

Smith, R. B. (1987). Aerial observations of the Yugoslavian Bora. *Journal of the Atmospheric* Sciences, *44*, 269–297.

Smith, R. B. (2002). Stratified flow over topography. In: R. Grimshaw (Ed.), *Environmental stratified flows* (pp. 119–159). Dordrecht, Netherlands: Kluwer.

Šoljan, V., Belušić, A., Šarović, K., Nimac, I., Brzaj, S., Suhin, J.,... Grisogono, B. (2018). Micro-scale properties of different bora types. *Atmosphere, 9*(4), 116–141. http://www.mdpi.com/2073-4433/9/4/116

Strauss, L., Serafin, S., & Grubišić, V. (2016). Atmospheric rotors and severe turbulence in a long deep valley. *Journal of the Atmospheric Sciences, 73*, 1481–1506.

Stull, R. B. (1988). *An introduction to boundary layer meteorology*. Dordrecht, Netherlands: Kluwer Academic.

Sun, J., Nappo, C. J., Mahrt, L., Belušić, D., Grisogono, B., Stauffer, D. R.,... Neff, W. D. (2015). Review of wave-turbulence interactions in the stable atmospheric boundary layer. *Reviews of Geophysics, 53*, 956–993. https://doi.org/10.1002/2015RG000487

Teixeira, M. A. C. (2014). The physics of orographic gravity wave drag. *Frontiers in Physics, 2*, 1–24.

Teixeira, M. A. C., Kirshbaum, D. J., Ólafsson, H., Sheridan, P. F., & Stiperski, I. (Eds.). (2016). *The atmosphere over mountainous regions*. Lausanne, Switzerland: Frontiers Media. http://doi.org/10.3389/978-2-88945-016-9

Teixeira, M. A. C., Miranda, P. M. A., & Cardoso, R. M. (2008). Asymptotic gravity wave drag expressions for non-hydrostatic rotating flow over a ridge. *Quarterly Journal of the Royal Meteorological Society, 134*, 271–276.

Thuillier, R. H., & Lappe, U. O. (1964). Wind and temperature profile characteristics from observations on a 1,400 ft tower. *Journal of Applied Meteorology, 3*, 299–306.

Von Kármán, T. (1948). Progress in the statistical theory of turbulence. *Proceedings of the National Academy of Sciences of the United States of America, 34*(11), 530–539.

Wyngaard, J. C. (2010). *Turbulence in the atmosphere*. Cambridge, UK: Cambridge University Press.

Yoshino, M. M. (1976). *Local wind bora.* Tokyo, Japan: University of Tokyo Press.
Zardi, D., & Whiteman, C. D. (2013). Diurnal mountain wind systems. In: F. K. Chow, S. F. J. de Wekker, & B. J. Snyder (Eds.), *Mountain weather research and forecasting* (pp. 35–119). Dordrecht, Netherlands: Springer.
Zhong, S., & Whiteman, C. D. (2008). Downslope flows on a low-angle slope and their interactions with valley inversions. Part II: Numerical modeling. *Journal of Applied Climatology, 47*, 2039–2057.

CHAPTER 6

FORECASTING OF TORNADOES AND DOWNBURSTS

Challenges, Prioritization, and Progress

DJORDJE ROMANIC

1. Introduction

WEATHER forecasting is the focal point of meteorological practice. Weather forecasts are important to the point that most countries have their own national weather agencies specialized at predicting the future state of the atmosphere over their limited geographic regions. In addition to the national and regional weather bureaus, global weather forecasts are currently produced by more than 10 large agencies. The benefits of reliable weather forecasts are many and span from human safety to economic implications in a wide range of sectors such as energy, tourism, transportation, agriculture, and urban planning.

Today's forecasts are very accurate at time scales of up to 5–7 days in advance. For instance, the precision of current 5- to 7-day forecasts is similar to that of a 1-day forecast from 50 years ago (Stern & Davidson 2015). These mid-range forecasts (i.e., 5–7 days) are usually accurate more than 80% of the time, but the accuracy drops as the length of the forecasting period increases. Forecasting involves solving a system of non-linear partial differential equations with given initial conditions using various computational methods and presenting these results to the public. The process of *computing* a weather forecast is hardly ever subjective but instead is profoundly objective and mathematically rigorous. However, in due honesty, there are times (approximately 20%) when weather forecasts are incorrect. We will learn in Section 2 and throughout this chapter, why this is the case.

Probably one of the biggest challenges in operational weather forecasting today is predicting the occurrence and physical characteristics of tornadoes and small-scale downbursts. Tornadoes are vigorously rotating columns of air that connect a thunderstorm cloud with the surface. Tornadic wind speeds can be extremely high and can even exceed 130 m s^{-1} near the surface. The destructive nature of tornadoes, evidenced by the damage they inflict on various structures, and the environment is used for their classification through the Enhanced Fujita (EF) scale (McDonald et al. 2006). Tornadoes are categorized from EF0- (weakest) to EF5-rated (strongest), with an associated wind speed interval for each of the categories. Downbursts, on the other hand, are strong downdrafts of cold air that emerge from thunderstorm clouds and spread out radially upon reaching the surface. Although downburst wind speeds are not as high as the winds in the strongest tornadoes, they still can reach peak velocities of more than 70 m s^{-1}: This velocity is equivalent to the wind speed in an EF3-rated tornado. While downbursts are not always as strong as tornadoes, they are much more frequent than tornadoes and therefore contribute to the bulk of reported structural damages in many regions around the globe. These high wind speeds in tornadoes and downbursts pose a threat to human safety, the environment, and structures. The interested reader is referred to Solari (2019) for further reading on wind–structure and wind–environment interactions, as well as pedestrian wind comfort.

This brief description of tornadoes and downbursts already suggests that the accurate forecasts of these high impact weather events is essential. Unfortunately, they are either impossible in some cases or very challenging and uncertain in other cases. The reasons are many, and this chapter presents and explores some of the obstacles that prevent accurate weather forecasts of tornadoes and thunderstorm downbursts. This chapter is intended as a high-level overview of various challenges related to tornado and downburst forecasts, progress, and prioritization that need to be addressed in future research on this subject. It is hoped that the material in this chapter is equally clear and interesting to the general public as well as to weather forecasting experts.

2. General Weather Forecasting

Depending on the delivery format, weather forecasting is an easy and challenging job at the same time. That is, forecasting can be highly qualitative or quantitative. Generally, the more quantitative (bound by values) a forecast is, the more useful it is to applications and the public, but the more difficult it is to make it reliable. To predict the future state of the atmosphere (e.g., the *exact* value of wind speeds and temperature 10 days in advance at some particular location), the forecaster would have to know in detail all physical processes that govern the weather globally and locally and would also need to know how to accurately represent these weather processes in a computer code. Moreover, this forecast would require that the current state of the atmosphere is perfectly known. Unfortunately, none of these statements is entirely correct. Our inability to know all physical phenomena and the precise laws that govern them is not surprising given the scales of atmospheric processes that range from subatomic (e.g., approximately 10^{-15} m) to planetary (long planetary waves, around 10^{7} m)—an interval that spans 22 orders of magnitude. This number is so vast that it is comparable to the estimated number of stars in the visible universe or, to put it in a different perspective, it is slightly smaller than the number of all atoms in our body.

The governing equations in weather forecasting are a system of several non-linear, partial differential equations (PDEs). Therefore, it is necessary to use numerical methods and various approximations to solve the equations and produce a weather forecast. The numerical models that have been developed to solve this system of equations through finite difference methods are called the *numerical weather prediction* (NWP) models. One of the best known and widely used NWP models in both research and operational weather forecasting is the *weather research and forecasting* (WRF[1]) model (Skamarock et al. 2008).

To solve the PDEs, the forecaster needs to provide a set of initial conditions and then carry out the numerical integration using a supercomputer. In weather forecasting, these initial conditions are meteorological measurements from weather stations, airplanes, ships, satellites, or any other measuring platform. Starting from the provided initial conditions, the forward-in-time integration of the PDEs is the weather forecast. Theoretically, it is possible to carry out this forward integration for 1 day, 88 days or even 1 million years, and, in principle, obtain the forecast for any period in advance. However, it is not possible to get reliable solutions to the PDEs for long integration periods because the system is mathematically *chaotic*. Chaotic systems exhibit sensitive dependence on initial conditions, which means that small changes in initial conditions will result in an arbitrary large spread of results over time. The system, however, is still deterministic, and the same initial conditions will always provide the same results at any moment in time; however, if the initial conditions are just slightly perturbed—no matter how tiny—the differences between the results will eventually become arbitrarily large (Figure 6.1). In addition to being sensitive to initial conditions, chaotic systems are also non-periodic, which means that the same solution will never begin to repeat regularly.

These properties of differential equations that describe our atmosphere were discovered by Lorenz (1963). Lorenz was running a numerical model that was internally saving results in memory with the precision of six digits after the decimal place. At one point, Lorenz wanted to redo one of his simulations, but he decided not to initiate this rerun from the beginning (as he did for the original simulation); instead he used the numbers halfway through the original simulation as the initial conditions for the rerun. However, in this process, Lorenz also rounded the numbers to three digits after the decimal. Later, he noticed that the results from the rerun were utterly different from the previously obtained original solution. In other words, Lorenz noticed that the solutions were extremely sensitive to the initial conditions.

We see now that weather forecasts are both deterministic and unpredictable. They are unpredictable because no one knows the initial conditions with perfect accuracy. First, the instruments used to measure temperature, wind, pressure, and other meteorological variables are not perfect, and they produce measuring errors. Second, various mathematical and physical approximations, such as the one shown in Figure 6.1, are unavoidable steps in solving PDEs. Consequently, the atmosphere becomes inherently unpredictable after some time (Figure 6.1). However, forecasts are also deterministic because if the integration of PDEs does start from the same initial conditions in multiple simulations, then the researcher will always arrive at the same results. With the current accuracy of numerical models and the quality of our measuring instruments, the validity range of weather forecasts in Figure 6.1 is up to 5–10 days in advance. To additionally minimize the uncertainties of weather forecasts, today's practice in most weather agencies is to run multiple models with multiple initial conditions that are all slightly perturbed from the original values. In this way, the level of sensitivity of weather forecasts to initial conditions can be quantified. When all simulations are completed, the most likely forecast is usually an ensemble average of these multiple weather prognoses.

The World Meteorological Organization (WMO) defines the following forecasting ranges. *Nowcasting* is the weather forecast 2 h in advance. An up to 12-h forecast is called a *very short-range weather forecast*, whereas the forecasts between 12 and 72 h (3 days) are known as *short-range weather forecasts*. *Medium-range weather forecasts* are up to 10 days. Forecasts beyond this period are called *extended-range weather forecasts* (up to 30 days), *long-range weather forecasts* (up to 2 years), and *climate forecasts* (beyond 2 years).

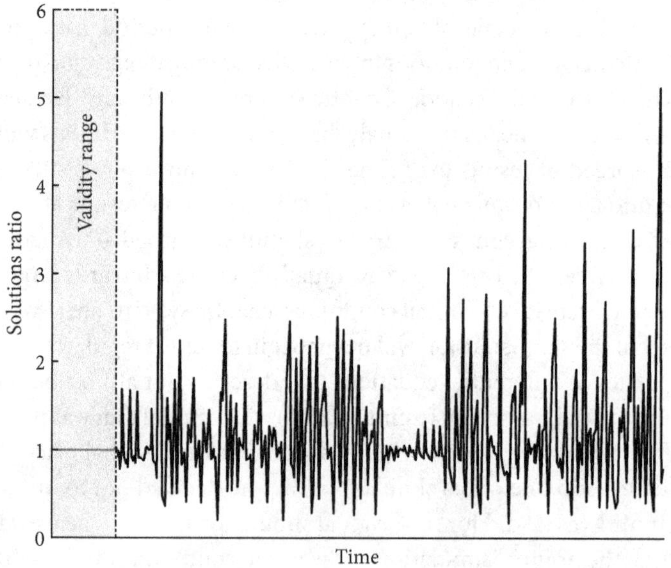

FIGURE 6.1 The ratio of two solutions to the system of differential equations from Lorenz (1963). One solution was initiated with the number $\pi = 3.142$, while the other solution was initiated with the number $\pi = 3.141593$. Since π is an irrational number that never ends, we will always have to make rounding at some point and thus introduce an error into the system. After some finite time, the error will grow and result in an arbitrary inaccurate solution of the system. In this case, the validity range schematically represents the time interval over which weather forecasts are reliable.

The last two in this classification are the most uncertain, and they are not given as an exact characterization of weather conditions but rather as the averaged weather parameters expressed as a departure from climate values for that season, year, decade, or some other long period. Even when formulated in this non-exact way, long-term and climate forecasts are more often than not very unreliable.

All considered, it is possible to distinguish between the intrinsic and practical predictability of weather (Melhauser & Zhang 2012). *Intrinsic predictability* is related to the chaotic nature of the PDEs and that even small errors in initial conditions and inaccurate approximations will eventually grow to be arbitrarily large (Lorenz 1963) (Figure 6.1). Because this challenge is an intrinsic mathematical property of the system, there is very little that can be done to change it. The *practical predictability* is what currently can be done to improve the quality of weather forecasts. These include better meteorological measurements, more accurate numerical approximations, and, most importantly, more complete knowledge of natural phenomena (e.g., tornadogenesis and downbursts). Another factor related to practical predictability is data assimilation methodologies, which are techniques that combine observational data with numerical model outputs to generate an optimal estimate of the evolving state of weather. From this point forward, the material in this chapter mostly focuses on the practical predictability of tornadoes and downbursts.

3. Tornado and Downburst Forecasting Challenges

So far, the discussion has not been strongly related to tornadoes and downbursts because it was essential to lay out the fundamentals of weather forecasting before tackling more complex problems. After establishing in the previous section what is possible and what is not possible in weather forecasting in general (e.g., Figure 6.1), we now discuss some of the complexities behind tornado and thunderstorm downburst forecasts. Hereafter, the term "tornado and downburst forecast" is abbreviated as TaDF, while the term "tornados and downbursts" is abbreviated as TaDs.

3.1. Numerical Weather Prediction Models

3.1.1. *Grid Resolution and Cloud-Resolving Modeling*

The first difficulty for TaDFs that deserves our attention—and one that is easy to grasp—is the relationship between the spatial sizes of TaDs and the size of computational grids used in numerical models. As we discussed in Section 2, the PDEs that describe our atmosphere are solved within a computational domain that has a finite mesh in both horizontal and vertical directions. In operational weather forecasting, grid size in the horizontal direction is usually between 3 and 5 km, whereas the vertical resolution spans from a few dozens to a few hundred meters in the first 1 km of the atmosphere. As the typical width of a tornado is around a few hundred meters, we notice that its horizontal size is much smaller than the model grid size. In simple words, the model "cannot see" the tornado (Figure 6.2). This spatiotemporal discretization of PDEs is a serious constraint set on TaDFs. It seems that the logical step here would be to make the grid finer, but this type of grid manipulation is not straightforward due to the limited computational power at our disposal. The finer the grid, the higher the accuracy of weather forecasts (at least in principle), but also the longer the time needed to produce them.

For example, *high-resolution rapid refresh* (HRRR) is a WRF-based model that is currently implemented at the National Centers for Environmental Prediction (NCEP). It is a real-time 3-km resolution, cloud-resolving, and convection-allowing atmospheric model that covers the entire United States (Benjamin et al. 2015). The model has 50 vertical levels and is initialized by 3-km grids, with 3-km radar data being assimilated every 15 min over a 1-h period (NOAA 2019). These short assimilation cycles add more details to the hourly data that were already assimilated from the 13 km radar-enhanced *rapid refresh* (RAP) model. The RAP is the North America-scale hourly updated assimilation and modeling system operational at NCEP.

Convection-allowing models (CAM) simulate deep convection, such as thunderstorm clouds, without convective parameterization. That is, they produce explicit forecasts of grid-scale precipitation through microphysics parameterization. The CAM approach requires the grid resolution in the horizontal direction to be less than approximately 4 km. On the other hand, *convection-resolving models* (CRM) simulate not only deep convection (as CAM), but also resolve individual convective cells without any convective parameterization involved. Research has shown that grid spacing for CRM forecasting must be less than at least 500 m, but more desirably less than 100 m.

Now a simple thought experiment is to increase the horizontal resolution of HRRR from 3 km (Figure 6.2) to 1 km and leave everything else the same. In this case, we have three times more grid points in the west-east and south-north directions. Hypothetically, if the discretization time step used in solving PDEs is the same as it was for the 3-km resolution, then the computer would require $3 \times 3 = 9$ times more computational time to deliver the forecast. However, this value is an underestimate of the real computational time due to the so-called *Courant-Friedrichs-Lewy (CFL) condition* (C):

$$C = \frac{|u_{max}|\Delta t}{\Delta x} \leq 1, \quad (1)$$

where u_{max} is the maximum speed the information can travel across Δx interval length during the Δt integration time step in the model. Therefore, Equation 1 links the space and time discretization to the maximum speed in the flow field. In other words, the time step (Δt) should be short enough so that it is quicker than the fastest traveling processes being simulated. Otherwise, the solutions will become numerically unstable and "blow up." Assuming $|u_{max}| = 30$ m s^{-1} and $\Delta x = 3$ km and 1 km, Equation 1 indicates that it is necessary to set $\Delta t = 100$ s and 33.3 s, respectively, in order for $C = 1$. Therefore, the preceding estimate of 3×3 needs to be adjusted to $3 \times 3 \times 3 = 27$ times more computational time to calculate the forecast using the finer grid resolution.

Let us now push this exercise to a hypothetical tornado-resolving simulation. In this case, the horizontal resolution in the model should be around 25 m in the horizontal direction, and we also need to keep in mind that tornado velocities can be very high. For example, we limit the maximum velocity to the upper range of EF3-rated tornados, which develop velocities up to 74 m s^{-1}. As discussed in Section 1, this value is also comparable to the maximum velocity in a downburst outflow. Covering the entire continental US excluding Alaska (i.e., the so-called CONUS run—CONtinental US) with this resolution and accounting for the preceding CFL criteria, the increase in the HRRR computational time from the current 3-km grid would be around $120 \times 120 \times 296 = 4{,}262{,}400$ times. In other words, if the model currently computes a 24-h forecast in, say, 1 h of real-time, it would require almost 500 years to calculate the new 24-h forecast on the finer grid. Stepping down from 3 km to 25 m resolution would also require more vertical levels to preserve the computational stability of the model, which would additionally increase already enormous computational time. Section 3.1.2 discusses how the computational time can be reduced if the high-resolution simulations are performed only over a limited geographic region (e.g., Tornado Alley in the United States) utilizing nesting and unstructured grids.

Unfortunately, more complications and factors than the grid resolution need to be accounted for in real (operational) forecasting. In this regard, the process of "quilting" in parallel computing is important. In simple language, *quilting* can be visualized as follows. Assuming N processors are working in parallel to calculate a forecast (i.e., computational domain is decomposed into a number of tiles, and each processor is responsible for one tile), then it is often not desirable that only one (or all) processor(s) collect the outputs from all other processors because of the varying computational times between different processing units, as well as several other technical/numerical constraints. Instead, it is usually more efficient to run the computations on, say, M ($M < N$) processors and have approximately $N - M$ processors handle quilting. Each quilting processor is responsible for supervising the progress of a group of computing processors. Instead of all N processors accessing the file system, only the quilting processors can transfer the results from the computing processors to the file system for writing. This technique minimizes waiting time between the processing units that finish their jobs and enables them to start new calculation. However, setting up quilting correctly requires a lot of experience, and, if not done in the right way, quilting can have the opposite effect and additionally increase the simulation time. Quilting is usually utilized if the domain is large, and the time taken to write an output is becoming significant in comparison to the time taken to integrate the model between output times. These issues are likely to be faced in TaDF that require fine grids and, therefore, large output files.

3.1.2. *Nesting and Unstructured Grids*

One way to tackle grid resolution constraints (Figure 6.2) is to use the concept of *nesting* (Figure 6.3). Instead of covering the entire computational domain with a fine mesh, many models have nesting options in which finer grids are embedded into coarser grids. Assuming A in Figure 6.3 is the area of interest for the TaDF, then it might be enough to construct the very high-resolution grid only above A and have coarser grids in the rest of the computational domain. While this can significantly reduce the model's computing time, there are some apparent issues with this approach. First, forecasts are still unreliable if TaDs form near the borders of A or if they form inside of A but subsequently leave this subdomain. Second, it is not possible to have an arbitrary large refinement ratio between the coarser and finer grids. For instance, the recommended grid resolution ratios in the WRF model are either 1:3 or 1:5 (Figure 6.3). In other words, if domain B has a horizontal grid resolution of 3 km, then A should have a resolution of either 1 km or 0.6 km (Figure 6.3). In other words, it is not recommended to jump immediately from 3-km to 25-m resolution; instead, a series of nests are needed to cascade down to 25-m resolution. Third, the finest domain should not be very small in the horizontal direction (e.g., a few km or less) due to the numerical instabilities that can arise, as well as the different damping constraints in the model that are present near the domain borders (the so-called *buffer zones*) and can contaminate the results in the center of the domain. Last, the number of vertical levels must be the same in all domains. While it is possible to manipulate the horizontal grid resolution by employing nesting

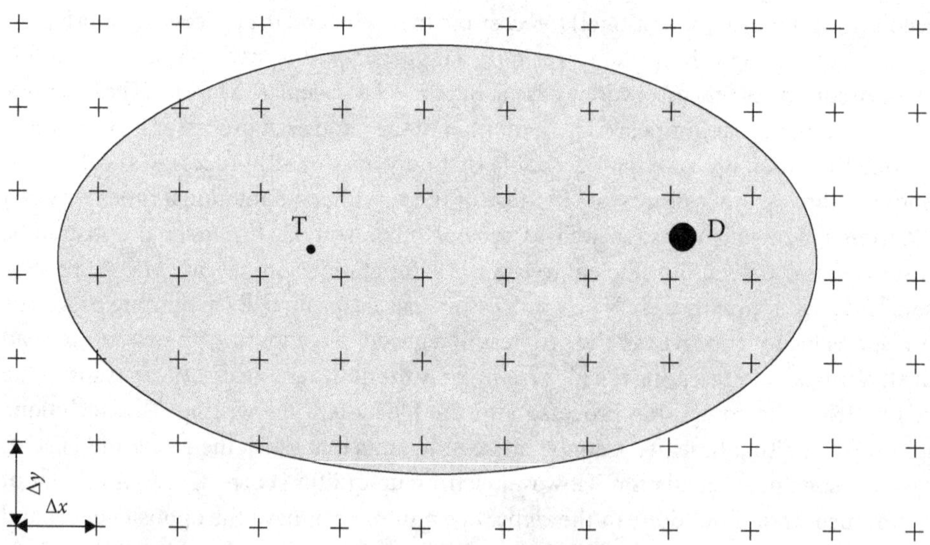

FIGURE 6.2 A 3-km grid is represented with black crosses ($\Delta x = \Delta y = 3$ km). The symbols T and D represent a 300-m and a 1,000-m wide tornado and downburst, respectively. The shaded ellipse is an idealized contour of a thunderstorm cloud. In this case, both tornado and downburst are subgrid processes that cannot be directly resolved on this coarse grid.

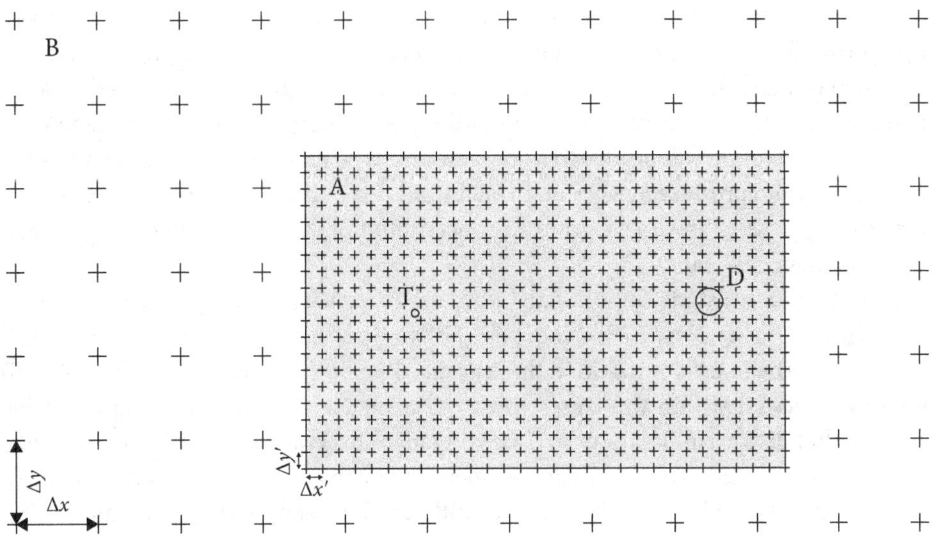

FIGURE 6.3 The concept of nesting in which the finer grid A is inserted into the coarser grid B. A nesting ratio of 1:5 was used to increase the grid resolution from B ($\Delta x = \Delta y = 3$ km) to A ($\Delta x' = \Delta y' = 0.6$ km). A tornado (T) with a width of 300 m is still a subgrid process, whereas the downburst (D) with a diameter of 1,000 m is underrepresented with only a few grid points.

methodologies, the same resolution in the vertical direction is needed in all domains. However, despite some of these limitations, nesting is commonly used in both operational weather forecasting and research. Nesting methods are well understood, and experienced forecasters usually know under which conditions nesting will significantly improve the quality of forecasts.

Returning to Figure 6.3, the finer nest (A) that is inside the coarser nest (B) is called the *child* of B (grid B is called the *parent* of A). Communication between child and parent nests can be one- or two-way. *One-way nesting* means that the information from the parent nest is fed into the child nest, but there is no feedback in the opposite direction. On the other hand, *two-way nesting* indicates that the information from the finer domain (i.e., child) is also returned to the coarser grid, where it makes an impact on computations. One-way nesting is typically more frequent than two-way nesting in operational forecasting. While there are many other characteristics of nesting that could be presented here, this short discussion should be enough for the reader to get the basic idea behind nesting methodology and how, in principle, it can be employed in TaDFs. The next paragraph provides a few examples of tornado forecasting using nesting.

Very recently, Sun et al. (2019) investigated the capabilities of WRF to simulate an EF4-rated supercell tornado that occurred on June 23, 2016, in Funing, China. The authors ran five nested forecast experiments with the horizontal grid spacing in the finest nest being 4,000 m, 1,333 m, 4,44 m, 1,48 m, and 49 m, respectively. While all simulations resolved the general characteristics of the observed supercell, the tornado did not develop until the grid spacing was 444m or less. Only the 49-m grid managed to replicate an EF3-rated tornado (Figure 6.4), which was still weaker than the observed EF4-rated tornado that occurred on that day. Since the strength of the simulated tornado increased with the increasing resolution of the grid, it is logical to expect that an even higher grid resolution was needed to resolve the observed EF4-rated tornado. In addition, the 444-m grid reproduced only a one-cell tornado-like vortex that was stripped of any fine-scale structure. The 49-m grid introduced multiple suction vortices (Figure 6.5) and better matched reality.

In another recent study, Snook et al. (2019) used an ensemble of 10 forecasts to simulate the May 20, 2013, Newcastle-Moore (US) EF5-rated tornado and its parent supercell (Figure 6.6). The finest 50-m grid in the Advanced Regional Prediction System (ARPS) model was nested within the larger 500-m grid that was initialized using ensemble Kalman filter data assimilation of surface and radar observations. While a tornado-like vortex was forecasted in all ensemble members, the intensity, trajectory, and duration of the simulated tornado were distinctively different between ensemble members. An example of the resolved tornado features from one of the ensemble members is shown in Figure 6.7. The ARPS model was also used by Schenkman et al. (2012) to simulate a tornadic mesovortex to investigate the process of tornadogenesis. The mesovortex was associated with a mesoscale convective system (MCS) that occurred over southwestern and central Oklahoma, on May 8–9, 2007. A 100-m grid was nested inside a larger 400-m grid, and the model successfully reproduced the mesovortex that created a tornado. Therefore, the models are capable of forecasting tornadoes, but the simulations require extensive modeling knowledge and fine computational grids. With the current computational power, the refinement level of these grids is not applicable in operational weather forecasting.

Another approach to improve TaDFs in operational weather forecasting might be the utilization of flexible and unstructured grids. A weather model that is built on an unstructured grid is the Model for Prediction Across Scales (MPAS)[2] (Skamarock et al. 2012), developed by Los Alamos National Laboratory (COSIM) and the National Centers for Atmospheric Research (NCAR) in the United States. This model is formulated on unstructured *Voronoi meshes*—formally known as *spherical centroidal Voronoi tesselations* (SCVTs)—which allow for the semi-uniform discretization of the sphere and local mesh refinements where needed (Figure 6.8). Volume elements in unstructured grids can be joined arbitrarily and not in an organized way, as in the example of the organized rectangular grid in Figure 6.3. Additionally, unstructured meshes can be generated to have smoothly varying mesh transitions between different domains (Figure 6.8). This methodology avoids abrupt transitions and buffer zones between adjacent nests, which is one of the issues associated with traditional one- or two-way nesting discussed earlier (Figure 6.3). Also, this type of meshing avoids the convergence of latitude-longitude tiles in polar regions that occurs in traditionally structured meshes. In various tests, MPAS has performed well in forecasting springtime thunderstorms over the US Great Plain, including the tornado that struck Moore, Oklahoma (Whitaker & Pegion 2015).

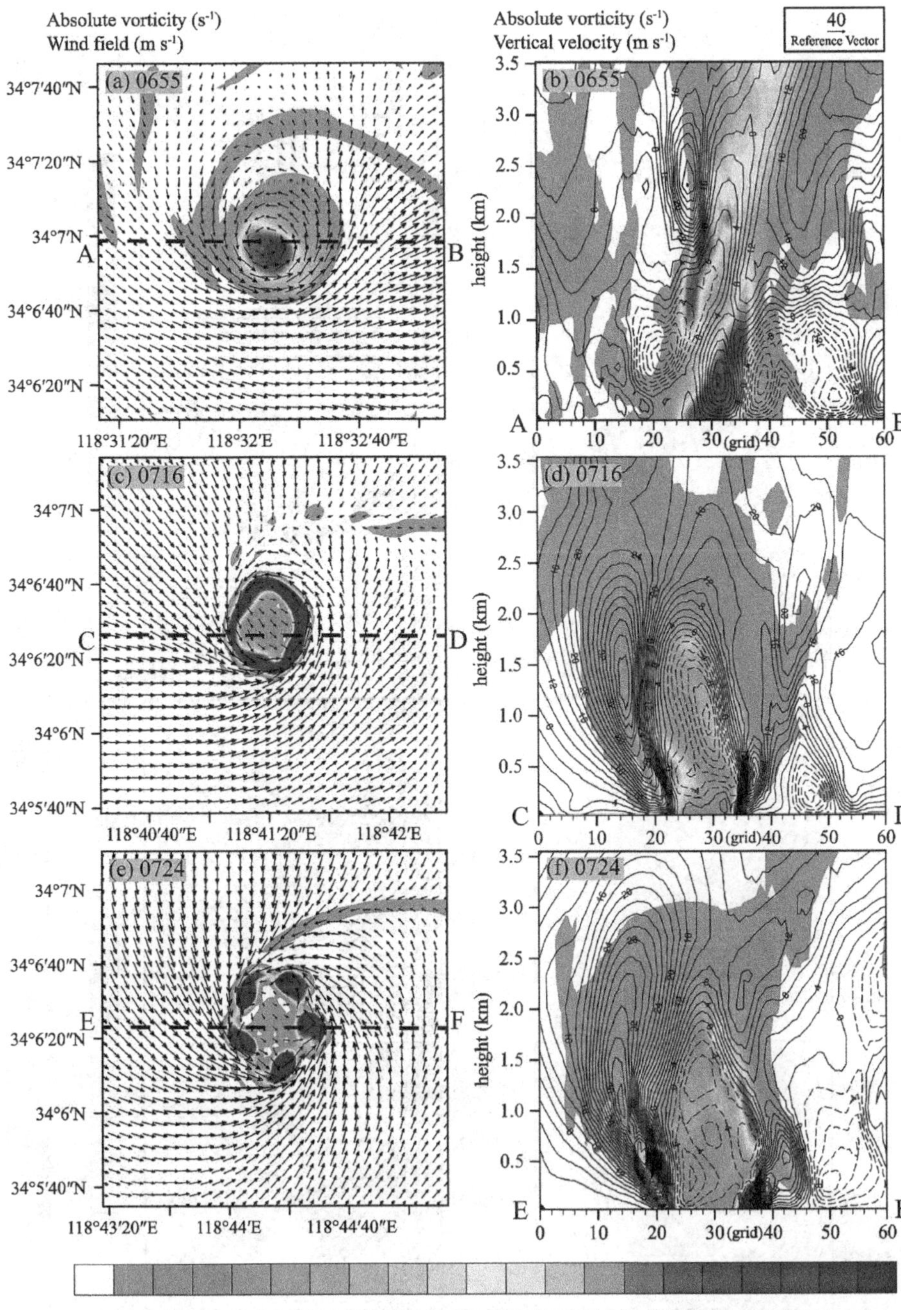

FIGURE 6.4 Left column: Horizontal cross-sections in the 49-m domain showing vertical vorticity and horizontal wind vectors at ~26 m above the ground. Right column: Vertical cross-sections of vertical vorticity and vertical velocity (contours every 2 m s^{-1} with negative contours dashed). The location of the cross-sections is marked as a dashed line in the left column. Each row represents a moment in time (indicated next to the subplot label).

Reproduced from Sun et al. (2019) with permission.

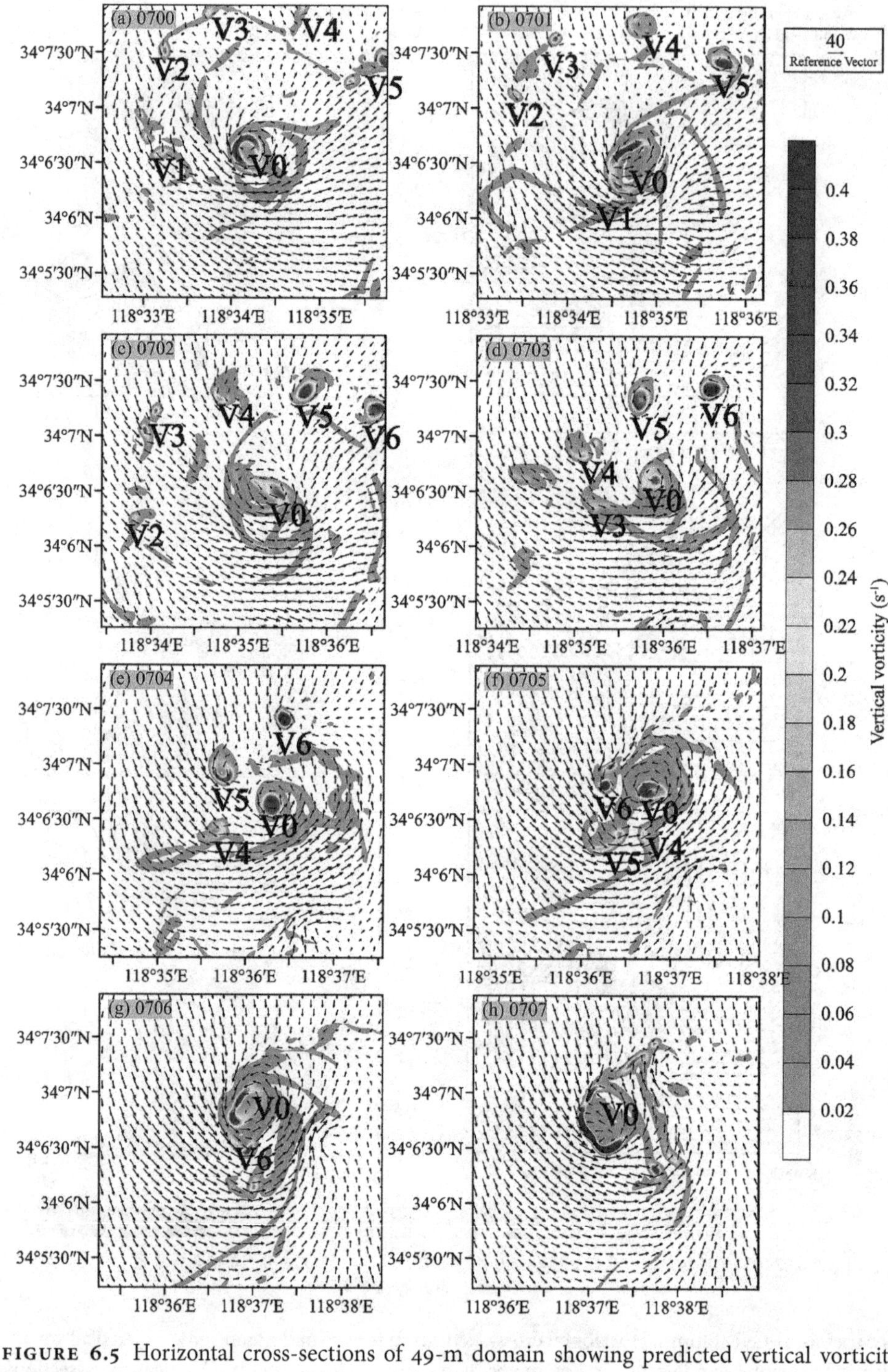

FIGURE 6.5 Horizontal cross-sections of 49-m domain showing predicted vertical vorticity and horizontal wind vectors at the first model level (26 m above ground) at (a) 0700, (b) 0701, (c) 0702, (d) 0703, (e) 0704, (f) 0705, (g) 0706, and (h) 0707 UTC (Coordinated Universal Time). The main tornado vortex is marked "V0," and sub-vortices revolving around the main vortex are marked "V1"–"V6."

Reproduced from Sun et al. (2019) with permission.

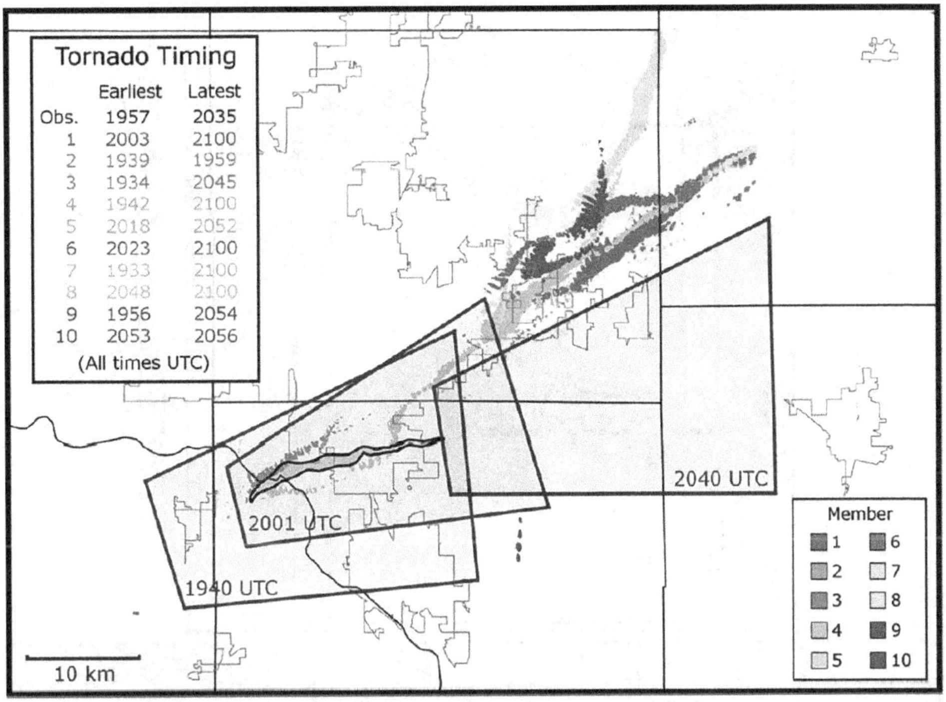

FIGURE 6.6 Swaths of wind speed exceeding the EF0-rated threshold (29 m s^{-1}) at the first model level above the surface for each of the 10 members of the 50-m ensemble (color-coded by ensemble member). Urban areas are outlined in gray, county boundaries are denoted by thin black lines, and the observed extent of EF0 or greater damage from the Newcastle–Moore tornado is indicated by a thick black contour. Tornado warnings that were issued between 19:30 and 21:00 UTC by the NWS Norman (labeled times) are plotted as dark red polygons. The earliest and latest times at which tornadoes were present for the observed tornado ("Obs.") and each ensemble member ("1"–"10") are listed in the upper-left inset.

Reproduced from Snook et al. (2019). Copyright American Meteorological Society. Used with permission.

However, unstructured grids are not flawless either. First, they require more information to be stored and retrieved than do structured grids. Some examples of the additional data required when using unstructured meshing are the neighbor connectivity list and element and node indices. Second, while most grid elements in MPAS are hexagons (six edges), pentagons (five edges), and septagons (seven edges) are sometimes present, too. This change of sizes and shapes of grid elements increases numerical approximation errors. Third, the unstructured meshes are prone to the data locality issue inside the computer's memory. This point is particularly important in parallel computing, which is employed in operational weather forecasting. In structured meshes, data for grid elements that are close geometrically inside the computational domain are also close in memory. Therefore, a smaller number of cache misses results in more efficient usage of memory bandwidth in structured meshes.

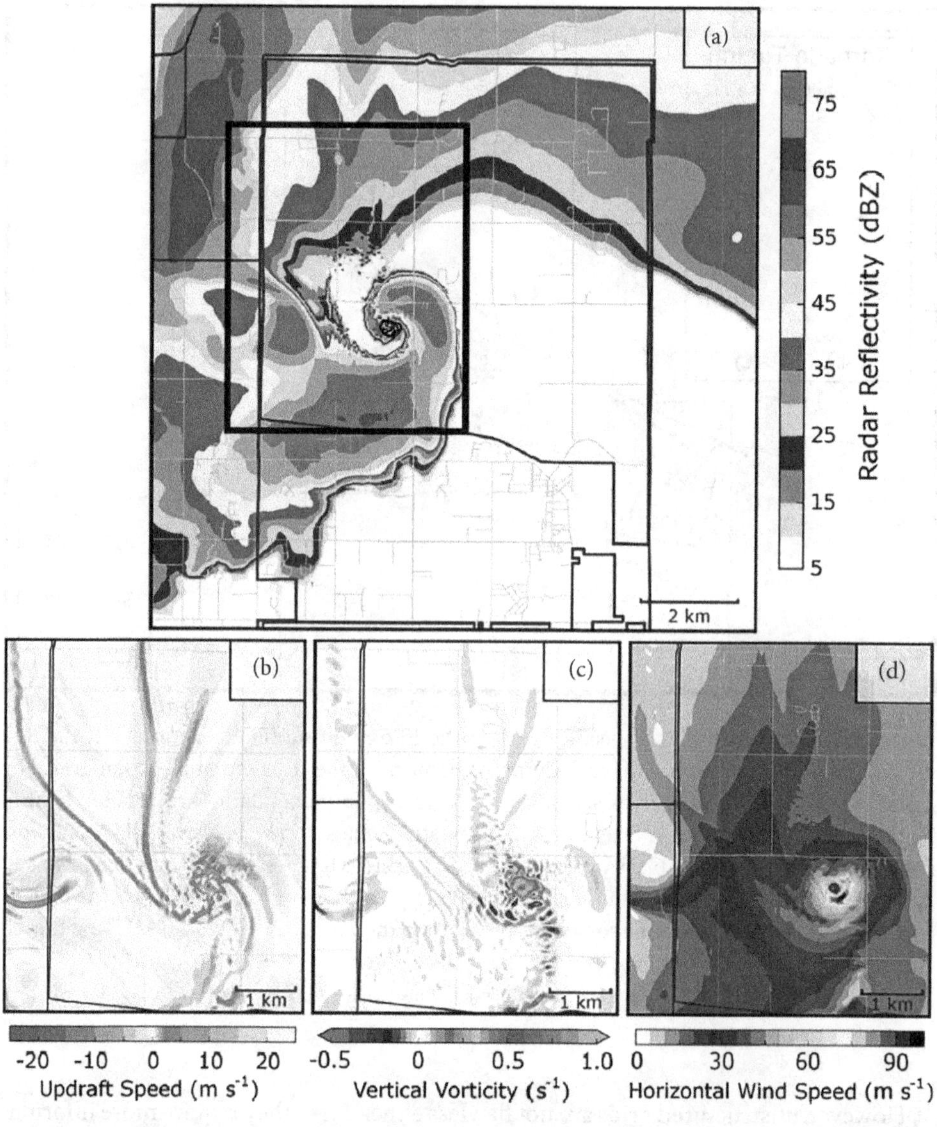

FIGURE 6.7 (a) Radar reflectivity, (b) vertical wind speed, (c) vertical vorticity, and (d) horizontal wind speed at 10 m above the surface in the region of tornadic circulation of ensemble member 1 at 20:40 UTC. The thick black box shown in (a) indicates the region shown in (b)–(d). City borders are indicated in dark purple, and roads are shown in light gray.

Reproduced from Snook et al. (2019). Copyright American Meteorological Society. Used with permission.

The Geophysical Fluid Dynamics Laboratory (GFDL) in the United States has developed the *finite-volume cubed-sphere dynamical core* (FV3[3]) scalable and variable-resolution model for both hydrostatic and non-hydrostatic atmospheric simulations (Lin 1997, 2004) (Figure 6.9). The model's discretization is guided by physical constraints as much as possible. The initial tests of the model's computational efficiency

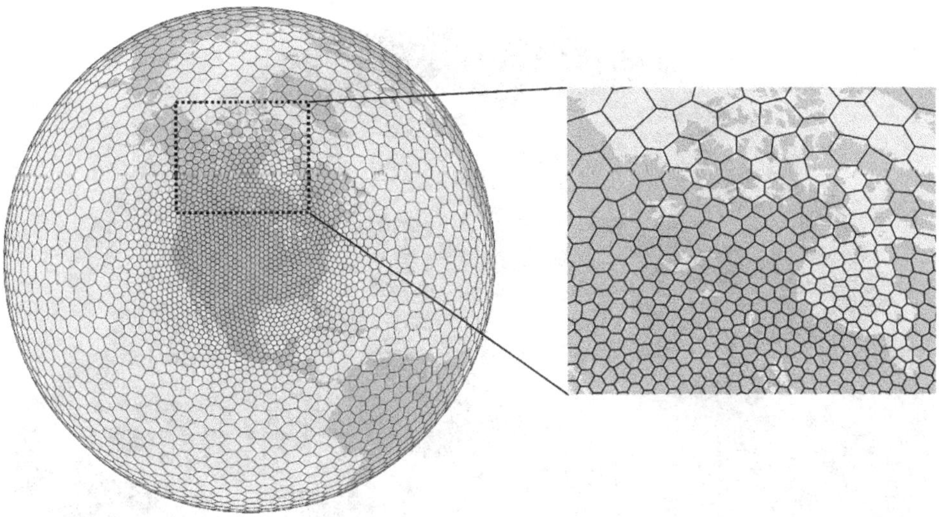

FIGURE 6.8 Unstructured Voronoi mesh in the Model for Prediction Across Scales (MPAS) model.

Courtesy of Bill Skamarock (COSIM and NCAR).

showed that FV3 is faster than most of its peers (Michalakes et al. 2015). FV3 will replace the dynamic core of the current Global Prediction System (GFS) due to its high fidelity, flexibility, computational efficiency, skillful forecasts, convective-scale simulations, and run-time switchable non-hydrostatic solver, as well as many other useful facets (Zhou et al. 2019). After implementing several physics schemes from WRF into FV3, Zhang et al. (2019) demonstrated that FV3 with an approximate 3.5-km convection-allowing grid that was two-way nested inside the global approximate 13-km grid had equivalent forecasting skills as the WRF model with similar resolution. Testing was conducted during the 2018 Spring Forecasting Experiment at the National Oceanic and Atmospheric Administration (NOAA)'s Hazardous Weather Testbed. Even if the entire simulation is at global scales, the stretching grid implemented in FV3 enables convection-resolving simulations in the areas of interest (Figure 6.9).

While the new-generation models, such as MPAS and FV3, as well as the well-established WRF, continue to improve the skills of severe weather forecasts associated with TaDs, there is still a large gap between the current operational weather forecasting and the explicit simulation of a tornado or small-scale downburst. The enormous computational expense of very high-resolution grids (e.g., ~25 m) coupled with our incomplete understanding of TaD formation and dynamics (see Section 3.3) presently prevent the explicit modeling of TaDs. Prioritization in this field needs to be given to further development of NWP models, such as MPAS, FV3, and WRF (HRRR), and their simulation across scales. Indeed, both MPAS and FV3 models have proved to be capable and efficient at modeling convective systems associated with severe TaDs across the continental United States. These two models were the finalists of the Next Generation Global Forecasting System (NGGFS) dynamic core evaluation that took place in 2016 (AVEC 2016).

(a)

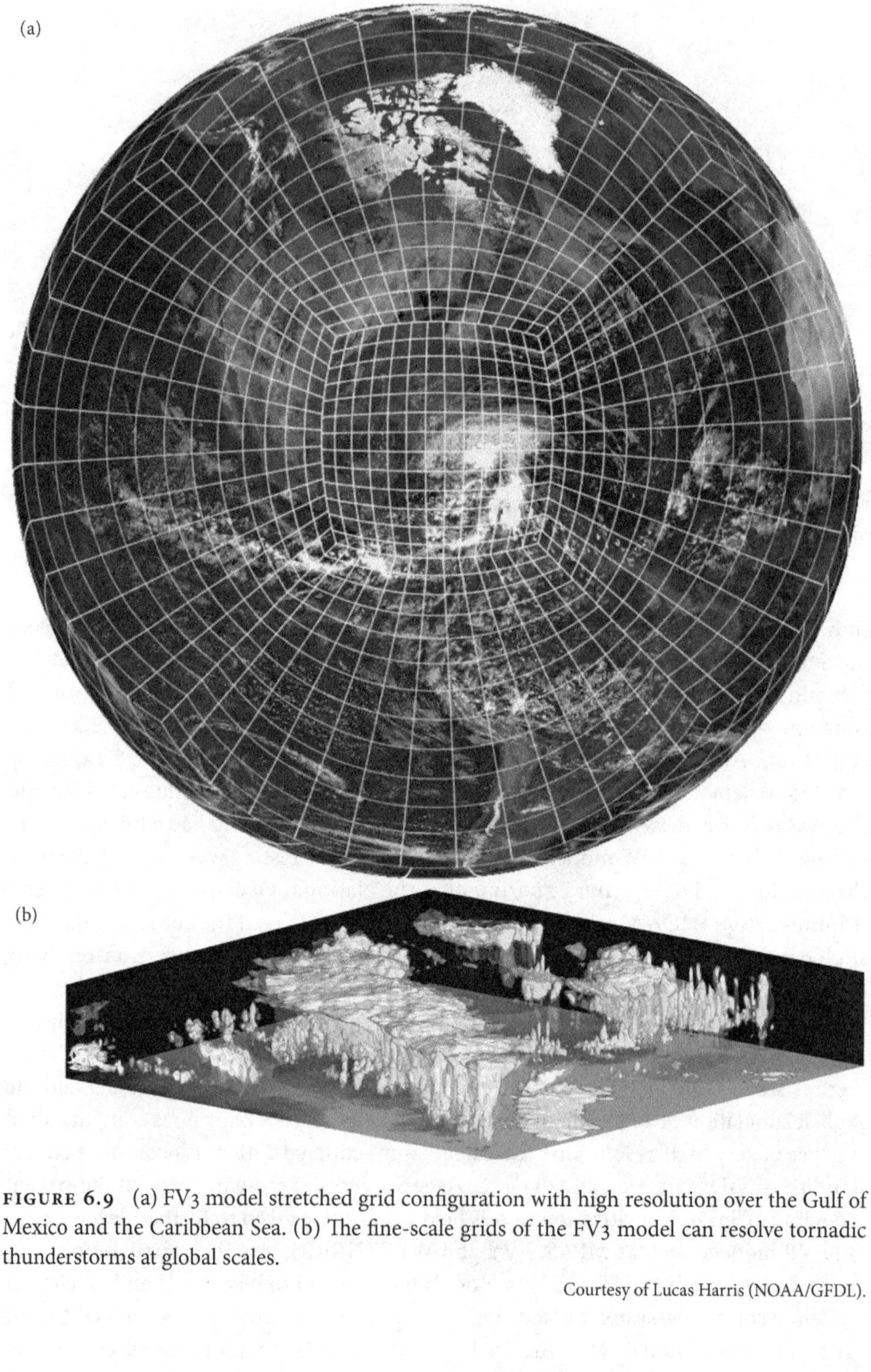

(b)

FIGURE 6.9 (a) FV3 model stretched grid configuration with high resolution over the Gulf of Mexico and the Caribbean Sea. (b) The fine-scale grids of the FV3 model can resolve tornadic thunderstorms at global scales.

Courtesy of Lucas Harris (NOAA/GFDL).

3.2. Warn-on-Forecast

Presently, in the United States, NOAA issues warnings for local severe weather only if they observe an early signal on radar or the weather hazard is already spotted. This approach provides the public with an average of 13 min advance notice before a tornado hits (NSSL 2015). A *warn-on-forecast* (WoF) is a NOAA research program developed to determine if the numerical weather prediction models with assimilated observations can be improved to the point that severe weather forecast warnings can be issued from model outputs (Stensrud et al. 2009). In WoF, data assimilation focuses on integrating Doppler radar observations of in-storm processes and satellite-retrieved products into the models that are using advanced microphysics schemes to model in-cloud processes explicitly. The WoF systems are focused on predicting specific weather hazards 30–60 min before their formation. WoF provides quantitative predictions in the form of probabilities of a weather hazard occurring and an estimate of the path that the hazard will likely take, and it also adapts to trends in the threat level based on new weather observations. As such, WoF is currently the foundation of the new *forecasting a continuum of environmental threats* (FACETs) concept, which serves as the delivery mechanism for WoF forecasts. FACETs provide user-specific and practical weather information about the hazard of interest.

Some of the areas of the ongoing research within the WoF framework are (NSSL 2015): (1) implementation of phased array radars and their capability to deliver more frequent updates than the current radar system, (2) testing of different assimilation techniques to quickly and accurately integrate radar products into models, (3) analyses of different ensembles of forecasting models that can be integrated into a single system that represents all (likely) outcomes for a given severe weather event, and (4) development of frameworks that will allow forecasters to quickly interpret model outputs and add value in generating the final forecast product. The present configuration of the WoF system consists of an ensemble of 36 members that are initialized within the Earth System Research Laboratory's HRRR Ensemble. The simulations are centered over expected areas of severe weather during the test day and cover a 900 × 900 km domain with 3-km grid spacing. Standard surface and remotely sensed observations that include radar reflectivity, radial velocity, and GOES-16[4] cloud water path are all assimilated into the WoF system every 15 min using the Gridpoint Statistical Interpolation (GSI)-Ensemble Kalman Filter[5] (EnKF) tool. The WoF forecasts are delivered using 18 ensemble members every 30 min and are up to 6 h in length, with results available every 5 min. An example[6] of WoF products is provided in Section 4.

3.3. Brief Theoretical Framework of Tornado and Downburst Formation

By reading only Sections 3.1 and 3.2, the reader could conclude that the only challenges for accurate TaDFs are the computational aspects (i.e., grid, meshing, computational time, and domains) and not our physical understanding of TaDs. However, this is not the case, and our understanding of TaDs formation and lifecycle is incomplete. In particular, the process of tornadogenesis is mostly unknown. Currently, there is no satisfactory theoretical framework that would deterministically predict the strength and path of a tornado based on some fundamental physical law(s) and local observations. The formation of multiple tornado outbreaks is even less understood. Therefore, we are currently limited to the probabilistic and qualitative approaches that provide a degree to which the atmospheric conditions are *prone* to forming TaDs (Section 4).

However, there is also a lot that we do know about TaD formation. From the climatological point of view, we have good knowledge of the regions and seasons in which TaDs form. Of course, data coverage could always be better, but the climatological aspects of TaD formation have advanced much further in comparison to our knowledge of the physics of TaD formation. The best availability of TaD data is for the continental United States, which receives approximately 1,000 tornadoes per year, usually occurring in March through May in the southern part of the United States and in summer in the northern states. Tornadoes mostly form between 3 and 9 PM local time. Globally, Canada sits at second place on the tornado occurrence list and receives around 100 tornadoes per year. Relative to its land area, interestingly, the United Kingdom has more tornadoes than any other country. Luckily for the Brits, most of these tornadoes are weak. Detailed climatological data on tornados and thunderstorms (e.g., Brooks et al. 2003) can subsequently be used to statistically model their frequency of occurrence, variability, and damages to structures through Monte Carlo simulations (e.g., Romanic et al. 2016; Refan et al. 2020). However, it should be clear that Monte Carlo modeling is by no means TaDF because the model only provides a statistical interpretation of historic data and not the future state of the atmosphere at any particular time.

The physical theories of TaD formation are much more complex and substantially less understood than the climatological picture of these phenomena. In very simple words, thunderstorms occur when warm and humid air near the ground is positioned below colder air. This temperature profile with height favors *convection*, and such weather conditions are known as *unstable* in meteorological terms. Tornadoes usually form in thunderstorms when instability is supplemented by intense variations of wind with height. This change of wind velocity with altitude is known as *wind shear*. The essential ingredients for tornado formation are the rising and spinning of warm and moist air accompanied by the sinking of cold air. On a tornado day, the lower atmosphere can be so moist that the thunderstorm cloud base is not more than about 500 m above the ground. Moreover, the wind shear becomes so intense that winds can intensify by more than 20 m s^{-1} between the surface and cloud base.

In this paragraph, we only outline the basic concepts of tornado formation, and the interested reader is referred to Markowski and Richardson (2010) and references therein for more details on this topic. The summary of tornado formation here follows the conceptual models described in Rotunno (1986) and Markowski and Richardson (2010, 2014). One of the prerequisites of tornadogenesis is the strong vertical vorticity $\left(\partial v / \partial x - \partial u / \partial y \right)$

close to the surface (u and v are the wind velocity components in the x and y directions, respectively). If the vertical vorticity near the ground preexists, then it can be further intensified by the vortex tube stretching due to the flow convergence in the lower atmosphere. On the other hand, tilting of the horizontal vorticity to the vertical vorticity or downward advection of the vertical vorticity from higher elevations is needed for tornado formation if there is no preexisting vertical vorticity close to the surface. Vorticity tilting due to convection alone is thought not to be enough for generating the adequate amount of vertical vorticity close to the surface because of the upward advection of the produced vorticity in the rising air. Therefore, the existence of a downdraft next to the updraft seems to be an essential requirement for tornado formation (if there is no preexisting strong vertical vorticity close to the surface). The importance of developed downdraft in tornadogenesis was also confirmed through different observational and numerical studies (Brandes 1978). Two dominant downdrafts in supercell thunderstorms, from which tornadoes usually form, are the *forward flank downdraft* (FFD) and the *rear flank downdraft* (RFD). For instance, numerical simulations have demonstrated that some of the air that enters a tornado is coming from the RFD (Wicker & Wilhelmson 1995). The strong wind shear in the lower atmosphere further contributes to enhancing the low-level streamwise vorticity. The subsequent tilting of the increased streamwise vorticity fosters a stronger mesocyclone.[7] The mesocyclone is characterized by low pressure in the center of rotation; the lower the pressure, the stronger the suction of the near-surface vorticity. So, where does the high relative humidity that was mentioned earlier fit in this description? The high relative humidity reduces evaporation (and, therefore, prevents cooling) in the radially advancing outflow from the impinging downdraft(s). Thus, the strong suction due to the pressure deficit in the mesocyclone, in combination with the air that spreads radially from the downdrafts, promotes intense upward accelerations and convergence in the lower atmosphere (Markowski & Richardson 2014). These dynamics seem to be the most important ingredients for tornado formation. However, some studies reported that only one-quarter or less of all mesocyclones detected by radars were associated with tornadoes (Trapp et al. 2005). Also, the interaction between the storm and the environment is not well understood, and this process can either augment or diminish further storm development (Markowski & Richardson 2010). The complexity and short spatiotemporal scales of all of these processes prevent their proper representation in NWP models, which puts an additional constraint on tornado forecasting.

Tornado simulations in cloud models are also challenging because of the sensitivity of tornado formation to initial conditions. That is, a slight change in initial conditions can result in the "hit" or "miss" of tornado formation. While the slight change of these parameters does not necessarily affect the formation of the whole supercell, it can significantly alter tornado formation and intensity. This pronounced sensitivity to initial conditions indicates that there are many aspects of tornadogenesis that we currently do not adequately understand. Observations also show that tornadoes sometimes form in nonsupercell clouds. See Markowski and Richardson (2010) and references therein for further reading on the formation of nonsupercell tornados.

Downburst formation is better understood than tornado formation. The contributing factors for downdraft formation and intensification are evaporation and melting of hydrometeors (e.g., raindrops, ice crystals, hail) and precipitation loading. The steady flow that can develop at the rear side of the storm entrains drier air into the middle and lower parts of the storm, which then promotes evaporation and melting. Downbursts can develop quickly, and their velocity increases as the air descend toward the ground. The downdraft will intensify further if the air underneath the storm has low relative humidity. The dry air under the cloud promotes evaporation, which then lowers the temperature in the downdraft and increases its negative buoyancy. Other contributing factors for downdraft formation—in particular, RFD formation—are the vertical perturbation of pressure gradients due to the gradients of vertical vorticity, as well as the pressure perturbations caused by vertical buoyancy variations (Markowski 2002). Once the downdraft hits the surface, it spreads radially outward in a starburst pattern, creating high-intensity winds near the ground. Strong winds near the boundary of the radially advancing outflow are often referred to as the *gust front*. An idealized model of a supercell thunderstorm with the usual location of a tornado, RFD, and FFD is provided in Figure 6.10.

The process of downburst formation is parameterized in numerical models because microphysics processes, such as evaporation and melting, must be parameterized. That is, weather forecasting models are not explicitly simulating any raindrop or an ice crystal in the atmosphere, but their presence is bulk represented through statistical distributions whose parameters depend on environmental conditions inside the cloud. The processes of entrainment of drier air into the cloud are also complicated to accurately model due to their small scales and significant dependency on turbulence. In addition, some microphysics processes are still not fully understood from the theoretical point of view. An example is the exact model of hail growth and mechanisms of electric charge separation in thunderstorms. This is not to say that we do not have any knowledge about these processes, but we simply note that, in some cases, the observational evidence contradicts existing theoretical models.

Prioritization in this field should be given to theoretical research on tornado formation and downburst outflow dynamics. However, this is easier said than done. The new generation of weather radars and high-resolution lidar measurements of thunderstorms coupled with numerical models using variational assimilation techniques are expected to further increase our theoretical knowledge of TaD formation and dynamics. Currently, meteorologists have effectively no way to inform the public on expected duration, detailed trajectory, or future intensity of a formed tornado and, to lesser a extent, a downburst. The gustiness and profiles of near-surface winds in TaD are also mostly unpredictable. The influence of surface roughness and orography on TaD formation and lifecycle also deserves more research.

When it comes to seasonal (and beyond) tornado forecasts, research suggests that there might be a link between multiyear cycles of El Niño, La Niña, and the Pacific Decadal Oscillation, on one hand, and the frequency of occurrence of tornadoes in the United States (Lee et al. 2012), on the other hand. This study reported that the transition

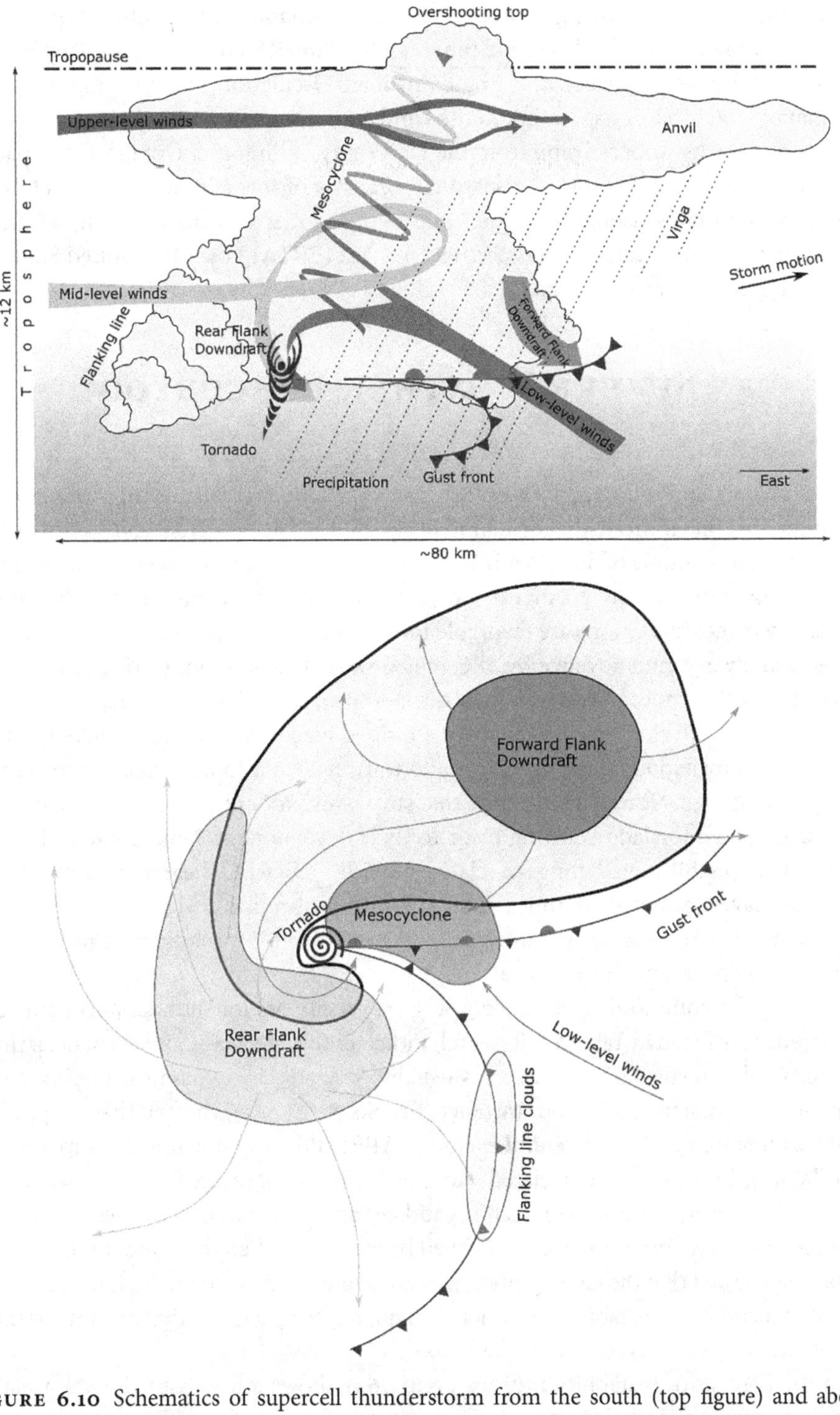

FIGURE 6.10 Schematics of supercell thunderstorm from the south (top figure) and above (bottom figure) (based on Lemon and Doswell 1979). The size of tornado vortex is not to scale with the rest of the cloud. This figure is an idealized schematic that serves to depict the primary dynamics in a supercell. The cloud dynamics of real supercells is significantly more complex than this.

of La Niña to El Niño during some spring seasons doubles the likelihood of a major tornado outbreak in the Midwest and the southern United States. Similarly, Tippett et al. (2015) found a link between the Madden-Julian Oscillation and El Niño-Southern Oscillation and hazardous weather in the United States. These and similar studies (e.g., recent research by Robert Trapp from the University of Illinois at Urbana-Champaign) provide the first insights into the seasonal forecasts of tornado occurrence. On time scales shorter than seasonal forecasts, Gensini et al. (2020) recently introduced a concept of the *extended-range tornado activity forecast* (ERTAF) over the United States for forecasting periods of 2–3 weeks.

4. Severe Weather Precursors

The NOAA Strom Prediction Center (SPC) successfully forecasts the high-risk convective outlooks[8] that indicate the formation of significant tornado events on a time scale of nowcasting to a couple of days in advance. This resource is one of the latest methods of monitoring and assessing predictions of TaDs and severe weather across the United States. If weather conditions are favorable for tornado development, a tornado *watch* is issued usually 4–8 h in advance by SPC meteorologists who monitor the weather 24/7 across the entire United States. Then, a tornado *warning* is issued by local meteorologists who are analyzing the weather over a limited geographic region if there is a high likelihood of tornado formation. A tornado warning can also be issued if a tornado is spotted in the area. Nowadays, it is rare that strong tornadoes form without the weather service issuing a tornado warning prior to their development. Numerical models are generally successful at predicting the relative humidity and wind shear in the atmospheric boundary layer (atmosphere in the first ~1 km above ground), which, as discussed in Section 3.3, are the crucial parameters for distinguishing tornadic and nontornadic supercell environments (Brooks et al. 2003).

Some of the commonly used meteorological precursors for thunderstorm and TaD occurrence are listed in Table 6.1. Research indicates that the development of deep thunderstorms that favor TaD formation is usually governed by dynamical forcing rather than pure thermal instability. For instance, Brooks et al. (2003) showed that the product of the *convective available potential energy* (CAPE) and 0–6 km wind shear magnitude (see Table 6.1) is positively correlated with the likelihood of severe thunderstorm development. However, none of the instability indices and parameters described in Table 6.1 tells the full story. Tornadoes were observed in situations when the values of some of the indices suggested that the environment was not prone to tornado formation. Such occasional inaccuracy of instability parameters is not surprising given that our understanding of tornadogenesis is incomplete (Section 3.3).

In the context of instability indices, Figure 6.11 shows a forecast of severe weather parameters from the WoF system (Section 3.2). The forecast was initiated at 00:00 UTC on May 3, 2018, and it is valid for 02:10 UTC on the same date. Because the model

Table 6.1 Some of the important instability indices used in TaDF

Index [unit]	Index description	Value	Value description
Level of free convection (LFC) [m]	The level at which a lifted parcel begins a free acceleration upward to the equilibrium level.	<2000	Tornadoes more likely
		<1000	Tornadoes very likely
Lifting condensation level (LCL) [m]	The level at which a parcel becomes saturated. A good estimate of cloud base height when parcels experience forced lifting.	500–800	Tornadoes very likely
Convective condensation level (CCL) [m]	The level at which condensation will occur if afternoon heating causes lifting parcels to reach saturation. CCL ≥ LCL.	500–800	Tornadoes very likely
Equilibrium level (EL) [m]	The level at which a lifted parcel becomes cooler than the environment and is no longer buoyant. A good indicator of thunderstorm anvil height.	Near tropopause	Deep thunderstorm development
Lapse rate (LR) [°C km^{-1}]	Rate of temperature change with height. Conditionally unstable air is unstable if it contains enough moisture.	~9.5	Unstable
		5.5–6	Stable
		6–9.5	Conditionally unstable
Surface-based convective available potential energy (SBCAPE) [J kg^{-1}]	A measure of instability over the depth of the troposphere. Related to updraft strength in thunderstorms.	<1,000	Weak instability
		100–2,500	Moderate instability
		2,501–4,000	Strong instability
		>4,000	Extreme instability
Convective inhibition (CIN) [J kg^{-1}]	Amount of energy that will prevent a parcel from lifting from the surface to the LFC. The CIN layer is also called the cap.	>0	No cap, weak convection possible
		0 to −20	Weak cap, single-cell thunderstorm possible
		−20 to −60	Moderate cap, best conditions for thunderstorm development
		<−60	Strong triggering/forcing mechanisms needed, thunderstorms unlikely.
Downdraft CAPE (DCAPE) [J kg^{-1}]	An estimate of the potential strength of rain-cooled downdrafts from thunderstorms.	>800	Moderate values, strong downdrafts possible
		>1,200	High values, strong downdrafts expected
0–6 km Wind shear vector [m s^{-1}]	Indicates the change in the wind throughout this height.	>18–20.5	Supercells common
		23–31	Tornadic supercell likely

(continued)

Table 6.1 Continued

Index [unit]	Index description	Value	Value description
0–1 km wind shear vector [m s^{-1}]	Indicates the change in the wind throughout this height.	>10	Tornados likely
Bulk Richardson number shear (BRN Shear) [m^2 s^{-2}]	Like the 0–6 km wind shear, except that the BRB Shear accounts for density changes with the height.	>35–40	Supercells probable
		>100	An increased threat of tornadoes from supercells
0–1 km Storm relative helicity (0–1SRH) [m^2 s^{-2}]	A measure of the potential for cyclonic updraft rotation in right-moving supercells in the first 0–1 km.	>100	An increased threat of tornadoes from supercells
0–3 km Storm relative helicity (0–3SRH) [m^2 s^{-2}]	Same as 0–1SRH, but in the first 0–3 km.	>250	An increased threat of tornadoes from supercells
Supercell composite parameter (SCP) [-]	An index that combines several of the above parameters to indicate the likelihood of supercells.	<2	Single- and multi-cells possible
		>2	Supercells likely
		~5	Tornadoes likely
Significant tornado parameter (STP) [-]	An index that combines several of the above parameters to indicate the likelihood of tornadoes from supercells.	>1	Most significant (EF2-rated and higher) tornadoes occurred when STP > 1
Energy-helicity index (EHI) [-]	Based on the assumption that storm rotation should be maximized when both CAPE and SRH are large.	>0.5–1	Associated with significant tornadoes in supercells
Vorticity generation parameter (VGP) [m s^{-2}]	An estimate of the rate of tilting and stretching of horizontal vorticity by a thunderstorm updraft.	>0.2	The increasing possibility of tornadic storms

This list is not exhaustive. The provided values and their ranges are indicative, and slightly different values can be found in the literature.

advanced 2 h and 10 min after initiation time, the forecast is just slightly exceeding the nowcasting range of 2 h. The prognostic variables in Figure 6.11 (i.e., composite reflectivity[9] and STP) are the product of the HRRR ensemble simulations that were centered on Oklahoma and Kansas. Figure 6.11 shows that the model predicts high values of composite reflectivity (exceeding 50 dBZ) near Norman in central Oklahoma. The verification of the forecast using a ground-based radar system is provided in Figure 6.12. The high fidelity of forecasted radar reflectivity can be attributed to the short forecasting period (nowcasting), continuous assimilation of new radar observations into the model, the ensemble modeling approach, high resolution of the HRRR ensemble members,

(a)

(b)

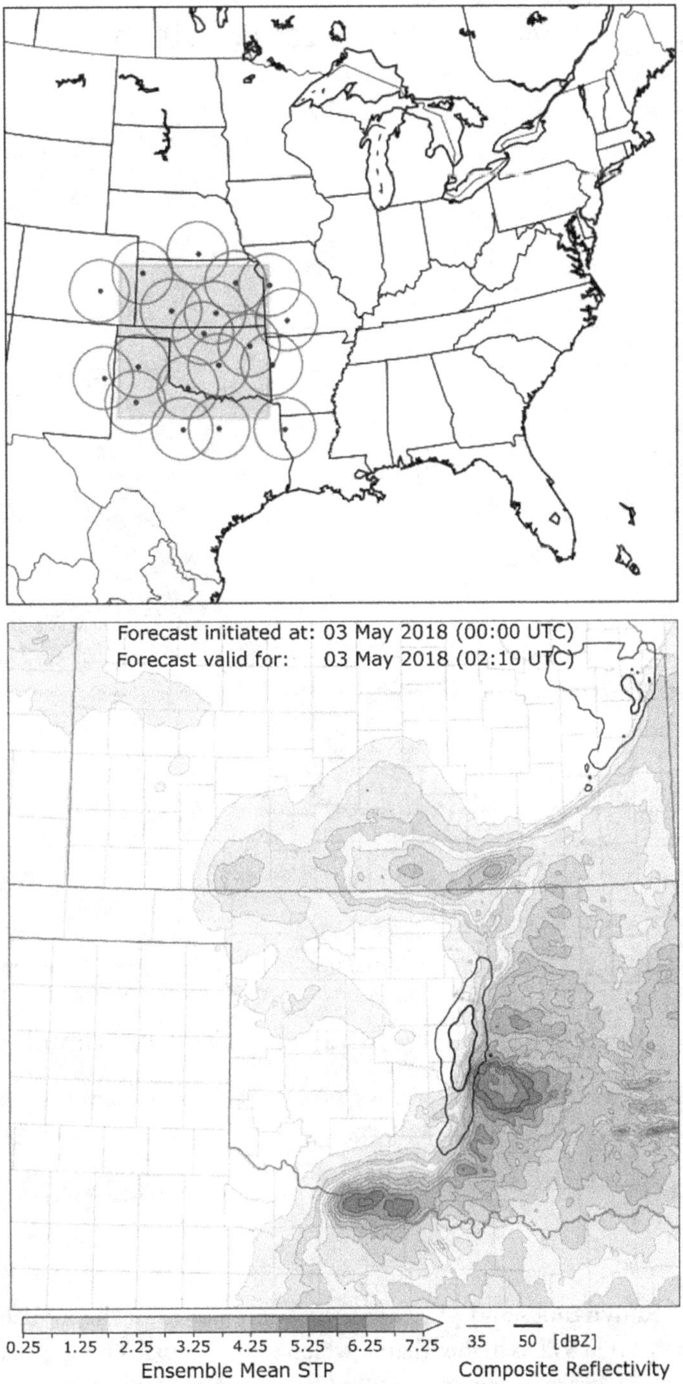

FIGURE 6.11 (a) 3-km high-resolution rapid refresh (HRRRE) background and nested warn-on-forecast (WoF) forecast verification sub-grid (green box). Radar locations in the WoF sub-grid shown as blue dots with 150-km range rings. (b) The WoF system forecast of the ensemble-mean significant tornado parameter (STP) (shaded regions) and probability matched mean composite reflectivity (black contours).

Data are available at https://wof.nssl.noaa.gov/realtime/

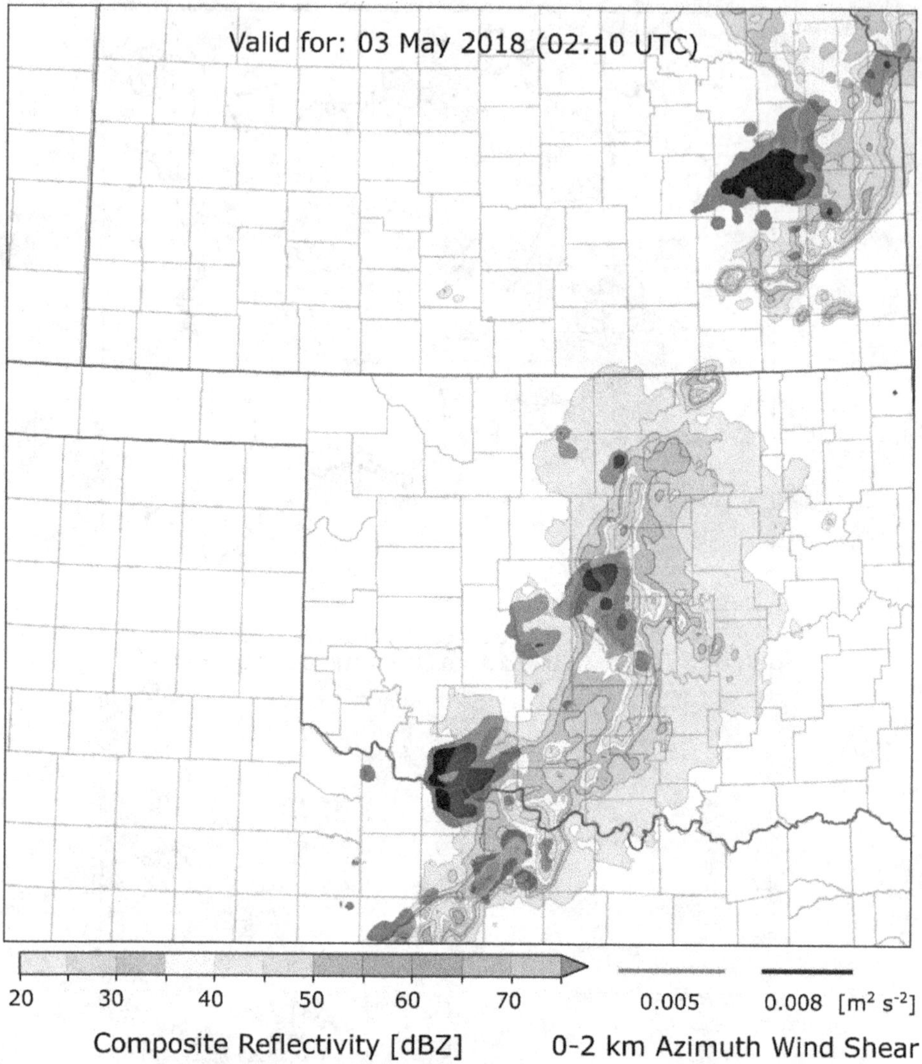

FIGURE 6.12 (a) Multi-radar/multi-sensor system (MRMS) verification of the warn-on-forecast (WoF) forecast of composite reflectivity in Figure 6.11. The observed 0–2 km azimuth wind shear also shown.

Data are available at https://wof.nssl.noaa.gov/realtime/

and advanced dynamics and physics schemes employed in the models. Figure 6.11 also shows that that STP is higher than 5, which indicates that strong tornadoes are very likely to develop. A tornado outbreak indeed happened on May 2–3, 2018, and it spawned 24 tornadoes across Kansas and Oklahoma on May 2. This example further demonstrates that we have advanced far in accurately predicting tornado-prone environments.

The WoF probabilistic concept of tornado forecasts is also depicted in Figure 6.13. The conceptual model of probabilistically forecasting TaD's occurrence and trajectory (the

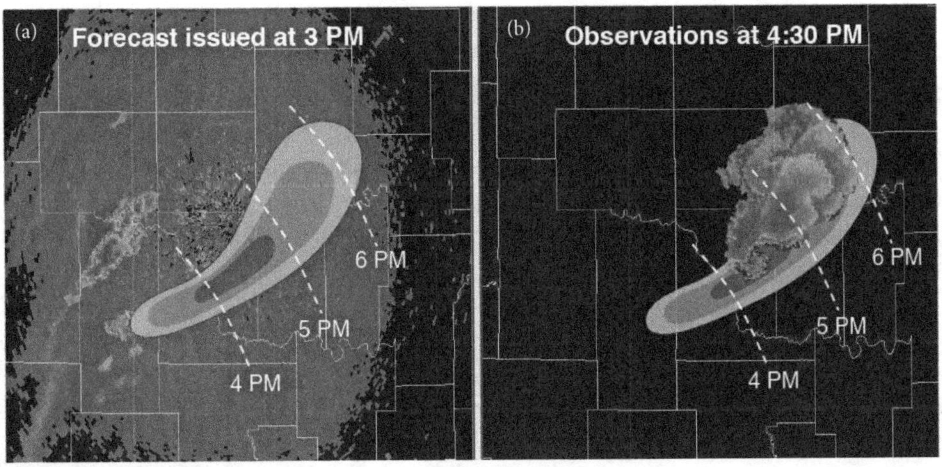

FIGURE 6.13 Radar and satellite observations of developing convection are assimilated into an ensemble of high-resolution numerical weather prediction models. This ensemble then issues a short-term (0–6 h) forecast that is used to create probabilistic guidance products for thunderstorm hazards; in this case, mesocyclone formation. If the ensemble forecast is accurate, the developing storms will intensify into supercells and move along the track of high predicted probabilities of mesocyclone formation. The blue zones represent different probabilities of the storm yielding a tornado.

Reproduced and modified from Stensrud et al. (2009). Copyright American Meteorological Society. Used with permission.

blue shaded region) is inspired by the forecasts of large-scale weather systems such as hurricanes. In principle, this type of forecasted TaD information in probabilistic format should be updated as frequently as possible (ideally with each new volume scan of radar observations) and used to govern warning decisions.

5. Summary and Conclusion

This chapter summarizes the current state of the art of TaDF. We first described the factors that prevent more reliable weather forecasts for long prognostic periods. The discretization of PDEs that describe atmospheric processes on a coarse grid is one of the main challenges for TaDF. This difficulty is due to the relatively coarse grid resolution (3 km or larger) in comparison to what is needed (~25–50 m) for the explicit modeling of TaDs. A simple increase of grid resolution is not a solution due to the limited computational power at our disposal. The effectiveness of nesting a finer grid inside a coarser grid was described in the context of TaDF. However, research in this field is still new and restricted to specific and idealized case studies.

WoF is one of the most recent TaDF attempts that combines rapid data assimilation of surface measurements, radar observations, and satellite data with high-resolution numerical models. The WoF system works toward issuing severe weather warnings based on model results supplemented with forecasters' supervision. Currently, the operational severe weather forecasts rely on the prediction of severe weather parameters and instability indices that are used as a proxy for the likelihood of a TaD occurrence.

The chapter also briefly summarized our current understanding of TaD formation. The goal was to familiarize the reader with the complexities of TaD development. While substantial progress has been made in this field over the past several decades, it is also necessary to recognize that one of the fundamental problems of TaDF is our inadequate understanding of TaD physics.

Acknowledgments

The author expresses his sincere gratitude to Ilija Jovičić and Dr. Miloš Lompar for providing many suggestions that substantially improved the quality of this work. The comments received from Prof. Horia Hangan are also highly appreciated. The author acknowledges partial support from the European Research Council (ERC) under the European Union's Horizon 2020 research and innovation program (grant agreement no. 741273) for the project THUNDERR—Detection, Simulation, Modelling, and Loading of Thunderstorm Outflows to Design Wind-Safer and Cost-Efficient Structures—awarded by an advanced grant (AdG) 2016.

Notes

1. The model is available at https://www2.mmm.ucar.edu/wrf/users/download/get_sources.html.
2. The model is available at: https://mpas-dev.github.io/.
3. The model description and development progress are available at https://www.gfdl.noaa.gov/fv3/.
4. GOES-16 is the first of the GOES-R series of Geostationary Operational Environmental Satellites operated by the National Aeronautics and Space Administration (NASA) and NOAA.
5. Read Zhu et al. (2013) and references therein for a technical description of the EnKF filter.
6. The WoF products are available at https://wof.nssl.noaa.gov/.
7. *Mesocyclone* is a large-scale updraft in supercell (3–10 km in diameter).
8. Convective outlooks can be accessed at https://www.spc.noaa.gov/products/outlook/.
9. *Composite reflectivity* is the overall maximum value of radar reflectivity that occurs in a vertical column at any altitude. *Radar reflectivity* is a measure of the efficiency of a target in intercepting and returning radio energy. In clouds, it depends on the size, shape, aspect ratio, and dielectric properties of the hydrometeors.

References

AVEC. (2016). *NGGPS Phase-2 Benchmarks and Software Evaluation* (Advanced Computing Evaluation Committee (AVEC) Report). National Energy Research Scientific Computing Center (NERSC), Berkeley, California, United States, 13.

Benjamin, S. G., Weygandt, S. S., Brown, J. M., Hu, M., Alexander, C. R., Smirnova, T. G., . . . Manikin, G. S. (2015). A North American hourly assimilation and model forecast cycle: The rapid refresh. *Monthly Weather Review*, *144*(4), 1669–1694. https://doi.org/10.1175/MWR-D-15-0242.1

Brandes, E. A. (1978). Mesocyclone evolution and tornadogenesis: Some observations. *Monthly Weather Review, 106*(7), 995–1011. https://doi.org/10.1175/1520-0493(1978)106<0995:MEATSO>2.0.CO;2

Brooks, H. E., Lee, J. W., & Craven, J. P. (2003). The spatial distribution of severe thunderstorm and tornado environments from global reanalysis data. *Atmospheric Research, 67–68*, 73–94. https://doi.org/10.1016/S0169-8095(03)00045-0

Gensini, V. A., Barrett, B. S., Allen, J. T., Gold, D., & Sirvatka, P. (2020). The Extended-Range Tornado Activity Forecast (ERTAF) Project. *Bulletin of the American Meteorological Society*. https://doi.org/10.1175/BAMS-D-19-0188.1

Lee, S-K., Atlas, R., Enfield, D., Wang, C., & Liu, H. (2012). Is there an optimal ENSO pattern that enhances large-scale atmospheric processes conducive to tornado outbreaks in the United States? *Journal of Climate, 26*(5), 1626–1642. https://doi.org/10.1175/JCLI-D-12-00128.1

Lemon, L. R., & Doswell, C. A. (1979). Severe thunderstorm evolution and mesocyclone structure as related to tornadogenesis. *Monthly Weather Review, 107*(9), 1184–1197. https://doi.org/10.1175/1520-0493(1979)107<1184:STEAMS>2.0.CO;2

Lin, S-J. (1997). A finite-volume integration method for computing pressure gradient force in general vertical coordinates. *Quarterly Journal of the Royal Meteorological Society, 123*(542), 1749–1762. https://doi.org/10.1002/qj.49712354214

Lin, S-J. (2004). A "vertically Lagrangian" finite-volume dynamical core for global models. *Monthly Weather Review, 132*(10), 2293–2307. https://doi.org/10.1175/1520-0493(2004)132<2293:AVLFDC>2.0.CO;2

Lorenz, E. N. (1963). Deterministic non-periodic flow. *Journal of the Atmospheric Sciences, 20*(2), 130–141. https://doi.org/10.1175/1520-0469(1963)020<0130:DNF>2.0.CO;2

Markowski, P., & Richardson, Y. (2010). *Mesoscale meteorology in mid-latitudes*. Chichester, UK: John Wiley.

Markowski, P., & Richardson, Y. (2014). What we know and don't know about tornado formation. *Physics Today, 67*(9), 26–31. https://doi.org/10.1063/PT.3.2514

Markowski, P. M. (2002). Hook echoes and rear-flank downdrafts: A review. *Monthly Weather Review, 130*(4), 852–876. https://doi.org/10.1175/1520-0493(2002)130<0852:HEARFD>2.0.CO;2

McDonald, J. R., Mehta, K. C., & Mani, S. (2006). *A recommendation for an enhanced Fujita Scale (EF-Scale)* (p. 111). Lubbock, TX: Wind Science and Engineering Center, Texas Tech University.

Melhauser, C., & Zhang, F. (2012). Practical and intrinsic predictability of severe and convective weather at the mesoscales. *Journal of the Atmospheric Sciences, 69*(11), 3350–3371 https://doi.org/10.1175/JAS-D-11-0315.1

Michalakes, J., Govett, M., Benson, R., Black, T., Juang, H., Reinecke, A., & Skamarock, B. (2015). *NGGPS level-1 benchmarks and software evaluation*. Advanced Computing Evaluation Committee (AVEC) Report (p. 22). Berkeley, CA: National Energy Research Scientific Computing Center (NERSC).

NOAA. (2019). *High-resolution rapid refresh (HRRR)*. Norman, OK: NOAA Research.

NSSL. (2015). *Warn-on-forecast fact sheet* (p. 2). Norman, OK: NOAA National Severe Storms Laboratory.

Refan, M., Romanic, D., Parvu, D., & Michel, G. (2020). Tornado loss model of Oklahoma and Kansas, United States, based on the historical tornado data and Monte Carlo simulation. *International Journal of Disaster Risk Reduction, 43*, 101369. https://doi.org/10.1016/j.ijdrr.2019.101369

Romanic, D., Refan, M., Wu, C-H., & Michel, G. (2016). Oklahoma tornado risk and variability: A statistical model. *International Journal of Disaster Risk Reduction, 16*, 19–32. https://doi.org/10.1016/j.ijdrr.2016.01.011

Rotunno, R. (1986). Tornadoes and tornadogenesis. In P. S. Ray (Ed.), *Mesoscale meteorology and forecasting* (pp. 415–436). Boston: American Meteorological Society.

Schenkman, A. D., Xue, M., & Shapiro, A. (2012). Tornadogenesis in a simulated mesovortex within a mesoscale convective system. *Journal of the Atmospheric Sciences, 69*(11), 3372–3390. https://doi.org/10.1175/JAS-D-12-038.1

Skamarock, W. C., Klemp, J. B., Duda, M. G., Fowler, L. D., Park, S-H., & Ringler, T. D. (2012). A multiscale nonhydrostatic atmospheric model using centroidal Voronoi tesselations and C-grid staggering. *Monthly Weather Review, 140*(9), 3090–3105. https://doi.org/10.1175/MWR-D-11-00215.1

Skamarock, W. C., Klemp, J. B., Dudhia, J., Gill, D. O., Barker, D. M., Duda, M., . . . Powers, J. G. (2008). *A description of the advanced research WRF Version 3*. NCAR technical note (p. 113). Boulder, CO: National Center for Atmospheric Research.

Snook, N., Xue, M., & Jung, Y. (2019). Tornado-resolving ensemble and probabilistic predictions of the 20 May 2013 Newcastle–Moore EF5 tornado. *Monthly Weather Review, 147*(4), 1215–1235. https://doi.org/10.1175/MWR-D-18-0236.1

Solari, G. (2019). *Wind science and engineering: Origins, developments, fundamentals and advancements*. Basel, Switzerland: Springer International.

Stensrud, D. J., Xue, M., Wicker, L. J., Kelleher, K. E., Foster, M. P., Schaefer, J. T., . . . Tuell, J. P. (2009). Convective-scale warn-on-forecast system. *Bulletin of the American Meteorological Society, 90*(10), 1487–1500. https://doi.org/10.1175/2009BAMS2795.1

Stern, H., & Davidson, N. E. (2015). Trends in the skill of weather prediction at lead times of 1–14 days. *Quarterly Journal of the Royal Meteorological Society, 141*(692), 2726–2736. https://doi.org/10.1002/qj.2559

Sun, Z., Xue, M., Zhu, K., & Zhou, B. (2019). Prediction of an EF4 supercell tornado in Funing, China: Resolution dependency of simulated tornadoes and their structures. *Atmospheric Research, 229*, 175–189. https://doi.org/10.1016/j.atmosres.2019.06.019

Tippett, M. K., Allen, J. T., Gensini, V. A., & Brooks, H. E. (2015). Climate and hazardous convective weather. *Current Climate Change Reports, 1*(2), 60–73. https://doi.org/10.1007/s40641-015-0006-6

Trapp, R. J., Stumpf, G. J., & Manross, K. L. (2005). A reassessment of the percentage of tornadic mesocyclones. *Weather and Forecasting, 20*(4), 680–687. https://doi.org/10.1175/WAF864.1

Whitaker, J., & Pegion, P. (2015). *HIWPP non-hydrostatic dynamical core tests: Results from 3-km, three-day simulations* (p. 29). Berkeley, CA: NOAA/ESRL/PSD.

Wicker, L. J., & Wilhelmson, R. B. (1995). Simulation and analysis of tornado development and decay within a three-dimensional supercell thunderstorm. *Journal of the Atmospheric Sciences, 52*(15), 2675–2703. https://doi.org/10.1175/1520-0469(1995)052<2675:SAAOTD>2.0.CO;2

Zhang, C., Xue, M., Supinie, T. A., Kong, F., Snook, N., Thomas, K. W., . . . Lin, S-J. (2019). How well does an FV3-based model predict precipitation at a convection-allowing resolution? Results from CAPS forecasts for the 2018 NOAA Hazardous Weather Test Bed with different physics combinations. *Geophysical Research Letters, 46*(6), 3523–3531. https://doi.org/10.1029/2018GL081702

Zhou, L., Lin S-J, Chen J-H, Harris, L. M., Chen, X., & Rees, S. L. (2019). Toward convective-scale prediction within the next generation global prediction system. *Bulletin of the American Meteorological Society, 100*(7), 1225–1243. https://doi.org/10.1175/BAMS-D-17-0246.1

Zhu, K., Pan, Y., Xue, M., Wang, X., Whitaker, J. S., Benjamin, S. G., Weygandt, S. S., & Hu, M. (2013). A regional GSI-based ensemble Kalman filter data assimilation system for the rapid refresh configuration: Testing at reduced resolution. *Monthly Weather Review, 141*(11), 4118–4139. https://doi.org/10.1175/MWR-D-13-00039.1

CHAPTER 7

POTENTIAL EFFECTS OF ANTHROPOGENIC CLIMATE CHANGE ON NON-SYNOPTIC WIND STORM HAZARDS

ROBERT J. TRAPP

1. Introduction

As reviewed in the chapter by Bluestein in this volume, one of the primary contributors to non-synoptic wind storms (NSWSs) is thunderstorm-induced phenomena such as tornadoes and downbursts. The intensity of these phenomena are intimately linked to the characteristics of their parent thunderstorm (or system of thunderstorms), which in turn depend to varying degrees on the larger scale meteorological (and physical–geographical) environment. To paraphrase L. F. Richardson's famous quote, "Little whirls feed on the velocity of big whirls."

The "big whirls"—here, the synoptic- to planetary-scale eddies—ultimately "feed" on the sun and its nonuniform heating of the earth's surface. Thus, any modification of that heating, be it natural or anthropogenic, has the potential to affect the intensity of the "little whirls." An analogous argument can be made regarding nonthunderstorm contributions to NSWSs.

This chapter begins with the background theory of how tornadoes and downbursts, specifically, relate to their larger scale environment (Section 2). Simple arguments are then made on how changes to the environment might result in changes in the intensity and other characteristics of tornadoes and downbursts. Next, in Section 3, analyses of historical data are examined to determine the existence of any long-term trends in such intensity/characteristics that might be indicative of an anthropogenic climate change (ACC) impact. A necessary digression from this examination is a discussion of the credibility of the historical data and thus trends. In Section 4, global (GCM) and regional climate model (RCM) simulations are consulted to determine if future impacts are to be expected; a discussion of the means by which these simulations are "downscaled" to thunderstorm (and smaller) scales is necessary here. Finally, Section 5 provides some insight into implications of the historical and projected trends on exposure and risk to NSWSs.

2. Background Theory

One means of distinguishing the different wind-generating processes in NSWSs is by the nature of the winds (i.e., rotational versus divergent), as well as by the vertical component of motion that is fundamental to the process. Tornadoes, which are well known as rotational phenomena, have a genesis that culminates in the rapid amplification ("stretching") of convective-storm–scale rotation by an *updraft*, or concentrated upward current of air. The divergent winds associated with downbursts necessarily involve a concentrated downward current of air, that is, a *downdraft*. The suggestion here is that one could predict the intensity and perhaps duration of the associated winds by identifying and quantifying the forcing(s) of such updrafts and downdrafts.

For example, these drafts are largely driven by (Archimedean) buoyancy, which can be assessed with the aid of vertical profiles of temperature (T) and humidity (q) within the draft's environment. A convenient way of summarizing the vertically dependent buoyancy is through a vertical integration of the profiles, as combined through the virtual temperature $T_v\left[=T\left(1+0.61q\right)\right]$:

$$\int g\frac{T_v'-\overline{T}_v}{\overline{T}_v}dz, \tag{1}$$

where g is standard gravity (= 9.81 m s^{-2}); the overbar is used to indicate the environment, as observed with a balloon-borne radiosonde, for example; and a prime indicates a "parcel," which has a specified vertical rate of change based on the parcel theory construct (e.g., Trapp, 2013, Chapter 5). For updraft assessment, the integration is taken from approximately the height of the assumed lower end of the updraft (cloud base) to the height of the updraft top (top of the tropopause). The integrated buoyancy in Equation 1 equates to the parameter known as *convective available potential energy* (CAPE; J kg^{-1}). It can be shown that the theoretical maximum updraft speed (w_{max}; m s^{-1}) attained over the depth of the tropospheric profile is

$$w_{max}=\sqrt{2\text{CAPE}}. \tag{2}$$

For tornadoes, we can make a further appeal to theory to relate CAPE, and thus w_{max}, to the maximum rotational winds (v_{max}; m s^{-1}). This is simply

$$v_{max} = w_{max} = \sqrt{2\text{CAPE}}. \tag{3}$$

This "thermodynamic speed limit" (e.g., Fiedler & Rotunno, 1986) is based on the premise that the hydrostatic pressure drop sustained (via mass conservation considerations) by a buoyant updraft will limit, and must be matched by, the pressure in the tornadic-vortex core (via cyclostrophic balance considerations). The thermodynamic speed limit is known to be exceeded in theoretically modeled, numerically simulated, and even observed tornadoes, and thus Equation 3 can be considered a conservative estimate. *Nevertheless,* Equation 3 *suggests that a systematic change in the magnitude of CAPE would affect the wind speeds in tornadoes.*

Unfortunately, the estimation and thus prediction/projection problem is more complicated than that because the wind-generating process in tornadoes also requires consideration of the means by which the convective-storm–scale rotation is generated. Although this does not appear to occur in a single, universal way, one can at least indirectly link the origin of rotation in strong tornadoes to the vertical profile of horizontal wind ($\vec{V} = u\hat{\imath} + v\hat{\jmath}$, where $\hat{\imath}$ and $\hat{\jmath}$ are the west–east and south–north basis vectors, respectively) in the environment. As with CAPE in Equation 1, it is convenient to summarize the vertical variation of $\vec{V}$ over a relevant depth using either a vertical derivative or vertical integral. An approximate vertical derivative is *bulk vertical wind shear,* which is simply the magnitude of the vector difference over some layer (e.g., the 0- to 6-km layer)

$$\left| \vec{\overline{V}}_{6\,\text{km}} - \vec{\overline{V}}_{0\,\text{km}} \right| = \text{S06}, \tag{4}$$

where, again, an overbar indicates the environment; bulk shear over the 0- to 1-km layer (S01) is also a relevant parameter for tornadoes. The vertical integral is an expression of *storm-relative environmental helicity* (SRH; m^2 s^{-2}), written as

$$\int \vec{\overline{V}}_{SR} \cdot \vec{\overline{\omega}} dz = \text{SRH}, \tag{5}$$

where SR indicates an environmental wind that is relative to the motion of the storm; $\vec{\overline{\omega}} = d\overline{v} / dz\hat{\imath} - d\overline{u} / dz\hat{\jmath}$, which represents the horizontal components of the environmental vorticity (hereinafter, horizontal vorticity; s^{-1}); and the integration limits are usually 0–3 km or 0–1 km. Notice that SRH is also considered to be a measure of vertical wind shear, albeit a vertically integrated one.

Parameters 4 and 5 both can be statistically related to tornado damage intensity (Enhanced Fujita Scale; EF) (Smith, Thompson, Dean, & Marsh, 2015), but a theoretical relationship like Equation 3 is less straightforward. Consider

$$\frac{\partial \zeta}{\partial t} \sim \vec{\overline{\omega}} \cdot \nabla w, \tag{6}$$

which isolates the generation of the vertical component of vorticity (hereinafter, vertical vorticity, ζ ; s^{-1}) due to the tilting of horizontal vorticity in horizontal gradients of vertical velocity (w) (see Trapp, 2013, Chapter 7). This can be approximated as

$$\sqrt{2\text{CAPE}} \times \text{S06} = \text{VGP}, \tag{7}$$

where VGP is the *vorticity generation parameter* (VGP) (Rasmussen & Blanchard, 1998). Strictly speaking, VGP explains the generation of storm-scale rotation at midlevels (~2–7 km above the ground), and thus additional arguments are needed to relate this parameter to the near-ground rotation that is required for tornado formation (Trapp, Marion, & Nesbitt, 2017). Nevertheless, the basic idea behind its use is that the stronger the storm-scale rotation, the stronger the tornado, per conservation of angular momentum/circulation (e.g., Trapp et al., 2017). *Thus,* Equation 7 *suggests that systematic changes in the magnitudes of CAPE and S06 would affect the wind speeds in tornadoes.*

Although the preceding focus on environmental wind shear and buoyancy was applied to tornadic winds, it turns out that these parameters also have relevance to the general class of damaging, convectively induced surface wind events to which downbursts belong. Following Corfidi, Coniglio, Cohen, and Mead (2015) (see also Mahoney & Lackmann, 2011), such damaging winds can be generated through "internal" and "external" mechanisms. Here, internally means that the surface winds arise primarily by convective-storm processes, including downdrafts, while externally implies that the surface winds are a result of the vertical transport of environmental winds (momentum) by convective-storm downdrafts, that is, "convective momentum transport" (Figure 7.1).

To be clear, both mechanisms require the existence of a convective storm, with a generalization that storms with more intense updrafts favor more intense downdrafts and thus stronger surface winds. The updraft–downdraft intensity relationship follows from the fact that precipitation particles are initially formed in rising air. Accordingly, rapidly rising air (i.e., stronger updrafts) more readily supports the growth of larger particles, which have larger fall speeds and can thus potentially generate stronger downdrafts; as discussed further in the chapter, downdraft strength is also modulated by the latent cooling associated with phase changes in precipitation. Implied here is a relevance of CAPE via Equation 2. But an additional relevant buoyancy parameter is downdraft CAPE (DCAPE; J kg^{-1}):

$$\text{DCAPE} = \int g \frac{-T_v^{\otimes} + \bar{\bar{T}}_v}{\bar{\bar{T}}_v} dz. \tag{8}$$

For integration over the lowest 2- to 3-km layer over the ground, DCAPE effectively provides an estimate of possible downdraft speeds attained as an air parcel descends through this layer to the ground. *Thus, systematic changes in the vertical profiles of environmental temperature and humidity, and more specifically, in the lapse rates of these variables, would affect downdraft intensity and associated surface winds.*

As with tornadoes, the intensity of convectively induced surface wind also depends in multiple ways on the vertical profile of environmental wind. Let us consider the external mechanism of convective momentum transport, which for the simplified case of a westerly wind component only (i.e., $\vec{V} = u\hat{i}$) can be expressed as

$$\frac{\partial u}{\partial t} \sim -w\frac{\partial \bar{u}}{\partial z}. \tag{9}$$

Equation 9 shows that the rate of westerly wind generation depends in part on the downdraft ($w < 0$) speed and in part on the vertical change of the environmental westerly wind. Wind generation is thus maximized when $\partial \bar{u} / \partial z$ is largest over the same levels where the downdraft is strongest, which is in the lower half of the troposphere, that is, the lowest ~5- to 6-km layer above the ground; this can be confirmed by integrating Equation 9 along the trajectory of a parcel descending to the ground in a downdraft (e.g., Figure 7.1). *The suggestion is that S06 and perhaps even S03 are relevant predictors of damaging winds; the further suggestion is that systematic changes to low- to midlevel environmental wind shear would affect damaging-wind intensity.*

This statement likewise applies to damaging winds that result from the downward transport of an internally generated *rear-inflow jet* (RIJ). The RIJ is one component of a highly coupled horizontal–vertical circulation driven fundamentally by the convective updrafts in a mesoscale convective system (MCS). The physical characteristics of the

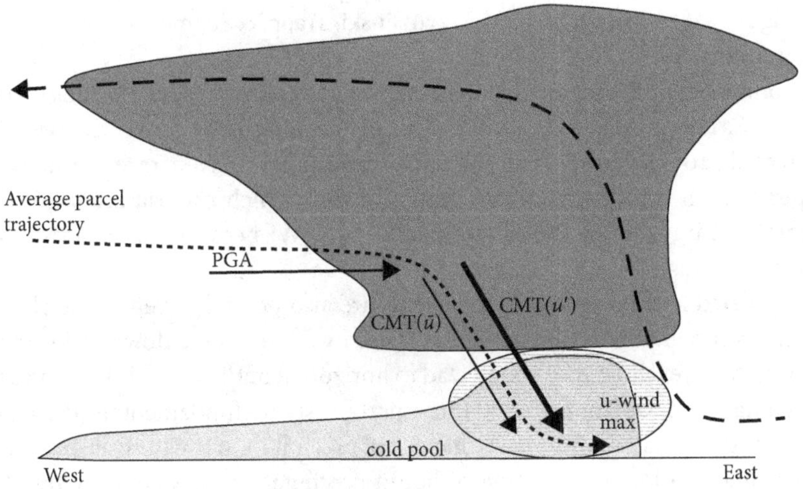

FIGURE 7.1 Schematic of convective momentum transport (CMT) within the vertical cross section of an idealized mesoscale convective system. Here, PGA indicates a pressure-gradient–driven rear-inflow jet, u is the west–east wind component, an overbar represents the environment, and a prime is a deviation from that environment (and thus generated by the storm). From Mahoney and Lackmann (2011).

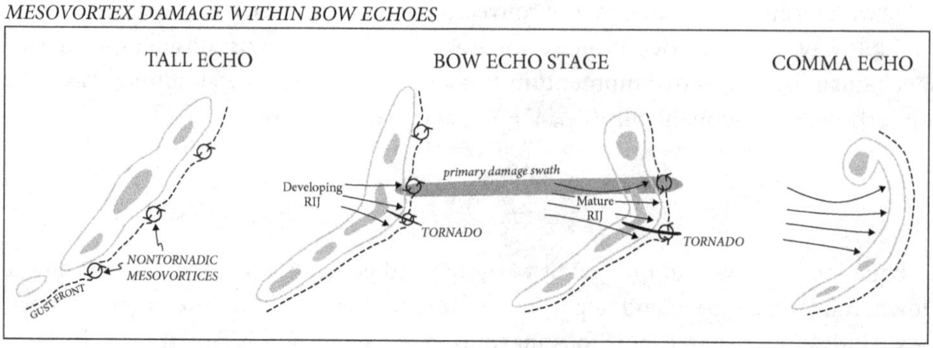

FIGURE 7.2 Schematic of damaging-wind generation by a rear-inflow jet (RIJ) and nontornadic mesovortices within a bowing mesoscale convective system (or "bow echo"). From Atkins et al. (2005).

updrafts are in turn highly dependent on the low-level wind shear (S03) and also CAPE, as has been shown theoretically and through numerical simulations by, for example, Rotunno, Klemp, and Weisman (1988). Sufficiently strong S03 is also necessary for the formation of near-ground "mesovortices" within MCSs (Trapp & Weisman, 2003; Weisman & Trapp, 2003). Mesovortices are distinct from and larger than tornadoes and cause swaths of damaging winds, which are in addition to the larger area of RIJ-induced winds (Figure 7.2) (Atkins, Bouchard, Przybylinski, Trapp, & Schmocker, 2005; Wheatley, Trapp, & Atkins, 2006).

The combined influence of RIJs and mesovortices in particularly long-lived and intense MCSs comprise *derechos*, which literally are long-lived convective windstorms (Corfidi et al., 2015; Johns & Hirt, 1987). Derechos are relatively rare (perhaps one to three per year in the United States, depending on which criteria are used; Corfidi et al., 2015) and arguably are the most impactful of NSWS because of their areal coverage of damaging winds.

More isolated and short-lived wind events are also possible, even in the absence of strong environmental winds/wind shear. These owe to extreme downdrafts or downbursts, which on reaching the ground lead to horizontal outflow winds known as *microbursts* (Fujita, 1981). Downbursts and thus microbursts are fundamentally driven by the latent cooling associated with phase changes in precipitation–melting and sublimation of frozen precipitation, evaporation of liquid precipitation, as well as by the drag or "loading" of the precipitation, as alluded to previously. The dominant effect depends on the details of the environmental temperature and humidity profiles, with relatively arid environments favoring latent cooling (and "dry microbursts"), and relatively humid environments favoring loading particularly from hail and large raindrops (and "wet microbursts"). Thus, as already noted, systematic changes in the character of temperature and humidity profiles would affect the wind speeds in microbursts.

3. Trends in Historical Data

Before examining historical data for the existence of long-term trends in the intensity/characteristics of NSWS phenomena, it is prudent to describe the data used to derive such trends. We focus here on wind.

Routine meteorological measurements of near-ground (10-m) wind in the atmosphere are made with some form of anemometer, which for US National Weather Service (NWS) systems have maximum operational speeds of 60–65 m s^{-1} (Sturgeon, 1999). Meteorological anemometers are typically deployed as part of (now-) automated weather stations, which also provide measurements of temperature, humidity, pressure, precipitation, and perhaps also visibility. Although these weather stations are often sited at airports, they are sufficiently distributed geospatially to form *synoptic-scale* networks (e.g., Trapp, 2013, Chapter 3) with station spacing of about 50–100 km in the US specialized *mesoscale* networks ("mesonets"), which have also been installed over some US states (e.g., Brock et al., 1994) and elsewhere around the world, and have finer station spacings. Nevertheless, long-term measurements of wind (and also of pressure, from which wind can be deduced) are only available on scales larger than those of tornadoes, downbursts, and other NSWS phenomena, meaning that fortuitous encounters are necessary for the phenomena on the smaller end of the NSWS spectrum (although see Smith et al., 2013, for an example application of these measurements).

Although imperfect, Doppler weather radar provides a source of wind information that is more comparable to the scales of these phenomena. Specifically, the information is in terms of the mean speed of raindrops, hailstones, and other microwave-signal scatterers along radar beams (which can be likened to spokes on a wheel). For the NEXRAD (WSR-88D) systems in the United States, the maximum unambiguous speed (also known as Doppler Nyquist velocity) is about 30 m s^{-1} (e.g., see Rauber & Nesbitt, 2018), but multiples of this speed can be detected and deduced (through a process known as dealiasing).

Unfortunately, for the purposes of acquiring near-ground wind speeds, Doppler weather radar has limitations: Radar beams are transmitted at nonzero elevation angles relative to the earth's curved surface, meaning that the heights of radar beams quickly increase with distance from the radar, and thus that sections of radar beams far from the radar have heights well above the ground. Sections of radar beams very near to the radar, on the other hand, have heights very near to the ground, meaning that the radar signal tends to be contaminated by ground targets (i.e., "clutter") and thus rendered useless; radar beams transmitted at (nearly) zero elevation angles would suffer this same fate. Thus, only within a relatively narrow range of distances from the radar are the heights of wind data directly relevant to the built and natural environment, although downward extrapolation is possible through the use of physical and statistical models. A roughly two-decade data record is now possible from NEXRAD, which is not exactly considered long term, but perhaps sufficient for some applications.

The final source of wind information, and the source used most often in convective-wind hazard analysis, is reports of wind or damage from human observers. In the United States, these solicited and unsolicited reports are compiled by the NWS and then archived by the National Oceanic and Atmospheric Administration (NOAA) National Centers for Environmental Information; elsewhere around the world, such as in Europe, the compilation of this information tends to be driven at a grassroots level by researchers. The US report database dates back to the 1950s and thus has a length adequate for climatological applications.

As discussed by Trapp and Brooks (2013) and many others, this database has a number of well-known caveats, however. The first are related to changes in population density, the introduction of videorecording and social media, and the general education of the observers about the phenomena. These in particular have introduced biases to the raw numbers of reported wind events. The second caveat relates to the use of damage to rate intensity, the change in the damage-scale rating system (from F to EF), and the reporting requirement that either a wind speed estimate be provided or damage type be indicated (see Edwards, Allen, & Carbin, 2018). For example, the variability in nonengineered building practices has led to biases in the damage-based intensity of tornadoes (e.g., see Marshall, 2002), as has the variability across NWS offices in how and when surveys of damage are conducted; the tendency for humans to overestimate wind have introduced biases into the perceived intensity of severe-wind events (e.g., Trapp, Wheatley, Atkins, Przybylinski, & Wolf, 2006). The report data must therefore be treated judiciously.

For tornadoes, one such judicious treatment is an evaluation of *tornado days* or, in other words, the number of days during some period that at least one tornado has occurred within a geospatial domain. The use of tornado days mitigates bias in overreporting, especially when a threshold on F/EF is applied. For example, Brooks, Carbin, and Marsh (2014) considered F/EF1+ tornado days over the contiguous United States (CONUS), and concluded that the number of F/EF1+ days per year had decreased at an average rate of about 1 day/year over the period 1990–2010 (Figure 7.3). Brooks et al. (2014) also concluded that the number of days per year of 30 or more F/EF1+ tornadoes (i.e., tornado-outbreak days) over the CONUS had increased to about 3 days/year over the period 2000–2010 (Figure 7.3). The simple storyline from Brooks et al., and subsequently from Elsner, Elsner, and Jagger (2015), is as follows: fewer tornado days per year but more tornadoes forming on those days.

A consequence of these opposite-signed trends is greater variability in tornado activity on annual as well as monthly timescales (Brooks et al., 2014). This variability can be alternatively quantified as the standard deviation of year-to-year differences in tornado reports. The resultant *volatility* (Tippett, 2014) can help identify when the tornado record appears little susceptible to the aforementioned reporting biases in reporting, such as during the 2000s. Indeed, Tippett (2014) found that tornado volatility had increased during the 2000s for weak as well as strong tornadoes, and that this increased volatility had a consistent signature in storm–environmental parameters. One implication of increased volatility, and the associated increase in tornado-event intermittency, is the possibility of reduced preparedness.

Preparedness may also be affected by an earlier start to the tornado season, as has been identified by Brooks et al. (2014). This is significant because such a trend in the annual tornado cycle is especially noticeable in F/EF3+ tornadoes (Trapp & Hoogewind, 2018, their Figure 7.2). The period of peak activity during the annual cycle has also been shifting toward earlier in the year (see also Long & Stoy, 2014): This is due in part to an overall reduction in tornado days during boreal summer months. Indeed, a monthly disaggregation of the tornado-day data by Trapp and Hoogewind (2018) showed that that the largest declines during the year have occurred in the month of July, thus explaining much of the annual tornado-day reduction noted by Brooks et al. (2014); this, incidentally, has been linked statistically and physically to declines in Arctic sea ice in July.

The July tornado-day reductions would seem to suggest a reduction in exposure. However, because July tornado activity in the United States is highest in the central-to-northern Great Plains, where population and population density are relatively low, the exposure has probably not changed significantly. Exposure to individual tornado events also depends on the size, intensity, and duration of the tornado, which from the perspective of tornado damage characteristics tend to be interrelated: On average, the more intense tornadoes have wider and longer damage tracks (Brooks, 2004; see also Elsner, Jagger, & Elsner, 2014). A natural question to ask is whether these characteristics have changed over time. Agee and Childs (2014) suggested that the (adjusted) mean track width for significant (F/EF2+) tornadoes has increased over the period 1953–2012. Elsner et al. (2014) reached a similar conclusion for mean track width as well as mean track length.

Proof that such trends have a meteorological basis await identification of a robust physical explanation. Here we refer to Trapp et al. (2017), who have presented evidence that tornado width and intensity—and perhaps even, by extension, tornado duration and track length—are controlled by the width of the updraft in the parent tornadic storm, and that the updraft width is highly dependent on the environmental vertical wind shear (e.g., S06 and SRH) and CAPE (see also Kirkpatrick, McCaul, & Cohen, 2009, and Marion & Trapp, 2019). In Section 4, we discuss how these controls may be expected to change in the future.

Although detection of trends in severe nontornadic winds is more problematic, owing especially to the allowance for (typically biased) human-estimated wind speeds to meet the 50-knot criterion (Edwards et al., 2018), new analyses of occurrence can shed light on exposure. It should be noted here that past severe-wind climatologies in the United States have been based (exclusively) on reports (e.g., Doswell, Brooks, & Kay, 2005). However, the geospatial details of these have been brought into question by Smith et al. (2013), who instead used wind gusts measured via the NWS Automated Surface Observing Systems (ASOS) and Automated Weather Observing Systems (AWOS). A significant difference between the report-based and measured-wind analyses is the

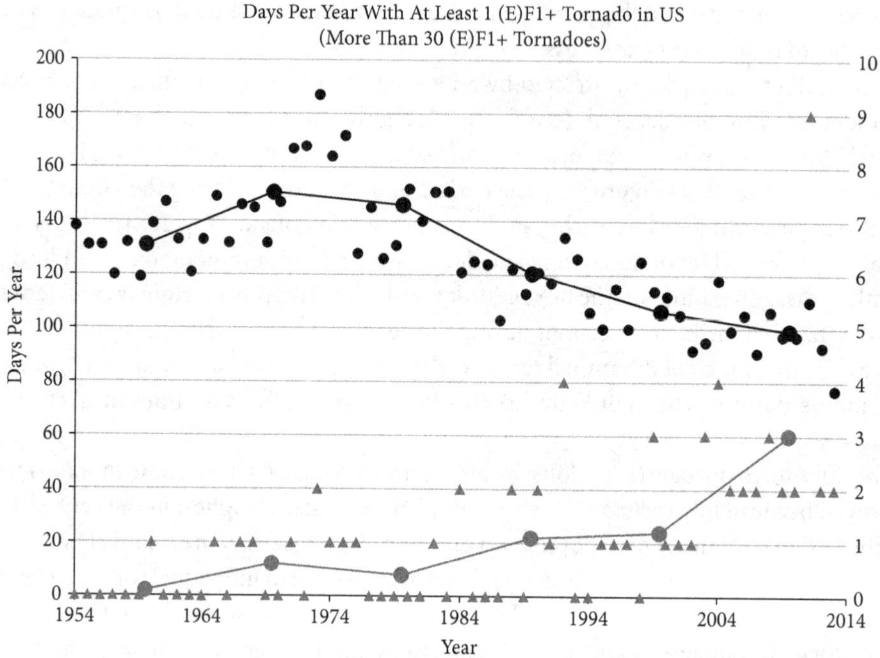

FIGURE 7.3 Annual tornado activity in the United States over the period 1955–2013. The small black circles show the number of days per year with at least one tornado rated F/EF1+, and the large black circles and line show the decadal mean line of such *tornado days*. The red triangles show the number of days per year with more than 30 tornadoes rated F/EF1+, and the red circles and line show the decadal mean of such tornado outbreaks. From Brooks et al. (2014). Reprinted with permission from AAAS.

geospatial location of the maximum severe-wind gust occurrence: the southern Appalachians from the reports versus the Great Plains from the measurements. This geolocation bias in the reports may be related to the relative (over)reliance on estimated gusts within the southeast United States (Edwards et al., 2018).

Nonetheless, if we accept the measured-gust occurrence as truth, then the convective-mode type attached to each severe wind event by Smith et al. (2013) provides additional information on the wind character. For example, "disorganized" convective storms (convective cells or clusters that did not meet quasilinear convective system [QLCS] or supercell criteria) accounted for 47 percent of the severe wind gusts and were most common in June–July–August. Compared to QLCS and supercell events, the disorganized storm events are likely to be more isolated and shorter lived (i.e., of microburst character). QLCSs accounted for 42 percent of the severe wind gusts and are likely to have longer and larger swaths of damaging winds; these were most common in May–June–July, from the Great Plains through the Midwest. Finally, supercells accounted for the remaining 11 percent of the events and tended to be most prevalent within the Great Plains in May–June–July. These events were most apt to be associated with significant wind gusts (e.g., speeds $\geq$ 65 knots).

4. Future Climate Projections

The goal of this section is to determine whether/how the temporal and geospatial distributions of the NSWSs revealed in Section 3 might change in the future owing to anthropogenic climate change (ACC). The underlying challenge to this goal is the temporal and spatial scales of NSWS: Because these are, by definition, subsynoptic scale, they are also below the nominal grid scale of typical GCMs and even many RCMs. This is especially the case with tornadoes and microbursts, which have length scales of ~O(100 m) to ~O(1-km), as compared to the ~O(100-km) and ~O(10-km) grid spacings of GCMs and RCMs, respectively. Fortunately, as demonstrated in Section 2, these NSWSs are strongly controlled by the temperature, humidity, and wind on the GCM and RCM scales. In other words, the concept of an "environment" can still be exploited, whereby certain behaviors are expected—such as tornado formation or lack thereof—given a heuristic range of environmental parameters (of CAPE, S06, convective inhibition [CIN], etc.).

The basic procedure established by Brooks, Lee, and Craven (2003), and then first implemented by Trapp et al. (2007) and Del Genio, Yao, and Jonas (2007), is to evaluate these environmental parameters at individual grid columns, just as they would be evaluated using the vertical profiles of temperature, humidity, and wind measured using radiosondes. Following Brooks et al. (2003), Trapp et al. (2007) combined CAPE and S06 into an *environmental proxy* for the Number of Days of SEVere thunderstorms (NDSEV). Specifically, where (at grid points) and when (on days)

$$\text{CAPE} \times \text{S06} > 10{,}000 \tag{10}$$

a severe thunderstorm (i.e., one that generates a tornado, straight-line winds with speeds ≥ 26 m s^{-1}, or hail with diameter ≥ 25 mm) is considered to be supported, and NDSEV is incremented. Variants of Equation 10 have subsequently been used (e.g., Seeley & Romps 2015), but they all make the necessary assumption that a convective storm has developed at that grid point and thus are able to realize the environmental CAPE and S06.

The study by Diffenbaugh, Scherer, and Trapp (2013) illustrates well the information possible from this environmental approach. Diffenbaugh et al. evaluated NDSEV as well as the individual parameters of CAPE and S06 using future (2070–2099) and historical (1970–1999) simulations from ten different GCMs from the Coupled Model Intercomparison Project phase 5 (CMIP5). Robust increases in CAPE were found throughout most of the United States during each season, as realized through an increase in the occurrence of very large (>3,000 J kg^{-1}) CAPE and decreases in the occurrence of moderate-to-small (<1,000 J kg^{-1}) CAPE (Figure 7.4). In contrast, S06 exhibited robust decreases throughout most of the United States during each season. This is consistent with the theoretical argument that the relatively more warming at high latitudes under

ACC will reduce the pole-to-equator temperature gradient and in turn reduce the vertical wind shear, as dictated by the thermal wind equation (e.g., see Trapp et al., 2007). When combined through Equation 10, however, the CAPE increases more than compensated for the S06 decreases to yield a mean increase in NDSEV of about a few days/season (Figure 7.4).

The projection under ACC is thus for an increase in the number of days supportive of severe-thunderstorm occurrence in the future. The updrafts within these future storms should tend to be stronger, given Equation 1 and the projected increases in CAPE. It is also possible to argue per Equations 3 and 7 that the tornadoes that form in these storms should be stronger. What limits this argument, however, is the uncertainty of whether or not the decreased S06 would still allow for the formation of supercell thunderstorms,

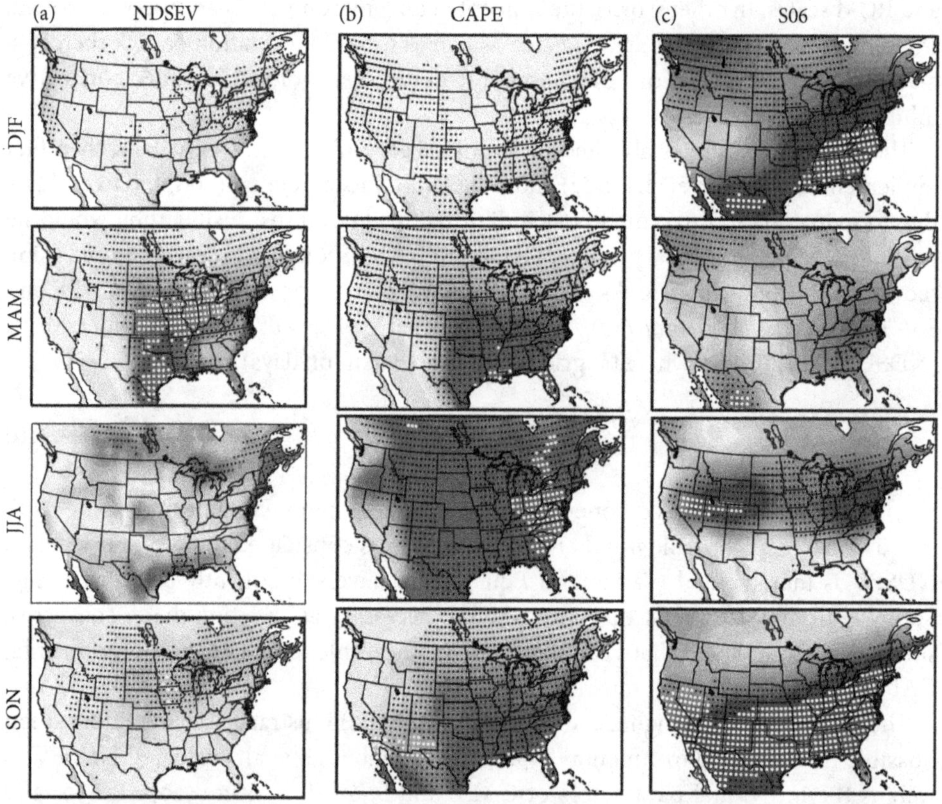

FIGURE 7.4 Mean seasonal change (future [2070–2099] minus historical [1970–1999]) in (A) NDSEV (days/season), (B) CAPE (J/kg), and (C) S06 (m/s) over each season (December–January–February; DJF), March–April–May (MAM), June–July–August (JJA), and September–October–November (SON). Black (gray) dots show locations where the ensemble signal exceeds one (two) standard deviations of the ensemble noise. From Diffenbaugh et al. (2013).

which are the primary generators of strong tornadoes. A closer inspection of the daily CAPE–So6 distribution shows that the decreases in So6 are mostly on days when CAPE is relatively low (Diffenbaugh et al., 2013). Tornadic supercells are known to form in low-CAPE/high-So6 environments, especially within the southeastern United States (Davis & Parker 2014), but tend not to yield the violent tornadoes often found in high-CAPE/high-So6 environments. Nonetheless, this uncertainty is still not removed completely because even a high-CAPE/high-So6 environment can support non-supercellular storms such as QLCSs (Smith, Thompson, Grams, Broyles, & Brooks, 2012; Thompson, Smith, Grams, Dean, & Broyles, 2012); as already discussed, these have wind hazards, but of a different character than supercells.

As an aside here, it would seem that one possible methodological approach is to use idealized numerical model experiments in which CAPE and vertical wind shear are systematically varied over an expected range in a future climate. We are aware of past attempts to do this (G. Lackmann, NC State University, personal communication, 2015); Table 7.1 summarizes some of our work (Marion & Trapp, 2019) adapted for this purpose and shows that the strongest tornado-like vortices occur with moderate CAPE and strong So6. While this provides valuable information in general, the information is somewhat limited by the fact that convection initiation (CI) is imposed and thus guaranteed in the idealized model; the means of initiating convection in the model also exerts some control on the convective morphology and thus hazard.

Indeed, these questions regarding convective morphology, and CI in general, have motivated the implementation of high-resolution *dynamical downscaling*, wherein the relatively coarse climate simulations from a GCM are used to drive regional models that can simulate the climate on computational grids that "allow" or "permit" the explicit representation of convective storms (e.g., Trapp, Smith, Grams, Dean, & Broyles, 2011; see also Prein et al., 2015). A typical "convection-permitting" model grid spacing is 4 km (e.g., see Kain et al., 2008), which is still too coarse to resolve a tornado, but sufficient to resolve a tornadic storm and other NSWSs.

At present, Hoogewind, Baldwin, and Trapp (2017) have generated the most comprehensive set of convection-permitting dynamical downscaling simulations. To quantify the spatiotemporal changes in severe thunderstorms, they analyzed grid-scale occurrences of upward vertical velocity (UVV) exceeding 22 m s^{-1}, a metric they determined corresponded well to gridded reports of severe thunderstorms over the historical period of 1971–2000. Their most basic conclusion is consistent with that from the environmental analyses, namely, that synthetic severe-thunderstorm occurrence over the future period of 2071–2100 is more frequent than over the historical period. However, the future increase is more muted in the dynamical downscaling simulations (Figure 7.5).

For example, a future mean increase of 7.3 (18.8) severe-thunderstorm days per year is projected for over the CONUS using the dynamical downscaling (environmental) approach; such a disparity varies regionally as well as seasonally. A future increase in the overall length of the severe-thunderstorm season is also projected, with enhanced severe-thunderstorm probability on both ends of the annual cycle. Finally, as expressed in terms of UVV and thus updraft speeds, a higher frequency of more intense severe

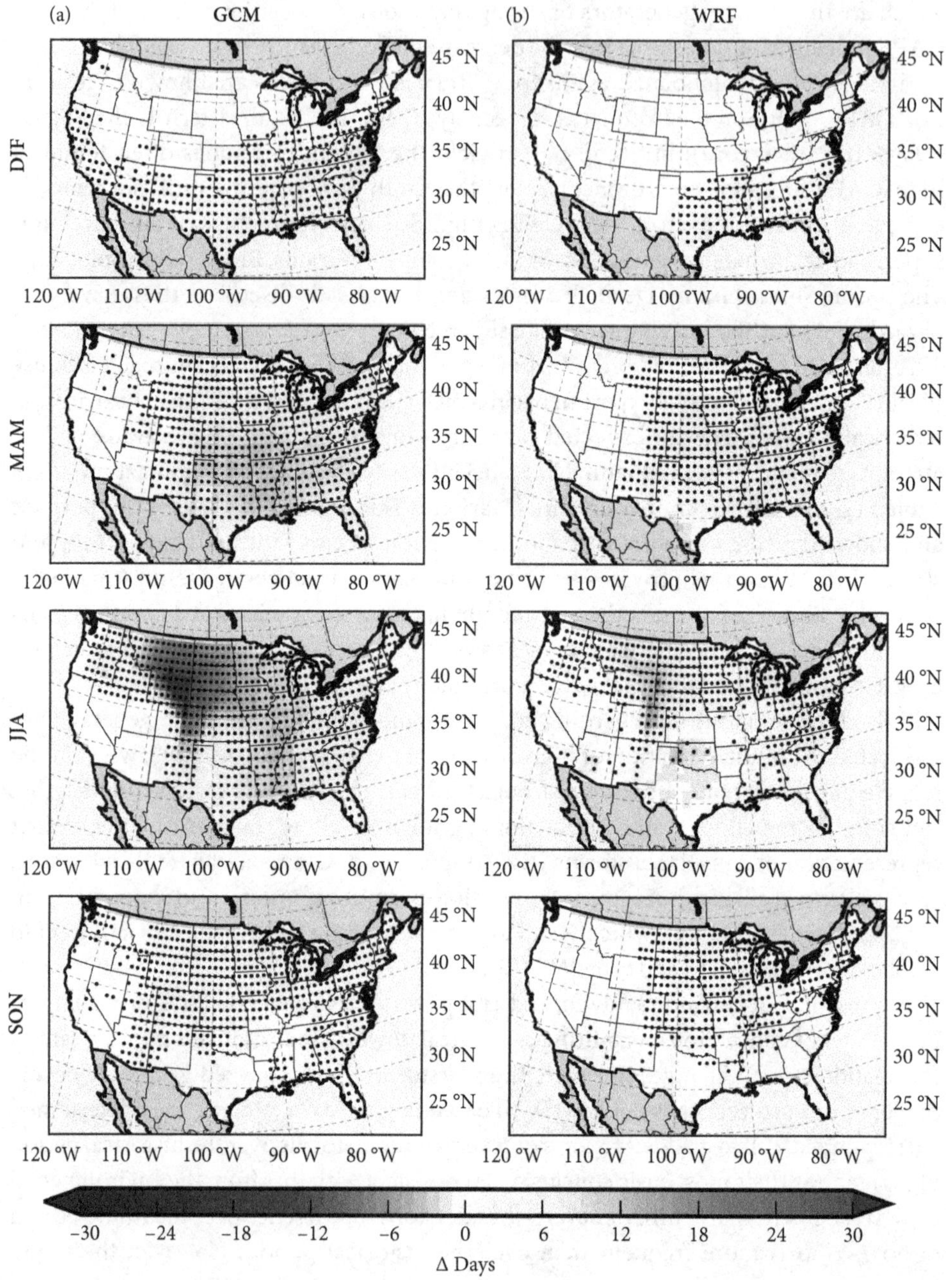

FIGURE 7.5 Mean seasonal change (future [2071–2100] minus historical [1971–2000]) in (a) NDSEV (days/season) as computed from the GFDL CM3 under CMIP5, and (b) synthetic severe-thunderstorm days (days/season) as computed from the high-resolution dynamical downscaling simulations using the Weather Research and Forecasting (WRF)model. The stippling shows locations where the distributions of seasonal means between these two periods are statistically significant from one another at the 95 percent confidence level. From Hoogewind et al. (2017).

thunderstorms is projected, especially over the central United States (northern Great Plains) over March–April–May (MAM) (June–July–August; JJA). In general terms, an increase in exposure to the severe-thunderstorm members of the NSWS family is projected for the United States under ACC.

To be clear, this conclusion—and the analogous one from the environmental approach—applies strictly to the general category of severe thunderstorms, which includes tornadic as well as nontornadic hazardous-wind occurrences. Efforts to extend this to specific hazards have thus far been more limited but are still possible. For example, consider the evaluation of *updraft helicity* (UH),

$$\mathrm{UH} = \int_{z=2km}^{z=5km} w\zeta\, dz, \tag{11}$$

which quantifies local (grid-scale) updraft rotation within the middle levels of convective storms. A rotating updraft is a supercell storm hallmark and also is generally considered to be a necessary condition for strong tornado formation. In forecasting operations with convection-permitting numerical prediction models, the use of a UH threshold (e.g., 50–75 $m^2\ s^2$) exhibits some skill in tornado prediction, especially when conditioned on the existence of tornado-favorable environmental parameters (e.g., Gallo, Clark, & Dembek, 2016). Here we evaluate a multivariate tornado-prediction parameter that includes a UH threshold—the Air Force Weather Agency tornado (AFWAtor) parameter (see Creighton et al., 2014)—using the Hoogewind et al. (2017) simulations to estimate tornado occurrence. As shown in Figure 7.6, the combined grid-scale and environmental analyses suggest future increases in tornado occurrences of a few days/season on average across the lower Mississippi Valley in MAM and in the

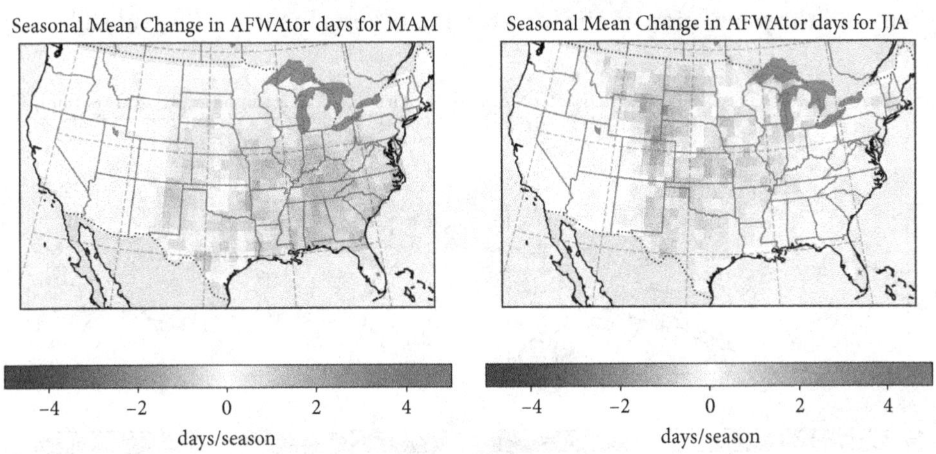

FIGURE 7.6 Mean seasonal change (future [2071–2100] minus historical [1971–2000]) in synthetic tornado days (days/season). Such tornado days are determined using exceedances of AFWAtor greater than 30 m s^{-1} (see Creighton et al., 2014) using the high-resolution dynamical downscaling simulations of Hoogewind et al. (2017).

northern Great Plains in JJA; decreases in occurrence of a few days/season are indicated in the central United States. Note that this preliminary estimation of tornado occurrence is consistent with the estimation of severe thunderstorms in general from these downscaling simulations (Figure 7.5).

A simpler procedure can be used to estimate convective-wind occurrence in the Hoogewind et al. (2017) simulations. Provided here is an evaluation of grid-scale magnitudes of 10-m horizontal winds, conditioned on the coexistence of a grid-scale convective cloud, as determined by simulated radar reflectivity exceeding 30 dBZ. These analyses indicate future increases in convective-wind occurrences of a few days/season on average over much of the central United States in MAM. In JJA, modest increases are indicated in the central United States, while larger increases are indicated in the southwestern United States through the northern Great Plains (Figure 7.7).

Unlike with the GCM-based projections shown in Figure 7.5, these downscaling-based projections do not include quantifications of solution robustness. This is because the considerable computational resources required for convection-permitting dynamical downscaling over large spatial and temporal domains so far have prevented the use of "ensembles," which are multiple realizations of the same model forecast but each with slightly different model conditions (e.g., see Chapter 4 in Trapp, 2013). Thus, the estimates/projections have been based on single realizations, using a single GCM driver and a single set of model physics (parameterizations of radiative transfer, turbulent mixing, cloud and precipitation processes, etc.).

As implemented, typical dynamical downscaling only allows indirect, statistical answers to basic questions such as whether ACC will cause tornadoes to be more intense. For this specific question, one might be able to conclude that the population of all storms in the past climate tended to exhibit weaker (or stronger) storm-scale rotation than does the population of all storms in the future climate. While valuable, this

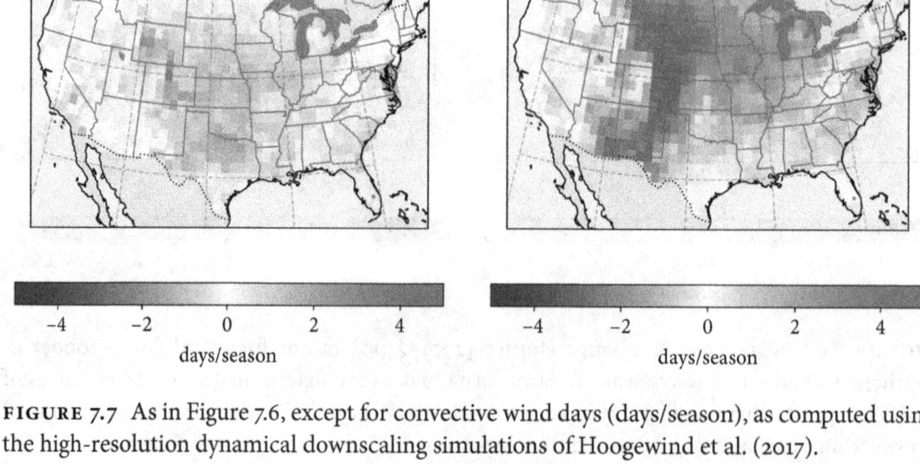

FIGURE 7.7 As in Figure 7.6, except for convective wind days (days/season), as computed using the high-resolution dynamical downscaling simulations of Hoogewind et al. (2017).

conclusion still does not say whether a tornadic storm from a past climate would be stronger (or weaker) if realized in a future climate and, if so, by how much. It also does not say whether the tornadic storm from a past climate would even be realized as a tornadic storm in a future climate.

The approach that does produce such answers is the event-level "pseudo-global-warming" (PGW) approach. As implemented by Lackmann (2013, 2015), Trapp and Hoogewind (2016), and more recently Carroll-Smith (2018), event-level PGW is a variant of, but complementary to, traditional dynamical downscaling. With this implementation, a specific convective-weather event (of a duration of one to several days) is simulated in both its true 4D (four-dimensional) environment and a 4D environment modified by a climate change delta (Δ; encompassing temperature, pressure/geopotential height, humidity, and wind). Because this approach is event level rather time period level, and thus involves relatively short time integrations, it provides a flexible framework for experimentation and examination of cause and effect in the relevant dynamical and microphysical processes.

Trapp and Hoogewind (2016) considered three tornadic supercell thunderstorms that occurred within the Great Plains of the United States during the month of May. An EF5 tornado formed on two of these days (May 4, 2007, and May 10, 2010), and an EF4 tornado on the other (May 19, 2013), implying that each event represented an extreme. The PGW Δ's were obtained as (three dimensional, 3D) differences from time means computed over May 1990–1999 and May 2090–2099. Thus, the temperature field used in the PGW simulations was

$$T(x,y,z,t) = T(x,y,z,t) + \Delta T \tag{12}$$

where $T(x,y,z,t)$ is the 3D, time-dependent forcing (from the North American Mesoscale Forecast System [NAM] analyses), and the climate change Δ in temperature is

$$\Delta T = \overline{T(x,y,z)}_{future} - \overline{T(x,y,z)}_{past}, \tag{13}$$

where the overbar indicates a time mean. Three separate Δ's were determined from three GCMs from the CMIP5 archive: Model for Interdisciplinary Research On Climate, version 5 (MIROC5), Geophysical Fluid Dynamics Laboratory Climate Model, version 3 (GFDL-CM3), and National Center for Atmospheric Research (NCAR) Community Climate System Model version 4 (CCSM4). Based on the analysis of Diffenbaugh et al. (2013), these GCMs provided a range in the historical frequency of severe-thunderstorm environments during the MAM season and thus provided a range in the possible realizations of the historical cases.

A significant finding by Trapp and Hoogewind (2016) is that PGW did not induce a change in the convective morphology, that is, the supercells simulated in their true, historical environments (the control [CTRL]) remained supercells under PGW. They did find, however, that PGW in some cases prevented initiation of the convective storms.

Table 7.1 Peak Near-Surface Wind Magnitude of Tornado-Like Vortices in Simulated Supercell Thunderstorm

CAPE (J kg^{-1})	S06 (m s^{-1})	Peak wind (m s^{-1})
1,000	21	31.99
1,000	28.5	34.41
2,000	21	36.06
2,000	28.5	42.90
2,000	36	48.91
3,000	28.5	39.08
3,000	36	43.56

The simulations were conducted over a range of environmental CAPE and S06. See Marion and Trapp (2019) for details on the model and experiments.

The storms that were able to initiate in the PGW environments ultimately evolved into relatively more intense supercells, with stronger mesocyclonic rotation. The suggestion is that these more intense supercells would have produced more intense tornadoes.

5. Implications for Exposure to NSWSs

Although comments regarding possible trends and future changes in exposure to NSWSs were made in Sections 3 and 4, these were directed mostly at the NSWS hazard. In this final section, the hazard information is combined with population and socioeconomic data to understand societal exposure. Although existing studies have focused mostly on tornadoes, the basic methodology can be generalized to other NSWSs.

There are different ways to quantify tornado exposure. In the approach of Strader, Ashley, Pingel, and Krmenec (2016), housing units (HUs) first were estimated using the spatially explicit regional growth model (SERGoM; Theobald, 2005); the SERGoM provides historical HU density on a high-resolution (100-m) grid over the CONUS. Future projections of HUs were estimated from a coupling of SERGoM with different integrated climate and land-use scenarios (Bierwagen et al., 2010). These gridded datasets were then combined with 10,000 Monte Carlo realizations of annual tornado-report data, in particular, the "tornado footprints" (damage path length by path width) of EF2+ tornadoes.

Strader et al. (2016) considered four US regions: the central Great Plains, high Great Plains, Midwest, and Midsouth. Although the central Great Plains region experiences the most significant tornadoes, and the Midwest has the greatest built-environment exposure, Strader et al. (2016) found the Midsouth region had the highest potential for tornado disaster. The projected increases in the built environment within the Midsouth

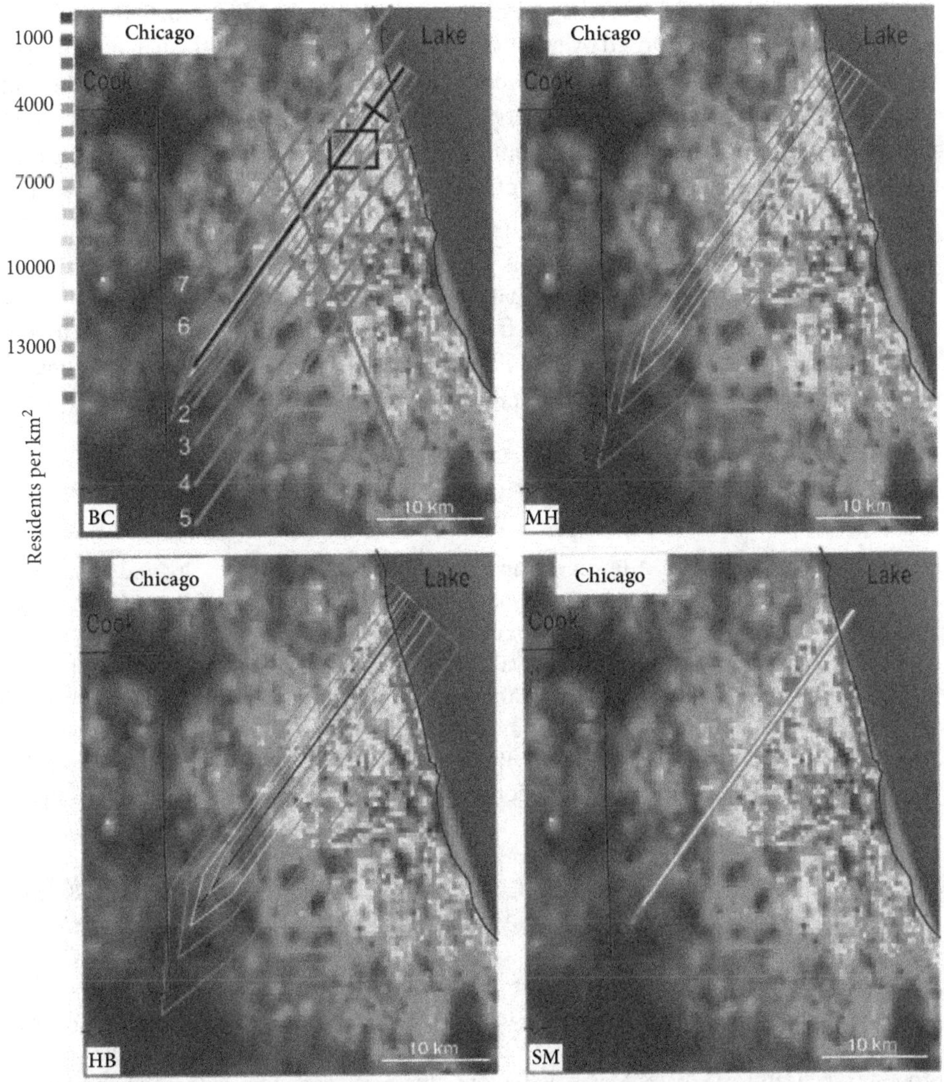

FIGURE 7.8 Hypothetical tracks of four different tornadoes through Chicago, Illinois. Colored lines represent peak winds of 120 (red), 102 (orange), 76 (yellow), 59 (green), and 43 m s^{-1} (cyan). From Wurman et al. (2007).

suggest even higher tornado disaster potential in the future; this basic conclusion generally applies to the other regions as well, albeit nonuniformly across space and time.

Akin to the event-level PGW modeling approach discussed in Section 4 is the hypothetical, event-occurrence exercise used, for example, by Rae and Stefkovich (2000) and Wurman, Robinson, Alexander, and Richardson (2007). This consists of the superimposition of real and idealized tornado tracks over different geographical areas and then an assessment of the hypothetically resultant damage and mortality. Wurman et al. (2007) used mobile radar–observed winds from two violent tornadoes with different wind-field distributions, as well as winds from "hybrid," idealized tornadoes. These wind fields were superimposed over densely populated urban centers, such as Chicago, Illinois (Figure 7.8), and then damage and mortality was estimated for each hypothetical track using US census block data. A large, violent tornado moving through a densely populated area of Chicago could potentially result in as many as 45,000 deaths and $40 billion in damage; with an assumption of better preparedness and response, the hypothetical death toll could still be as high as 4,500.

Such dire predictions are based on current-day populations and current-climate tornado sizes and intensities. Future changes in population are expected, while future changes in tornado characteristics and frequency are possible. Comparable statements about other NSWSs seem reasonable. Further risk modeling is necessary to convert ongoing and future NSWS research to actionable plans to mitigate NSWS impact.

References

Agee, E., & Childs, S. (2014). Adjustments in tornado counts, F-scale intensity, and path width for assessing significant tornado destruction. *Journal of Applied Meteorology and Climatology, 53*, 1494–1505. doi:10.1175/JAMC-D-13-0235.1

Atkins, N. T., Bouchard, C. S., Przybylinski, R. W., Trapp, R. J., & Schmocker, G. (2005). Damaging surface wind mechanisms within the 10 June 2003 Saint Louis bow echo during BAMEX. *Monthly Weather Review, 133*, 2275–2296. http://dx.doi.org/10.1175/MWR2973.1

Bierwagen, B. G., Theobald, D. M., Pyke, C. R., Choate, A., Groth, P., Thomas, J. V., & Morefield, P. (2010). National housing and impervious surface scenarios for integrated climate impact assessments. *Proceedings of the National Academy of Sciences of the United States of America, 107*, 20887 LP–20892. http://www.pnas.org/content/107/49/20887.abstract

Brock, F. V., Crawford, K. C., Elliott, R. L., Cuperus, G. W., Stadler, S. J., Johnson, H. L., & Eilts, M. D. (1994) The Oklahoma mesonet: A technical overview. *Journal of Atmospheric and Oceanic Technology, 12*, 5–19. https://doi.org/10.1175/1520-0426(1995)012%3C0005:TOMATO%3E2.0.CO;2

Brooks, H. E. (2004). On the relationship of tornado path length and width to intensity. *Weather Forecasting, 19*, 310–319. doi:10.1175/1520-0434(2004)019<0310:OTROTP>2.0.CO;2

Brooks, H. E., Carbin, G. W., & Marsh, P. T. (2014). Increased variability of tornado occurrence in the United States. *Science 80*(346), 349–352. http://science.sciencemag.org/content/346/6207/349.abstract

Brooks, H. E., Lee, J. W., & Craven, J. P. (2003). The spatial distribution of severe thunderstorm and tornado environments from global reanalysis data. *Atmospheric Research, 67–68*, 73–94.

Carroll-Smith, D. (2018). *"If it happened in . . .": A pseudo-global warming assessment of tropical cyclone tornadoes.* (Ph.D. dissertation). University of Illinois at Urbana Champaign, p. 100. Retrieved from http://hdl.handle.net/2142/101477.

Corfidi, S. F., Coniglio, M. C., Cohen, A. E., & Mead, C. M. (2015). A proposed revision to the definition of "Derecho." *Bulletin of the American Meteorological Society, 97*, 935–949. doi:10.1175/BAMS-D-14-00254.1

Creighton, G., Kuchera, E., Adams-Selin, R. D., McCormick, J., Rentschler, S. & Wickard, B. (2014). *AFWA Diagnostics in WRF*. http://www2.mmm.ucar.edu/wrf/users/docs/AFWA_Diagnostics_in_WRF.pdf

Davis, J. M., & Parker, M. D. (2014). Radar climatology of tornadic and nontornadic vortices in high-shear, low-CAPE environments in the mid-Atlantic and southeastern United States. *Weather Forecasting, 29*, 828–853. http://dx.doi.org/10.1175/WAF-D-13-00127.1

Del Genio, A. D., Yao, M.-S., & Jonas, J. (2007). Will moist convection be stronger in a warmer climate? *Geophysical Research Letters, 34*, L16703. doi:10.1029/2007GL030525

Diffenbaugh, N. S., Scherer, M., & Trapp, R. J. (2013). Robust increases in severe thunderstorm environments in response to greenhouse forcing. *Proceedings of the National Academy of Sciences of the United States of America, 110*, 16361–16366. doi:10.1073/pnas.1307758110

Doswell, C. A., Brooks, H. E., & Kay, M. P. (2005). Climatological estimates of daily local nontornadic severe thunderstorm probability for the United States. *Weather Forecasting, 20*, 577–595. doi:10.1175/WAF866.1

Edwards, R., Allen, J. T., & Carbin, G. W. (2018, August). Reliability and climatological impacts of convective wind estimations. *Journal of Applied Meteorology and Climatology*, 1825–1845. https://doi.org/10.1175/JAMC-D-17-0306.1

Elsner, J. B., Jagger, T. H., & Elsner, I. J. (2014). Tornado intensity estimated from damage path dimensions. *PLoS One, 9*, e107571. https://doi.org/10.1371/journal.pone.0107571

Elsner, J. B., Elsner, S. C., & Jagger, T. H. (2015). The increasing efficiency of tornado days in the United States. *Climate Dynamics, 45*, 651–659. doi:10.1007/s00382-014-2277-3

Fiedler, B. H., & Rotunno, R. (1986). A theory for the maximum windspeeds in tornado-like vortices. *Journal of Atmospheric Sciences, 43*, 2328–2340. doi:10.1175/1520-0469(1986)043<2328:ATOTMW>2.0.CO;2

Fujita, T. T. (1981). Tornadoes and downbursts in the context of generalized planetary scales. *Journal of Atmospheric Sciences, 38*, 1511–1534. doi:10.1175/1520-0469(1981)038<1511:TADITC>2.0.CO;2

Gallo, B. T., Clark, A. J., & Dembek, S. R. (2016). Forecasting tornadoes using convection-permitting ensembles. *Weather Forecasting, 31*, 273–295. doi:10.1175/WAF-D-15-0134.1

Hoogewind, K. A., Baldwin, M. E., & Trapp, R. J. (2017). The impact of climate change on hazardous convective weather in the United States: Insight from high-resolution dynamical downscaling. *Journal of Climate, 30*, 10081–10100. doi:10.1175/JCLI-D-16-0885.1

Johns, R. H., & Hirt, W. D. (1987). Derechos: Widespread convectively induced windstorms. *Weather Forecasting, 2*, 32–49. doi:10.1175/1520-0434(1987)002<0032:DWCIW>2.0.CO;2

Kain, J. S., Weiss, S. J., Bright, D. R., Baldwin, M. E., Levitt, J. J., Carbin, G. W., . . . Thomas, K. W. (2008). Some practical considerations regarding horizontal resolution in the first generation of operational convection-allowing NWP. *Weather Forecasting, 23*, 931–952. doi:10.1175/WAF2007106.1

Kirkpatrick, C., McCaul, E. W., Jr., & Cohen, C. (2009). Variability of updraft and downdraft characteristics in a large parameter space study of convective storms. *Monthly Weather Review, 137*, 1550–1561. doi:10.1175/2008MWR2703.1

Lackmann, G. M. (2013). The south-central US Flood of May 2010: Present and future. *Journal of Climate, 26*, 4688–4709. doi:10.1175/JCLI-D-12-00392.1

Lackmann, G. M. (2015). Hurricane Sandy before 1900 and after 2100. *Bulletin of the American Meteorological Society, 96*, 547–560. doi:10.1175/BAMS-D-14-00123.1

Long, J. A., & Stoy, P. C. (2014). Peak tornado activity is occurring earlier in the heart of "Tornado Alley." *Geophysical Research Letters, 41*, 6259–6264. doi:10.1002/2014GL061385

Mahoney, K. M., & Lackmann, G. M. (2011). The sensitivity of momentum transport and severe surface winds to environmental moisture in idealized simulations of a mesoscale convective system. *Monthly Weather Review, 139*, 1352–1369. doi:10.1175/2010MWR3468.1

Marion, G. R., & Trapp, R. (2019). The dynamical coupling of convective updrafts, downdrafts, & cold pools in simulated supercell thunderstorms. *Journal of Geophysical Research: Atmospheres, 124*, 664–683. https://doi.org/10.1029/2018JD029055

Marshall, T. P. (2002). Tornado damage survey at Moore, Oklahoma. *Weather Forecasting, 17*, 582–598. doi:10.1175/1520-0434(2002)017<0582:TDSAMO>2.0.CO;2

Prein, A., Langhans, W., Fosser, G., Ferrone, A., Ban, N., Goergen, K. . . . Leung, R. (2015). A review on regional convection-permitting climate modeling: Demonstrations, prospects, and challenges. *Reviews of Geophysics, 53*, 323–361. doi:10.1002/2014RG000475

Rae, S., & Stefkovich, J. (2000, September). The tornado damage risk assessment predicting the impact of a big outbreak in Dallas–Fort Worth, Texas. *Preprints, 20th Conference on Severe Local Storms*, American Meteorological Society, 295–296.

Rasmussen, E. N., & Blanchard, D. O. (1998). A baseline climatology of sounding-derived supercell and tornado forecast parameters. *Weather Forecasting, 13*, 1148–1164. doi:10.1175/1520-0434(1998)013<1148:ABCOSD>2.0.CO;2

Rauber, R. M., & Nesbitt, S. W. (2018). *Radar meteorology, an introduction*. Wiley Blackwell, p. 461.

Rotunno, R., Klemp, J. B., & Weisman, M. L. (1988). A theory for strong, long-lived squall lines. *Journal of Atmospheric Sciences, 45*, 463–485. Doi:10.1175/1520-0469(1988)045<0463:ATFSLL>2.0.CO;2

Seeley, J. T., & Romps, D. M. (2015). The effect of global warming on severe thunderstorms in the United States. *Journal of Climate, 28*, 2443–2458. Doi:10.1175/JCLI-D-14-00382.1

Smith, B. T., Castellanos, T. E., Winters, A. C., Mead, C. M., Dean, A. R., & Thompson, R. L. (2013). Measured severe convective wind climatology and associated convective modes of thunderstorms in the contiguous United States, 2003–09. *Weather Forecasting, 28*, 229–236. Doi:10.1175/WAF-D-12-00096.1

Smith, B. T., Thompson, R. L., Dean, A. R., & Marsh, P. T. (2015). Diagnosing the conditional probability of tornado damage rating using environmental and radar attributes. *Weather Forecasting, 30*, 914–932. Doi:10.1175/WAF-D-14-00122.1

Smith, B. T., Thompson, R. L., Grams, J. S., Broyles, C., & Brooks, H. E. (2012) Convective modes for significant severe thunderstorms in the contiguous United States. Part I: Storm classification and climatology. *Weather Forecasting, 27*, 1114–1135. Doi:10.1175/WAF-D-11-00115.1

Strader, S. M., Ashley, W. S., Pingel, T. J., & Krmenec, A. J. (2016). Observed and projected changes in United States tornado exposure. *Weather, Climate, and Society, 9*, 109–123. Doi:10.1175/WCAS-D-16-0041.1

Sturgeon, M. (1999, January). A wind tunnel acceptance test method for sonic anemometers. *Preprints, 15th International Conference on Interactive Information and Processing Systems (IIPS) for Meteorology, Oceanography, and Hydrology*, American Meteorological Society.

Theobald, D. M. (2005). Landscape patterns of exurban growth in the USA from 1980 to 2020. *Ecology and Society*, 10. Retrieved from http://www.ecologyandsociety.org/vol10/iss1/art32/.

Thompson, R. L., Smith, B. T., Grams, J. S., Dean, A. R., & Broyles, C. (2012). Convective modes for significant severe thunderstorms in the contiguous United States. Part II: Supercell and QLCS tornado environments. *Weather Forecasting, 27*, 1136–1154. Doi:10.1175/WAF-D-11-00116.1

Tippett, M. K. (2014). Changing volatility of US annual tornado reports. *Geophysical Research Letters, 41*, 6956–6961. http://doi.wiley.com/10.1002/2014GL061347

Trapp, R. J. (2013). *Mesoscale-convective processes in the atmosphere*. Cambridge, UK: Cambridge University Press.

Trapp, R. J., & Brooks, H. E. (2013). Regional characterization of tornado activity. *Journal of Appied Meteorology and Climatology*, 52, 654–659. Doi:10.1175/JAMC-D-12-0173.1

Trapp, R. J., Diffenbaugh, N. S., Brooks, H. E., Baldwin, M. E., Robinson, E. D., & Pal, J. S. (2007). Changes in severe thunderstorm environment frequency during the twenty-first century caused by anthropogenically enhanced global radiative forcing. *Proceedings of the National Academy of Sciences of the United States of America*, *104*, 19719–19723. Doi:10.1073/pnas.0705494104

Trapp, R. J., & Hoogewind, K. A. (2016). The realization of extreme tornadic storm events under future anthropogenic climate change. *Journal of Climate*, 29, 5251–5265. Doi:10.1175/JCLI-D-15-0623.1.

Trapp, R. J., & Hoogewind, K. A. (2018). Exploring a possible connection between U.S. Tornado activity and Arctic sea ice. *npj Climate and Atmospheric Science*, *1*, 14. doi:10.1038/s41612-018-0025-9.

Trapp, R. J., Marion, G. R., & Nesbitt, S. W. (2017). The regulation of tornado intensity by updraft area. *Journal of Atmospheric Sciences*, *74*, 4199–4211. https://doi.org/10.1175/JAS-D-16-0331.1

Trapp, R. J., Robinson, E., Baldwin, M., Diffenbaugh, N., & Schwedler, B. J. (2011). Regional climate of hazardous convective weather through high-resolution dynamical downscaling. *Climate Dynamics*, *37*, 677–688. doi:10.1007/s00382-010-0826-y

Trapp, R. J., & Weisman, M. L. (2003). Low-level mesovortices within squall lines and bow echoes. Part II: Their genesis and implications. *Monthly Weather Review*, *131*, 2804–2823. doi:10.1175/1520-0493(2003)131<2804:LMWSLA>2.0.CO;2

Trapp, R. J., Wheatley, D. M., Atkins, N. T., Przybylinski, R. W., & Wolf, R. (2006). Buyer beware: Some words of caution on the use of severe wind reports in postevent assessment and research. *Weather Forecasting*, *21*, 408–415. doi:10.1175/WAF925.1

Weisman, M. L., & Trapp, R. J. (2003). Low-level mesovortices within squall lines and bow echoes. Part I: Overview and dependence on environmental shear. *Monthly Weather Review*, *131*, 2779–2803. doi:10.1175/1520-0493(2003)131<2779:LMWSLA>2.0.CO;2

Wheatley, D. M., Trapp, R. J., & Atkins, N. T. (2006). Radar and damage analysis of severe bow echoes observed during BAMEX. *Monthly Weather Review*, *134*, 791–806. doi:10.1175/MWR3100.1

Wurman, J., Robinson, P., Alexander, C., & Richardson, Y. (2007). Low-level winds in tornadoes and potential catastrophic tornado impacts in urban areas. *Bulletin of the American Meteorological Society*, *88*, 31–46. doi:10.1175/BAMS-88-1-31

SECTION 2

IN SITU DATA ACQUISITION FOR NSWS

CHAPTER 8

MEASURING THE ENGINEERING-RELEVANT ASPECTS OF NON-SYNOPTIC WIND HAZARDS

JOHN L. SCHROEDER

1. Introduction

Historically, the near-surface flow fields within all wind storms have been assumed by engineers to be reasonably well represented by synoptic wind systems (SWSs). SWSs are driven by large-scale pressure gradients and result in long-lasting, widespread, moderate wind speed events, which provide an easy target for historical measurement campaigns. Indeed, over 80 years ago, Sherlock and Stout (1937) measured the wind structure in two winter storms in Michigan using an array of instrumented towers and then produced diagrams of horizontal and vertical wind structure. Numerous other tower-based wind measurement campaigns, at times coupled with complementary wind load information, have since been acquired at other sites around the world (e.g., Levitan & Mehta, 1992a, 1992b). Given the inherent stationarity of SWSs, straightforward calculation of traditional wind engineering properties typically follow, and the results provide a convenient blueprint for extrapolation to higher wind speeds, replication in wind tunnels, and codification into engineering design.

However, for many locations inland from the coastline, thunderstorms and tornadoes represent the wind hazards of concern. These events have been termed non-synoptic wind systems (NSWSs) by the community. While the forcing mechanism

(Smith et al., 2013) and structure (Gunter & Schroeder, 2015) of an individual NSWS event vary, the most significant portion of these events is often very small spatially and embedded within a broader phenomenon (e.g., a rear-flank downdraft surge embedded within the broader rear-flank downdraft of a supercell thunderstorm). While the recognition of the importance of these NSWS events is increasing, their engineering-relevant characteristics are still largely unknown (Lombardo, Smith, Schroeder, & Mehta, 2014). Of course, given their transient and localized character, thunderstorm- and tornado-generated winds are challenging to measure. Given their inherent nonstationarity, they can also be difficult to analyze using traditional statistical or spectral analysis (Kwon & Kareem, 2009) approaches.

Obviously, hurricanes pose a significant wind hazard (Congressional Budget Office, 2016) threat to coastal communities. While smaller in scale than SWSs, hurricanes are still large-scale thermodynamically driven circulations that can impact a region for days. While storm dynamics provide a mechanism for a compressed boundary layer and supergradient, low-level, jet-like features in the inner core (Kepert, 2001; Kepert & Wang, 2001), analysis of previous measurements indicated the near-surface wind structure associated with the broader circulation remains relatively well behaved (e.g., Giammanco, Schroeder, & Powell, 2012; Vickery & Skerlj, 2005). Even within the eyewall region of historically sampled storms, significant perturbations from established norms are relatively rare (Schroeder, Paulsen-Edwards, & Giammanco, 2009). Yet, even with a growing inventory of research-grade measurements made by university-led mobile tower programs over the past 20 years, the vast majority of these hurricane wind measurements are still well below design level.

By definition, NSWS would also include small-scale phenomena that are embedded within the broader hurricane circulation. For instance, satellite and operational radar measurements have previously identified shear-induced features termed *mesovortices* within the eyewall of hurricanes. These features have been highlighted as a potential means by which some of the most devastating wind damage occurs (e.g., Willoughby & Black, 1996). Higher resolution mobile research radar has also identified mesovortices and tornado-scale vortices in the eyewall (e.g., Wurman & Koshiba, 2018), and mobile tower measurements have recently been acquired from some of these same events (Fernández-Cabán et al., 2019). These recent measurements indicate there could be significant differences in the structure of these localized events relative to nominal SWS assumptions. These features are also largely unaccounted for by the current wind field reconstruction methods (Powell, Houston, Amat, & Morisseau-Leroy, 1998; Vickery, Lin, Skerlj, Twisdale, & Huang, 2003), which results in an increase in the uncertainty when constructing a wind field of record using the established techniques.

The application of SWS wind structure to describe all design-level wind events remains questionable, while the current engineering-based description of many NSWS phenomena remains limited. Both of these factors create significant uncertainty in the resultant engineering design process. While next-generation wind tunnels (Standohar-Alfano, Estes, Johnston, Morrison, & Brown-Giammanco, 2017); newly advanced tornado simulators (e.g., Haan, Sarkar, & Gallus, 2008; Refan & Hangan, 2016; Tang, Changda,

Wu, Zuo, & James, 2018); and high-resolution modeling efforts (e.g., Lewellen, Lewellen, & Sykes, 1997; Natarajan & Hangan, 2012) offer opportunities to advance our engineering understanding of NSWS flow fields and the wind loads they produce, the fundamental gap in knowledge regarding the near-surface engineering-relevant aspects of NSWSs stymies progress. Therefore, while difficult to acquire, there is an essential need to make engineering-relevant measurements from NSWSs.

2. Previous Wind Storm Measurements and Limitations

Measurement efforts in wind storms have been ongoing for decades, but the usefulness of the acquired information to advance engineering design and modeling efforts is often limited.

2.1. Hurricanes

In hurricanes, reconnaissance aircraft have been successfully used for decades to operationally monitor storm intensity, location, and movement over the open ocean up to landfall. Subsequent integration of the GPS dropwindsonde platform (Hock & Franklin, 1999) expanded our knowledge of the variation in wind speeds with height within the offshore hurricane boundary layer, while the incorporation of the stepped-frequency microwave radiometer (SFMR) (Uhlhorn & Black, 2003) significantly bolstered offshore near-surface mean wind field estimates. At landfall, the national network of S-band Next-Generation Weather Radars (NEXRAD) has provided elevated coverage of the hurricane's precipitation and radial velocity fields. While all of these platforms have proven exceptionally valuable from the perspective of an operational meteorologist, they provide only modest insights into detailed hurricane wind fields near the ground overland. Even as late as the mid-1990s, there remained a dearth of high-quality overland wind measurements from the near-surface regions of hurricanes to assist hurricane boundary layer characterization and advance engineering design.

Since the late 1990s, the prestorm placement of mobile research radars (Wurman & Winslow, 1998) and instrumented towers (Poss, 2000; Schroeder & Smith, 2003) has become common for most landfalling hurricanes of consequence in the United States. Additionally, the development of state (e.g., Kimball, Mulekar, Cummings, & Stamates, 2010) and privately operated (e.g., WeatherFlow's Coastal Mesonet) mesonets has further bolstered the inventory of quality measurements available along the coast and inland. Collectively, these measurements and the accompanying metadata provide a relatively robust basis to assess the mean wind field of the broader circulation and provide support wind field reconstruction efforts.

The mobile instrumented tower and research radar programs have the advantage of offering high-resolution measurements of value to the engineering community. This information has provided a basis for engineering-relevant studies of the hurricane boundary layer and associated turbulence (e.g., Schroeder et al., 2009; Yu, Chowdhury, & Masters, 2008); documentation of the structure and evolution of smaller scale hurricane boundary layer wind features (Lorsolo, Schroeder, Dodge, & Marks, 2008; Wurman & Winslow, 1998); and the capture of design-level wind events associated with eyewall mesovortices (Fernández-Cabán et al., 2019; Wurman & Koshiba, 2018). Still, high-resolution measurements from the most devastating areas (e.g., mesovortices) within significant hurricanes remain rare, and the collective support to fund integrated high-quality measurements geared toward comprehensive engineering interests remains elusive. Instead, ad hoc coordination between researchers has been the norm for collecting these invaluable data the past two decades.

2.2. Thunderstorms and Tornadoes

Considerable effort has been expended to acquire research-relevant measurements within thunderstorms and tornadoes to advance atmospheric understanding and predictive capabilities. For example, VORTEX (1994–1995), IHOP (2002), BAMEX (2003), VORTEX 2 (2009–2010), PECAN (2015), and VORTEX SE (2016–2018) are just a few examples of large-scale, multi-institutional, collaborative, field measurement campaigns. These large-scale efforts have typically employed a fleet of instrumentation, which has included research radars, portable weather stations, mobile mesonets, instrumented drones, portable sounding systems, and so on to acquire integrated measurements and advance our knowledge regarding storm evolution and structure, convective initiation, tornado genesis and structure, and more. This knowledge has directly led to documented forecast improvements and enhanced lead time for the public issue of severe weather watches and warnings and undoubtedly has saved many lives. However, these same efforts have led to only minimal amounts of near-surface, high-resolution measurements of value to advancing engineering design.

Fujita (1978, 1981) and Hjelmfelt (1988) studied thunderstorm downdrafts and outflows from an aviation safety perspective, but detailed engineering studies of thunderstorm outflow-generated NSWSs have been limited. Full-scale measurements acquired using stationary and mobile instrumented towers (Lombardo et al., 2014; Orwig & Schroeder, 2007) and high-resolution mobile research radars (Gunter & Schroeder, 2015; Gunter, Schroeder, Weiss, & Bruning, 2017) have yielded limited insights. Collectively, these measurements have provided a glimpse of the near-surface vertical and horizontal wind structure of thunderstorm outflows, as well as a few examples of high-resolution wind speed and direction time histories from damaging NSWS events. However, these limited measurements also indicate that a significant amount of variability can exist between different events, which is not surprising given their varied forcing mechanisms

(Smith et al., 2013), and differences in event age and surface characteristics at the time of sampling. Given the limited effort and small sample size, the near-surface structure and evolution of thunderstorm-generated wind events remain under characterized.

While wind engineering–oriented measurement campaigns in thunderstorms have been limited, direct in situ measurements of wind speeds within tornadoes are exceedingly rare. However, at least six fixed mesonet stations in Oklahoma and Texas have been impacted by tornados since their installation. While of interest, maximum wind speeds in all cases were below 151 mph (Burgett, 2007; Mersereau, 2018), and a 5-minute reporting interval was used to generate the wind speed and direction statistics, limiting the value of these records. Karstens, Samaras, Lee, and Finley (2010) placed hardened in-situ tornado pressure recorders (HITPRs) in the path of tornadoes from 2002 to 2008. These probes, combined with unintended mobile mesonet vehicle interactions with tornadoes during the same experiment, provided pressure deficit information from nine tornadoes and near-surface wind speed measurements from three tornadoes. While limited in context, the pressure measurements are perhaps the most valuable, in situ, engineering-relevant measurements of tornadoes acquired to date, as the surface pressure deficit time histories provide a basis for replication in laboratory and numerical simulations.

Given the difficulty of making in situ measurements, mobile research radars have been the instrument of choice for documenting the wind fields within tornadoes (e.g., Snyder & Bluestein, 2014; Wurman & Alexander, 2005). However, in some cases the acquired information is located too high above ground level to directly assist engineers. Also, unlike storm-scale analysis efforts within thunderstorms, employing dual-Doppler synthesis to resolve smaller scales of motion of value to engineers within the core of a tornado is troublesome for several reasons (Wurman & Koshiba, 2010). Hence, without coupling basic assumptions about the flow field, most tornado-scale radar studies are limited to the evaluation of the one-dimensional radial velocity magnitude, structure, and evolution at vertical levels where measurements are acquired. These studies have proven exceptionally valuable for estimating wind field profiles across tornado vortices; however, a comprehensive description of the low-level turbulent wind field and near-ground vertical changes in wind speed within a tornado remains elusive.

3. The NSWS Challenge

While the need to inform engineering design and validate modeling efforts is clear, the acquisition of measurements to comprehensively serve the engineering community is not without difficulty. There are atmospheric- and engineering-based challenges associated with making such measurements at relevant scale. Coupling these challenges with the inherent limitations of available measurement technologies and techniques exacerbates the difficulty of the task.

3.1. Atmospheric-Based Challenges

The NSWSs are nominally transient, quick-hitting events, often lasting just minutes at a particular location, which complicates their measurement. This quick-hitting nature is illustrated via the wind speed time history shown in Figure 8.1. In this case, extreme winds were generated by a passing supercell rear-flank downdraft. At the measurement location, the entire event lasted a total of approximately 10 minutes, while the strongest embedded surge lasted only 3 minutes. In addition to their temporal brevity, NSWS events are relatively small in spatial size. These events may be embedded within a larger scaled weather system (e.g., a hurricane or larger scale thunderstorm), and the portion of the event that meets or exceeds design level is often localized (i.e., < 10 km^2). Figure 8.2 shows the spatial representation of an embedded wind speed surge within the rear-flank downdraft of a supercell thunderstorm as revealed by dual-Doppler synthesis of research radar measurements. While the parent supercell thunderstorm itself was relatively large, the area of interest (with wind speeds greater than 40 m/s) was relatively small and contained within the broader rear-flank downdraft. Tornadoes also maintain small spatial and temporal footprints. Brooks (2004) indicated that the average area impacted by a tornado (as defined by track width and length) is less than 1 km^2, and tornadoes are often on the ground (https://www.spc.noaa.gov/faq/tornado/) for less than 10 minutes.

Beyond the difficulty associated with scale, NSWSs can also be relatively violent. The wind hazard of interest is often coupled with other hazards, such as storm surge, freshwater flooding, lightning, or hail. These hazards pose a direct threat to instrumentation

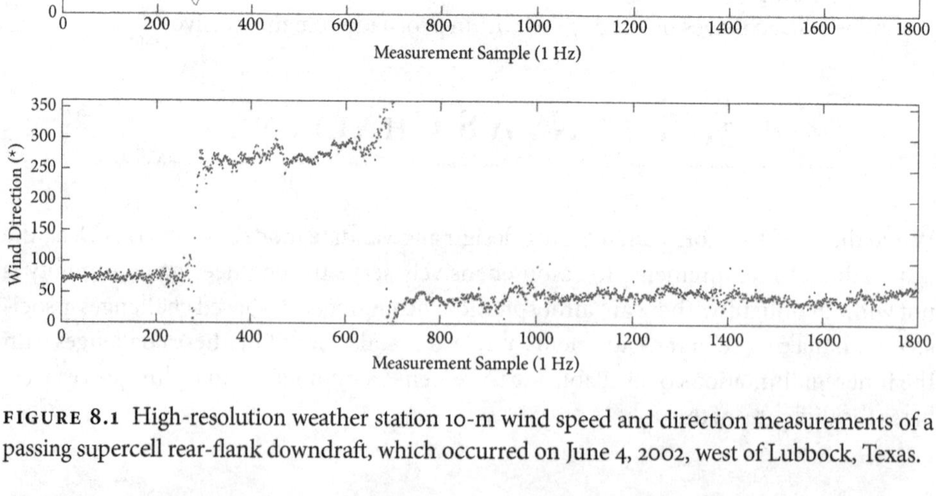

FIGURE 8.1 High-resolution weather station 10-m wind speed and direction measurements of a passing supercell rear-flank downdraft, which occurred on June 4, 2002, west of Lubbock, Texas.

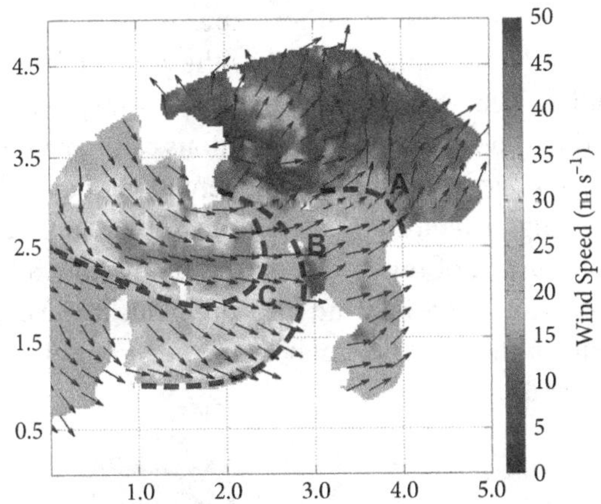

FIGURE 8.2 Dual-Doppler analysis of the wind field within the rear-flank downdraft of a supercell thunderstorm near Dumas, Texas, on May 18, 2010. Subjective analysis of embedded rear-flank downdraft surges was used for identification. The 1-km grid can be used to evaluate the spatial context of the embedded events. Figure adapted from Skinner et al. (2014).

and people in the field. Further, the measurement capability of some technologies can be compromised in many commonly found NSWS conditions. For example, in situ wind sensors can be damaged by large hail, while radar signal returns may be limited by attenuation in heavy rain or hail. Control of exposed radar antennas may also be lost in high-wind conditions. These additional hazards must be accounted for when planning the logistics of a particular field campaign.

3.2. Engineering-Based Challenges

There also exist challenges related to acquiring relevant four-dimensional wind information at sufficient spatial and temporal scales to assist engineers. Ideally, spatial resolutions of tens of meters and temporal resolutions of seconds (or even subseconds) would be available to comprehensively serve engineering interests. However, atmospheric science–based studies using remote sensing technologies typically offer gridded data sets with spatial resolutions of 100–1,000+ m and temporal resolutions of 1 to 10 minutes. Tower-based research measurements often yield high temporal resolution time histories (e.g., sampling rates of 1 Hz or higher) but are limited to a few specific points spatially.

Adding to the measurement challenge, the near-surface region of the atmosphere is most critical for engineers, yet this same region is only a subset of the typical domain of interest to an atmospheric scientist. Near-surface wind measurements are often difficult to measure via remote sensing technologies as terrain, trees, and buildings provide blockage that limits coverage and also introduces ground clutter contamination into the data. Hence, many remote sensing–based measurement campaigns are often focused above the near-surface region due to either experimental design or ground-based constraints. This constraint fundamentally limits the ability of these measurements to assist engineering studies. Even when instrumented towers are placed in the field to collect high temporal resolution time histories, complications can remain if proper documentation is not acquired. For instance, careful documentation of each deployment site and its immediate surroundings is required to standardize any acquired measurements and fully understand the influence of roughness on the acquired wind speed records. The same complication exists for low-level wind measurements made using research radars, where spatial variability in the wind field may be forced by variability in the underlying surface roughness or terrain.

3.3. Measurement Technologies and Approaches

Historically, the wind engineering community has largely been dependent on various types of anemometry as the basis for providing information about the wind field. However, new technologies and approaches are available and starting to make contributions at relevant scales.

3.3.1. *Weather Stations*

A significant number of near-surface weather stations exist across many areas. Examples in the United States include the Automated Surface Observing System (ASOS), state-wide mesonets (Brock et al., 1995; Kimball et al., 2010; Schroeder et al., 2005), and relevant private networks, such as that offered by WeatherFlow, Incorporated (e.g., the Hurricane Network). Combined, these networks and others represent a significant number of measurement systems that commonly sample NSWSs near the surface. In general, these fixed-measurement assets provide mean wind speeds/directions and gust wind speeds with reporting intervals on the order of 1 to 5 minutes. Many of these stations directly contribute to the nowcasting needs of operational meteorologists and provide input for extreme value analysis of NSWS events (Lombardo, 2012). However, most stations are not equipped to provide high-resolution time histories, which would be of significant value to wind engineers. While these surface stations provide great coverage for many locations within the United States, they are not concentrated enough to resolve micro- to meso-gamma scales of motion.

The university-led mobile tower programs (Figure 8.3), such as the towers placed by the University of Florida in hurricane-related research projects, are an exception to this general rule. These mobile facilities are often delivered on demand by researchers to provide high temporal resolution and, in some cases, multilevel/multitower measurements from extreme wind events. To enhance deployment speed, provide larger deployment arrays, and provide a dense observation network, smaller, rapidly deployable measurement systems such as Texas Tech University's StickNet (Weiss & Schroeder, 2008) platform were developed. These platforms enhance the ability to quickly place an array of instruments across a localized region in the path of a transient NSWS event to

FIGURE 8.3A. Portable weather stations: Florida Coastal Monitoring Program's Tower 3 deployed for Hurricane Michael (2018).

Photo credit: Dr. Forrest Masters, University of Florida.

provide enhanced data coverage. However, wind information is typically only available from one level (usually within 3 m of ground level).

3.3.2. *Remote Sensing*

Remote sensing technologies have successfully been used to extract information from NSWSs for decades. A remote sensing instrument does not have to be located within the event itself to acquire measurements, but rather nearby, with the event within range of the instrument. This separation enables extraction of information from even the most violent events while limiting the risk to the instrument or individual(s) deploying it. However, the usefulness of such information from an engineering perspective is largely dependent on the spatial and temporal resolution of the measurements and the vertical elevation(s) at which they are available.

3.3.2.1. LIDAR

Lidar offers a narrow "pencil" beam that does not expand in the azimuthal dimension with range. This characteristic is beneficial to mitigating ground clutter contamination when acquiring measurements from the lowest levels of the atmosphere, and the

FIGURE 8.3B. Portable weather stations: Texas Tech University's StickNet platform being deployed during the VORTEX2 project.

Photo credit: Texas Tech University.

crossbeam resolution of the measurements with range is preserved. However, the use of lidar in NSWSs is not practical in most cases due to the inability of the technology to penetrate precipitation.

While there could be isolated cases where information could be extracted in low-precipitation environments, the scan speed of the technology also compromises measurement value from an engineering perspective. For example, a scanning lidar typically scans at 1–3°/s, which means a single horizontal plan position indicator (PPI) scan with a width of 60° would take from 20 to 60 seconds to acquire. Attempting to objectively analyze multiple scans of lidar data into a three-dimensional volume at temporal scales relevant for wind engineering applications would be futile in many cases. In modest winds, boundary layer structures (e.g., a gust) would enter one side of the measurement domain and exit the other side before the acquisition of a single volume is complete.

Maximum range for most lidars is also limited to about 3–6 km in many environments, and range resolution is typically at least 60 m (i.e., the along-beam dimension of the pencil beam). Hence, lidar's ability to address engineering-based NSWS questions is minimal, and its only contribution is likely limited to making radial velocity measurements of a feature of interest (e.g., a tornado) using a very narrow sector (e.g., 20°) just above the surface in a few select environments.

3.3.2.2. WEATHER SURVEILLANCE RADAR

Historical field campaigns in severe convective storms have often been headlined by weather radars. Pulsed Doppler weather radar returns a power-weighted estimate of the radial velocity across the measurement volume. However, the largest and most abundant scatterers within the measurement volume will make the most significant contributions to the estimate, and in some cases (e.g., large debris found within a tornado) the motion of these scatterers may not represent the actual wind field (Snyder & Bluestein, 2014). Regardless, radar provides a proven technology to acquire measurements from even the most violent NSWSs; therefore, it is worthy to examine how it can contribute to wind engineering applications.

From an operational perspective, the national network of 160 weather surveillance radars owned by the National Oceanic and Atmospheric Administration (NOAA) in the United States provides outstanding coverage to assist weather forecasters with nowcasting duties. These permanently sited S-band radars and their 8.5-m antennas provide a beam width of 0.925° and range resolution of 250 m. Information is nominally conveyed in "superresolution" at 0.5° or a legacy resolution of 1.0°. A base scan is usually acquired at a 0.5° elevation angle, while the remainder of the volume is acquired at higher tilt angles over the course of approximately a 5- to 10-minute period. While a powerful resource for providing weather surveillance, this operations-based radar network is constrained from serving the engineering community. The relatively wide beam width limits the ability to extract relevant scales of motion, while the minimum elevation angle of 0.5° and broader operationally driven volume coverage pattern minimizes the time focused on the lowest levels of the atmosphere. The relatively slow revisit times also limit the ability to track and examine the advection of smaller scales of motion, and the distance between radar sites negates any chance of synthesizing dual-Doppler measurements at heights or scales of value to engineers.

3.3.2.3. RESEARCH RADARS

Luckily, there is an expanding and diverse inventory of research weather radars (Bluestein et al., 2014) that is available for studying events such as NSWSs. These research radars have led the way in studying severe storms for decades and continue to contribute today with new and novel measurements. Without being handcuffed by operational duties, these instruments can be focused on the task at hand via customized scanning strategies dedicated to the measurement of specific phenomena at low levels. The fact that many are truck mounted provides an additional dimension of freedom to position the instrument within reasonable range of an event of interest. This flexibility also mitigates attenuation that is prevalent at shorter wavelengths in precipitation and enhances azimuthal resolution (through minimizing the range to the event of interest). Multiple research radars can be deployed and operated in a collaborative fashion to acquire coordinated measurements and provide a basis for multi-Doppler synthesis and extraction of the full wind speed and direction. If the distance between the radars is limited, the resulting dual-Doppler analysis can be of significant value to engineering analyses.

The usefulness of the acquired information to the engineering community is largely dictated by the underlying limitations in spatiotemporal resolution and the ability to extract information from the lowest levels of the atmosphere. Most research radars have wavelengths that range from 5 cm (C band) to 3 mm (W band). Because many are affixed to trucks as shown in Figure 8.4, their antenna size is constrained by roadway regulations, but they nominally yield beam widths between 0.18° and 1.5°, which subsequently yields measurement volume horizontal cross sections of 15.7–130.9 m at 5 km and 47.1–392.7 m at 15 km. Range resolution also varies. For conventional pulsed radars, the range resolution is a function of the pulse duration, but for radars that integrate pulse compression technology, the range resolution is a function of the inverse bandwidth. Common range resolutions exist between 10 and 250 m. Scan speeds also vary significantly from 5° to 180°/s. This diversity, along with the advent of phased array technology, provides a significant difference in how quickly a two- or three-dimensional volume of space can be scanned and a significant diversity in effective resolution (taking into account beam width, scan speed, and integration time for each averaged radial velocity measurement).

FIGURE 8.4A. Mobile truck mounted research radars, including a C-band SMART Radar (SR3) deployed on the shoulder of the Cape Fear River bridge in North Carolina for Hurricane Florence (2018) deployment.

Photo credit: Dr. Michael Biggerstaff, Oklahoma University.

FIGURE 8.4B. An X-band RaXPOL radar deployed at White Sands Missile Range in New Mexico.

Photo credit: Dr. Michael Biggerstaff, Oklahoma University.

FIGURE 8.4C. A TTUKa radar intercepting a supercell thunderstorm.

Photo credit: Texas Tech University.

For extracting engineering-relevant information, enhanced spatiotemporal resolution is paramount. Hence, higher resolution radars with faster scan speeds are better positioned to contribute to engineering studies. Physical size constraints favor smaller wavelength radars whose smaller beam widths also enable extraction of information from regions closer to the ground. Of course, these same radars are most vulnerable to attenuation issues. Regardless, any radar would need to be placed relatively close to the NSWS event of interest to limit the impact of beam spreading, reduce the above-ground height of the measurements, and help mitigate attenuation.

3.4. Other Relevant Technologies

Other, less common, technologies have also made contributions, but they also maintain limitations that must be understood.

3.4.1. *High-Resolution Pressure Deficit Measurements*

Karstens et al. (2010) documented pressure measurements collected from beneath tornadoes during the 2002–2008 time period. In this study, one or two HITPR probes (Samaras & Lee, 2004) were placed in the anticipated path of tornadoes to measure high temporal resolution pressure deficits. In several cases, information acquired by the HITPR probes was coupled with research radar or mobile mesonet information from the same or nearby locations. While limited, this information has provided significant value to those attempting to numerically or experimentally model tornadic wind fields.

3.4.2. *Moving Measurement Platforms*

Mobile mesonets (i.e., a vehicle rack-mounted instrument suite) (Straka, Rasmussen, & Fredrickson, 1996) and instrumented drones offer two methods to maneuver in situ measurement platforms in and around thunderstorms. While these measurement systems have proven useful for documenting thermodynamic and kinematic gradients near thunderstorms, their usefulness to extract engineering-relevant scales is questionable. First, the motion and location of the observing platform must be successfully known with sufficient detail to acquire relevant information from a moving platform. Second, there are added concerns regarding the influence of changing roughness or terrain on any acquired wind measurements. While these issues may not be of concern to an atmospheric scientist seeking documentation of a frontal boundary or the position of the rear-flank gust front, they are of concern to a wind engineer. These issues complicate the use of these technologies to document the mean wind field and makes reliable extraction of any turbulence quantities unlikely. If a mobile mesonet is stationary, then some of these concerns are mitigated (assuming proper metadata are collected). In this case, a mobile Mesonet performs equivalently to an instrumented tower, except it provides the added advantage of being able to be deployed just before or even during the arrival of the most relevant portion of the event.

4. Equating Measurement Paradigms

Measurement technologies and approaches continue to evolve, but the ability to translate information from one measurement paradigm to another is important and not always straightforward. Traditional wind engineering analyses have almost exclusively been based on the evaluation of high temporal resolution time histories acquired with in situ anemometry, yet radar is the predominant technology currently used to extract storm-scale information from NSWS events.

Anemometers, when combined with a data acquisition system, typically produce a digitized record of the two- or three-dimensional wind flow, such as that shown in Figure 8.1. The dynamic response characteristics of a given sensor can vary considerably, but if research-grade measurements are available, the extraction of micro- and turbulence-scale information of value to engineers is possible. However, instrumented towers are typically limited in height (usually 10 m or less) and nominally provide only a single measurement point. Unless there are multiple levels of information available or there is an array of towers placed, it is impossible to assess lateral or vertical correlations/coherency or horizontal/vertical gradients in the wind field, which are also important for engineering applications.

In contrast, a weather radar is typically employed to spatially map the precipitation and radial velocity fields across a two-dimensional or three-dimensional domain. Individual estimates of radial velocity for a given resolution volume (Figure 8.5) are acquired "instantaneously" (at least for the purposes of this discussion), but they represent a spatial average, not a point value. The exact time of the acquisition within the broader experimental domain is controlled by the implemented scanning strategy and

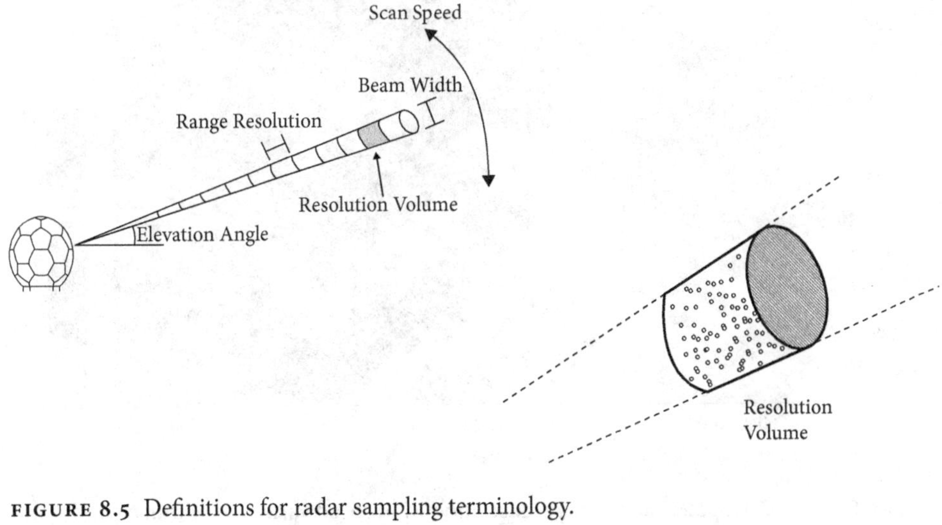

FIGURE 8.5 Definitions for radar sampling terminology.

technology constraints (e.g., scan speed). Using a repetitive scanning strategy across a two- or three-dimensional experimental domain results in periodic measurements for an individual resolution volume. These revisit times are typically on the order of 1–10 minutes for most atmospheric science-based studies using a mechanically scanned radar. However, rapid scan or phased array radar technologies can reduce this timeline by an order of magnitude, while lidars would typically require more than an order of magnitude more time to provide equivalent data coverage for a given set of azimuthal and elevation angles.

Beyond the spatiotemporal constraints of the underlying measurements, the technique often used to study storm-scale motions (e.g., dual-Doppler synthesis) combines coordinated measurements acquired using multiple radars to estimate the horizontal (and at times vertical) wind fields across a portion of the experimental domain (Miller & Strauch, 1974). While these methods have successfully estimated larger scales of motion, their applicability to reproduce smaller scales of motion has only more recently been explored in more detail. Hirth and Schroeder (2013) used information from the two high-resolution research radars to explore the horizontal complex flows found within and surrounding wind farms, such as that shown in Figure 8.6. In these cases, the experimental domains were typically limited to less than 100 km^2, the revisit times were approximately 60 s, and the wind speeds remained below 20 m/s. The result provided

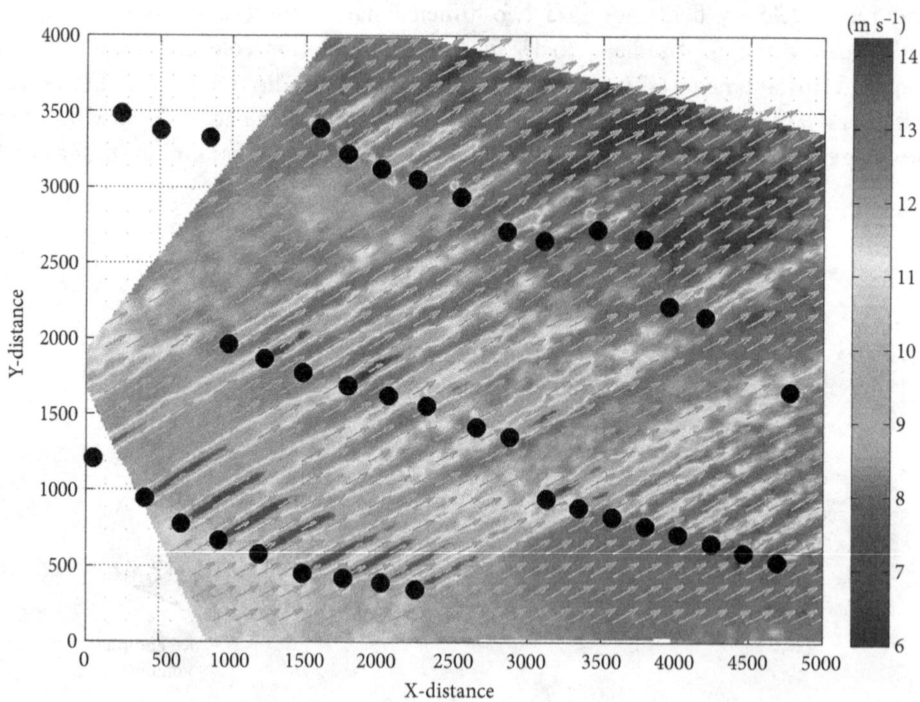

FIGURE 8.6 Dual-Doppler radar image illustrating wind speed variations within a wind plant. Wind turbine locations are identified by the black dots.

three-dimensional snapshots of the wind throughout the rotor-swept layer every minute and illustrated the ability to track larger scale gusts and lulls moving through the domain.

Traditional multi-Doppler synthesis techniques attempt to account for advection of the flow field between radar revisit times, but storm-scale studies are concerned with broader regions rather than tracking resolved boundary layer structures. Hence, corrections are often based on storm translation during the measurement period rather than the translation of the embedded boundary layer flow itself. To account for this fundamental difference, Duncan, Hirth, and Schroeder (2019) used the same measurement techniques as Hirth and Schroeder (2013) but modified the advection correction schemes to account for the translation of boundary layer flow features. A subsequent space-to-time conversion technique resulted in the extraction of wind speed and direction time histories for various locations within the experimental domain. Reporting intervals for such information were on the order of every few seconds. Validation of the methodology relative to an instrumented tower has been successful in low-to-moderate wind speeds, but the applicability of the methodology to higher wind speed events is unknown. For NSWS flow fields that maintain significant velocities, localized discontinuities, or significant curvature, the task of extracting smaller scales of motion becomes considerably more challenging.

5. The Measurement Path Forward

Future engineering-oriented measurement campaigns will likely make use of both traditional anemometry and remote sensing technologies to document the characteristics of NSWSs. For remote sensing platforms that maintain excellent spatiotemporal resolution, extraction of some engineering-relevant measurements appears to be within reach. Yet, more work is needed to bridge the gap between the different measurement paradigms and provide more effective translation of the subsequent results to the engineering community. Hence, additional studies, even outside of NSWSs, are needed to augment our current knowledge. These efforts include how to extract baseline estimates of turbulence intensity, integral scales, and so on from radar-acquired measurements. Validation of the results with those based on traditional analyses employing research-grade anemometry will be required. Extrapolation of the developed techniques to environments with higher wind speeds is also required, as well as developing an understanding of the limitations at various ranges (i.e., azimuthal resolutions). With appropriate techniques in place for paradigm translation and a better understanding of the inherent limitations, mobile research radars can become a more effective tool for engineers.

One method to expand engineering use of measurements collected from traditional weather stations is to retrofit existing stations to enable higher sampling rates during periods of high winds or for specific events (at the discretion of the network operator). These high-resolution measurements could then be separated from the operational data stream and set aside for study. A good target for this sort of adaptation is the state-oriented mesonets (in contrast to the national network of ASOS) given their flexibility to incorporate such a change. This investment would continue to provide new value as additional events are sampled each year. However, an individual measurement location and height can only provide limited information and do not provide comprehensive four-dimensional, engineering-relevant information.

The development of highly instrumented sites is another option. In this case, a relatively dense array of multilayered measurements could be deployed at a select few sites. Optimally, this approach would integrate information from both weather stations and remote sensing technologies/techniques. High-resolution multi-Doppler information targeting the lowest levels of the atmosphere could be acquired and compared to the underlying instrumented towers. Mobile instrumentation (both weather stations and radars) could also be deployed within the array to provide additional resolution and value for targeted campaigns. While a highly instrumented site has the potential to successfully measure high-quality comprehensive data sets, one would have to "wait" for events of interest to occur, and developing such a site would require significant capital outlay.

Given the transient and localized nature of most NSWS design-level events, a concerted effort to "chase" high-end events over a multiyear time period would also seem like a reasonable approach. The experimental domain would need be focused not only where NSWS events are prevalent, but also where extraction of low-level information is feasible. Hence, operations in the regions where terrains or trees are prevalent would preclude quality data acquisition using existing technologies.

Assuming differences in the measurement paradigms can be sufficiently addressed, high-resolution mobile radars appear the best positioned tool to contribute relevant measurements to meet engineering needs in the near term. Platforms that offer mobility, high spatiotemporal resolution, and the ability to extract information from the low levels will provide the most significant value. Given the spatial constraints of the problem, the quest for engineering-relevant, four-dimensional wind fields would favor higher resolution radars at W- to X-band frequencies deployed relatively close to the event. Any of these wavelengths can be compromised by attenuation, but given the measurement domain of interest, is usually limited in range (~10 km is more than adequate), these frequencies are still appropriate for many NSWS cases. Assuming the underlying spatial resolution is supportive of extracting relevant scales of motion, phased array or rapid scan technologies are also valuable given the reduction in revisit times enables better tracking of wind structure through the domain. However, the cost of these technologies may be prohibitive.

Initial results indicate multi-Doppler synthesis techniques may provide significant value to the engineering community for some thunderstorm cases. However, these multi-Doppler approaches will need to be completed using enhanced objective analysis and advection correction techniques to preserve the relevant scales of motion, which are different from those traditionally employed by the atmospheric science community to extract storm-scale motions. Regardless, favorable deployment scenarios would include placing two or more research radars separated by 5–10 km in areas with minimal/no terrain or obstructions and then using rapid and coordinated scanning of the lowest portion of the atmosphere to create a basis for the subsequent analysis.

For violent tornadoes, placing traditional weather sensors in their path and having them survive and yield relevant data seem less feasible. Even if platforms are hardened, deployment timelines must be exceptionally short to support fortuitous placement within the tornado itself. Historic HITPR deployments have shown value and success. Relative to previous experiments, a significant increase in the number of HITPR deployments for a given storm/tornado could drastically increase the amount of information available to engineers and could also allow a wider measurement array to be placed at an earlier time. This change from historical efforts would increase the likelihood of success and simultaneously alleviate safety concerns.

Finally, the smaller scaled NSWS features (e.g., eyewall mesovortices) found within the broader hurricane circulation pose an especially challenging measurement case. The NSWS features themselves are localized, yet they are embedded within a challenging environment of elevated wind, storm surge, and freshwater flooding. Historical deployments of mobile research radars and towers targeting these features have had some success, but there are always safety concerns for manned instrumentation embedded within the hazard itself. One way to alter this approach is to deploy mobile radar systems that are built to operate independently of direct human interaction. Even without cell service to a given site, the system can be designed to operate autonomously, withstand the wind, and collect targeted information from the lowest levels of the atmosphere. It seems appropriate to consider evolving current research radar technology to incorporate these fundamental capabilities.

While new technologies and techniques will almost assuredly alter the steps forward, there is a considerable amount of work that can be done with existing technologies to advance our understanding of the engineering-relevant characteristics of NSWS. Given the varying forcing mechanisms, stability regimes, exposures, and even terrain conditions associated with each unique event, no one data set will answer the broad spectrum of open scientific questions. Rather a prolonged and diverse approach needs to be set into motion. Much like the historical projects led by the atmospheric science community, a targeted field campaign merging the most appropriate technologies to support measurement of low-level, small scales of motion within NSWSs needs to be commissioned and properly supported. Without such a targeted effort, the wind engineering community will continue to see small-scale, ad hoc research efforts that struggle to provide even the most basic answers to long-standing questions.

References

Bluestein, H. B., Rauber, R. M., Burgess, D. W., Albrecht, B., Ellis, S. M., Richardson, Y. P., . . . Weckworth, T. M. (2014). Radar in atmospheric sciences and related research: Current systems, emerging technology, and future needs. *Bulletin of the American Meteorological Society*, 95, 1850–1861.

Brock, F. V., Crawford, K. C., Elliott, R. L., Cuperus, G. W., Stadler, S. J., Johnson, H. L., & Eilts, M. D. (1995). The Oklahoma mesonet: A technical overview. *Journal of Atmospheric and Oceanic Technology, 12*, 5–19.

Brooks, H. E. (2004). On the relationship of tornado path length and width to intensity. *Weather and Forecasting, 19*, 310–319.

Burgett, W. (2007). McLean tornado summary report: 28 March 2007. Retrieved from http://www.mesonet.ttu.edu/temp/McLean_tornado_28March2007/mclean_summary.html

Congressional Budget Office. (2016). Potential increases in hurricane damage in the United States: implications for the federal budget. Retrieved from https://www.cbo.gov/sites/default/files/114th-congress-2015-2016/reports/51518-hurricane-damage-onecol.pdf

Duncan, J. B., Jr., Hirth, B. D., & Schroeder, J. L. (2019). Enhanced estimation of boundary layer advective properties to improve space-to-time conversion processes for wind energy applications. *Journal of Wind Energy*. Advance online publication. doi:10.1002/we.2350

Fernández-Cabán, P. L., Alford, A. A., Bell, M. J., Biggerstaff, M. I., Carrie, G. D., Hirth, B., . . . Masters, F. J. (2019, May). Observing Hurricane Harvey's eyewall at landfall. *Bulletin of the American Meteorological Society*. doi:10.1175/BAMS-D-17-0237.1

Fujita, T. T. (1978). *Manual of downburst identification* (SMRP Research Paper no. 156). Chicago, IL: Department of Geophysical Science, University of Chicago.

Fujita, T. T. (1981). Tornadoes and downbursts in the context of generalized planetary scales. *Journal of the Atmospheric Sciences, 38*, 1511–1534.

Giammanco, I. M., Schroeder, J. L., & Powell, M. D. (2012). Observed characteristics of tropical cyclone vertical wind profiles. *Wind and Structures, 15*, 65–86.

Gunter, W. S., & Schroeder, J. L. (2015). High-resolution full-scale measurements of thunderstorm outflow winds. *Journal of Wind Engineering and Industrial Aerodynamics, 138*, 13–26.

Gunter, W. S., Schroeder, J. L., Weiss, C. C., & Bruning, E. C. (2017). Surface measurements of the 5 June 2013 damaging thunderstorm wind event near Pep, Texas. *Journal of Wind and Structures, 24*, 185–204.

Haan, F. L., Sarkar, P. P., & Gallus, W. A. (2008). Design, construction and performance of a large tornado simulator for wind engineering applications. *Engineering Structures, 30*, 1146–1159.

Hirth, B. D., & Schroeder, J. L. (2013). Documenting wind speed and power deficits behind a utility-scale wind turbine. *Journal of Applied Meteorology and. Climatology, 52*, 39–46.

Hjelmfelt, M. R. (1988). Structure and life cycle of microburst outflows observed in Colorado. *Journal of Applied Meteorology, 27*, 900–927.

Hock, T. F., & Franklin, J. L. (1999). The NCAR GPS dropwindsonde. *Bulletin of the American Meteorological Society, 80*, 407–420.

Karstens, C. D., Samaras, T. M., Lee, B. D., & Finley, C. A. (2010). Near-ground pressure and wind measurements in tornadoes. *Monthly Weather Review, 138*, 2570–2588.

Kepert, J. D. (2001). The dynamics of boundary layer jets within the tropical cyclone core. Part I: Linear theory. *Journal of the Atmospheric Sciences, 58*, 2469–2484.

Kepert, J. D., & Wang, Y. (2001). The dynamics of boundary layer jets within the tropical cyclone core. Part II: Nonlinear enhancement. *Journal of the Atmospheric Sciences, 58*, 2485–2501.

Kimball, S. K., Mulekar, M. S., Cummings, S., & Stamates, J. (2010). The University of South Alabama Mesonet and Coastal Observing System: A technical and statistical overview. *Journal of Atmospheric and Oceanic Technology, 27*, 1417–1439.

Kwon, D., & Kareem, A. (2009). Gustfront factor: new framework for wind load effects on structures. *Journal of Structural Engineering, 135*, 717–732.

Levitan, M. L., & Mehta, K. C. (1992a). Texas Tech field experiments for wind loads part 1: Building and pressure measuring system. *Journal of Wind Engineering and Industrial Aerodynamics, 43*, 1565–1576.

Levitan, M. L., & Mehta, K. C. (1992b). Texas Tech field experiments for wind loads part II: Meteorological instrumentation and terrain parameters, *Journal of Wind Engineering and Industrial Aerodynamics, 43*, 1577–1588.

Lewellen, W. S., Lewellen D. C., & Sykes, R. I. (1997). Large-eddy simulation of a tornado's interaction with the surface. *Journal of the Atmospheric Sciences, 54*, 581–605.

Lombardo, F. T. (2012). Improved wind speed estimation for wind engineering applications. *Journal of Wind Engineering and Industrial Aerodynamics*, 104–106, 278–284.

Lombardo, F. T., Smith, D., Schroeder, J. L., & Mehta, K. (2014). Thunderstorm characteristics of importance to wind engineering. *Journal of Wind Engineering and Industrial Aerodynamics, 125*, 121–132.

Lorsolo, S., Schroeder, J. L., Dodge, P., & Marks, F. (2008). An observational study of hurricane boundary layer small-scale features. *Monthly Weather Review, 136*, 2871–2893.

Mersereau, D. (2018). Tornadoes in Oklahoma have directly hit four weather stations in the past seven years. *Forbes.com*. Retrieved from https://www.forbes.com/sites/dennismersereau/2018/08/30/tornadoes-in-oklahoma-have-directly-hit-four-weather-stations-in-the-past-seven-years

Miller, L. J., & Strauch, R. G. (1974). A dual Doppler radar method for the determination of wind velocities within precipitating weather systems. *Remote Sensing of Environment, 3*, 219–235.

Natarajan, D., & Hangan, H. (2012). Large eddy simulations of translation and surface roughness effects on tornado like vortices. *Journal of Wind Engineering and Industrial Aerodynamics, 104–106*, 577–584.

Orwig, K. D., & Schroeder, J. L. (2007). Near-surface wind characteristics of extreme thunderstorm outflows. *Journal of Wind Engineering and Industrial Aerodynamics, 95*, 565–584.

Poss, D. B. (2000). *Design and evaluation of a mobile wind instrumentation tower for hurricane wind measurements* (Master's thesis). Clemson University, Clemson, SC.

Powell, M. D., Houston, S. H., Amat, L. R., & Morisseau-Leroy, N. (1998). The HRD real-time hurricane wind analysis system. *Journal of Wind Engineering and Industrial Aerodynamics, 77/78*, 53–64.

Refan, M., & Hangan, H. (2016). Characterization of tornado-like flow fields in a new model scale wind testing chamber. *Journal of Wind Engineering and Industrial Aerodynamics, 151*, 107–121.

Samaras, T. M., & Lee, J. J. (2004, January). Pressure measurements within a large tornado. Preprints, Eighth Symposium on Integrated Observing and Assimilation Systems for Atmosphere, Oceans and Land Surface, Seattle, WA, American Meteorological Society, 4.9.

Schroeder, J. L., Burgett, W. S., Haynie, K. B., Sonmez, I., Skwira, G. D., Doggett, A. L., & Lipe, J. W. (2005). The West Texas Mesonet: A technical overview. *Journal of Atmospheric and Oceanic Technology, 22*, 211–222.

Schroeder, J. L., Paulsen-Edwards, B., & Giammanco, I. M. (2009). Observed tropical cyclone wind flow characteristics. *Journal of Wind and Structures, 12*, 347–379.

Schroeder, J. L., & Smith, D. A. (2003). Hurricane Bonnie wind flow characteristics. *Journal of Wind Engineering and Industrial Aerodynamics, 91*, 767–789.

Sherlock, R. H., & Stout, M. B. (1937). Wind structure in winter storms. *Journal of the Aerospace Sciences, 5*, 53–61.

Skinner, P., Weiss, C., French, M., Bluestein, H., Markowski, P., & Richardson, Y. (2014). VORTEX2 observations of a low-level mesocyclone with multiple internal rear-flank downdraft momentum surges in the 18 May 2010 Dumas, Texas, supercell. *Monthly Weather Review, 142*, 2935–2960.

Smith, B. T., Castellanos, T. E., Winters, A. C., Mead, C. M., Dean, A. R., & Thompson, R. L. (2013). Measured severe convective wind climatology and associated convective modes of thunderstorms in the contiguous United States, 2003–09. *Weather and Forecasting, 28*, 229–236.

Snyder, J. C., & Bluestein, H. B. (2014). Some considerations for the use of high-resolution mobile radar data in tornado intensity determination. *Weather and Forecasting, 29*, 799–827.

Standohar-Alfano, C. D., Estes, H., Johnston, T., Morrison, M. J., & Brown-Giammanco, T. M. (2017). Reducing losses from wind-related natural perils: Research at the IBHS Research Center. *Frontiers in Built Environment, 3*. doi:10.3389/fbuil.2017.00009

Straka, J. M., Rasmussen, E. N., & Fredrickson, S. E. (1996). A mobile mesonet for finescale meteorological observations. *Journal of Atmospheric and Oceanic Technology, 13*, 921–936.

Tang, Z., Changda, F., Wu, L., Zuo, D., & James, D. L. (2018). Characteristics of tornado-like vortices simulated in a large-scale Ward-type simulator. *Boundary-Layer Meteorology, 166*, 327–350.

Uhlhorn, E. W., & Black, P. G. (2003). Verification of remotely sensed sea surface wind in hurricanes. *Journal of Atmospheric and Oceanic Technology, 20*, 99–116.

Vickery, P. J., Lin, J. X., Skerlj, P. F., Twisdale, L. A., & Huang, K. (2003). HAZUS-MH hurricane model methodology. I: Hurricane hazard, terrain and wind load modeling. *Natural Hazards Review, 7*, 82–93.

Vickery, P. J., & Skerlj, P. F. (2005). Hurricane gust factors revisited. *Journal of Structural Engineering, 131*, 825–832.

Weiss, C. C., & Schroeder, J. L. (2008). StickNet: A new portable rapidly deployable surface observation system. *Bulletin of the American Meteorological Society, 89*, 1502–1503.

Willoughby, H. E., & Black, P. G. (1996). Hurricane Andrew in Florida: Dynamics of a disaster. *Bulletin of the American Meteorological Society, 77*, 543–549.

Wurman, J., & Alexander, C. R. (2005). The 30 May 1998 Spencer, South Dakota, storm. Part II: comparison of observed damage and radar-derived winds in the tornadoes. *Monthly Weather Review, 133*, 97–119.

Wurman, J., & Koshiba, K. (2010). Finescale single- and dual-Doppler analysis of tornado intensification, maintenance, and dissipation in the Orleans, Nebraska, supercell. *Monthly Weather Review, 146*, 713–722.

Wurman, J., & Koshiba, K. (2018). The role of small-scale vortices in enhancing surface winds and damage in Hurricane Harvey (2017). *Monthly Weather Review, 146*, 713–722.

Wurman, J., & Winslow, J. (1998). Intense sub-kilometer-scale boundary layer rolls observed in Hurricane Fran. *Science, 280*, 555–557.

Yu, B., Chowdhury, A. G., & Masters, F. J. (2008). Hurricane power spectra, co-spectra and integral length scales. *Bound-Layer Meteorology, 129*, 411–430.

CHAPTER 9

NON-SYNOPTIC WINDS DATA BASIS

UWE ULBRICH AND EDMUND P. MEREDITH

What are the best ways of creating non-synoptic wind system (NSWS) data basis? Are current data basis and practice satisfactory? What are future ways of better generating, managing, and using NSWS-related data basis?

The imagination of an ideal data basis is a spatially dense, long-term wind measurement network that is homogeneous and provides a perfect basis for a statistical estimation of extreme events. As such a data basis is not achievable, we explore some possibilities that may serve different usage scenarios.

1. Local Wind Measurements

Local wind measurements carried out with anemometers of different construction type provide time series of the local wind climate and thus a basis of knowledge of that climate. The measurements are influenced by a number of different factors, including the synoptic and non-synoptic wind types occurring at a specific site. With respect to the usability of such time series, a number of factors must be taken into account (cf. World Meteorological Organization [WMO], 2003).

1.1. Quality/Homogeneity of the Data

Proper maintenance of the instruments used is a key issue in local wind measurements. A lack of such maintenance can lead to misinterpretation. Corrosion and frictional effects affecting mechanical instruments like standard anemometers can falsely suggest a decline of wind speeds. As another example, a deviation from the vertical setup of the mast used for the measurements can lead to biases in both the measured wind speed and the direction. Problems in data quality can be induced by the electrical parts of the instruments, which translate the primary measurements into digital information. While these examples underline why local measurement sites must be carefully maintained, there are also risks to the data quality from inhomogeneities caused by replacement or repair of the measurement devices. Thus, suitable metadata should be available in order to enable the user to check for such inhomogeneities.

In the field of non-synoptic winds, sudden changes of wind speeds and directions can be an important feature, which should be reflected in the data in order to allow for its assessment. Available station data are, however, not stored at high frequencies, so that they represent only part of the temporal spectrum of winds. The internal averaging implemented by some instrumental devices can have its origin in the inertia of the instrument, which prevents an accurate reflection of rapidly (range of seconds) changing wind speeds in the data.

1.2. Representativeness of the Data

Even with optimum quality of local wind measurements, the usability of data for an assessment of non-synoptic winds may be strongly limited by the fact that they are point measurements in the atmosphere. These point measurements are not always representative of the wind conditions in their surroundings. For wind, the systematic deviations of wind climate away from the measurement location in both horizontal and vertical dimensions is a key issue. The rules for synoptic stations imposed by the WMO and national weather services demand that wind-measuring synoptic stations should be designed to represent the wind climate in their vicinity. This requirement is clearly not fulfilled for the vertical dimension. Surface winds are strongly dependent on the height at which the measurements are carried out. The standard height is 10 m above ground, with a well-kept grass lawn of sufficient spatial extension beneath. The background to this requirement is the influence of surface roughness, which is a key factor in the (idealized) logarithmic vertical wind profile in the atmospheric boundary layer. Measurements at a different height than 10 m above the ground have systematically lower or higher wind speeds than those at the standard level. With respect to mean winds (i.e., not local phenomena like gusts), standard methods are available to estimate the values at the reference height, assuming the surface roughness can be quantified from available metadata about the site (e.g., Wan, Wang, & Swail, 2010). Even then, the measured values cannot simply be transferred to locations with other than standard conditions at the ground (e.g., different from the standard roughness of the surface). Even worse, there are often additional factors like buildings, trees, or wind farms in the vicinity of a station. They usually impose biases in wind directions and speeds and can also influence attributes like gustiness. For these reasons, wind-measuring stations should be sufficiently distant from the contamination sources (i.e., regions where microscale processes are a major factor for local winds, rendering the idealized logarithmic wind profile erroneous or producing unrepresentative extremes). It should also be noted that trees and buildings in the vicinity of a station impose direction-dependent biases. Such effects can be further complicated due to the annual cycle of climate, as the changing presence of tree leaves will induce a changing influence of air-flow biases. Like the aforementioned inhomogeneities induced by maintenance and replacement of instruments, growth of trees and tree cutting are important sources of inhomogeneity.

While the aforementioned idealized logarithmic wind profile is typically a useful tool for deriving wind speeds at various heights within the atmospheric boundary layer, the theoretical assumptions underlying the approach (i.e., neutral buoyancy within the boundary layer) are not always applicable. In particular for NSWSs (e.g., those associated with convective activity), the assumption of neutral atmospheric stability may be inappropriate. Rather, the occurrence of convection is related to the occurrence of local instability in the atmosphere. The presence of buoyancy within the mixed layer—particularly in the surface layer—can strongly modulate the vertical wind profile, requiring the more complex Monin–Obukhov similarity theory to estimate wind profiles (Monin & Obukhov, 1954).

Thus, the local wind measurements are very restricted in their representativeness, and one should be aware of variability in all directions, which can often lead to large differences within a short distance. An approach to overcome this problem is the concentration of many local measurement devices in a small but spatially extended test site. Such installations have been set up at several sites, sometimes called mesonets. The collection of measurements within mesonets can provide information on the horizontal changes of wind climate at the specific site. Given a synchronous measurement at the different locations, these can even provide information on the structure of the non-synoptic winds at the site. Such experimental sites have also been set up in order to investigate the vertical structure of non-synoptic (extreme) winds (e.g., making use of an available radio transmission tower) (e.g., Clobes, Willecke, & Peil, 2011).

Another issue that is relevant for local measurements is the temporal representativeness of the data sample, which may be limited by a combination of insufficiently long records and interannual variability and, indeed, climate change. Specific phenomena or extreme events change their likelihood with the large-scale atmospheric conditions, which are not distributed equally in time but rather may occur more often in some years and less in others. Thus, the risk of errors in estimating frequencies and magnitudes of the occurrence of certain wind events may arise if the analysis period is not sufficiently long to fully represent natural interannual atmospheric fluctuations. As a result, the WMO has set standards for estimating climatologies: data should be collected over a time span of at least 30 years. Additionally, if possible, a so-called climate normal period of a certain range of years, such as 1961–1990, should be used; this provides a standard reference period for estimating return levels and event frequencies. In times of anthropogenic climate change, the underlying assumption of a stable climate within and beyond the 30-year period may no longer be adequate, as systematic shifts in the climate away from the reference period climate and toward warmer conditions will likely have a range of impacts on winds; in such cases, guidance may be sought from climate models.

2. Remote Sensing Techniques

Remote sensing techniques measure wind in a range around the (usually fixed) location of the device. With respect to the setup of a data basis for non-synoptic data, a clear advantage of the remote sensing techniques is that they provide information over a beam extending radially from the device. A main disadvantage of such systems is that their basis is the Doppler effect, and thus only the wind components toward or away from the instrument can be captured. Other challenges of these measurements are similar to those of local point measurements, such as questions of homogeneity and representativeness of the data. In order to extend the measurements of winds from data within a single direction toward two or even three dimensions, some of the available systems include a scanning procedure so that winds in ranges of angles (horizontally and height) are evaluated. In order to avoid a significant change of the wind field's structures during a scan, the temporal offset between neighboring directions should be small. The question of what a significant change is depends, of course, on the wind field structures investigated. This is related to the general conflict between the desire to cover a large air volume (including larger distances from the instrument) and to have a high spatial and temporal resolution of the measured data. A LIDAR (light detection and ranging) system, for example, will measure winds within a small beam, which approximately does not broaden with distance. Thus, the volume of air investigated in terms of wind speed remains very small within the reach of the beam. Other instruments allow for a broadening of the beam, so that fine structures of the wind field become less detectable with increasing distance. The respective devices, Doppler radars, for example, typically incorporate a scanning mode.

With respect to the completeness of data obtained from remote sensing methodologies, it must be borne in mind that they measure winds via backscattering. They thus require that aerosol or cloud droplets are flowing with the air surrounding them. This imposes two limitations on these measurements. First, with insufficient aerosol content, the scattered signal may be too weak, so that no interpretation is possible due to a low signal-to-noise ratio. In particular, the data on winds at larger distances from the device become unmeasurable. Second, with an excess of those particles a certain distance within the beam, there is not enough power of the backscattered signal to measure wind speeds behind the particle accumulation areas. Thus, the extension of the measurements in the beam will essentially vary. There is therefore a risk of data gaps, and it must be ensured that such gaps do not occur systematically in conjunction with the type of phenomena being assessed.

3. Numerical Models

Numerical models have been constructed to provide information on meteorological parameters, including winds, on a three-dimensional grid. While weather forecasts are a basic application of these models, they can also be used to estimate the three-dimensional state of the atmosphere at different times on the basis of the available measurements, including those from remote sensing. The combination of model estimates (available on a three-dimensional grid at the model's computational time steps) and measurements called data assimilation leads to the so-called analysis products, which are often taken as reality and thus as a suitable basis for an evaluation of observed structures and developments, such as those of non-synoptic wind structures. There are, however, a number of restrictions preventing a direct application in this area.

The basic numerical weather forecast models must be global in order to include all types of interactions in Earth's atmospheric system. Current operational weather forecast models have a spatial grid resolution of about 10 km and thus are not able to represent many non-synoptic wind phenomena. As fine-scale structures are not present, their interactions with the large scale and resulting feedbacks are approximated by the so-called physical parameterization schemes. These schemes make use of empirical–statistical relationships. Their output represents an average estimate out of the possible range of scale interactions between resolved and unresolved processes. Particular situations and extremes may thus be poorly represented (Volosciuk, Maraun, Semenov, & Park, 2015).

Thus, the usability of analysis products depends on if and how well non-synoptic wind features are incorporated. It should be noted that this is a question of not only spatial resolution, but also the inclusion of mechanisms in the atmosphere or at the surface that are required for the generation of non-synoptic wind features. It is also important to note that the analysis will not necessarily incorporate a feature even if it is present in the primary data available for the analysis. The assimilation process will remove many features that cannot be properly resolved at the model's resolution.

The quality of an analysis product depends on the measurements incorporated, the assimilation (which shall also minimize influences of measurement errors), and the numerical model used. Changes in this system lead to inhomogeneities in the data. While effects of increasing data availability (in particular from satellites) can hardly be avoided in a reasonable way, effects from the frequent changes in the forecasting model introduced in its operational use are commonly excluded by the production of the so-called reanalysis, which makes use of a fixed version of the numerical model.

For reasons of limited computational resources, it is not possible to have global models at an extremely fine grid spacing. The importance of smaller scale variability and features has led to the development of dynamical downscaling methods. Here, a higher resolution regional model receives lateral boundary conditions from a lower resolution global model, allowing it to develop fine-scale features within its interior (Feser, Rockel, von Storch, Winterfeldt, & Zahn, 2011; Rummukainen, 2010). This so-called nesting approach can have multiple steps, with an even higher resolution regional domain nested within the first dynamical downscaling step. Nesting can also be two way, with the global and regional models running in parallel, and the fine-scale fields of the regional model feeding back into the global model (e.g., Lorenz & Jacob, 2005).

With the nesting approach, more and more physical processes can be explicitly resolved and the reliance on (often imperfect) parametrization schemes thus reduced. As an example, the use of convection-permitting regional models (grid spacing < 4 km) has become common in recent years (Prein et al., 2015), allowing deep convective processes to be explicitly resolved and offering marked improvement in the representation of convective winds (e.g., updrafts, downdrafts, and cold-pool outflows) (Hentgen, Ban, Kröner, Leutwyler, & Schär, 2019; Meredith, Maraun, Semenov, & Park, 2015), as well as extreme wind gusts in general (Kendon et al., 2017). Regional models have recently also been used to produce regional reanalyses e.g., over Europe and Germany, in order to permit the incorporation of features on this scale into the analysis product.

In a similar way as the ensemble approach in operational weather forecasting, analysis products can be used as a basis for exploring different possible weather developments on the fine resolution under the given large-scale conditions (e.g., Noyelle, Ulbrich, Becker, & Meredith, 2019; Pardowitz et al., 2016). This provides an important broadening of the basis for investigations, provided that the features under investigation can be properly produced by the modeling system. This, again, requires a check of the model outcomes against available direct observations as the atmosphere in any numerical model can only be a limited approximation of reality.

Even convection-permitting regional models will, however, be unable to explicitly represent all wind-related phenomena of interest (e.g., subgrid-scale orographic flows and highly localized extreme gusts). Spatial resolutions of the order of 100 m or less are required to resolve such fine-scale features. For these purposes, as in many engineering applications, large-eddy simulations (LESs) are used. Using LES models, the majority of turbulent kinetic energy of various non-synoptic wind phenomena can be accounted for by resolved eddies and a very fine-scale representation of three-dimensional wind fields produced. Example applications include modeling of gusts in urban environments (e.g., Ahmad, Inagaki, Kanda, Onodera, & Aoki, 2017) and tornadoes (e.g., Nolan, Dahl, Bryan, & Rotunno, 2017). Certain global weather models can also be run regionally in an LES configuration, such as the Advanced Research Weather Research and Forecasting (AR-WRF; Skamarock et al., 2019) and Icosohedral Nonhydrostatic (ICON; Dipankar et al., 2015) models, though, due to the computational expense involved, the simulation domains must be kept small. As with regional models, the results of LES simulations are dependent on the large-scale boundary conditions provided. A proper sample of such driving boundary conditions is required in order to gain reliable statistics of non-synoptic winds from this modeling approach.

REFERENCES

Ahmad, N. H., Inagaki, A., Kanda, M., Onodera, M., & Aoki, T. (2017). Large-eddy simulation of the gust index in an urban area using the lattice Boltzmann method. *Boundary-Layer Meteorology, 163*, 447–467. doi:10.1007/s10546-017-0233-6

Clobes, M., Willecke, A., & Peil, U. (2011). Shape-dependent characteristics of full-scale wind profiles. *Journal of Wind Engineering and Industrial Aerodynamics, 99*(9), 919–930.

Dipankar, A., Stevens, B., Heinze, R., Moseley, C., Zängl, G., Giorgetta, M., & Brdar, S. (2015). Large eddy simulation using the general circulation model ICON. *Journal of Advances in Modeling Earth Systems, 7*, 963–986. doi:10.1002/2015MS000431

Feser, F., Rockel, B., von Storch, H., Winterfeldt, J., & Zahn, M. (2011). Regional climate models add value to global model data: A review and selected examples. *Bulletin of the American Meteorological Society, 92*(9), 1181–1192. doi:10.1175/2011BAMS3061.1

Hentgen, L., Ban, N., Kröner, N., Leutwyler, D., & Schär, C. (2019). Clouds in convection-resolving climate simulations over Europe. *Journal of Geophysical Research: Atmospheres, 124*, 3849–3870. doi:10.1029/2018JD030150

Kendon, E. J., Ban, N., Roberts, N. M., Fowler, H. J., Roberts, M. J., Chan, S. C., . . . Wilkinson, J. M. (2017). Do convection-permitting regional climate models improve projections of future precipitation change? *Bulletin of the American Meteorological Society, 98*(1), 79–93. doi:10.1175/BAMS-D-15-0004.1

Lorenz, P., & Jacob, D. (2005). Influence of regional scale information on the global circulation: A two-way nesting climate simulation. *Geophysical Research Letters, 32*, L18706. doi:10.1029/2005GL023351

Meredith, E. P., Maraun, D., Semenov, V. A., & Park, W. (2015). Evidence for added value of convection-permitting models for studying changes in extreme precipitation. *Journal of Geophysical Research: Atmospheres, 120*, 12,500–512,513. doi:10.1002/2015JD024238

Monin, A. S., & Obukhov, A. M. (1954). Basic laws of turbulent mixing in the surface layer of the atmosphere. *Contributions of the Geophysical Institute of the Academy of Sciences of the USSR, 151*(163), e187.

Nolan, D. S., Dahl, N. A., Bryan, G. H., & Rotunno, R. (2017). Tornado vortex structure, intensity, and surface wind gusts in large-eddy simulations with fully developed turbulence. *Journal of the Atmospheric Sciences, 74*(5), 1573–1597.

Noyelle, R., Ulbrich, U., Becker, N., & Meredith, E. P. (2019). Assessing the impact of sea surface temperatures on a simulated medicane using ensemble simulations. *Natural Hazards and Earth System Sciences, 19*, 941–955. doi:10.5194/nhess-19-941-2019

Pardowitz, T., Befort, T. J., Leckebusch, G. C., & Ulbrich, U. (2016). Estimating uncertainties from high resolution simulations of extreme wind storms and consequences for impacts. *Meteorologische Zeitschrift, 25*, 531–541.

Prein, A. F., Langhans, W., Fosser, G., Ferrone, A., Ban, N., Goergen, K., . . . Leung, R. (2015). A review on regional convection-permitting climate modeling: Demonstrations, prospects, and challenges. *Reviews of Geophysics, 53*, 323–361. doi:10.1002/2014RG000475

Rummukainen, M. (2010). State-of-the-art with regional climate models. *Wiley Interdisciplinary Reviews: Climate Change, 1*(1), 82–96. doi:10.1002/wcc.008

Skamarock, W. C., Klemp, J. B., Dudhia, J., Gill, D. O., Liu, Z., Berner, J., . . . Huang, X.-Y. (2019). *A description of the advanced research WRF version 4* (No. NCAR/TN-556+STR). doi:10.5065/1dfh-6p97

Volosciuk, C., Maraun, D., Semenov, V. A., & Park, W. (2015). Extreme precipitation in an atmosphere general circulation model: Impact of horizontal and vertical model resolutions. *Journal of Climate*, *28*, 1184–1205.

Wan, H., Wang, X. L., & Swail, V. R (2010). Homogenization and trend analysis of Canadian near-surface wind speeds. *Journal of Climate*, 23, 1209–1225.

World Meteorological Organization (WMO). (2003). *Guidelines on climate observation networks and systems* (WMO/TD No. 1185). Contributions by N. Plummer, T. Allsopp, J. A. Lopez. Geneva, Switzerland: World Meteorological Organization WCDMP Guidelines.

SECTION 3

MODELING METHODS FOR NSWS

CHAPTER 10

ANALYTICAL AND SEMI-EMPIRICAL MODELS OF TORNADOES AND DOWNBURSTS

DJORDJE ROMANIC AND HORIA HANGAN

1. Introduction

Experimental and numerical simulations of non-synoptic winds are many times complemented by analytical and semi-empirical models. These models are inexpensive to run and produce mostly mean flow fields that can be further used in several types of applications spanning from risk analysis, guidelines, and codes to estimating further effects such as debris flight passes and damage and losses to buildings and structures.

Herein we present a review of most of these models for tornadic and downburst flows. Some are developcns in the Navier-Stokes equations and searching for exact but many times inviscid solutions occasionally complemented by boundary layer equations to take into account surface effects. Other use simple superposition of generic, canonical flows for which the individual solutions are known. These solutions are then ensembled by empirical or semi-empirical fitting procedures. Very few of these models try to address turbulent or fluctuating flow fields. All models have a series of constants that are fitted against experiments or numerical simulations.

We have made an effort to present these models in an organized manner and to underline their assumptions and limitations.

2. Tornadoes

The flow field in a three-dimensional stationary tornado vortex consists of tangential, radial, and axial velocity components if the flow is presented in the cylindrical coordinate system. Because tangential velocity is the dominant component in most cases, the most basic one-dimensional tornado vortex models neglect the radial and axial components.

The governing equations for tornadic flows are the Naiver-Stokes equations in cylindrical coordinates (Figure 10.1):

$$\frac{1}{r}\frac{\partial(ru_r)}{\partial r}+\frac{1}{r}\frac{\partial u_\theta}{\partial \theta}+\frac{\partial u_z}{\partial z}=0 \tag{1}$$

$$\begin{aligned}\frac{\partial u_r}{\partial t}+u_r\frac{\partial u_r}{\partial r}+\frac{u_\theta}{r}\frac{\partial u_r}{\partial \theta}-\frac{u_\theta^2}{r}+u_z\frac{\partial u_r}{\partial z}=-\frac{1}{\rho}\frac{\partial p}{\partial r}+g_r+\upsilon\Bigg(&\frac{1}{r}\frac{\partial}{\partial r}\left(r\frac{\partial u_r}{\partial r}\right)-\frac{u_r}{r^2}\\&+\frac{1}{r^2}\frac{\partial^2 u_r}{\partial \theta^2}-\frac{2}{r^2}\frac{\partial u_\theta}{\partial \theta}+\frac{\partial^2 u_r}{\partial z^2}\Bigg)\end{aligned} \tag{2}$$

$$\begin{aligned}\frac{\partial u_\theta}{\partial t}+u_r\frac{\partial u_\theta}{\partial r}+\frac{u_\theta}{r}\frac{\partial u_\theta}{\partial \theta}-\frac{u_\theta u_r}{r}+u_z\frac{\partial u_\theta}{\partial z}=-\frac{1}{\rho r}\frac{\partial p}{\partial \theta}+g_\theta+\upsilon\Bigg(&\frac{1}{r}\frac{\partial}{\partial r}\left(r\frac{\partial u_\theta}{\partial r}\right)-\frac{u_\theta}{r^2}\\&+\frac{1}{r^2}\frac{\partial^2 u_\theta}{\partial \theta^2}-\frac{2}{r^2}\frac{\partial u_r}{\partial \theta}+\frac{\partial^2 u_\theta}{\partial z^2}\Bigg)\end{aligned} \tag{3}$$

$$\begin{aligned}\frac{\partial u_z}{\partial t}+u_r\frac{\partial u_z}{\partial r}+\frac{u_\theta}{r}\frac{\partial u_z}{\partial \theta}+u_z\frac{\partial u_z}{\partial z}=-\frac{1}{\rho}\frac{\partial p}{\partial z}+g_z+\upsilon\Bigg(&\frac{1}{r}\frac{\partial}{\partial r}\left(r\frac{\partial u_z}{\partial r}\right)\\&+\frac{1}{r^2}\frac{\partial^2 u_z}{\partial \theta^2}-\frac{2}{r^2}\frac{\partial u_r}{\partial \theta}+\frac{\partial^2 u_z}{\partial z^2}\Bigg),\end{aligned} \tag{4}$$

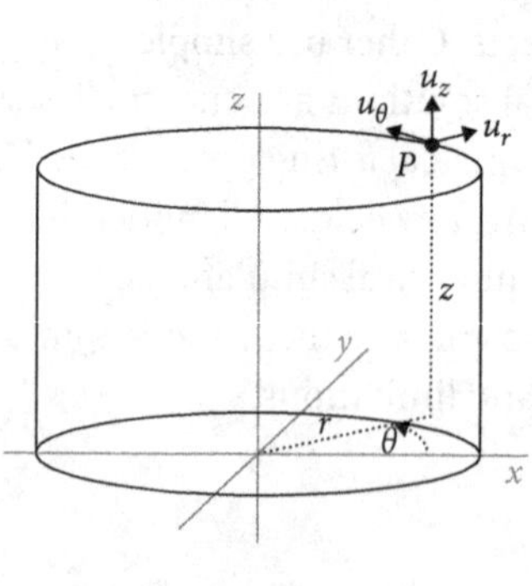

FIGURE 10.1 Cylindrical coordinate system and flow field nomenclature at a point P in the flow. Cartesian coordinates are in gray.

where t is the time; p is the static pressure; ρ is the density of air; υ is the kinematic viscosity, and g_r, g_θ, and g_z are three components of gravitational acceleration. See Table 10.1 for the complete list of all symbols used in this chapter.

In this system of equations, Equation (1) is the mass continuity equation for an incompressible fluid. Equations (2)–(4) are the radial (u_r), tangential (u_θ), and vertical (u_z);

Table 10.1 Nomenclature

Variable	Description
c_1, c_2	Coefficients in the Oseguera-Bowles-Vicroy downburst model
c_{ae1}, c_{ae2}, c_k, c_l, c_e, c_ζ, c_δ, c_ε, c_κ, c_μ	Coefficients in Abd-Elaal downburst model
c_{brs}	Shape coefficient in Burgers-Rott and Sullivan tornado models
c_m, c_η, c_χ	Coefficients in Li downburst model
c_v	Shape coefficient in Vatistas tornado model
c_{w1}, c_{w2}, c_{w3}	Shape coefficient in Wood and White tornado model
c_β, c_γ	Parameters in Li downburst model
d, d'	Distance from real and image doublet sheets, respectively, to a point P in downburst outflow
d_v	Viscous core diameter in the ring vortex model
f_{kw}	Function that accounts for velocity fluctuations outside the core region in Kuo-Wen tornado model
f_f, f_g, f_p, f_q	Shaping functions in Oseguera-Bowles downburst model
g_r, g_θ, g_z	Components of the acceleration due to gravity in cylindrical coordinates
h_s	Average size of roughness elements in Mason downburst model
j, N	Summation counter and the total number of points in the sum
k	Constant in the doublet model of a downburst and flow rate multiplier in Xu-Hangan-Yu model
k_{fr}	Flow rate multiplier
l_{in}, l_{out}	In- and out-of-boundary-layer functions in Mason downburst model
r_1, r_2, k_r	Smallest and farthest distance of a point P to vortex rings in the vortex ring downburst model. k_r is a combination of r_1 and r_2.
r_c	Tornado core radius
r_d	Core radius of the ring in the downburst ring model
r_h	Horizontal distance from the vortex centerline to a point P in downburst outflow
r_0	Doublet sheet radius
$r_{r.m}$, $z_{r.m}$, $u_{r.m}$	Radius and height, respectively, of the maximum radial velocity $(u_{r.m})$ in Baker tornado model

(continued)

Table 10.1 Continued

Variable	Description
r_v, z_v	Radial radius and height of the vortex ring in the downburst vortex ring model
r_m, r_t	Time-independent and time-dependent radius to the maximum horizontal velocity in a downburst outflow, respectively
s_1, s_2	Dummy variables of integration
t	Time
u_{10}	Mean wind speed at 10 m above the surface
u_{max}	Maximum horizontal velocity in a downburst
u_r, u_θ, u_z	Radial, tangential, and vertical velocity components, respectively, in tornado and downburst flow fields
u_x, u_y, u_z	Velocity components in a Cartesian coordinate system
u_{trans}	Downburst translation velocity
$u_{z,max}$	Maximum (descending) velocity in a downburst downdraft
$u_{\theta,m}$	Maximum tangential velocity
$U_{\theta,fluc}$, $U_{\theta,sub-vortex}$, $U_{\theta,wandering}$	Instantaneous fluctuating velocity, fluctuations attributed to coherent structures, and fluctuations attributed to vortex wandering, respectively, in Karami-Hangan tornado model
v_v	Downdraft velocity in the center of the vortex ring
w_0	Vertical velocity shaping function in Fujita tornado model
w_{jet}	Centerline velocity in impinging jet
x, y, z	Cartesian coordinates with the origin in tornado vortex center or downburst center
z^*	Characteristic height outside boundary layer in Oseguera-Bowles downburst models
z_c	Normalized height of the lower part of downburst flow in Mason model
z_h	Outflow depth in Oseguera-Bowles downburst model
$\boldsymbol{e}(\boldsymbol{p})$	Vector of differences between observed $(\boldsymbol{v}_o)$ and modeled $(\boldsymbol{v}_m)$ velocities for the given configuration of model parameters $(\boldsymbol{p})$ in Schultz downburst model
$\boldsymbol{v}_o$, $\boldsymbol{v}_m$	Vectors of observed and modeled velocities, respectively, in Schultz downburst model
$\boldsymbol{p}$	Vector of model parameters in Schultz model
$\boldsymbol{B}$	Matrix of weighing coefficients in the minimization problem
D	Downburst downdraft diameter
F_f	Shaping function in Fujita tornado model
F_u, F_w	Empirical functions in Abd-Elaal downburst model
H	Doublet sheet height
H_{in}	Inflow layer depth in Fujita tornado model
J	Cost function in minimization problems

Variable	Description
O, P	Coordinate system origin and an arbitrary point in tornado or downburst flow fields, respectively
PDF	Probability density function
R	Downdraft radius in downburst models
R_c	Radial length scale in Holmes and Oliver downburst model
R_f	Roughness factor in Mason downburst model
R_{in}, R_{out}	Inner and outer radii, respectively, in Fujita tornado model
R_{ref}, $u_{r,ref}$	Vortex reference radius and reference radial velocity in Xu and Hangan tornado model
S_r	Radial swirl ratio in Xu and Hangan tornado vortex model
T	Time scale constant in Holmes and Oliver downburst model
X, Y, Z	Coordinates of vortex ring center in a Cartesian system that does not have origin In the vortex center
α	Impinging jet tilt angle
α_{in}	Inflow angle in Fujita tornado model
α_r	Power-law coefficient in the velocity profile of atmospheric boundary layer winds
γ	Shape coefficient in Baker model
δ, δ_∞	Boundary layer height and its value at infinity, respectively, in Kuo-Wen tornado model
δ_{li}	Characteristic boundary layer thickness in the downburst model of Li
$\boldsymbol{\delta}$	Characteristic height inside boundary layer in Oseguera-Bowles downburst models
λ	Velocity scaling factor in Oseguera-Bowles downburst model
ξ, η, ζ	Coordinates of doublet singularity
ϖ	Downburst age scale in Abd-Elaal downburst model
ρ	Density of air
σ, σ_m	Doublet strength per unit area and the maximum doublet strength, respectively
ς	Function in Abd-Elaal downburst model
υ, υ_e	Kinematic viscosity and eddy viscosity of air, respectively
ϕ	Function variable of an integral
Γ, Γ_∞	Vortex circulation and its value at infinity, respectively
Π	Downburst intensity factor
ψ	Stream function
Y	Velocity damping factor in Schultz downburst model
Ψ_r, Ψ_α, Ψ	Scaling factors for roughness and jet tilt angle, and their sum, respectively, in Mason downburst model

see Figure 10.1) momentum equations, respectively, that are the direct consequence of the second Newton's law, with the forces per unit mass included on the right-hand side of the equations. Indeed, most of the tornado vortex models reviewed in this chapter can be derived by neglecting some terms and further manipulating other terms in the preceding equations. All solutions discussed in this chapter are steady state $(\partial/\partial t = \mathbf{0})$. It is only very recently that attempts have been made to simulate the fluctuating flow field.

The vertical and radial velocity components are disregarded in one-dimensional tornado vortex models. This assumption removes Equations (2) and (4) from the problem. In addition, most of the models consider inviscid fluid $(\upsilon = \mathbf{0})$ while neglecting body forces $(g_r = g_\theta = g_z = \mathbf{0})$, which significantly simplifies the preceding system of equations.

2.1. The Evolution of Rankine Vortex Model

2.1.1. *The RV Model*

The simplest and probably the most widely known analytical tornado vortex model is the Rankine vortex model based on the original work of Rankine (1858). This model represents an inviscid steady-state vortex whose core rotates as a rigid body, and the outer regions follow the potential flow with the tangential velocity (u_θ) being inversely proportional to the radial distance (r) from the vortex center. Formally, the RV model is expressed as

$$u_\theta = \begin{cases} \dfrac{u_{\theta,m} r}{r_c}, & r \le r_c \\ \dfrac{u_{\theta,m} r_c}{r}, & r > r_c, \end{cases} \tag{5}$$

where $u_{\theta,m}$ is the maximum tangential velocity observed at the core radius r_c. This model is also sometimes also called the *combined RV model* due to the combination of a rigid body and potential flow rotation.

2.1.2. *The MRV Model*

The preceding oversimplification of the inner region of a tornado vortex by the rigid-body rotation in the case of Ranking model is adjusted in the *Modified Rankine Vortex* (MRV) model as

$$u_\theta = \frac{r\Gamma_\infty}{\pi\left(r^2 + r_c^2\right)}, \tag{6}$$

where $\Gamma_\infty = 2\pi r_c u_{\theta,m}$ is the maximum vortex strength. Figure 10.2 shows tangential velocity profiles in the radial direction in both RV and MRV models. Notice that the MRV model provides a smooth transition between the inner and outer regions in the flow.

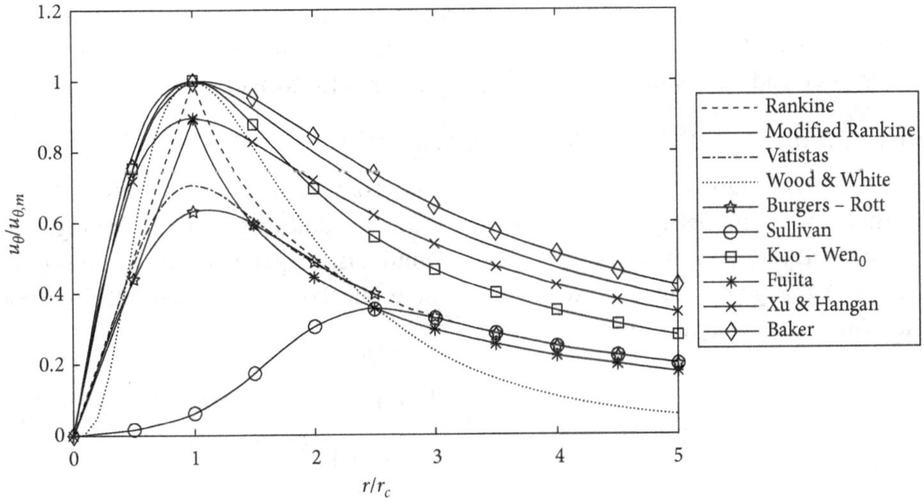

FIGURE 10.2 Radial profiles of tangential velocity calculated using different analytical models of tornadoes. Apart from Rankine and modified Rankine vortex models, the shapes of all other profiles can be manipulated depending on the choice of model parameters. The normalization is in respect to $u_{\theta,m}$ and r_c of the RV model.

2.1.3. *Modifications to MRV Model by Vatistas*

Vatistas et al. (1991) introduced a non-dimensional modulating factor (c_v) that governs the velocity profiles near the core radius (r_c). The core radius is, once again, defined as the radius of the maximum tangential velocity $(u_{\theta,m})$. Similar to the MRV, this modification provides a smooth transition between the inner and outer regions in the flow. In this model, tangential velocity is expressed as

$$u_\theta = u_{\theta,m} \frac{r/r_c}{\left(1+\left(r/r_c\right)^{2c_v}\right)^{1/c_v}} \tag{7}$$

Values $c_v < 1$ should be omitted because of singularities that occur along the vertical axis that goes through the vortex center. The radial profile of u_θ in the Vatistas et al. (1991) model for $c_v = 2$ is depicted in Figure 10.2. The value of $c_v = 2$ gives a similar result to the tangential flow field of Burgers-Rott model (Section 2.2).

Using the θ-momentum equation (Equation 3) in the simplified form due to the axisymmetric and steady-state flow, and also invoking the continuity equation (Equation 1), the other two velocity components can be expressed as

$$u_r = -2\left(c_v + 1\right)\frac{\upsilon_e}{r_c}\frac{\left(r/r_c\right)^{2c_v - 1}}{1+\left(r/r_c\right)^{2c_v}} \tag{8}$$

$$u_z = 4c_v\left(c_v + 1\right)\frac{\upsilon_e z}{r_c^2}\frac{\left(r/r_c\right)^{2(c_v - 1)}}{\left(1+\left(r/r_c\right)^{2c_v}\right)^2}. \tag{9}$$

Here, υ_e is the eddy viscosity that is used instead of kinematic molecular viscosity of air. Therefore, eddy viscosity with the common values between 1 m^2 s^{-1} and 10 m^2 s^{-1} should be considered as an additional shaping factor in the tornado velocity profiles.

2.1.4. *Modifications to Vatistas's Model by Wood and White*

Wood and White (2011) further built up on the model of Vatistas et al. (1991) by developing a tornado vortex model of tangential velocities that, in addition to maximum tangential velocity and radius of maximum tangential wind, also depends on three power-law exponents that govern different segments of the profile. The proposed model in Wood and White (2011) is in the form

$$u_\theta = c_{w1}^{\ c_{w2}} u_{\theta,m} \frac{\left(r/r_c\right)^{c_{w3}}}{\left(c_{w1} - c_{w3} + c_{w3}\left(r/r_c\right)^{c_{w1}/c_{w2}}\right)^{c_{w2}}}, \tag{10}$$

where c_{w1}, c_{w2}, and c_{w3} are the model-shaping parameters that satisfy $1 \leq c_{w3} < c_{w1}$ and $c_{w2} > 0$. While the model assumes the same form as Vatistas et al.'s (1991) model (Equation 7), the additional parameters introduce more flexibility of the model to fit real observations. The parameter c_{w3} governs the linearity or non-linearity of the inner region. It is easy to demonstrate that $c_{w3} = 1$ corresponds to the rigid-body rotation that was introduced in the original RV model. The parameter c_{w1}, on the other hand, controls the outer segment of the profile (i.e., $r > r_m$); namely, as c_{w1} increases, the decline of u_θ becomes steeper. Last, c_{w2} governs the shape of the profile near $r = r_m$. Figure 10.2 shows the radial profile of tangential velocity from this tornado vortex model ($c_{w1} = c_{w2} = 10$, $c_{w3} = 5$).

2.2. Burgers-Rott Model

This model is widely used in both wind engineering and meteorological communities for representing tornado-like vortices in numerical models, physical experiments, and the real atmosphere. Burgers-Rott model (Burgers 1948; Rott 1958) is a steady-state model with a constant viscosity throughout the flow and with the following assumption: (1) tangential velocity is only a function of radial distance from the vortex center, (2) vertical velocity is linearly dependent on the height, and (3) radial velocity is dependent on both radial distance from the vortex center as well as the height above the ground. With these assumptions, this model is an exact solution of the Navier-Stokes equations (Equations 1–4):

$$u_r = -c_{brs} r \tag{11}$$

$$u_\theta = \frac{\Gamma_\infty}{2\pi r}\left(1 - e^{-c_{brs} r^2 / 2\upsilon_e}\right) \tag{12}$$

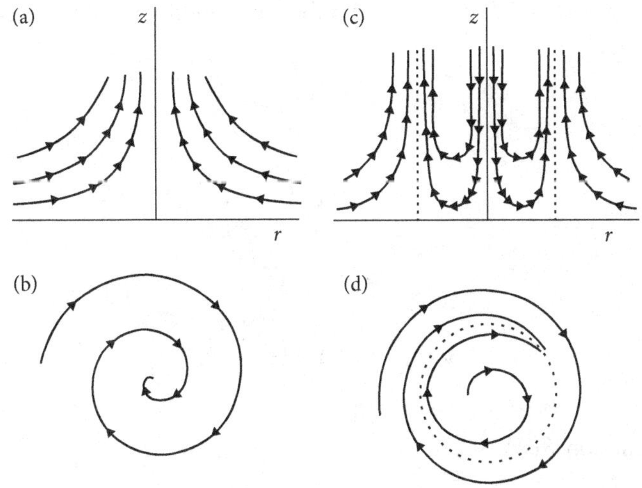

FIGURE 10.3 Schematics of projections of streamlines onto a vertical plane through the vortex axis and horizontal plane. (a, b) One-cell vortex (e.g., Burgers-Rott model) and (c, d) two-cell vortex (e.g., Sullivan model).

$$u_z = 2c_{brs}z, \tag{13}$$

where c_{brs} is the velocity gradient [s^{-1}] or also called the strength constant. Equation (11) shows that $c_{brs} > 0$ results in u_r being directed toward the vortex center. It is also easy to demonstrate that the model produces unrealistic results of u_r and u_z as $r \to \infty$ and $z \to \infty$, respectively. Therefore, the model always needs to be used with caution. The parameter c_{brs} can be expressed in terms of υ_e and r_c as

$$c_{brs} = \frac{2\upsilon_e}{r_c^2}. \tag{14}$$

The tangential velocity profile along the radial direction obtained from the Burgers-Rott model is portrayed in Figure 10.2. Notice that the Burgers-Rott model produces the maximum tangential velocity that is 64% of the maximum velocity in the RV and the MRV models. Last, this model belongs to the family of single-cell vortices (Figure 10.3).

2.3. Sullivan Model

Sullivan (1959) developed a two-cell (Figure 10.3) steady-state tornado model as an exact solution of simplified Navier-Stokes equations (Equations 1–4). While the assumptions are mostly similar to those in Burgers-Rott model, Sullivan (1959) allowed the vertical velocity to depend on both height and radial distance from the vortex center. Also similar to Burgers-Rott model, the static pressure is only a function of height and radial distance. With these assumptions, the velocity components of Sullivan model are

$$u_r = -c_{brs} r + \frac{6\upsilon_e}{r}\left(1 - \mathrm{e}^{-\frac{c_{brs} r^2}{2\upsilon_e}}\right) \tag{15}$$

$$u_\theta = \frac{\Gamma_\infty}{2\pi r} \frac{S\left(\frac{c_{brs} r^2}{2\upsilon_e}\right)}{S(\infty)} \tag{16}$$

$$u_z = 2c_{brs} z\left(1 - 3\mathrm{e}^{-\frac{c_{brs} r^2}{2\upsilon_e}}\right), \tag{17}$$

where the function $S(\phi)$ is

$$S(\phi) = \int_0^\phi \mathrm{e}^{\left(-s_1 + 3\int_0^{s_1}\left(1-\mathrm{e}^{-s_2}\right)\frac{ds_2}{s_2}\right)} ds_1, \tag{18}$$

where ϕ is the function argument and s_1 and s_2 are dummy variables. Similar to the Burgers-Rott model, the Sullivan model also produces unrealistic values of u_r and u_z for the limiting cases of r and z. Notice that $S\left(c_{brs} r^2 / 2\upsilon_e\right)/S(\infty)$ goes to 1 as $r \to \infty$, which also means that u_θ approaches the potential flow solution of the RV model when $r \to \infty$. Figure 10.2 shows the radial profile of tangential velocity from this model. Notice that the Sullivan model produces a peak velocity that is about 35% of the RV model, and, in addition to the velocity peak discrepancy, r_c is also displaced to the right in comparison to the RV model.

2.4. Kuo-Wen Model

Another three-dimensional tornado vortex model was proposed by Kuo (1971). This model has an advantage over the other models presented here because it explicitly accounts for the boundary layer effects in a tornado vortex. Similar to the common assumption in all other models, this steady-state model also assumes that the pressure only depends on r and z. In Kuo's model, the highest concentration of vorticity is in the core region, and a near-zero vorticity is found outside the core (i.e., a common feature of all Rankine-like vortices).

The vertical and radial distributions of velocities are obtained by separately solving two non-linear differential equations that Kuo (1971) derived from Equations (1)–(4) and that govern the flow inside and above the boundary layer. The region inside the core is characterized by an Ekman-like boundary layer with oscillatory distribution of velocities in the vertical direction. The outer region possesses the typical boundary layer velocity profile, with the velocities approaching their reference values without any osculation. Kuo's (1971) solution also establishes a weak downward motion in the outer region and a strong upward current in the inner region of a tornado vortex. It was demonstrated that the strongest u_z is always inside $r < r_c$ and at the location where the boundary layer growth is the most rapid. Depending on the boundary condition used in the numerical solutions of non-linear equations, Kuo (1971) also showed that the surface inflow angles are in the approximate range 10°–30° in the outer region.

While mathematically rigorous, Kuo's (1971) model is also very complex, and this system of two non-linear differential equations does not possess a closed-form solution of any set of initial conditions. Therefore, the model was later simplified by Wen's (1975) proposing regression equations that fit numerical solutions of Kuo's (1971) non-linear differential equations.

The boundary layer's thickest $\delta = \delta(r)$ that is now a function of r is given as

$$\delta = \delta_\infty \left(1 - e^{-0.5\left(\frac{r}{r_{m,0}}\right)^2} \right), \tag{19}$$

where δ_∞ is the boundary layer thickness far away from the vortex (i.e., theoretically at infinity) and $r_{m,0}$ is the radius of the maximum u_θ above the boundary layer (the second subscript "0" will indicate variables that represent the flow in the region above the boundary layer).

In this model, the velocity components have different expressions in the regions above and inside the boundary layer. The three velocity components above the boundary layer are defined as

$$u_{r,0} = 0 \tag{20}$$

$$u_{\theta,0} = 1.4 u_{\theta,m,0} \frac{r_{m,0}}{r} \left(1 - e^{-1.256\left(\frac{r}{r_{m,0}}\right)^2} \right) \tag{21}$$

$$u_{z,0} = 93 u_{\theta,m,0} \left(\frac{r}{r_{m,0}} \right)^3 e^{-5\left(\frac{r}{r_{m,0}}\right)}. \tag{22}$$

Inside the boundary layer, the velocity components are given by

$$u_{r,z} = 0.672 u_{\theta,0} e^{-\pi\frac{z}{\delta}} \sin\left(\left(f_{kw} + 1 \right) \pi \frac{z}{\delta} \right) \tag{23}$$

$$u_{\theta,z} = u_{\theta,m,0} \left(1 - e^{-\pi\frac{z}{\delta}} \cos\left(2\pi f_{kw} \frac{z}{\delta} \right) \right) \tag{24}$$

$$u_{z,z} = u_{z,0} \left(1 - e^{-\pi\frac{z}{\delta}} \cos\left(2\pi f_{kw} \frac{z}{\delta} \right) \right), \tag{25}$$

where f_{kw} is the function that depicts the velocity fluctuations outside the core region:

$$f_{kw} = 1.2 e^{-0.8\left(\frac{r}{r_{m,0}}\right)^4}. \tag{26}$$

Notice that the second subscript "z" in Equations (23)–(25) indicates that the profiles represent the flow field inside the boundary layer, and the profiles are height-dependent. Figure 10.2 shows the radial profile of tangential velocity above the boundary layer (Equation 21) from this model. The radial velocity profiles inside the boundary layer are height-dependent, and the interested reader is referred to Kuo (1971) and Kim and Matsui (2017) for more details.

2.5. Fujita Model

This three-dimensional model developed by Fujita (1978) seems to be less represented in the literature than, for instance, the Burgers-Rott or Sullivan models. Fujita's model was primarily developed for wind engineering applications in which, according to Fujita (1978), the parameters that govern tornado damages on structures are (1) outer core radius (Figure 10.4), (2) tornado translation velocity, (3) maximum tangential velocity, (4) maximum pressure drop, and (5) maximum rate of pressure change (the last two points are not discussed in this review). In this model, the core rotates as a rigid body and the vertical velocity is zero. An updraft is distributed along the outer region. Also, the inflow toward the tornado center is confined to a narrow layer called the *inflow layer*. The maximum radial and tangential velocities are located around the core radius. A schematic of Fujita's vortex that is characterized by inner and cores is shown in Figure 10.4.

The relationship between the inner (R_{in}) and outer (R_{out}) cores is

$$R_{in} = R_{out}\left(0.9 - 0.7e^{-0.005R_{out}}\right). \tag{27}$$

The depth of the inflow layer (H_{in}) is

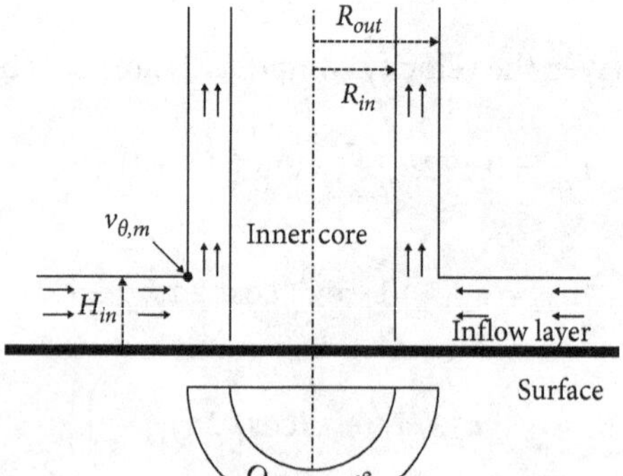

FIGURE 10.4 Inner and outer cores of the Fujita's vortex. The inner core rotates like a sold disk and is free of any vertical velocity.

$$H_{in} = 0.55R_{out}\left(1-\left(0.9-0.7e^{-0.005R_{out}}\right)^2\right). \tag{28}$$

Tangential velocity depends on radius and height in the following manner:

$$u_\theta = u_{\theta,m}F_f(r)F_f(z), \tag{29}$$

where

$$F_f(r) = \begin{cases} \dfrac{r}{R_{out}}, & r < R_{out} \\ \dfrac{R_{out}}{r}, & r \geq R_{out} \end{cases} \tag{30}$$

$$F_f(z) = \begin{cases} \left(\dfrac{z}{H_{in}}\right)^{\frac{1}{6}}, & z < H_{in} \\ e^{-0.03\left(\frac{z}{H_{in}}-1\right)}, & z \geq H_{in}. \end{cases} \tag{31}$$

Notice that the function $F_f(r)$ resembles the RV model (Section 2.1.1). The radial velocity is directly proportional to tangential velocity and inflow angle (α_{in}):

$$u_r = u_\theta \tan(\alpha_{in}), \tag{32}$$

where

$$\tan(\alpha_{in}) = \begin{cases} 0, & r < R_{in} \\ \dfrac{\tan(\alpha_{in,0})}{1-\left(0.9-0.7e^{-0.005R_{out}}\right)^2}\left(1-\left(\dfrac{R_{out}}{r}\right)^2\left(0.9-0.7e^{-0.005R_{out}}\right)^2\right), & R_{in} \leq r < R_{out} \\ \tan(\alpha_{in,0}), & r \geq R_{out} \end{cases} \tag{33}$$

$$\tan(\alpha_{in,o}) == \begin{cases} -0.75\left(1-\left(\dfrac{z}{H_{in}}\right)^{1.5}\right), & z < H_{in} \\ 0.0217\left(1-e^{-0.03\left(\frac{z}{H_{in}}-1\right)}\right), & z \geq H_{in}. \end{cases} \tag{34}$$

The reference inflow angle $(\alpha_{in,o})$ is the value of α_{in} at the outer core radius. Last, the vertical velocity is given as

$$\tan(\alpha_{in}) = \begin{cases} 0, & r < R_{in} \\ u_{\theta,m}w_0, & R_{in} \leq r < R_{out}, \\ 0, & r \geq R_{out} \end{cases} \tag{35}$$

where

$$w_0 = \begin{cases} 0.044\left(16\left(\dfrac{z}{H_{in}}\right)^{\frac{7}{6}} - 7\left(\dfrac{z}{H_{in}}\right)^{\frac{8}{3}}\right), & z < H_{in} \\ 0.398\left(2\mathrm{e}^{-0.03\left(\frac{z}{H_{in}}-1\right)} - \mathrm{e}^{-0.06\left(\frac{z}{H_{in}}-1\right)}\right), & z \geq H_{in} \end{cases} . \tag{36}$$

The tangential velocity component from Fujita's model is plotted in Figure 10.2 for comparison against other analytical models discussed in this chapter. For a unit value of $u_{\theta,m}$ and $z = 0.5H_{in}$, the model produces results that have shapes similar to the RV model. The damping of $u_{\theta,m}$ in respect to the RV case is due to the $F_f(z)$ function in the domain $z < H_{in}$ (Equation 31). That is, if $z = H_{in}$, the model is equal to the RV vortex. See Kim and Matsui (2017) for more details on the three-dimensionality of the flow field in the Fujita model.

2.6. Xu and Hangan Model

Xu and Hangan (2009) modeled a tornado-like vortex as a free narrow jet solution combined with a modified Rankine vortex. These two flow parts are connected through the swirl ratio. In their study, swirl ratio is unconventionally defined as the ratio of tangential and radial velocities. Traditionally, this parameter represents the ratio of the maximum tangential velocity and the mean updraft velocity in the vortex. In contrast to the Burgers-Rott (Section 2.2), Sullivan (Section 2.3), and Kuo-Wen (Section 2.4) models, this model is not a direct solution of Navier-Stokes equations. The model has been tested against experimental data for a swirl ratio of 0.28 only.

The three velocity components are

$$u_r = u_{r,ref} R_{ref}\left(\left(\frac{1}{r} + \frac{2r}{z^2}\right)\mathrm{e}^{-\left(\frac{r}{z}\right)^2} - \frac{1}{r}\right) \tag{37}$$

$$u_\theta = u_{r,ref} R_{ref} 2S_r\left(\left(\frac{r_c}{R_{ref}}\right)^2 + 1\right)\frac{r}{r_c^2 + r^2} \tag{38}$$

$$u_z = u_{r,ref} R_{ref} \frac{2}{z}\mathrm{e}^{-\left(\frac{r}{z}\right)^2}, \tag{39}$$

where R_{ref} is the vortex reference radius (i.e., inlet of the Ward-type vortex chamber) and $u_{r,ref}$ is the radial influx velocity at R_{ref} and S_r is the radial swirl ratio defined at the inlet:

$$S_r = \frac{u_{\theta,0}}{2u_{r,0}} \tag{40}$$

that was fixed (0.28) in Xu and Hangan (2009). The second subscript "o" indicates that the velocities are defined at the inlet. The radial profile of tangential velocity from this model is shown in Figure 10.2.

2.7. Baker Model

Baker and Sterling (2017) recently proposed another three-dimensional steady-state and inviscid tornado model in their study of tornado debris flights. In their model, all three velocity components, as well as the static pressure, are dependent on r and z. While their originally one-cell vortex model can in principle be tuned to produce the velocity profiles of a two-cell vortex, we limit the discussion here to the original proposition of a one-cell vortex model in Baker and Sterling (2017). Namely, the extension to a two-cell vortex is not an exact solution of Euler equations (i.e., Navier-Stokes equations with the preceding assumptions).

The single-cell vortex represented by this model has the following velocity components:

$$u_r = u_{r,m} \frac{-4\dfrac{r}{r_{r,m}}\dfrac{z}{z_{r,m}}}{\left(1+\left(\dfrac{r}{r_{rm}}\right)^2\right)\left(1+\left(\dfrac{z}{z_{rm}}\right)^2\right)} \tag{41}$$

$$u_\theta = u_{r,m} \frac{\dfrac{2S_r}{\ln(2)}\dfrac{r}{r_{r,m}}\ln\left(1+\left(\dfrac{z}{z_{r,m}}\right)^2\right)}{\left(1+\left(\dfrac{r}{r_{rm}}\right)^2\right)} \tag{42}$$

$$u_z = u_{r,m} \frac{4\dfrac{z_{r,m}}{r_{r,m}}\ln\left(1+\left(\dfrac{z}{z_{r,m}}\right)^2\right)}{\left(1+\left(\dfrac{r}{r_{rm}}\right)^2\right)^2}, \tag{43}$$

where $u_{r,m}$ is the maximum radial velocity, and $r_{r,m}$ and $z_{r,m}$ are the radius and height of $u_{r,m}$. Also, S_r is the radial definition of the swirl ratio [the ratio of $u_\theta\left(r_{r,m}, z_{r,m}\right)$ and $u_r(r_{r,m}, z_{r,m})$ at the reference height]. According to Baker and Sterling (2017), this model produces more realistic profiles of radial velocity, which makes it more suitable for the study of lifting and flying debris in tornadic flows. However, a drawback of this model is in Equations (42) and (43) that show u_θ and u_z are unbounded with height. Therefore, the model needs to be applied with caution at higher elevations above the surface. Similar shortcomings are observed in the Burger-Rott model (Section 2.2), but only for u_z. The radial profile of tangential velocity in Baker's model is shown in Figure 10.2 (for $r_{r,m} = 1.1r_c$, where r_c is the core radius defined as the radius of the maximum tangential velocity). If $r_{r,m} = r_c$, and the radial swirl ratio is 1, the model reduces to the MRV vortex. This model produces higher values of tangential velocity in the outer regions than all other models considered in this chapter.

2.8. Modeling of the Fluctuating Flow Field

To date no attempts have been made to model the instantaneous flow field of tornado-like vortices (TLVs). Most recently a simplified large-scale fluctuations models has been developed by Karami and Hangan based on their coherent structure analysis of TLVs (Karami et al. 2019).

2.8.1. *Karami and Hangan Large Scale Fluctuations Model*

Karami et al. (2018) used the results of a modal analysis for TLVs (Karami et al. 2019) to propose for the first time a preliminary model that includes a large-scale fluctuating flow field. These large-scale fluctuations are based on vortex wandering motion and sub-vortex dynamics (coherent structures). The fluctuating tangential velocity (the only one considered as being dominant) can be decomposed into three components:

$$U_{\theta,fluc} = U_{\theta,sub-vortex} + U_{\theta,wandering} \tag{44}$$

where $U_{\theta,fluc}$ is the instantaneous fluctuating velocity, $U_{\theta,sub-vortex}$ is the fluctuations attributed to coherent structures (sub-vortex dynamics), and $U_{\theta,wandering}$ is the fluctuations attributed to vortex wandering motion. Before vortex breakdown, only $U_{\theta,wandering}$ is significant, and $U_{\theta,sub-vortex}$ is negligible, whereas after vortex breakdown only $U_{\theta,sub-vortex}$ is significant, and $U_{\theta,wandering}$ is negligible (Karami et al. 2019).

Before vortex breakdown, the effects of vortex wandering motions are implemented into the original Burgers-Rott model using a convolution integral as follows (Devenport et al. 1996):

$$U_{\theta,corrected}(x,y) = \int_{-\infty}^{+\infty}\int_{-\infty}^{+\infty} PDF_{wandering\ motion}\, U_{\theta,uncorrected}(x,y)\,dxdy, \tag{45}$$

where $U_{\theta,uncorrected}(x,y)$ is the original Burgers-Rott model and $PDF_{wandering\ motion}$ is the probability density function (PDF) of the random displacement of the vortex with respect to its mean center, which is assumed to be a Gaussian joint PDF (Ashton et al. 2019).

After vortex breakdown only the sub-vortex dynamics is modeled, consisting of a recirculation bubble at the center and a single or double spiral structure rotating around the bubble. This vortex dynamics is simplified to a system of generic but representative vortices that are further used to derive the resulting fluctuating vorticity and velocity fields (Karami et al. 2019).

3. Downbursts

This section starts the survey of analytical and empirical models of downbursts by examining some of the models that are based on impinging or plain jet flow. This category of models is the most common in downburst research. The review then proceeds with the ring-vortex models that were initially developed for the specific purpose of increasing the safety of aircraft to downburst outflows. The gravity current models, based on the principle of negative buoyancy of downdraft in respect to the ambient atmosphere, are not considered in this chapter. As discussed in Section 1, thermodynamic effects on tornadic and downburst flows also are not considered in this chapter.

3.1. Jet-like Models

3.1.1. *The Evolution of the OBV Model*

While the phenomenon of downbursts was scientifically studied in atmospheric sciences since the 1940s, some of the first analytical models of downburst outflows were proposed much later. We start this review by examining the original model of Oseguera and Bowles (1988)—the so-called OB model—and how it has been modified and improved throughout the years.

3.1.1.1. THE OB MODEL

Assuming an axisymmetric and stationary downburst outflow without any inclination of the downdraft, the incompressible mass continuity equation in the cylindrical coordinate system (see Equation 1) reads

$$\frac{\partial u_r}{\partial r} + \frac{\partial u_z}{\partial z} + \frac{u_r}{r} = 0, \tag{46}$$

where u_r and u_z are the mean velocity components in the radial (r) and vertical (z) directions, respectively (see Table 10.1 for the full list of all symbols). The radial distance is measured from the downburst center (e.g., see Figure 10.1). The functions

$$u_z = f_g\left(r^2\right) f_q\left(z\right) \tag{47}$$

$$u_r = \frac{f_f\left(r^2\right)}{2} f_p\left(z\right) \tag{48}$$

are a solution to Equation (46), provided that

$$f_r'\left(r^2\right)=\frac{\lambda}{2}f_g\left(r^2\right) \tag{49}$$

$$f_q'\left(z\right)=-\lambda f_p\left(z\right), \tag{50}$$

where λ [in s^{-1}] is the scaling factor. The shapes of f_f, f_g, f_q, and f_p functions were chosen to satisfy some of the observed characteristics of downburst outflows, such as the strong dependency of the outflow velocity field in radial and height directions. Moreover, the shaping functions had to replicate the well-known nose-like velocity profile with height. Oseguera and Bowles (1988) adopted a couple of shaping functions from the Terminal Area Simulation System (TASS) numerical model results, as

$$f_g\left(r^2\right)=e^{-\left(\frac{r}{R}\right)^2} \tag{51}$$

$$f_p\left(z\right)=e^{-\frac{z}{z*}}-e^{-\frac{z}{\varepsilon}}, \tag{52}$$

where R is the downdraft radius, ε is the characteristic height inside the boundary layer, and z^* is the characteristic height outside of the boundary layer (i.e., the altitude at which $u_r=0.5u_{\max}$, where $u_{\max}$ is the maximum horizontal velocity). After obtaining the functions f_f and f_q by integrating Equations (49) and (50), the two velocity components are given as

$$u_r=\frac{\lambda R^2}{2r}\left(1-e^{-\left(\frac{r}{R}\right)^2}\right)\left(e^{-\frac{z}{z*}}-e^{-\frac{z}{\varepsilon}}\right) \tag{53}$$

$$u_z=\lambda e^{-\left(\frac{r}{R}\right)^2}\left(\varepsilon\left(e^{-\frac{z}{\varepsilon}}-1\right)-z*\left(e^{-\frac{z}{z*}}-1\right)\right). \tag{54}$$

After manipulating the above expressions, Oseguera and Bowles (1988) further showed that a ratio of the maximum horizontal velocity $\left(u_{\max}\right)$ and the maximum downdraft velocity $\left(u_{z,max}\right)$ can be expressed as

$$\frac{u_{max}}{u_{z,max}}=\frac{0.2357R}{z*\left(e^{-\left(\frac{z_h}{z*}\right)}-0.92\right)}, \tag{55}$$

where z_h is the outflow depth.

3.1.1.2. THE OBV MODEL

While the results reported in Oseguera and Bowles (1988) showed a proper matching with full-scale and laboratory data, Vicroy (1991) and Vicroy (1992) noted that the function $f(r)$ was markedly different from the observations. Vicroy (1992) therefore proposed the following shaping function:

$$f_f(r) = \frac{\lambda r}{2} \mathrm{e}^{\frac{2-\left(\frac{r}{r_m}\right)^{2\alpha}}{2\alpha}}, \tag{56}$$

where α is the function coefficients and r_m is the radius to the maximum horizontal velocity.

After implementing the new shaping function into the OB equations and replacing few variables in the OB model with different parameters, the new model—the so-called OBV model—assumes the following velocity profiles:

$$u_r = \frac{\lambda r}{2}\left(\mathrm{e}^{-c_1\frac{z}{z_m}} - \mathrm{e}^{-c_2\frac{z}{z_m}}\right)\mathrm{e}^{\frac{2-\left(\frac{r}{r_m}\right)^{2\alpha}}{2\alpha}} \tag{57}$$

$$u_z = -\lambda\left(\frac{z_m}{c_1}\left(\mathrm{e}^{-c_1\frac{z}{z_m}} - 1\right) - \frac{z_m}{c_2}\left(\mathrm{e}^{-c_2\frac{z}{z_m}} - 1\right)\right)\left(1 - \frac{1}{2}\left(\frac{r}{r_m}\right)^{2\alpha}\right)\mathrm{e}^{\frac{2-\left(\frac{r}{r_m}\right)^{2\alpha}}{2\alpha}}. \tag{58}$$

where z_m is the height of the maximum horizontal velocity, and $c_1 = -0.15$ and $c_2 = -3.2175$ are model coefficients. The OBV model is easier to use than the OB model because of the reduction of model parameters. Namely, the OBV model is fully defined in terms of r_m and z_m, the model coefficient $\alpha(\approx 2)$, and the scale factor λ that governs the strength of the outflow wind field. The parameters r_m and z_m are more intuitive and probably easier to obtain from full-scale data than the inside and outside boundary-layer characteristic heights in the original OB model.

3.1.1.3. MODIFICATIONS TO THE OBV MODEL BY CHAY

The OBV model was later modified by Chay et al. (2006) by manipulating the terms that account for storm intensity and the height of the maximum wind speed in the outflow. First, the model was extended to the time domain by introducing the intensity factor (Π) that accounts for storm intensity as

$$\Pi = \begin{cases} \frac{t}{5}, & 0 \le t \le 5 \\ \mathrm{e}^{-\frac{t-5}{11.542}}, & t > 5 \end{cases} \tag{59}$$

where t is the time (in minutes). The shape of this function was proposed after examining the full-scale Doppler radar data of several downbursts from Wilson et al. (1984) and Hjelmfelt (1988). Second, Chay et al. (2006) expressed λ in terms of $u_{\max}$ and r_m that simplifies the implementation of the model in practical applications; that is:

$$\lambda = \frac{2u_{max}}{r_m}\left(\frac{1}{e^{\frac{1}{2\alpha}}\left(e^{c_1} - e^{c_2}\right)}\right). \tag{60}$$

Furthermore, the translation velocity of the thunderstorm cloud $\left(u_{\text{trans}}\right)$ was added as a vector sum to the axisymmetric velocity field from the original OBV analytical model. Moreover, the radius at which the maximum outflow velocity occurs $\left(r_m\right)$ was replaced by the time-dependent variable $\left(r_t\right)$, but the exact shape of this function was not proposed. Similarly, the height at which the maximum outflow velocity occurs $\left(z_m\right)$, which was assumed to be constant throughout the outflow in the OBV model, was substituted by z_r that accounts for the radial dependency of z_m. After combining all these changes, the modified OBV model in Chay et al. (2006) was written in the following form:

$$u_r = \Pi \frac{u_{\max r}}{r_t} \frac{e^{c_1 \frac{z}{z_r}} - e^{c_2 \frac{z}{z_r}}}{e^{c_1} - e^{c_2}} e^{\frac{1-\left(\frac{r}{r_m}\right)^{2\alpha}}{2\alpha}} + u_{\text{trans}} \tag{61}$$

$$u_z = -2\Pi \frac{u_{\max z_m}}{r_t} \frac{\frac{1}{c_1}\left(e^{-c_1 \frac{z}{z_r}} - 1\right) - \frac{1}{c_2}\left(e^{-c_2 \frac{z}{z_r}} - 1\right)}{e^{c_1} - e^{c_2}} \left(1 - \frac{1}{2}\left(\frac{r}{r_m}\right)^{2\alpha}\right) e^{\frac{1-\left(\frac{r}{r_m}\right)^{2\alpha}}{2\alpha}}. \tag{62}$$

The vector addition of u_{trans} to axisymmetric outflow makes the modified OBV model asymmetric, whereas the Π term formulates the model in the time domain. Following the work of Chen and Letchford (2004), Chay et al. (2006) also added turbulent fluctuations to mean velocities of the modified OBV model using the evolutionary power spectral density method.

3.1.1.4. MODIFICATIONS TO THE OBV MODEL BY LI

Li et al. (2012) further modified the OBV model by accounting for the non-linear growth of the downburst outflow boundary layer. This modification subsequently affects the location of the maximum horizontal wind speed in the outflow. The OBV model was reformed by proposing a new vertical shaping function for radial velocity

$$f_p(z) = \left(\frac{z}{z_m}\right)^{c_\gamma} e^{\gamma\left(1-\frac{z}{z_m}\right)}. \tag{63}$$

and a new horizontal shaping function for radial velocity

$$f_f(r) = \begin{cases} \frac{r}{r_m}, & r \le r_m \\ e^{-\left(\frac{r-r_m}{R_c}\right)^{c_\beta}}, & r > r_m \end{cases}, \tag{64}$$

where c_γ and $c_\beta > 1$ are two non-dimensional parameters and R_c is the radial length scale adopted from Holmes and Oliver (2000). R_c is defined as the distance to the high-pressure ring in the downburst outflow. After applying $f_f(r)$ to the continuity equations, a new vertical shaping function of the horizontal wind was derived as

$$f_g\left(r^2\right)=\begin{cases} 1, & r\le r_m \\ \dfrac{e^{-\left(\frac{r-r_m}{R_c}\right)^{c_\beta}}}{2r}-\dfrac{c_\beta\left(\dfrac{r-r_m}{R_c}\right)^{c_\beta-1} e^{-\left(\frac{r-r_m}{R_c}\right)^{c_\beta}}}{2R_c}, & r>r_m. \end{cases} \tag{65}$$

Non-linear growth of an impinging jet boundary layer thickness was also modeled through the implementation of shaping functions that depict the relationship between $z_m(r)$ and $\delta_{li}(r)$, where δ_{li} is the characteristic boundary layer thickens, expressed in the following implicit form:

$$\frac{e^{c_1\frac{\delta_{li}}{z_m}}-e^{c_2\frac{\delta_{li}}{z_m}}}{e^{c_1}-e^{c_2}}=0.5. \tag{66}$$

Similarly, a non-linear relationship between r_m and z was expressed as

$$\frac{r_m}{D}=c_\eta-\frac{z}{c_m}e^{\frac{1-\left(\frac{z}{c_m}\right)^{2c_\chi}}{2c_\chi}}, \tag{67}$$

where $c_\chi=-0.0363$, $c_m=0.0078D$, $c_\eta=2.01$ are the model parameters, and $D=2R$ is the impinging jet diameter. Since these shaping functions, as well as the proposed coefficient values, are heavily based on computational fluid dynamics simulations and wind tunnel data, further model verifications are desirable against full-scale downbursts. Last, Li et al. (2012) also developed a new vertical shaping function for the OBV model, but the resulting expression could not be expressed in terms of a closed-form solution and thus requires numerical integrations.

3.1.1.5. MODIFICATIONS TO THE OBV MODEL BY ABD-ELAAL

After noticing a small discrepancy between the vertical shaping function in Li et al. (2012) (Equation 63) and available data, Abd-Elaal et al. (2013) proposed a new shaping function in the form

$$f_p(z)=\left(\frac{z}{z_m}\right)^{c_{ae2}-1}e^{c_{ae1}\left(\frac{z}{z_m}\right)^{c_{ae2}}}, \tag{68}$$

where c_{ae1} and c_{ae2} are non-dimensional parameters. This function enables the derivation of the vertical speed shaping function, $f_q(z)$, in a closed-form that was not possible in Li et al. (2012). That is,

$$f_q(z)=-\lambda\frac{z_m}{c_{ae1}c_{ae2}}\left(e^{c_{ae1}\left(\frac{z}{z_m}\right)^{c_{ae2}}}-1\right). \tag{69}$$

Abd-Elaal et al. (2013) also noticed that $f_g\left(r^2\right)=1$ (i.e., fixed constant) for $r\le r_m$ (Equation 65) in the model of Li et al. (2012) did not properly match some of the available field data. Therefore, Abd-Elaal et al. (2013) introduced

$$f_f(r)=\frac{\lambda r}{2}\left(\mathrm{e}^{-\left(2c_\zeta-\left(\frac{c_\delta r^2}{r_m^2}\right)^{c_\zeta}\right)^2}+c_\varepsilon \mathrm{e}^{-c_\kappa\left(\frac{r}{r_m}\right)^{2c_\mu}}\right). \tag{70}$$

where c_ζ, c_δ, c_ε, c_κ, and c_μ are the non-dimensional parameters. The corresponding $f_g(r^2)$ is in the form

$$f_g(r^2)=\left(1+2c_\zeta c_\mu\left(2c_\zeta-\varsigma\right)\right)\mathrm{e}^{-(2c_\zeta-\varsigma)^2}+c_\varepsilon \mathrm{e}^{-c_\kappa\left(\frac{r}{r_m}\right)^{2c_\mu}}\left(1-c_\kappa c_\mu\left(\frac{r}{r_m}\right)^{2c_\mu}\right) \tag{71}$$

where

$$\varsigma=\left(\frac{c_\delta r^2}{r_m^2}\right)^{c_\zeta}. \tag{72}$$

The implementation of the new shaping functions into the OBV model results in the following expressions for the horizontal and vertical velocity components in their respective order:

$$u_r=\frac{\lambda r}{2}\left(e^{-(2c_\zeta-\varsigma)^2}+c_\varepsilon \mathrm{e}^{-c_\kappa\left(\frac{r}{r_m}\right)^{2c_\mu}}\right)\left(\frac{z}{z_m}\right)^{c_{ae2}-1}\mathrm{e}^{c_{ae1}\left(\frac{z}{z_m}\right)^{c_{ae2}}} \tag{73}$$

$$u_z=-\lambda\left(\left(1+2c_\zeta\varsigma\left(2c_\zeta-\varsigma\right)\right)\mathrm{e}^{-(2c_\zeta-\varsigma)^2}+c_\varepsilon \mathrm{e}^{-c_\kappa\left(\frac{r}{r_m}\right)^{2c_\mu}}\left(1-c_\kappa c_\mu\left(\frac{r}{r_m}\right)^{2c_\mu}\right)\right)$$
$$\frac{z_m}{c_{ae1}c_{ae2}}\left(\mathrm{e}^{c_{ae1}\left(\frac{z}{z_m}\right)^{c_{ae2}}}-1\right). \tag{74}$$

Similar to Li et al. (2012), Abd-Elaal et al. (2013) also studied the non-linear growth of the downburst boundary using numerical and experimental data. They proposed the following dependency of the height of the maximum horizontal velocity on r:

$$z_m(r)=z_m\left(\frac{1.1+7\left(\ln\frac{r}{D}-0.2\right)^2}{\frac{r}{D}}\right), \tag{75}$$

and the following profile of the radial distance of the maximum horizontal wind speed:

$$r_m(z)=c_k D-\frac{z}{c_l}\mathrm{e}^{-\left(\frac{1-\left(\frac{z}{c_l D}\right)^{-c_e}}{c_e}\right)}. \tag{76}$$

Here, c_k, c_l, and c_e are non-dimensional parameters obtained from fitting the experimental data. It is crucial to note that all shaping functions in the preceding discussion, including the original OB model, are empirical functions determined on either limited full-scale or experimental data. That is, the functions are not an explicate solution of governing physical law. Therefore, the models always demonstrate satisfactory matching against experimental data that were used for calibration of shaping functions. Due to the empirical nature of shaping functions (usually derived from numerical or wind tunnel data), the application of the equations to a new set of data that were not used in the model's development tends to diminish the model's performance significantly. Also, this model suffers from too many empirical coefficients that severely limit its applicability to real data.

Last, we describe the proposed improvements to the original OBV model by Abd-Elaal et al. (2014). Similar to Chay et al. (2006), Abd-Elaal et al. (2014) extended the OBV model to the time domain and added the storm translation speed to the otherwise axisymmetric downburst outflow. The functions that depict the temporal behavior of downburst outflow were obtained based on numerical detached eddy simulations. By fitting the numerical results, the authors proposed mathematical functions that describe the spatiotemporal evolution of the outflow (equations not shown here). The storm translation velocity was kept constant and described using the power law:

$$u_{\mathrm{trans}} = u_{10}\left(\frac{z}{10}\right)^{\alpha_r}, \tag{77}$$

where u_{10} is the mean wind speed at 10 m above the ground and α_r is the power-law coefficient that depends on surface roughness and atmospheric stability. However, as Abd-Elaal et al. (2014) also noticed in their study, this formula is most likely inaccurate in describing storm translation for numerous reasons, the obvious one being the absence of wind direction dependency on the height and inapplicability to winds above the atmospheric boundary layer, which predominantly govern cloud translation.

After implementing the preceding modifications in mathematical form, the new OBV model in Abd-Elaal et al. (2014) was cast in the following form:

$$u_r(t,r,z,\varpi) = u_r(r,z)\,F_u(t,r,\varpi) + u_{\mathrm{trans}} \tag{78}$$

$$u_z(t,r,z,\varpi) = \begin{cases} u_z(r,z)\,F_w(t,r,\varpi), & r < D \\ 17w_{\mathrm{jet}}\left(1 - e^{-0.3\frac{Z}{D}}\right)F_w(t,r,\varpi), & r \geq D \end{cases} \tag{79}$$

where w_{jet} is the centerline jet velocity. The functions $F_u(t,r,\varpi)$ and $F_w(t,r,\varpi)$ are empirically derived from numerical experiments (expressions not presented here), and ϖ is the so-called scale age for downburst outflow (in minutes). The age of downburst outflow in this context corresponds to the period of the impinging flow from the inlet jet and not the total time of downburst duration in, for example, anemometer records.

This model was compared against several full-scale downburst events, and the results showed good performance for the model. However, a more comprehensive verification of all analytical models would be in terms of the entire downburst flow field reconstruction/comparison and not focused only on matching a point measurement. That is, a comparison of the reconstructed flow field from an analytical model against Doppler radar measurements would demonstrate the accuracy of the model to resolve the entire flow field in four dimensions (i.e., both space and time).

3.1.2. *Plain Wall Jet Model of Verhoff and Impinging Jet Model of Wood*

Numerous impinging jet and plain wall jet experiments in the field of experimental fluid mechanics have been performed over the past several decades to study both mean and turbulent flow properties. The research on impinging jets has been motivated by many applications of these types of flows in aerospace engineering, combustion engines, mechanical engineering, and hydraulics, to name a few. However, the resulting flow fields from these experiments resemble some of the observed features of real downbursts in the atmosphere, such as the existence of toroidal vortices that travel outward from the downdraft center and the nose-shaped velocity profile with height. Because a comprehensive review of all significant impinging jet and plain wall jet studies is practically impossible, we present here a model developed by Verhoff (1963) in his experimental study of plain wall jets because this model was later used to construct a new empirical model of downbursts proposed by Wood et al. (2001).

While the studies of Sigalla (1958), Schwarz and Cosart (1961), and Myers et al. (1963) have all confirmed the similarity of velocity profiles in the plain wall jets that was initially found in the experimental work of Förthmann (1936), it was Verhoff (1963) who formulated an empirical equation that correctly described the data (Rajaratnam 1976):

$$\frac{u_r}{u_m} = 1.48\left(\frac{z}{z*}\right)^{\frac{1}{7}}\left(1-\operatorname{erf}\left(0.68\frac{z}{z*}\right)\right). \tag{80}$$

Here, u_r is the radial velocity of the plain wall jet, u_m is the maximum value of u_r, and z^* is the height where $u = 0.5u_m$ (the same quantity was used in the formulation of the OB model in Section 3.1.1.1). Also, erf is the error function, defined as

$$\operatorname{erf}(\phi) = \frac{2}{\sqrt{\pi}}\int_0^{\phi} e^{-s_1^2}\, ds_1, \tag{81}$$

where ϕ is a random variable, and s_1 is a dummy variable. Although some books present Equation (81) without the term $2/\sqrt{\pi}$, the function represents twice the integral of the Gaussian distribution with zero mean and a variance of ½.

A model similar to Equation (80) was later proposed by Wood et al. (2001) in their physical and numerical study of impinging jets with application to downbursts, that is:

$$\frac{u_r}{u_m} = 1.55\left(\frac{z}{z*}\right)^{\frac{1}{6}}\left(1-\operatorname{erf}\left(0.70\frac{z}{z*}\right)\right). \tag{82}$$

Wood et al. (2001) have demonstrated that the model accurately fitted their experimental and numerical data. This empirical model is today widely used in wind engineering studies that focus on downburst outflows and their interaction with structures. It is also important to note that the self-similarity of the radial velocity profiles does not exist until $r/D>3$, but is closely achieved for $r/D>1.5$ (Knowles and Myszko 1998), where r and D are the radial distance from the downdraft center and downdraft diameter, respectively. However, the maximum velocities in the jet outflows are observed at $1.0<r/D<1.5$.

3.1.3. *Empirical Model of an Inclined Jet with Surface Roughness by Mason*

This model was developed as an extension of the model by Wood et al. (2001) (Section 3.1.2) to account for the absence of self-similarity of velocity profiles for $r/D<1.5$. In addition, the model attempts to account for surface roughness and the inclination of the impinging jet, thereby resembling more closely real downburst downdrafts that are usually tilted and not perfectly vertical.

First, the influence of surface roughness is empirically modeled using the following function that was derived based on physical experiments in a wind tunnel:

$$\frac{z_m}{z^*}=\frac{R_f}{15}\left(2+\mathrm{e}^{-\left(0.45\left(\frac{r}{D}\right)\right)^4}\right)\left(1-\mathrm{e}^{-\left(0.95\left(\frac{r}{D}\right)\right)^3}\right), \tag{83}$$

where the roughness factor $\left(R_f\right)$ is

$$R_f=100\left(\frac{h_s}{D}\right)+1 \tag{84}$$

and h_s is the average dimension of surface roughness features.

The introduced modification in Equation (83) still does not ensure the self-similarity of the profiles close to the impingement zone, and, therefore, additional corrections are needed. Two additional terms that are added are the so-called "in-boundary layer" and "out-of boundary layer," terms l_{in} and l_{out}, respectively. This terminology follows the work of Oseguera and Bowles (1988) (Section 3.1.1.1), and the terms refer to the regions $z<z_m$ and $z>z_m$, respectively. Empirically, l_{in} was found to be

$$l_{in}=1-\left(\frac{1}{R_f}\right)^2\left(1-\frac{\min\left(\frac{z}{z^*},\frac{z_m}{z^*}\right)}{\frac{z_m}{z^*}}\right), \tag{85}$$

where the min function ensures that the correction is only applied within the in-boundary layer region. The authors reported that l_{in} is mostly independent of the tilt angle, radial position in the flow, and surface roughness. However, the combination of downdraft inclination and surface roughness requires the introduction of the parameter $1/R_f$.

Once again, using the empirical fitting of impinging jet data from physical experiments, l_{out} was represented as

$$l_{out} = \Psi \cdot \cos\left(\frac{\pi}{2}\left(\frac{\min\left(\left|\frac{z}{z^*} - z_c\right|, 1 - z_c\right)}{1 - z_c}\right)\right)^{\frac{3}{2}}, \tag{86}$$

where z_c is the normalized height of the mid-point in the region $z_m/z^* < z/z^* < 1$ that is calculated as

$$z_c = \frac{\frac{z_m}{z^*} + 1}{2}. \tag{87}$$

The scaling factor (Ψ) is calculated as $\Psi = \Psi_r + \Psi_\alpha$ (where α is the tilting angle of the downdraft in radians) as

$$\Psi_r = 0.065e^{-\left(2\left(\frac{r}{D}\right)-1\right)^5}. \tag{88}$$

$$\Psi_\alpha = 0.1\theta^{0.45}e^{-\left(\frac{\frac{r}{D}}{1.1+0.8\theta}\right)^5}. \tag{89}$$

Both Ψ_r and Ψ_α tend to zero as the flow field reaches self-similarity. Also, a closer inspection of Equation (86) shows that l_{out} is only applied up to the level $z/z^* = 1$.

The implementation of these empirical terms to the model of Wood et al. (2001) (Section 3.1.2) results in the final expression of the non-dimensional velocity profile:

$$\frac{u_r}{u_m} = 1.55 l_{in}\left(\frac{z}{z^*}\right)^{\frac{1}{6}}\left(1 - \mathrm{erf}\left(0.70\frac{z}{z^*}\right)\right) + l_{out}. \tag{90}$$

The scaling of a roughness parameter in this and similar models deserves further research due to the Reynolds number's dependency between the velocity and surface roughness.

3.1.4. *Holmes and Oliver Model*

The model presented in this section is an empirical model based on single and multiple Doppler radar measurements of downbursts in Colorado, in the United States (Hjelmfelt 1988). Downburst field measurements were collected during the Joint Airport Weather Studies (JAWS) project (Fujita 1981) and were synthesized by Wilson et al. (1984) and Hjelmfelt (1988) in two seminal papers that presented some of the essential features of downburst outflows. In figure 12 of Hjelmfelt (1988), the author depicted the radial profile of radial velocities measured along the azimuth of the maximum outflow velocity and demonstrated that the wall jet experiments are accurate at reproducing the velocity profile for $r < r_m$. The matching is less accurate in the outer region of the outflow where the wall jet profile overestimates the field measurements.

By combining Hjelmfelt's (1988) observation with the potential flow solution of an impinging jet (for $r < r_m$), Holmes and Oliver (2000) proposed an empirical model of downburst outflow in the form

$$u_r = \begin{cases} u_m \dfrac{r}{r_m} e^{-\frac{t}{T}}, & r < r_m \\ u_m e^{-\left(\frac{r-r_m}{R_c}\right)^2} e^{-\frac{t}{T}}, & r \geq r_m. \end{cases} \tag{91}$$

Here, u_r is the radial velocity, u_m is the maximum radial velocity, r is the radial distance measured from downburst center, r_m is the radius of u_m, and R_c is the radial length scale. The factor $e^{-t/T}$ extends the model to the time domain and accounts for downburst decay with time (t). The time $t = 0$ is at the moment when the downburst reaches peak intensity, and T is the time scaling constant. Holmes and Oliver (2000) compared their model against the Andrews Air Force Base downburst that occurred on 1 August 1983 and showed a good match between the anemometer measurements and the proposed model. In this comparison, the translation speed of the cloud was added as a simple vector addition to u_r, similar to the approach implemented in other expressions from the OBV family of models. Last, the model does not account for the change of wind speed with height.

3.1.4.1. XU- HANGAN-YU MODEL

This model provides an exact inviscid analytical solution for a family of impinging jets applicable to round, planar, or annular jets and combined with boundary layer empirical solutions applicable to laminar and turbulent boundary layers (Xu et al. 2008). The inflow (W_∞) for this family of impinging jets is expressed as

$$-W_\infty = e^{\frac{-r_*^2}{k}}, \tag{92}$$

where k is the flow rate multiplier and $r_* = r/R$ is the non-dimensional radial distance. This model extended the analytical formulation of Lee et al. (1998), which provided for the first time an analytical solution for a simple $(k = 1)$ Gaussian impinging jet. The extended model of Xu, Hangan and Yu gives the following radial and axial velocity solutions:

$$u_*(r_*, z_*) = r_* e^{-r_*^2/k} \sum_{n=1}^{\infty} c_n L_n^1\left(\frac{2r_*^2}{k}\right) \sqrt{8n/k} e^{-\sqrt{8n/k} z_*} \tag{93}$$

$$w_*(r_*, z_*) = -e^{-r_*^2/k} + 2\left(1 - \frac{r_*^2}{k}\right) e^{-r_*^2/k} \sum_{n=1}^{\infty} c_n L_n^1\left(\frac{2r_*^2}{k}\right) e^{-\sqrt{8n/k} z_*} \tag{94}$$

$u_\star = u/u_m$, $w_\star = w/w_m$ are the non-dimensional radial and vertical velocities (u_m and w_m are their maximum values at r_m and jet centerline, respectively), $z_\star$ is the non-dimensional height (normalized to half-width of plain jet slot), and L_n^1 are the associated Laguerre polynomials. The coefficients c_n are expressed as

$$c_n = \frac{\int_0^\infty \frac{k}{2}\left(1-e^{-r_\star^2/k}\right)L_n^1\left(\frac{2r_\star^2}{k}\right)e^{-r_\star^2/k}\, r_\star dr_\star}{\int_0^\infty r_\star^2 e^{-r_\star^2/k}\left(L_n^1\left(\frac{2r_\star^2}{k}\right)\right)^2 r_\star dr_\star} \tag{95}$$

Within the boundary layer $(z<\delta)$, a polynomial approximation by Homman is used in conjunction with empirical solutions for laminar or turbulent boundary layer depth $\delta(r)$ via

$$u(r,\eta) = \left(-1.43\eta^4 + 5.39\eta^3 - 7.45\eta^2 + 4.49\eta\right)u_m, \tag{96}$$

where $\eta = z/\delta$ and u_m is the maximum velocity.

This model showed good agreement with experiments by Landreth and Adrian (1990) in terms of flow and surface pressure fields for both laminar and turbulent boundary layers, as well as for different Reynolds numbers.

3.2. Ring-Vortex Models

While the OB model (Oseguera and Bowles 1988) was formulated utilizing the boundary layer theory and stagnation flow principles, that model does not account for the existence of single or multiple annular vortex rings that descend from higher elevations together with the descending air current. These ring vortices develop due to Kelvin-Helmholtz instability that acts in the shear layer between the downdraft and the surrounding ambient air. Once the downdraft hits the surface, the vortices continue to spread radially outward from the downdraft in a starburst pattern. This section summarizes several analytical models of downbursts that account for the existence of ring vortices.

3.2.1. *Doublet Model of Zhu and Etkin*

Zhu and Etkin (1985) proposed a three-dimensional model of downbursts based on a circular doublet sheet of changing intensity (Figure 10.5). The model considers a doublet distribution of strength $\sigma(\xi,\eta)$ per unit area on the surface S at height H. In the three-dimensional Cartesian space (x,y,z), this doublet configuration (Figure 10.5) generates the following velocity components at a point $P(x,y,z)$:

$$u_x = -\frac{3}{4\pi}\iint_S \sigma(\xi,\eta)\frac{(z-H)(x-\xi)}{d^5}d\xi d\eta \tag{97}$$

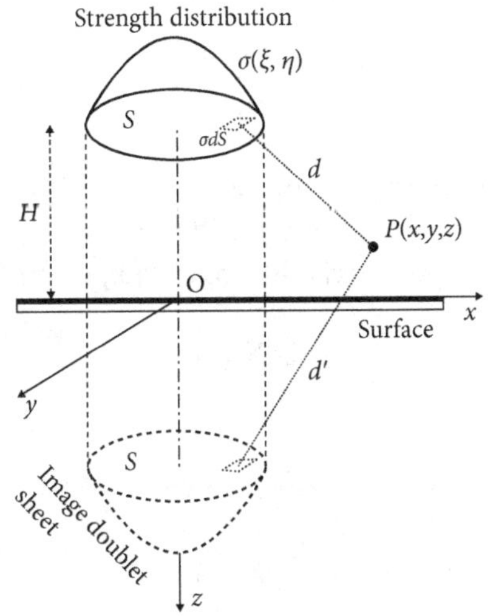

FIGURE 10.5 The doublet sheet that is above the surface is mirrored by a second doublet sheet below the surface level, with an opposite sign to satisfy the impermeable boundary condition at the surface. Therefore, for the image doublet, H is replaced by $-H$ and σ by $-\sigma$ in Equations (97)–(99).

$$u_y = -\frac{3}{4\pi}\iint_S \sigma(\xi,\eta)\frac{(z-H)(y-\xi)}{d^5}d\xi d\eta \tag{98}$$

$$u_z = -\frac{1}{4\pi}\iint_S \sigma(\xi,\eta)\left(\frac{3(z-H)^2}{d^5}-\frac{1}{r^5}\right)d\xi d\eta, \tag{99}$$

where $d^2 = (x-\xi)^2 + (y-\eta)^2 + (z-H)^2$ and (ξ,η) are coordinates of a singularity. However, instead of a single doublet, as described earlier, the authors further used the linear superposition principle of potential flows. They proposed a series of uniform concentric doublet sheets of different radii and strengths to represent better the distribution of observed velocities in an idealized downburst flow field.

Their trial-and-error tests showed that the best strength distribution of the doublet is in the form

$$\sigma(r) = \sigma_m\left(1-\sin\left(\frac{d}{r_0}\right)\right), \tag{100}$$

where r_0 is the radius of the doublet sheet and σ_m is the maximum strength. This distribution of doublet sheet strengths is then approximated as a superposition of N different constant-strength doublets as

$$\sigma(r) = \sum_{j=1}^{N}\sigma_j, \tag{101}$$

where each doublet is in the form

$$\sigma_i = k\cos\left(\frac{\pi}{2}\right)\left(\frac{r_i}{r_0}\right) \tag{102}$$

and $k = \sigma_m / N$ is a constant that can be fine-tuned to replicate the desired downburst intensity. The incorporation of this superposition method into the model induces the new downburst velocity components at the point $P(x, y, z)$ as

$$u_x = \sum_{i=1}^{N} \sigma_i u_{xo_i} \tag{103}$$

$$u_y = \sum_{i=1}^{N} \sigma_i u_{yo_i} \tag{104}$$

$$u_z = \sum_{i=1}^{N} \sigma_i u_{zo_i}, \tag{105}$$

where each of u_{xo_i}, u_{yo_i}, and u_{zo_i} can be evaluated using Equations (97)–(99), and the subscript o is used to indicate the distance from the origin (see Figure 10.5). While this model is elegantly formulated, the integration of Equations (97)–(99) results in cumbersome expressions (see appendix in Zhu and Etkin 1985) that are additionally complicated due to their summation over N terms. These complications were relaxed in the model of Ivan (1986) that is presented in the following section.

3.2.2. *Ring-Vortex Model of Ivan*

To account for annular vortex dynamics, Ivan (1986) developed a ring-vortex model with application to aircraft flight simulations. Indeed, the early research on downbursts, including the previously discussed model of Zhu and Etkin (1985), was mostly motivated by numerous aircraft accidents across the United States that were caused by downburst outflows. Ivan's (1986) mathematical model of downbursts is based on the concept of two ring vortices and a stream function for a ring-vortex downburst. Instead of a cosine radial distribution of the doublet singularity intensity that required complicated integrations in the model of Zhu and Etkin (1985), the formulation of Ivan (1986), which relies on the stream function of the ring-vortex, enables the derivation of an elegant closed-form solution of the velocity flow field.

Consider an isolated three-dimensional ring vortex with the vortex radius r_v and the toroidal core radius r_d situated at height z_v above the surface (Figure 10.6). The downward velocity in the vortex core is v_v and, moreover, the flow is symmetric around the vertical axis (z). The stream function at a point $P(x, y, z)$ in the flow of a single vortex ring can be expressed as (Lamb 1945)

$$\psi = -\frac{\Gamma}{2\pi}(r_1 + r_2)\big(F(k_r) - E(k_r)\big), \tag{106}$$

where Γ is the circulation, r_1 and r_2 represent the closest and farthest distances of the point P from the ring vortex, respectively, and $F(k_r) - E(k_r)$ is a combination of complete elliptic integrals of the modulus $k_r = (r_2 - r_1)/(r_2 + r_1)$.

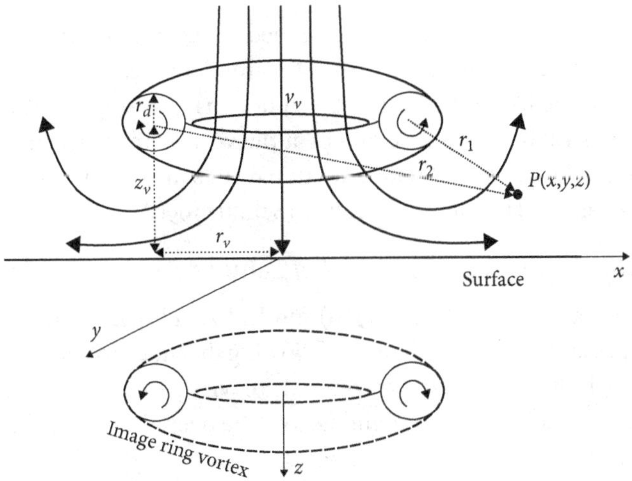

FIGURE 10.6 Geometric details and flow field in the vortex-ring model. Notice the conceptual similarity with the doublet sheet model in Figure 10.5.

In the range of $0 \le k_r^2 \le 1$, the following approximation is very accurate:

$$F(k_r) - E(k_r) \approx \frac{0.788 k_r^2}{0.25 + 0.75\sqrt{1 - k_r^2}}. \tag{107}$$

Ivan (1986) further showed that Γ can be expressed in terms of centerline downdraft velocity (v_v) as

$$\Gamma = \frac{2 v_v r_v}{1 - \dfrac{1}{\left(1 + \left(\dfrac{2 z_v}{r_v}\right)^2\right)^{1.5}}}. \tag{108}$$

Accounting for the mirror (see Figure 10.6) vortex that is needed to satisfy the boundary condition at the surface and knowing the closed-form expression for Ã from Equation (108), the resulting stream function at the point P can be expressed as

$$\psi = -\frac{\Gamma}{2\pi}\left(r_{1,r} + r_{2,r}\right)\frac{0.788 k_{r,r}^2}{0.25 + 0.75\sqrt{1 - k_r^2}} - \left(r_{1,i} + r_{2,i}\right)\frac{0.788 k_{r,i}^2}{0.25 + 0.75\sqrt{1 - k_i^2}}, \tag{109}$$

where the second subscripts "r" and "i" stand for the real and image vortex rings, respectively. The radial and vertical velocity components can be derived from the stream function as

$$u_r = \frac{1}{r}\frac{\partial \psi}{\partial z} \tag{110}$$

$$u_z = -\frac{1}{r}\frac{\partial \psi}{\partial r}. \tag{111}$$

The values of u_r and u_z are then obtained using numerical differentiation of Equations (110) and (111). The results from this model nicely resembled the primary vortex observed in the JAWS field project (Fujita 1981). It is important to note that the preceding model is valid everywhere except at the vortex ring filament and the central axis. In both cases, the numerical differentiation results in a singularity. In this case, the velocities can be replaced by their limit values (Schultz 1990).

3.2.3. *A Multiple Ring-Vortex Model of Schultz*

This model is an extension of Ivan's (1986) model to multiple vortex rings, with each vortex being represented as a ring filament following the modified potential flow theory. In this model, Schultz (1990) also accounted for viscosity effects at the core of the primary vortex using a velocity damping factor. The overall wind field is obtained as a summation of the wind fields of all vortex rings.

In the Cartesian coordinate system with the origin at (X,Y,Z), the velocity components at a point $P(x,y,z)$ can be expressed as

$$u_x = \frac{x-X}{r_h^2}\frac{\partial \psi}{\partial z} \tag{112}$$

$$u_y = \frac{y-Y}{r_h^2}\frac{\partial \psi}{\partial z} \tag{113}$$

$$u_z = -\frac{1}{r_h^2}\frac{\partial \psi}{\partial r}, \tag{114}$$

where $r_h = \sqrt{(x-X)^2 + (y-Y)^2}$ is the distance of P from the central axis of the vortex ring, and ψ is the stream function defined as in Ivan (1986). As discussed in Section 3.2.2, Equations (112)–(114) are ill-defined along the centerline of the vortex ring as well as on the vortex filament. The singularity along the central axis is removed by forcing the find field to its limit value, that is:

$$u_x = 0 \tag{115}$$

$$u_y = 0 \tag{116}$$

$$u_z = \frac{\Gamma r_v^2}{2}\left(\left(r_v^2 + (Z+z)^2\right)^{-\frac{3}{2}} - \left(r_v^2 + (Z-z)^2\right)^{-\frac{3}{2}}\right), \tag{117}$$

where Γ is the circulation as defined in Section 3.2.2, and r_v is the vortex ring radius.

The singularity that occurs in the vortex core is handled by introducing a term that modifies the otherwise purely potential flow model. The potential vortex filament is replaced with a viscous core of finite dimensions in which the radial velocity distribution follows the pattern of a spreading line vortex in a viscous fluid. This approach distributes the vorticity away from its high concentration at the vortex filament while still preserving the total circulation in the ring. Also, a wind field far from the ring is not affected by this modification because vorticity is confined to the close vicinity of the vortex filament.

An empirically derived velocity damping factor (Y) that ensures a continuous wind field everywhere in the flow is expressed as

$$\mathrm{Y}=1-\mathrm{e}^{\left(\frac{\left(\frac{r_1}{r_v}\right)^2}{\left(0.4215\frac{d_v}{r_v}+0.0822\left(\frac{d_v}{r_v}\right)^2-0.0969\left(\frac{d_v}{r_v}\right)^3\right)^2}\right)}, \tag{118}$$

where d_v is the filament viscose core diameter and r_1 is the shortest distance from P to the filament (see Section 3.2.2). Therefore, this model is governed by the following six parameters $\left(X,Y,Z,r_v,d_v,\Gamma\right)^{\mathrm{T}}$, where $(\cdot)^{\mathrm{T}}$ is the transpose operator.

The number of rings is arbitrary and, in principle, should be chosen in a way that matches the observed downburst to be reconstructed. Addition of each ring adds six more parameters (i.e., $\left(X,Y,Z,r_v,d_v,\Gamma\right)^{\mathrm{T}}$) to the model, and their estimates are obtained by minimizing the squared error cost function J:

$$J=\frac{1}{N}\sum_{j=1}^{N}\boldsymbol{e}_j^{\mathrm{T}}(\boldsymbol{p})\boldsymbol{B}\boldsymbol{e}_j(\boldsymbol{p}). \tag{119}$$

Here, N is the number of vortex pair rings, $\boldsymbol{B}$ is the matrix of weighing coefficients, $\boldsymbol{e}_j(\boldsymbol{p})=\boldsymbol{v}_o-\boldsymbol{v}_m$ is the difference between the observed $(\boldsymbol{v}_o)$ and modeled $(\boldsymbol{v}_m)$ wind velocity at the point i (and a given time instant), and $\boldsymbol{p}$ is the vector of model parameters. Schultz (1990) used the Newton-Raphson method to minimize J after the initial guess of model parameters. This model was later used by Zhao and Bryson (1990) to calculate the optimal path of an aircraft flying through a downburst. The model was also improved by Jesson and Sterling (2018) who added the secondary vortex that develops at the leading edge of the primary vortex.

4. Conclusion

We have brought together an exhaustive number of analytical and semi-empirical models related to tornadic and downburst wind systems and presented them in an organized way, stressing their assumptions and main outcomes as well as their limitations. Most of these models only characterize the mean velocity fields. While their applicability is important mostly for risk-related analysis, these models need to be further tested against experimental or numerical simulations. In the future they should also be complemented with translation, surface roughness, and mostly turbulence characteristics, which are all important aspects.

REFERENCES

Abd-Elaal, E.-S., Mills, J. E., & Ma, X. (2013). An analytical model for simulating steady state flows of downburst. *Journal of Wind Engineering and Industrial Aerodynamics*, *115*, 53–64. doi:10.1016/j.jweia.2013.01.005

Abd-Elaal, E.-S., Mills, J. E., & Ma, X. (2014). Empirical models for predicting unsteady-state downburst wind speeds. *Journal of Wind Engineering and Industrial Aerodynamics*, *129*, 49–63. doi:10.1016/j.jweia.2014.03.011

Ashton, R., Refan, M., Iungo, G. V., & Hangan, H. (2019). Wandering corrections from PIV measurements of tornado-like vortices. *Journal of Wind Engineering and Industrial Aerodynamics*, *189*, 163–172. doi:10.1016/j.jweia.2019.02.010

Baker, C. J., & Sterling, M. (2017). Modelling wind fields and debris flight in tornadoes. *Journal of Wind Engineering and Industrial Aerodynamics*, *168*, 312–321. doi:10.1016/j.jweia.2017.06.017

Burgers, J. M. (1948). A mathematical model illustrating the theory of turbulence. In R. Von Mises and T. Von Kármán (Eds.), *Advances in applied mechanics* (pp. 171–199). Amsterdam, Netherlands: Elsevier.

Chay, M. T., Albermani, F., & Wilson, R. (2006). Numerical and analytical simulation of downburst wind loads. *Engineering Structures*, *28*(2), 240–254. doi:10.1016/j.engstruct.2005.07.007

Chen, L., & Letchford, C. W. (2004). A deterministic–stochastic hybrid model of downbursts and its impact on a cantilevered structure. *Engineering Structures*, *26*(5), 619–629. doi:10.1016/j.engstruct.2003.12.009

Devenport, W. J., Rife, M. C., Liapis, S. I., & Follin, G. J. (1996). The structure and development of a wing-tip vortex. *Journal of Fluid Mechanics*, *312*, 67–106. doi:10.1017/S0022112096001929

Förthmann, E. (1936). *Turbulent jet expansion* (Technical memorandum). Washington, DC: NACA.

Fujita, T. T. (1978). *Workbook of tornadoes and high winds for engineering applications*. Department of the Geophysical Sciences. Chicago: University of Chicago Press.

Fujita, T. T. (1981). Tornadoes and downbursts in the context of generalized planetary scales. *Journal of the Atmospheric Sciences*, *38*(8), 1511–1534. doi:10.1175/1520-0469(1981)038<1511:TADITC>2.0.CO;2

Hjelmfelt, M. R. (1988). Structure and life cycle of microburst outflows observed in Colorado. *Journal of Applied Meteorology*, *27*(8), 900–927. doi:10.1175/1520-0450(1988)027<0900:SALCOM>2.0.CO;2

Holmes, J. D., & Oliver, S. E. (2000). An empirical model of a downburst. *Engineering Structures*, *22*(9), 1167–1172. doi:10.1016/S0141-0296(99)00058-9

Ivan, M. (1986). A ring-vortex downburst model for flight simulations. *Journal of Aircraft*, *23*(3), 232–236. doi:10.2514/3.45294

Jesson, M., & Sterling, M. (2018). A simple vortex model of a thunderstorm downburst—A parametric evaluation. *Journal of Wind Engineering and Industrial Aerodynamics*, *174*, 1–9. doi:10.1016/j.jweia.2017.12.001

Karami, M., Hangan, H., Carassale, L., & Peerhossaini, H. (2019). Coherent structures in tornado-like vortices. *Physics of Fluids*, *31*(8), 085118. doi:10.1063/1.5111530

Karami, M., Hangan, H., Carassale, L., & Refan, M. (2018). *An approximate model for large-scale fluctuations in tornado-like vortices*. Paper presented at the XV Conference of the Italian Association for Wind Engineering (In-Vento 2018). Napoli, Italy.

Kim, Y. C., & Matsui, M. (2017). Analytical and empirical models of tornado vortices: A comparative study. *Journal of Wind Engineering and Industrial Aerodynamics, 171*, 230–247. doi:10.1016/j.jweia.2017.10.009

Knowles, K., & Myszko, M. (1998). Turbulence measurements in radial wall-jets. *Experimental Thermal and Fluid Science, 17*(1), 71–78. doi:10.1016/S0894-1777(97)10051-6

Kuo, H. L. (1971). Axisymmetric flows in the boundary layer of a maintained vortex. *Journal of the Atmospheric Sciences, 28*(1), 20–41. doi:10.1175/1520-0469(1971)028<0020:AFITBL>2.0.CO;2

Lamb, S. H. (1945). *Hydrodynamics*. New York: Dover.

Landreth, C. C., & Adrian R. J. (1990). Impingement of a low Reynolds number turbulent circular jet onto a flat plate at normal incidence. *Experiments in Fluids, 9*(1–2), 74–84. doi:10.1007/BF00575338

Lee, J. A., Burggraf, O. R., & Conlisk, A. T. (1998). On the impulsive blocking of a vortex-jet. *Journal of Fluid Mechanics, 369*, 301–331. doi:10.1017/S0022112098001815

Li, C., Li, Q. S., Xiao, Y. Q., & Ou, J. P. (2012). A revised empirical model and CFD simulations for 3D axisymmetric steady-state flows of downbursts and impinging jets. *Journal of Wind Engineering and Industrial Aerodynamics, 102*, 48–60. doi:10.1016/j.jweia.2011.12.004

Myers, G. E., Schauer, J. J., & Eustis, R. H. (1963). Plane turbulent wall jet flow development and friction factor. *Journal of Basic Engineering, 85*(1), 47–53. doi:10.1115/1.3656534

Oseguera, R. M., & Bowles, R. L. (1988). *A simple, analytic 3-dimensional downburst model based on boundary layer stagnation flow* (Technical memorandum no. 100632). Hampton, VA: NASA Langley Research Center.

Rajaratnam, N. (1976). *Turbulent jets*. Amsterdam, Netherlands: Elsevier.

Rankine, W. J. M. (1858). *A manual of applied mechanics*. R. Griffin.

Rott, N. (1958). On the viscous core of a line vortex. *Zeitschrift für angewandte Mathematik und Physik ZAMP, 9*(5), 543–553. doi:10.1007/BF02424773

Schultz, T. A. (1990). Multiple vortex ring model of the DFW microburst. *Journal of Aircraft, 27*(2), 163–168. doi:10.2514/3.45913

Schwarz, W. H., & Cosart, WP. (1961). The two-dimensional turbulent wall-jet. *Journal of Fluid Mechanics, 10*(4), 481–495. doi:10.1017/S0022112061000299

Sigalla, A. (1958). Measurements of skin friction in a plane turbulent wall jet. *The Aeronautical Journal, 62*(576), 873–877. doi:10.1017/S0368393100070231

Sullivan, R. D. (1959). A two-cell vortex solution of the Navier-Stokes equations. *Journal of the Aerospace Sciences, 26*(11), 767–768. doi:10.2514/8.8303

Vatistas, G. H., Kozel, V., & Mih, W. C. (1991). A simpler model for concentrated vortices. *Experiments in Fluids, 11*(1), 73–76. doi:10.1007/BF00198434

Verhoff, A. (1963). *The two-dimensional turbulent wall jet with and without an external stream* (Report no. 626). Princeton, NJ: Princeton University Press.

Vicroy, D. D. (1991). *A simple, analytical, axisymmetric microburst model for downdraft estimation* (Technical memorandum no. 104053). Hampton, VA: NASA Langley Research Center.

Vicroy, D. D. (1992). Assessment of microburst models for downdraft estimation. *Journal of Aircraft, 29*(6), 1043–1048. doi:10.2514/3.46282

Wen, Y-K. (1975). Dynamic tornadic wind loads on tall buildings. *Journal of the Structural Division. ASCE, 101*(1), 169–185.

Wilson, J. W., Roberts, R. D., Kessinger, C., & McCarthy, J. (1984). Microburst wind structure and evaluation of Doppler radar for airport wind shear detection. *Journal of Climate and Applied Meteorology, 23*(6), 898–915. doi:10.1175/1520-0450(1984)023<0898:MWSAEO>2.0.CO;2

Wood, G. S., Kwok, K. C. S., Motteram, N. A., & Fletcher, D. F. (2001). Physical and numerical modelling of thunderstorm downbursts. *Journal of Wind Engineering and Industrial Aerodynamics*, *89*(6), 535–552. doi:10.1016/S0167-6105(00)00090-8

Wood, V. T., & White, L. W. (2011). A new parametric model of vortex tangential-wind profiles: Development, testing, and verification. *Journal of the Atmospheric Sciences*, *68*(5), 990–1006. doi:10.1175/2011JAS3588.1

Xu, Z., & Hangan, H. (2009). An inviscid solution for modeling of tornadolike vortices. *Journal of Applied Mechanics*, *76*(3), 031011-031011–5. doi:10.1115/1.3063632

Xu, Z., Hangan, H., & Yu, P. (2008). Analytical solutions for a family of Gaussian impinging jets. *Journal of Applied Mechanics*, *75*(2). doi:10.1115/1.2775502

Zhao, Y., & Bryson, A. E. (1990). Optimal paths through downbursts. *Journal of Guidance, Control, and Dynamics*, *13*(5), 813–818. doi:10.2514/3.25406

Zhu, S., & Etkin, B. (1985). Model of the wind field in a downburst. *Journal of Aircraft*, *22*(7), 595–601. doi:10.2514/3.45171

CHAPTER 11

MESOSCALE, MICROSCALE, AND NUMERICAL MODELS

Limitations and Future Evolution

PATRICK HAWBECKER

1. Introduction

Non-synoptic wind storms (NSWSs), including thunderstorm outflow, downbursts, and tornadoes, are capable of producing widespread structural damage and destruction. In a 2017 report by Munich Re, it was estimated that nearly 60 percent of the insured losses (a figure totaling $14 billion) within the United States were caused by severe thunderstorm damage in 2016. Further, the annual average loss from severe convective storms was estimated to be $11.23 billion based on Property Claim Services from a Willis Re 2017 report. This large amount of loss is in part not only due to the extreme wind produced by these NSWS events, but also because of the lack of design guidelines or code implementations for these catastrophic events (Orwig & Schroeder, 2007).

In order to generate these guidelines, it is imperative to have an accurate representation of the flow field during NSWS events through observations, physical models, or numerical models. This is very difficult to obtain with routine observations alone. Typical routine observational platforms include point measurements from surface stations and mesonets, vertical profiles from radiosondes (weather balloons) and meteorological towers, and volumetric data from remote sensing equipment such as radars and lidars. These platforms are extremely valuable in characterizing the near-surface winds and large-scale properties of the parent storms yet fail to obtain complete spatial coverage of the small-scale structures, including tornadoes and downbursts. Highly advanced remote sensing equipment available today, such as mobile radars (Weiss, Schroeder, Guynes, Skinner, & Beck, 2009; Wurman, Straka, Rasmussen, Randall, & Zahrai, 1997), allows for these events to be characterized at much higher fidelity than in years past and has allowed for observations that were not previously recorded. These data are invaluable for analysis and model validation; however, it remains very difficult, if not impossible, to capture the full wind field associated with NSWS events at low levels to be utilized within the wind engineering models due to noise and ground clutter (Oye, Mueller, & Smith, 1995; Rhee & Lombardo, 2018). Physical simulations can complement this surface layer given that they are calibrated on full-scale data (Refan & Hangan, 2018). However, like microscale computational fluid dynamics (CFD) models, physical simulations are generally limited in their general physics as well as in scale or Reynolds number.

Historically, one of the most common means for deducing the near-surface wind speeds from these storms was from observations of the damage that occurred. The Fujita, or F, Scale from Fujita (1971, 1981) introduced a way to relate structural damage to the wind speeds needed to inflict the damage based on design codes. This rating system became widely used to rate the intensity of tornadoes and downbursts and was eventually updated to the Enhanced Fujita, or EF, Scale (McDonald & Mehta, 2006). While the EF Scale is still actively in use, several studies have pointed out how indirect estimations of extreme winds based on these damage–wind relationships can be imprecise and uncertain (Doswell, Brooks, & Dotzek, 2009; Doswell & Burgess, 1988; Marshall, 2002; Reynolds, 1971), making other methods critical to determine the actual wind fields in NSWS events. Due to this, high-resolution numerical modeling is proving to be a very useful tool that is capable of serving as a surrogate to observational data (Dahl, Nolan, Bryan, & Rotunno, 2017) for many applications.

Numerical model results provide a prediction of the wind fields by solving the equations of motion. They can provide a full volume of gridded data with several diagnostic variables not easily measured in the real atmosphere. The method for solving the governing equations can help to determine the appropriate application for the solver with Reynolds-averaged Navier–Stokes (RANS) models typically utilized for mesoscale purposes (and sometimes high-fidelity microscale simulations for engineering purposes), and large-eddy simulation (LES) models typically utilized for microscale purposes (see Section 2). Applications of these numerical models range from simulating the parent storm and mesoscale processes for forecasting, dynamics inspection, and sensitivity experiments (Section 3) to simulating the flow field of NSWSs at high fidelity for turbulence characterization (Section 4). The types of microscale solvers discussed here are separated into categories consisting of idealized source models in which microphysical processes are excluded and those that consider microphysics, known as cloud models. The uses and limitations of each model are discussed, followed by a summary of open research questions and future directions involving the numerical modeling of NSWS events in Section 5.

2. RANS AND LES TECHNIQUES

The Navier–Stokes equation describes the momentum balance for fluids and is one of the principle equations that are solved within numerical models of atmospheric flows. Turbulent systems require mass and energy balances as well (De Visscher, 2013, Ch. 10). However, for the explanation of the differences between RANS and LES techniques, only the Navier–Stokes equations are discussed here.

Directly solving the Navier–Stokes equations (what is known as direct numerical simulation, DNS) is presently impossible for applications of NSWS. Utilizing DNS would include solving for the flow across all scales from the Kolmogorov scale (on the order of 1 mm) to the mesoscale (on the order of 10 km) or even synoptic scales (on the order of 1,000 km or more). As shown by De Visscher (2013, Ch. 10), even considering an example with five orders of magnitude in each direction, one would need a numerical grid with 10^{15} nodes and a time step smaller than 0.001 second, resulting in over 100 million time steps to run the code for 1 day of simulation. Thus, it is necessary to break the problem into the scales to be resolved and the scales to be parameterized.

Two of the most common ways in which the scales can be separated are (1) through applying the Reynolds-averaging equations on all variables in the Navier–Stokes equations to remove the time-varying perturbations (the RANS method) and (2) through filtering each variable in the Navier–Stokes equations to remove the subgrid scale signal (the LES method). The fundamental difference between these two techniques is that RANS considers flows in which the ratio of the energy containing turbulence length scale (l) to horizontal grid spacing (Δ) is small ($\Delta \gg l$); thus, none of the turbulence is resolved. LES considers flows in which l/Δ is large; thus, the grid spacing is sufficiently small in order to resolve the energy-containing eddies (Wyngaard, 2004).

2.1. RANS Modeling

The RANS technique consists of separating the time average and turbulent portions of each variable, which reduces the Navier–Stokes equations to mean components (which can be directly solved) and viscous and turbulent stresses (which can be parameterized by the mean values) (De Visscher, 2013, Ch. 10). This technique carries the assumption that the scale of the turbulent perturbations is much smaller than the scale of the mean flow. Consider the following form of the RANS equations from Fröhlich and Von Terzi (2008):

$$\frac{\partial \langle u_i \rangle}{\partial t} + \frac{\partial \langle u_i \rangle \langle u_j \rangle}{\partial x_j} + \frac{\partial \langle p \rangle}{\partial x_i} = \nu \frac{\partial^2 \langle u_i \rangle}{\partial x_j^2} - \frac{\partial \tau_{ij}^{RANS}}{\partial x_j} \tag{1}$$

where angle brackets denote the Reynolds averaging of a variable.

This method of closing the equation set produces a very efficient solution to the governing equations and is valid assuming the grid spacing is larger than the largest turbulent length scales of the flow (Wyngaard, 2004). However, the solution produces, by definition, average values at each grid cell with no turbulent fluctuations (De Visscher, 2013, Ch. 10). In order to close this equation set, the variances and covariances (τ_{ij}^{RANS}) need to be parameterized. Parameterizations for these terms typically range from a first-order closure (resolving only the mean components and parameterizing all variances and covariances) to a 1.5-order closure (taking advantage of the relationship between turbulent kinetic energy and energy dissipation rate in order to calculate some second-order moments prognostically and parameterizing the rest) (De Visscher, 2013, Ch. 10). This allows all turbulent fluctuations to be described by physical quantities as shown by Fröhlich and Von Terzi (2008), where in the K-ϵ model,

$$\tau_{ij}^{RANS} = f\left(\frac{\partial\langle u_i\rangle}{\partial x_i}, \kappa, \varepsilon, C\right) \tag{2}$$

Here, C is a model constant, κ is the turbulent kinetic energy, and the turbulent dissipation rate is denoted by ϵ. Utilizing a closure method such as this, the solution is in fact a steady-state solution but can be unsteady by providing unsteady boundary conditions or source terms (Fröhlich & Von Terzi, 2008). To be said another way, many turbulent features within outflow winds, tornadoes, and other NSWS events that may be important for structural analysis would not show up within a RANS solution (De Visscher, 2013, Ch. 10). For simulations of the main storm structure or a mesoscale simulation of a large-scale NSWS event, this may be reasonable. In high-fidelity simulations, however, the laminar flow field (in the sense that turbulent fluctuations are not present) will be a highly simplified solution and may underpredict or neglect small-scale processes that can impact the large-scale motions.

Within RANS models that are utilized for meteorological purposes, it is necessary to include equations for potential temperature, pressure, as well as moisture and hydrometeor variables (e.g., cloud water, rain, hail, etc.). The microphysical processes will augment the velocity and temperature fields, which is necessary to consider for modeling phenomena such as downbursts, for which the cooling through phase changes and drag from the hydrometeors are the main processes in generating the downward momentum. This adds to the computational expense of the model framework by including more state variables that need to be solved. However, when meeting the RANS assumptions and keeping Δ large (with respect to l), the number of grids is typically small enough such that mesoscale models can still be run rather easily in a predictive or near–real-time mode over entire countries even with the increased number of equations.

2.2. LES Modeling

With LES, a filter is utilized to separate the largest energy-containing eddies in the flow field of interest from the smaller, dissipative eddies. The equation set for LES looks very similar to that of Equation 1 except that the equations represent a spatial filtered solution where the turbulent motions are filtered at some length scale, rather than a time-averaged solution that removes all turbulence (De Visscher, 2013, Ch. 10). Fröhlich and Von Terzi (2008) showed the filtered Navier–Stokes equations as

$$\frac{\partial \overline{u_i}}{\partial t} + \frac{\partial \overline{u_i}\,\overline{u_j}}{\partial x_j} + \frac{\partial \overline{p}}{\partial x_i} = \nu \frac{\partial^2 \overline{u_i}}{\partial x_j^2} - \frac{\partial \tau_{ij}^{LES}}{\partial x_j} \tag{3}$$

where the overbar denotes the spatial filtering. This filter can be any size larger than the model grid size, although simply using the model grid size as the filter size is a viable approach. It was shown by Fröhlich and Von Terzi (2008) that closure within an LES model, known as the subgrid scale (SGS) model, solves the turbulent stresses as a function of the filter size Δ, which can be shown in the Smagorinsky model relationship:

$$\tau_{ij}^{LES} = f\left(\frac{\partial \overline{u_i}}{\partial x_j}, \Delta, C\right) \tag{4}$$

For cloud modeling, it is common to have an equation for the diffusion coefficient of momentum to be a function of some length scale that is dependent on the grid or filter length multiplied by the square root of the subgrid turbulent kinetic energy (Wilhelmson & Wicker, 2001).

What is produced is no longer a time-averaged quantity that is being solved for; the Leonard and cross-term stresses are no longer equal to unity. That is, $\overline{\overline{u_i}} \neq \overline{u_i}$ when utilizing the filter. The inclusion of these terms provides unsteady data, such as fluid–structure coupling, which may be necessary for engineering purposes (Fröhlich & Von Terzi, 2008) and in high-fidelity simulations of NSWS events.

Utilizing this technique is computationally more expensive than RANS; thus, it is common for the omission of moisture from the model altogether. Instead, again considering downbursts as an example, the changes to the temperature and velocity fields can be parameterized as a heat sink rather than directly solving for the hydrometeor fields and the latent heating process involved in order to generate downward momentum.

3. Mesoscale Modeling of NSWS

Mesoscale modeling typically considers the simulations of real events in case study analysis, sensitivity experiments, or within a forecasting capacity. NSWS events such as tornadoes and microbursts cannot be simulated directly from mesoscale models because, by definition, these simulations are run at too coarse of a spatiotemporal resolution (with respect to the scales of tornadoes and microbursts). The main focus within mesoscale simulations is on the parent storm and kilometer-scale features, which would include the larger NSWS events such as macrobursts and derechos. Output is a grid average, which can be easily seen from the governing equations of a typical mesoscale model and thus will lack in any turbulent characteristics. However, the mean profiles that are generated will include the effects of storm dynamics and microphysical processes.

Utilizing mesoscale model data for wind engineering purposes as a whole is not common. There exist studies of the mesoscale effects of the urban canopy Masson (2006), wind energy forecasting improvement with mesoscale ensembles Deppe, Gallus, and Takle (2013), and generating engineering models based on mesoscale output (Xue, Li, Song, Chen, & Wang, 2017), but very few directly connect mesoscale simulations of NSWS events for engineering purposes. This is due to the inherent differences between the two fields, where mesoscale simulations are considering much larger scales than within engineering studies and the direct downscaling of mesoscale information is not a straightforward task. Further, there are several parameterizations within mesoscale models that have large impacts on the wind fields produced by NSWS events that are not typically found within engineering models.

3.1. Parameterization Schemes

There are several processes involved within NSWS events that the mesoscale model does not resolve explicitly and necessitates a parameterization scheme in order to account for the effects on the flow fields. These processes include everything from radiative transfer to the modeling of clouds and cloud microphysics. However, in regard to the wind engineering community, the main parameterization schemes of interest are the land–surface model (LSM), the surface layer scheme, the boundary layer scheme, and, to a lesser extent, the microphysics scheme. A basic understanding of the purpose of these parameterization schemes is provided here with a focus on the sensitivity of the wind field within NSWS events to the scheme.

3.1.1. *Land–Surface Model and Surface Layer Scheme*

In a typical mesoscale model, the lower boundary conditions are specified within the LSM by defining the type of land use category at each model grid, which in turn relates to a surface roughness length, soil moisture, and albedo. The roughness length, or drag, specifications are communicated from the LSM to the surface layer scheme in order to calculate the bulk exchange coefficients and friction velocities. These are then are then utilized to determine the fluxes of momentum, heat, and water vapor between the surface and atmosphere (F. Chen, Janjić, & Mitchell, 1997). Surface layer schemes based on Monin–Obukhov similarity theory (MOST; Monin & Obukhov, 1954) have been extensively utilized within mesoscale models, with the empirically derived stability functions (which are used to calculate the surface fluxes) varying between schemes (Prasad, Srinivas, Naidu, Baskaran, & Venkatraman, 2016). It is with MOST that the logarithmic profile for wind speed can be described in the surface layer, which is paramount within the wind engineering community for structural design (Stull, 1988). Altogether, the LSM and surface layer scheme provide the lower boundary condition for the vertical transport that is calculated within the boundary layer parameterization.

3.1.2. *Boundary Layer Parameterization*

Planetary boundary layer (PBL) parameterizations are utilized within mesoscale models to predict the momentum and scalar fluxes generated by mixing from atmospheric eddies. Many of the PBL parameterizations used within meteorological models are one dimensional and only consider the vertical flux from these eddies and assume horizontal homogeneity. However, a three-dimensional PBL parameterization was recently implemented within the weather research and forecasting (WRF) model based on the turbulence model from Mellor and Yamada (1982) (Kosovic et al., 2016).

Atmospheric eddies are generally confined within the PBL; however, they can exist within all regions of the atmosphere. For example, when a large thunderstorm reaches the tropopause, turbulent structures, or eddies, will exist throughout the storm. Thus, the PBL scheme is applied on each grid cell throughout the atmospheric column (Skamarock et al., 2008) and is the method for turbulence closure within atmospheric RANS models. As mentioned previously, the wind field within mesoscale simulations will be representative of the mean flow field where the effects of turbulence are accounted for by the PBL parameterization. It is clear from this that the PBL parameterization can have large impacts on storm structure and dynamics (Hawbecker, Basu, & Manuel, 2017; Stensrud, 2009; Stensrud & Weiss, 2002) as well as the resulting wind field at low levels that is important within the wind engineering community (Hawbecker et al., 2017).

3.1.3. *Microphysics Parameterization*

Microphysics parameterization schemes simulate moisture processes within the numerical model, including cloud and hydrometeor development and the resulting latent heating and cooling. These processes will then alter the flow field by generating positive or negative buoyancy (from phase changes) and increasing drag induced by the hydrometeor field and when coupled with the radiation scheme, can cause scattering, reflections, shadowing, and absorption. The number of processes that are accounted for within the microphysics scheme will vary between very minimal and very complex. Within the Kessler scheme (Kessler, 1969), only the mixing ratio of water vapor, cloud water, and rain are calculated, and the number concentration of the hydrometeors is assumed to follow a defined curve. The Milbrandt–Yau triple-moment scheme (Milbrandt & Yau, 2005) directly solves for the mixing ratio, number concentration, and radar reflectivity for seven hydrometeor types from cloud water to hail. Generally, the simple microphysics parameterizations have been developed for coarse mesoscale model simulations, while the complex schemes were developed for high-resolution cloud model simulations. As increases in computational power allow for higher resolution simulations, it may become more appropriate for the complex microphysics schemes to be employed (Skamarock et al., 2008). That being said, with increasing complexity comes increased computational expense in memory to store new variables as well as in the number of calculations performed as several new equations are introduced for the phase changes of hydrometeors.

Simulating NSWS events such as downbursts is highly dependent on the microphysics parameterization as it has been estimated that roughly 80 percent of the downward velocity in downbursts can be attributed to latent cooling from hydrometeor phase changes (Houze, 1993). Additionally, it has been found through idealized single-column studies that ice phases are required to generate strong downbursts (Srivastava, 1987). Mesoscale simulations of a downburst-producing thunderstorm show that increasing the complexity of the microphysics parameterization produces a similar result in that the outflow winds are increased once additional ice phase hydrometeors are considered along with the calculation of the drop size distribution (Hawbecker et al., 2017). This same principle has been shown to affect mesocyclone and tornado track accuracy due to their dependence on cold pool development (Dawson, Xue, Milbrandt, & Shapiro, 2015; Dawson, Xue, Milbrandt, & Yau, 2010). Although this is discussed generally within a mesoscale framework, the same arguments and impacts will be applicable in microscale, or cloud modeling, studies.

3.2. Applications

Considering the various NSWS events, a mesoscale model is mostly used for the simulation of the larger scale phenomena, such as bow echoes (Xu, Xue, & Wang, 2015), derechos (Fierro et al., 2014; Toll, Männik, Luhamaa, & Rõõm, 2015), and the parent storms of tornadoes and microbursts, but not the tornadoes or microbursts themselves. The simulations are often of historically significant cases or forecasts and thus can be directly compared to observations in order to validate the model performance. Coupled with observations, mesoscale model climatologies can produce risk maps for areas of severe wind and tornadoes. In the case of wind energy, wind resource and risk assessment can be done with mesoscale modeling, though microscale terrain features can play a significant role in local winds and must be considered. Further, the adjustments to the parameterization schemes can often help to determine which physical processes are most important in the event of interest. Hawbecker et al. (2017) showed the sensitivity of the low-level wind field to the microphysics parameterization in a severe wind event that damaged several commercial wind turbines.

It is also possible to extract environmental variables and soundings from the mesoscale simulations to be used as the boundary and initial conditions within microscale solvers. In axisymmetric, rotational models for idealized tornado simulations, the updraft velocity and rotational speed that are used to drive the model can be extracted from mesoscale simulations of a specific case study. Yao et al. (2018) extracted vertical profiles of wind speed, temperature, and humidity from a mesoscale model with 3-km grid spacing to initialize and drive the boundaries of a cloud model simulation of a thunderstorm producing an EF4 tornado. This approach has benefits over simply using observed atmospheric soundings to initialize the microscale simulations due to the larger spatial coverage of available simulated soundings and the fact that the mesoscale sounding will be a horizontally averaged value (over the model grid) with height. Observed soundings are point measurements and may not be appropriate to assign as the background state of the entire simulation domain as instabilities and turbulence are included within the data. When observed soundings are used, it is common to first stabilize the profiles in order to eliminate any layer instabilities. This is not an issue when mesoscale models are employed for initial and boundary conditions. It is also possible to get closer to the storm environment with a mesoscale model as opposed to with observations. This is not to say that utilizing mesoscale results to drive idealized microscale solvers is always the best approach, but mesoscale models offer the unique ability to predict mean environmental variables and provide a spatiotemporal coverage that is not possible with observations.

At a lower level, several relationships or metrics have been developed to relate environmental characteristics to expected low-level wind speeds for tornadoes and downbursts. This can provide an operational advantage for NSWS events where mesoscale models can provide an estimate to what the maximum wind speed may be, although it will not be explicitly resolved. McCann (1994) developed a relationship through a machine learning algorithm between the atmospheric profiles of temperature and water vapor mixing ratio (which are commonly measured in atmospheric soundings and predicted by mesoscale models) and the maximum outflow wind speed that was observed. Another approach considered using downdraft mass flux as a proxy for small-scale downbursts within a mesoscale simulation (De Meutter et al., 2015). Das, Das, Chowdhury, and Karmakar (2016) and Avgoustoglou et al. (2018) used several metrics to give an idea of the tornadic potential of simulated storms. These metrics, such as storm-relative (environmental) helicity (SRH or SREH, respectively; see Thompson, Mead, & Edwards, 2007), bulk Richardson number shear (BRNSHR), or energy helicity index (EHI), which is a function of convective available potential energy (CAPE) and SRH (Das et al., 2016), can then be utilized to predict the intensity of the tornado.

One of the more recent developments made possible by advances in computational power is the ability to nest from mesoscale simulations directly to microscale. This approach has only recently become possible; thus, there are not many examples of it in the literature. One example for numerical modeling of a NSWS event is that of a tornadic supercell (Mashiko, 2016), where a series of three one-way nested domains, from a 1-km grid spacing mesoscale simulation to a 50-m grid spacing LES, are utilized to simulate one of the most destructive tornadoes recorded in Japan. Elaboration on this approach is provided in Section 4.

By taking advantage of the fact that mesoscale output is a horizontally averaged value, it is possible to take the output and superimpose turbulence in order to produce a reasonable estimate of the full wind field. Although mesoscale models were not utilized for the studies, Nguyen, Manuel, and Veers (2011) and Nguyen and Manuel (2017) investigated the influence of downburst winds on wind turbine loads by adding stochastic turbulence onto the mean wind fields produced by analytical models. The same principle can be applied with output from mesoscale modeling; however, significant downscaling would be necessary.

3.3. Limitations

In terms of the applicability of mesoscale model simulations of NSWS events for wind engineering purposes, several limitations exist. Some of these limitations can be improved over time; however, some limitations of the mesoscale model are inherent and cannot be addressed through mesoscale models alone. Mesoscale, or RANS, models will never provide turbulent structures in the solution, such as tornadoes, shedding vortices around structures, and small circulations within downburst winds. Thus, for many engineering purposes, turbulence would need to be applied retroactively. That being said, the dearth of complete data sets at high spatiotemporal resolution for NSWS events has made it difficult to determine the characteristics of the turbulent wind fields. More observational efforts such as those of Gunter and Schroeder (2015) are needed in order to provide the necessary information for realistic stochastic turbulence to be applied to the mean wind fields. If high spatiotemporal resolution is desired, mesoscale models are simply not the best approach.

Within mesoscale models, the parameterization schemes, themselves, have scales of applicability. For example, the assumptions within the PBL parameterization are commonly assumed to no longer apply for $\Delta_{x,y}$ less than 1 km. Unlike the typical PBL parameterization used within engineering RANS models, the vast majority of PBL parameterizations in meteorological mesoscale models are one dimensional (as opposed to three dimensional in engineering models) due to the relatively cheap computational cost. This may be appropriate for large-scale studies; however, as the simulations increase in fidelity and complexity, it is not clear if this is the best approach (see Kosovic et al., 2016, for an example of a three-dimensional PBL parameterization within the WRF model). Additionally, many of the more simplistic microphysics parameterization schemes were developed with large-scale mesoscale modeling or global models in mind, which would not accurately represent many NSWS events. Meanwhile, many of the complex schemes were intended to be used within cloud-resolving models and thus may not be appropriate for general mesoscale simulations but may be beneficial for the simulation of NSWS events.

Observationally speaking, there is an abundance of measurements within the surface layer as it is the layer of the atmosphere in which we spend the vast majority of our time (Stull, 1988), and the surface layer schemes have been tested within a variety of conditions, including stable, unstable, neutral, and complex terrain (Prasad et al., 2016). However, for NSWS events such as tornadoes and downbursts, the observations are limited, and the surface layer parameterizations are not well substantiated. Further, as these events (especially tornadoes) are smaller in scale than the typical mesoscale model grid, the model's near-surface wind speeds and surface fluxes that are produced within these events may not be representative.

Last, from a fundamental prospective, mesoscale models are attempting to capture all processes involved within the atmosphere, which makes it difficult to isolate processes and attribute the cause and effect of certain phenomena or features. Idealized simulations at small scales will typically allow for a more easily manageable framework and can produce a clearer picture to determine the origin of such features. Certainly, if one desires high-fidelity results, the large spatial extent of the NSWS events requires large simulation domains, which in turn make simulations computationally expensive. Further, the governing equations within mesoscale, or RANS, models are not applicable once the horizontal grid spacing decreases past about 1 km. When numerical modeling gets to this resolution and higher, a different modeling method is required.

4. Microscale Modeling of NSWS

Simulating NSWS events with microscale, or LES, models allows for the possibility of the complex turbulent structures involved within tornadoes and thunderstorm outflow dynamics to be resolved. Horizontal grid spacings on the order of 100 m or smaller are recommended for storm modeling as grid spacings of 1 km—the grid spacing utilized in the earliest storm models—produce results where the subgrid turbulent kinetic energy (TKE) is excessive and do not faithfully capture the kinetic energy spectrum (Bryan, Wyngaard, & Fritsch, 2003). In other words, due to these microscale and cloud models utilizing LES methods, the grid spacing must be such that $\Delta \ll l$, where l is the turbulence length scale of the NSWS event. Further, as discussed previously, the spatial extent of the full NSWS event can be rather large; thus, simulating the full scale of the event at high resolution is extremely computationally expensive, but not impossible (see Section 4.3). For this reason, in the engineering community, the parent storm, or dynamic processes of the parent storm, is often simplified as source terms within the model in order to decrease the domain size requirements set by the extent of the parent storm. It is also common to simplify the governing equations by neglecting microphysical processes, allowing more computational power to be devoted to increasing the spatial resolution. This type of idealized setting is very common within microscale modeling of NSWS events where topography and processes such as radiation or microphysics may also be neglected. By neglecting processes such as these, the simulations are run in a very controlled setting in which specific processes can be studied and analyzed without interference from additional influences.

The initial and boundary conditions for microscale simulations of NSWS events are typically considered to be horizontally homogeneous and specified by a single vertical profile of wind speed, temperature, and, for models that consider moisture, water vapor mixing ratio or relative humidity. This profile can be theoretical or based on composite atmospheric soundings (see Weisman & Klemp, 1982, and Wakimoto, 1985) or can be specified directly from observational data such as radiosondes or vertical scanning lidars. By utilizing an observed profile, one can attempt to re-create an observed storm and can compare to observations but must take care to stabilize the profile such that there are no full layers of instability within the domain. While the majority of studies considered this profile to remain fixed with time, there are techniques available to allow the mean profile to be temporally varying. One such technique known as base-state substitution (BSS) replaces the original background mean profile with a new profile, or series of profiles, over a user-specified time while retaining the perturbations that have already developed (Letkewicz, French, & Parker, 2013).

It is also very common for the environment in NSWS simulations to be laminar, or containing no ambient turbulence. This type of cold starting of simulations has been shown to be very problematic for LES of convective events where the near-surface vertical wind profiles are unrealistic and contain excessive vertical shear with a nonfree slip lower boundary condition (Markowski & Bryan, 2016). Including realistic ambient turbulence could require running the simulations without the NSWS events (either the full storm or the simplified source) for several hours before initializing the event or using a technique to insert turbulence beforehand. This produces an additional computational burden on these simulations and makes the results more convoluted by including the effects of ambient turbulence on the NSWS wind fields. However, without turbulence in the model, the validity of LES comes into question (Markowski & Bryan, 2016).

The LES models are commonly utilized to simulate NSWS events such as downbursts (see Section 4.1) and tornadoes (see Section 4.2). While the same model framework may be utilized to study these events, the model setup and source terms for these are vastly different.

4.1. Cooling Source Models

Downburst winds in the atmosphere are generated through the combination of cooling by hydrometeor phase changes and the drag induced on the air by falling hydrometeors. Houze (1993) estimated that roughly 80 percent of the downward momentum within downbursts can be attributed to latent cooling. Thus, a common way for downbursts to be simulated is by including an elevated cooling source (or heat sink) within the model domain to simulate this process while neglecting hydrometeor drag. The cooling source is typically defined by a cosine-squared function of a specified radius that increases to a certain value at the cooling source center over a given time, first utilized by Anderson, Orf, and Straka (1992). The source produces a column of cold (relative to the ambient environment) air that descends toward the surface and spreads out laterally, generating downburst wind patterns along with the associated structures, such as the ring vortex. This ring vortex is a circular tube of strong horizontal vorticity that is generated at the interface of the cold outflow and the ambient environmental air (see Fujita, 1985).

Domain sizes are typically constrained in the vertical to be well above the specified boundary layer height to leave room for a damping region for vertically propagating waves and in the horizontal by the size of the downburst being modeled. This generally produces domain sizes of O(10) km in the horizontal directions and O(1) km in the vertical. Thus, considering a grid spacing O(10) m, the computational grid will consist of roughly 100 million nodes that will be solved for lengths of time on the order of 10 minutes. Within axisymmetric models, this domain size can be decreased substantially; however, this removes any possible deviations in symmetry when observed downbursts are often asymmetric. Due to the simplicity of the governing equations within idealized simulations such as this, it is possible to generate fairly high spatiotemporal resolution simulations of idealized downbursts with current supercomputers. These numerical results can then be utilized in lieu of the analytical models and laboratory experiments that have historically been utilized to estimate the loads generated by NSWS events on structures such as buildings (X. Chen, 2008; Sengupta, Haan, Sarkar, & Balaramudu, 2008); electricity transmission towers (Chay, Albermani, & Wilson, 2006; Savory, Parke, Zeinoddini, Toy, & Disney, 2001); wind turbines (Nguyen et al., 2011; Nguyen & Manuel, 2015); and wind farms (Nguyen & Manuel, 2013).

A similar modeling framework is utilized for the study of gust fronts as first done by Droegemeier and Wilhelmson (1987); a cooling source was located on a lateral boundary in order to simulate the propagation of a density current. Fujita (1980, 1981) showed how downbursts and gust fronts, or density currents, differ, though the general principles behind these events are similar enough that the modeling framework has few differences. The same assumptions for the cooling source are made where latent cooling from microphysical processes is what generates the dense air; however, the model itself does not need to solve for the microphysics. Because this type of simulation does not include the descent of the cold air, the vertical extent of the domain can be decreased.

4.2. Tornado Models

In order to simulate tornadoes and tornadogenesis at high spatiotemporal resolution, axisymmetric and three-dimensional vortex models have commonly been employed (Wilhelmson & Wicker, 2001). Though most of the initial tornado simulations were conducted with axisymmetric models, Fiedler (1998) found large differences between the axisymmetric simulations and three-dimensional vortex models, with the latter more closely representing observations. With this vortex model setup, the tornado is simulated without explicitly representing the parent storm and associated processes (Wilhelmson & Wicker, 2001). Vortex models are the numerical response to the Ward chamber where tornadoes are simulated in a laboratory setting, with a combination of fans to generate vertical motion and rudders to generate environmental vorticity (Ward, 1972). The numerical technique, known as the Fiedler chamber (Fiedler, 1995; Trapp & Fiedler, 1995), utilizes a constant buoyancy source at the top of the numerical domain with rotating environmental winds in order to emulate the updraft of the parent thunderstorm and the rotation in the atmosphere, respectively (Bluestein, 2017). These types of simulations allow for high spatiotemporal resolution for the tornado and, most importantly, for engineering purposes, of the near-surface wind field. With this technique, suction vortices, which have been theorized to be able to have wind speeds twice that of the parent vortex (Fujita, 1971), are able to be simulated, which increases the representativeness of the simulated tornado to reality (Fiedler, 1998). Lewellen (1993) found that the grid spacing needed for the flow to be relatively insensitive to the turbulent mixing parameters was around 5 m within the tornado and around 10 m throughout the rest of the domain. High-resolution requirements such as this make full-storm simulations of tornadoes incredibly difficult. However, they are well within the realm of possibility for Fiedler chamber simulations. Vortex models are actively used within the research community and are growing in complexity and realism (see, e.g., Bryan, Dahl, Nolan, & Rotunno, 2017). That being said, this framework generates the tornado without the consideration of the rear-flank downdraft and surface cold pool, two important factors within tornadogenesis (Davies-Jones & Brooks, 1993; Trapp & Fiedler, 1995).

Utilizing both a cold bubble (somewhat similar to that used in the downburst simulations in Section 4.1) and a warm cylinder of air, the downdraft and updraft of the parent storm can be represented, respectively (Trapp and Fiedler, 1995). The intersection of these two features generates a realistic representation of the updraft and downdraft couplet and contains most of the physics currently believed to be necessary for tornadogenesis within a supercell thunderstorm (Wilhelmson and Wicker, 2001). Studies with this model framework were able to investigate the impacts of model parameters such as the downdraft intensity and updraft translation speed on tornadogenesis, concluding that tornadogenesis is highly sensitive to each parameter and is confined to a limited region of the parameter space (Trapp and Fiedler, 1995). This modeling framework generally requires larger domains than that of the vortex model, thus, utilizing this framework makes it more difficult to achieve the high spatiotemporal resolution. By utilizing this framework, however, one gets closer to simulating the full spectrum of physical processes within a tornado and during tornadogenesis.

4.3. Storm-Scale Models

Full storm-scale model simulations at spatiotemporal resolution that can be utilized within the engineering community are extremely rare within the published scientific community due to the massive size of the problem at hand (see Orf, Kantor, & Savory, 2012, and Orf, Wilhelmson, & Wicker, 2016, for examples). The domain extent of storm-scale simulation largely depends on the phenomenon of interest. For example, with single-cell convection for downburst or tornado studies, a common domain size is O(10) km, whereas for squall line studies, the lateral extent of the domain could expand into several hundred kilometers. It is common to utilize radiative lateral boundary conditions within these simulations; however, it is important to keep the storm fully contained within the simulation domain such that interactions with this boundary are limited. For periodic domains in infinite squall line simulations, the domain limitations can be smaller relative to when utilizing radiative boundaries. Additionally, as there are large vertical motions, excitement of gravity waves is possible and may contaminate the solution; thus, wave-damping regions or radiative boundaries are typically considered (Wilhelmson & Wicker, 2001). These boundary conditions, while necessary, make it such that a large portion of the computational domain is not acceptable for analysis due to numerical contamination.

Bryan et al. (2003) showed that the solutions for convective simulations do not converge even down to grid spacings of 125 m. This results in the necessary grid spacing to be around 100 m or smaller for convective modeling. Last, thunderstorm lifetimes are on the order of tens of minutes to multiple hours, with the features of interest (e.g., tornadoes and downbursts) having timescales that are much shorter. It is from this that the numerical expense of these problems can be realized where large domains with small grid sizes are run for hours and outputting data at minute, or subminute, intervals. In order to decrease the numerical grid size, mesh refinement or nesting is possible such that a higher resolution mesh is generated at a single location or following the storm (Skamarock & Klemp, 1993).

The majority of the parameterizations utilized in the previously discussed simulation methods apply to storm-scale modeling. However, when considering the simulation of the full storm, it is no longer reasonable to exclude microphysical processes. The principles of microphysics parameterizations as discussed in Section 3.1.3 apply directly to cloud models as well. Many of the more advanced microphysics schemes have been developed within cloud models and are specifically intended to be utilized in high-resolution, storm-scale modeling.

On of the most significant differences between the methodology for full-storm modeling and idealized simulations of any NSWS phenomenon is that the parent storm is triggered, rather than triggering the phenomenon, itself. The ambient environment is prescribed in such a way that the developed storm will then produce the tornado, downburst, or other NSWS phenomenon of interest. The model is initialized by prescribing a vertical profile of wind speed, temperature, and water vapor mixing ratio or relative humidity, as discussed previously, depicting an environment that is favorable for the development of the NSWS event of interest.

The simulations often start with a laminar environment (i.e., no resolved turbulence), where convection is initiated by emulating a thermal through a warm bubble within the boundary layer to generate positive buoyancy (Carpenter, Droegemeier, & Blyth, 1998a; Weisman, Skamarock, & Klemp, 1997). It is common to adjust the mixing ratio within the warm bubble region due to the increase in temperature effectively causing drying of the parcel. Other initialization mechanisms exist, such as constant flux of heat (i.e., a continuous warm bubble) and forced convergence, in which the wind field perturbations are prescribed such that there is convergence at low levels of the boundary layer. Each of these methods attempts to re-create physical processes that are found to generate storms in the atmosphere, though each is sensitive to the physical specifications of the trigger mechanism (Carpenter et al., 1998a; Loftus, Weber, & Doswell, 2008). For example, with the warm bubble approach, the size and intensity of the bubble are seen to directly relate to the size and intensity of the simulated storm (Carpenter et al., 1998a). Due to this, it is important for the initial convection to be analyzed with caution and, assuming a quasi–steady-state solution is achieved, only analyze the simulation after the initial convection. Brooks (1992) argued that even after reaching a quasi–steady state, the solution has a dependence on the initial convection. Thus, a lot of care must be taken in the initialization of convection in order to decrease the impact of the trigger on the end result.

4.4. Limitations

Within the wind engineering community, microscale modeling of NSWS events may be necessary in order to accurately determine the effect of these events on structures such as buildings, transmission towers, and wind turbines. For some applications, it may be sufficient to utilize simple analytical or empirical models, although if turbulent structures are of interest, then these techniques may not suffice. However, there are several limitations to the microscale modeling techniques described that are inherent to the framework itself and in practice.

Within each microscale modeling technique discussed, the domain extents are rather large with respect to the grid spacing required to resolve the features of interest. In order for the LES solution to reproduce the largest eddies in the inertial subrange, the spatial resolution for events such as tornadoes is on the order of 1 m (Bluestein, 2017; Lewellen, 1993). With domain sizes on the order of 10 km, this resolves features spanning four orders of magnitude and can generate a very convoluted result, making it difficult to separate the processes and mechanisms that are active in the simulation. Aside from the computational cost of simply running these simulations, the analysis and visualization of these large data sets becomes exceptionally difficult, and new techniques are being developed regularly in order to manage the file sizes (Orf et al., 2016).

Another limitation lies within the generation of the NSWS event within the model. In order to generate the phenomenon of interest, each of these simulation techniques requires a trigger or source. The sensitivity of the flow field to the specifications of this source, whether it is a warm bubble to trigger convection or a cold bubble to mimic the latent cooling of downbursts, has been shown to be very large, and careful consideration must be taken with these parameters during the experimental design. For many studies, the amplitude of the warm bubble, cooling source, and so on have been chosen based on previous studies or modified through trial and error in order to obtain the desired result.

Within the downburst community, the cooling source that is utilized is more or less universal and based on that of Anderson et al. (1992). This cooling source, however, is only truly representative of dry downbursts that are most common in the high plains of the United States and areas with relatively low humidity. In dry downbursts, all of the evaporation and latent cooling occurs before the downburst reaches the surface. For wet downbursts or microbursts, which are common in the southeastern United States and regions with high humidity, latent microphysical processes will continue throughout the entire downburst duration. No cooling source method has been developed for this downburst type.

Last, microscale simulations of NSWS events are idealized in nature; however, many of the simplifications that have been used historically were due to a lack of computational resources to consider all of the processes at play. As computational power has increased, it is now possible to revisit the assumptions and simplifications that were previously utilized and determine if the simulations can be made more realistic by including more sophisticated parameterizations or including realistic lower boundary conditions. Many of the seminal microscale simulations of tornadoes focused on tornadogenesis and did not consider friction on the lower boundary, rendering the data near the surface useless for the wind engineering community (Bluestein, 2017). As discussed by Bluestein (2017), the role of surface friction and the lower boundary conditions in tornadic studies is not able to be ignored for wind engineering purposes, and it is important to simulate the effects more realistically. Finally, the observed flow field for downbursts, tornadoes, and other NSWS events has been shown to be asymmetric for both the NSWS event and the environment; however, the use of axisymmetric models remains prevalent within the research community. For certain applications, this modeling technique suffices; for engineering purposes, however, these asymmetries may prove important in the design and load modeling of structures.

5. Areas for Improvement and Open Research Topics

Increasing the representativeness of numerical simulations of NSWS events is an active area of research within the meteorological and engineering communities. However, one of the major deficiencies for numerical modeling at high fidelity is that there exists a scant amount of complete observational data sets that can be utilized to validate the model results. Validating the turbulence characteristics of tornado or downburst wind fields is essential for the engineering community in order to develop design codes or low-order models of these events. The same can be said for parameterization validation within mesoscale modeling for such events. For example, the vast majority of PBL and surface layer parameterizations have been developed from, and validated against, simple convective, neutral, and stable boundary layer observations. It remains to be seen how well these parameterization schemes perform, or how they can be improved, for simulations of NSWS events where stable flow and convection can be in constant interaction.

Additionally, a major aspect of LES modeling has been largely ignored within the severe storms and engineering communities: ambient turbulence. Until recently, very few studies have utilized numerical simulations with realistic turbulence in the ambient environment, electing, instead, to simulate the events in a laminar environment. Markowski and Bryan (2016) highlighted the dangers to such an approach where near-surface wind fields—which are most important to the wind engineering community—are shown to be unrealistic when simulated in a laminar environment. Several studies have already begun to include ambient turbulence (see Bryan et al., 2017; Hawbecker, Basu, & Manuel, 2018; and Nowotarski & Markowski, 2016) and have highlighted the importance of such in NSWS simulations. This step helps to bridge the gap between the highly idealized simulations and reality, which in turn allows for potential model validation with observations to be more robust.

Microscale modeling of NSWS events has also shown that the wind field can be very sensitive to the trigger or source characteristics. Due to this, it may be important to develop more robust initialization strategies. Improvements to the cooling source for downburst simulations may be possible by tracking the downburst and associated cooling within storm-scale simulations in order to develop a more realistic spatially and temporally varying source. The warm bubble, which is used to represent a rising thermal that initiates deep convection, is typically utilized to trigger convection within storm-scale models and may also be improved through investigation of the convective boundary layer within LES simulations driven by radiative heating. Studies that have triggered deep convection through surface heating have been conducted previously (Balaji & Clark, 1988; Carpenter, Droegemeier, & Blyth, 1998a, 1998b, 1998c), though not for the purpose of trigger mechanism improvement. By improving the trigger, though the sensitivity will likely still exist, the range of sizes and intensities of the warm bubble can be limited within a range of possibilities as opposed to the current methods of adaptation from past studies or trial and error.

Overall, because the problem of simulating NSWS events for engineering purposes is so large, one of the major near-term goals is to develop robust mesoscale-to-microscale coupling strategies. With this approach, the mesoscale features that are important driving factors within NSWSs can be included in the microscale simulation in order to simulate realistic events. There exist several strategies for this mesoscale-to-microscale coupling. The mesoscale model can simply be scaled down from several kilometers to several meters through nesting without having to change the model framework (Mashiko, 2016). This method is promising as the information on the microscale domain, or domains, can then provide information back to the mesoscale domains in what is known as two-way nesting. A disadvantage to this method is that as the grid spacing gets smaller, some of the parameterizations will become inappropriate and will need to be changed. For example, it is common to turn off the cumulus parameterization at around 4 km. For other parameterizations, such as the PBL parameterization, it is less clear where the parameterization should be turned off or changed to something appropriate for LES methods. Because of this, scale-aware parameterization schemes are being developed such that the scales of applicability are simply built into the scheme, and the contribution will decrease as fidelity increases. Improvements of scale-aware and dynamic parameterization schemes may greatly improve NSWS simulations in both mesoscale and microscale frameworks.

Another option to couple the mesoscale and microscale is to run a microscale solver that is driven by source terms, or at the boundaries, by mesoscale results in an offline modeling framework. With this approach, an idealized microscale simulation can be influenced by the mesoscale simulation in the large scales but does not need to simulate the full extent of the NSWS event with the microscale solver. Further, it is not necessary to utilize the same model for both the microscale and mesoscale simulations with this approach. In many instances, it may be advantageous to utilize a full-physics atmospheric model for the mesoscale simulation but then switch to a more idealized LES model for the microscale simulation.

With these approaches, however, several obstacles must be dealt with. When the mesoscale flow enters the microscale domain, it does so without resolved turbulence at the scale of the microscale grid. Because of this, it may become necessary to perturb the flow at the boundaries such that the fetch (the region at the inflow side of the domain that is without resolved turbulence) is decreased (see, for example, Mirocha, Kosović, & Kirkil, 2014 and Muñoz-Esparza, Kosović, Mirocha, & van Beeck, 2014). It is unclear regarding what perturbation strategies should be utilized when it is a storm entering the boundary. Additionally, when transitioning from mesoscale to microscale, careful consideration must be made in the mapping of the mesoscale data to the microscale domain when terrain is considered and when the model physics or assumptions differ. This applies to the boundary conditions as well as during the initialization process. With effective methods for coupling mesoscale to microscale simulations, realistic simulations of NSWSs can be achieved, which will advance our understanding of these events and help to develop proper engineering regulations.

While this is not an exhaustive set of areas of future research and limitations, future research targeting these and other developments within numerical modeling of NSWS events will benefit the wind engineering community by providing realistic data sets to develop structural standards and regulations, as well as determine the impacts of such events on structures.

Acknowledgments

My work has, thus far, been supported through the National Science Foundation, North Carolina State University, and the National Renewable Energy Laboratory. I would like to thank my collaborators on current and previous projects that have contributed to many of the ideas and concepts presented within this chapter.

References

Anderson, J. R., Orf, L. G., & Straka, J. M. (1992). A 3-D model system for simulating thunderstorm microburst outflows. *Meteorology and Atmospheric Physics*, 49, 125–131.

Avgoustoglou, E., Matsangouras, I., Pytharoulis, I., Kamperakis, N., Mylonas, M., Nastos, P., & Bluestein, H. (2018). Numerical modeling analysis of the mesoscale environment conducive to two tornado events using the COSMO. Gr model over Greece. *Atmospheric Research*, *208*, 148–155.

Balaji, V., & Clark, T. L. (1988). Scale selection in locally forced convective fields and the initiation of deep cumulus. *Journal of the Atmospheric Sciences*, 45, 3188–3211.

Bluestein, H. B. (2017, September). Tornadoes and their parent convective storms. *Oxford Handbooks Online*, pp. 1–58. doi:10.1093/oxfordhb/9780190699420.013.15

Brooks, H. E. (1992). *Operational implications of the sensitivity of modelled thunderstorms to thermal perturbations*. Presented at the Fourth Workshop on Operational Meteorology (pp. 398–407), Whistler, Canada.

Bryan, G. H., Dahl, N. A., Nolan, D. S., & Rotunno, R. (2017). An eddy injection method for large-eddy simulations of tornado-like vortices. *Monthly Weather Review*, 145, 1937–1961.

Bryan, G. H., Wyngaard, J. C., & Fritsch, J. M. (2003). Resolution requirements for the simulation of deep moist convection. *Monthly Weather Review, 131*, 2394–2416.

Carpenter, R. L., Jr., Droegemeier, K. K., & Blyth, A. M. (1998a). Entrainment and detrainment in numerically simulated cumulus congestus clouds. Part I: General results. *Journal of the Atmospheric Sciences, 55*, 3417–3432.

Carpenter, R. L., Jr., Droegemeier, K. K., & Blyth, A. M. (1998b). Entrainment and detrainment in numerically simulated cumulus congestus clouds. Part II: Cloud budgets, *Journal of the Atmospheric Sciences, 55*, 3433–3439.

Carpenter, R. L., Jr., Droegemeier, K. K., & Blyth, A. M. (1998c). Entrainment and detrainment in numerically simulated cumulus congestus clouds. Part III: Parcel analysis. *Journal of the Atmospheric Sciences, 55*, 3440–3455.

Chay, M. T., Albermani, F., & Wilson, R. (2006). Numerical and analytical simulation of downburst wind loads. *Engineering Structures, 28*, 240–254.

Chen, F., Janjić, Z., & Mitchell, K. (1997). Impact of atmospheric surface-layer parameterizations in the new land-surface scheme of the NCEP mesoscale Eta model. *Boundary-Layer Meteorology, 85*, 391–421. doi:10.1023/A:1000531001463

Chen, X. (2008). Analysis of alongwind tall building response to transient nonstationary winds. *Journal of Structural Engineering, 134*, 782–791.

Dahl, N. A., Nolan, D. S., Bryan, G. H., & Rotunno, R. (2017). Using high-resolution simulations to quantify underestimates of tornado intensity from in situ observations. *Monthly Weather Review, 145*, 1963–1982.

Das, M. K., Das, S., Chowdhury, M. A. M., & Karmakar, S. (2016). Simulation of tornado over Brahmanbaria on 22 March 2013 using Doppler weather radar and WRF model, Geomatics. *Natural Hazards and Risk, 7*, 1577–1599.

Davies-Jones, R., & Brooks, H. (1993). Mesocyclogenesis from a theoretical perspective. In C. Church, D. Burgess, C. Doswell, & R. Davies-Jones (Eds.), *The tornado: Its structure, dynamics, prediction, and hazards* (Geophysical Monograph Series, Vol. 79, pp. 105–114). Washington, DC: American Geophysical Union.

Dawson, D. T., Xue, M., Milbrandt, J. A., & Shapiro, A. (2015). Sensitivity of real-data simulations of the 3 May 1999 Oklahoma City tornadic supercell and associated tornadoes to multimoment microphysics. Part I: Storm-and tornado-scale numerical forecasts. *Monthly Weather Review, 143*, 2241–2265.

Dawson, D. T., Xue, M., Milbrandt, J. A., & Yau, M. (2010). Comparison of evaporation and cold pool development between single-moment and multimoment bulk microphysics schemes in idealized simulations of tornadic thunderstorms. *Monthly Weather Review, 138*, 1152–1171.

De Meutter, P., Gerard, L., Smet, G., Hamid, K., Hamdi, R., Degrauwe, D., & Termonia, P. (2015). Predicting small-scale, short-lived downbursts: Case study with the NWP limited-area ALARO model for the Pukkelpop thunderstorm. *Monthly Weather Review, 143*, 742–756.

Deppe, A. J., Gallus, W. A., Jr., & Takle, E. S. (2013). A WRF ensemble for improved wind speed forecasts at turbine height. *Weather and Forecasting, 28*, 212–228.

De Visscher, A. (2013). *Air dispersion modeling: foundations and applications*. New York, NY: Wiley.

Doswell, C. A., III, Brooks, H. E., & Dotzek, N. (2009). On the implementation of the enhanced Fujita Scale in the USA. *Atmospheric Research, 93*, 554–563.

Doswell, C. A., III, & Burgess, D. W. (1988). On some issues of United States tornado climatology. *Monthly Weather Review, 116*, 495–501.

Droegemeier, K. K., & Wilhelmson, R. B. (1987). Numerical simulation of thunderstorm outflow dynamics. Part I: Outflow sensitivity experiments and turbulence dynamics. *Journal of the Atmospheric Sciences, 44*, 1180–1210.

Facts + Statistics: U.S. catastrophes. (2017). © 2017 Munich Re, NatCatSERVICE; Property Claim Services (PCS®)*, a Verisk Analytics® business. As of February 2017. Retrieved from http://www.iii.org/fact statistic/catastrophes us

Fiedler, B. H. (1998). Wind-speed limits in numerically simulated tornadoes with suction vortices. Quarterly *Journal of the Royal Meteorological Society, 124*, 2377–2392.

Fiedler, B. H. (1995). On modelling tornadoes in isolation from the parent storm. *Atmosphere-Ocean, 33*, 501–512, 1995.

Fierro, A. O., Gao, J., Ziegler, C. L., Mansell, E. R., MacGorman, D. R., & Dembek, S. R. (2014). Evaluation of a cloud-scale lightning data assimilation technique and a 3DVAR method for the analysis and short-term forecast of the 29 June 2012 derecho event. *Monthly Weather Review, 142*, 183–202.

Fröhlich, J., & Von Terzi, D. (2008). Hybrid LES/RANS methods for the simulation of turbulent flows. *Progress in Aerospace Sciences, 44*, 349–377.

Fujita, T. T. (1971). *Proposed characterization of tornadoes and hurricanes by area and intensity* (Technical Report SMRP Research Paper 91). Chicago, IL: Department of the Geophysical Sciences, University of Chicago.

Fujita, T. T. (1980, April). *Downbursts and microbursts—An aviation hazard [downdrafts beneath thunderstorms]*. Presented at the Conference on Radar Meteorology, pp. 94–101, Miami Beach, FL.

Fujita, T. T. (1981). Tornadoes and downbursts in the context of generalized planetary scales. *Journal of the Atmospheric Sciences, 38*, 1511–1534.

Fujita, T. T. (1985). *The downburst* (SMRP Research Paper Number 210). Chicago, IL: Department of the Geophysical Sciences, University of Chicago.

Gunter, W. S., & Schroeder, J. L. (2015). High-resolution full-scale measurements of thunderstorm outflow winds. *Journal of Wind Engineering and Industrial Aerodynamics, 138*, 13–26.

Hawbecker, P., Basu, S., & Manuel, L. (2017). Realistic simulations of the July 1, 2011 severe wind event over the Buffalo Ridge Wind Farm. *Wind Energy, 20*, 1803–1822. doi:10.1002/we.2122.

Hawbecker, P., Basu, S., & Manuel, L. (2018). Investigating the impact of atmospheric stability on thunderstorm outflow winds and turbulence. *Wind Energy Science*, 3, 203, 2018.

Houze, R. A., Jr. (1993). *Cloud dynamics* (Vol. 53). New York, NY: Academic Press.

Kessler, E. (1969). On the distribution and continuity of water substance in atmospheric circulations. In *On the distribution and continuity of water substance in atmospheric circulations* (Meteorological Monographs, Vol. 32, pp. 1–84). Boston, MA: American Meteorological Society.

Kosovic, B., Jiménez, P., Haupt, S., Olson, J., Bao, J., Grell, E., & Kenyon, J. (2016). A three-dimensional PBL parameterization for high-resolution mesoscale simulation over heterogeneous and complex terrain. *AGU Fall Meeting Abstracts*, Abstract A21I-05.

Letkewicz, C. E., French, A. J., & Parker, M. D. (2013). Base-state substitution: An idealized modeling technique for approximating environmental variability. *Monthly Weather Review, 141*, 3062–3086.

Lewellen, W. (1993). Tornado vortex theory. In C. Church, D. Burgess, C. Doswell, & R. Davies-Jones (Eds.), *The tornado: Its structure, dynamics, prediction, and hazards* (Geophysical Monograph Series, Vol. 79, pp. 19–39). Washington, DC: American Geophysical Union.

Loftus, A. M., Weber, D. B., & Doswell, C. A., III. (2008). Parameterized mesoscale forcing mechanisms for initiating numerically simulated isolated multicellular convection. *Monthly Weather Review, 136*, 2408–2421.
Markowski, P. M., & Bryan, G. H. (2016). LES of laminar flow in the PBL: A potential problem for convective storm simulations. *Monthly Weather Review, 144*, 1841–1850.
Marshall, T. P. (2002). Tornado damage survey at Moore, Oklahoma. *Weather and Forecasting, 17*, 582–598.
Mashiko, W. (2016). A numerical study of the 6 May 2012 Tsukuba City supercell tornado. Part I: Vorticity sources of low-level and midlevel mesocyclones. *Monthly Weather Review, 144*, 1069–1092.
Masson, V. (2006). Urban surface modeling and the meso-scale impact of cities. *Theoretical and Applied Climatology, 84*, 35–45.
McCann, D. W. (1994). WINDEX—A new index for forecasting microburst potential. *Weather and Forecasting, 9*, 532–541.
McDonald, J. R., & Mehta, K. C. (2006). *A recommendation for an Enhanced Fujita Scale (EF-Scale)*. Lubbock, TX: Wind Science and Engineering Center, Texas Tech University.
Mellor, G. L., & Yamada, T. (1982). Development of a turbulence closure model for geophysical fluid problems. *Reviews of Geophysics, 20*, 851–875.
Milbrandt, J., & Yau, M. (2005). A multimoment bulk microphysics parameterization. Part II: A proposed three-moment closure and scheme description. *Journal of the Atmospheric Sciences, 62*, 3065–3081.
Mirocha, J., Kosović, B., & Kirkil, G. (2014). Resolved turbulence characteristics in large-eddy simulations nested within mesoscale simulations using the Weather Research and Forecasting Model.Â *Monthly Weather Review*,Â *142*, 806–831.
Monin, A., & Obukhov, A. (1954). Basic laws of turbulent mixing in the surface layer of the atmosphere. Contrib. Geophys. *Inst. Acad. Sci. USSR, 151*, e187.
Muñoz-Esparza, D., Kosović, B., Mirocha, J., & van Beeck, J. (2014). Bridging the transition from mesoscale to microscale turbulence in numerical weather prediction models. *Boundary-layer meteorology, 153*, 409–440.
Nguyen, H. H., & Manuel, L. (2013). Thunderstorm downburst risks to wind farms. *Journal of Renewable and Sustainable Energy, 5*, 013120.
Nguyen, H. H., & Manuel, L. (2015). Extreme and fatigue loads on wind turbines during thunderstorm downbursts: The influence of alternative turbulence models. *Journal of Renewable and Sustainable Energy, 7*, 013 102.
Nguyen, H. H., & Manuel, L. (2017, January). *On simulation of a rear-flank downburst with non-stationary turbulence and its influence on wind turbine extreme loads*. Presented at the 35th Wind Energy Symposium, p. 0682, Grapevine, TX.
Nguyen, H. H., Manuel, L., & Veers, P. S. (2011). Wind turbine loads during simulated thunderstorm microbursts, *Journal of Renewable and Sustainable Energy, 3*, 053 104.
Nowotarski, C. J., & Markowski, P. M. (2016). Modifications to the near-storm environment induced by simulated supercell thunderstorms. *Monthly Weather Review, 144*, 273–293.
Orf, L., Kantor, E., & Savory, E. (2012). Simulation of a downburst-producing thunderstorm using a very high-resolution three-dimensional cloud model. *Journal of Wind Engineering and Industrial Aerodynamics, 104–106*, 547–557.
Orf, L., Wilhelmson, R., & Wicker, L. (2016). Visualization of a simulated long-track EF5 tornado embedded within a supercell thunderstorm. *Parallel Computing, 55*, 28–34.
Orwig, K. D., & Schroeder, J. L. (2007). Near-surface wind characteristics of extreme thunderstorm outflows. *Journal of Wind Engineering and Industrial Aerodynamics, 95*, 565–584.

Oye, R., Mueller, C., & Smith, S. (1995, October). *Software for radar translation, visualization, editing, and interpolation.* In Preprints, 27th Conference on Radar Meteorology, Vail, CO, American Meteorological Society, pp. 359–361.

Prasad, K. H., Srinivas, C. V., Naidu, C. V., Baskaran, R., & Venkatraman, B. (2016). Assessment of surface layer parameterizations in ARW using micro-meteorological observations from a tropical station. *Meteorological Applications*, 23, 191–208.

Refan, M., & Hangan, H. (2018). Near surface experimental exploration of tornado vortices. *Journal of Wind Engineering and Industrial Aerodynamics*, 175, 120–135.

Reynolds, G. W. (1971). *Complications in estimating the magnitudes of tornado forces from damage analysis.* In Seventh Conference on Severe Local Storms, Amer. Meteor. Soc., Kansas City, MO, pp. 5–7.

Rhee, D. M., & Lombardo, F. T. (2018). Improved near-surface wind speed characterization using damage patterns. *Journal of Wind Engineering and Industrial Aerodynamics*, *180*, 288–297.

Savory, E., Parke, G. A. R., Zeinoddini, M., Toy, N., & Disney, P. (2001). Modelling of tornado and microburst-induced wind loading and failure of a lattice transmission tower. *Engineering Structures*, 23, 365–375.

Sengupta, A., Haan, F. L., Sarkar, P. P., & Balaramudu, V. (2008). Transient loads on buildings in microburst and tornado winds. *Journal of Wind Engineering and Industrial Aerodynamics*, *96*, 2173–2187.

Skamarock, W. C., & Klemp, J. B. (1993). Adaptive grid refinement for two-dimensional and three-dimensional nonhydrostatic atmospheric flow. *Monthly Weather Review*, *121*, 788–804.

Skamarock, W. C., Klemp, J. B., Dudhia, J., Gill, D. O., Barker, D. M., Duda, M. G., .Â .Â . Powers, J. G. (2008). *A description of the Advanced Research WRF Version 3* (Technical Report NCAR/TN-457+STR). Boulder, CO: National Center for Atmospheric Research.

Srivastava, R. C. (1987). A model of intense downdrafts driven by the melting and evaporation of precipitation. *Journal of the Atmospheric Sciences*, 44, 1752–1773.

Stensrud, D. J. (2009). *Parameterization schemes: keys to understanding numerical weather prediction models.* Cambridge, UK: Cambridge University Press.

Stensrud, D. J., & Weiss, S. J. (2002). Mesoscale model ensemble forecasts of the 3 May 1999 tornado outbreak. *Weather and Forecasting*, *17*, 526–543.

Stull, R. B. (1988). *An introduction to boundary layer meteorology.* Dordrecht, Netherlands: Kluwer Academic.

Thompson, R. L., Mead, C. M., & Edwards, R. (2007). Effective storm-relative helicity and bulk shear in supercell thunderstorm environments. *Weather and Forecasting*, *22*, 102–115.

Toll, V., Männik, A., Luhamaa, A., & Rõõm, R. (2015). Hindcast experiments of the derecho in Estonia on 08 August, 2010: Modelling derecho with NWP model HARMONIE. *Atmospheric Research*, *158*, 179–191.

Trapp, J. R., & Fiedler, B. H. (1995). Tornado-like vortexgenesis in a simplified numerical model. *Journal of the Atmospheric Sciences*, 52, 3757–3778.

Vermeire, B. C., Orf, L. G., & Savory, E. (2011). Improved modelling of downburst outflows for wind engineering applications using a cooling source approach. *Journal of Wind Engineering and Industrial Aerodynamics*, 99, 801–814.

Wakimoto, R. M. (1985). Forecasting dry microburst activity over the high plains. *Monthly Weather Review*, *113*, 1131–1143.

Ward, N. B. (1972). The exploration of certain features of tornado dynamics using a laboratory model. *Journal of the Atmospheric Sciences*, 29, 1194–1204.

Weisman, M. L., & Klemp, J. B. (1982). The dependence of numerically simulated convective storms on vertical wind shear and buoyancy. *Monthly Weather Review, 110*, 504–520.

Weisman, M. L., Skamarock, W. C., & Klemp, J. B. (1997). The resolution dependence of explicitly modeled convective systems. *Monthly Weather Review, 125*, 527–548.

Weiss, C. C., Schroeder, J. L., Guynes, J., Skinner, P., & Beck, J. (2009). *The TTUKA mobile Doppler radar: Coordinated radar and in situ measurements of supercell thunderstorms during project VORTEX2*. In Preprints, 34th Conference on Radar Meteorology, Williamsburg, VA, American Meteorological Society, p. 11B.2.

Wilhelmson, R. B., & Wicker, L. J. (2001, November). Numerical modeling of severe local storms. *Severe Convective Storms*, 123–166.

Willis Re: Thunderstorms Pose as Much Property Risk as Hurricanes. (2017). Retrieved from http://extremeweather.columbia.edu/2017/03/23/willis-re-thunderstorms-pose-as-much-property-risk-as-hurricanes/

Wurman, J., Straka, J., Rasmussen, E., Randall, M., & Zahrai, A. (1997). Design and deployment of a portable, pencil-beam, pulsed, 3-cm Doppler radar. *Journal of Atmospheric and Oceanic Technology, 14*, 1502–1512.

Wyngaard, J. C. (2004). Toward numerical modeling in the "Terra Incognita." *Journal of the Atmospheric Sciences, 61*, 1816–1826.

Xu, X., Xue, M., & Wang, Y. (2015). Mesovortices within the 8 May 2009 bow echo over the central United States: Analyses of the characteristics and evolution based on Doppler radar observations and a high-resolution model simulation. *Monthly Weather Review, 143*, 2266–2290.

Xue, L., Li, Y., Song, L., Chen, W., & Wang, B. (2017). A WRF-based engineering wind field model for tropical cyclones and its applications. *Natural Hazards, 87*, 1735–1750.

Yao, D., Xue, H., Yin, J., Sun, J., Liang, X., & Guo, J. (2018). Investigation into the formation, structure, and evolution of an EF4 tornado in East China using a high-resolution numerical simulation. *Journal of Meteorological Research, 32*, 157–171.

CHAPTER 12

NOVEL PHYSICAL SIMULATORS FOR NON-SYNOPTIC WIND SYSTEMS

Limitations and Future Evolution

HORIA HANGAN, MARYAM REFAN,
PARTHA P. SARKAR, AND DELONG ZUO

1. Introduction

The effects of wind on buildings and structures has been traditionally addressed through physical simulations of synoptic, atmospheric boundary layer (ABL) winds in boundary layer wind tunnels. Given the fact that non-synoptic wind systems (NSWS) such as tornadoes and downbursts are three-dimensional, time-dependent, translating, and non-stationary wind events, these systems cannot be reproduced in typical boundary layer wind tunnels. As a result, a new generation of physical simulators has emerged in the past decades. These new simulators have evolved from the research-oriented tornado vortex chambers (TVCs) or impinging jet installations simulating downburst flows toward larger, more complex, and many time multifan simulators. Herein, we attempt to review and analyze the main NSWS simulators and their capabilities to reproduce generic full-scale events.

2. Tornado Simulators

2.1. Flow Parameters

Any physical flow simulator is governed by a set of geometric, kinematic, and dynamic non-dimensional parameters. For tornado simulators these are the aspect ratio, *a*; the swirl ratio, S; and the radial Reynolds number, Re_r. Early laboratory (Ward 1972; Church et al. 1979; Snow 1982) and numerical studies (Lewellen 1962; Davies-Jones 1973) have shown that the complex flow structure of tornado-like vortices depends primarily on the swirl ratio, which is ultimately a ratio between the initial tangential and radial velocities in the flow. Swirl ratio, S, is defined as:

$$S = \frac{r_o \Gamma}{2Q'h} \tag{1}$$

where r_o is a characteristic radius usually corresponding to that of the updraft region, Λ is the circulation, Q' is the volumetric flow rate per unit inflow (or inlet) height, and h is the inflow height. Swirl ratio is an important tornado-vortex parameter which characterizes the vortex breakdown and the transition to multiple vortices (Davies-Jones 1973; Rotunno 1979). Controlling the variation of the swirl ratio is therefore important for tornado simulators and results in the evolution of tornado-like vortices. This evolution starts with a thin laminar core for very small swirls, followed by a turbulent vortex breakdown aloft for increased small swirls, touchdown for moderate swirls, and a two-celled vortex followed by multiple vortices for high swirls.

The other dimensionless parameters governing the laboratory simulation of tornado-like vortices are aspect ratio and radial Reynolds number. Aspect ratio, a, is defined as the ratio of the inflow height to the updraft radius, which is less than 1 for a real tornado (Church et al. 1979), and the radial Reynolds number defined as:

$$Re_r = \frac{Q'}{2\pi v} \tag{2}$$

where v is the dynamic viscosity of the fluid.

2.2. Tornado Simulators

2.2.1. *Tornado Vortex Chambers*

Laboratory simulations of tornado-like vortices began in the early 1970s, with a focus on examining the nature and structure of vortex flows. They offered the advantage of controlled experimental conditions and repeatability of experiments. The simulations were initially conducted in TVCs similar to that shown in Figure 12.1 (Wan & Chang 1972; Ward 1972; Davies-Jones 1973; Church et al. 1977, 1979; Rotunno 1979; Lund & Snow 1993) which enabled an independent control of the flow rate and the tangential velocity components by means of a fan at the top and a swirling mechanism at the bottom. However, due to the presence of the swirling mechanism, instrumentation access to the flow region of interest was difficult.

Following the early studies, several simpler laboratory apparatuses were built by various researchers to specifically measure and analyze pressure and forces on objects immersed in the vortex flow (Matsui & Tamura 2005; Haan Jr et al. 2008; Mishra et al. 2008; Hashemi Tari et al. 2010; Hangan 2014). Some of the most recent

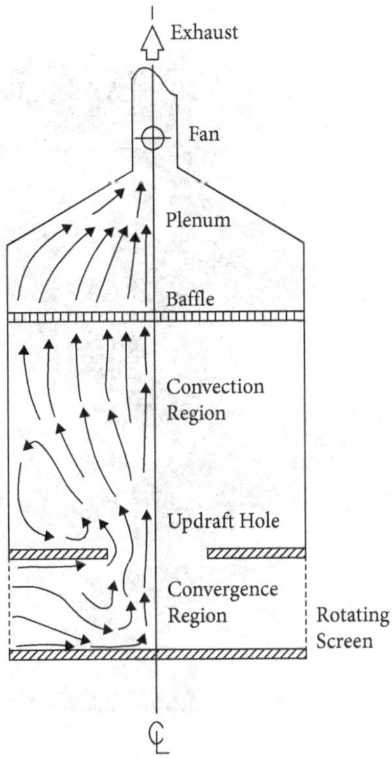

FIGURE 12.1 Schematic of the major components of the Ward-type TVC (Church et al. 1979).

simulators developed by the wind engineering community are briefly described in the following sections.

2.2.2. *Iowa State University (ISU) Tornado Simulator*

The ISU Tornado Simulator (see Figure 12.2) was designed (Sarkar et al. 2003) and constructed (2005) to meet two important requirements: first, to accommodate physical models of civil engineering structures of reasonable size for the purpose of assessing tornado-induced wind loads on them, and, second, to be able to translate along the ground plane for a realistic simulation of a natural tornado. A schematic illustration of the ISU Tornado Simulator is shown in Figure 12.2. In this facility (see Haan et al. 2008 for details), a stationary or a translating tornado-like vortex with 0.46–1.12 m maximum core diameter (distance between maximum tangential speeds on both sides of tornado center) and with maximum tangential velocities ranging from 6.9 to 14.5 m s^{-1} can be generated. It simulates a tornado-like vortex by producing a strong region of updraft surrounded by a spinning tube of air that descends toward the ground plane like the rear-flank downdraft (RFD) in a real tornado. Updraft is generated with a 1.83 m diameter fan driven by a motor (37 kW or 50 hp) mounted vertically inside a circular duct that is concentric with respect to a 0.30-meter wide downflow circular duct of 5.49 m in

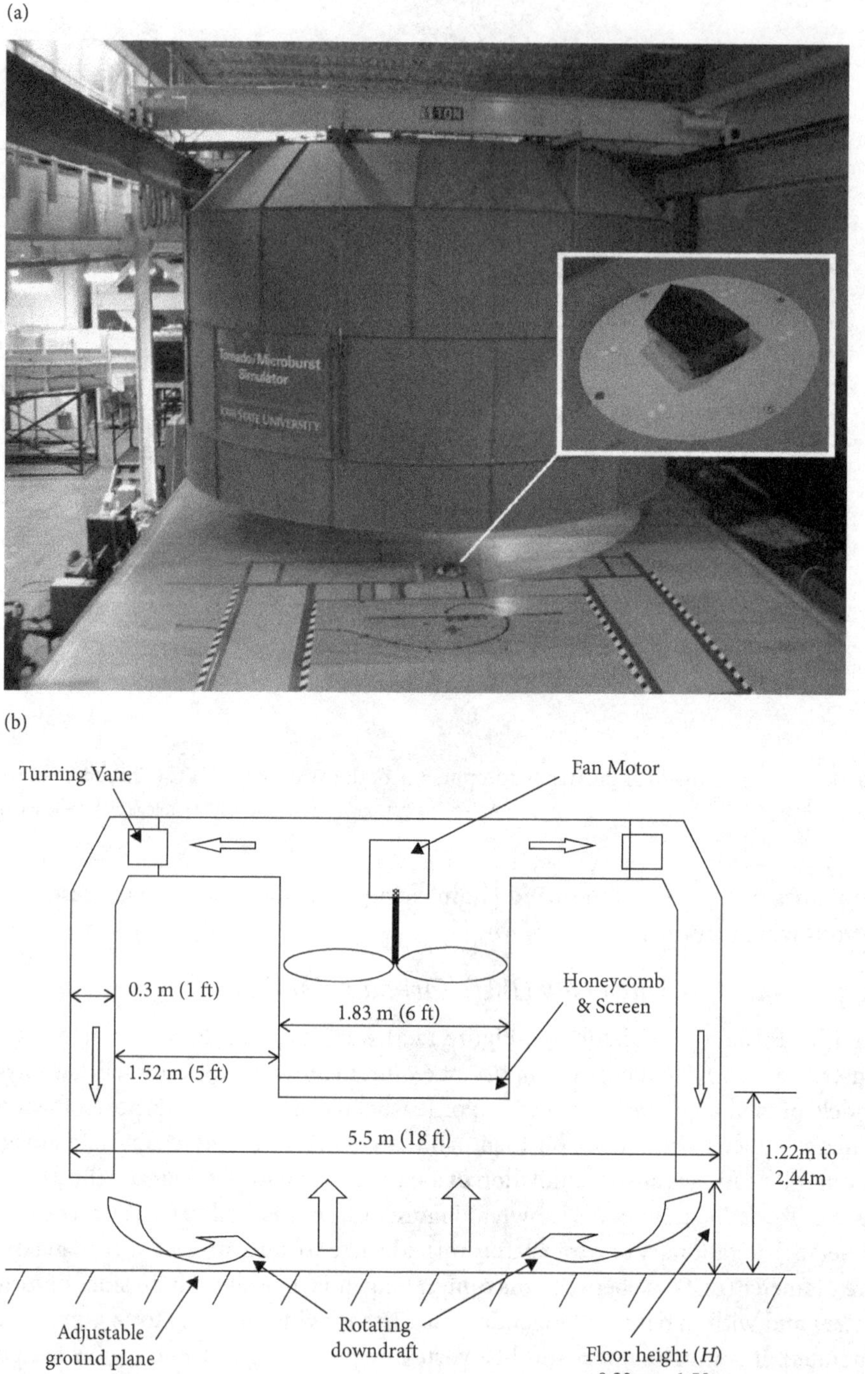

FIGURE 12.2 Schematic illustration and photo of the ISU tornado (Haan Jr et al. 2008).

Diagram: Copyright Elsevier 2008. Photo by Partha P. Sarkar.

diameter. The fan sucks air upward to simulate an updraft which passes through a series of screens and a honeycomb that tries to eliminate effects of the fan on the upstream flow. The upstream flow turns into a horizontal duct comprised of two spaced circular plates, then flows radially outward at the top of the simulator. The flow gains angular momentum after passing through a series of equally spaced vanes (hinged flat plates) placed along the outer periphery of the circular plates at a fixed angle. The fan duct and downflow duct are suspended from an overhead crane (5-ton) so that they can move above a 10.4-meter long by 6.1-meter wide ground plane. The tornado that forms moves along the ground plane as the entire fan/downdraft-producing mechanism translates. It generates vortices that can translate horizontally at speeds up to 0.61 m s^{-1}. The circulation in the flow is controlled by varying the angle of the vanes with respect to the radial direction, the flow speed is controlled by varying the power/speed of the fan, and the aspect ratio of the vortex is controlled by varying the gap between the ground plane and the downflow duct.

A 1:3 scaled Mini ISU Tornado Simulator version of the ISU Tornado Simulator was built to overcome the difficulty of visualizing the full vortex in the ISU Tornado Simulator where the size of the vortex generated is too large for the field of view of the particle image velocimetry (PIV) system to make a high-resolution measurement. This 1:3 scaled tornado simulator is used to study detailed vortex structures near the ground plane.

2.2.3. *Western Small-Scale Simulator*

Hashemi-Tari et al. (2010) developed a modified version of the ISU Tornado Simulator at Western and, for the first time, characterized both the mean and turbulent flow field for a range of swirl ratios using PIV measurements. A transparent floor panel and ducts were used to enable PIV measurements inside the simulator.

2.2.4. *TTU Simulator*

A novel simulator was built at Texas Tech University (TTU) in 2001 (Wang et al. 2001). It had the ability to independently control the flow rate via an updraft suction fan and the tangential motion of the inflow through 16 slotted jets distributed around the inlet circumference to generate swirl. The simulator was later modified by Mishra et al. (2008) to provide easy access to the simulation chamber for measurement purposes and facilitate experiments for wind engineering applications. Figure 12.3 displays the simulator resulting from modification (TTU vortex simulator II). The updraft radius was fixed at 0.19 m and the inflow height was variable between 0.06 and 0.19 m. A vortex blower at the base of the simulator is used to regulate the flow from each slot jet and thereby adjust the swirl ratio.

In 2008, a large-scale Ward-type simulator, VorTECH, was commissioned at Texas Tech University. With a chamber of 10.2 m in diameter, this simulator is capable of generating vortex cores of larger than 0.5 m in diameter and mean tangential velocities in excess of 15 m/s. The chamber can be configured to be 1–2 m in height, depending on the desired aspect ratio for the simulation, and the diameter of the updraft hole is 4 m in

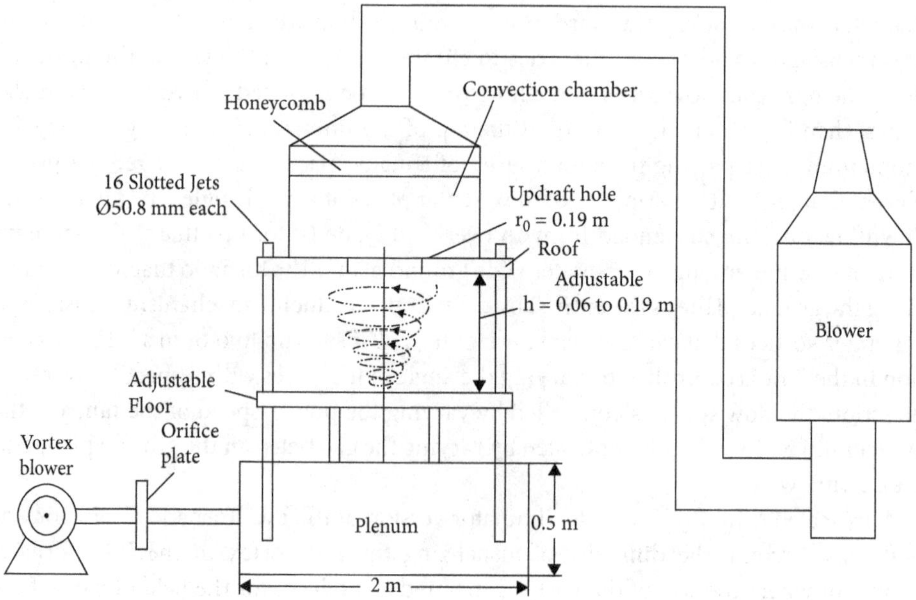

FIGURE 12.3 Schematic drawing of the second generation vortex simulator at Texas Tech University (Mishra et al. 2008).

diameter. Eight suction fans at the top of the simulator can provide up to 75 m^3/s of updraft, while 64 vanes around the periphery can be adjusted to change the direction of the inflow and, consequently, the structure of the vortex. With a recent upgrade, the center 4.2 m section of the floor of VorTECH can translate at constant speeds up to 1.5 m/s over a distance of at least 5m. This enables the study of tornadic loading on structures with relative motion between the tornado and the structures. Figure 12.4 shows a cross section of TTU's VorTECH. One important feature of this simulator is that the control room sits beneath the vortex chamber, which allows for easy access to flow field and pressure measurements.

2.2.5. *Wind Engineering, Energy and Environment (WindEEE) Dome*

WindEEE (Hangan 2014) is a unique large, hexagonal wind testing chamber, or "wind dome," of 25 m inner diameter and 40 m outer diameter (including the return circuit, see Figure 12.5). By using a system of 100 dynamic fans on the peripheral walls coupled with 6 larger fans at the ceiling level, WindEEE can produce a variety of wind systems including 4 m diameter translating tornado vortices and downbursts as well as a variety of dynamically shear flows.

The WindEEE dome passed through conceptual and engineering design phases, and construction was finished in September 2013 with commissioning in September 2014. A 1/11 scale model of the dome, Model WindEEE Dome (MWD), was designed and used to complete all validations tests for the WindEEE Dome.

(a)

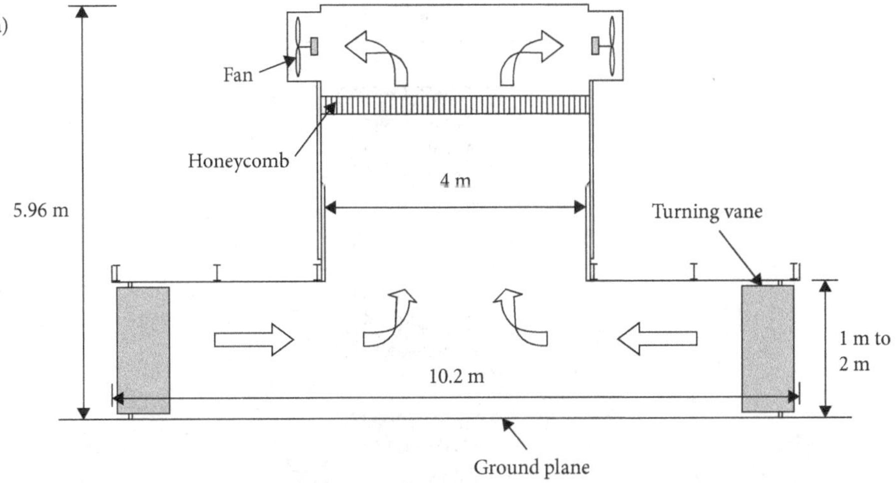

(b)

FIGURE 12.4 TTU VorTECH. Cross-section of VorTECH at Texas Tech University (Tang et al. 2018).

Diagram: Copyright Springer Nature 2017. Photo by Zuo Delong.

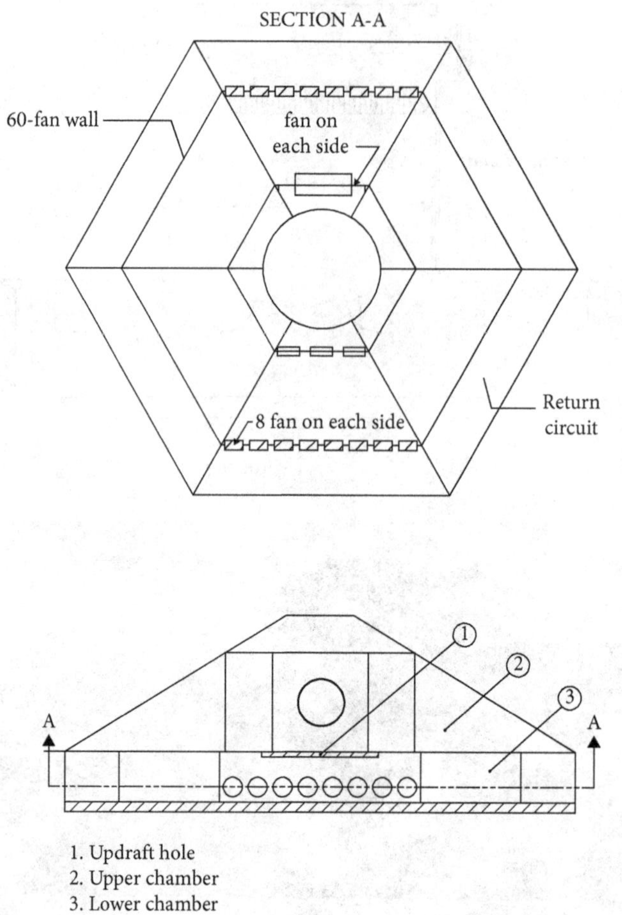

FIGURE 12.5 Cross-section of WindEEE Dome.

Photo by WindEEE Research Institute, Western University.

There are two possible configurations for generating tornado-like vortices inside the dome: Mode A, using top fans to provide updraft and periphery vanes at a given angle to generate swirl and, Mode B, running top fans and periphery fans as a source of suction and inflow, respectively, while still using vanes to control the swirl.

In Mode A, WindEEE generates more intense and smaller tornado vortices with maximum tangential velocities ranging in between 8 and 16 m/s and diameters varying between 0.6 and 1.2 m. In Mode B, WindEEE generates less intense and larger tornadoes with maximum tangential velocities around 7.5 m/s and diameters around 2.4 m. Both Mode A (Refan & Hangan 2018) and Mode B tornadoes (Ashrafi et al. 2018) have been validated in terms of vortex structure and have been scaled against full-scale tornado events ranging between EF0 to EF3. Mode A produces velocity scales of approximately 1/3 and geometric scales of between 1/150 to 1/200, while Mode B produces velocity scales of approximately 1/6 and geometric scales up to 1/50. WindEEE tornadoes can be translated at speeds just below 2 m/s using a guillotine system (Figure 12.6).

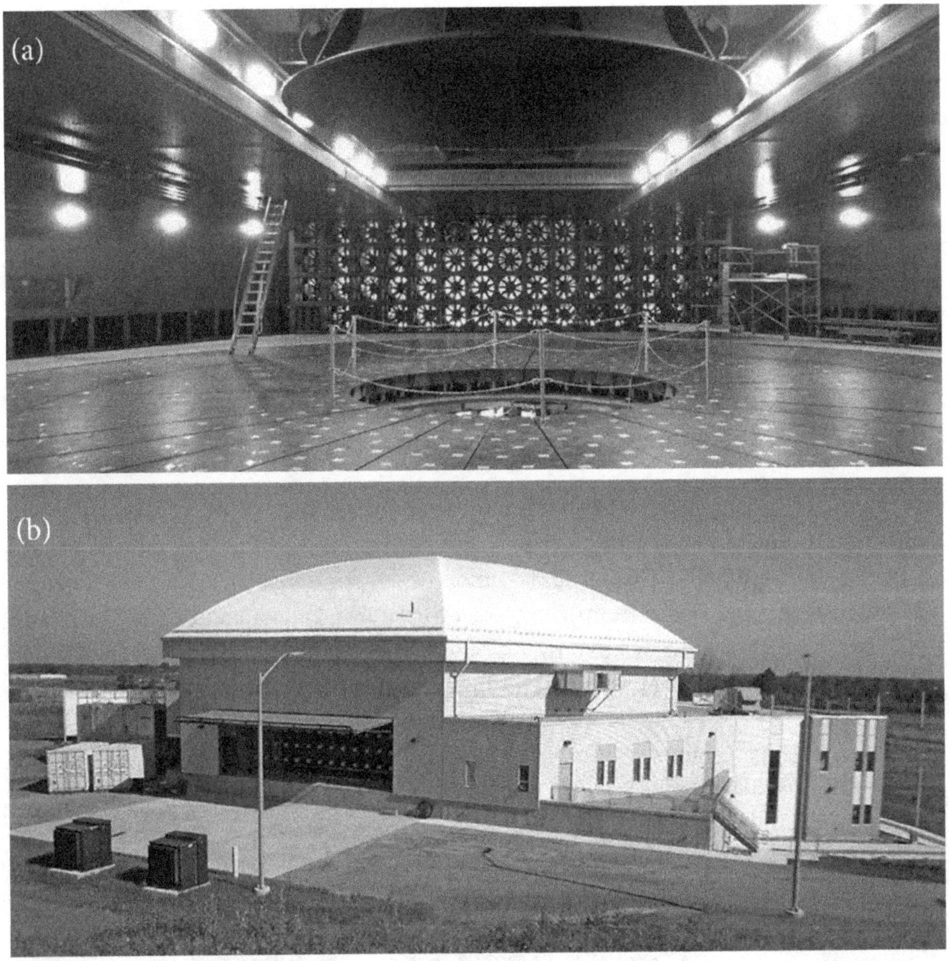

FIGURE 12.6 WindEEE Dome: (a) Inner test chamber; (b) External view.
Photos by WindEEE Research Institute, Western University.

2.3. Real Tornado Versus Simulated Vortex

Typical characteristics of tornadoes in nature are compared with those of laboratory simulated vortices and listed in Table 12.1.

As the radial Reynolds number of a real tornado is many orders of magnitude larger than those of tornado-like vortices simulated in laboratories, it can be concluded that dynamic scaling requirements are not satisfied. However, many studies (e.g.; Ward 1972; Davies-Jones 1973; Jischke & Parang 1974; Church et al. 1979; Refan & Hangan 2018; Tang et al. 2018) have shown that, for a given geometry of the simulator and for a smooth surface, if the radial Reynolds number is large enough (3 × 105) to ensure turbulent flow, the core radius and transition from a single vortex to multiple vortices are independent of radial Reynolds number and are strongly a function of swirl ratio.

Table 12.1 Dimensionless groups and translation speed for a real tornado and simulated tornado-like vortices

	Likely Atmospheric Range (Church et al. 1979)	Purdue (Church et al. 1979)	ISU (Haan et al. 2008)	Model WindEEE Dome	WindEEE Dome
Aspect ratio	0.2–1	0.2–3	0.25–1.68	0.35–1	0.35–1
Swirl ratio	0.05–2	0.01–27.5	0.08–1.14	0.1–1.3	TBD
Reynolds number	10^9– 10^{11}	4.1×10^3 1.2×10^5	3×10^5	8.4×10^4	7.5×10^6
Translation speed (mph)	22–40	Not applicable	1.35	0.56	4.5

Unlike ABL flows, where the maximum velocity is linked to one coordinate (the depth of the boundary layer), the position of the maximum tangential velocity in tornado vortices is linked to both a core radius (r_c) and a corresponding height(z_{max}). That generates two geometric scales with respect to real tornadoes. Hangan and Kim (2008) and Refan et al. (2014) have shown that by varying the swirl ratio of the simulation, the two geometric scales intersect at a swirl ratio that matches the real and simulated tornado vortex structures. This matching procedure made use of a first tornado database for wind engineering, Refan et al. (2017), to determine the simulated tornado vortex scale with respect to real tornadoes and to produce a possible linkage between real tornadoes' EF scale and simulated swirl ratios. This procedure can be adapted and applied to any physical or numerical simulations (Figure 12.7).

3. Downburst Simulators

Similar to the case of tornado simulators, several downburst simulators have emerged to reproduce downburst events in a laboratory environment. Some of these downburst simulators are listed here. Although these simulators vary in size, shape, fan configurations, and wind speeds, all employ radial impinging jets to simulate downbursts. Sometimes the tornado and downburst simulations are just different flow scenarios in the same facility.

3.1. ISU Microburst Simulator

The ISU Microburst Simulator was designed to produce an impinging jet or a downdraft flow through an axial fan driven by a step motor. The vertically mounted metal duct that houses the motor and fan can automatically translate horizontally on two parallel rails to simulate a translating microburst on a ground plane. A honeycomb and screen were

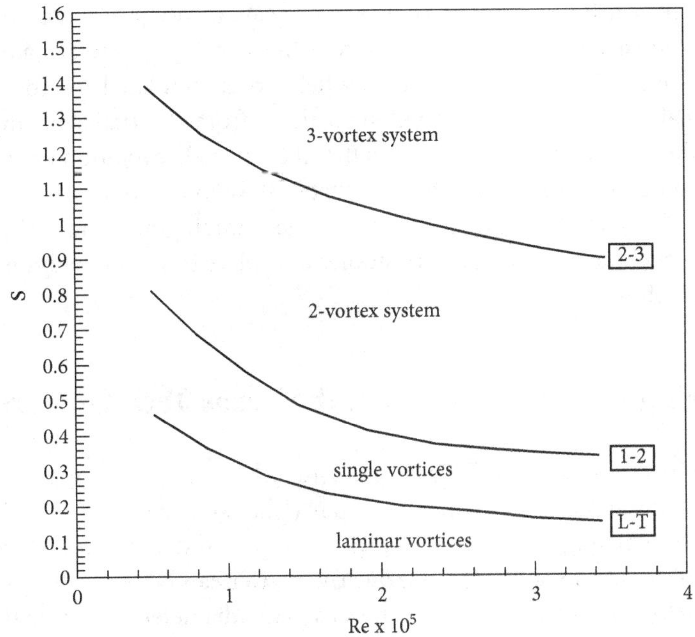

FIGURE 12.7 Critical (transition) swirl ratio as a function of radial Reynolds number; L T → laminar to turbulent, 1–2 → Single cell to two-cell vortices, 2–3 → two-cell vortices to three-cell vortices (Hashemi Tari, 2010).

placed near the nozzle exit to produce a uniform jet flow. The first form of this simulator that was used to study wind and static pressure profiles in a stationary microburst (Sengupta 2007; Sengupta & Sarkar 2008) and wind loads on buildings in a translating microburst (Sengupta et al. 2008) had a 3:1 area contraction following the fan and leading up to a nozzle diameter (D) of 203 mm to make the issuing jet uniform with a maximum velocity of 10 m s^{-1}. The simulator could translate with a maximum speed of 0.225 m s^{-1} using a mechanism operated with a garage door opener (0.37 kW or 0.5 hp). The distance (H) of the ground plane or impinging platform (2.44 × 1.22 m) from the nozzle could be varied from 203 mm to 826 mm using a mechanical jack allowing H/D to vary from 1 to 4.1, which falls into the microburst range in nature (H/D = 0.75–7.5). Comparing this with a downburst event recorded at Andrews Air Force Base (AFB) (Fujita 1985) yielded a geometric scale of about 1:2600 of the simulated microburst at ISU (Sengupta 2007). The second form of this simulator that was used to study effects of microburst wind on buildings (Zhang et al. 2013a, 2013b, 2014a, 2014b) and a wind turbine (Zhang et al. 2015) had an exhaust nozzle diameter of 610 mm. A geometric scale of approximately 1:650 was estimated by comparing this nozzle diameter to a small-size microburst of 400 m diameter in nature (Hjelmfelt 1988).

The commonly used technique for laboratory simulation, the steady impinging jet, provides an average flow field with a reasonable radial-velocity profile at the critical location (max velocity location), but it lacks time-dependent information which can prove to be critical for building design. An alternate laboratory model, the *transient impinging jet model*, was developed to improve the laboratory simulation technique at ISU. Comparison of time-dependent flow fields of a simulated microburst from a computational "cooling-source model" with those from the transient impinging jet model and those from the field has shown that the flow field evolution predicted by the transient impinging jet model is similar to that of the cooling source model with some differences (Zhang et al. 2013a). Thus, the transient impinging jet model provides an alternative model for future laboratory studies to explore transient loads on civil structures in a microburst.

3.2. Moving Jet Wind Tunnel at the Texas Tech University

The Moving Jet Wind Tunnel at TTU is a centrifugal blower-based downburst simulator where the maximum size of jet diameter is 0.51 m (Chay and Letchford 2002) (Figure 8b). Here, a 5.6 kW centrifugal blower running at 3,450 rpm produces a flow which impinges on a test surface 0.87 m above the nozzle. The blower can be translated at a constant velocity of up to 2 m/s. Mean velocity of the jet at half a diameter away from the jet outlet is 10 m/s with 4% turbulence intensity (Letchford & Chay 2002). Geometric scale of the downburst produced here was about 1:3000 to 1:3500 based on the observations of the full-scale downburst events reported in Hjelmfelt (1988).

3.3. Downburst Simulator at the University of Birmingham

The downburst simulator at the University of Birmingham employs a flap control mechanism to release pressured flow with a jet diameter of 1 m (McConville et al. 2009). This downburst simulator is adaptable to three different heights, 1, 1.5, and 2 m between the test surface and the bottom of the nozzle. Nine axial flow fans generate the flow, with a 10 mm deep honeycomb sheet placed below to reduce the turbulence of the impinging jet. The mean jet flow speed can be varied between 6 and 16 m/s. Here, translation of downburst is simulated using a movable platform with a translation speed of 3 m/s. The length scales of the simulated downbursts are around 1:700 and 1:1000 based on the downburst events recorded at Andrews AFB on August 1, 1983, reported in Fujita (1985).

3.4. Impinging Jet Simulator at Western University

The flow in this simulator was generated using fully developed horizontal pipe flows impinging on a vertical aluminum plate (Xu & Hangan 2008; see Figure 8c). A settling chamber and a machined bellmouth was used at the entrance of the pipe to condition the flow. Diameter of the horizontal pipe (D_j) is 3.8 cm. The distance between the pipe exit and the impinging surface (H_j) can be varied so that a ratio, H_j/D_j, of from 1 to 4 can

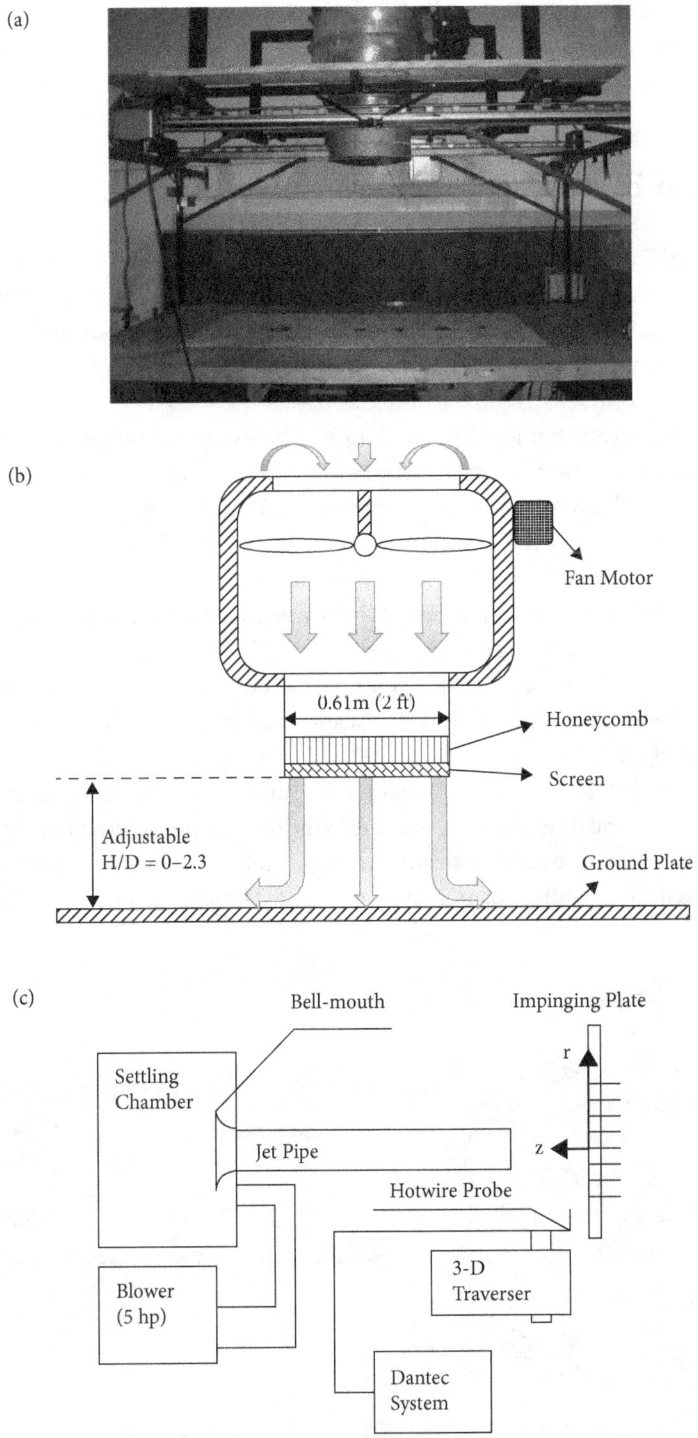

FIGURE 12.8 Downburst simulators at: (a) ISU (Zhang et al. 2013a); (b) TTU (Chay and Letchford, 2002); and (c) Western University.

be achieved. A grid can also be installed in the horizontal pipe at various axial distances to generate different inflow turbulence intensities. The simulator could produce both steady-state impinging jets as well as, for the first time, dynamic impinging jets.

3.5. WindEEE Dome at Western University

The WindEEE Dome at Western University is the state-of-the-art large scale wind testing chamber where a wide variety of atmospheric flows (e.g., tornado, thunderstorm downburst, ABL) can be simulated at various scales and intensities (Hangan 2014; Hangan et al. 2017). In this chapter, the focus is on WindEEE's capabilities to produce downburst flows only.

WindEEE has a test chamber of 25 m diameter, a hexagonal footprint, with a total 106 fans, of which 100 fans are set up on peripheral walls and 6 other larger fans are situated above the testing chamber ceiling. One hundred fans on the peripheral wall are distributed as 8 fans on each of 5 walls, and another 60 fans are installed on 1 wall in a matrix of 15 fans per row × 4 rows. Each of the 100 fans in the test chamber has a diameter of 0.8 m, with a nominal power of 25 kW whereas the 6 larger fans on the upper chamber have a diameter of 2 m each with a nominal power of 220 kW. The air from the upper plenum is fed to the test chamber by a circular opening with a bellmouth and mechanically operated louvers. In the dynamic mode, it takes approximately 2 s to fully open the louvres from the closed position. The diameter of the circular opening can be varied from 1.6 to 4.5 m using annular rings of different diameters. The height of the test chamber (H) is 3.8 m. A schematic of the downburst flow at the WindEEE Dome is shown in Figure 12.9.

The simulator can also act as a steady impinging jet, which has been demonstrated to reproduce well intermediate or gust type of events (Jubayer et al. 2019). To simulate a

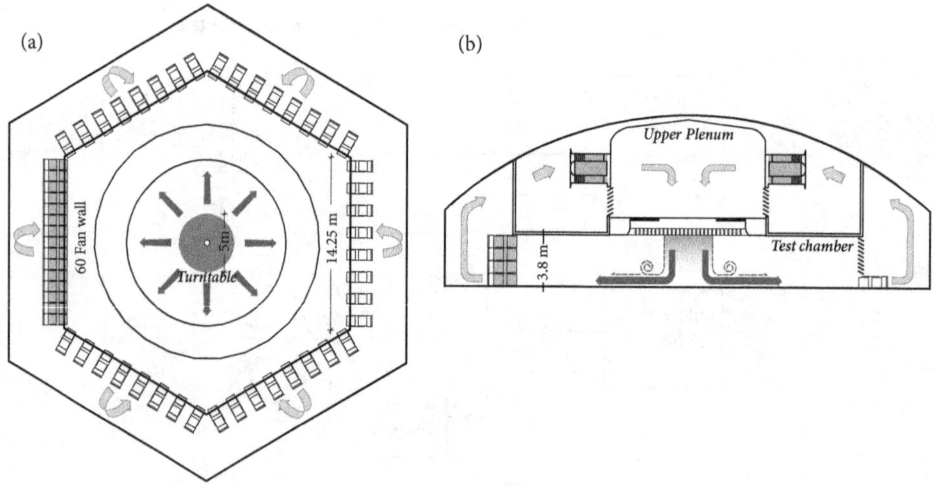

FIGURE 12.9 Schematic of the downburst flow simulation at WindEEE. Horizontal section (a), vertical section (b).

downburst event, the six fans in the upper plenum are run with the louvers closed to develop pressure. With a buildup of pressure in the upper chamber, louvre positions are changed to nearly open to create the sudden downfall of pressurized air that slowly reaches to the test chamber ground surface and produces a downburst. Downbursts produced in WindEEE have been scaled against several real events (e.g., Burlando et al. 2017, 2018) using several procedures including size, ramp-up time, or statistical approaches. Based on these results, the range of geometric scales covered by WindEEE is between 1/100 and 1/1,000, and the range of velocity scales is typically between ½ to ¼.

Similar to the case of tornado simulations, the downbursts can be translated to a maximum speed of 2 m/s by using a guillotine system that translates the bellmouth. The louvres at the bellmouth level can dynamically open to different angles, thus allowing the simulation of inclined downburst events. In addition, and uniquely so, WindEEE can combine the impinging jet mode with a boundary layer mode by simultaneously using the 60-fan wall to create the ABL and the upper fans to generate the downburst (Romanic et al. 2019). By combining the action of these two systems, WindEEE can generate the ramp-up in downburst events which corresponds to the effects of environmental winds on the downburst.

References

Ashrafi, A., Romanic, D., Chowdhury, J., & Hangan H. (2018, September 26–27). *Producing 1/100 and larger scale tornadoes in a wind simulator.* Conference presentation. Tornado Hazard Wind Assessment and Reduction Symposium (THWARTS 2018), Champaign, IL, United States.

Burlando, M., Romanic, D., Hangan, H., & Solari, G. (2018, September 9–12). *Wind tunnel experimentation on stationary downbursts at WindEEE.* XV Conference of the Italian Association for Wind Engineering (IN-VENTO 2018). Napoli, Italy.

Burlando, M., Romanic, D., Solari, G., Hangan, H., & Zhang, S. (2017). Field data analysis and weather scenario of a downburst event in Livorno, Italy on October 1, 2012. *Monthly Weather Review/American Meteorological Society, 145*, 3507–3527. https://doi.org/10.1175/MWR-D-17-0018.1

Chay, M. T., & Letchford, C. W. (2002). Pressure distributions on a cube in a simulated thunderstorm downburst—Part A: Stationary downburst observations. *Journal of Wind Engineering and Industrial Aerodynamics, 90*, 711–732. https://doi.org/10.1016/S0167-6105(02)00158-7

Church, C. R., Snow, J. T., & Agee, E. M. (1977). Tornado vortex simulation at Purdue University. *Bulletin of the American Meteorological Society, 58*, 900–908.

Church, C. R., Snow, J. T., Baker, G. L., & Agee, E. M. (1979). Characteristics of tornado-like vortices as a function of swirl ratio: A laboratory investigation. *Journal of Atmospheric Sciences, 36*, 1155–1176.

Davies-Jones, R. P. (1973). The dependence of core radius on swirl ratio in a tornado simulator. *Journal of Atmospheric Sciences, 30*, 1427–1430.

Fujita, T. T. (1985). *The downburst: Microburst and macroburst.* Chicago, IL: University of Chicago Press.

Haan Jr, F. L., Sarkar, P. P., & Gallus, W. A. (2008). Design, construction and performance of a large tornado simulator for wind engineering applications. *Engineering Structures, 30*, 1146–1159.

Hangan, H. (2014). The Wind Engineering Energy and Environment (WindEEE) Dome at Western University, Canada. *Wind Engineers, JAWE, 39*(4), 350–351.

Hangan, H., & Kim, J. (2008). Swirl ratio effects on tornado vortices in relation to the Fujita scale. *Wind and Structures, 11*(4), 291–302.

Hangan, H., Refan, M., Jubayer, C., Romanic, D., Parvu, D., LoTufo, J., & Costache, A. (2017). Novel techniques in wind engineering. *Journal of Wind Engineering and Industrial Aerodynamics, 171*, 12–33. https://doi.org/10.1016/j.jweia.2017.09.010

Hashemi Tari, P., Gurka, R., & Hangan, H. (2010). Experimental investigation of a tornado-like vortex dynamics with swirl ratio: The mean and turbulent flow fields. *Journal of Wind Engineering and Industrial Aerodynamics, 98*, 936–944.

Hjelmfelt, M. R. (1988). Structure and life cycle of microburst outflows observed in Colorado. *Journal of Applied Meteorology, 27*, 900–927. https://doi.org/10.1175/1520-0450(1988)027<0900:SALCOM>2.0.CO;2

Jischke, M. C., & Parang, M. (1974). Properties of simulated tornado-like vortices. *Journal of Atmospheric Sciences, 31*, 506–512.

Jubayer, C., Romanic, D., & Hangan H. (2019). Aerodynamic loading of a typical low rise building for an experimental stationary and non-Gaussian impinging jet. *Wind and Structures, 28*, 315–329.

Letchford, C. W., & Chay, M. T. (2002). Pressure distributions on a cube in a simulated thunderstorm downburst. Part B: Moving downburst observations. *Journal of Wind Engineering and Industrial Aerodynamics, 90*, 733–753. https://doi.org/10.1016/S0167-6105(02)00163-0

Lewellen, W. S. (1962). A solution of three dimensional vortex flows with strong circulation. *Journal of Fluid Mechanics, 14*, 420–432.

Lund, D. E., & Snow, J. T. (1993). Laser doppler velocimetry measurements in tornado like vortices. In C. Church, D. Burgess, C. Doswell, and R. Davies-Jone (Eds.), *The tornado: Its structure, dynamics, prediction, and hazards*. Washington, DC: American Geophysical Union.

Matsui, M., & Tamura, Y. (2005). Characteristics of tornado-like flow as functions of swirl ratio and surface roughness. *Academic Reports, Faculty of Engineering, Tokyo Institute of Polytechnics (Japan), 28*(1), 113–119.

McConville, A. C., Sterling, M., & Baker, C. J. (2009). The physical simulation of thunderstorm downbursts using an impinging jet. *Wind and Structures, 12*, 133–149. https://doi.org/10.12989/was.2009.12.2.133

Mishra, A. R., James, D. L., & Letchford, C. W. (2008). Physical simulation of a single-celled tornado-like vortex, Part B: Wind loading on a cubical model. *Journal of Wind Engineering and Industrial Aerodynamics, 96*(8), 1258–1273. https://doi.org/10.1016/j.jweia.2008.02.027

Refan, M., & Hangan, H. (2018). Near surface experimental exploration of tornado vortices. *Journal of Fluids Engineering, 175*, 120–135. https://doi.org/10.1016/j.jweia.2018.01.042

Refan, M., Hangan, H., & Wurman, J. (2014). Reproducing tornadoes in laboratory using proper scaling. *Journal of Wind Engineering and Industrial Aerodynamics, 135*, 136–148.

Refan, M., Hangan, H., Wurman, J., & Kosiba, K. (2017). Doppler radar-derived wind field of several tornadoes with application to engineering simulations. *Engineering Structures, 148*, 509–521.

Romanic, D., LoTufo J., & Hangan, H. (2019). Transient behavior in impinging jets in crossflow with application to downburst flows. *Journal of Wind Engineering and Industrial Aerodynamics, 184*, 209–227.

Rotunno, R. A. (1979). A study in tornado-like vortex dynamics. *Journal of Atmospheric Sciences, 36*, 140–155.

Sarkar, P. P., R. Kardell, F. Haan, & W. Gallus (2003). Design of ISU tornado vortex simulator [conference presentation]. *Proceedings of the Eleventh International Conference on Wind Engineering*. Lubbock, Texas, June 2–5, 2003.

Sengupta, A. (2007). *Study of microburst-induced wind flow and its effects on cube shaped buildings using numerical and experimental simulations of an impinging jet* [unpublished PhD dissertation]. Iowa State University, Ames, Iowa.

Sengupta, A., Haan, F. L., Sarkar, P. P., & Balaramudu, V. (2008). Transient loads on buildings in microburst and tornado winds. *Journal of Wind Engineering and Industrial Aerodynamics, 96*, 2173–2187. https://doi.org/10.1016/j.jweia.2008.02.050

Sengupta, A., & Sarkar, P. P. (2008). Experimental measurement and numerical simulation of an impinging jet with application to thunderstorm microburst winds. *Journal of Wind Engineering and Industrial Aerodynamics, 96*, 345–365. https://doi.org/10.1016/j.jweia.2007.09.001

Snow, J. T. (1982). A review of recent advances in tornado vortex dynamics. *Reviews of Geophysics, 20*, 953–964.

Tang, Z., Feng, C., Wu, L., Zuo, D., & James, D. L. (2018). Characteristics of tornado-like vortices simulated in a large-scale ward-type simulator. *Boundary-Layer Meteorology*, 166, 327–350.

Wan, C. A., & Chang, C. C. (1972). Measurement of the velocity field in a simulated tornado-like vortex using a three dimensional velocity probe. *Journal of Atmospheric Sciences, 29*, 116 127.

Wang, J., James, D. L., Letchford, C. W., Peterson, C., & Snow, J. T. (2001). Development of a prototype tornado simulator for the assessment of fluid-structure interaction [conference paper]. *Proceedings of the 1st American Conference on Wind Engineering*, Clemson University, South Carolina. June 4–6, 2001.

Ward, N. B. (1972). The exploration of certain features of tornado dynamics using a laboratory model. *Journal of Atmospheric Sciences, 29*, 1194–1204.

Xu, Z., & Hangan, H. (2008). Scale, boundary and inlet condition effects on impinging jets. *Journal of Wind Engineering and Industrial Aerodynamics, 96*, 2383–2402. https://doi.org/10.1016/j.jweia.2008.04.002

Zhang, Y., Hu, H., & Sarkar, P. P. (2013a). Modeling of microburst outflows using impinging jet and cooling source approaches and their comparison, *Engineering Structures, 56*, 779–793. http://dx.doi.org/10.1016/j.engstruct.2013.06.003

Zhang, Y., Hu, H., & Sarkar, P. P. (2014b). Comparison of microburst-wind loads on low-rise structures of various geometric shapes. *Journal of Wind Engineering and Industrial Aerodynamics, 133*, 181–190. http://dx.doi.org/10.1016/j.jweia.2014.06.012

Zhang, Y., Sarkar, P. P., & Hu, H. (2014a). An experimental study on wind loads acting on a high-rise building model induced by microburst-like winds. *Journal of Fluids and Structures* 50, 547–564. http://dx.doi.org/10.1016/j.jfluidstructs.2014.07.010

Zhang, Y., Sarkar, P. P., & Hu, H. (2013b). An experimental study of flow fields and wind loads on gable-roof building models in microburst-like wind. *Experiments in Fluids*, 54, 1511. https://doi.org/10.1007/s00348-013-1511-9

Zhang, Y., Sarkar, P., & Hu, H. (2015). An experimental investigation on the characteristics of fluid-structure interactions of a wind turbine model sited in microburst-like winds. *Journal of Fluids and Structures*, 57(8), 206–218. http://dx.doi.org/10.1016/j.jfluidstructs.2015.06.016

Further Reading

Abbott, P. L. (2004). *Natural disasters*. Boston, MA: McGraw Hill.

Alahyari, A., & Longmire, E. K. (1994). Particle image velocimetry in a variable density flow: Application to a dynamically evolving microburst. *Experiments in Fluids, 17*, 434–440. https://doi.org/10.1007/BF01877047

Ashton, A., Leduc, M., & Boodoo, S. (2010). *The day the skies rained tornadoes* [Powerpoint Slides]. http://www.yorku.ca/glomw10/Tues_March_23_AM/Ashton_Leduc_August%20 20th_Talk%20One.ppt

Bakke, P. (1957). An experimental investigation of a wall jet. *Journal of Fluid Mechanics* 2, 467–472. https://doi.org/10.1017/S0022112057000270

Bledsoe, B. (2009). VORTEX 2 (Tornado Research Project) National tornado experiment to begin in May. KKTV. Gray Television.

Bluestein, H. B., & Pazmany, A. L. (2000). Observations of tornadoes and other convective phenomena with a mobile, 3-mm wavelength, Doppler radar: The spring 1999 field experiment. *Bulletin of the American Meteorological Society, 81*(12), 2939–2952.

Burlando, M., De Gaetano, P., Repetto, M. P., & Solari, G. (2014). *Wind and waves numerical forecasting for safety access to port areas: The "Wind, Ports, and Sea" project* [Conference presentation]. The 6th International Symposium on Computational Wind Engineering, Hamburg, Germany.

Charlton, R. B., Kachman, B. M., & Wojtiw, L. (1995). Urban hailstorms: A view from Alberta. *Natural Hazards, 12*, 29–75.

Chay, M. T., Wilson, R., & Albermani, F. (2008). Gust occurrence in simulated non-stationary winds. *Journal of Wind Engineering and Industrial Aerodynamics, 96*, 2161–2172. https://doi.org/10.1016/j.jweia.2008.02.059

Cheng, V. Y. S., Arhonditsis, G. B., Sills, D. M. L., Auld, H., Shephard, M. W., Gough, W. A., & Klaassen, J. (2013). Probability of tornado occurrence across Canada. *Journal of Climate, 26*(23), 9415–9428. https://doi.org/10.1175/JCLI-D-13-00093.1

Choi, E. C., & Hidayat, F. A. (2002). Dynamic response of structures to thunderstorm winds. *Progress in Structural Engineering and Materials, 4*, 408–416. https://doi.org/10.1002/pse.132

Egentowich, J. M., Kaplan, M. L., Lin, Y.-L., & Riordan, A. J. (2000a). Mesoscale simulations of dynamical factors discriminating between a tornado outbreak and non-event over the southeast US Part II: 48-6 hour precursors. *Meteorology and Atmospheric Physics, 74*(1–4), 159–187. https://doi.org/10.1007/s007030070031

Egentowich, J. M., Kaplan, M. L., Lin, Y.-L., & Riordan, A. J. (2000b). Mesoscale simulations of dynamical factors discriminating between a tornado outbreak and non-event over the southeast US Part III: 6 hour precursors. *Meteorology and Atmospheric Physics, 74* (1–4), 189–215. https://doi.org/10.1007/s007030070032

Etkin, D., Brun, S. E., Shabbar, A., & Joe, P. (2001). Tornado climatology of Canada revisited: Tornado activity during different phases of ENSO. *International Journal of Climatology, 21* (8), 915–38. https://doi.org/10.1002/joc.654

Fierro, A. O., Mansell, E. R., Ziegler, C. L., & MacGorman, D. R. (2012). Application of a lightning data assimilation technique in the WRF-ARW model at cloud-resolving scales for the tornado outbreak of 24 May 2011. *Monthly Weather Review, 140*(8), 2609–2627. https://doi.org/10.1175/MWR-D-11-00299.1

Fujita, T. T. (1974). Jumbo tornado outbreak of 3 April 1974. *Weatherwise, 27*, 116–126. https://doi.org/10.1080/00431672.1974.9931693

Fujita, T. T. (1978). *Manual of downburst identification for Project NIMROD* (SMRP Research Paper 156).

Fujita, T. (1990). Downbursts: Meteorological features and wind field characteristics. *The Sixth U. S. National Conference on Wind Engineering, Journal of Wind Engineering and Industrial Aerodynamics 36*, 75–86. https://doi.org/10.1016/0167-6105(90)90294-M

Grazulis, T. P. (1991). *Significant tornadoes, 1880–1989*. Vol.1. Discussion and Analysis. Environmental Films.

Haan Jr, F. L., Balaramudu, V., & Sarkar, P. (2010). Tornado-induced wind loads on a low-rise building. *Journal of Structural Engineering, 136*(1), 106–116. https://doi.org/10.1061/(ASCE)ST.1943-541X.0000093

Henjes, K. (1997). Isotropic and anisotropic correlations in turbulent wind speed data. *Boundary Layer Meteorology, 84*, 149–167. https://doi.org/10.1023/A:1000386716341

Hjelmfelt, M. R. (1987). The microbursts of 22 June 1982 in JAWS. *Journal of the Atmospheric Sciences, 44*, 1646–1665. https://doi.org/10.1175/1520-0469(1987)044<1646:TMOJIJ>2.0.CO;2

Holmes, J. D., Hangan, H. M., Schroeder, J. L., Letchford, C. W., & Orwig, K. D. (2008). A forensic study of the Lubbock-Reese downdraft of 2002. *Wind and Structures, 11*, 137–152. https://doi.org/10.12989/was.2008.11.2.137

Killick, R., Fearnhead, P., & Eckley, I. A. (2012). Optimal detection of changepoints with a linear computational cost. *Journal of the American Statistical Association, 107*, 1590–1598. https://doi.org/10.1080/01621459.2012.737745

Kim, J., & Hangan, H. (2007). Numerical simulations of impinging jets with application to downbursts. *Journal of Wind Engineering and Industrial Aerodynamics, 95*, 279–298. https://doi.org/10.1016/j.jweia.2006.07.002

Lavielle, M. (2005). Using penalized contrasts for the change-point problem. *Signal Process, 85*, 1501–1510. https://doi.org/10.1016/j.sigpro.2005.01.012

Lee, W. C., & Wurman, J. (2005). Diagnosed three-dimensional axisymmetric structure of the Mulhall Tornado on 3 May 1999. *American Meteorological Society, 62*, 2373–2393.

Letchford, C., & Illidge, G. C. (1999). Turbulence and topographic effects in simulated thunderstorm downdrafts by wind tunnel jet. *Wind Engineering in the 21st Century, 1–3*(1), 1907–1912.

Litta, A. J., Mohanty, U. C., & Bhan, S. C. (2010). Numerical simulation of a tornado over Ludhiana (India) using WRF-NMM model. *Meteorological Applications, 17*(1), 64–75. https://doi.org/10.1002/met.162

Litta, A. J., Mohanty, U. C., Prasad, K. S., Mohapatra, M., Tyagi, A., & Sahu, S. C. (2012). Simulation of tornado over Orissa (India) on March 31, 2009, using WRF–NMM model. *Natural Hazards, 61*(3), 1219–1242. https://doi.org/10.1007/s11069-011-9979-1

Lombardo, F. T., Smith, D. A., Schroeder, J. L., & Mehta, K. C. (2014). Thunderstorm characteristics of importance to wind engineering. *Journal of Wind Engineering and Industrial Aerodynamics, 125*, 121–132. https://doi.org/10.1016/j.jweia.2013.12.004

Lundgren, T. S., Yao, J., & Mansour, N. N. (1992). Microburst modelling and scaling. *Journal of Fluid Mechanics, 239*, 461–488. https://doi.org/10.1017/S002211209200449X

Mason, M. S., Letchford, C. W., & James, D. L. (2005). Pulsed wall jet simulation of a stationary thunderstorm downburst, Part A: Physical structure and flow field characterization. *Journal of Wind Engineering and Industrial Aerodynamics, 93*, 557–580. https://doi.org/10.1016/j.jweia.2005.05.006

Matsangouras, I. T., Nastos, P. T., & Pytharoulis, I. (2011). Synoptic-mesoscale analysis and numerical modeling of a tornado event on 12 February 2010 in northern Greece.

International Journal of Advanced Science and Research, 6(July), 187–194. https://doi.org/10.5194/asr-6-187-2011

McCullough, M., Kwon, D. K., Kareem, A., & Wang, L. (2014). Efficacy of averaging interval for nonstationary winds. *Journal of Engineering Mechanics, 140*, 1–19. https://doi.org/10.1061/(ASCE)EM.1943-7889.0000641

Nasir, Z., & Bitsuamlak, G. (2014). *Similarities and differences among tornadic and synoptic flow induced loads on a bluff-body* (conference presentation). EMI 2014 Engineering Mechanics Institute Conference, Hamilton, Canada.

Natarajan, D. (2011). *Numerical simulation of tornado-like vortices* [unpublished PhD dissertation]. University of Western Ontario, London, Ontario, Canada.

Natarajan, D., & Hangan, H. (2012). Large eddy simulations of translation and surface roughness effects on tornado-like vortices. *Journal of Wind Engineering and Industrial Aerodynamics, 104–106*, 577–584.

Oseguera, R. M. B. (1988). *A simple, analytic 3-dimensional downburst model based on boundary layer stagnation flow*. NASA Technical Memorandum. Washington, DC: NASA.

Panneer Selvam, R., & Holmes, J. D. (1992). Numerical simulation of thunderstorm downdrafts. *Journal of Wind Engineering and Industrial Aerodynamics*, Special Issue, *8th International Conference on Wind Engineering, 44*, 2817–2825. https://doi.org/10.1016/0167-6105(92)90076-M

Pielke, R. A. Sr. (1998). *Hurricanes: Their nature and impact on society*. New York: Wiley.

Refan, M. (2014). *Physical simulation of tornado-like vortices* [unpublished PhD dissertation]. University of Western Ontario, London, Ontario, Canada.

Selvam, R. P., & Millett, P. C. (2003). Computer modeling of tornado forces on buildings. *Wind and Structures, 6*(3), 209–220.

Sengupta, A., Haan, F. L., Sarkar, P. P., & Balaramudu, V. (2006). *Transient loads on buildings in microburst and tornado winds* [conference presentation]. The Fourth International Symposium on Computational Wind Engineering (CWE2006), Yokohama, Japan.

Shafer, C. M., Mercer, A. E., Doswell, C. A, Richman, M. B, & Leslie, L. M. (2009). Evaluation of WRF forecasts of tornadic and nontornadic outbreaks when initialized with synoptic-scaleinput. *Monthly Weather Review, 137*(4), 1250–1271. https://doi.org/10.1175/2008MWR2597.1

Shafer, C. M., Mercer, A. E., Leslie, L. M., Richman, M. B., & Doswell, C. A. (2010). Evaluation of WRF model simulations of tornadic and nontornadic outbreaks occurring in the Spring and Fall. *Monthly Weather Review, 138*(11), 4098–4119. https://doi.org/10.1175/2010MWR3269.1

Solari, G., & Piccardo, G. (2001). Probabilistic 3-D turbulence modeling for gust buffeting of structures. *Probabilistic Engineering Mechanics, 16*, 73–86. https://doi.org/10.1016/S0266-8920(00)00010-2

Solari, G., Burlando, M., De Gaetano, P., & Gaetano, M. P. (2015). Characteristics of thunderstorms relevant to the wind loading of structures. *Wind and Structures, 20*, 763–791. https://doi.org/10.12989/was.2015.20.6.763

Solari, G., Repetto, M. P., Burlando, M., De Gaetano, P., Pizzo, M., Tizzi, M., & Parodi, M. (2012). The wind forecast for safety management of port areas. *Journal of Wind Engineering and Industrial Aerodynamics, 104*, 266–277. https://doi.org/10.1016/j.jweia.2012.03.029

Stensrud, D. J., & Weiss, S. J. (2002). Mesoscale model ensemble forecasts of the 3 May 1999 tornado outbreak. *Weather and Forecasting, 17*(3), 526–543. https://doi.org/10.1175/1520-0434(2002)017<0526:MMEFOT>2.0.CO;2

Wilson, J. W., Roberts, R. D., Kessinger, C., & McCarthy, J. (1984). Microburst wind structure and evaluation of doppler radar for airport wind shear detection. *Journal of Applied*

Meteorology and Climatology, *23*, 898–915. https://doi.org/10.1175/1520-0450(1984)023<0898:MWSAEO>2.0.CO;2

Wolfson, M. M. (1988). Characteristics of microbursts in the continental United States. *The Lincoln Lab Journal*, *1*(1).

Wood, G. S., Kwok, K. C. S., Motteram, N. A., & Fletcher, D. F. (2001). Physical and numerical modelling of thunderstorm downbursts. *Journal of Wind Engineering and Industrial Aerodynamics*, *89*, 535–552. https://doi.org/10.1016/S0167-6105(00)00090-8

Wurman, J. (2002, June). The multiple-vortex structure of a tornado. *American Meteorological Society*, 473–505.

Wurman, J., & Alexander, C. R. (2005). The 30 May 1998 Spencer, South Dakota, storm. Part II: Comparison of observed damage and radar-derived winds in the tornadoes. *Monthly Weather Review*, *133*, 97–119.

Wurman, J., & Gill, S. (2000). Fine-scale radar observations of the Dimmitt, Texas (2 June 1995) tornado. *Monthly Weather Review*, *128*(7), 2135–2164.

Yao, J., & Lundgren, T. S. (1996). Experimental investigation of microbursts. *Experiments in Fluids*, *21*, 17–25. https://doi.org/10.1007/BF00204631

CHAPTER 13

RELATION BETWEEN FULL-SCALE AND MODEL DATA

HORIA HANGAN, MARYAM REFAN, AND DJORDJE ROMANIC

1. Introduction

When a flow phenomenon is reproduced through physical or numerical simulations, a group of nondimensional parameters need to be the same in full scale (i.e., prototype) and the simulation (i.e., model). These parameters can usually be reduced to a geometric, a kinematic, and a dynamic parameter. The geometric parameter is the ratio between two main geometric dimensions in the flow; the kinematic parameter involves a ratio of two defining velocities in the flow, whereas the dynamic parameter involves a ratio between two dominant forces in the flow, and it is most of the time represented by the Reynolds number. The geometric and the dynamic similarities can be fulfilled in most cases, while the Reynolds number is usually much smaller in simulations compared to reality; therefore, some evidence of Reynolds number independency above a certain critical value should be provided.

For synoptic winds simulated as boundary layer flows over a variety of rough surfaces, the mean geometric similarity is based on the ratio between the depth of the boundary layer and a dimension related to the object tested, usually the height. The scaling between reality and model is typically based on the ratio between the depth of the two boundary layers, which can be interpreted as a gradient height (z_g) and which corresponds to the position where 95 percent or 98 percent of the maximum velocity is attained.

For non-synoptic wind fields such as tornadoes and thunderstorm downbursts, things are complicated by the three-dimensionality of the flow (which involves two length parameters for the position of the maximum velocity) or by the fact that these phenomena occur in nature over a range of sizes and shapes; therefore, a statistical rather than deterministic scaling approach is more appropriate. Here, we treat the scaling of tornadoes first, followed by the scaling for downbursts.

2. Tornado Scaling

Despite the significant number of analytical, experimental, and numerical studies and advances in measurement methods, the investigation of the tornado loading on buildings and structures has been limited by the definition of a proper scaling between simulated and real tornadoes. In order to conclude that a simulated tornado-like vortex is a valid representation of tornado flow in nature, it is important that the geometric, kinematic, and dynamic similitudes are analyzed.

The difficulty with the case of tornadic flows originates in the definition of the main nondimensional number governing tornado-like flows, that is, the swirl ratio (S). This important parameter is defined based on the geometry and boundaries of a simulator and is location dependent. Therefore, it is somewhat subjective to calculate the swirl ratio for a real tornado as there is no clear definition of inlet/outlet boundary conditions in a field tornado. An alternative approach was suggested by Hangan and Kim (2008) in which they compared simulated velocity data at various swirl ratios with that of Doppler radar full-scale data from the F4-rated tornado of Spencer, South Dakota, on May 30, 1998. Hangan and Kim (2008) showed that a simulated tornado-like vortex with $S = 2$ best represents the Spencer tornado (see Figure 13.1). Although their results were promising, the approach was not validated until later due to the shortage of full-scale velocity data from tornadoes of various structures and intensities.

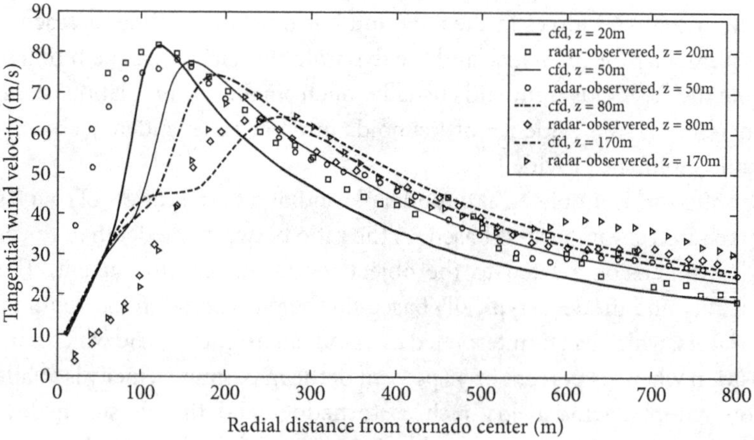

FIGURE 13.1 Matched tangential wind velocity profiles from Doppler radar data and CFD (computational fluid dynamics) (Hangan & Kim, 2008).

By following a similar approach, Refan, Hangan, and Wurman (2014) successfully compared laboratory simulated tornado data with full-scale Doppler radar data of various tornadoes and estimated the scaling ratios of the simulations as well as the swirl ratio of the real tornado.

The scaling process developed by Refan et al. (2014) constitutes a framework that can be applied to both experimental and numerical tornado datasets. While the scaling process is therefore independent of the tornado simulator design or the boundary conditions of the simulations, the results in terms of scaling will differ from one physical simulator to another or between numerical simulations. Once the scaling and the range of validity (over swirl ratios) of the simulators are known, the simulated tornadoes are comparable with full-scale tornadoes. The main requirements are simulation data over a wide range of swirl ratios as well as the availability of full-scale data of tornadoes with different structures and intensities. This full-scale database is presently under expansion.

2.1. Length and Velocity Scale Ratios

There are various geometric lengths in a simulator, such as updraft, inflow, and inner chamber height, as well as the core radius (r_c) and the height corresponding to the maximum tangential wind speed (z_{max}). Among these lengths, only the two last ones are flow related and potentially measurable in a real tornado: r_c and z_{max}. Therefore, two length scale ratios can be defined as the ratios between full-scale Doppler radar (index *D*) and simulation (index *S*) data: $r_{c,D} / r_{c,S}$ and $z_{max,D} / z_{max,S}$. The velocity scale of tornado simulation needs to be identified independent of the radial Reynolds number condition. As suggested by Refan et al. (2014), the ratio between the overall maximum tangential velocity of a real tornado and that of a simulated one ($V_{tan,max,D} / V_{tan,max,S}$) is selected to calculate the velocity length scales for each simulated tornado.

2.2. Matching Process

Establishing a relationship between simulated and real tornadoes is the main focus of this section. In order to achieve this, a set of experimental and full-scale data from tornadoes of various intensities and characteristics is required. The following are the main steps of the matching process:

1. Collect simulation (experimental/numerical) wind velocity data for tornado vortices at different swirl ratios.
2. For each swirl ratio, determine the overall maximum tangential velocity of the simulated tornado over various heights and radii, $V_{tan,max} = V_{tan}\left(r_{c,max}, z_{max}\right)$, and then compare it with that of the full-scale measurements. This way, the velocity scaling could be approximated.

3. For every given swirl ratio, determine the core radius and the height corresponding to the overall maximum tangential velocity for the simulated vortex ($r_{c,max,S}$ and $z_{max,S}$, respectively) and compare them to their counterparts in the natural tornado ($r_{c,max,D}$ and $z_{max,D}$, respectively). This provides the two length scale ratios ($r_{c,max,D} / r_{c,max,S}$ and $z_{max,D} / z_{max,S}$).
4. Plot the two length scales as a function of the swirl ratio. Since in proper fluid mechanics simulations the length scale must be a single value, it is expected that the two length scale ratios converge toward one value at a certain swirl ratio. This is a key condition that, if satisfied, may then be used to relate swirl ratio to the Fujita scale and therefore model tornado-like vortices (experimental or numerical) to full-scale tornadoes.
5. Two trends may be obtained: (1) the two length scales intersect at a certain swirl ratio or (2) the two length scale ratios show converging behavior toward a range of swirl ratios.
6. Based on the length scale ratio variation with swirl ratio, the following matching procedure is applied:
   (i) If there is a range of swirl ratios (rather than a single value) over which convergence/intersection occurs, the chosen swirl ratio is based on the vital structure of the tornado (i.e., single-celled, two-celled tornado, etc.).
   (ii) If there is a range of convergence that is consistent with the structure of the real vortex, the experimental results are scaled up using length scales corresponding to that range of swirl ratios, and the radial profiles of the tangential velocities at various heights are compared to the ones extracted from the full-scale data. The length scale results in the most accurate estimation of the maximum tangential velocity, and the corresponding core radius is then selected to represent the geometric scaling of the simulation. This point-by-point procedure is applied to all the full-scale (Doppler radar) tornado volumes available.
   (iii) If the difference between the two length scale ratios at the convergence is significant, the priority is given to the length scale determined using $r_{c,max,D} / r_{c,max,S}$.
7. Once the swirl ratio of the full-scale data and the length and velocity scale ratios of the simulation are determined the tangential velocity profiles of the simulation can be scaled up and compared with the full-scale data.

2.3. Example of Scaling Practice

Refan et al. (2014) applied the scaling from the model WindEEE (Wind Engineering, Energy, and Environment) Dome at Western University and full-scale Doppler radar data collected at that time. The Model WindEEE Dome (MWD) is a model of the WindEEE Dome scaled at 1/11; it was built to validate the conceptual design of WindEEE Dome. By using a system of 100 dynamic fans on the six peripheral walls coupled with 18 larger fans at the ceiling level, MWD can produce any type of wind system, including

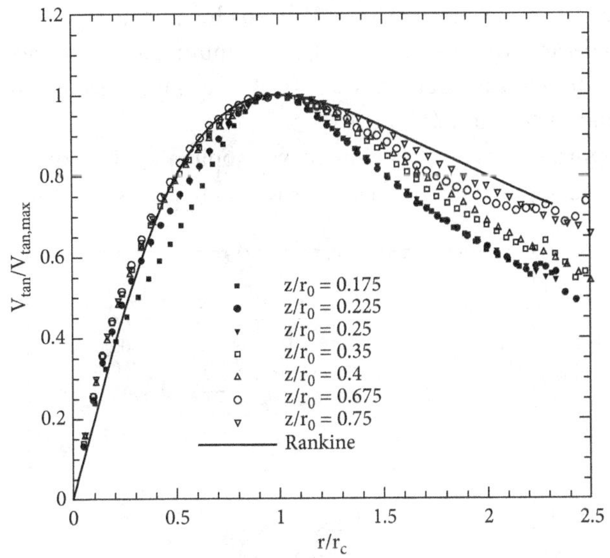

FIGURE 13.2 Radial profiles of the normalized tangential velocity at various heights for $S = 0.57$. Reproduced from Refan (2014).

translating tornadoes and downbursts as well as a variety of time-dependent shear flows. See Hangan (2014), Refan (2014), and Refan et al. (2014) for details of the simulator. Tornado-like vortices can be generated in this simulator using top fans as the source of suction and the adjustable periphery vanes at the lower chamber to control the swirl.

Tornado-like vortices of various structures and intensities were simulated in the MWD. Particle image velocimetry (PIV) was employed to characterize the velocity field of tornado-like vortices for various swirl ratios. The PIV system measured the mean velocity field of the simulated tornado-like vortices at eight swirl ratios and at eight different heights above the surface. Figure 13.2 shows time and azimuthally averaged tangential velocity (herein after called tangential velocities) variations with radius at various heights and for $S = 0.57$. The maximum tangential velocity and the corresponding radius (i.e., core radius) at each height are used to normalize the velocity and radius, respectively. Results are also compared with the Rankine vortex model.

Subsequently, full-scale Doppler radar data from various tornado events, provided by the Center for Severe Weather Research (CSWR), were analyzed. The ground-based velocity track display (GBVTD) method was employed to extract an axisymmetric three-dimensional structure of the parent vortex, mainly tangential velocity profiles at various heights from single-Doppler radar data of the Spencer, South Dakota, 1998 (F4); Stockton, Kansas, 2005 (F1); Clairemont, Texas, 2005 (F0); Happy, Texas, 2007 (EF0); and Goshen County (LaGrange), Wyoming, 2009 (EF2) tornadoes. Each tornado was studied at various instants of its life cycle. In total, nine volumes of data were analyzed with this mathematical method. The term *volume* refers to one complete radar scan of the tornado from regions very close to the ground to hundreds of meters aloft. As an example, Figure 13.3 shows the tangential velocity variation with radius retrieved for volume 2 of the Happy, Texas, tornado at various heights using the GBVTD method. Velocities and radii are normalized using the maximum tangential velocity and the core radius corresponding to each height, respectively. Results are again compared with the modified Rankine vortex model.

The scaling approach briefed previously was applied to the experimental and full-scale data collected. Figure 13.4 displays the two length scales' variation with swirl ratio

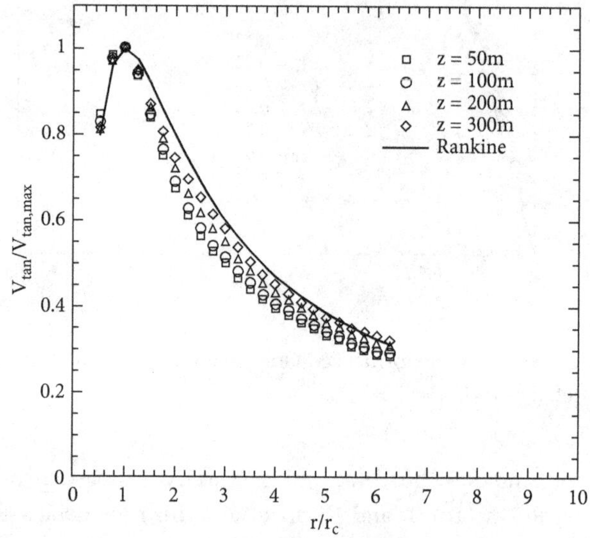

FIGURE 13.3 Radial profiles of the tangential velocity at different heights compared with the modified Rankine vortex model for Happy, Texas, 2007 (EF0). Reproduced from Refan (2014).

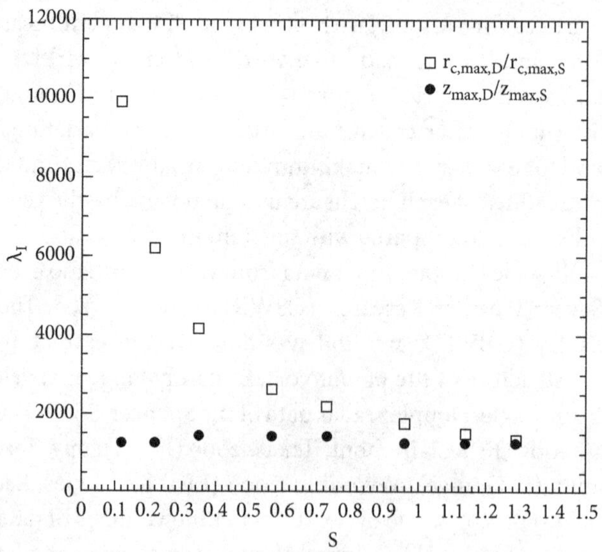

FIGURE 13.4 Geometric scaling ratio as a function of swirl ratio for the Happy, Texas, 2007 tornado from 02:03:20 to 02:04:17 UTC. Reproduced from Refan (2014).

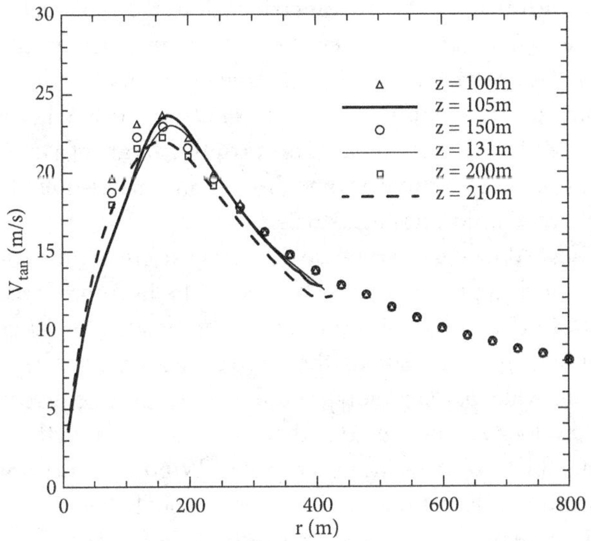

FIGURE 13.5 Comparison between simulated (solid lines) and full-scale (symbols) tangential velocity profiles at various heights for the Happy, Texas, 2007 tornado (from 02:03:20 to 02:04:17 UTC) after applying the velocity and length scale ratios. Reproduced from Refan (2014).

for one volume of the Happy, Texas, 2007, tornado from 02:03:20 to 02:04:17 UTC (coordinated universal time). It is observed that as the swirl ratio increases, the two length scales demonstrate a convergence behavior over a range of swirl ratios. Based on the GBVTD-retrieved velocity fields, this volume of the Happy, Texas, 2007, tornado is at the touch-down stage. As a result, the length scale associated with $S = 0.6$ was chosen for this event.

Finally the velocity Λ scale ($\Lambda_v = 1/2.75$) and the length scale ($\Lambda_l = 1/2{,}627$) for volume 4 of the Happy, Texas, 2007 tornado can be used to scale up the experimental tangential velocity profiles and then compare them with the full-scale data. Figure 13.5 shows that, overall, radial profiles of the tangential velocity obtained through PIV measurements match very well with the full-scale measurements at different heights. The agreement is observed inside and outside of the tornado core region as well as for the core radius and the maximum tangential velocity. This practice was performed for other volumes of data and resulted in good agreement between simulations of full-scale data.

3. Downburst Scaling

The first attempts to scale downburst were based on an artificial geometric scaling between the "diameter" of the full-scale downburst and that of the impinging jet simulating it (Chay & Letchford, 2002; Jesson & Sterling, 2018; Kim & Hangan, 2007; Mason & Wood, 2005; McConville, Sterling, & Baker, 2009). Alternative approaches used a velocity scale as a starting point complemented afterward by the use of a time scale and followed by the determination of the geometric scale as the product of these two scales. The velocity scale was based on either the maximum velocity in the time record (McConville et al., 2009) or the use of peak or gust factors (Junayed, Jubayer, Parvu, Romanic, Hangan, 2019). The time scales are usually determined as a measure of the ramp-up time to the peak velocity or based on an optimization of a moving average time (Junayed et al., 2019). Other approaches and discussions on downburst scaling were also proposed (Jesson, Lombardo, Sterling, & Baker, 2019).

However, full-scale downburst events span over a variety of time and velocity scales; therefore, a statistical approach to scaling seems to be more adequate. Romanic, Nicolini, Hangan, Burland, and Giovanni (2019) proposed a novel approach to scale downbursts in impinging jet simulations; the approach was based on matching the statistical properties of full-scale (*p*, i.e., prototype) and model (*m*) velocity records of measured wind speeds. All *p* velocity records used in their study, as well as in this overview of their work, are from the full-scale measuring campaigns Wind and Ports (Solari et al., 2012) and Wind, Ports and Sea (Repetto et al., 2018) carried out between 2009 and 2015 along the Mediterranean coast of Italy. The *m* velocity records were obtained from the downburst simulations in the WindEEE Dome (Hangan, 2010; Hangan et al., 2017) at Western University in Canada. Here, we review and comment on their methodology.

3.1. Statistical Scaling of Downbursts—Matching Process

The statistical scaling approach in Romanic, Nicolini, et al. (2019) starts by decomposing the instantaneous wind speed (v) of a downburst event into the slowly varying moving mean ($\overline{v}$) and the residual fluctuations (v') (Burlando, Romanić, Solari, Hangan, & Zhang, 2017; Holmes, Hangan, Schroeder, Letchford, & Orwig, 2008; Solari, Burlando, De Gaetano, & Repetto, 2015). The moving average window for the *p* velocity records is fixed at a value of $T_p = 30$ s. This decomposition procedure further enables the formulation of γ functions that depict the main signature of a transient downburst outflow through velocity ramp-up and slowdown after the maximum wind speed is reached ($\overline{v}_{max}$; Solari, De Gaetano, & Repetto, 2015):

$$\gamma(t) = \frac{\overline{v}(t)}{\overline{v}_{max}}, \tag{1}$$

where $\overline{v}_{max}$ is the maximum of $\overline{v}(t)$. Romanic, Nicolini, et al. (2019) proposed the following relation for the mean velocity scale (M_v) between *p* and *m* velocity records:

$$M_v = \frac{\overline{v}_{max,p}\gamma_p(t_p)}{\overline{v}_{max,m}\gamma_m(t_m)} \tag{2}$$

Further on, an iterative method was proposed to obtain the proper value of T_m:

1. For the *m* velocity record, assume a value of averaging window $T_m = T_{m,n}$.
2. For the chosen $T_{m,n}$ from the previous step, obtain $\overline{v}_{m,n}(t_{m,n})$ and $\gamma_{m,n}(t_{m,n})$ records.

3. Evaluate the mean velocity scale, $M_{v,n}\left(T_{m,n}\right)$.
4. The proper M_v is the value that provides the lowest root-mean-square error (RMSE) between the two records of $\gamma_p\left(t_p\right)$ and $\gamma_{m,n}\left(t_{m,n}\right)$.
5. In the last iterative step, determine the value of $T_{m,P}$ (the subscript P stands for "Proper") that resulted in the lowest value of $\text{RMSE}_P\left(T_{m,P}\right)$ and consequently compute $M_{v,P}$.

Last, the length scales are calculated from the proper time and velocity scales obtained using the iterative procedure outlined.

3.2. Scaling Results and Applications

Romanic, Nicolini, et al. (2019) reported the $\gamma_m\left(t_m\right)$ functions obtained from a database of 1,400 downburst records from the WindEEE Dome simulator. This set of wind data currently represents the largest database of experimentally produced downburst-like outflows. At the same time, Zhang, Solari, De Gaetano, Burlando, and Repetto (2018) analyzed the shapes of γ_p records of 277 downburst wind speed time series measured along the Italian coast in the Mediterranean. The resulting γ functions of both m and p events are depicted in Figure 13.6. The ensemble mean γ_p in Figure 6a (thick line) resembles the main signature of a downburst event through a sudden increase (also known as ramp-up) and decrease of the mean wind speed.

The time scales are obtained by the best match between γ functions of p and m downbursts. The shape of $\gamma_m\left(t_m\right)$ depends on the choice of the averaging period T_m in such a way that γ_m is steeper for smaller T_m and tends to flatten out with increasing T_m. Finding

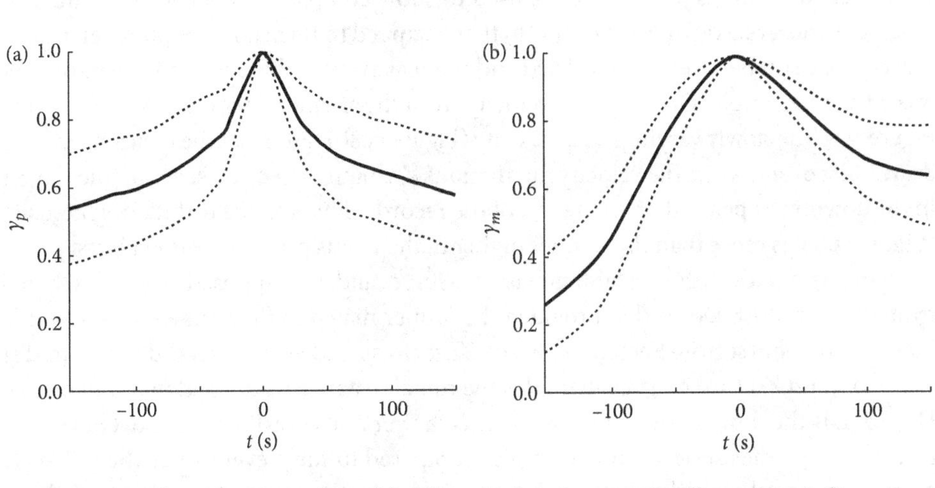

FIGURE 13.6 The γ functions from (a) full-scale downburst events from the Mediterranean (modified after Zhang et al., 2018, and Romanic, Nicolini, et al., 2019) and (b) the WindEEE Dome downbursts (modified after Romanic, Nicolini, et al., 2019). The full thick line represents the mean γ function, and the dotted lines are ±1 standard deviation around the mean.

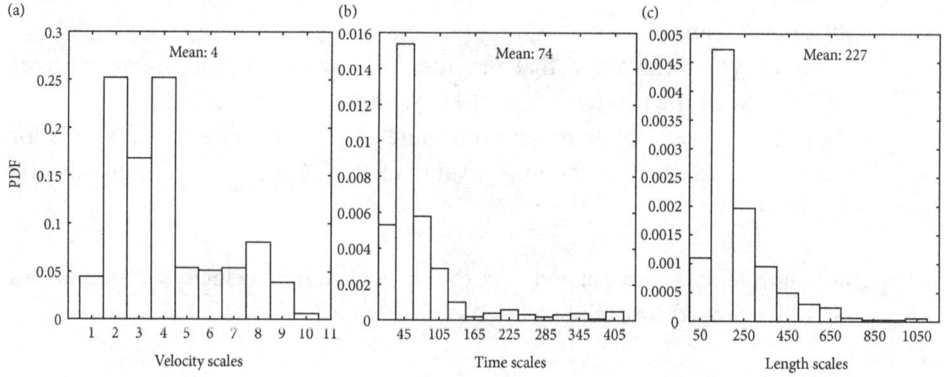

FIGURE 13.7 The WindEEE Dome downburst scales: (a) velocity (mean flow), (b) time, and (c) length scales (modified after Romanic, Nicolini, et al., 2019).

the proper averaging window $T_{m,P}$ is usually reduced to the problem of determining the value of T_m that gives the similar slope between the tails of γ_m and γ_p. Then, the velocity scales are obtained as the ratio of slowly varying velocity records that are evaluated using the averaging windows of $T_{m,P}$ and T_p = 30 s for the model and full-scale data, respectively. Last, the combination of these two scales provides the proper geometric scales.

Histograms of kinematic and geometric scales between *p* and *m* downburst records are shown in Figure 13.7. The velocity scales of *m* downburst in the WindEEE Dome and *p* events based on the Mediterranean database are typically between 1:2 and 1:4. The time scales are usually around 1:75. The combination of these two scales results in length scales on the order of 1:230 on average, but smaller length scales (i.e., 1:400 or below) are also fairly common.

The scaling process just detailed is used to show comparisons between a simulated (*m*) event and a real downburst event (*p*) that occurred in Livorno on September 4, 2011. The comparison between the modeled and the real events is based on one anemometer record as an example. Decomposed velocity from the anemometer LI.05 is presented in Figure 13.8. The slowly varying mean velocity (Figure 13.8b) resembles the typical shape of a downburst event, while the velocity fluctuations v'_p (Figure 13.8c) are several times larger in the downburst peaks than in the rest of the record. However, the turbulence intensity (Figure 13.8e) is more than three times higher in the winds prior to the downburst peak.

Figure 13.9 shows the instantaneous measured and decomposed velocity records from the WindEEE Dome that provided the proper match to the *p* time series from the Livorno downburst from September 4, 2011. The presented velocity record is captured at R/D = 0.2 and Z/D = 0.08438, where *R* is the radial distance from the downdraft center, D = 3.2 m is the downdraft diameter, and *Z* is the vertical coordinate. The overall structure of the *m* times series is much simpler compared to the *p* event since the full-scale events are complicated by factors such the background atmospheric boundary layer (ABL) winds, thunderstorm cloud motion, surface roughness changes, and atmospheric stability, to name just a few. We also observe that the level of turbulence intensity

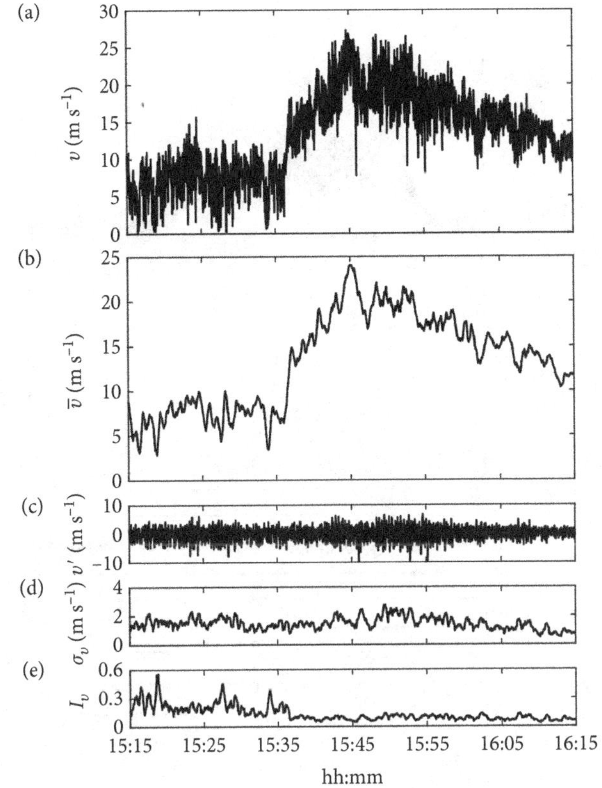

FIGURE 13.8 Decomposed wind velocity of the p downburst that occurred in Livorno on September 4, 2011, for the anemometer LI.05 (modified after Nicolini, 2018).

in the m record (Figure 9e) is higher than in the p event (Figure 8e) during the downburst peak.

The reduced turbulence fluctuations ($\tilde{v}'$) and the associated histograms and spectral plots for the analyzed p and m velocity records are shown in Figure 13.10. The most evident difference between the two records is the quasi-deterministic ramp-up segment in the m event (Figure 13.10d; centered at around 2 s into the time series). While both histograms of $\tilde{v}'$ are fairly Gaussian, the stronger non-Gaussianity is observed in the m record, probably due to the same quasi-deterministic ramp-up feature of the event. The empirical histogram of the m record is positively skewed, while the p event is characterized with a negatively skewed histogram of $\tilde{v}'$. Both $\tilde{v}'$ records are characterized by small deviations from the −5/3 slope in the inertial subrange of turbulence.

The obtained γ functions (Figure 13.11) and the scaling procedure outlined in Section 3.1 are further used to evaluate the mean flow velocity scale (M_v) and the length scale (Λ_l) in Figure 13.12. As a validation of this scaling methodology, Figure 13.13 shows the predicted height of the LI.05 anemometer in Livorno that recorded the investigated p velocity record. The computed anemometer height from the obtained length scale is

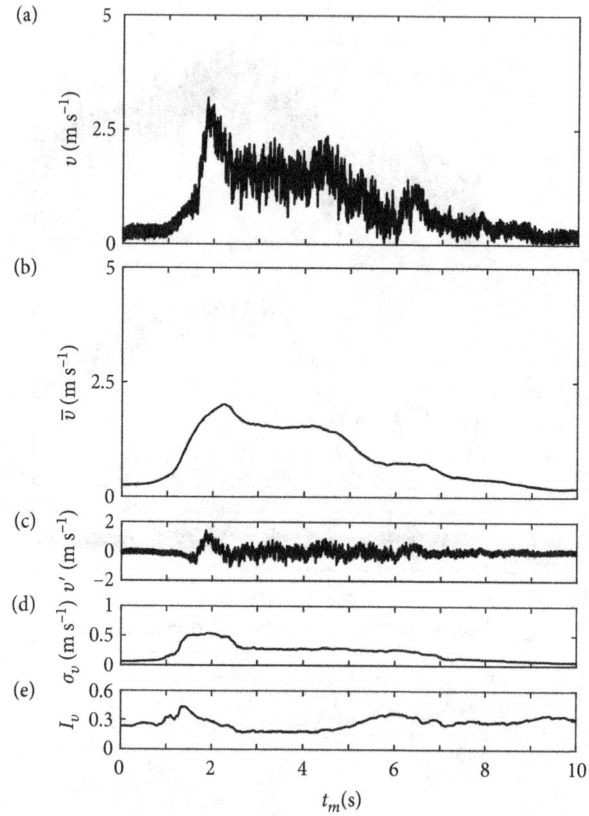

FIGURE 13.9 Decomposed wind velocity of the m downburst from the WindEEE Dome that showed the best match with the p record in Figure 13.8. The velocity record obtained at $R/D = 0.2$ and $Z/D = 0.08438$ (modified after Nicolini, 2018).

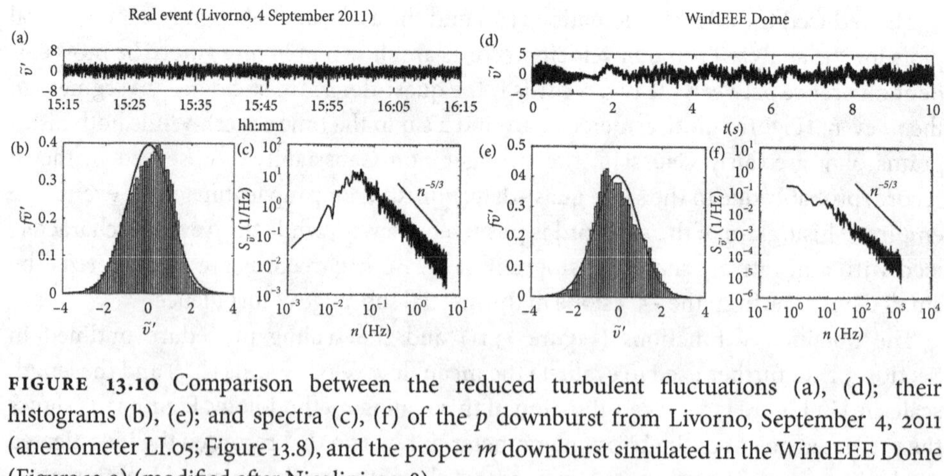

FIGURE 13.10 Comparison between the reduced turbulent fluctuations (a), (d); their histograms (b), (e); and spectra (c), (f) of the p downburst from Livorno, September 4, 2011 (anemometer LI.05; Figure 13.8), and the proper m downburst simulated in the WindEEE Dome (Figure 13.9) (modified after Nicolini, 2018).

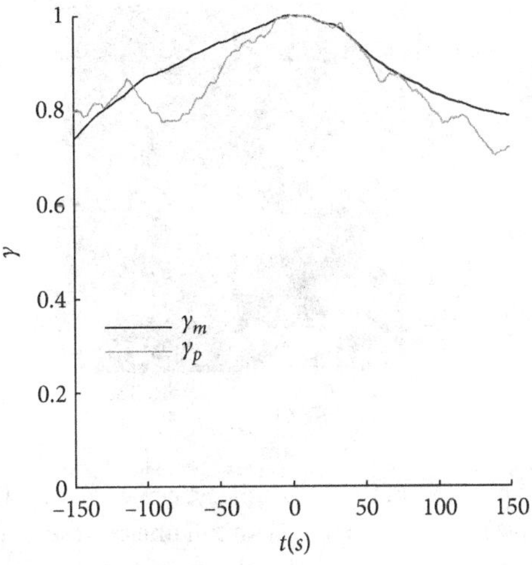

FIGURE 13.11 The γ functions for the p downburst from Livorno, September 4, 2011 (anemometer LI.05; Figure 13.8), and the proper m downburst simulated in the WindEEE Dome (Figure 13.9) (modified after Nicolini, 2018).

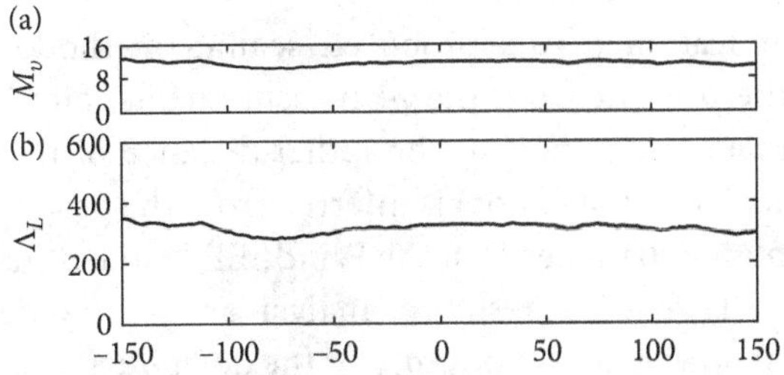

FIGURE 13.12 The obtained scales between the WindEEE Dome downburst (Figure 13.9) and the Livorno, September 4, 2011 (anemometer LI.05; Figure 13.8), events: (a) mean velocity scales and (b) length scales (modified after Nicolini, 2018).

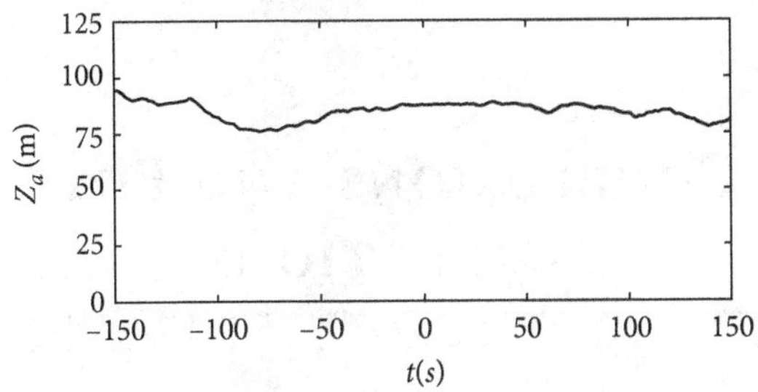

FIGURE 13.13 Evaluated height of the anemometer LI.05 (Z_a) in the Livorno, September 4, 2011, event using the calculated length scale from the WindEEE Dome in Figure 13.12. The real anemometer height is 75 m above ground level (modified after Nicolini, 2018).

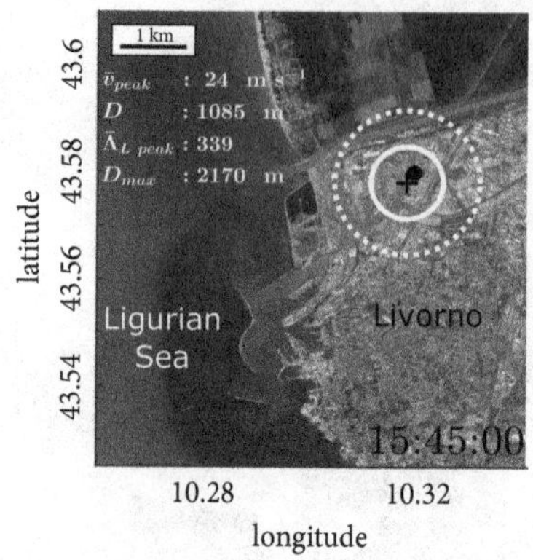

FIGURE 13.14 Estimated position of the *p* downburst center at 15:45 UTC (×) and the location of the LI.05 anemometer (●). The white circle with full line represents the estimated diameter of the *p* downdraft, while the white dashed circle is the estimated diameter of the maximum radial velocity (D_{max}) in the *p* outflow of the September 4, 2011, Livorno downburst (modified after Nicolini, 2018).

approximately 80 m above ground, while the real height of the LI.05 anemometer is at 75 m above ground (Solari et al., 2012).

Last, another application of the proposed scaling methodology is the partial reconstruction of the *p* event from the *m* measurements. Similar to the previous prediction of the anemometer height, the radial distance of the downburst center from the measurement location (LI.05) is inferred from the proper *m* velocity record. The employed downburst diameter *D* in the WindEEE Dome is then used to estimate the diameter of the real event. The results of analysis are portrayed in Figure 13.14. The estimated downburst downdraft is 1,085 m, and the downburst touchdown center was approximately 220 m southwest from the LI.05 anemometer. The azimuthal direction of the downburst from the anemometer was estimated from the wind direction record (not shown) of this event. This thunderstorm downburst cloud—similar to the majority of downbursts in this region (Burlando, Zhang, & Solari, 2018)—was formed over the sea and then moved over land. The downburst location in Figure 13.14 is shown at the moment of the maximum velocity in the slowly varying wind speed record ($\bar{v}_p$ = 24 m s^{-1}; Figure 8b).

4. Conclusions and Future Directions

This chapter introduces recent results aiming to compare full-scale data and simulations of tornadoes and to establish a framework (Refan et al., 2014) through which simulations of tornado-like vortices (both physical and numerical) can be compared to full-scale data. It is shown that the tornado geometric scaling involves two length scales corresponding to the height and the radius corresponding to the maximum tangential velocity. When plotted against a varying swirl ratio, the two scales merge to a common scaling value. This frame for establishing a tornado scaling can be applied for any numerical and physical simulations, and it allows for proper translation of results from model to full-scale results.

The chapter also introduces the last results related to the scaling of downburst events. A statistical approach proposed by Romanic, Nicolini, et al. (2019) compares the γ functions (defined as the ratio between the slowly varying mean velocity and its maximum value) between modeled and full-scale downburst events. The best match between γ_p and γ_m is obtained through an iterative procedure that finds the best averaging window of the modeled data (p stands for "prototype" and m stands for "model").

This scaling method was then tested on 1,400 m downburst-like records experimentally simulated in the WindEEE Dome (Hangan et al., 2017) at Western University in Canada and on 17 p records from the Mediterranean in Italy measured under the scope of the European project Wind and Ports (Solari et al., 2012). The same scaling method can then be used to correlate the position and spatial extent of real and simulated events.

While the presented methodologies for scaling tornadoes and downbursts are quite different, they both focus on the main flow field. Once more full-scale and simulation data become available for the fluctuating velocity fields, it is important to incorporate turbulence scaling in the two frameworks. In addition, the current methodologies of downburst scaling will have to be modified or new methodologies developed in order to account for physically produced downburst outflows embedded into ABL winds (Romanic, LoTufo, & Hangan 2019a).

References

Burlando, M., Romanić, D., Solari, G., Hangan, H., & Zhang, S. (2017). Field data analysis and weather scenario of a downburst event in Livorno, Italy, on 1 October 2012. *Monthly Weather Review, 145*(9), 3507–3527. doi:10.1175/MWR-D-17-0018.1.

Burlando, M., Zhang, S., & Solari, G. (2018). Monitoring, cataloguing, and weather scenarios of thunderstorm outflows in the northern Mediterranean. *Natural Hazards and Earth System Sciences, 18*(9), 2309–2330. doi:10.5194/nhess-18-2309-2018.

Chay, M. T., & Letchford, C. W. (2002). Pressure distributions on a cube in a simulated thunderstorm downburst—Part A: Stationary downburst observations. *Journal of Wind Engineering and Industrial Aerodynamics, 90*(7), 711–732. doi:10.1016/S0167-6105(02)00158-7.

Hangan, H. (2010). Current and future directions for wind research at western: A new quantum leap in wind research through the wind engineering, energy and environment (WindEEE) Dome. *Wind Engineers, JAWE, 35*(4), 277–281. doi:10.5359/jawe.35.277.

Hangan, H. (2014). The wind engineering energy and environment (WindEEE) Dome at Western University, Canada. *Wind Engineers, JAWE, 39*(4), 350–351.

Hangan, H., & Kim, J. (2008). Swirl ratio effects on tornado vortices in relation to the Fujita scale. *Wind and Structures, 11*(4), 291–302.

Hangan, H., Refan, M., Jubayer, C., Romanic, D., Parvu, D., LoTufo, J., & Costache, A. (2017). Novel techniques in wind engineering. *Journal of Wind Engineering and Industrial Aerodynamics, 171*, 12–33. doi:10.1016/j.jweia.2017.09.010.

Holmes, J. D., Hangan, H. M., Schroeder, J. L., Letchford, C. W., & Orwig, K. D. (2008). A forensic study of the Lubbock-Reese downdraft of 2002. *Wind and Structures, 11*(2), 137–152. doi:10.12989/was.2008.11.2.137.

Jesson, M., Lombardo, F. T., Sterling, M., & Baker, C. (2019). The physical simulation of a transient, downburst-like event—How complex does it need to be? *Journal of Wind Engineering and Industrial Aerodynamics, 189*, 135–150. doi:10.1016/j.jweia.2019.03.021.

Jesson, M., & Sterling, M. (2018). A simple vortex model of a thunderstorm downburst—A parametric evaluation. *Journal of Wind Engineering and Industrial Aerodynamics, 174*, 1–9. doi:10.1016/j.jweia.2017.12.001.

Junayed, C., Jubayer, C., Parvu, D., Romanic, D., & Hangan, H. (2019). Flow field dynamics of large-scale experimentally produced downburst flows. *Journal of Wind Engineering and Industrial Aerodynamics, 188*, 61–79. doi:10.1016/j.jweia.2019.02.008.

Kim, J., & Hangan, H. (2007). Numerical simulations of impinging jets with application to downbursts. *Journal of Wind Engineering and Industrial Aerodynamics, 95*(4), 279–298. doi:10.1016/j.jweia.2006.07.002.

Mason, M. S., & Wood, G. S. (2005). Influence of jet inclination on structural loading in an experimentally simulated microburst. paper presented at the 6th Asia-Pacific Conference on Wind Engineering. Seoul, South Korea.

McConville, A. C., Sterling, M., & Baker, C. J. (2009). The physical simulation of thunderstorm downbursts using an impinging jet. *Wind and Structures, 12*(2), 133–149.

Nicolini, E. (2018). *A new approach for downburst scaling in experimentally produced downburst-like impinging jets* (Unpublished master's thesis). University of Genoa, Italy.

Refan, M. (2014). *Physical simulation of Tornado-like vortices* (PhD dissertation). University of Western Ontario, London, Ontario, Canada.

Refan, M., Hangan, H., & Wurman, J. (2014). Reproducing Tornadoes in Laboratory Using Proper Scaling. *Journal of Wind Engineering and Industrial Aerodynamics, 135*, 136–148.

Repetto, M. P., Burlando, M., Solari, G., De Gaetano, P., Pizzo, M., & Tizzi, M. (2018). A web-based GIS platform for the safe management and risk assessment of complex structural and infrastructural systems exposed to wind. *Advances in Engineering Software, 117*, 29–45. doi:10.1016/j.advengsoft.2017.03.002.

Romanic, D., LoTufo, J., & Hangan, H. (2019). Transient behavior in impinging jets in cross-flow with application to downburst flows. *Journal of Wind Engineering and Industrial Aerodynamics, 184*, 209–227. doi:10.1016/j.jweia.2018.11.020.

Romanic, D., Nicolini, E., Hangan, H., Burlando, M., & Giovanni, S. (2019). A novel approach to scaling experimentally produced downburst-like impinging jet outflows. *Journal of Wind Engineering and Industrial Aerodynamics*, In review.

Solari, G., Burlando, M., De Gaetano, P., & Repetto, M. P. (2015). Characteristics of thunderstorms relevant to the wind loading of structures. *Wind and Structures, 20*(6), 763–791. doi:10.12989/was.2015.20.6.763.

Solari, G., De Gaetano, P., & Repetto, M. P. (2015). Thunderstorm response spectrum: Fundamentals and case study. *Journal of Wind Engineering and Industrial Aerodynamics, 143*, 62–77. doi:10.1016/j.jweia.2015.04.009.

Solari, G., Repetto, M. P., Burlando, M., De Gaetano, P., Pizzo, M., Tizzi, M., & Parodi, M. (2012). The wind forecast for safety management of port areas. *Journal of Wind Engineering and Industrial Aerodynamics, 104–106*: 266–277. doi:10.1016/j.jweia.2012.03.029.

Zhang, S., Solari, G., De Gaetano, P., Burlando, M., & Repetto, M. P. (2018). A refined analysis of thunderstorm outflow characteristics relevant to the wind loading of structures. *Probabilistic Engineering Mechanics, 54*: 9–24. doi:10.1016/j.probengmech.2017.06.003.

CHAPTER 14

OTHER MODELS: DATA-DRIVEN APPROACHES FOR NON-STATIONARY NON-SYNOPTIC WINDS

DAE KUN KWON AND AHSAN KAREEM

1. Introduction

A data-driven approach has gained its popularity in many fields mainly due to the rapid advances in measurement and monitoring systems that enable the collection of long-term data. In the context of the treatment of time-dependent non-stationary wind events such as hurricanes and typhoons; thunderstorms and downbursts, etc., data-driven approaches can work in terms of characterization, modeling, and simulation. Indeed, any empirical models can be regarded as a data-driven approach because those models are usually derived from numerical, experimental, or full-scale data. From this viewpoint, a data-driven model differs from mechanics-based models, which are generally formulated on the basic mechanics involved. An example data-driven modeling in non-stationary wind events is to establish an empirical non-stationary wind spectrum such as an evolutionary wind spectrum (e.g., Chen and Letchford 2004a, 2004b; Chay et al. 2006; Wang 2007; Kareem 2008; Kwon and Kareem 2009, 2013a, 2019; Kwon et al. 2012; Hu et al. 2013, 2017a; Zhou et al. 2015; Huang et al. 2015; Solari 2016; Solari et al. 2017; Solari and De Gaetano 2018; Wang et al. 2016, 2018, 2019; Zhao et al. 2019; Kareem et al. 2019). Data-driven simulation approaches that utilize the instantaneous features of available data can offer a straightforward and statistically meaningful approach to non-stationary simulation (e.g., Wang et al. 2002; Wang and Kareem 2004; Wang 2007; Kareem 2008; Wang et al. 2013, 2014; McCullough 2016, Peng et al. 2017, Hu et al. 2017b,

Huang et al. 2020). In addition, it is very important to extract or decompose time-dependent components from non-stationary data—*detrending*—to better understand non-stationary wind characteristics (e.g., McCullough et al. 2014; McCullough 2016).

A major drawback of such data-driven approaches inevitably involves the paucity of data. To date, only limited non-synoptic wind datasets are available due to the inherent nature of time dependency, which often hinders the full-scale capture of such events. More attempts at full-scale capture of these events may lead to better identifying time dependency and wind characteristics such as gust factor (GF), turbulence intensity (TI), turbulence integral length scale (TILS), etc. Another potential issue is the reliability of measured data and potential biases. For example, Solari et al. (2015a) reported the characteristics of measured full-scale thunderstorms and downbursts datasets that showed significant spatial variability among extreme events and their measurement locations. Utilizing the simulation-based approach, one could design structures for a prescribed event and its statistical emulations to form a suite of time histories. A time-domain approach may then be pursued to estimate load effects, and statistical averages may be derived based on these simulations. Akin to an earthquake, one may utilize intensity scaling of non-synoptic winds for design purposes to select a most probable event or one associated with a given return period.

Assuming that a large set of reliable datasets is available, the very first choice for codification is to extract salient features from the datasets then fit empirical model(s) for conventional code and standard formats. However, in light of uncertainties, a data-driven design involving uncertainty quantification may offer a better option to probabilistically evaluate non-synoptic wind effects on structures. For a better description of such effects, it is desirable to identify the probabilistic and statistical information of wind characteristic parameters from datasets such as statistical moments (mean, standard deviation, etc.) and probability distributions. Then, this information involving uncertainties associated with parameters is taken into account in a design procedure similar to the treatment used in codes and standards.

This chapter reviews data-driven approaches employed in the field of non-stationary non-synoptic winds for their modeling, simulation, and wind characteristics perspective. Note that because of data availability, wind events such as hurricanes and typhoons and thunderstorms and downbursts are mainly discussed here.

2. Identification of Time-Dependent Feature

2.1. Non-Stationary Wind Velocity Model

In a stationary wind velocity model, the velocity is typically decomposed into mean and fluctuation wind components, which may not be valid for non-stationary non-synoptic winds such as thunderstorms and downbursts; hurricanes, typhoons, and cyclones, etc. Thus, a non-stationary model has generally been employed instead (e.g., Chen and Letchford 2004a, 2004b; Wang and Kareem 2004; Xu et al, 2005; Chay et al. 2006; Kwon and Kareem 2009, 2013a; 2019; Kwon et al. 2012; Hu et al. 2013; Solari et al. 2015b; Solari 2016; Solari and De Gaetano 2018; Wang et al. 2016, 2018, 2019; Kareem et al. 2019).

$$U(z,t) = \bar{U}(z,t) + u(z,t) \tag{1}$$

where z and t are the spatial and temporal variables, respectively; $U(z, t)$ is the non-stationary wind; and $\bar{U}(z,t)$ is the time-varying mean component of non-stationary wind. The mean wind speed varies continuously in time in an attempt to better represent the trends and capture embedded time-varying frequency components; $u(z,t)$ is the fluctuating component of non-stationary wind, which may also be non-stationary and may be treated as a product of a non-stationary standard deviation [$\sigma(z,t)$] and zero-mean Gaussian process [$\bar{u}(z,t)$] with unit variance for simplicity.

$$u(z,t) = \sigma(z,t)\bar{u}(z,t) \tag{2}$$

An intrinsic roadblock in this model is the presence of time-varying features and spatiotemporal dependency. One way to overcome this complexity, a decomposition in space (z) and time (t), may be introduced.

$$\begin{aligned} \bar{U}(z,t) &= \bar{U}(z)\cdot\bar{U}(t) \\ \sigma(z,t) &= \sigma(z)\cdot\sigma(t) \end{aligned} \tag{3}$$

where $\bar{U}(z)$ is the vertical profile of non-stationary wind; and $\bar{U}(t)$ is the (normalized) temporal variation of non-stationary wind. Indeed, Equation (3) implies that all time functions along the heights are fully correlated without a phase difference, which is an ideal condition. However, recent studies reported that, in some cases, spatiotemporal dependence has been noted for the cases of thunderstorms and downbursts (Lombardo et al. 2014; Scott Gunter and Schroeder 2015). For the fluctuating component, the decoupling in Equation (3) may be alleviated with the introduction of a coherence function, but, for the time-varying mean component, it may yield an overestimation strongly depending on the selection of the vertical profile of non-stationary winds [$\bar{U}(z)$]. The time function [$\bar{U}(t)$] is generally treated in a normalized analytical format for thunderstorm and downburst cases (e.g., Holmes and Oliver 2000; Chen and Letchford 2004a, 2004b; Chay et al. 2006; Kwon and Kareem 2009, 2013a, 2019; Solari 2016; Solari and De Gaetano 2018; Kareem et al. 2019); that is, the maximum amplitude of the time function is equal to unity and wind velocities being described in the vertical profile, whereas, for hurricanes and typhoons, it generally relies on a data-based approach without establishing a specific analytical model (i.e., detrending time-varying mean component from measured data; see, e.g., Hu et al. 2013; Wang et al. 2018; Kareem et al. 2019).

The vertical profile of thunderstorm and downburst winds is critical in evaluating the wind effects on structures, which is different from that of boundary layer winds bearing an inverted velocity profile with its maxima near the ground, thus giving it a nose-shaped profile (e.g., Fujita 1985; Hjelmfelt 1988). Several analytical/empirical models to describe the vertical profile of thunderstorm and downburst winds have been proposed based on limited full-scale measurements, wind tunnel experiments, or computational fluid dynamics (CFD) (e.g., Oseguera and Bowles 1988; Vicroy 1991, 1992; Wood et al. 2001; Chay et al. 2006). For the case of hurricane/typhoon winds, the wind profile usually shows a logarithmic increase to a maximum (e.g., near 500 m), then a gradual decrease with height after reaching the maximum, for which semi- and analytical models have been proposed because of model parameters such as eddy viscosity, etc. (e.g., Meng et al. 1995, 1997; Franklin et al. 2003; Powell et al. 2003; Vickery et al. 2009; Giammanco et al. 2013; Li et al. 2015c; Fang et al. 2018).

Note that although this separable type description may be an adequate representation of the velocity field for codification, certainly this model may be modified should additional data suggest an alternative description.

2.2. Detrending Non-Stationary Winds

It is a natural fit to use time-frequency tools for time-dependent non-stationary non-synoptic winds. With the development of time-frequency analysis tools such as wavelet transform (WT), empirical mode decomposition (EMD), blind source separation (BSS), etc., it is possible to separate slowly varying (low-frequency) trends from the quickly varying (high-frequency) components of the signal typically observed in non-synoptic winds (e.g., Farge 1992; Gurley and Kareem 1999; Hayashi 1994; Huang et al. 1998, 1999; Kareem and Kijewski 2002; Rilling et al. 2003; Wang and Kareem 2004; Xu et al. 2005; Wang et al. 2013, 2014; McCullough et al. 2014; Guo 2015; Guo and Kareem 2016; Kareem et al. 2019). In this section, an analysis framework that uses three different approaches (i.e., WT, EMD, and BSS) to extract the time-varying mean wind speed is discussed.

2.2.1. *Non-Synoptic Wind Datasets*

Two non-stationary wind speed datasets are utilized throughout this chapter. One dataset was measured during the passage of the 2002 hurricane known as Hurricane Lili by the Florida Coastal Monitoring Project (FCMP 2011) and consists of 19 consecutive hours of data sampled at 0.33 Hz. The data were recorded at Tower 3 in Lydia, Louisiana, which is east of the hurricane trajectory. The other dataset is a rear-flank downdraft (RFD; i.e., downburst data) obtained from the 2002 thunderstorm outflow field experiment by the Department of Atmospheric Science of Texas Tech University. The data were measured from Tower 4 at a height of 10 meters, and the sampling rate and period are 1.0 Hz and 30 minutes, respectively (Gast and Schroeder 2003; Orwig and Schroeder 2007). The time histories of the two wind speed datasets, shown in

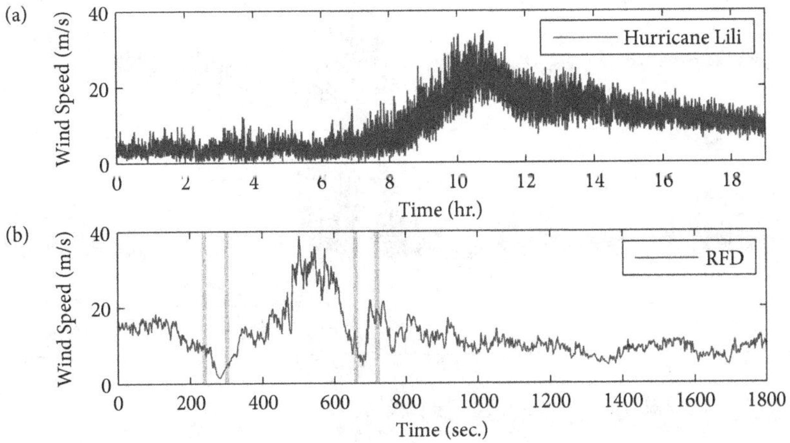

FIGURE 14.1 Time histories of two non-stationary wind datasets: (a) Hurricane Lili and (b) rear-flank downdraft (RFD).

From McCullough et al. (2014).

Figure 14.1, have somewhat different non-stationary characteristics. The RFD data in Figure 1b shows more rapid changes in wind speed over shorter periods than does the Hurricane Lili data in Figure 1a. In addition, it should be noted that two regions of the RFD data, 240~302 seconds and 661~720 seconds, denoted by vertical dotted lines, indicate sudden changes in wind direction, which may produce erroneous statistical values (Orwig and Schroeder 2007). These two non-stationary datasets will be mainly used throughout subsequent sections of this chapter.

2.2.2. *Wavelet Analysis*

Wavelet analysis is a convenient tool to obtain the time-frequency information of a process, and it has found several engineering applications in recent years. The analysis uses a set of basis functions that are simply dilations and translations of a parent wavelet function to decompose a broadband signal into a collection of successively more band-limited components (Kareem and Kijewski 2002). At each level, the low-frequency output of the previous level is decomposed into adjacent high- and low-frequency sub-bands by a pair of high-pass and low-pass filters. For example, the Daubechies wavelet of order 15 within the discrete wavelet transform (DWT) algorithm was utilized to demonstrate the decomposition of the Hurricane Lili signal using DWT at six levels with one approximation (Figure 14.2). As the number of levels increases, the frequency components filtered out become lower and the approximation better represents the time-varying mean.

The approximation shown in Figure 14.2 corresponds to the 1-hour time-varying mean. To obtain the target mean information, the original signal may need to be resampled before wavelet analysis. For example, to get the 10-minute (600-sec.) mean using a seven-level decomposition (six levels and one approximation), the data need to be resampled such that the sampling frequency is $2^7/600$ Hz (128/600 Hz). Figure 14.3 shows a

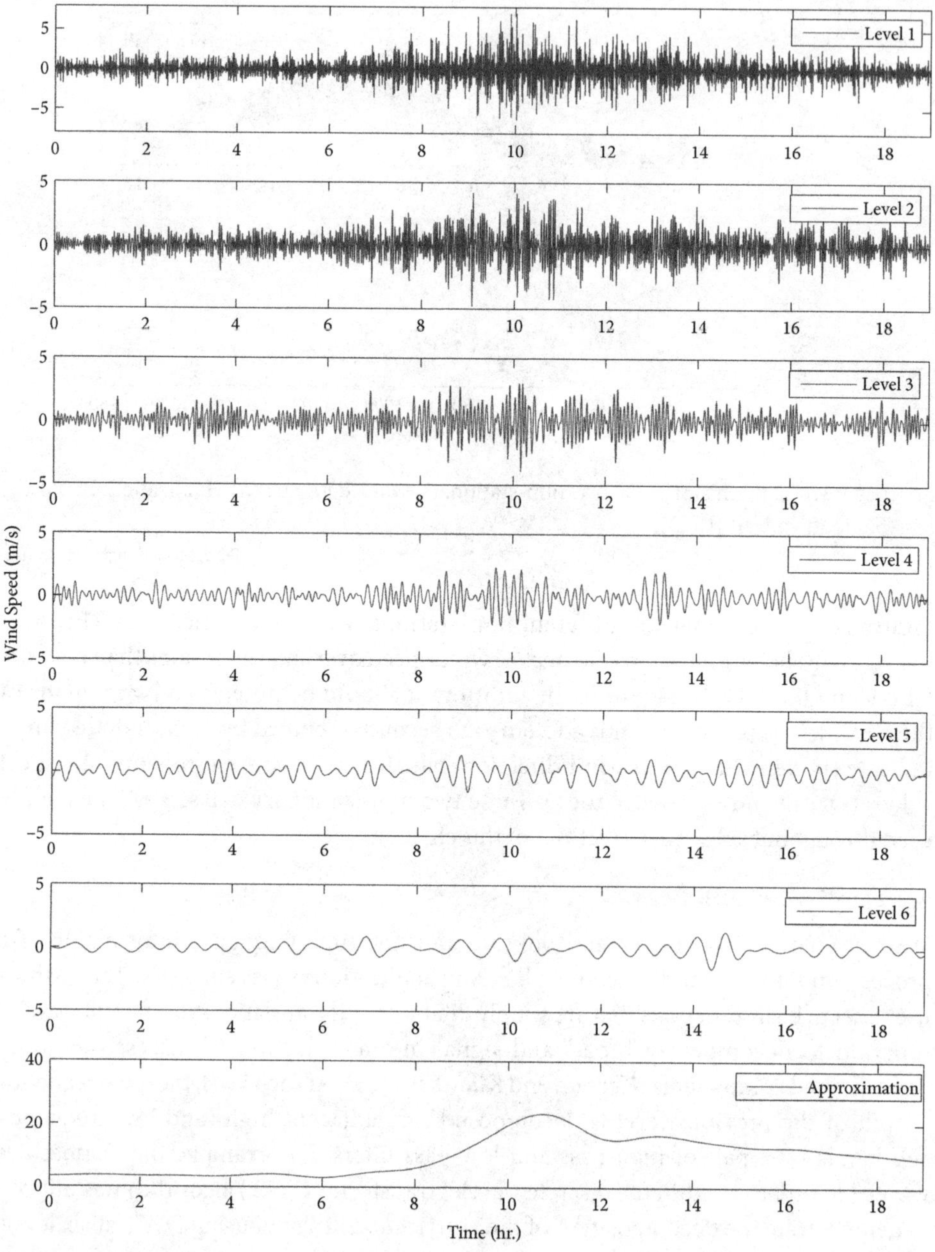

FIGURE 14.2 Discrete wavelet transform decomposition of the Hurricane Lili data for 1-hour averaging interval size.

From McCullough et al. (2014).

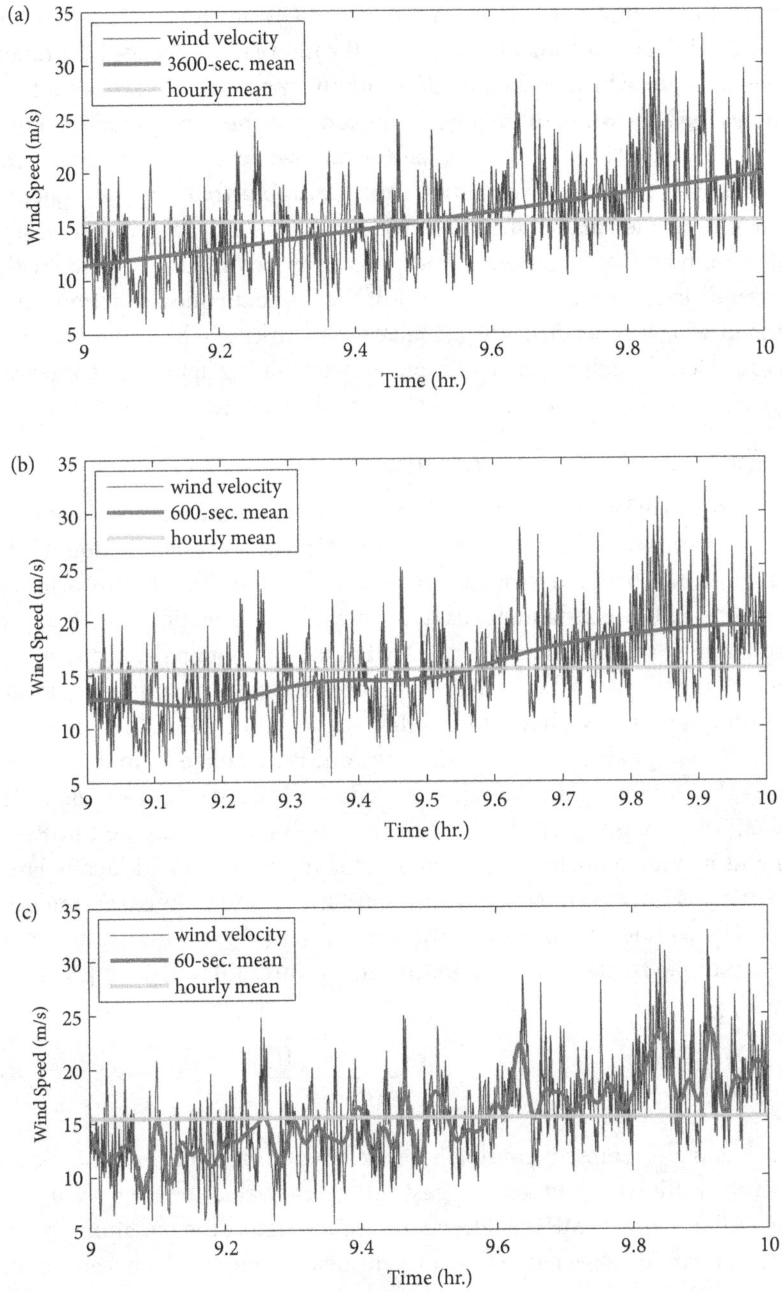

FIGURE 14.3 Comparison of constant hourly mean with time-varying mean estimated using discrete wavelet transform (DWT) for the ninth hour of the Hurricane Lili data: (a) time-varying 1-hour mean; (b) time-varying 10-minute mean; (c) time-varying 1-minute mean.

From McCullough et al. (2014).

comparison of the constant hourly mean with the time-varying 1-hour mean (Figure 3a), 10-minute mean (Figure 3b), and 1-minute mean (Figure 3c) for the ninth hour of the Hurricane Lili data. Therefore, the choice of the interval size dictates the resampling of the necessary data, which, in conjunction with the specification of the number of levels, determines the bandwidth of frequencies included in the approximation. For the same number of decomposition levels, smaller interval sizes lead to a larger bandwidth of frequencies included in the approximation function, as illustrated by the higher frequency components included in the time-varying mean. Conversely, larger interval sizes lead to a smaller bandwidth of frequencies, leading to an approximation function with only the lowest frequencies. The exact bandwidth of thc approximation function can be easily determined using wavelet theory (e.g., Mallat 1998; Gurley and Kareem 1999; McCullough et al. 2014; McCullough 2016). This topic of determining appropriate interval size for non-synoptic winds deserves more study and will be further discussed in later sections.

2.2.3. *Empirical Mode Decomposition*

EMD, introduced in Huang et al. (1998, 1999), also facilitates the analysis of a non-stationary signal by decomposing the signal into a series of intrinsic mode functions (IMFs) and a residue. The IMFs, believed to be band-limited, allow the Hilbert transform (HT) operation, which yields the instantaneous frequency (IF) information. In this approach, the same resampling procedure used with DWT is used to obtain data consistent with the different averaging interval sizes. Next, a procedure known as *sifting* uses local maxima and minima together with a cubic spline fitting to produce the upper and lower envelope of the signal. The mean of the envelope is extracted from the signal, and the residue is then subjected to the same procedure until specific stopping criteria have been met. The stopping criteria used in this analysis are presented in Rilling et al. (2003) and include globally small fluctuations in the mean and locally large excursions. Sifting is iterated until an evaluation function for some fraction of the total duration $(1 - \alpha)$ is less than the first threshold $[\sigma(t) < \theta_1]$ and the evaluation function for the remaining fraction is less than the second threshold $[\sigma(t) < \theta_2]$. The evaluation function is given as:

$$\sigma(t) = \left| \frac{e_{\max}(t) + e_{\min}(t)}{e_{\max}(t) - e_{\min}(t)} \right| \tag{4}$$

where $e_{\max}$ and $e_{\min}$ represent envelopes defined by the local maxima and local minima, respectively. Rilling et al. (2003) suggest setting the values to $\alpha \approx 0.05$, $\theta_1 \approx 0.05$, and $\theta_2 \approx 10\theta_1$. After the first IMF is subtracted from the signal, the remaining is treated as a new time series and subjected to the same sifting procedure, which leads to the second IMF. The EMD procedure continues until no more meaningful IMFs are obtained. The remaining residue represents the time-varying mean, similar to the approximation shown for the DWT decomposition in Figure 14.2.

Note that Kijewski-Correa and Kareem (2007) have attempted to compare the performance between continuous WT and EMD with HT in extracting signals embedded in both high and low noise levels. They reported that whereas both approaches can capture the instantaneous frequency in a mean sense, regardless of the noise level, signal extraction and reconstruction from the empirical bases of EMD with HT can be problematic as noise levels increase, thus wavelet transforms might provide a more reliable alternative for such analyses. More recent advances in EMD formulation may overcome this disadvantage.

2.2.4. *Blind Source Separation*

The BSS methods have shown great potential in separating both stationary and non-stationary signals. There have been several studies on different BSS algorithms since the 1980s, and they are widely used in many disciplines, such as biomedical signal processing, voice recognition, image reconstruction, financial data analysis, vibration problems, etc. Several elegant algorithms emerged in the electrical engineering field, such as independent component analysis (ICA) (Hyvärinen et al. 2001), algorithm for multiple unknown signals extraction (AMUSE) (Tong et al. 1990), second-order blind identification (SOBI) (Belouchrani et al. 1997), time-frequency blind source separation (TFBSS) based on spatial time-frequency distribution (STFD) (Belouchrani and Amin 1998; Guo 2015; Guo and Kareem 2016), etc.

A significant feature of BSS methods is indicated by the term "blind," which means that only general assumptions about sources (e.g., their statistical independence or uncorrelation) are needed, while the knowledge of the statistical distribution of the sources is not required. Due to this appealing feature, various BSS methods have been used in the field of output-only modal analysis/identification problems, where measurement of input is difficult or impossible (Kerschen et al. 2007; Poncelet et al. 2007; Zhou and Chelidze 2007; McNeill and Zimmerman 2008, 2010; Hazra et al. 2010; Hazra and Narasimhan, 2010; Hazra et al. 2012; Yang and Nagarajaiah 2013; Guo 2015; Guo and Kareem 2016).

To address the challenges caused by non-stationarity, time-frequency blind source separation (TFBSS) methods using spatial time-frequency distribution (STFD) matrices were reported by Guo (2015) and Guo and Kareem (2016); these were developed in the electronic engineering field (e.g., Belouchrani and Amin, 1998; Zhang and Amin, 2000). This kind of method separates non-stationary sources with different spectral components in the time-frequency plane by selectively using the effective information in single auto-term regions where only one source has energy. By excluding overlapped signatures of different modal responses, which will contaminate the effective information, these methods can solve the aforementioned problem of decreased quality of separation caused by non-stationarity. In addition, the methods are adaptive for non-stationary signals in the sense that different analysis regions can be selected for different source signals according to their different energy distributions in the time-frequency plane.

Note that BSS methods are potentially useful but, thus far, have rarely been explored for the application of non-synoptic winds.

3. Modeling Fluctuating Component of Non-Stationary Winds

3.1. Spectral Model: Evolutionary Power Spectrum

For non-stationary winds, the slowly varying mean wind component observed in the non-stationary wind (Equation 1) is usually treated as a deterministic part, while the non-stationary fluctuating wind component, $u(t)$ (Equation 3) is non-deterministic (stochastic) and thus often modeled as an evolutionary process (Priestley 1965, 1967).

$$u(t) = \int_{-\infty}^{+\infty} A(\omega,t)e^{i\omega t}dZ(\omega) \tag{5}$$

where $A(\omega, t)$ is a modulating function that is assumed to be slowly varying with time, and $Z(\omega)$ is a zero-mean Gaussian process. The corresponding EPS of $u(t)$ $[G(\omega,t)]$ can then be expressed as

$$G(\omega,t) = \left|A(\omega,t)\right|^2 S(\omega) \tag{6}$$

where $S(\omega)$ is the power spectrum of the zero-mean Gaussian process, $Z(\omega)$.

Estimation of the EPS from sample time histories has been extensively investigated, resulting in methods such as windowed Fourier transform (e.g., Priestley 1965, 1967), the time-varying correlation function-based method (e.g., Benowitz et al. 2015), the time-varying autoregressive (AR) model (e.g., Chen 2005), and wavelet-based methods (e.g., Spanos and Failla 2004; Wang and Kareem 2004; Huang and Chen 2009; Zhou et al. 2015; Huang et al. 2015; Wang L. et al. 2013, 2014; Wang H. et al. 2016, 2018). However, all of these methods face an inherent difficulty in practice: the inability to establish the ensemble average of EPS because sample time histories of non-stationary events are rare, and only one sample time history is usually available (Wang L. et al. 2014). The lack of an ensemble average may lead to large random errors in the ESPD. Therefore, all these estimation methods should be used with careful consideration of uncertainties, even if special procedures have been incorporated to alleviate random error (Kareem et al. 2019).

Alternatively, parametric modeling of EPS of non-stationary tropical storms and downburst winds has been attempted based on limited full-scale data. Observations reported in the literature (Chen and Letchford 2006; Huang and Chen 2009; Hu et al. 2013; Huang et al. 2015; Wang H. et al. 2016; Li et al. 2017) have suggested that, in many cases, wind spectrum models of stationary winds may be extended to describe the characteristics of non-stationary winds by plugging in a time dimension to describe the time-dependent turbulence spectral contents; this is based on either a Karman or Kaimal spectrum (e.g., Hu and Xu 2013; Li et al. 2015a; Li et al. 2017; Wang H. et al. 2018). The Karman and Kaimal formats, respectively, are given here.

$$\frac{S_{uu}(\omega,t)}{\sigma_u^2(t)} = \frac{1}{2\pi}\frac{\dfrac{4L_u(t)}{\bar{U}(t)}}{\left[1+70.8\left(\dfrac{\omega L_u(t)}{2\pi\bar{U}(t)}\right)^2\right]^{\frac{5}{6}}}$$

$$\frac{S_{uu}(\omega,t)}{u_*^2(t)} = \frac{1}{4\pi}\frac{\dfrac{200z}{\bar{U}(t)}}{\left[1+50\left(\dfrac{\omega z}{2\pi\bar{U}(t)}\right)\right]^{\frac{5}{3}}} \tag{7}$$

where subscript u is the longitudinal (alongwind) direction; $S_{uu}(\omega)$ is the longitudinal EPS; ω is the circular frequency [rad]; $\sigma_u(t)$ is the time-varying standard deviation of fluctuating component of non-stationary wind; $u_*(t)$ is the time-varying friction velocity; z is the height; $L_u(t)$ is the time-varying TILS; and $\bar{U}(t)$ is the time-varying mean wind velocity.

This type of fully non-stationary model may be further simplified as an amplitude modulation function (ignoring the frequency component in the modulating function to be a time-related function only) for simplifying the overall computation of non-stationary structural response, deriving a closed-form solution, etc. (e.g., Chen and Letchford 2006; Kwon and Kareem 2009, 2012, 2013a, 2019; Huang et al. 2015; Solari 2016). Note that Huang et al. (2015) compared a fully non-stationary model with a simplified model for thunderstorm and downburst data and reported that the performance of a simplified model was comparable to that of a fully non-stationary model.

The EPS-based model also allows the incorporation of other essential characteristics of non-stationary winds. For example, Li et al. (2015b) recently reported a conceptual spectral model of typhoon winds that was based on theoretical consideration and verified by full-scale measurements. It included the presence of additional high-frequency subranges beyond the three subranges that have typically been modeled in wind power spectrum models (Figure 14.4). Hu et al. (2017a) and Tao and Wang (2019) considered the time-dependent boundary of the inertial subrange—especially in the high-frequency range of EPS, encompassing the conceptual spectral model by Li et al. (2015b)—because the fundamental frequency of large civil structures may fall within that range.

3.2. Equivalent Wind Spectrum

Recently, Solari (2016) and Solari and De Gaetano (2018) have introduced an approach for the design of thunderstorm by focusing on downburst events, referred to as the thunderstorm response spectrum technique (TRST). The TRST is adopted from the concept of the response spectrum being widely used in the field of earthquake engineering

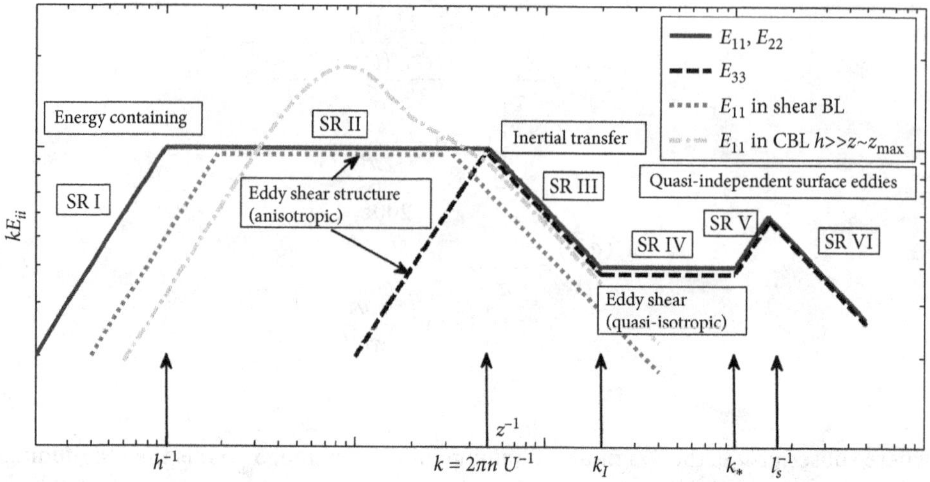

FIGURE 14.4 Schematic diagram of the sub-ranges (SR) of the turbulent velocity spectra in the tropical cyclone boundary layer.

From Li et al. (2015b).

as well as blast design. As a part of their framework, the concept of the equivalent wind spectrum technique (EWST), which was originally introduced for stationary winds (Solari 1988; Piccardo and Solari 1998), was extended for thunderstorms and downbursts. The EWST is a method that drastically reduces the computational burden for evaluating the wind-induced response to stationary synoptic winds. It consists of replacing the actual turbulent fluctuation, as a random function of time and space, by an equivalent turbulent fluctuation as a random function of time identically coherent in space. This quantity is defined in such a way that the aerodynamic admittance associated with it matches, in the best possible way, the actual turbulence field (Solari 2016; Solari and De Gaetano 2018).

For the application of the ESWT to non-stationary thunderstorm outflows, a reduced equivalent wind velocity was introduced in terms of the reduced size factor, time, and the equivalent wind velocity invoking the EWS technique. The calculation of the equivalent wind velocity (v_{eq}) uses a rather complex procedure using datasets, thus it is a data-based approach. Solari (2016) suggested three different methods depending on the use of dimensional/nondimensional parameters, although the underlying procedure is similar in each. Although this technique was only applied to thunderstorm and downburst events, it may be plausible to apply it to other non-synoptic winds such as hurricanes, typhoons, etc.

An inherent shortcoming in a data-based approach like the ESWT is the lack of representative datasets. However, it is worth noting that full-scale spatial information regarding thunderstorms and downbursts, such as vertical profile, TI, etc., is also being measured through a LiDAR system in the range of 40–250 meter heights and at a sampling rate of 1 Hz in the Wind, Ports, and Sea project (Repetto et al. 2018). It is anticipated that such data will provide better knowledge of spatial information, which has been mostly limited in past studies concerning thunderstorm and downburst events.

3.3. Probability Distribution: Statistical Modeling

The definition of statistical modeling is quite broad, and even analytical models derived through full-scale data of non-stationary winds may also be regarded as statistical modeling to some extent. In addition, extracting slowly time-varying mean winds from non-stationary winds (detrending) may also be considered as regression analysis in statistics, which is treated as statistical modeling. Thus, only the probability distribution of a stochastic non-stationary fluctuating wind component is discussed in this section.

A widely utilized approach in the sense of an evolutionary process (Equations 5 and 6) is to assume that the evolutionary wind spectrum is, as discussed earlier, a product of a modulating function and a zero-mean Gaussian process (Equation 6). Since the modulating function is also assumed to be slowly varying and treated as deterministic to some extents, especially introducing an amplitude modulating function, most of the non-synoptic wind studies in the past have implicitly assumed Gaussian distribution due to its convenience and simplicity in many respects such as modeling, simulation, deriving analytical expression, etc. Certainly, it is a daunting task to identify an appropriate probability distribution given the fact that limited full-scale data for non-stationary non-synoptic winds prevent us from reaching a consensus in the wind research community.

It is worth noting that McCullough et al. (2014) investigated the Gaussian assumption for fluctuating components of non-stationary winds with Hurricane Lili and RFD data and found that a double exponential distribution (also known as a *Laplace distribution*, see Equation 8) was a better fit than the Gaussian distribution using several fixed averaging intervals (e.g., 1-hr., 10-min., etc.).

$$p(x) = \frac{1}{\sigma\sqrt{2}} \exp\left(-\frac{\sqrt{2}\left|x-\mu\right|}{\sigma} \right) \tag{8}$$

In fact, the probability distribution of the fluctuating components approaches the double exponential distribution, an observation that has been made in other cases (e.g., Ward and Boland 2007; Nazir and Bouffard 2012). Nonetheless, McCullough et al. (2014) also reported that while the fluctuating component of the entire dataset may deviate from Gaussian, the assumption of a Gaussian or Gaussian-like distribution within each segment (time interval size/duration of analysis) may remain valid if segmentation and detrending are performed properly.

One of the interests in defining the probability distribution is to estimate extremes, which are often expressed as the *gust* or *peak factor* for codification purpose (e.g., wind speeds, structural response, etc.). For stationary winds, the *Davenport peak factor* (Poisson based approximation) (Davenport 1967) is widely adopted in most codes and standards, although work on it is still under way for non-stationary winds. It is worth noting that Michaelov et al. (2001) extended the Davenport peak factor in terms of equivalent time-dependent parameters such as equivalent time duration, time-dependent standard deviation, zero-crossing rate, etc. and expressed it in a modified Gumbel distribution format.

$$g_{ns} = \sqrt{2\ln\left[v_0(t)T_{eq}\right]} + \frac{0.5772}{\sqrt{2\ln\left[v_0(t)T_{eq}\right]}} \quad (9)$$

where g_{ns} is the non-stationary peak factor; $v_0(t)$ is the time-dependent zero-crossing rate; and T_{eq} is the equivalent time duration. Note that Michaelov et al. (2001) also reported other form by including the Vanmarcke approach (Vanmarcke 1975).

Note that the underlying concept employed in Michaelov et al. (2001) has been widely utilized for estimation extremes of wind speeds and structural response (e.g., Kwon and Kareem 2009, 2013a, 2019; Kwon et al. 2012; Hu and Xu 2014; Huang et al. 2013; Chen 2015; Huang and Gu 2019; Kareem et al. 2019). Kwon and Kareem (2019) also reported a possible simplification of Equation (8) for codification purposes under the assumption of a half-sine time function with a long pulse duration in thunderstorms and downbursts. Huang et al. (2018) recently attempted the non-stationary statistical modeling of extreme wind speed, fitting data to a generalized extreme value distribution in terms of generalized maximum likelihood approach.

4. Time-Series Simulation of Non-Stationary Winds

The most popular simulation approach concerning the fluctuating component of non-stationary winds is based on prescribed analytical models describing time-varying features and coherence functions and relying on the evolutionary process mentioned earlier. In this approach, a zero-mean Gaussian process in the non-stationary fluctuating component (Equations 1 and 2) is simulated just as stationary winds are simulated in terms of a simulation scheme, and the spectral representation method (e.g., Shinozuka and Deodatis 1991) is widely adopted for this purpose due to its simplicity and reasonably fast computation. Then, an analytically described amplitude modulating function (e.g., Equation 6) is multiplied with the simulated results to represent non-stationary fluctuating winds (e.g., Chen and Letchford 2004; Chay et al. 2006; Kareem 2008; Kwon and Kareem 2013a; Li et al. 2017; Peng et al. 2017, Hu et al. 2017b, Bao and Li 2019; Wang et al. 2019; Kareem et al. 2019; Huang et al. 2020). Certainly, one of the critical issues involving this simulation of random processes is the need for a theoretical model for the physical process being simulated (Wang et al. 2013).

With those developments, a move toward simulation methods that incorporate time-frequency analysis and modeling tools can be seen. For example, Wang et al. (2002), Spanos and Failla (2004), and Liang et al. (2007) utilized wavelet-based methods to determine the EPSs of non-stationary processes. Gurley and Kareem (1999) used a wavelet transform to model ground motion records as a sum of modulated stationary processes centered at narrow-banded frequencies, similar to Der Kiureghian and Crempien (1989). Iyama and Kuwamura (1999) applied wavelet transforms into the analysis and simulation of earthquake data from the viewpoint of energy input. Kwon et al. (2007) combined wavelet decomposition with AR models, modeling each frequency band separately. An alternative way to track the time-varying frequency content of a signal uses EMD together with the HT (e.g., Wen and Gu 2004). While a host of novel methods to simulate non-stationary random processes have emerged, few have emphasized the physical interpretation of the instantaneous information obtained from non-stationary random processes.

Data-driven simulation approaches that utilize the instantaneous features of available data offer a straightforward and statistically meaningful approach to non-stationary simulation (e.g., Wen and Gu 2004; Wang and Kareem 2004; Wang 2007; Kareem 2008; Wang et al. 2013, 2014). The procedure introduced herein (Wang 2007; Wang et al. 2013, 2014) involves the tandem implementation of wavelet and Hilbert transforms and does not invoke assumptions that may be inappropriate (e.g., functional form and piecewise stationarity). This method relies on the property that the discrete wavelet decomposition of a non-stationary process can be treated as mono-component processes that meet the basic requirements of the HT. Thus, the instantaneous amplitude and frequency information are known following the transformation, revealing a great deal about the energy fluctuations at different frequencies (Buresti et al. 2004). This information can then be used within a simulation framework to generate additional realizations from the same process. This approach is extendable to the simulation of multivariate random processes utilizing proper orthogonal decomposition (POD).

For demonstration, RFD data measured from four towers at 10-meter height and their simulated data are shown in Figure 14.5, showing the similitude of the rapid variation of wind velocities in the simulation. The comparison of the Hilbert contour spectrum, as shown in Figure 14.6, further attests that the time-varying characteristics of the measurement at Tower 3 is captured and preserved during the simulation process, with high energy concentrating on the frequency range up to 0.2 Hz during 600–800 seconds. From the comparisons of the Hilbert spectrum of other measurement components, similar conclusions were reached. More detailed information and theoretical backgrounds can be found in Wang (2007) and Wang et al. (2013, 2014).

5. Wind Characteristics: Averaging Interval

The conventional approach to deal with the characteristics of atmospheric boundary-layer winds has been a stationary wind model since Davenport (1961, 1967) proposed the quasi-static and strip theory for computing wind forces on structures. This approach

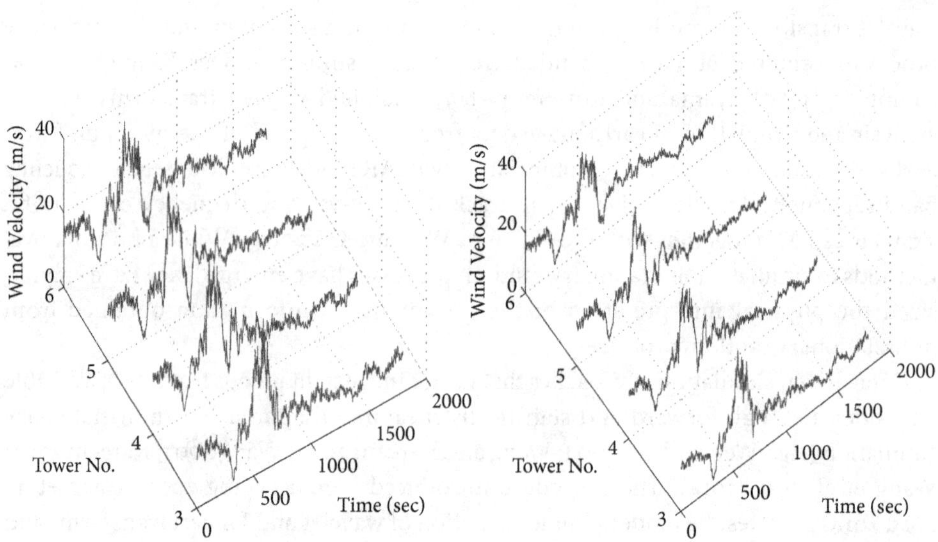

FIGURE 14.5 Simulation of rear-flank downdraft (RFD) data recorded from four different towers at 10-meter height (left: measured data; right: simulated results).

From Wang (2007).

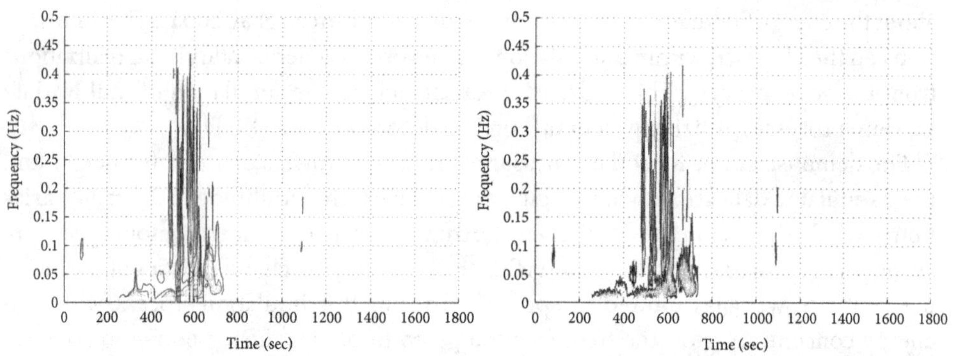

FIGURE 14.6 Hilbert contour spectrum of downburst at 10-meter height for Tower 3 (left: measured data; right: simulated results).

From Wang (2007).

assumes that the fluctuating component (random turbulence) of boundary layer winds can be treated as a zero-mean stationary Gaussian random process by removing a constant mean from the wind data. Based on this stationary wind model, most developments in wind engineering have used a fixed averaging interval (FAI) to study boundary layer winds in several categories, such as modeling, simulation, GF, and TI, due to its simplicity. For example, 10-minute mean wind speeds (Architectural Institute of Japan [AIJ] 2004), 1-hour mean wind speeds (NBCC 2005), and 3-second peak gust wind speeds (American Society of Civil Engineers [ASCE] 2016) have been adopted in most major international codes and standards as basic wind speeds using FAI (e.g., Kijewski and Kareem 1998; Solari and Kareem 1998; Zhou et al. 2002; Kwon and Kareem 2013b). Non-stationarity of data during hurricanes and typhoons or thunderstorms and downbursts, however, often precludes utilization of conventional analysis tools and makes the accurate estimation of turbulent wind characteristics during these events difficult.

5.1. Fixed Averaging Interval for Stationary and Non-Stationary Winds

The turbulence statistics on which this chapter will focus are GF, TI, and TILS. Each definition of the turbulence statistics under stationary winds, which is defined based on FAI, is reviewed here. The GF is generally defined as the ratio of the peak gust speed (e.g., 3-sec. gust in ASCE 7-16) to the mean wind speed (e.g., 1-hr., 10-min.), as shown in Table 14.1, where $\hat{U}_T(t)$ and $\bar{U}_T$ are the gust and the mean wind speeds over the user-defined interval, T, respectively. The TI is given by the ratio of the standard deviation of the fluctuating wind speed component to the mean wind speed (Table 14.1), where σ_{uT} is the standard deviation of the fluctuating component over the user-defined interval T. The TILS attempts to quantify the dominant gust sizes in the wind event. Its definition is shown in Table 14.1, where $R_u(\tau)$ is the autocorrelation of the fluctuating component. To overcome issues with spurious periodicities at high lags, the autocorrelation is truncated at the first zero-crossing (Iyengar and Farell 2001).

As mentioned earlier, non-stationary winds usually show time-varying features in their mean and fluctuating wind components. Thus, conventional definitions based on the assumption of stationarity may not be valid for non-stationary winds. Accordingly, new definitions will be required for non-stationary winds even if conventional FAI is used. After a time-varying mean wind speed $[\bar{U}(t)]$ has been identified using time-frequency tools such as wavelet analysis (e.g., DWT), EMD, or BSS as mentioned earlier, the fluctuating wind speed $[u(t)]$ can be acquired by subtracting the time-varying mean wind speed from the measured wind speed time history, $U(t)$. Since a time-varying mean is utilized in this approach, the turbulence statistics need to be modified to be consistent with the statistics defined by the traditional approaches in Table 14.1. Several definitions considering time-varying feature of non-stationary winds have been reported in past pieces of literature (e.g., Choi and Hidayat 2002; Choi 2004; Wang and Kareem 2004; Xu et al. 2005; Orwig and Schroeder 2007; Wang 2007; Wang et al. 2013; McCullough et al. 2014; Lombardo et al. 2014; Solari et al. 2015a, 2015b; Huang et al. 2015; Wang et al. 2016; McCullough 2016; Huang et al. 2018; Huang and Gu 2019), but this chapter presents the definitions reported in McCullough et al. (2014).

The revised GF using FAI is estimated by the maximum ratio of the peak gust wind speeds to the corresponding time-varying mean, as expressed in Table 14.1,

Table 14.1 Comparison of the definition of wind parameters for averaging interval schemes

	FAI-S (stationary)	FAI-NS (non-stationary)	VAI[1]
GF	$GF_T = \dfrac{\max\left[\hat{U}_T(t)\right]}{\bar{U}_T}$	$GF_T' = \max\left[\dfrac{\hat{U}_T(t)}{\bar{U}_T(t)}\right]$	$GF_{\delta T}' = \max\left[\dfrac{\hat{U}_{\delta T}(t)}{\bar{U}_{\delta T}(t)}\right]$
TI	$TI_T = \dfrac{\sigma_{uT}}{\bar{U}_T}$	$TI_T' = E\left[\dfrac{\sigma_{u'T}}{\bar{U}_T(t)}\right]$	$TI_{\delta T}' = E\left[\dfrac{\sigma_{u'\delta T}}{\bar{U}_{\delta T}(t)}\right]$
TILS	$L_u = \bar{U}\int_0^{\infty}\dfrac{R_u(\tau)}{\sigma_u^2}d\tau$	$L_u' = E\left[\bar{U}(t)\int_0^{\infty}\dfrac{R_{u'}(\tau)}{\sigma_{u'}^2}d\tau\right]$	$L_{u\delta T}' = E\left[\bar{U}_{\delta T}(t)\int_0^{\infty}\dfrac{R_{u'\delta T}(\tau)}{\sigma_{u'\delta T}^2}d\tau\right]$

* GF, gust factor; TI, turbulence intensity; TILS, turbulence integral length scale.
[1] Following the expressions defined in McCullough et al. (2014).

where $\hat{U}_T(t)$ and $\bar{U}_T(t)$ are the time-varying gust and mean wind speeds, respectively, over the user-defined interval, T. In addition, the TI of non-stationary winds is defined as the expected value of the standard deviation of the fluctuating component over the time-varying mean for the interval T, as shown in Table 14.1, where E[] denotes the expected value and $\sigma_{u'T}$ represents the standard deviation of the fluctuating wind speed for the time interval T. Finally, the TILS is written in Table 14.1, where $R_{u'}(\tau)$ and $\sigma_{u'}^2$ are the autocorrelation and variance of the fluctuating component after the time-varying mean is removed, respectively.

It is worth noting that the turbulence statistics for non-stationary winds depends on a user-defined time interval (T) in the FAI, which also differs from stationary events. For example, widely adopted (but not consensus) time interval values for FAI are 1 minute for hurricanes and typhoons, and 30–60 seconds for thunderstorms and downbursts. For strong non-stationary winds (e.g., transient events), rapid changes in wind speed and direction may be lost using FAI methods since the interval sizes cannot be varied to capture shorter or longer events. To overcome those limitations, the concept of a variable averaging interval (VAI) was instead proposed by McCullough et al. (2014) as an alternative approach to conventional FAI methods.

5.2. Variable Averaging Interval: Data-Driven Approach

The VAI approach introduced here determines requisite averaging blocks (windows) based on the measured non-stationary wind data itself. Similar techniques, such as variable-interval time-averaging (VITA), have been used in the past for the detection of abrupt changes and sampling in turbulent boundary layers with some experimental success (e.g., Blackwelder and Kaplan 1976; Subramanian et al. 1982). The VITA technique was developed as a way to concentrate on a local region in space and modify the averaging time to be compatible with the time scale of the phenomenon of interest.

Similar variable segmentation schemes are necessary that are more suitable for winds associated with hurricanes and typhoons, thunderstorms and downbursts, etc.

The focus of the VAI schemes is to find the most appropriate method of data segmentation to determine the statistical characteristics of the non-stationary winds. An example of the variable segmentation that results when using a VAI scheme for the Hurricane Lili and RFD data is shown in Figure 14.7. Four schemes were introduced and compared to evaluate their suitability for application within the newly proposed VAI approach in McCullough et al. (2014). It is important to note that these schemes were not developed for the purpose of VAI; the authors have extended their original concepts to be utilized within a VAI approach with proper assumptions. The following four segmentation schemes are considered: time-dependent memory method (TDMM) (e.g., Treviño 1999; Treviño and Andreas 2000), t-test method (e.g., Ang and Tang 1975), confidence interval method (e.g., NIST/SEMATECH 2003), and penalized contrast (PC) method (Lavielle 2005). The PC method utilized several different examples of contrast functions based on the characteristics that are of interest. The different contrast functions detect changes in the mean (PC-mean), variance (PC-variance), mean and variance together (PC-MV), cumulative distribution function (PC-CDF), and power spectral density (PC-PSD).

The traditional definitions for the turbulence statistics once again need to be updated for use with VAI methods. Due to the nature of the VAI, the revised definitions should be compatible with variable segment sizes and different detrending schemes. Several

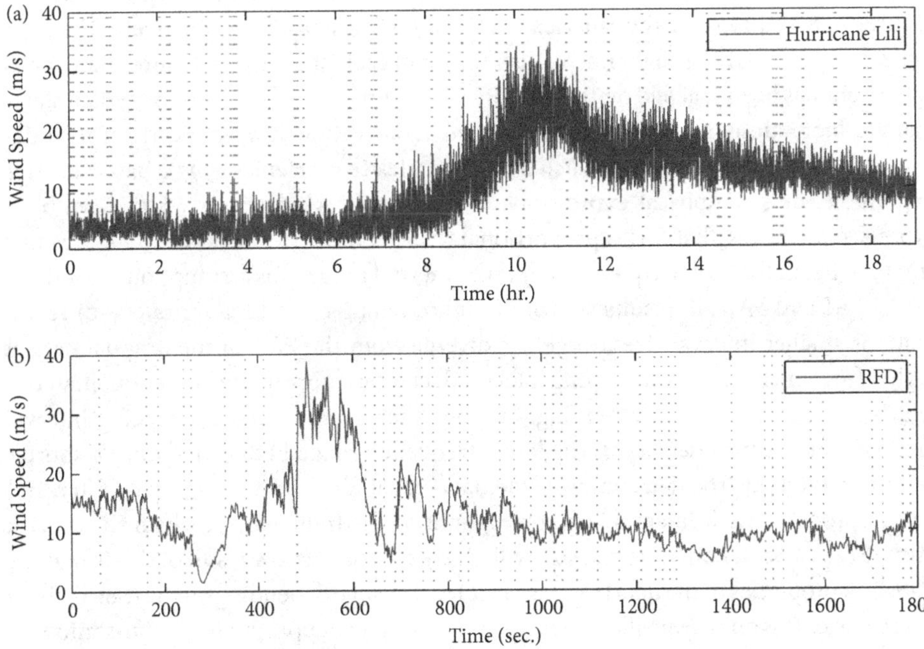

FIGURE 14.7 Illustration of variable averaging interval (VAI) for Hurricane Lili data (a) and the rear-flank downdraft (RFD) data (b).

From McCullough et al. (2014).

detrending schemes may be used within each segment (e.g., linear, quadratic, cubic, etc.). The proposed new definitions follow readily from those determined for the FAI for non-stationary (FAI-NS) schemes summarized in Table 14.1, which feature a time-varying mean. The new GF definition for VAI takes into account the variable segment size and the time-varying mean determined by the detrending method chosen, as shown in Table 14.1, where subscript δT is a variable averaging interval. The TI and TILS for VAI are defined in similarly to the definitions used in FAI-NS, also shown in Table 14.1. Detailed theoretical background about the four VAI schemes is omitted for the sake of brevity but can be found in McCullough et al. (2014).

5.3. Evaluation and Comparison of Averaging Interval Schemes

Three different schemes have been introduced in this section: the traditional FAI approach, the alternative FAI-NS using DWT (FAI-NS-DWT) or EMD (FAI-NS-EMD), and the VAI approach. Taking into account the record lengths of given datasets, for all of the FAI schemes, interval sizes of 1 hour, 10 minutes, and 1 minute are used for the Hurricane Lili data and 5 minute, 2 minute, and 1 minute for the RFD data.

5.3.1. *Power Spectral Density*

The PSDs of the fluctuating components using each of the schemes can be found in Figures 14.8 and 14.9 for the Hurricane Lili and RFD datasets, respectively. The $f^{(-5/3)}$ frequency power law (straight line) is also shown in each of the figures. Almost universal agreement has been reached on the usefulness of Kolmogorov's hypothesis, which states that the fluctuations at the high-frequency end of the turbulent wind spectrum are independent of external conditions such as boundaries (e.g., Henjes 1997). Based on this hypothesis, most empirical expressions of fluctuating wind spectra, $S(f)$, have been expressed assuming that $S(f)$ is proportional to $f^{(-5/3)}$. It can be observed in Figures 14.8 and 14.9 that in the high frequency range, i.e., inertial range, this assumption is valid.

The FAI and FAI-NS results shown in Figure 14.8(a–c) and Figure 14.9(a–c) reveal that, for smaller interval sizes, the PSDs diverge from the PSD of the original data at higher frequencies. As the averaging interval decreases, more of the low-frequency content is filtered out due to detrending. On the other hand, a longer averaging interval preserves its spectral density in the lower frequency range better than does a shorter averaging interval. The detrending of the data essentially acts as a high-pass filter, and the reciprocal of each interval size corresponds to the frequency at which the spectra diverge (i.e., a 60-second interval size will diverge at 1/60 Hz). A result of this is that the statistical properties of the data (e.g., variance) will change depending on the user-defined interval size. This once again illustrates why FAI may not be appropriate for non-stationary winds.

Figures 14.8(d) and 24.9(d) illustrate the PSD results using VAI for the Hurricane Lili and RFD datasets, respectively. From the viewpoint that a desirable detrend preserves

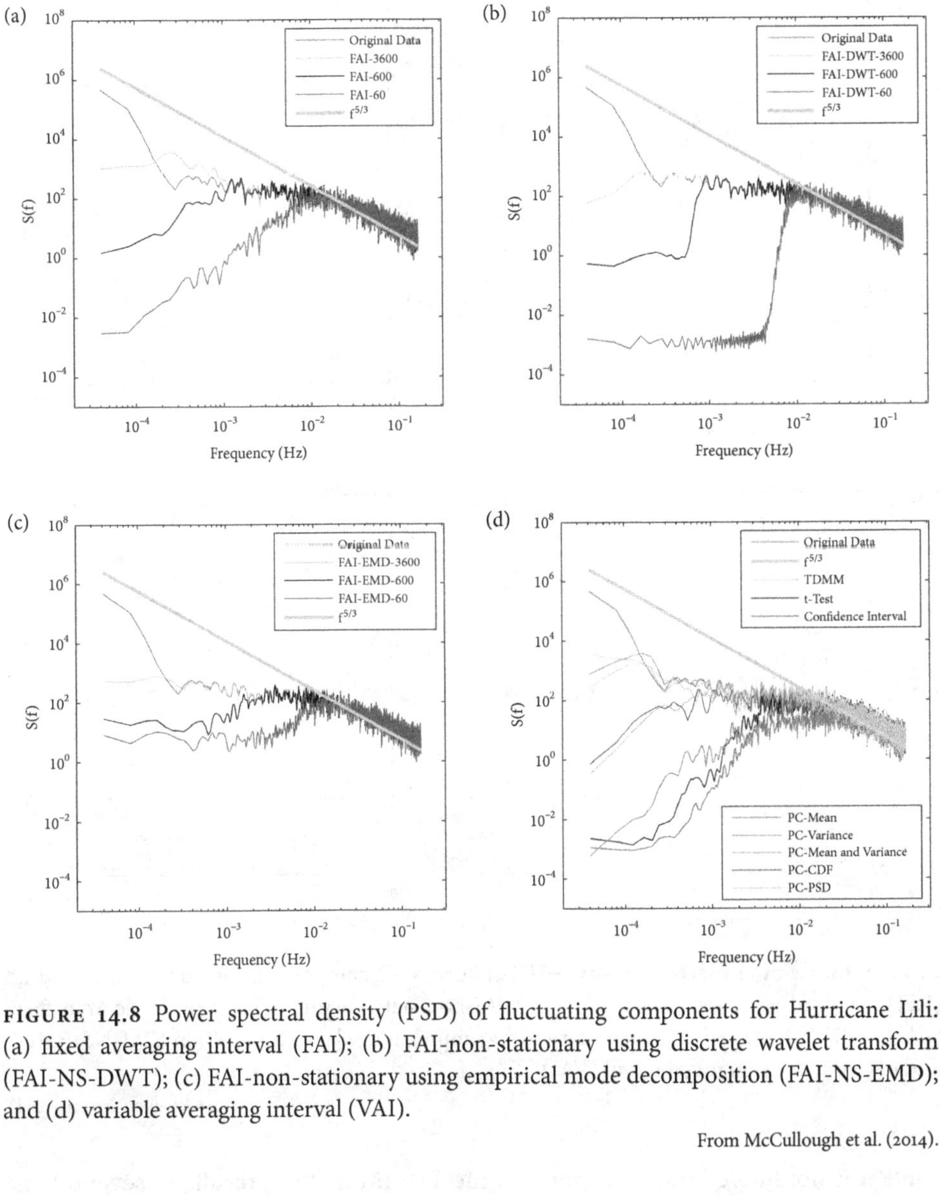

FIGURE 14.8 Power spectral density (PSD) of fluctuating components for Hurricane Lili: (a) fixed averaging interval (FAI); (b) FAI-non-stationary using discrete wavelet transform (FAI-NS-DWT); (c) FAI-non-stationary using empirical mode decomposition (FAI-NS-EMD); and (d) variable averaging interval (VAI).

From McCullough et al. (2014).

the PSD in the high-frequency range, the PC-mean method is readily identified as unsuitable for the Hurricane Lili data in Figure 14.8(d). The RFD results in Figure 14.9(d) confirm that the PC-mean method is not able to preserve the PSD in the high-frequency range, but the PC-mean and variance (PC-MV) method also shows poor performance. These results can be explained, in part, by looking at the averaging interval sizes for the different methods in Tables 14.2 and 14.3. PC-mean results in a small average interval size for both Hurricane Lili and RFD, leading to divergence in the higher frequency. While the average interval sizes for the TDMM and confidence interval schemes are also small, these schemes are segmented and detrended such that the high-frequency

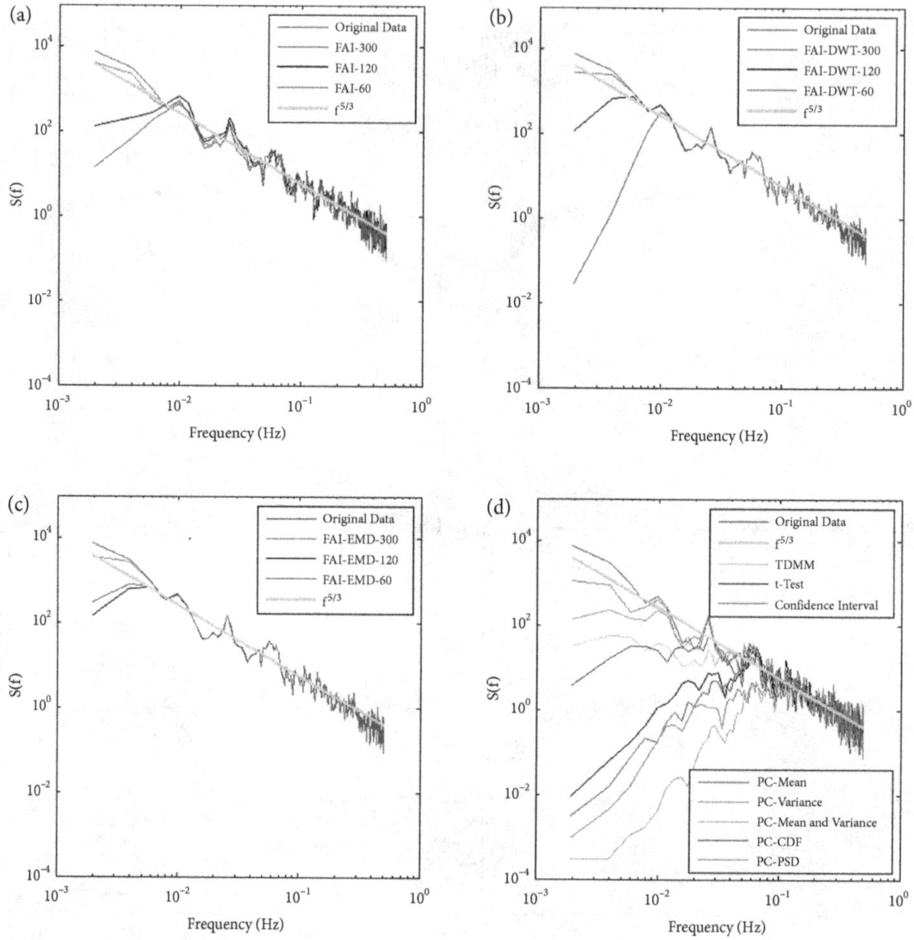

FIGURE 14.9 Power spectral density (PSD) of fluctuating components for rear-flank downdraft (RFD): (a) fixed averaging interval (FAI); (b) FAI-non-stationary using discrete wavelet transform (FAI-NS-DWT); (c) FAI-non-stationary using empirical mode decomposition (FAI-NS-EMD); and (d) variable averaging interval (VAI).

From McCullough et al. (2014).

content is not filtered out. For example, the TDMM method results in several large segments, with many smaller segments in between, allowing for the preservation of the PSD in the lower frequency range. The PC-mean scheme, however, results in small segments throughout the full length of the dataset, resulting in the filtering out of high-frequency content.

5.3.2. *Gust Factor*

For the Hurricane Lili data, Table 14.2 summarizes a variety of the statistics for the portion of the time history that constitutes the event where "gust factor" is denoted as GF in Tables 14.2 and 14.3. It is observed for the FAI schemes in Table 14.2 that longer averaging intervals correspond to larger mean GFs than do shorter averaging intervals. This is a result of the effect of smaller mean wind speeds in longer averaging intervals. For the same interval length in the FAI schemes, all have similar mean and maximum GF results and small SDs, again showing the appropriateness of the alternative turbulence statistic expressions in Table 14.1. The relative agreement between the traditional FAI schemes and the FAI-NS schemes shows that an appropriate fixed interval size may be able to properly capture the characteristics of the slowly varying non-stationary data. Looking at the VAI schemes in Table 14.2, the PC-mean method is characterized by small mean GFs as a result of small interval sizes, as discussed previously in the discussion of PSD. It is also apparent that the VAI methods with the larger interval sizes are characterized by larger mean GFs, particularly the PC-variance and PC-PSD schemes. The t-test, confidence interval, PC-MV, PC-CDF, and PC-PSD methods have a maximum GF of around 1.7 at the beginning of the event. The mean values vary as a result of the various mean interval sizes for each scheme. The t-test and confidence interval schemes, which have small mean interval sizes, have mean GFs similar to those of the FAI and FAI-NS schemes with a 1-minute interval size. The PC-variance and PC-PSD schemes result in mean GFS similar to the 1-hour FAI and FAI-NS schemes. Finally, the PC-M&V scheme has a mean GF result similar to the 10-minute interval schemes.

Considering the RFD event (Table 14.3), only the 2- and 1-minute interval sizes will be used for the FAI schemes because the 5-minute interval size is not able to successfully characterize the short event. The maximum and mean GF results for the FAI scheme show larger variations than those for Hurricane Lili; this is likely a result of the event being quite short and characterized by large changes in wind speed amplitude over small time periods. The maximum GFs range from 1.36 to 1.69, and the mean GFs range from 1.30 to 1.62. Again, the smaller interval size leads to smaller mean GFs; however, other trends for the FAI schemes are difficult to discern. Looking at the VAI methods, the PC-variance method has one interval that encompasses the entire event, leading to a very high GF, and TDMM has very small interval sizes. The PC-mean, t-test, and confidence interval methods show good agreement for the mean and maximum GFs. For reference, Orwig and Schroeder (2007) obtained a maximum GF of 1.48 during RFD using a 120-second FAI with a detrending scheme developed for non-stationary data. The PC-MV and PC-PSD methods yield maximum GFs of 1.44 and 1.60, respectively. While the locations of the maximums do not coincide, the PC-PSD method does yield a GF of 1.44 at the same location of the PC-M&V maximum. The confidence interval, t-test, PC-mean, and PC-CDF methods yield maximums at the same approximate location, which coincides with the location of the maximum found by Orwig and Schroeder (2007). It is apparent that the traditional FAI scheme is not effective for characterizing the GF in a non-stationary event with rapid changes in wind speed. The FAI-NS schemes with an appropriately chosen interval size (~60 sec) and the VAI schemes give comparable results.

5.3.3. *Turbulence Intensity*

The overall trends for TI were similar to those for GFs given in Table 14.2 for the Hurricane Lili data, where "turbulence intensity" is denoted as TI in Tables 14.2 and 14.3.

The FAI and FAI-NS schemes with the same interval size resulted in similar mean and maximum TIs. Once again, the maxima for all interval sizes occurred at the beginning of the event (i.e., in the same location). For the FAI schemes, longer averaging intervals showed larger mean TIs than did the shorter averaging intervals. For the VAI schemes, the PC-mean method had small mean and maximum TIs because of small interval sizes. The mean and maximum TI results for the t-test and confidence interval methods showed good agreement and occurred in the same location (i.e., at the beginning of the event). Similarly, the maxima for the PC-MV, PC-CDF, and PC-PSD methods showed good agreement and occurred at the beginning of the event. The RFD results in Table 14.3 again show larger variation for the TI results than do the results for the Hurricane Lili data. The FAI schemes were once again characterized by erroneously high statistical values as a result of the inability of a constant mean to effectively detrend data characterized by rapid changes in wind speed.

The FAI-NS schemes, however, yielded mean and maximum TIs that were similar, with peaks occurring at the same locations as the peaks for the GF. Looking at the VAI schemes, the PC-variance scheme leads to only one interval with unrealistically high TIs. The t-test and confidence interval methods show good agreement for the mean and maximum TIs. For reference, Orwig and Schroeder (2007) obtained a maximum TI of 20% during RFD. The maximum TIs estimated using the PC-mean, PC-CDF, and PCPSD methods were 19.6%, 26.4%, and 19.8%, respectively, and occurred at the location of the maxima for the t-test and confidence interval schemes.

5.3.4. *Turbulence Integral Length Scale*

The results for the TILS showed much larger variation than the GF and TI results. Inherently, the length scale is a much more difficult statistic to estimate properly, especially in non-stationary winds. The large variation in length scale estimates due to the interval size used for detrending has been noted in the literature (Yu et al. 2008). The interval sizes and detrending schemes used affect the autocorrelations, which affect the length scale estimations. Also, autocorrelation often lacks a smooth decay due to the short segments of data. Therefore, trends for length scale will be discussed with an understanding that the values have large variability, as shown in Tables 14.2 and 14.3. Looking first at the FAI results for the Hurricane Lili event in Table 14.2, it is apparent that a smaller interval size results in smaller length scales because of the decrease in autocorrelation. The length scale for the 1-hour FAI is extremely large, which can be explained by the serial correlation that remains in the data even after detrending. The results for the 10- and 1-minute interval sizes were similar. For the VAI schemes, the length scales were directly related to their autocorrelations. The PC-mean method is characterized by small intervals and excessive detrending, leading to a small length scale. The length scale results for the t-test and confidence interval methods show good agreement between them and with the 1-minute FAI and FAI-NS length scales. The PC-MV and PC-CDF methods were characterized by larger length scales of 129.3 and 98.8 meters, respectively. The length scales reported in Yu et al. (2008) for the Hurricane Lili data ranged from 67 to 189 meters depending on the interval size used, ranging from

Table 14.2 Summary of statistics for the Hurricane Lili event

Scheme	Average interval size (sec)[a]	Mean GF	GF STD	Max GF	Mean TI (%)	TI STD (%)	Max TI (%)	Turbulence integral length scale (m)
FAI	3,600	1.54	0.05	1.63	17.2	1.4	19.1	955.7
FAI	600	1.44	0.08	1.76	16.1	2.0	22.7	116.8
FAI	60	1.26	0.08	1.62	13.9	3.8	30.3	64.3
FAI-NS-DWT	3,600	1.54	0.06	1.65	16.6	1.3	19.0	234.9
FAI-NS-DWT	600	1.44	0.08	1.73	16.0	2.0	22.2	119.8
FAI-NS-DWT	60	1.24	0.08	1.65	13.1	3.6	30.8	61.3
FAI-NS-EMD	3,600	1.54	0.05	1.66	16.6	1.3	18.6	233.8
FAI-NS-EMD	600	1.43	0.08	1.72	15.9	2.0	22.1	112.0
FAI-NS-EMD	60	1.25	0.09	1.66	13.6	3.7	30.5	74.5
TDMM	476.5	1.31	0.15	1.57	16.7	3.9	30.2	1956.8
t-Test	119.1	1.29	0.10	1.76	13.8	3.4	26.9	65.3
CI	95.0	1.27	0.09	1.65	13.2	3.4	26.6	61.9
PC-Mean	62.1	1.16	0.08	1.46	9.1	3.4	16.7	35.5
PC-Variance	6,480	1.57	0.06	1.63	16.4	1.4	17.8	431.3
PC-MV	2,700	1.48	0.09	1.69	16.2	2.3	21.8	129.3
PC-CDF	830.8	1.37	0.14	1.69	13.6	3.7	22.0	98.8
PC-PSD	5,400	1.57	0.09	1.69	17.3	2.7	21.8	319.8

[a] Mean values of all averaging intervals for variable averaging interval (VAI) schemes.
GF, gust factor; TI, turbulence intensity; STD, standard deviation.

5 minute to 1 hour. For the RFD data in Table 14.3, the length scales for the FAI schemes are generally much larger than those for the VAI schemes, except the PC-variance method. As discussed previously, the FAI schemes maintain nonzero serial autocorrelation longer than many of the VAI schemes. Excluding the 2-minute FAI and FAI-NS-DWT results, the remaining FAI schemes resulted in similar length scales. The VAI schemes, however, resulted in much lower length scales because of their autocorrelation coefficients more quickly approaching zero.

Throughout the investigation of turbulence statistics, it is difficult to directly compare the VAI results with previous studies because of the differences between FAI and VAI schemes. It was evident, however, that some VAI schemes showed better performance in the case of Hurricane Lili, while others were better for use with the RFD data. This observation may be a result of the different characteristics of non-stationarity in the Lili and RFD events, as well as the different concepts for each VAI scheme. In addition, the TILS is extremely sensitive to the length of the data, sampling rate, and method of detrending. Therefore, this statistic is not ideal for comparing the various averaging

Table 14.3 Summary of statistics for RFD event

Scheme	Average interval size (sec)[a]	Mean GF	GF STD	Max GF	Mean TI (%)	TI STD (%)	Turbulence integral length scale (m)
FAI	120	1.62	0.02	1.64	30.8	3.5	368.6
FAI	60	1.43	0.16	1.69	20.6	11.8	119.2
FAI-NS-DWT	120	1.41	0.17	1.59	20.0	4.4	155.7
FAI-NS-DWT	60	1.30	0.07	1.38	16.0	4.4	100.9
FAI-NS-EMD	120	1.34	0.05	1.38	18.1	1.4	119.2
FAI-NS-EMD	60	1.30	0.05	1.36	17.6	5.1	121.2
TDMM	11.2	1.14	0.11	1.36	8.2	3.1	125.2
t-Test	18.9	1.16	0.09	1.31	8.2	4.2	28.3
CI	16.3	1.13	0.07	1.33	7.4	3.6	21.8
PC-Mean	14.9	1.12	0.07	1.31	7.1	3.6	15.7
PC-Variance	358	1.93	-	1.93	40.0	-	489.5
PC-MV	51.1	1.24	0.14	1.44	9.4	4.2	47.1
PC-CDF	32.5	1.24	0.17	1.71	10.7	6.1	36.2
PC-PSD	89.5	1.43	0.14	1.60	15.1	4.0	60.1

[a] Mean values of all averaging intervals for variable averaging interval (VAI) schemes.
GF, gust factor; TI, turbulence intensity; STD, standard deviation.

interval schemes. Robust schemes with broad applicability are necessary for the effective implementation of the VAI approach.

6. Conclusion

For several decades, studies have been focused on identifying the characteristics of non-stationary non-synoptic winds using data from experiments, numerical simulations, and full-scale measurements. Yet a definitive design consensus for practice has not emerged, partially due to a lack of consistency in the definition of wind characteristics among studies in conjunction with the scarcity of available data needed to corroborate. This is further compounded by the difficulty in dealing with non-stationarity data, which feature time dependency. It is hoped that the rapid advance in monitoring systems and their deployments at fixed and mobile sites will enable us to collect long-term data. This would enable data-driven approaches to offer a promising alternative avenue for better characterization, modeling, and simulation.

Acknowledgments

The authors gratefully acknowledge the support in part by the NSF Grant CMMI-1562244.

References

Ang, A. H.-S., and Tang, W. H. (1975). *Probability concepts in engineering planning and design: Volume I*. New York: Wiley.

American Society of Civil Engineers (ASCE). (2016). *Minimum design loads and associated criteria for buildings and other structures: ASCE 7-16*. Reston, VA: American Society of Civil Engineers.

Architectural Institute of Japan (AIJ). (2004). *RLB recommendations for loads on buildings*. Tokyo: Structural Standards Committee, Architectural Institute of Japan.

Bao, X., and Li, C. (2019). Fast simulation of non-stationary wind velocity based on time-frequency interpolation. *Journal of Wind Engineering and Industrial Aerodynamics*, *193*, 103982.

Belouchrani, A., Abed-Meraim, K., Cardoso, J.-F., and Moulines, E. (1997). A blind source separation technique using second-order statistics. *IEEE Transactions on Signal Processing*, *45*, 434–444.

Belouchrani, A., and Amin, M. G. (1998). Blind source separation based on time-frequency signal representations. *IEEE Transactions on Signal Processing*, *46*, 2888–2897.

Benowitz, B. A., Shields, M. D., and Deodatis, G. (2015). Determining evolutionary spectra from non-stationary autocorrelation functions, *Probabilistic Engineering Mechanics*, 41, 73–88.

Blackwelder, R. F., and Kaplan, R. E. (1976). On the wall structure of the turbulent boundary layer. *Journal of Fluid Mechanics*, *76*, 89–112.

Buresti, G., Lombardi, G., and Bellazzini, J. (2004). On the analysis of fluctuating velocity signals through methods based on the wavelet and Hilbert transforms. *Chaos, Solitons and Fractals*, *20*, 149–158.

Chay, M., Albermani, F., and Wilson, R. (2006). Numerical and analytical simulation of downburst wind loads. *Engineering Structures*, *28*(2), 240–254.

Chen, L., and Letchford, C. W. (2004a). A deterministic–stochastic hybrid model of downbursts and its impact on a cantilevered structure. *Engineering Structures*, *26*(5), 619–629.

Chen, L., and Letchford, C. W. (2004b). Parametric study on the along-wind response of the CAARC building to downbursts in the time domain. *Journal of Wind Engineering and Industrial Aerodynamics*, *92*(9), 703–724.

Chen, L. (2005). Vector time-varying autoregressive (TVAR) models and their application to downburst wind speeds. Ph.D. Dissertation, Lubbock, TX: Texas Tech University.

Chen, X. (2015). Analysis of multimode coupled buffeting response of long-span bridges to nonstationary winds with force parameters from stationary wind. *Journal of Structural Engineering, ASCE*, *141*(4), 04014131.

Choi, E. C. C., and Hidayat, F. A. (2002). Dynamic response of structures to thunderstorm winds. *Progress in Structural Engineering and Materials*, *4*(4), 408–416.

Choi, E. C. C. (2004). Field measurement and experimental study of wind speed profile during thunderstorms. *Journal of Wind Engineering and Industrial Aerodynamics*, *92*(3–4), 275–290.

Davenport, A. G. (1961). The application of statistical concepts to the wind loading of structures. *Proceedings of the Institution of Civil Engineers, 19*(4), 449–472.

Davenport, A. G. (1967). Gust loading factors. *Journal of Structural Division, ASCE, 93*(3), 11–34.

Der Kiureghian, A., and Crempien, J. (1989). An evolutionary model for earthquake ground motion. *Structural Safety, 6*, 235–246.

Fang, G., Zhao, L., Cao, S., Ge, Y., and Pang, W. (2018). A novel analytical model for wind field simulation under typhoon boundary layer considering multi-field correlation and height-dependency. *Journal of Wind Engineering and Industrial Aerodynamics, 175*, 77–89.

Farge, M. (1992). Wavelet transforms and their applications to turbulence. *Annual Review of Fluid Mechanics, 24*(1), 395–458.

FCMP. (2011). Florida Coastal Monitoring Program. http://fcmp.ce.ufl.edu.

Franklin, J. L., Black, M. L., and Valde, K. (2003). GPS dropwindsonde wind profiles in hurricanes and their operational implications. *Weather and Forecasting, 18*(1), 32–44.

Fujita, T. T. (1985). *The Downburst*. SMRP Research Paper No. 210, Chicago, IL: University of Chicago.

Gast, K. D., and Schroeder, J. L. (2003). Supercell rear-flank downdraft as sampled in the 2002 thunderstorm outflow experiment. In *Proceedings of the 11th International Conference on Wind Engineering* (CD-ROM). Lubbock, TX, Paper No. C08-2.

Giammanco, I. M., Schroeder, J. L., and Powell, M. D. (2013). GPS dropwindsonde and WSR-88D observations of tropical cyclone vertical wind profiles and their characteristics. *Weather and Forecasting, 28*(1), 77–99.

Guo, Y. (2015). *Nonstationary system identification techniques*. PhD dissertation. University of Notre Dame.

Guo, Y., and Kareem, A. (2016). System identification of nonstationary data using time-frequency blind source separation, *Journal of Sound and Vibration, 371*, 110–131.

Gurley, K. R., and Kareem, A. (1999). Applications of wavelet transforms in earthquake, wind and ocean engineering. *Engineering Structures, 21*, 149–167.

Hayashi, T. (1994). An analysis of wind velocity fluctuations in the atmospheric surface layer using an orthogonal wavelet transform. *Boundary Layer Meteorology, 70*, 307–326.

Hazra, B., and Narasimhan, S. (2010). Wavelet-based blind identification of the UCLA Factor building using ambient and earthquake responses. *Smart Materials and Structures, 19*, 1–10.

Hazra, B., Roffel, A. J., Narasimhan, S., and Pandey, M. D. (2010). Modified cross-correlation method for the blind identification of structures. *Journal of Engineering Mechanics, ASCE, 136*(7), 889–897.

Hazra, B., Sadhu, A., Roffel, A. J., and Narasimhan, S. (2012). Hybrid time-frequency blind source separation towards ambient system identification of structures. *Computer-Aided Civil and Infrastructure Engineering, 27*, 314–332.

Henjes, K. (1997). Isotropic and anisotropic correlations in turbulent wind speed data. *Boundary-Layer Meteorology, 84*(1), 149–167.

Hjelmfelt M. R. (1988). Structure and life cycle of microburst outflows observed in Colorado. *Journal of Applied Meteorology, 27*(8), 900–927.

Holmes, J. D., and Oliver, S. E. (2000). An empirical model of a downburst. *Engineering Structures, 22*(9), 1167–1172.

Hu, L., and Xu, Y. L. (2014). Extreme value of typhoon-induced nonstationary buffeting response of long-span bridges. *Probabilistic Engineering Mechanics, 36*, 19–27.

Hu, L., Xu, Y.-L., and Huang, W.-F. (2013). Typhoon-induced non-stationary buffeting response of long-span bridges in complex terrain. *Engineering Structures, 57*, 406–415.

Hu, L., Xu, Y. L., Zhu, Q., Guo, A., and Kareem, A. (2017a). Tropical storm-induced buffeting response of long-span bridges: Enhanced nonstationary buffeting force model. *Journal of Structural Engineering, ASCE, 143*(6), 04017027.

Hu, L., Xu, Z., Xu, Y.-L., Li, L., and Kareem, A. (2017b). Error analysis of spatially-varying seismic ground motion simulation by spectral representation method. *Journal of Engineering Mechanics, ASCE, 143*(9), 04017083.

Huang, G., and Chen, X. (2009). Wavelets-based estimation of multivariate evolutionary spectra and its application to nonstationary downburst winds. *Engineering Structures, 31*, 976–989.

Huang, G., Chen, X., Liao, H., and Li, M. (2013). Predicting tall building response to nonstationary winds using multiple wind speed samples. *Wind and Structures, 17*, 227–244.

Huang, G., Peng, L., Kareem, A., Song, C. (2020). Data-driven simulation of multivariate nonstationary winds: A hybrid multivariate empirical mode decomposition and spectral representation method. *Journal of Wind Engineering and Industrial Aerodynamics, 197*, 104073.

Huang, G., Zheng, H., Xu, Y.-L., and Li, Y. (2015). Spectrum models for nonstationary extreme winds. *Journal of Structural Engineering, ASCE, 141*(10), 04015010.

Huang, N. E., Shen, Z., and Long, S. R. (1999). A new view of nonlinear water waves: The Hilbert spectrum. *Annual Review of Fluid Mechanics, 31*(1), 417–457.

Huang, N. E., Shen, Z., Long, S. R., Wu, M. C., Shih, H. H., Zheng, Q., . . . Liu, H. H. (1998). The empirical mode decomposition and the Hilbert spectrum for nonlinear and non-stationary time series analysis. *Proceedings of the Royal Society of London A, 454*, 903–995.

Huang, M., Li. Q., Xu, H., Lou, W., and Lin, N. (2018). Non-stationary statistical modeling of extreme wind speed series with exposure correction. *Wind and Structures, 26*(3), 129–146.

Huang, Z., and Gu, M. (2019). Estimation of peak factor and gust factor of nonstationary wind speed. *Journal of Wind Engineering and Industrial Aerodynamics, 193*, 103953.

Hyvärinen, A., Karhunen, J., and Oja, E. (2001). *Independent component analysis*. New York: Wiley.

Iyama, J., and Kuwamura, H. (1999). Application of wavelets to analysis and simulation of earthquake motions. *Earthquake Engineering and Structural Dynamics, 28*, 255–272.

Iyengar, A. K. S., and Farell, C. (2001). Experimental issues in atmospheric boundary layer simulations: Roughness length and integral length scale determination. *Journal of Wind Engineering and Industrial Aerodynamics, 89*(11–12), 1059–1080.

Kareem, A. (2008). Numerical simulation of wind effects: A probabilistic perspective. *Journal of Wind Engineering and Industrial Aerodynamics, 96*, 1472–1497.

Kareem, A., Hu, L., Guo, Y., and Kwon, D. K. (2019). Generalized wind loading chain: Time-frequency modeling framework for nonstationary wind effects on structures. *Journal of Structural Engineering, ASCE, 145*(10), 04019092.

Kareem, A., and Kijewski, T. (2002). Time-frequency analysis of wind effects on structures. *Journal of Wind Engineering and Industrial Aerodynamics, 90*(12–15), 1435–1452.

Kerschen, G., Poncelet, F., and Golinval, J. C. (2007). Physical interpretation of independent component analysis in structural dynamics. *Mechanical Systems and Signal Processing, 21*, 1561–1575.

Kijewski, T., and Kareem, A. (1998). Dynamic wind effects: A comparative study of provisions in codes and standards with wind tunnel data. *Wind and Structures, 1*(1), 77–109.

Kijewski-Correa, T., and Kareem, A. (2007). Performance of wavelet transform and empirical mode decomposition in extracting signals embedded in noise. *Journal of Engineering Mechanics, ASCE, 133*(7), 849–852.

Kwon, D., and Kareem, A. (2009). Gust front factor: New framework for wind load effects on structures. *Journal of Structural Engineering, ASCE, 135*(6), 717–732.

Kwon, D., and Kareem, A. (2013a). Gust front factor: New framework for wind load effects on structures. *Engineering Structures, 48*, 635–644.

Kwon, D. K., and Kareem, A. (2013b). Comparative study of major international wind codes and standards for wind effects on tall buildings. *Engineering Structures, 51*, 23–35.

Kwon, D. K., and Kareem, A. (2019). Towards codification of thunderstorm/downburst using gust front factor: model-based and data-driven perspectives. *Engineering Structures*, 199, 109608.

Kwon, D., Kareem, A., Butler, K. (2012). Gust front loading effects on wind turbine tower systems. *Journal of Wind Engineering and Industrial Aerodynamics, 104–106*, 109–115.

Kwon, H.-H., Lall, U., and Khalil, A. (2007). Stochastic simulation model for nonstationary time series using an autoregressive wavelet decomposition: Applications to rainfall and temperature. *Water Resources Research, 43*, W05407.

Lavielle, M. (2005). Using penalized contrasts for the change-point problem. *Signal Processing, 85*(8), 1501–1510.

Li, J., Li, C., He, L., and Shen, J. (2015a). Extended modulating functions for simulation of wind velocities with weak and strong nonstationarity. *Renewable Energy, 83*, 384–397.

Li, L., Kareem, A., Hunt, J., Xiao, Y., Zhou, C., and Song, L. (2015b). Turbulence spectra for boundary-layer winds in Tropical cyclones: A conceptual framework and field measurements at coastlines. *Boundary-Layer Meteorology, 154*, 243–263.

Li, L., Kareem, A., Xiao, Y., Song, L., and Zhou, C. (2015c). A comparative study of field measurements of the turbulence characteristics of typhoon and hurricane winds. *Journal of Wind Engineering and Industrial Aerodynamics, 140*, 49–66.

Li, Y., Togbenou, K., Xiang, H., and Chen, N. (2017). Simulation of non-stationary wind velocity field on bridges based on Taylor series. *Journal of Wind Engineering and Industrial Aerodynamics, 169*, 117–127.

Liang, J., Chaudhuri, S. R., and Shinozuka, M. (2007). Simulation of nonstationary stochastic processes by spectral representation. *Journal of Engineering Mechanics, ASCE, 133*, 616–627.

Lombardo, F. T., Smith, D. A., Schroeder, J. L., and Mehta, K. C. (2014). Thunderstorm characteristics of importance to wind engineering. *Journal of Wind Engineering and Industrial Aerodynamics, 125*, 121–132.

Mallat, S. (1998). *A wavelet tour of signal processing*. San Diego, CA: Academic Press.

McCullough, M. (2016). *Data-driven models: From data to knowledge for extreme events*. PhD dissertation. University of Notre Dame.

McCullough, M., Kwon, D. K., Kareem, A., and Wang, L. (2014). Efficacy of averaging interval for nonstationary winds. *Journal of Engineering Mechanics, ASCE, 140*(1), 1–19.

McNeill, S. I., and Zimmerman, D. C. (2008). A framework for blind modal identification using joint approximate diagonalization. *Mechanical Systems and Signal Processing, 22*, 1526–1548.

McNeill, S. I., and Zimmerman, D. C. (2010). Relating independent components to free vibration modal responses. *Shock and Vibration, 17*, 161–170.

Meng, Y., Matsui, M., Hibi, K. (1995). An analytical model for simulation of the wind field in a typhoon boundary layer. *Journal of Wind Engineering and Industrial Aerodynamics, 56*(2–3), 291–310.

Meng, Y., Matsui, M., Hibi, K. (1997). A numerical study of the wind field in a typhoon boundary layer. *Journal of Wind Engineering and Industrial Aerodynamics, 67–68*(97), 437–448.

Michaelov, G., Lutes, L. D., and Sarkani, S. (2001). Extreme value of response to nonstationary excitation. *Journal of Engineering Mechanics, ASCE, 127*(4), 352–363.

Nazir, M. S., and Bouffard, F. (2012). Intra-hour wind power characteristics for flexible operations. In *Proceedings of Power and Energy Society General Meeting*, IEEE, New York, 1–8.

NBCC. (2005). National Building Code of Canada. Ottawa, Ontario: Institute for Research in Construction (IRC).

NIST/SEMATECH. (2003). e-Handbook of statistical methods. http://www.itl.nist.gov/div898/handbook/ (April, 2020).

Oseguera, R. M., and Bowles, R. L. (1988). A simple, analytical 3-dimentional downburst model based on boundary layer stagnation flow. *NASA Technical Memorandum*, 100632.

Orwig, K. D., and Schroeder, J. L. (2007). Near-surface wind characteristics of extreme thunderstorm outflows. *Journal of Wind Engineering and Industrial Aerodynamics, 95*(7), 565–584.

Peng, L., Huang, G., Chen, X., and Kareem, A. (2017). Simulation of multivariate nonstationary wind velocity process: Hybrid of stochastic wave and proper orthogonal decomposition approach. *Journal of Engineering Mechanics, ASCE, 143*(9), 04017064.

Piccardo, G., and Solari, G. (1998). Generalized equivalent spectrum technique. *Wind and Structures, 1*(2), 161–74.

Poncelet, F., Kerschen, G., Golinval, J. C., and Verhelst, D. (2007). Output-only modal analysis using blind source separation techniques. *Mechanical Systems and Signal Processing, 21*, 2335–2358.

Powell, M. D., Vickery, P. J., and Reinhold, T. A. (2003). Reduced drag coefficient for high wind speeds in tropical cyclones. *Nature, 422*(6929), 279.

Priestley, M. B. (1965). Evolutionary spectra and non-stationary processes. *Journal of the Royal Statistical Society. Series B (Methodological), 27*(2), 204–237.

Priestley, M. B. (1967). Power spectral analysis of nonstationary random process. *J. Sound and Vibration, 6*(1), 86–97.

Repetto, M. P., Burlando, M., Solari, G., De Gaetano, P., Pizzo, M., and Tizzi, M. (2018). A web-based GIS platform for the safe management and risk assessment of complex structural and infrastructural systems exposed to wind. *Advances in Engineering Software, 117*, 29–45.

Rilling, G., Flandrin, P., and Gonçalvès, P. (2003). On empirical mode decomposition and its algorithms. In *IEEE-EURASIP Workshop on Nonlinear Signal and Image Processing*. Grado, Italy.

Scott Gunter, W., and Schroeder, J. L. (2015). High-resolution full-scale measurements of thunderstorm outflow winds. *Journal of Wind Engineering and Industrial Aerodynamics, 138*, 13–26.

Shinozuka, M., and Deodatis, G. (1991). Simulation of stochastic processes by spectral representation. *Applied Mechanics Reviews, 44*(4), 191–204.

Solari, G. (2016). Thunderstorm response spectrum technique: Theory and applications. *Engineering Structures, 108*, 28–46.

Solari, G. (1988). Equivalent wind spectrum technique: Theory and applications. *Journal of Structural Engineering, ASCE, 114*(6), 1303–23.

Solari, G., Burlando, M., De Gaetano, P., and Repetto, M. P. (2015a). Characteristics of thunderstorms relevant to the wind loading of structures. *Wind and Structures, 20*(6), 763–791.

Solari, G., De Gaetano, P., and Repetto, M. P. (2015b). Thunderstorm response spectrum: Fundamentals and case study. *Journal of Wind Engineering and Industrial Aerodynamics, 143*, 62–77.

Solari, G., and De Gaetano, P. (2018). Dynamic response of structures to thunderstorm outflows: Response spectrum technique vs time-domain analysis. *Engineering Structures, 176*, 188–207.

Solari, G., and Kareem, A. (1998). On the formulation of ASCE 7–95 gust effect factor. *Journal of Wind Engineering and Industrial Aerodynamics, 77–78*, 673–684.

Solari, G., Rainisio, D., and De Gaetano, P. (2017). Hybrid simulation of thunderstorm outflows and wind-excited response of structures. *Meccanica*, 52(13), 3197–3220.

Spanos, P., and Failla, G. (2004). Evolutionary spectra estimation using wavelets. *Journal of Engineering Mechanics, ASCE, 130*, 952–960.

Subramanian, C. S., Rajagopalan, S., Antonia, R. A., and Chambers, A. J. (1982). Comparison of conditional sampling and averaging techniques in a turbulent boundary layer. *Journal of Fluid Mechanics, 123*, 335–362.

Tao, T., and Wang, H. (2019). Modelling of longitudinal evolutionary power spectral density of typhoon winds considering high-frequency subrange, *Journal of Wind Engineering and Industrial Aerodynamics*, 193, 103957.

Tong, L., Soon, V. C., Huang, Y.-F., and Liu, R.-W. (1990). AMUSE: A new blind identification algorithm. In *Proceedings of the IEEE International Symposium on Circuits and Systems*, 1784–1787.

Treviño, G. (1999). *Spectrum analysis method and apparatus*. Washington, DC: US Patent Office.

Treviño, G., and Andreas, E. L. (2000). Averaging intervals for spectral analysis of nonstationary turbulence. *Boundary-Layer Meteorology*, 95(2), 231–247.

Vanmarcke, E. H. (1975). On the distribution of the first-passage time for normal stationary processes. *Journal of Applied Mechanics, 42*, 215–220.

Vickery, P. J., Masters, F. J., Powell, M. D., and Wadhera, D. (2009). Hurricane hazard modeling: The past, present, and future. *Journal of Wind Engineering and Industrial Aerodynamics*, 97(7), 392–405.

Vicroy, D. D. (1991). A simple, analytical, axisymmetric microburst model for downdraft estimation. *NASA Technical Memorandum*, 104053.

Vicroy, D. D. (1992). Assessment of microburst models for downdraft estimation. *Journal of Aircraft, AIAA*, 29(6), 1043–1048.

Wang, J., Fan, L., Qian, S., and Zhou, J. (2002). Simulations of non-stationary frequency content and its importance to seismic assessment of structures. *Earthquake Engineering and Structural Dynamics, 31*, 993–1005.

Wang, L. (2007). *Stochastic modeling and simulation of transient events*. PhD dissertation. University of Notre Dame.

Wang, L., and Kareem, A. (2004). Modeling of non-stationary winds in gust fronts. In In *Proceedings of the 9th ASCE Joint Specialty Conference on Probabilistic Mechanics and Structural Reliability* (CD-ROM), PMC2004. Albuquerque, NM.

Wang, L., McCullough, M., and Kareem, A. (2013). A data-driven approach for simulation of full-scale downburst wind speeds. *Journal of Wind Engineering and Industrial Aerodynamics, 123*, 171–190.

Wang, L., McCullough, M., and Kareem, A. (2014). Modeling and simulation of nonstationary processes utilizing wavelet and Hilbert transforms. *Journal of Engineering Mechanics, ASCE, 140*(2), 345–360.

Wang, H., Wu, T., Tao, T., Li, A., and Kareem, A. (2016). Measurements and analysis of non-stationary wind characteristics at Sutong Bridge in Typhoon Damrey. *Journal of Wind Engineering and Industrial Aerodynamics, 123*, 100–106.

Wang, H., Xu, Z., Feng, D., and Tao, T. (2019). Non-stationary turbulent wind field simulation of bridge deck using non-negative matrix factorization. *Journal of Wind Engineering and Industrial Aerodynamics, 188*, 235–246.

Wang, H., Xu, Z., Wu, T., and Mao, J. (2018). Evolutionary power spectral density of recorded typhoons at Sutong Bridge using harmonic wavelets. *Journal of Wind Engineering and Industrial Aerodynamics, 177*, 197–212.

Ward, K., and Boland, J. (2007). Modelling the volatility in wind farm output. In *Proceedings of International Congress on Modelling and Simulation (MODSIM)*, Modelling and Simulation Society of Australia and New Zealand, Christchurch, New Zealand, 1858–1863.

Wen, Y. K., and Gu, P. (2004). Description and simulation of nonstationary processes based on Hilbert spectra. *Journal of Engineering Mechanics, ASCE, 130*, 942–951.

Wood, G. S., Kwok, K. C. S., Motteram, N. A., and Fletcher, D. F. (2001). Physical and numerical modelling of thunderstorm downbursts. *Journal of Wind Engineering and Industrial Aerodynamics, 89*(6), 535–552.

Xu, Y. L., Chen, J., and Hui, M. C. H. (2005). Comparison of stationary and non-stationary wind models for turbulent winds over a complex terrain. In *Proceedings of the 9th International Conference on Structural Safety and Reliability* (*ICOSSAR*). Rome, Italy.

Yang, Y., and Nagarajaiah, S. (2013). Time-frequency blind source separation using independent component analysis for output-only modal identification of highly damped structures. *Journal of Structural Engineering, ASCE, 139*(10), 1780–1793.

Yu, B., Chowdhury, A. G., and Masters, F. J. (2008). Hurricane wind power spectra, cospectra, and integral length scales. *Boundary Layer Meteorology, 129*(3), 411–430.

Zhang, Y., and Amin, M. G. (2000). Blind separation of sources based on their time-frequency signatures. In *Proceedings of the IEEE International Conference on Acoustics, Speech, and Signal Processing*, 3132–3135.

Zhao, N., Huang, G., Yang, Q., Zhou, X., and Kareem, A. (2019). Fast convolution integration-based nonstationary response analysis of linear and nonlinear structures with nonproportional damping. *Journal of Engineering Mechanics, ASCE, 145*(8), 04019053.

Zhou, W., and Chelidze, D. (2007). Blind source separation based vibration mode identification. *Mechanical Systems and Signal Processing, 21*, 3072–3087.

Zhou, Y., Kijewski, T., and Kareem, A. (2002). Along-wind load effects on tall buildings: Comparative study of major international codes and standards, *Journal of Structural Engineering, ASCE, 128*(6), 788–796.

Zhou, G.-D., Ding, Y.-L., and Li, A.-Q. (2015). Evolutionary spectra estimation of field measurement typhoon processes using wavelets. *Mathematical Problems in Engineering*, 945203.

SECTION 4

STRUCTURAL AND OTHER IMPACTS OF NSWS

CHAPTER 15

AERODYNAMIC LOADING DUE TO NON-SYNOPTIC WIND SYSTEMS

JUBAYER CHOWDHURY AND TENG WU

1. Introduction

Synoptic winds are associated with large-scale meteorological systems (e.g., high- and low-pressure systems, cyclones, hurricanes) and can last for days, blowing over hundreds of kilometers. It is the wind that is responsible for the everyday weather. Synoptic winds are statistically stationary and make up the atmospheric boundary layer (ABL). The ABL forms as the wind comes in contact with the ground surface. On the other hand, non-synoptic wind systems are statistically nonstationary and are local phenomena. Non-synoptic wind systems are short lived, sometimes lasting only a few minutes. Examples of non-synoptic wind systems are tornadoes and downbursts.

In a mixed climate where different weather phenomena such as extra tropical cyclones, hurricanes, downbursts, and tornadoes coexist, it is important to separate the events for homogeneous analysis. During the period of 2010–2015, wind measurements were taken along the coast of the Tyrrhenian Sea in Italy under the research programs "Wind and Ports" (Solari et al., 2012), and "Wind, Ports, and Sea" (Burlando, De Gaetano, Repetto, & Solari, 2014) to separate different wind events through statistical analysis. Upon analyzing the extreme wind speed records from the weather stations along the coast of the Tyrrhenian Sea, De Gaetano, Repetto, Repetto, and Solari (2014) divided the records in three groups: (1) stationary Gaussian (e.g., ABL), (2) nonstationary non-Gaussian (e.g., downbursts), and (3) nonstationary Gaussian (e.g., gust fronts). The third group, nonstationary Gaussian, is also known as intermediate events (Jubayer, Romanic, & Hangan, 2019a; Kasperski, 2002). Research and knowledge of the intermediate events are very limited and need further attention.

Seismic and wind loads are the two major lateral loads that need to be carefully considered in the design of structures. Extreme wind events, whether synoptic or non-synoptic, have inflicted excessive structural damage and heavy casualties. Structures experience static wind loads in the form of drag and lift forces as well as overturning moment, which may result in torsional divergence (Simiu & Scanlan, 1996). In addition to the static wind loads, a structure might also encounter dynamic wind load if the natural frequency of the structure is lower than about 1 Hz (Holmes, 2015). Different forms of dynamic response due to wind are galloping (single-degree-of-freedom large-amplitude oscillation); flutter (two-degrees of freedom coupled motion); and vortex lock-in (vibration due to vortex shedding) (Simiu & Scanlan, 1996).

2. Current Building Codes and Standards

The American Society of Civil Engineers (ASCE) building code (ASCE 7-16) (ASCE, 2017) did not consider tornadoes or any other non-synoptic winds in the main wind load provisions, mostly because of their low probability of occurrence. However, in the commentary segment, the standard has included a tornado wind speed map for tornado safe room design obtained from Federal Emergency Management Agency (FEMA, 2008). In addition to the map, tornado factors for enclosed and partially enclosed buildings are provided to be used with the GC_p obtained for straight-line synoptic winds. Here, G is the gust effect factor, and C_p is a non-dimensional pressure parameter, the pressure coefficient. The tornado factors result in higher loads for tornadic flow by considering the higher internal pressure and higher design wind speed compared to ABL winds. The standard explicitly mentions that the tornado factor has a considerable level of uncertainty in the development of the tornado design loads. Given the significant differences in the flow characteristics between tornado and ABL flows, one might question the validity of using the same GC_p as the ABL for evaluating tornadic load since the pressure distributions on the building surfaces would be different between tornadoes and ABL (Haan, Balaramudu, & Sarkar, 2010; Razavi & Sarkar, 2018; Sarkar, Haan, Balaramudu, & Sengupta, 2006). The standard does not provide any provision for downbursts.

National Building Code of Canada (NBCC 2015; Canadian Commission on Building and Fire Codes, 2015) also does not provide any guidelines for tornado-resilient building design. In an earlier version of the code (NBCC 2010; Canadian Commission on Building and Fire Codes, 2012), it was stipulated that other than exceptional buildings, such as nuclear power plants, it is not economically feasible to design buildings for tornadoes. The code also pointed out that, in situations where 90 percent of the occupants were killed or injured in tornadoes, the anchorage of house floors into the foundation and the anchorage of roofs to the concrete wall did not satisfy the requirements of building construction. A map of tornado-prone regions in Canada were included NBCC 2020 (Canadian Commission on Building and Fire Codes, 2012), which outlines three different zones: regions prone to significant tornadoes, regions prone to tornadoes, and regions where tornadoes are possible. Eurocode (EN 1991-1-4; European Committee for Standardization, 2005) and Australia/New Zealand standards (AS/NZS 1170.2:2011; Joint Standards Australia/Standards New Zealand Committee, 2015) do not provide any guidance for tornadoes or any other non-synoptic winds.

Building codes around the world hold a key position in ensuring best practice in engineering-based design and construction. Therefore, it is of utmost importance that the building codes and standards gradually address designing structures for non-synoptic winds, providing a common and harmonized rule for structural design for such wind systems.

3. Flow Characteristics of Non-Synoptic Winds

Within the traditional analysis framework, approaching wind fields are usually simplified as a steady flow part, which is characterized by the mean wind speed and profile, and the zero-mean fluctuating flow part, which is characterized by statistical properties such as turbulence intensity and integral scale, gust factor, peak factor, probability distribution, and power spectrum (Davenport, 1961; Harris, 1970). However, for a typical non-synoptic wind process, there is a trend (time-varying mean values) that needs to be extracted from the nonstationary wind data. An accurate, reliable analysis framework of flow structure and turbulence characteristics in the nonstationary wind fields is essential to understand the nature of non-synoptic winds. More importantly, it is the fundamental building block for accurate modeling of extreme winds.

One possible approach to extract the time-varying mean values is based on time–frequency analysis tools such as empirical model decomposition (EMD) (Huang et al., 1998; Xu & Chen, 2004) and wavelet techniques (Olhede & Walden, 2004; L. Wang, McCullough, & Kareem, 2014). For nonstationary wind processes, the definition of statistical properties needs to be updated accordingly so that the time-varying mean values could be naturally integrated. A nonstationary wind model, in which the constant mean wind speed is replaced by time-varying mean values, is an effective tool for this purpose (J. Chen, Hui, & Xu, 2007).

Nonstationary wind data are typically characterized by not only the time-varying mean values but also the time-dependent amplitudes, which may be modeled in simple cases by an envelope function (H. Wang, Wu, Tao, Li, & Kareem, 2016; Tao,

H. Wang, & Wu, 2016). In some cases, these features may be frequency dependent, which influences the power spectral density (PSD). In the development of evolutionary spectral theory, Priestley (1965) preserved the physical interpretation of a spectrum and the conventional definition of frequency. Hence, the evolutionary power spectral density (EPSD) has been widely implemented since its introduction. The EPSD of wind speed time history could represent the distribution of the local turbulent energy in both the time and frequency domains.

A wavelet-based scheme has been used to effectively estimate the EPSD (Spanos & Failla, 2004) for alleviating the resolution issue (optimal adjustment of the resolutions in both time and frequency). High simulation accuracy can be obtained based on the assumption that the time-varying rate of modulation function of the spectrum is very slow with respect to the duration of the selected wavelet (Spanos & Failla, 2004; H. Wang, Xu, Wu, & Mao, 2018). Another effective approach to characterize the nonstationary winds is to utilize the Hilbert spectrum together with the discrete time wavelet decomposition technique (Hilbert-wavelet scheme) (e.g., Buresti, Lombardi, & Bellazzini, 2004). The target Hilbert spectrum for nonstationary simulation is calculated based on the recorded wind samples during non-synoptic events (H. Wang & Wu, 2018; H. Wang & Wu, 2020; L. Wang et al., 2014; Wu, 2015).

Over the years, several wind-testing chambers have emerged to simulate non-synoptic winds, primarily tornadoes and downbursts. Some of the notable non-synoptic wind simulators are the tornado/microburst simulator at Iowa State University (ISU) in the United States (Haan et al., 2010); VorTECH simulator of the National Wind Institute at Texas Tech University, also in the United States (C. Letchford, Levitz, & James, 2015); downburst simulator at the University of Birmingham, United Kingdom (McConville et al., 2009); and the Wind Engineering, Energy, and Environment (WindEEE) Dome at Western University, Canada (Hangan, 2014; Hangan et al., 2017). These simulators can produce model tornadoes replicating Enhanced Fujita 0 (EF0) to EF3 Scale full-scale tornadoes. Wide ranges of model downburst sizes and wind speeds are also reported in the literature from the non-synoptic wind simulators listed above. For example, the model downbursts in the WindEEE Dome can reach up to a 33-m/s wind speed for a jet diameter of 3.2 m (Junayed et al., 2019).

One of the major challenges associated in simulating non-synoptic winds is to obtain the scaling ratios (length, velocity, and time) between full-scale events and laboratory simulations. This is mostly because the characteristics of the full-scale non-synoptic events can vary quite significantly between events, which means one particular laboratory simulation can result in varieties of scales if compared with different events. In addition, there seems to be a lack of consensus among researchers on how to properly scale non-synoptic wind events. Ample recording of full-scale events from all around the world, which has its own challenges due to the uncertainty with these events, is one of the key elements to obtain an appropriate scaling method. Accurate scaling is also an essential prerequisite to obtain design wind loads for structures from wind tunnel simulations.

4. Non-Synoptic Wind Loads

The assurance of structural safety and reliability under non-synoptic winds requires accurate modeling of wind-induced effects due to transient tornadoes and downbursts (Kareem, & Wu, 2013). In this section, an overview of non-synoptic wind loads with discussions on how these loads compare with synoptic wind loads is presented.

4.1. Tornadoes

The strength of a tornado is determined by damage surveys in its aftermath. By carefully analyzing the degree of the structural damage, a range of wind speeds are associated with the tornado. Tornado rating is done using the EF scale, from EF0 to EF5, EF5 being the strongest (McDonald, Mehta, Smith, & Womble, 2009). According to the National Oceanic and Atmospheric Administration (NOAA), approximately 97 percent of the tornadoes recorded over the period of 1991–2010 in the United States were rated EF2 or weaker. This statistic implies that designing buildings to withstand tornadoes with up to an EF2 strength could significantly reduce the probability of damages and inflicted losses. Of importance is that, given that the EF rating is solely based on the degree of damage, an EF5-equivalent tornado going over a barren land would be rated as EF0, meaning it is highly likely that some of the tornadoes are rated lower than their actual strength.

To estimate the tornadic wind loads, testing structures in the physical tornado simulators is probably the most reliable way used to date due to the complexities associated with the flow–structure interactions. One of the earliest studies on tornadic wind loads on a small building was by Chang (1971). The building model was placed in the modified Ward simulator (Ward, 1972), and it was reported that the mean surface pressures were significantly higher in swirling flow than straight-line ABL flow. The higher load is an outcome of the combined effect of suction and dynamic pressures. Jischke and Light (1983) observed similar behavior of tornadic wind loads on a small building as Chang (1971) observed. In addition, they also reported that the swirl in the flow significantly altered the wind loads on the rectangular building model compared to ABL flow. Results strongly suggested that the maximum wind speed, building location with respect to the tornado vortex and the building orientation were the most important parameters in estimating the critical wind loads.

Several buildings, varying in size and shape, were tested in the tornado simulator at ISU. Sengupta, Haan, Sarkar, and Balaramudu (2008) estimated tornadic wind loads on a cubic (geometric scale 1:100) and a tall building (geometric scale 1:500) and compared the loads between two building orientations (0° and 45°), two tornado core radii (0.30 and 0.53 m), and stationary and translating tornadoes. Except for the stationary tornado case with the smaller core radius and 0° building orientation, the tall building experienced higher uplift than the cubic building, with a maximum difference of about

50 percent. Translating tornadoes also produced higher uplift force than stationary tornadoes. For the cubic building in general, the tornado with the smaller core radius and 0° building orientation resulted in higher uplift forces than larger core radius and 45° building orientation. For the tall building, the observation is quite the opposite, where the larger core radius and 45° building orientation generated higher uplift compared to the smaller core radius and 0° building orientation. This clearly indicates that the building size is also a critical parameter in estimating tornadic wind loads.

Razavi and Sarkar (2018) also performed a rigorous parametric study for tornadic wind loads on a 1:200 scaled low-rise gable roof (roof slope 35°) building to investigate the influence of swirl ratio (0.05 and 0.22); translation speed (0.15, 0.30, 0.50 m/s); building's lateral position with respect to the translation path (17 lateral locations for 0.05 swirl ratio and 21 locations for 0.22 swirl ratio); and building orientation (12 orientations at 15° intervals). Overall, a lower swirl ratio generated higher wind loads than the higher swirl ratio. Peak uplift increased with translation speed for a lower swirl ratio, but for a higher swirl ratio the change in peak uplift with translation speed was minimal. For the higher swirl ratio case, a building's lateral location close to the core radius on both sides on the vortex resulted in higher peak uplift compared to other locations. For both swirl ratios in general, building orientations within ± 30° resulted in higher peak uplift, where 0° orientation is the case in which the roof ridge is parallel to the translation path.

Sarkar et al. (2006) evaluated tornadic wind loads on a 1:500 geometrically scaled tall building and found that tornadoes with strength F2 or higher would exceed the minimum design load set by ASCE 7-02 by a factor of 1.8 or higher. A 1:100 scaled gable roof low-rise building with roof slope of 35° was tested by Haan et al. (2010), and it was found that the peak external uplift force coefficients exceeded the minimum design wind loads in ASCE 7-05 by factors between 1.8 and 3.2. Roueche, Prevatt, Haan, and Datin (2015) utilized the pressure data from Haan et al. (2010) and observed that, for roof-to-wall connections, peak shear forces were 1.8 times stronger in tornadoes compared to the ASCE 7-10 building standard. In a recent study, Jubayer, Vickery, Hangan, and Banik (2019b) investigated the tornadic wind loads on two of the gable roof (roof slope 4.76°) low-rise buildings from the National Institute of Standards and Technology (NIST) database (Ho, Surry, Morrish, & Kopp, 2005) in the WindEEE Dome. Preliminary results showed that a for swirl ratio of 0.76, the component and cladding loads were significantly higher (a factor of 3.5) than the design loads prescribed in the ASCE 7-16 (ASCE, 2017) building code, especially in the middle of the roof (Zone 1′) and middle of the wall (Zone 4) zones (Figure 1a and 1b, respectively). Results presented in Figure 15.1 are for two different building orientations (0° and 45°, here 0° being a roof ridge normal to the translation path) and at two different lateral locations from the translation path (core radius, 2 × core radius). The tornado factor (TF) included in the commentary section of ASCE 7-16 is also plotted for comparison. Note that the experiments reported by Jubayer et al. (2019b) were not exhaustive and did not envelope all load scenarios.

Presently, the clear consensus among the studies investigating tornadic wind loads on buildings is that the peak wind loads are considerably higher in tornadic flow than ABL flow. Published literature reported factors as high as 3.2 (Haan et al., 2010) and 3.5

(Jubayer et al., 2019b) for tornadic wind load when compared with ASCE design wind loads. In addition, previous studies have reported the following parameters to be influential with regard to peak wind loads on buildings in tornadic flow:

- Swirl ratio
- Building location (relative to the path of translation)
- Building dimension (plan, height, roof slope, etc.)
- Building orientation (relative to the path of translation)
- Stationarity of the vortex (stationary or translating)
- Core radius (mainly the ratio between core radius and building size)
- Translation speed of the vortex

The slender bridges in the coastline area are prone to suffering from the tornadoes spawned by typhoons/hurricanes (e.g., Fujita, Watanabe, Tsuchiya, & Shimada, 1972; Hill, Malkin, & Schulz, 1966). Also, there is a high risk for bridges built in mountainous areas, due to the complex topography (Li, Tang, Cai, & Liao, 2010), to be affected by a severe convective storm that may initiate tornado-type winds. In 2003, the Kinzua Viaduct over the Kinzua Creek gorge in Pennsylvania in the United States was struck by a tornado, and a large portion of the bridge collapsed (Leech, 2005). The mean wind velocity profile of a tornado is characterized as transient in time and horizontally nonuniform in space, together with the intensive vertical wind flow (updraft).

However, there are few studies of tornado loads and their effects on slender bridges. A. Chen, Liu, and Zhou (2005) presented a preliminary study of bridge performance subject to a simulated tornado event. The results obtained showed that the response of a long-span bridge under tornado is significant. In their study, the unsteady wind–structure interaction effects were not taken into account. To consider the wind–structure interaction effects on the bridge response under tornadic winds, the motion-induced loads should be well characterized.

Hao and Wu (2016) investigated the nonuniform, intensive vertical wind velocity and transient natures of tornado events and their effects on the global behavior of a long-span bridge. Besides the aerostatic behaviors under the tornado-like wind-induced loads, the motion-induced effects were modeled using the semiempirical model with a two-dimensional (2D) indicial response function (Hao & Wu, 2017). Both nonlinear aerostatic analysis and linear aeroelastic analysis in time domain were conducted based on a 3D finite-element model to investigate the bridge performance under the most unfavorable tornado pattern considering wind–structure interactions. Their results indicated that the nonuniform, intensive vertical wind velocity and transient features associated with the tornado events were very important to consider in the simulation of tornado loads on long-span bridges.

Recently, J. Cao, Ren, Cao, and Ge (2019) measured the wind pressure on a streamlined bridge deck using a tornado vortex simulator. Significant discrepancies were noted between tornado-induced wind loads and results from conventional boundary-layer wind tunnels.

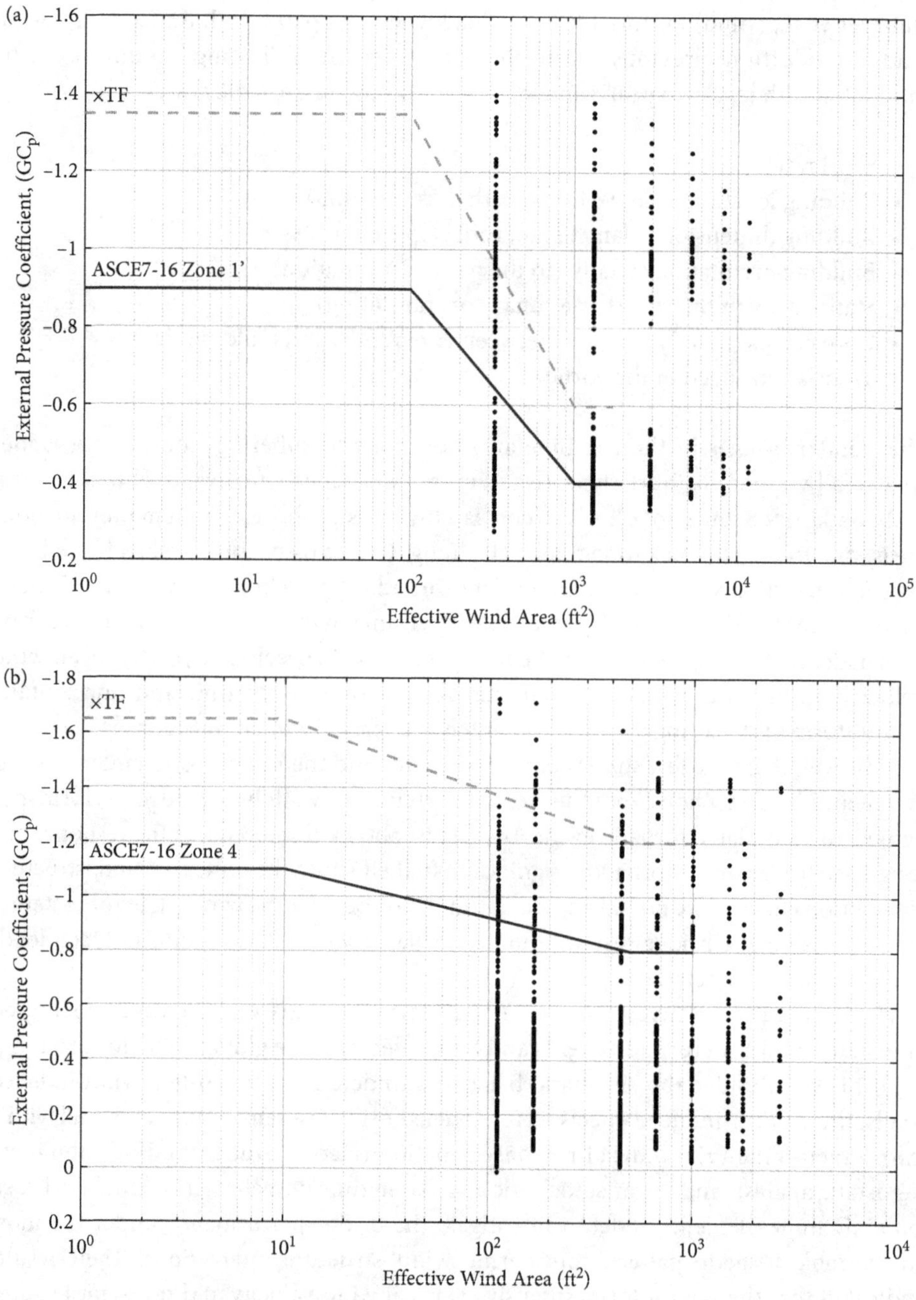

FIGURE 15.1 Comparison of component and cladding loads on low-rise building between tornado and ASCE 7-16 (Jubayer et al., 2019b).

Tornadoes are one of the major weather-related events that cause severe damage to transmission lines every year. Research focusing on tornadic loads on transmission lines did not get much traction until the 1990s. One of the earliest studies that estimated tornadic wind loads on transmission lines was by Ishac and White (1995). Using an analytical approach, a weight increase of less than 3 percent of the existing tower design was proposed to account for tornado loads for up to F2 tornadoes. Savory, Parke, Zeinoddini, Toy, and Disney (2001) performed dynamic structural analysis to study the response of a lattice tower for one tornado and one downburst event. The analysis predicted higher loads for tornado than downburst, resulting in shear failure of the tower for the tornado event. A similar observation was reported by Hamada and El Damatty (2011), where for an F2 tornado, the loads on transmission lines were found to be higher than for the downburst and ABL wind.

In a recent study, Alipour, Sarkar, Dikshit, Razavi, and Jafari (2020) employed aerodynamic force coefficients from conventional wind tunnels with empirically developed tornado wind speed profiles to obtain tornadic loads on transmission lines. Laboratory simulations of a tornado with a model transmission tower were also performed to validate the results. A similar variation of shear force coefficients was found between the proposed method and the validation experiment. This study points toward using conventional wind tunnels for estimating tornadic wind loads on transmission lines. However, more studies are needed to affirm the previous statement.

4.2. Downbursts

Thunderstorm downbursts produce strong wind gusts very close to the ground and therefore can be detrimental for ground-mounted structures. Thunderstorm wind dictates the design wind speed in the noncoastal areas in many parts of the world. Due to the distinct flow characteristics of downbursts from the ABL, researchers have been conducting studies to estimate wind loads on buildings and other structures in downburst-like flows. To estimate wind loads in downburst flows, different methods have been utilized by the researchers to model the flow in physical simulators or numerical simulations. These methods include the cooling source model, which uses negatively buoyant thermodynamic cooling from a predefined source, radial impinging jet, and slotted wall jet. Radial impinging jets are most widely used to physically and numerically simulate downburst flows. In the current section, focus is given to the downburst wind loads obtained using radial impinging jets.

Using numerical simulations, Kim et al. (2007) evaluated wind loads on a tall building in downburst-like impinging jet flows. Both shear and base moment ratios were found to be higher in the downburst when compared with ABL winds. In a two-part study, pressure distributions on a cube at a scale of 1:3,000 were estimated using stationary and translating radial impinging jets by Chay and Letchford (2002) and Letchford and Chay (2002). Translating downbursts, which is more realistic, produced higher wind loads compared to stationary downbursts. Pressure distributions on the cube varied significantly with the relative position of the cube from the downdraft. Maximum pressure was observed when the cube was one diameter of the jet away from the center of the jet. Sarkar et al. (2006) studied the downburst wind loads on a 1:500 scaled tall building in the ISU simulator and compared the loads with ASCE 7-02 standard (ASCE, 2002). Although the force and moment coefficients were lower in downbursts compared to the ASCE 7-02 standard, it was concluded that the downbursts produced biaxial bending and twisting moment that are not usually considered for design in ASCE 7-02 (2002).

However, for low-rise gable roof buildings, Zhang, Sarkar, and Hu (2013) reported higher surface pressures in downburst flows compared to the ASCE 7-05 standard (ASCE 2006). Similar to the observations by Chay and Letchford (2002), Zhang et al. (2013) also reported that surface pressures are sensitive to the relative position of the building from the jet. Using large-eddy simulation of impinging jet, Iida and Uematsu (2019) studied a rectangular building model and found the pressure coefficients were higher in the downburst compared to the ABL. As rightly pointed out by Iida and Uematsu (2019), proper definition of pressure coefficients is essential for meaningful comparison of wind loads for these two vastly different systems (downburst and ABL).

While the design of the long-span bridges in thunderstorm-prone sites could employ a high extreme wind speed derived from the mixed climates (Cook, Harris, & Whiting, 2003), most of the current considerations are limited to the straight-line winds of the neutral ABL (e.g., B. Cao & Sarkar, 2015). The thunderstorm downbursts generate non-synoptic winds presenting time-varying mean values and nonstationary turbulence. It is unlikely that the downburst winds will result in aerodynamic instabilities (e.g., flutter or vortex-induced vibration) of bridges due to their transient nature and hence a lack of "buildup" time for structural dynamics. On the other hand, the long-span bridge may experience significant forced vibration (loss of serviceability) resulting from the transient wind loading during a downburst event.

To well characterize the transient bridge aerodynamics under non-synoptic downburst winds, Hao and Wu (2017) extended the 1D indicial response function in the widely used semiempirical, linearized analysis framework of bridge aerodynamics under stationary winds to a 2D case by introducing an additional timescale in the wind–bridge interaction system. Furthermore, Hao and Wu (2018) carried out the time domain buffeting analysis of a long-span suspension bridge under the traveling downburst by utilizing the computational fluid dynamics–computational structure dynamics hybrid methodology to investigate the transient and nonstationary effects on the non-synoptic wind-induced structural dynamic response. Their numerical results suggested that the significant transient mean wind velocities of the downburst dominated the bridge dynamic performance; however, the gust-induced aerodynamic loads could not be ignored due to their large contribution to the response.

Downbursts are often responsible for widespread damage to transmission lines. The majority of the studies employed numerical simulations to estimate downburst wind loads on transmission lines. Using a finite element model, Shehata, El Damatty, and Savory (2005) investigated transmission lines in downburst flow and found that the downburst wind loading was significantly higher (as high as 304 percent) compared to straight-line winds. It was also reported that the wind loads are sensitive to the relative location of the transmission line from the downdraft. The ratio between the length of the transmission line span and downburst diameter is also a governing parameter in downburst wind loading (Shehata & El Damatty, 2007). Darwish (2010) reported that the forces on a guyed transmission line were increased by 15–20 percent due to the downburst turbulence component.

Recently, an aeroelastic model of a transmission line at 1:50 scale was tested by Elawady et al. (2017) at the WindEEE Dome simulator. This study is believed to be the first of its kind where an aeroelastic model of a downburst was tested in a physical downburst simulator. The study found that the response of the tower and conductor was mainly background, and quasi-steady analysis could be appropriate for obtaining downburst wind loads on transmission lines. Most of the previous studies on transmission lines looked into stationary downbursts, and adding translation might cause a difference in transmission line response under downburst wind. A comprehensive review of downburst wind loads on transmission lines can be found in the work of Abd-Elaal, Mills, and Ma. (2018).

5. Future Direction

Research concerning wind loads on structures due to non-synoptic winds is still in the early stage. Model-scale physical experiments, numerical simulations, and analytical models have helped us to gain some valuable insights on the wind–structure interactions and corresponding loads for non-synoptic wind systems. However, in many instances the results were not validated as the full-scale records are limited due to the extreme uncertainty associated with these events. Field measurements of non-synoptic wind events are also immensely important for proper scaling of these events in physical and numerical simulators.

Previous studies have also shown some disparity when it comes to wind loads on structures in non-synoptic winds, which can only be resolved with further credible studies of this topic. The pressure coefficient, which is one of the most important non-dimensional parameters in calculating wind loads, might need to be revisited when it comes to reference static and dynamic pressures. A wide range of choices of reference static and dynamic pressures can be found in the literature for non-synoptic wind events, and this has resulted in differences in wind load comparisons between non-synoptic winds and ABL. A proper choice of references for the pressure coefficient is also needed for meaningful comparisons of non-synoptic wind loads with straight-line ABL winds.

Finally, building codes and standards around the world need to provide some sort of concrete guidelines on how to design buildings and structures for non-synoptic winds. For small structures and residential low-rise buildings, non-synoptic winds probably would not govern the design of wind loads due to the high uncertainty associated with these events. Whereas the guidance in the building codes would improve the resiliency of structures to non-synoptic winds, the provisions would most likely not make the structures indestructible to these events.

References

Abd-Elaal, E.-S., Mills, J., & Ma, X. (2018). A review of transmission line systems under downburst wind loads. *Journal of Wind Engineering and Industrial Aerodynamics, 179*, 503–513. https://doi.org/10.1016/j.jweia.2018.07.004

Alipour, A., Sarkar, P., Dikshit, S., Razavi, A., & Jafari, M. (2020). Analytical approach to characterize tornado-induced loads on lattice structures. *Journal of Structural Engineering, 146*, 04020108. https://doi.org/10.1061/(ASCE)ST.1943-541X.0002660

American Society of Civil Engineers. (ASCE). (2002). *Minimum design loads for buildings and other structures, standards (ASCE 7-02).* Reston, VA: Author.

American Society of Civil Engineers (ASCE). (2006). *Minimum design loads for buildings and other structures, standards (ASCE 7-05).* Reston, VA: Author.

American Society of Civil Engineers (ASCE). (2017). *Minimum design loads and associated criteria for buildings and other structures, standards (ASCE 7-16).* Reston, VA: Author.

Buresti, G., Lombardi, G., & Bellazzini, J. (2004). On the analysis of fluctuating velocity signals through methods based on the wavelet and Hilbert transforms. *Chaos Solitons Fractals, 20*(1), 149–158.

Burlando, M., De Gaetano, P., Repetto, M. P., & Solari, G. (2014, June). *Wind and waves numerical forecasting for safety access to port areas: the "Wind, Ports, and Sea" project.* Presented at the 6th International Symposium on Computational Wind Engineering, Hamburg, Germany.

Canadian Commission on Building and Fire Codes. (2012). *National building code of Canada 2010 (NBCC 2010).* National Research Council Canada. Retrieved from http://www.nrc-cnrc.gc.ca/eng/publications/codes_centre/2010_national_building_code.html

Canadian Commission on Building and Fire Codes. (2015). *National building code of Canada 2015 (NBCC 2015).* National Research Council Canada. http://www.nrc-cnrc.gc.ca/eng/publications/codes_centre/2015_national_building_code.html

Cao, B., & Sarkar, P. P. (2015). Numerical simulation of dynamic response of a long-span bridge to assess its vulnerability to non-synoptic wind. *Engineering Structures, 84*, 67–75.

Cao, J., Ren, S., Cao, S., & Ge, Y. (2019). Physical simulations on wind loading characteristics of streamlined bridge decks under tornado-like vortices. *Journal of Wind Engineering and Industrial Aerodynamics, 189*, 56–70.

Chang, C. C. (1971). *Tornado wind effects on buildings and structures with laboratory simulation.* Presented at the 3rd International Conference on Wind Effects on Buildings and Structures, September 6–11, Tokyo, Japan.

Chay, M. T., & Letchford, C. W. (2002). Pressure distributions on a cube in a simulated thunderstorm downburst—Part A: Stationary downburst observations. *Journal of Wind Engineering and Industrial Aerodynamics, 90*, 711–732. https://doi.org/10.1016/S0167-6105(02)00158-7

Chen, A., Liu, Z., & Zhou, Z. (2005). Tornado effects on large span cable-stayed bridges [In Chinese]. *Journal of Tongji University (Natural Science), 33*(5), 569.

Chen, J., Hui, M. C. H., & Xu, Y. (2007). A comparative study of stationary and non-stationary wind models using field measurements. *Boundary-Layer Meteorology, 122*, 105–121.

Cook, N. J., Harris, R., & Whiting, R. (2003). Extreme wind speeds in mixed climates revisited. *Journal of Wind Engineering and Industrial Aerodynamics, 91*(3), 403–422.

Darwish, M. (2010). Characteristics and design of downburst loaded transmission lines. PhD thesis, Western University, London, ON, Canada.

Davenport, A. G. (1961). The spectrum of horizontal gustiness near the ground in high winds. *Quarterly Journal of the Royal Meteorological Society, 87*(372), 194–211.

De Gaetano, P., Repetto, M. P., Repetto, T., & Solari, G. (2014). Separation and classification of extreme wind events from anemometric records. *Journal of Wind Engineering and Industrial Aerodynamics, 126*, 132–143. https://doi.org/10.1016/j.jweia.2014.01.006.

Elawady, A., Aboshosha, H., El Damatty, A., Bitsuamlak, G., Hangan, H., & Elatar, A. (2017). Aero-elastic testing of multi-spanned transmission line subjected to downbursts. *Journal of Wind Engineering and Industrial Aerodynamics, 169*, 194–216. https://doi.org/10.1016/j.jweia.2017.07.010

European Committee for Standardization. (2005). *Eurocode 1: Actions on structures—Part 1-4: General actions—Wind actions (EN 1991-1-4)*. Brussels: European Committee for Standardization.

Federal Emergency Management Agency (FEMA). (2008). *Safe rooms for tornadoes and hurricanes: Guidance for community and residential safe rooms.* https://www.fema.gov/fema-p-361-safe-rooms-tornadoes-and-hurricanes-guidance-community-and-residential-safe-rooms

Fujita, T. T., Watanabe, K., Tsuchiya, K., & Shimada, M. (1972). Typhoon-associated tornadoes in Japan and new evidence of suction vortices in a tornado near Tokyo. *Journal of the Meteorological Society of Japan* (2nd ser.), *50*(5), 431–453.

Haan, F. L., Balaramudu, V. K., & Sarkar, P. P. (2010). Tornado-induced wind loads on a low-rise building. *Journal of Structural Engineering, 136*, 106–116. https://doi.org/10.1061/(ASCE)ST.1943-541X.0000093

Hamada, A., & El Damatty, A. A. (2011). Behaviour of guyed transmission line structures under tornado wind loading. *Computers and Structures, 89*, 986–1003. https://doi.org/10.1016/j.compstruc.2011.01.015

Hangan, H. (2014). The Wind Engineering Energy and Environment (WindEEE) dome at Western University, Canada. *Wind Engineers, JAWE, 39*, 350–351. https://doi.org/10.5359/jawe.39.350

Hangan, H., Refan, M., Jubayer, C., Romanic, D., Parvu, D., LoTufo, J., & Costache, A. (2017). Novel techniques in wind engineering. *Journal of Wind Engineering and Industrial Aerodynamics, 171*, 12–33. https://doi.org/10.1016/j.jweia.2017.09.010

Hao, J., & Wu, T. (2016, August–September). *Tornado-induced effects on aerostatic and aeroelastic behaviors of long-span bridge.* Presented at the Proceedings of the 2016 World Congress on Advances in Civil, Environmental, and Materials Research, Jeju Island, Korea.

Hao, J., & Wu, T. (2017). Nonsynoptic wind-induced transient effects on linear bridge aerodynamics. *Journal of Engineering Mechanics, 143*(9), 04017092.

Hao, J., & Wu, T. (2018). Downburst-induced transient response of a long-span bridge: A CFD-CSD-based hybrid approach. *Journal of Wind Engineering and Industrial Aerodynamics, 179*, 273–286.

Harris, R. I. (1970). The nature of the wind. In *Proceedings of the Seminar on the Modern Design of Wind Sensitive Structures* (pp. 29–55). London, UK: Institution of Civil Engineers.

Hill, E., Malkin, W., & Schulz, W., Jr. (1966). Tornadoes associated with cyclones of tropical origin-practical features. *Journal of Applied Meteorology, 5*(6), 745–763.

Ho, T. C. E., Surry, D., Morrish, D., & Kopp, G. A., (2005). The UWO contribution to the NIST aerodynamic database for wind loads on low buildings: Part 1. Archiving format and basic aerodynamic data. *Journal of Wind Engineering and Industrial Aerodynamics, 93*, 1–30. https://doi.org/10.1016/j.jweia.2004.07.006

Holmes, J. D. (2015). *Wind loading of structures* (3rd ed.). Boca Raton, FL: CRC Press.

Huang, N. E., Shen, Z., Long, S. R., Wu, M. C., Shih, H. H., Zheng, Q., . . . Liu, H. H. (1998). The empirical mode decomposition and the Hilbert spectrum for nonlinear and non-stationary time series analysis. *Proceedings of the Royal Society of London A: Mathematical, Physical and Engineering Science, 454*, 903–995.

Iida, Y., & Uematsu, Y. (2019). Numerical study of wind loads on buildings induced by downbursts. *Journal of Wind Engineering and Industrial Aerodynamics, 191*, 103–116. https://doi.org/10.1016/j.jweia.2019.05.018

Ishac, M. F., & White, H. B. (1995). Effect of tornado loads on transmission lines. *IEEE Transactions on Power Delivery, 10*, 445–451. https://doi.org/10.1109/61.368368

Jischke, M. C., & Light, B. D. (1983). Laboratory simulation of tornadic wind loads on a rectangular model structure. *Journal of Wind Engineering and Industrial Aerodynamics, 13*, 371–382. https://doi.org/10.1016/0167-6105(83)90157-5

Joint Standards Australia/Standards New Zealand Committee. (2015). *Structural design actions—Part 2: Wind actions (AS/NZS 1170.2:2011)*. Retrieved from http://shop.standards.co.nz/catalog/1170.2%3A2011%28AS%7CNZS%29/view

Jubayer, C., Romanic, D., & Hangan, H. (2019a). Aerodynamic loading of a typical low-rise building for an experimental stationary and non-Gaussian impinging jet. *Wind Structures, 28*, 315–329. https://doi.org/10.12989/was.2019.28.5.315

Jubayer, C., Vickery, P., Hangan, H., & Banik, S. (2019b). *Wind loads on low-rise buildings in tornado-like vortices*. Presented at the Joint Canadian Society for Mechanical Engineering and CFD Society of Canada International Congress, June 2–5, London, Ontario, Canada.

Junayed, C., Jubayer, C., Parvu, D., Romanic, D., & Hangan, H. (2019). Flow field dynamics of large-scale experimentally produced downburst flows. *Journal of Wind Engineering and Industrial Aerodynamics, 188*, 61–79. https://doi.org/10.1016/j.jweia.2019.02.008

Kareem, A., & Wu, T. (2013). Wind-Induced Effects on Bluff Bodies in Turbulent Flows: Nonstationary, Non-Gaussian and Nonlinear Features. *Journal of Wind Engineering and Industrial Aerodynamics, 122*, 21–37.

Kasperski, M. (2002). A new wind zone map of Germany. *Journal of Wind Engineering and Industrial Aerodynamics, 90*, 1271–1287. https://doi.org/10.1016/S0167-6105(02)00257-X

Kim, J., Hangan, H., & Ho, T. C. E. (2007). Downburst versus boundary layer induced wind loads for tall buildings. *Wind and Structures, 10*(5), 481–494.

Leech, T. (2005). The collapse of the Kinzua viaduct: A combination of design oversight and material fatigue left a century-old railroad bridge vulnerable to an F-1 tornado. *American Scientist, 93*(4), 348–353.

Letchford, C., Levitz, B., & James, D. (2015). Internal pressure dynamics in simulated tornadoes. In Ingraffea, N. & Libby, M. (Eds.), *Proceedings of Structural Congress 2015*. Reston, VA: American Society of Civil Engineers. https://doi.org/10.1061/9780784479117.233

Letchford, C. W., & Chay, M. T. (2002). Pressure distributions on a cube in a simulated thunderstorm downburst. Part B: Moving downburst observations. *Journal of Wind Engineering and Industrial Aerodynamics, 90*, 733–753. https://doi.org/10.1016/S0167-6105(02)00163-0

Li, Y., Tang, K., Cai, X., & Liao, H. (2010). Integrated wind speed standard for long-span bridges over deep-cutting gorge. *Journal of Southwest Jiaotong University, 45*(2), 167–173.

McConville, A. C., Sterling, M., & Baker, C. J. (2009). The physical simulation of thunderstorm downbursts using an impinging jet. *Wind Structures*, *12*, 133–149. https://doi.org/10.12989/was.2009.12.2.133

McDonald, J. R., Mehta, K. C., Smith, D. A., & Womble, J. A. (2009, November). The Enhanced Fujita Scale: Development and implementation. In Chen, S., de Leon, A. D., Dolhon, A. M., Drerup, M. J., & Parfitt, M. K. (Eds.), *Proceedings of Forensic Engineering* (pp. 719–728). Reston, VA: American Society of Civil Engineers. https://doi.org/10.1061/41082(362)73

Olhede, S., & Walden, A. T. (2004). The Hilbert spectrum via wavelet projections. *Proceedings of the Royal Society of London A*, *460*, 955–975.

Priestley, M. B. (1965). Evolutionary spectra and non-stationary processes. *Journal of the Royal Statistical Society, Series B (Methodological)*, *27*, 204–237.

Razavi, A., & Sarkar, P. (2018). Tornado-induced wind loads on a low-rise building: Influence of swirl ratio, translation speed and building parameters. *Engineering Structures*, *167*, 1–12. https://doi.org/10.1016/j.engstruct.2018.03.020

Roueche, D. B., Prevatt, D. O., Haan, F. L., & Datin, P. L. (2015). An estimate of tornado loads on a wood-frame building using database-assisted design methodology. *Journal of Wind Engineering and Industrial Aerodynamics*, *138*, 27–35. https://doi.org/10.1016/j.jweia.2014.11.011

Sarkar, P. P., Haan, F. L., Jr., Balaramudu, V., & Sengupta, A. (2006). Laboratory simulation of tornado and microburst to assess wind loads on buildings. In Brad Cross (Ed.), *Proceedings of Structures Congress 2006*. Reston, VA: American Society of Civil Engineers. https://doi.org/10.1061/40889(201)11

Savory, E., Parke, G. A. R., Zeinoddini, M., Toy, N., & Disney, P. (2001). Modelling of tornado and microburst-induced wind loading and failure of a lattice transmission tower. *Engineering Structures*, *23*, 365–375. https://doi.org/10.1016/S0141-0296(00)00045-6

Sengupta, A., Haan, F. L., Sarkar, P. P., & Balaramudu, V. (2008). Transient loads on buildings in microburst and tornado winds. Proceedings of the 4th International Symposium on Computational Wind Engineering (CWE2006). *Journal of Wind Engineering and Industrial Aerodynamics*, *96*, 2173–2187. https://doi.org/10.1016/j.jweia.2008.02.050

Shehata, A. Y., & El Damatty, A. A. (2007). Behaviour of guyed transmission line structures under downburst wind loading. *Wind Structures*, *10*, 249–268. https://doi.org/10.12989/was.2007.10.3.249

Shehata, A. Y., El Damatty, A. A., & Savory, E. (2005). Finite element modeling of transmission line under downburst wind loading. *Finite Elements in Analysis and Design*, *42*, 71–89. https://doi.org/10.1016/j.finel.2005.05.005

Simiu, E., & Scanlan, R. H. (1996). *Winds effects on structures: Fundamentals and applications to design*. New York, NY: Wiley.

Solari, G., Repetto, M. P., Burlando, M., De Gaetano, P., Pizzo, M., Tizzi, M., & Parodi, M. (2012). The wind forecast for safety management of port areas. Proceedings of the 13th International Conference on Wind Engineering. *Journal of Wind Engineering and Industrial Aerodynamics*, *104*, 266–277. https://doi.org/10.1016/j.jweia.2012.03.029

Spanos, P. D., & Failla, G. (2004). Evolutionary spectra estimation using wavelets. *ASCE Journal of Engineering Mechanics*, *130*(8), 952–960.

Tao, T., Wang, H., & Wu, T. (2016). Comparative Study of the Wind Characteristics of a Strong Wind Event Based on Stationary and Nonstationary Models. *ASCE Journal of Structural Engineering*, *143*(5), 04016230.

Wang, H., & Wu, T. (2018). Hilbert-wavelet-based nonstationary wind field simulation: A multi-scale spatial correlated scheme. *ASCE Journal of Engineering Mechanics*, *144*(8), 04018063.

Wang, H. & Wu, T. (2020). Time-Varying Multiscale Spatial Correlation: Its Simulation and Application to Wind Loading of Structures. *ASCE Journal of Structural Engineering, 146*(7), 04020138.

Wang, H., Wu, T., Tao, T., Li, A., & Kareem, A. (2016). Measurements and analysis of non-stationary wind characteristics at Sutong Bridge in Typhoon Damrey. *Journal of Wind Engineering and Industrial Aerodynamics, 151*, 100–106.

Wang, H., Xu, Z., Wu, T., & Mao, J. (2018). Evolutionary power spectral density of recorded typhoons at Sutong Bridge using harmonic wavelets. *Journal of Wind Engineering and Industrial Aerodynamics, 177*, 197–212.

Wang, L., McCullough, M., & Kareem, A. (2014). Modeling and simulation of nonstationary processes utilizing wavelet and Hilbert transforms. *ASCE Journal of Engineering Mechanics, 140*(2), 345–360.

Ward, N. B. (1972). The exploration of certain features of tornado dynamics using a laboratory model. *Journal of Atmospheric Science, 29*, 1194–1204. https://doi.org/10.1175/1520-0469(1972)029<1194:TEOCFO>2.0.CO;2

Wu, T. (2015). Simulation of nonstationary wind velocity field utilizing multi-scale spatial correlation nested Hilbert-wavelet scheme. In Loredo-Souza, A. M. (Ed.), *Proceedings of 14th International Conference on Wind Engineering* (ICWE14, Porto Alegre, Brazil). Porto Alegre, Brazil: Techno Press.

Xu, Y., & Chen, J., (2004). Characterizing nonstationary wind speed using empirical mode decomposition. *ASCE Journal of Structural Engineering, 130*(6), 912–920.

Zhang, Y., Sarkar, P., & Hu, H. (2013). An experimental study of flow fields and wind loads on gable-roof building models in microburst-like wind. *Experiments in Fluids, 54*, 1–21. https://doi.org/10.1007/s00348-013-1511-9

CHAPTER 16

EXTREMELY GUSTY WINDS FROM A VIEWPOINT OF AERODYNAMIC FORCE

JUNJI MAEDA, TAKASHI TAKEUCHI,
ERIKO TOMOKIYO, AND YUKIO TAMURA

1. Introduction

A gust of wind as a meteorological phenomenon means a sudden increase in the wind; however, the clear criteria for the time and wind speed are not defined. Is it possible to define the criteria quantitatively in more detail? To represent a gusty wind numerically, a continuous record of an instantaneous wind speed is required. At present, the instantaneous wind speed at many observatory stations in Japan is a 3-s averaged value of a propeller anemometer output signal scanned in 4 Hz (Japan Meteorology Agency [JMA], 2007). The maximum value of 1- or 3-s averaged wind speed is smaller than the maximum value of the scanned value at 4 Hz, but the maximum value of 1-s average is closer to that obtained from 4-Hz scanning. An anemometer with higher response and scanning performance for wind fluctuations can record more accurate instantaneous wind speed values. Figures 16.1 to 16.3, which show the time history of an instantaneous wind speed, amplify this discussion. Although their observation locations are different, each figure displays a 24-h history of the 1-s average of the signal of a propeller anemometer scanned at 4 Hz and installed at roughly 30 m above the ground surface. The upper graph in each figure shows the wind speed, while the lower graph shows the wind direction, compared against time on the horizontal axis. In the captions, the "NeWMeK (Network for Wind Measurement in Kyushu) No." is the name of the observation network used and its observation point number, the details of which are described in Section 3.

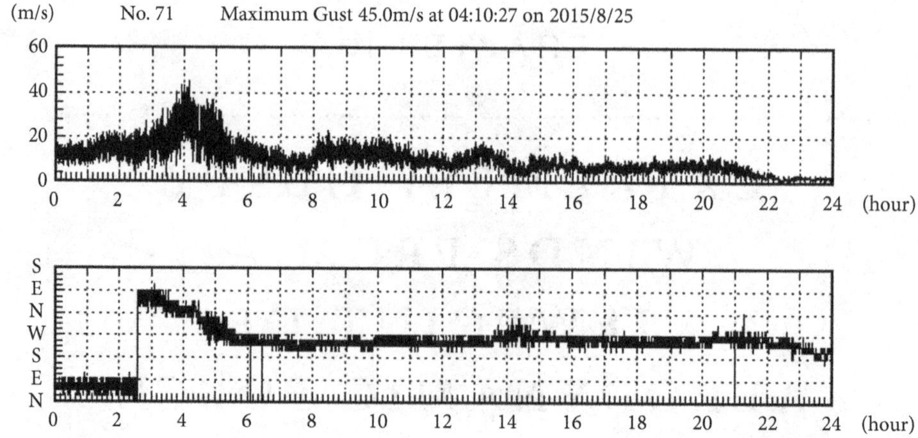

FIGURE 16.1 Example of wind record during a typhoon, NeWMeK No. 71 (Typhoon Goni, Kumamoto Prefecture, 2015).

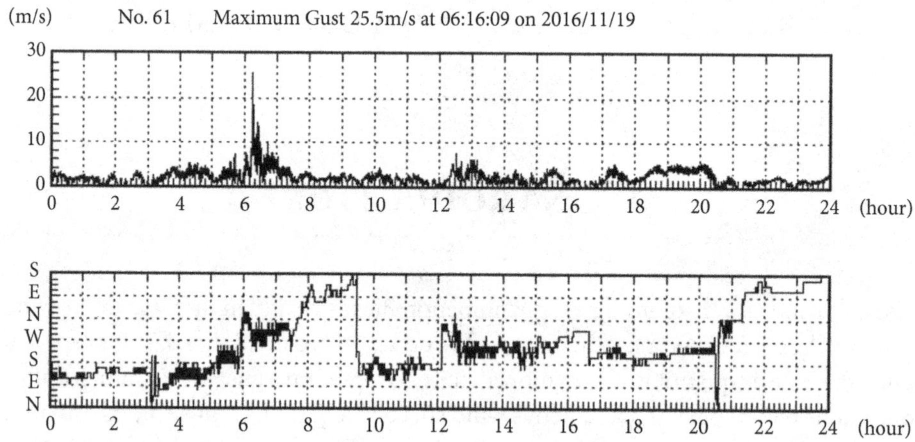

FIGURE 16.2 Example of wind record during a downburst, NeWMeK No. 61 (Kumamoto Prefecture, 2016).

Figure 16.1 shows an example of observations during a typhoon traveling at a speed of about 40 km/h. The time-change patterns of wind speed, wind direction, and the intensity of turbulence differ depending on the distance between the center position of the typhoon and the observation point, as well as the moving speed or surrounding circumstances. However, the average wind speed changes gradually, and a quasi-steady state forms for a period of time between several minutes to several tens of minutes.

Figures 16.2 and 16.3 are examples of observations of a sudden change in wind speed due to non-synoptic local wind not due to the presence of a typhoon. In the 24-h display of these figures, the existence of "spike-type gusts" indicates a sudden change in wind speed, and Figures 16.2 and 16.3 are due to a downburst and a gust front, respectively. Even in Figure 16.1, however, in which the wind speed appears to increase or decrease

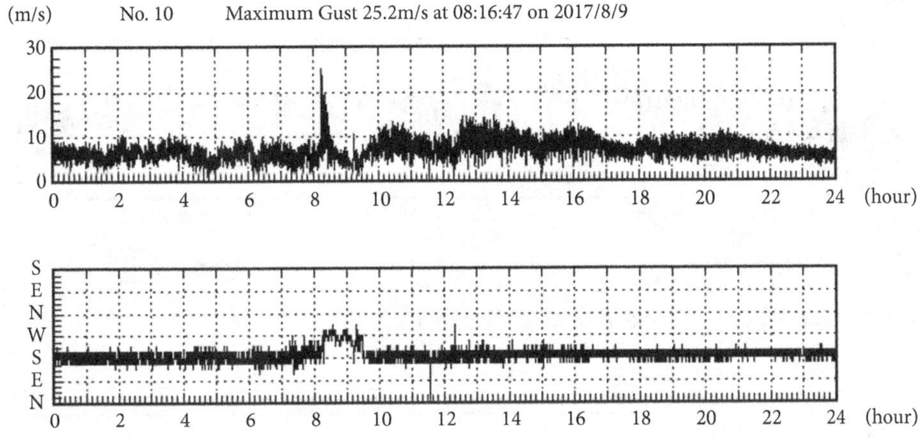

FIGURE 16.3 Example of wind record during a gust front, NeWMeK No. 10 (Fukuoka Prefecture, 2017).

gently, the presence of sharp and large wind speed changes can be observed as shown in Figure 16.4, which focuses an hour before the maximum wind speed. Such a large wind fluctuation exceeds the range where the fluctuation of wind speed can be regarded as a quasi-steady random process, so it is suggested that a larger gust occurred even during the high wind speed. As shown in Figures 16.5 and 16.6, which focus on the time axis before and after the spike-type gusts shown in Figures 16.2 and 16.3, not only the difference between the continuous wind speed values should be considered but also the time difference.

The wind speed difference and time difference that define a large gusty wind depend strongly on the width (average time) of the time domain defining the instantaneous wind speed. Therefore, from the viewpoint of weather observation, it is possible to judge various gusty winds against sequential records of instantaneous wind speed according to the performance of digital equipment and resolution of output signals.

In the field of fluid dynamics, it is understood that the body's fluid force in a steady flow changes into a different phenomenon when encountering a strong unsteady flow, and related early studies have used tank experiments. Taneda (Taneda, 1972) created the situation in which an elliptical cylinder was subjected to a step-function-like flow field by moving the cylinder in a rest state to a constant speed over a short period of time in a water tank and showed that the lift force of the cylinder just after beginning to move was much larger than in the steady flow. As for the reason, Taneda referred to the appearance of the potential flow immediately after the start of movement. Taneda's experimental results were expected to contribute to the accuracy verification of a numerical fluid analysis method under development at the time. Such early studies on strong unsteady fluid forces using tank experiments reported the existence of an unsteady fluid force greater than a steady fluid force (Sarpkaya, 1966).

One of the pioneering findings that focuses on the magnitude of the inertial force caused by a sudden change in the flow velocity over a short period of time is *Morrison's*

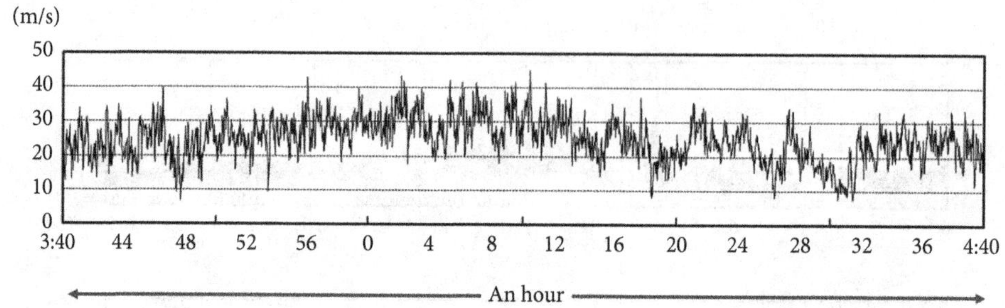

FIGURE 16.4 Expansion of the time axis of Figure 16.1.

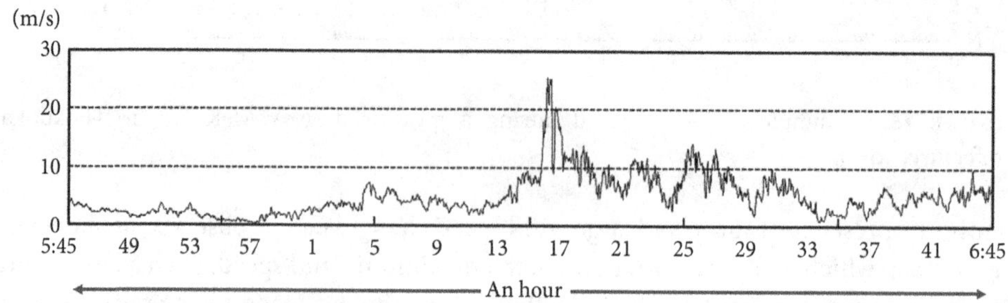

FIGURE 16.5 Expansion of time zone 05:45 to 06:45 in Figure 16.2.

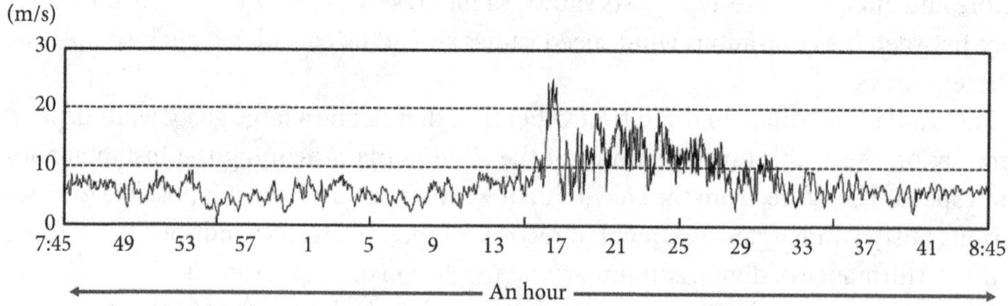

FIGURE 16.6 Expansion of time zone 07:45 to 08:45 in Figure 16.3.

formula, representing the wave force acting on a cylindrical object; the equation has been widely used in coastal engineering since the 1950s (Morison, O'Brien, Johnson, & Schaaf, 1950). Nomura reported that the numerical results corresponded well to the Morrison equation, in a range of low gust wind speeds, in which the component proportioned to the acceleration of wind speed played a large role in the drag fluctuation (Nomura, 2009, 2010; Nomura, Kitamura, & Kitagawa, 2000). It was added that the component proportional to the square of wind speed became dominant as the wind speed increased.

Matsumoto and colleagues (Shiraishi, Matsumoto, & Kitagawa, 1982; Matsumoto, Shimamura, Maeda, Shirato, Yagi, Kawashima, Hori, & Hashimoto, 2007) applied the aerodynamic indicial function, which is notable as one of the earliest studies and represents the transient characteristics of aerodynamic force acting on a bluff body, to the gusty winds and their wind tunnel experiment using a special device. They indicated that the wind force acting on two-dimensional rectangular columns and cylinders increased the steady wind force by about 1.2 times against the step function gusts that rise in a short time under some conditions. They named it an *overshoot phenomenon of wind force*.

A severe gusty wind when a tornado passes shows strong nonstationary disturbance as a non-synoptic local wind. Therefore, it has been known that information on the aerodynamic characteristics of an object in such as a boundary-layer wind tunnel flow cannot be applied to a tornadic load. For example, Wen (1975) pointed out the specificity of the tornado load as follows (p. 169):

I. For airtight structures, the suction force due to the sudden pressure drop at the core of the tornado becomes at least as equally important as the drag force.
II. The wind changes speed and direction rapidly. The large pressure gradient and the high acceleration in such unsteady flows will undoubtedly produce a very large inertia force. For this reason the conventional wind force formula, based on the quasi-steady assumption (inertia force not considered), may not be applicable; the effect of inertia force, therefore, needs to be examined.
III. A typical damaging tornado has a core width of several hundred feet and an average translation speed of about 40 mph (18 m/s), thus the duration of the intense wind action on a structure is very brief (a matter of seconds).

The gust loading factor theory (Davenport, 1967) is widely used in wind load design, using the aerodynamic characteristics of the structures in the wind velocity fluctuation field based on stationary stochastic process theory. However, a strong nonsteady gust such as of a tornado and downburst has no definition for the intensity of turbulence, the scale of turbulence, and the energy distribution in a frequency range. This is because these are based on the mean wind speed and require some average of time and space. Solari et al. (Solari, Burlando, De Gaetano, & Repetto, 2015; Zhang, Solari, De Gaetano, Burlando, & Repetto, 2018) investigated the statistical properties of thunderstorm outflow phenomena recorded by a wind-monitoring network system with 31 ultrasonic anemometers. They revealed the statistical properties of strongly transient fluctuation in wind speed, but not the effects of such a sudden change of wind speed on the aerodynamic admittance function of structures.

The next section on a "gusty wind force" further develops Matsumoto's research using our unique wind tunnel device (Shiraishi et al., 1982; Matsumoto et al., 2007). Note that a sudden change in wind generated by the tunnel is different from a real gust. Nevertheless, although limited, the ability to control the rise time of the wind speed plays the most important role in investigating the overshoot force due to the suddenly changing wind speed. Through such experiments, the overshoot wind force exceeding the steady wind force due to the sudden increase in wind speed will be clarified from the aerodynamic point of view, so the unsteady wind force, affected by the size of an object, with other factors can be explained. In addition, the extent to which current computational fluid dynamics (CFD) technology can reproduce the results of an actual gust of wind is shown and how it can contribute to the various investigations of gusty wind force is discussed.

In the Section 3, for an example of a specified structure, the information on the occurrence of a gusty wind that might have generated an overshoot wind force under a strong wind during a typhoon and non-typhoon is analyzed using the NeWMeK records. Also, how well a wind speed record of 3-s averaged value can capture gusts that seem to induce overshoot wind force is investigated.

2. Characteristics of Unsteady Wind Force on a Body Subjected to a Short-Rise-Time Gust

2.1. Gust Generation Method Using a Small Wind Tunnel

This section introduces both the use of a wind tunnel to generate a short-rise-time gust and the characteristics of an unsteady wind force on certain bodies with basic shapes in such a short-rise-time gust. Here, *short rise time* refers to a rapid change in wind speed that causes unsteady wind forces, which differ from that under a quasi-steady wind. In this section, focusing on the rise time that causes an overshoot wind force in the quantitative definition of a strong unsteady wind, the values of rise time of a gust in experiments and CFD simulations were selected. The unsteady flow around each body was simulated using CFD software, and then the mechanism generating the unsteady wind force was investigated from the simulated flow pattern.

We used a suction-based Eiffel-type wind tunnel at Kyushu University. Figure 16.7 shows a schematic of our test system (Takeuchi & Maeda, 2012). The cross-sectional area of the working space was 1.5 by 1.5 m, and the available length was 3 m. A step-function-like gust was generated by the following procedure using rows of flat blades to generate a pulsating flow: First, we shut the front blade rows, as shown in Figure 16.7, to keep the working space calm while the fan began rotating. Concurrently, we opened the top and bottom blade rows to prevent negative pressure in the suction chamber. Second, we quickly opened all the front blade rows to 90° and simultaneously closed the top and bottom blade rows. This caused an immediate flow into the working space, thus generating a step-function-like gust.

The rise time of the wind flow could be varied from as little as 0.2 s up to 5 s by adjusting the rotation speed of the blade rows. Wind tunnel tests were performed using the procedure, and the wind forces acting on a specimen were measured simultaneously with the wind speed in the working space. Figure 16.8 shows an example of the time

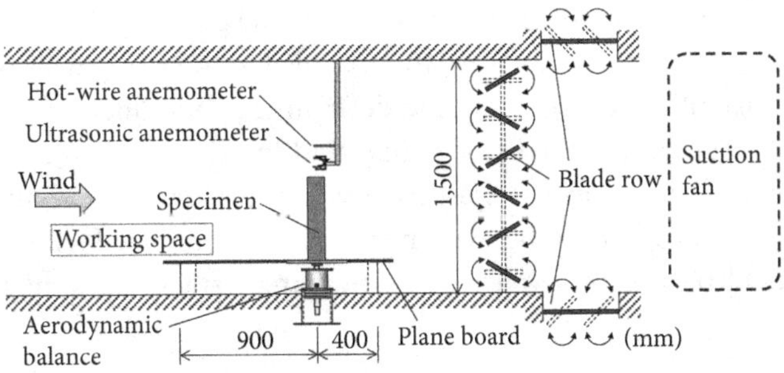

FIGURE 16.7 Schematic of the wind tunnel.

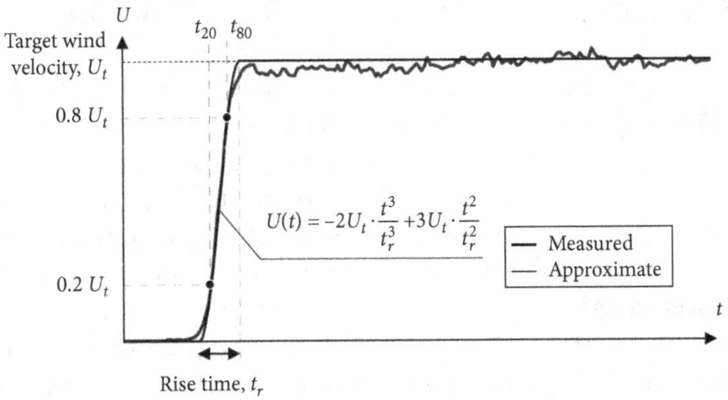

FIGURE 16.8 Time evolution of the wind velocity resulting from a step-function-like gust. (From Takeuchi & Maeda, 2012. Reproduced with permission.)

evolution of the wind velocity generated by controlling the blade rows (Takeuchi & Maeda, 2012). The gust rise time t_r was defined as the time required for an approaching wind to reach the target wind velocity U_t. The average wind speed measured in the steady flow after opening the blade rows was used as the target wind velocity. In this study, the rise time was determined according to the following procedure: First, we obtained U_t from a time evolution of the measured wind velocity. Second, we determined t_{20} and t_{80}, the times required for the wind velocity to increase to 20 percent and 80 percent of U_t, respectively. We assumed the following fit equation for the rising wind velocity:

$$U(t) = -2U_t\left(\frac{t}{t_r}\right)^3 + 3U_t\left(\frac{t}{t_r}\right)^2. \tag{1}$$

When this function passed through the two points, t_{20} and t_{80}, t_r was determined by the following equation:

$$t_r = 2.348\left(t_{80} - t_{20}\right). \tag{2}$$

The value obtained by Equation 2 was thus used as the gust rise time t_r. The reason for using t_{20} and t_{80} was that they were easy to recognize in the rising wind speed. For this section, we consistently used Equation 1 to determine t_r, but different expressions were needed depending on the pattern of the rising wind speed.

The wind forces acting on the specimen were measured using an aerodynamic balance, and the wind speed in the working space was confirmed with a hot-wire anemometer and an ultrasonic anemometer. The measured data were scanned at a frequency of 1,000 Hz.

2.2. Distribution of Wind Speed and Gust Rise Time in Wind Tunnel Generating a Short-Rise-Time Gust

We investigated some features of short-rise-time gusts generated by the procedure in Section 2.1. Figure 16.9 shows time evolutions of the wind velocity measured at the specimen position when the blade rows were opened in 0.2, 0.8, and 1.4 s with $U_t = 2.0$ m/s (Takeuchi & Maeda, 2012). Each wind velocity started to rise simultaneously with the opening of the blade rows. The gust rise times obtained by Equation 2, however, were 0.16, 0.72, and 1.27 s, respectively, which were slightly smaller than the durations of opening the blade rows.

Figure 16.10 shows time evolutions of the wind velocity measured at the specimen position when the blade rows were opened in 0.2 s with U_t = 3.4, 4.9, and 10.0 m/s (Takeuchi & Maeda, 2012). The gust rise times for Figures 10a and 10b were 0.14 and 0.15 s, respectively. The gust rise time for Figure 10c, however, was 0.38 s, which was longer than the duration of opening the blade rows.

This result shows that the gust rise time could exceed the opening time of the blade rows when the target wind velocity was high. For this section, therefore, we dealt with a parameter range that could achieve both the specified rise time and the target wind speed, which was held to 7 m/s or less. Incidentally, it is thought the disturbance in wind velocity seen around 1 s in Figure 10c was caused by the vibration of the device controlling the blade rows.

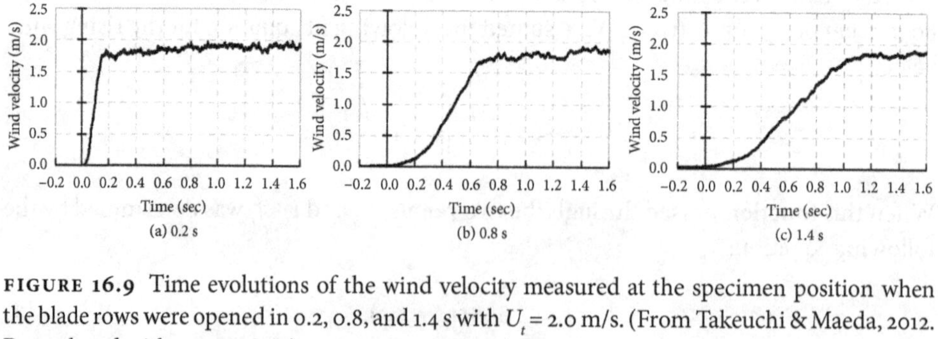

FIGURE 16.9 Time evolutions of the wind velocity measured at the specimen position when the blade rows were opened in 0.2, 0.8, and 1.4 s with $U_t = 2.0$ m/s. (From Takeuchi & Maeda, 2012. Reproduced with permission.)

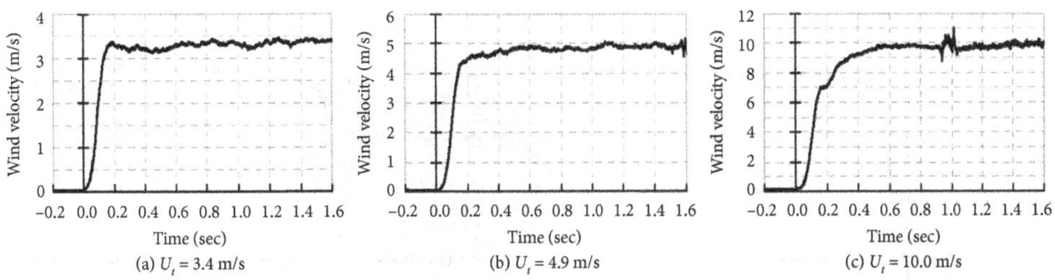

FIGURE 16.10 Time evolutions of the wind velocity measured at the specimen position when the blade rows were opened in 0.2 s with U_t = 3.4, 4.9, and 10.0 m/s. (From Takeuchi & Maeda, 2012. Reproduced with permission.)

Figure 16.11 shows time evolutions of the wind velocity at five different points in the wind tunnel at the specimen position when the blade rows were opened in 0.2 s with U_t = 5.0 m/s (Takeuchi & Maeda, 2012). The reached wind velocity slightly varied with the measurement position. The gust rise time, however, was 0.15 to 0.16 s regardless of the position. At point c2 close to the floor, the wind velocity was heavily disturbed from approximately 0.2 s after the wind velocity had risen. Until then, however, all measurement points showed similar rises in the wind speed. For any point, the results confirmed that the wind speed rose within the opening time of the blade rows, and the rise time was almost the same for all points.

Next, we verified the simultaneity of the generated gust rising at different points in the wind tunnel. Figure 16.12 shows comparisons of the time evolutions of the wind velocity measured at different points when the blade rows were opened in 0.2 s with U_t = 4.0 m/s (Takeuchi, Maeda, & Kawashita, 2008). Figure 12a shows a comparison for points A and B near the aerodynamic balance, which indicates that the velocities rose simultaneously with a short-rise-time gust. Figure 16.12b shows a comparison for point A and point C, located 1,000 mm windward of point A, which indicates the same result. Thus, we found that the wind velocity in the wind tunnel overall rose simultaneously with a short-rise-time gust.

This is almost the same situation as in the experiment of Taneda (1972), in which an elliptical cylinder began to move abruptly. This situation in which the wind speed rapidly changes around a body is not necessarily similar to a gust of natural wind around a large structure. If the rapidly changing wind flow has a spatial scale relatively larger than the size of the structure, however, then the unsteady wind force due to such a gust is considered close to the situation in this experiment.

2.3. Characteristics of Unsteady Wind Force on Basic-Shaped Bodies Subjected to a Short-Rise-Time Gust and Effect of Non-Dimensional Rise Time

The specimens were circular cylinders and square prisms 500 mm in height, as shown in Figure 16.13 (Takeuchi & Maeda, 2012). The cylinders had diameters of 60 and 106 mm,

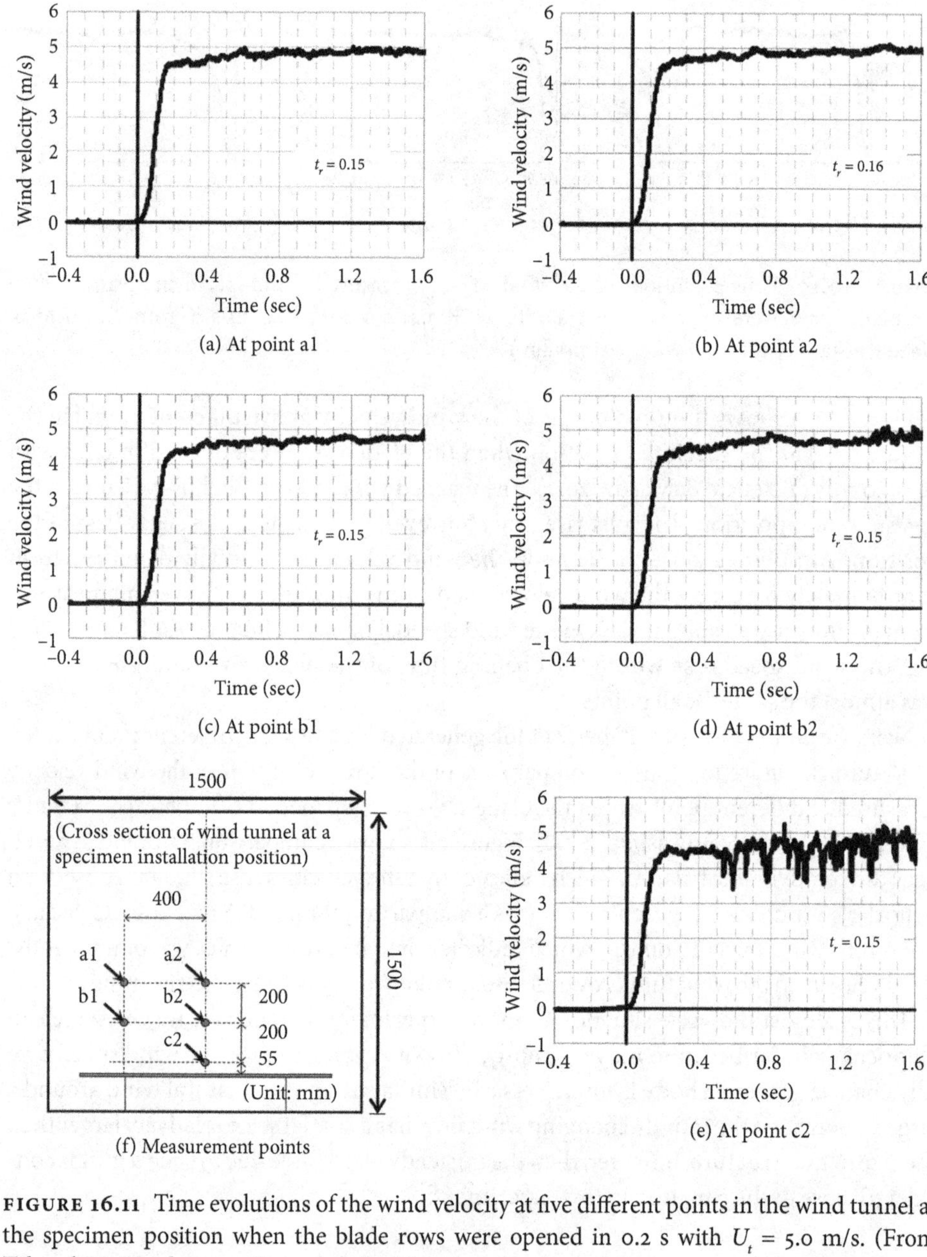

FIGURE 16.11 Time evolutions of the wind velocity at five different points in the wind tunnel at the specimen position when the blade rows were opened in 0.2 s with U_t = 5.0 m/s. (From Takeuchi & Maeda, 2012. Reproduced with permission.)

while the prisms had side lengths of 80 and 120 mm. The specimens were fixed to the aerodynamic balance, and the drag acting on them was measured. The square prisms were fixed with 0° and 45° angles of attack. Here, the angle of attack was defined as 0° when the prism was fixed to the balance with two sides parallel to the wind direction and two sides perpendicular.

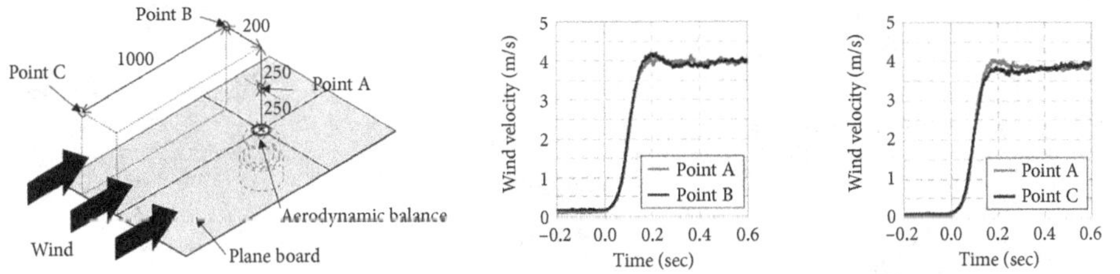

(a) Positions of measurement points (b) Comparison between points A and B (c) Comparison between points A and C

FIGURE 16.12 Comparisons of the time evolutions of the wind velocity measured at points A, B, and C (U_t = 4.0 m/s, t_r = 0.15 s). (From Takeuchi & Maeda, 2013. Copyright Elsevier 2013).

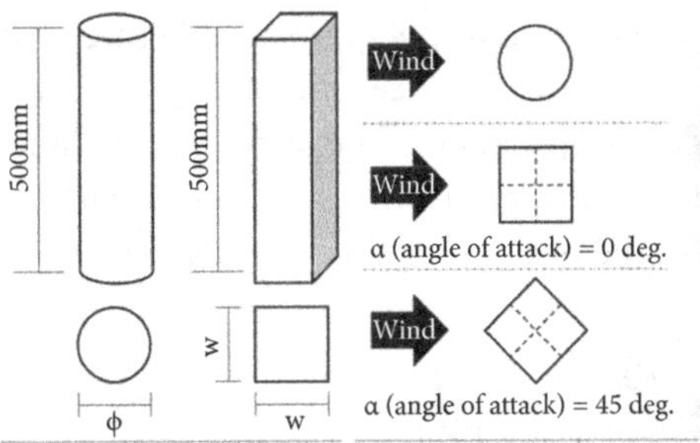

Specimen	Diameter (mm)	Specimen	Side length (mm)
ϕ 60-Cylinder	60	80mm-Square-prism	80
ϕ 100-Cylinder	106	120mm-Square-prism	120

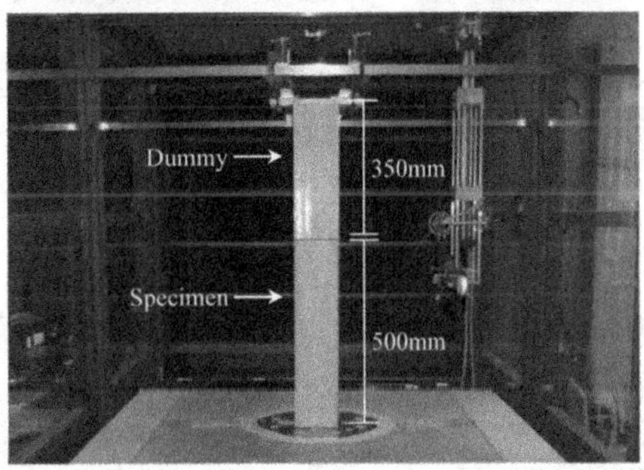

(a) Specimen shapes (b) Photograph of installation

FIGURE 16.13 Specimen shapes and installation. (From Takeuchi & Maeda, 2012. Reproduced with permission.)

For this testing, we did not evaluate the lateral force on the specimens, as they were all symmetrical with respect to the wind direction axis passing through their centers; therefore, the lateral force was near zero. Additionally, a dummy 350 mm in height was placed above the specimen to achieve a two-dimensional flow around it, as shown in Figure 16.13b. The dummy was placed several millimeters from the specimen so as not to measure the wind force on the dummy.

Figure 16.14 shows time evolutions of the drag on the ϕ60-cylinder specimen and the 80-mm square-prism specimen at both angles of attack when the wind velocity reached 2.0 m/s in rise times of 0.15 and 1.3 s (Takeuchi & Maeda, 2012). The figures display the time axis so that the wind velocity began to rise at 0 s, and they show a 2-s time range from the measured data.

In the case of the 0.15-s rise time, we confirmed the overshoot phenomenon of the aerodynamic forces, which reached values much larger than the steady-state drag values. The overshoot phenomenon reached a peak immediately before the wind velocity reached the target velocity. It then decreased immediately and gradually approached a steady state.

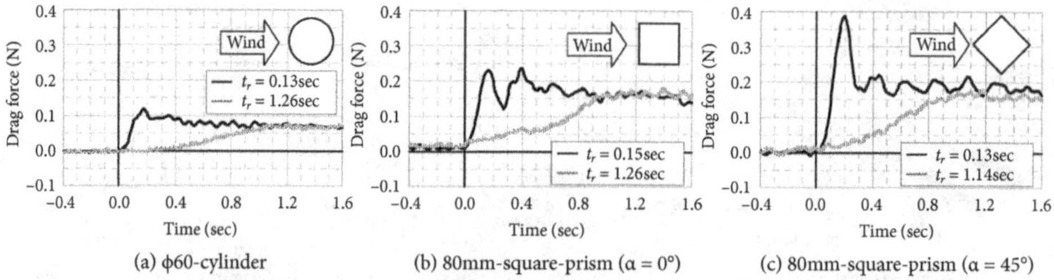

FIGURE 16.14 Time evolutions of the drag on various specimens (U_t = 2.0 m/s, t_r = 0.15, 1.3 s). (From Takeuchi & Maeda, 2012. Reproduced with permission.)

To clarify the impact of the rise time on the overshoot phenomenon, Figure 16.15 shows the relationship between the peak value of the wind force and the rise time for a target wind velocity of 2.0 m/s with three different specimens. The figure also includes the amplification coefficient of the wind force, denoted as the overshoot coefficient. We defined the overshoot coefficient as the ratio of the peak value to the steady-state value of the wind force:

$$\text{overshoot coefficient} = \frac{Peak\ wind\ force}{Steady\ wind\ force}, \tag{3}$$

where the peak wind force was defined as the actual maximum force, and the steady wind force was defined as the mean force over 20 s of steady flow. Note that Figure 16.15 does not show the overshoot coefficient for rise times over 1.14 s because we could not obtain a peak value higher than the steady-state value.

Both the peak value and the overshoot coefficient for each specimen decreased with increasing rise time. For target wind velocities other than 2.0 m/s, the peak wind force and overshoot coefficient showed the same tendency to decrease with increasing rise time.

Figure 16.16 shows the peak values of the wind force and the overshoot coefficients for various target wind velocities with a rise time of 0.15 s (Takeuchi & Maeda, 2012). The peak values clearly increased with the target wind velocity, whereas the overshoot coefficients decreased.

Figure 16.17 shows the effect of the specimen cross section on the non-dimensional time evolution of the ratio of the drag to the steady-state drag for a wind velocity of 2.0 m/s and rise time of 0.15 s (Takeuchi & Maeda, 2012). The non-dimensional time t' was defined by the following equation in reference to Taneda (1972):

$$t' = \frac{U_t \cdot t}{d}, \tag{4}$$

where U_t is the target wind velocity, t is the time, and d is a reference length defined as the streamwise length of the specimen.

The non-dimensional time taken for the drag to achieve a peak was shorter for a large cross section than for a small cross section. The magnitude of overshoot phenomenon increased with the cross section of the specimen.

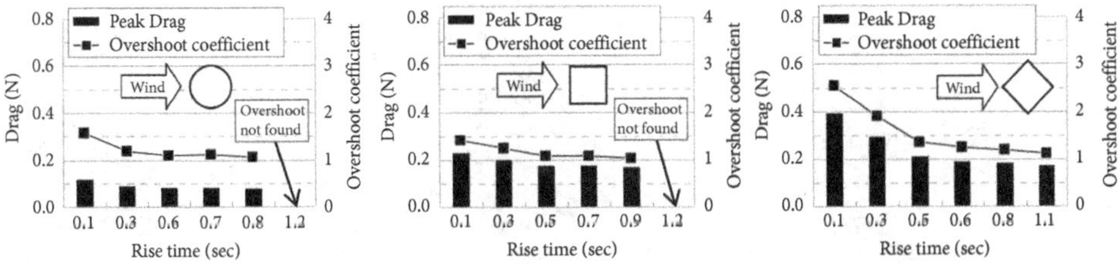

(a) ϕ60-cylinder (b) 80mm-square-prism (α = 0°) (c) 80mm-square-prism (α = 45°)

FIGURE 16.15 Peak wind force values and overshoot coefficients with respect to the rise time (U_t = 2.0 m/s). (From Takeuchi & Maeda, 2012. Reproduced with permission.)

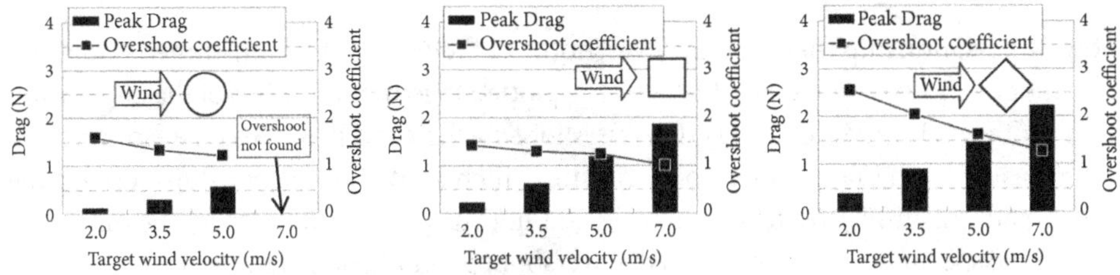

(a) ϕ60-cylinder (b) 80mm-square-prism (α = 0°) (c) 80mm-square-prism (α = 45°)

FIGURE 16.16 Peak wind force values and overshoot coefficients with respect to the target wind velocity (t_r = 0.15 s). (From Takeuchi & Maeda, 2012. Reproduced with permission.)

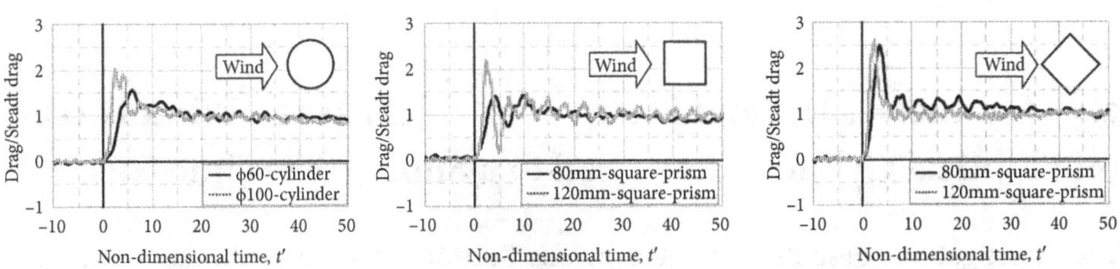

(a) ϕ60-cylinder (b) 80mm-square-prism (α = 0°) (c) 80mm-square-prism (α = 45°)

FIGURE 16.17 Non-dimensional time history data (U_t = 2.0 m/s, t_r = 0.15 s). (From Takeuchi & Maeda, 2012. Reproduced with permission.)

Taneda (1972) reported that at the initial stage just after an elliptical cylinder has started to move, the convection and viscous terms in the Navier-Stokes equation can be negligible because Ut/d and $\nu t/d^2$ are very small. A potential flow then forms, and the lift on the elliptical cylinder becomes large.

We previously reported that the overshoot phenomenon depends on the non-dimensional rise time t_r', defined by Equation 5 (Takeuchi & Maeda, 2012):

$$t_r' = \frac{U_t \cdot t_r}{d}, \tag{5}$$

where U_t is the target wind velocity, t_r is the rise time, and d is the reference length defined as the streamwise length of the specimen. To clarify this tendency, Figure 16.18

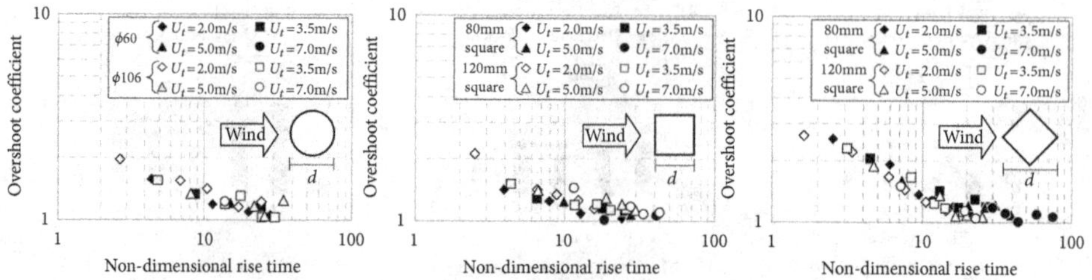

(a) ϕ60-cylinder (b) 80mm-square-prism (α = 0°) (c) 80mm-square-prism (α = 45°)

FIGURE 16.18 Relationship between the overshoot coefficient and non-dimensional rise time. (From Takeuchi & Maeda, 2012. Reproduced with permission.)

shows the relationships between the overshoot coefficient and the non-dimensional rise time under various conditions, which were all plotted together regardless of the target wind velocity (Takeuchi & Maeda, 2012). The overshoot coefficients all followed the same pattern of decreasing as the non-dimensional rise time increased. The relationships among the plotted patterns in Figure 18a through 18c did not make much difference, showing little effect of the specimen cross section on the overshoot phenomenon of the wind force.

These results indicate the possibility of estimating the non–steady-state drag acting on an actual structure subjected to an actual gust by using the relationship between the overshoot coefficient of the drag and the non-dimensional rise time obtained by wind gust testing.

2.4. Characteristics and Mechanism of Unsteady Wind Force on Elliptical Cylinder Subjected to a Short-Rise-Time Gust

The investigations described in the previous Section 2.3 were for bodies with basic shapes and a symmetrical cross section, and the overshoot phenomenon occurred only for the drag. Taneda (1972) reported, however, that an overshoot phenomenon of the lateral force occurs for a body with an asymmetrical cross section, like an elliptical cylinder with an angle of attack other than 0°. For this section, we thus investigated the characteristics of an unsteady wind force on such a cylinder subjected to a short-rise-time gust generated by a wind tunnel (Takeuchi et al., 2008).

Figure 16.19 shows the wind tunnel and the specimens, which were two elliptical cylinders with an axial ratio of 2:1 and a height of 500 mm. We refer to the specimens with a major axis of 70 and 140 mm as the "small cylinder" and "large cylinder," respectively. The specimens were fixed to the aerodynamic balance with an angle of attack α of 45°, and the drag and lateral force acting on them were measured at a frequency of 1,000 Hz.

Figure 16.20 shows the time evolutions of the wind velocity, drag, and lateral force for the large-cylinder specimen with the wind velocity reaching 2.0 m/s in a rise time of 0.15 s (Takeuchi et al., 2008). We confirmed the overshoot phenomenon for the lateral force as well as the drag and found that the peak value of the lateral force was larger than that of the drag.

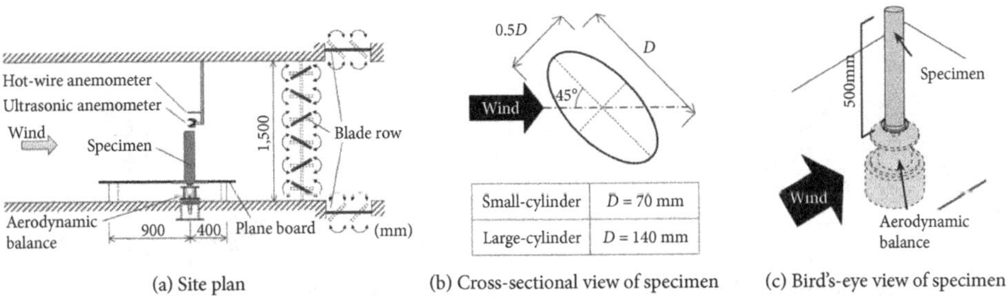

FIGURE 16.19 Wind tunnel site plan and specimens. (From Takeuchi, Maeda, Hayata, & Kawashita, 2009. Reproduced with permission.)

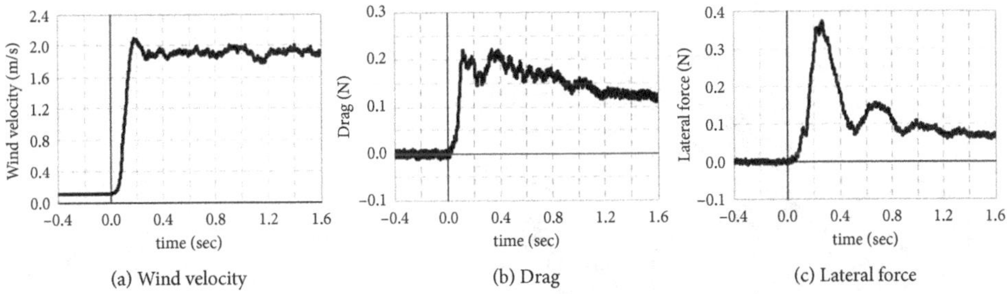

FIGURE 16.20 Time history data for the large-cylinder specimen (D = 140 mm, U_s = 0.0 m/s, U_t = 2.0 m/s, t_r = 0.12 s). (From Takeuchi et al., 2009. Reproduced with permission.)

Figure 16.21 shows the relationship between the overshoot coefficient and the non-dimensional rise time (Takeuchi et al., 2008). As in Figure 16.18, the overshoot coefficients were all plotted together regardless of the target wind velocity. The results indicated a strong relationship between the overshoot phenomenon and the non-dimensional rise time. The overshoot coefficient tended to increase as the non-dimensional rise time decreased, regardless of the cylinder cross section.

In this section, the test results are all for bodies subjected to a gust rising in a short time from calm. In contrast, Takeuchi et al. (2008) investigated the unsteady wind force on an elliptical cylinder subjected to a gust rising in a short time from a steady flow. They confirmed that the overshoot phenomenon also occurred for such a gust rising from a steady flow. They also reported that the overshoot phenomenon was more significant in the case of a gust with a smaller ratio of the initial wind velocity to the target wind velocity.

2.5. CFD Simulation of Flow Around an Elliptical Cylinder Subjected to a Step-Function-Like Gust

The flow around a body under a step-function-like gust has been investigated by CFD simulations with the aim of clarifying the mechanism generating the unsteady wind forces. We simulated the flow around an elliptical cylinder subjected to a step-function-like

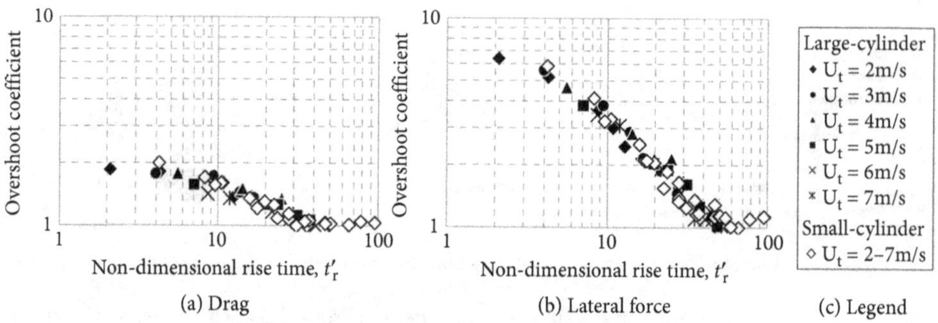

FIGURE 16.21 Relationship between the overshoot coefficient and the non-dimensional rise time (U_s/U_t = 0.0). (From Takeuchi & Maeda, 2013. Copyright Elsevier 2013.)

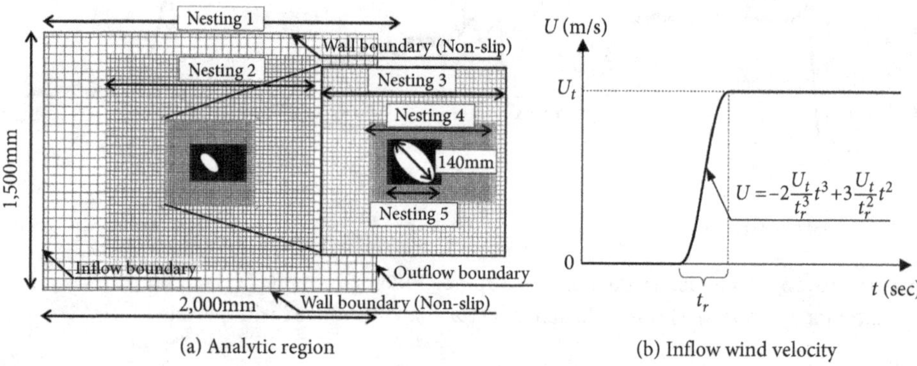

FIGURE 16.22 Analytic region and inflow wind velocity for CFD analysis. (From Takeuchi et al., 2014. Reproduced with permission.)

gust with CFD software (CDAJ) by using a k-ϵ model, as shown in Figure 16.22. A two-dimensional incompressible flow was assumed, and the differential equations were discretized by the finite volume method. The monotone advection and reconstruction scheme (MARS) was applied to the convective term. The pressure and velocity were coupled by the pressure implicit with splitting of operators (PISO) method. The standard k-ϵ model was used for the turbulence model. The analytic region was nested, and the model's boundary conditions were set up as shown in Figure 16.22a (Takeuchi, Maeda, & Otsubo, 2014). We modeled the inflow wind velocity on the short-rise-time wind velocity measured in the wind tunnel tests, shown in Figure 16.22b. The simulation, as an example, was conducted with the aims of confirming how much the experimental results obtained under the limited range of short-rise-time gusts could be reproduced with the CFD procedures and investigating the expandability of the non-dimensional time. However, the effects of the selection of a turbulence model on the reproducibility of the overshoot phenomenon in the field of an actual gusty wind have not been sufficiently clarified, and the verification is necessaryin future works.

Figure 16.23 shows comparisons between the CFD and experimental results for the drag and lateral force with a target wind velocity of 2.0 m/s and a rise time of 0.15 s (Takeuchi et al., 2014). For the CFD results, the wind force was converted to the force

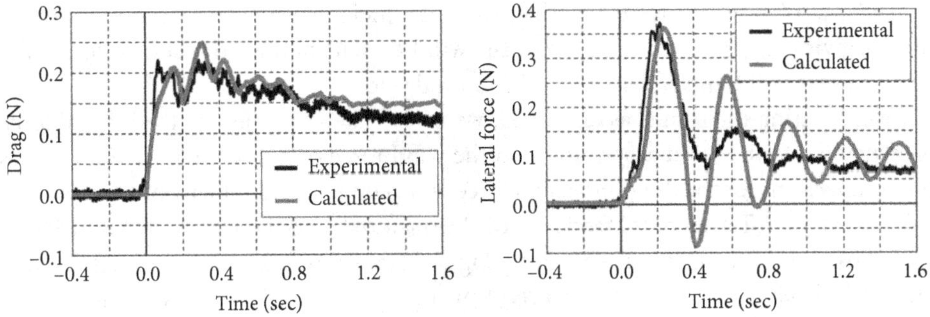

(a) Drag (b) Lateral force

FIGURE 16.23 Comparisons between the calculated and experimental results (U_t = 2.0 m/s, t_r = 0.2 s). (From Takeuchi et al., 2014. Reproduced with permission.)

acting on an elliptical cylinder with a length of 500 mm. We confirmed that the overshoot phenomenon occurred in both the calculated and the experimental results. The peak wind force values obtained by CFD matched well with the experimental values. For lateral force peaks after the second peak, however, the CFD results gave larger values than those for the experimental results. It would appear that this is due to the assumption of two-dimensional flow.

2.6. Effect of Inertial Force Proportional to Flow Acceleration on Unsteady Wind Force Acting on an Elliptical Cylinder Subjected to a Short-Rise-Time Gust

Several studies (Morison et al., 1950; Nomura et al., 2000) have pointed out that the inertial force proportional to the flow acceleration of a short-rise-time gust amplifies the drag on a body. We thus investigated the effect of the inertial force on the wind force acting on an elliptical cylinder subjected to such a gust. The drag and lateral force on the cylinder were obtained by the following equations based on the Morison equation:

$$\text{Drag} = \frac{1}{2}\rho d_y h U^2 C_D + \rho \frac{\pi}{4} d_y^2 h \frac{dU}{dt} C_{MD}, \tag{6}$$

$$\text{Lateral force} = \frac{1}{2}\rho d_y h U^2 C_L + \rho \frac{\pi}{4} d_y^2 h \frac{dU}{dt} C_{ML}. \tag{7}$$

Here, ρ is the air density (kg/m^3), d_y is the frontal projected width of the cylinder (m), h is the height of the cylinder (m), C_D is the drag coefficient under a steady flow, C_L is the lateral force coefficient under a steady flow, C_{MD} and C_{ML} are the respective inertial coefficients (or mass coefficients), U is the wind velocity, and dU/dt is the wind acceleration.

We compared the wind force calculated using these quasi–steady-state equations with an added inertial term proportional to the wind acceleration and the wind force obtained by CFD. The drag coefficient C_D and lateral force coefficient C_L, were obtained from the wind tunnel tests and CFD simulation in a steady flow, giving a C_D of approximately

1.0 and a C_L of approximately 0.6. The inertial coefficients were derived from potential theory, giving $C_{MD} = 1.8$ and $C_{ML} = 0.6$. The wind velocity and wind acceleration for the equations were obtained from the modeled wind velocity.

By comparing the results as described, we investigated the effect of the inertial force proportional to the wind acceleration on the wind force overshoot. Figure 16.24 shows the comparisons for the drag and lateral force with a target wind velocity of 2.0 m/s and a rise time of 0.2 s (Takeuchi & Maeda, 2010). The peak values obtained by CFD were larger than those calculated from the equations. Figure 16.25 shows the same comparisons for a target wind velocity of 1.0 m/s and a rise time of 0.02 s (Takeuchi & Maeda, 2010). The respective peak values of the simulated drag and lateral force corresponded well with those of the calculated drag and lateral force. In this case, the overshoot phenomena of the drag and lateral force were strongly influenced by the inertial force.

Figure 16.26 shows the relationship between the overshoot coefficient and the non-dimensional rise time for the drag and lateral force (Takeuchi & Maeda, 2010). The solid lines represent the results calculated from Equations 6 and 7, while the black squares and gray rhombuses represent the experimental and CFD results, respectively.

With a non-dimensional rise time of about 10, the overshoot coefficients given by the equations were smaller than those from the experimental or CFD results. In this case, the inertial force had only a small effect on the overshoot phenomenon, which was therefore caused by other factors. With a shorter non-dimensional rise time, the overshoot coefficients given by the equations were larger. The overshoot coefficients from the CFD results corresponded with those from the equations for a non-dimensional rise time of less than 3 for the drag and of less than 0.6 for the lateral force. Therefore, the inertial force strongly influenced the overshoot phenomenon when the non-dimensional rise time was relatively small.

In addition, the circle in Figure 26b shows the first peak from Taneda's testing (Taneda, 1972), which measured the unsteady lift acting on an elliptical cylinder rapidly started at a 45° angle of attack by using a water tank. The non-dimensional rise time and overshoot coefficient were obtained from the text and a figure in Taneda (1972) and were

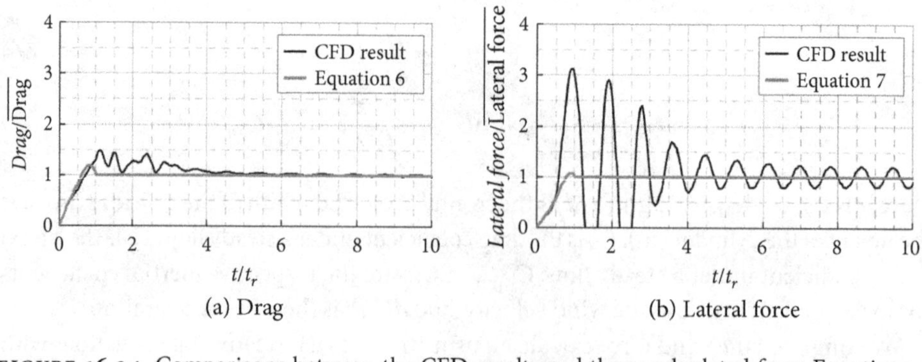

FIGURE 16.24 Comparisons between the CFD results and those calculated from Equations 6 and 7 (U_t = 2.0 m/s, t_r = 0.2 s, D = 70 mm). (From Takeuchi & Maeda, 2010. Reproduced with permission.)

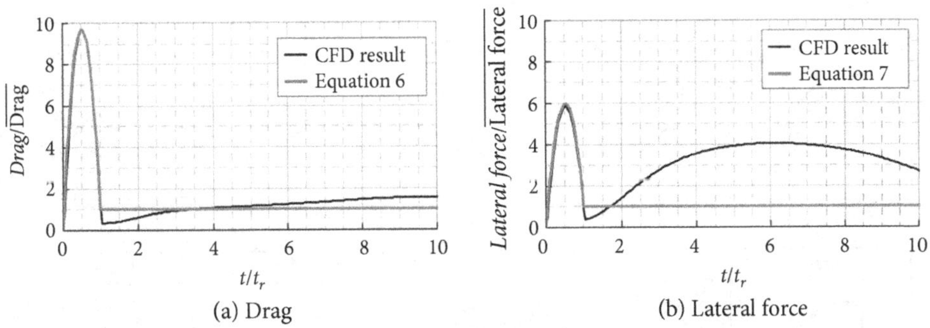

FIGURE 16.25 Comparisons between the CFD results and those calculated from Equations 6 and 7 (U_t = 1.0 m/s, t_r = 0.02 s, D = 70 mm). (From Takeuchi & Maeda, 2010. Reproduced with permission.)

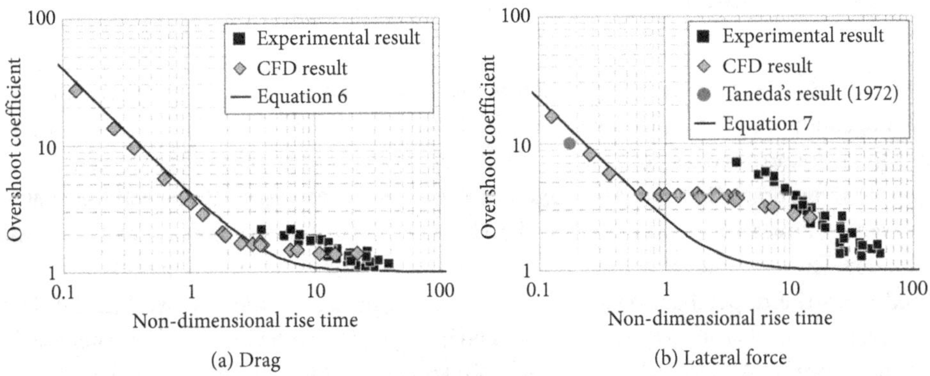

FIGURE 16.26 Relationship between the overshoot coefficient and non-dimensional rise time. (From Takeuchi & Maeda, 2010. Reproduced with permission.)

about 0.2 and 10, respectively. Because the circle appears near the solid line expressing the effect of the inertial force, it seems that the first peak from Taneda's testing was strongly influenced by the inertial force.

2.7. Transition Process of Flow Around an Elliptical Cylinder Subjected to a Short-Rise-Time Gust

In the previous Section 2.6, we showed that the overshoot phenomenon was affected by the inertial force when the non-dimensional rise time was very short. When a large wind force acts on a real structure subjected to a short-rise-time gust, however, the non-dimensional rise time depends on the size of the structure but is assumed to be about 1 to 10. In this case, the overshoot phenomenon is influenced by factors other than the inertial force. In this section, we thus investigate the transition of the flow around an elliptical cylinder subjected to a short-rise-time gust, as simulated by CFD.

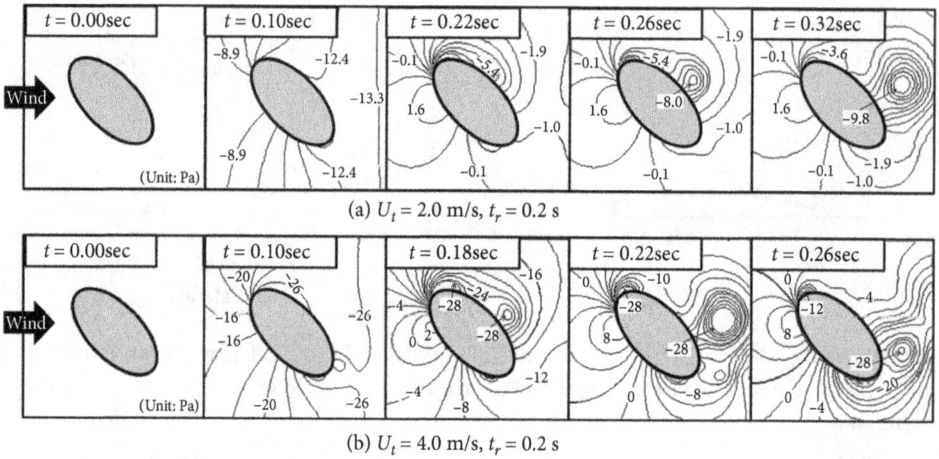

FIGURE 16.27 Isobaric maps (U_t = 2.0, 4.0 m/s; t_r = 0.2 s).

Figure 16.27 shows CFD-simulated isobaric maps of the flow around an elliptical cylinder subjected to a short-rise-time gust in the cases of U_t = 2.0 m/s and t_r = 0.2 s, and U_t = 4.0 m/s and t_r = 0.2 s, respectively (Takeuchi et al., 2014). Each set of maps shows the pattern at five points in time. At t = 0.10 s in Figure 16.27a, the flow pattern was like that in a potential flow. By t = 0.22 s, a vortex had been generated on the upper back face of the cylinder. The lateral force achieved a peak at around this time, and then the vortex was shed from the upper back face. The cylinder was pulled by the vortex, and the drag achieved a peak at around t = 0.32 s. As shown in Figure 16.27b, a vortex was also generated on the upper back face with U_t = 4.0 m/s. In this case, however, the vortex was shed at t = 0.18 s, compared to t = 0.22 s in the previous case. A vortex generated on the leeward upper face is apparently shed more rapidly in the case of a gust with a higher target wind velocity. Therefore, it appears that unsteady vortices are generated on a body subjected to a short-rise-time gust, which causes the overshoot phenomenon.

3. Gust Occurrence in Certain Wind Records from the View of an Overshoot Wind Force

3.1. Example of an Overshoot Wind Force Extracted from Records of Wind Speeds in a Certain Real Observation Network

The existence of overshoot wind force on an object is shown in Section 2, although it is only demonstrated in the modeled suddenly changed wind condition, which differs from an actual condition. The study showed that the occurrence of overshoot wind force is affected by not only rise time and velocity of gust but also the size of the object. "Non-dimensional rise time," which is calculated from the rise time and velocity of the gust and the size of an object, has been used to arrange the results of the experiments. It was then found that larger wind forces occur in a sudden wind change condition with non-dimensional rise times of 20 to 100 than those wind forces that occur in a static condition, though the object size affects these results. These findings show that overshoot wind forces do not always occur in gusts of step function type or spike type in a non-synoptic wind condition, as indicated in the introduction, but also occur when there is the possibility of overshoot wind forces occurring in synoptic wind conditions caused by a typhoon. As an example, Figure 16.28 shows a large wind speed increase observed in the time range when wind speeds became large.

Figure 16.29 shows the time duration of wind speed for 1 min, including 19:11 indicated in Figure 16.28 (Ishida et al., 2000). In the duration, a sudden wind speed change can be seen to occur with a wind speed difference U_t of 22.9 m/s and rise time t_r of 4 s. Consider the true example of a train car derailed on the leeward 500 m away from an observation point in the West Ward, Fukuoka City, on March 19, 1998. The car was 3 m wide, and the axis of it was at a right angle to the gust direction.

Let us consider this situation using non-dimensional rise time. Since wind speed increased with a wind speed difference U_t of 22.9 m/s and the rise time t_r of 4 s, non-dimensional rise time t_r' is 30.5 given that the representative length was a width D of 3 m for the train car. Takeuchi et al. (2008) have indicated the relationship between non-dimensional rise time and overshoot coefficient of wind force in Figures 30a and 30b when gusts rising from a calm condition affect a train-like object where non-dimensional rise time $t_r' = (U_t t_r)/D$, U_t is wind speed, t_r is rise time, D is the representative length of an object, and the overshoot coefficient of wind force is a ratio of peak wind force and static wind force. If the gusts rise from a calm condition, the coefficient of wind force is

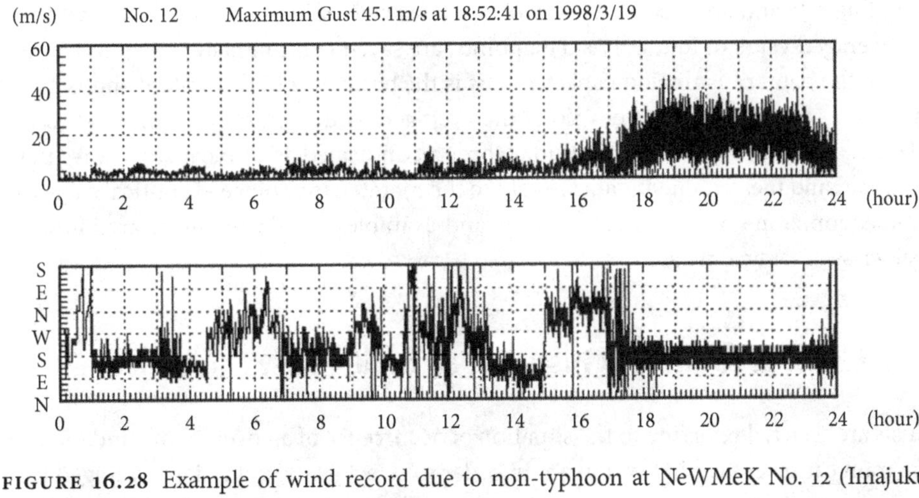

FIGURE 16.28 Example of wind record due to non-typhoon at NeWMeK No. 12 (Imajuku, West Ward, Fukuoka City), March 19, 1998.

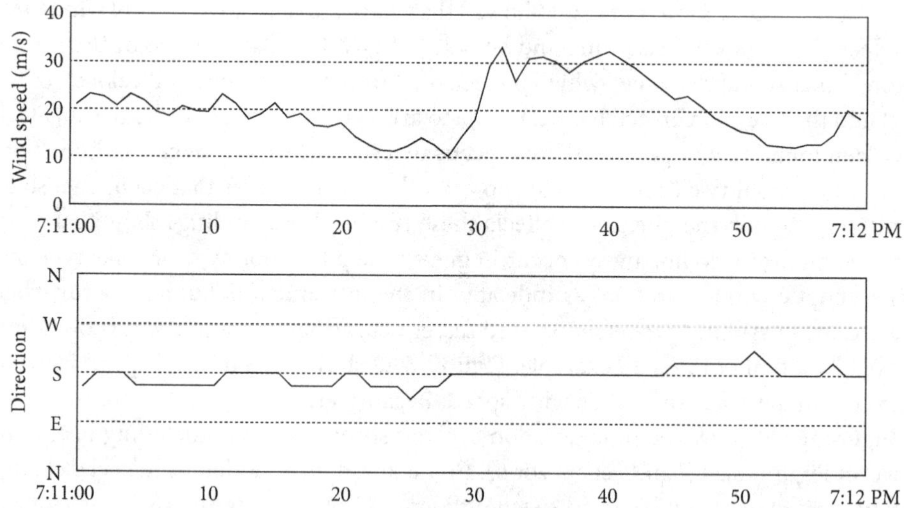

FIGURE 16.29 Example of suddenly changing wind speed (for 1 min) at NeWMeK No. 12 (Imajuku, West Ward, Fukuoka City), March 19, 1998.

1.2 in the case of drag force and 1.9 in the case of lift force. Thus, wind forces greater than the static wind force will occur under gusts.

These gusts, which are expected to generate the overshoot wind force on an object roughly the same size as a train car, were discovered by chance in the high-resolution observation. The tip of the gusts is just seen with the high-resolution data.

All durations of wind speed were data recorded by NeWMeK (Tomokiyo, Maeda, Ishida, & Imamura, 2004), a wide-area network system intended to measure wind speed and direction, segmenting the Kyushu area into high-density arrays. As of March 2007, the system had 123 observation points at intervals of approximately 20 km throughout the Kyushu area. An anemometer positioned at the top of each transmission tower was designed as an observation point. Both wind speed and direction are scanned at a 4-Hz sampling rate and are recorded every 0.25 s. However, only 1-s averaged data, which is the averaged value of four 0.25-s data points, are saved continuously due to the limitation of the data transmission capacity. This is the basic data of the system and is called "1-s averaged data." The system saves the averaged value and maximum wind record (data in each 0.25 s) for 10 min. Wind records are indicated as a time history wave in 1-s intervals, and measurement data are sent to the maintenance office via optical cables. In this section, a method of selecting gusts and examples of selected gusts are indicated using 1-sec averaged data.

3.2. Criteria to Select Gusts Based on a "Gust Effect Scale"

Gusts are selected referring to the situation of occurrence of an overshoot wind force on the train car discussed in Section 3.1. In order to select gusts, a standard averaged time

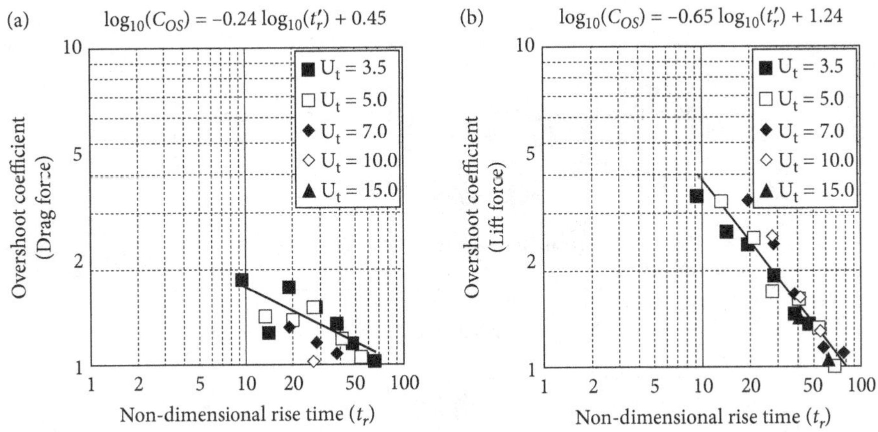

FIGURE 16.30 Relationship between non-dimensional rise time and overshoot coefficient: (a) in the case of drag force; (b) in the case of lift force.

and wind speed difference are required for a period during which gusts occur. In this study, the averaged time has been decided as 6 s based on the moving speed of a tornado in Nobeoka City, Miyazaki, Japan, in 2006 (Committee of Wind Disaster Study, 2007). Several criteria have been chosen in order to select the observed gusts when the train car was derailed in the West Ward, Fukuoka City, in 1998. Wind speeds have been corrected using a power law with a power index of 1/11 (Okauchi, Ito, & Miyata, 1977). Gusts have been selected based on the change in wind speed to refer to short-time-rising gusts used in the wind tunnel experiments. Additionally, the change of wind direction has been used to investigate the statistical properties of gusts.

Based on this basic policy, gusts have been selected from 1-s averaged wind data that were measured from 2009 to 2013 at 123 NeWMeK observation points using the procedure and criteria in the flow chart shown in Figure 16.31 (Tomokiyo, Maeda, Nogata, & Takeuchi, 2012; Tomokiyo & Maeda, 2016).

The "gust effect scale" divided by an object size indicates the "non-dimensional rise time." In what follows, it is assumed how many gusts are included in the records of NeWMeK. The gusts considered are defined as winds capable of causing overshoot wind force on an object roughly the width of a train car, where the train car body axis is perpendicular to the gust direction.

Gusts that can cause overshoot wind forces on a real-scale object of train car shape were selected based on gust wind tunnel tests on a train-shaped specimen and were conducted by Takeuchi et al. (2008). Referring to Takeuchi et al. (2008), the relationship between the overshoot coefficients, O_{SD} and O_{SL}, and non-dimensional rise time t_r' of the drag and lift forces on train-shaped objects under gusts rising from a calm condition can be approximated using the equations that follow. Here, $t_r' = (U_t t_r)/D$, t_r is the rise time of the wind speed, U_t is the target wind speed of the experiment, and D is the representative length of the object.

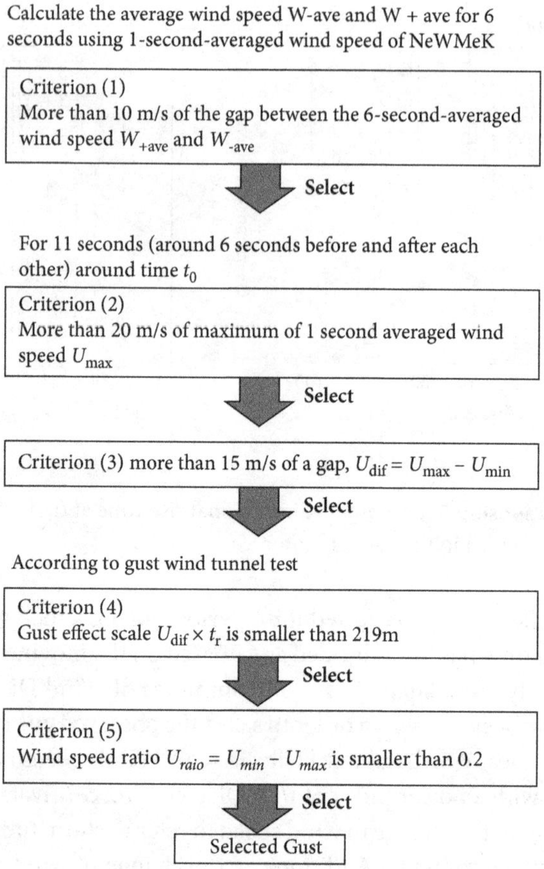

FIGURE 16.31 Flow chart for selecting gusts.

$$O_{SD} = 2.8 \frac{1}{t_r'^{\,0.24}} \tag{8a}$$

$$O_{SL} = 17.3 \frac{1}{t_r'^{\,0.65}} \tag{8b}$$

According to the formulas of Equations 8a and 8b, the non-dimensional rise time t_r' is about 73 when O_{SD} equals 1, and the t_r' is about 80 when O_{SL} equals 1. Therefore, in the case that the non-dimensionless rise time is 73 or less, it is assumed that an overshoot wind force is caused on the train-shaped object. If a representative length D of an object is 3 m (the width of the train car specified in Section 3.1), an overshoot wind force is caused when $U_t \times t_r$ is smaller than 219 m. Therefore, replacing the target wind speed U_t of the experiment with the difference of the observation wind speed U_{dif}, gusts that cause overshoot wind force on a train car are selected when the gust effect scale $U_{dif} \times t_r$ is 219 m or less. This is criterion (4) in Figure 16.31.

Although gusts rising from calm conditions are assumed in Equations 8a and 8b, most actual observed gusts suddenly rise from a certain wind speed condition. Takeuchi and Maeda (2012) have reported gust tunnel experiments for an elliptical column under gusts rising from a certain condition. According to the results, the smaller the ratio of the initial wind speed to the target wind speed (hereinafter referred to as the wind speed ratio), the larger the overshoot wind force is. If the wind speed ratio is less than 0.2, it has been reported that overshoot phenomenon is almost the same as that under a gust rising from calm conditions. Although the report is concerned with the results for elliptic columns rather than train-shaped objects and considers that Equations 8a and 8b can be applicable with the wind speed ratio of 0.2 or less. Thus, when the wind speed ratio of observations $U_{ratio} = U_{min}/U_{max}$ is less than or equal to 0.2, the condition is a criterion (5) in Figure 16.31. The criterion assumes the condition for selecting gusts that cause overshoot wind force on a train car. *Selected gusts* are the gusts with rise time selected by satisfying the five conditions shown in Figure 16.31.

3.3. Occurrence of Selected Gusts Focusing on Gust Effect Scale and Example of the Selected Gusts

Gusts focusing on the gust effect scale and the wind speed ratio of gust, as indicated in Figure 16.31, were selected from the 1-s averaged wind speed record in NeWMeK over 5 years from 2009 to 2013 (1,825 days). As described previously, it has been said that the overshoot phenomenon tends to occur when the non-dimensional rise time is smaller and the wind speed ratio is smaller. So, the maximum wind speed U_{max}, the minimum value U_{min}, the difference between wind speeds U_{dif}, and the rise time of wind speed t_r were obtained. For this, t_r is the difference between t_{max}, at which the maximum of 1-s averaged wind speed U_{max} occurred, and t_{min}, at which the minimum value U_{min} occurred. The gust effect scale $U_{\mathrm{dif}} \times t_r$ and the wind speed ratio U_{ratio} are also calculated. Noda, Hisanobu, Waki, and Nagao (2012) indicated that non-dimensional acceleration of wind speed affected overshoot phenomenon. Therefore, the wind acceleration a_r (= U_{dif}/t_r) was obtained also. The wind direction difference was obtained, and D_{dif} is the gap of the wind direction (16 direction classifications) when the maximum and the minimum wind speed were recorded.

A total of 1,298 gusts were selected from 123 observations of NeWMeK in the 5 years from 2009 to 2013. Between 2010 and 2013, there were 21 typhoons that caused 623 selected gusts over 31 days. On the other hand, 675 selected gusts that derived from strong winds other than typhoons were selected from records for 110 days. Excluding typhoon situations, 235 selected gusts occurred in 42 days. Almost all of the gusts were caused by a so-called bomb cyclone, a low-pressure system developed over 1 hPa in an hour. Table 16.1 shows the mode of each parameter. There were no significant differences between the maximum and minimum wind speeds, the gap of wind speed and the gust effect scale caused by typhoon and non-typhoon situations. The statistical characteristics of the gusts such as the wind speed rise time, wind speed ratio, and wind direction gap are almost the same.

Let us consider representative examples of selected gusts. Focusing on the gust effect scale, which affects the overshoot phenomena among meteorological conditions, four

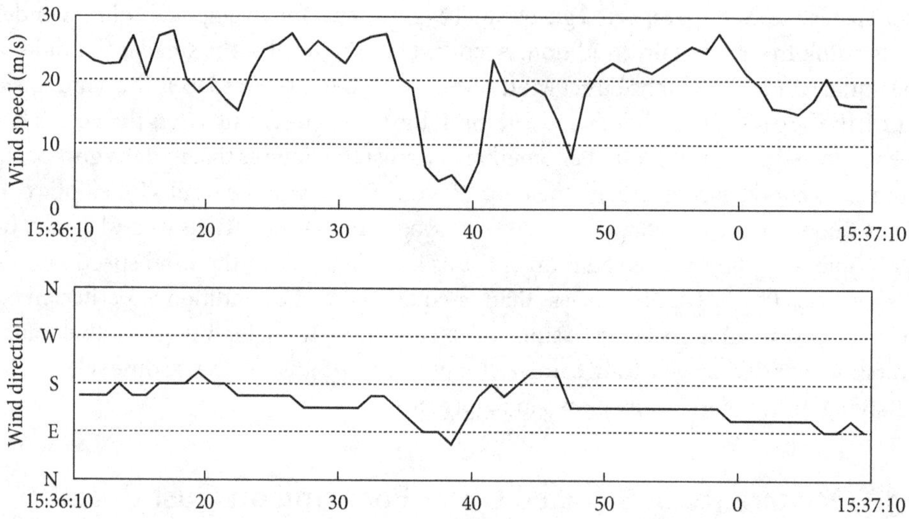

FIGURE 16.32 Example of gusts during a typhoon (October 8, 2013, NeWMeK No. 41, anemometer height 36 m).

cases of gust durations are shown in Figures 16.32 to 16.35. It can be seen that selected gusts showed a smaller gust effect scale. However, height correction did not perform to wind speed records in these figures. These figures indicate 1-min duration of 1-s averaged wind speed, including peak gusts. The examples of gusts derived from typhoons are shown in Figures 16.32 and 16.33, those caused by bomb cyclone are shown in Figure 16.34, while the examples during other strong winds are shown in Figure 16.35. As shown in Figure 16.33, a step-function-like gust arose from almost calm conditions during strong winds and continued for a long time, similar to typhoons and bomb cyclones. In addition, as shown in Figure 16.32, there was a case where winds calmed down momentarily during continuous strong winds. Figure 16.34 shows a case in which wind speeds changed instantaneously under strong winds because of strong turbulence intensity. On the other hand, Figure 16.35 is an example of wind due to non-typhoon-like seasonal wind. The average speed while seasonal wind continues is often not quite as large as the average speed during a typhoon or bomb cyclone. As shown in these figures, there were spike-type gusts that returned to low speed after wind speeds increased suddenly under calm conditions. All of these records seem to be for gusts that can cause overshoot wind force on train cars.

3.4. Occurrence of Selected Gusts Focusing on 3-s Averaged Peak Gusts and 10-min-Averaged Wind Speed

In order to analyze gusts with a sudden change in wind speed, ground observations using high-response anemometers at high-resolution locations are required. However,

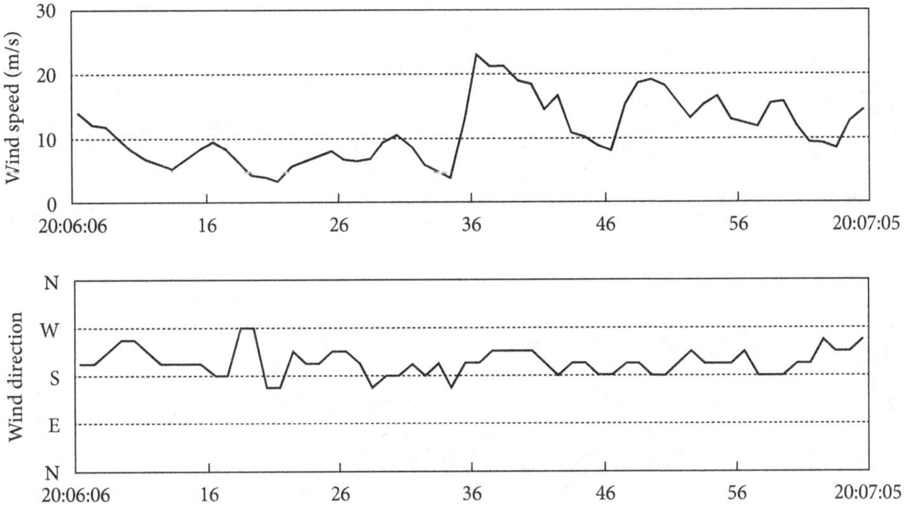

FIGURE 16.33 Example of gust during typhoon (October 8, 2013, NeWMeK No. 15, anemometer height 16.4 m).

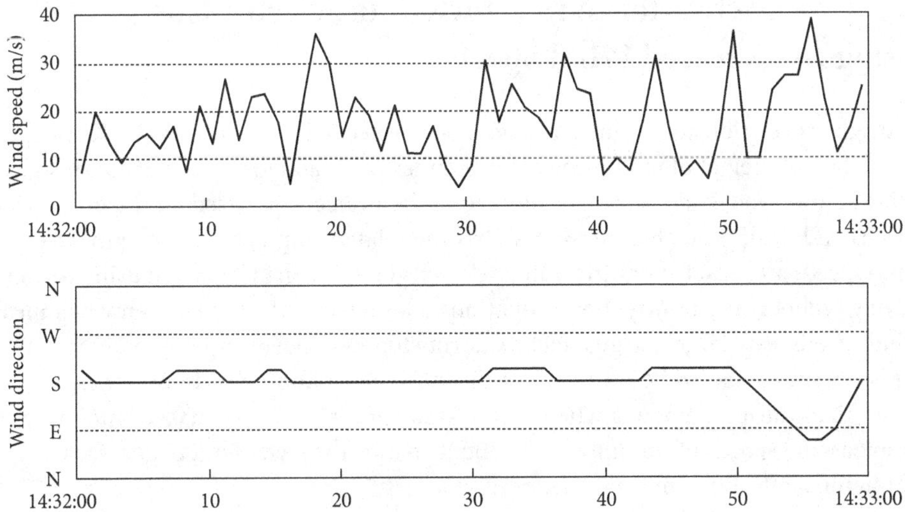

FIGURE 16.34 Example of gust during bomb cyclone (March 14, 2013, NeWMeK No. 102, anemometer height 39 m).

there are not as many wind observations that record 1-s averaged speed as described previously. Can we manage to use 3-s averaged peak gusts in 10 min to investigate "selected gusts"? The 10-min averaged wind speeds and peak gusts (3-s averaged value) corresponding to the JMA records were calculated from NeWMeK data. The relationship between selected gusts based on 1-s averaged value and the calculated 10-min records was then examined, focusing on 10 min including selected gusts.

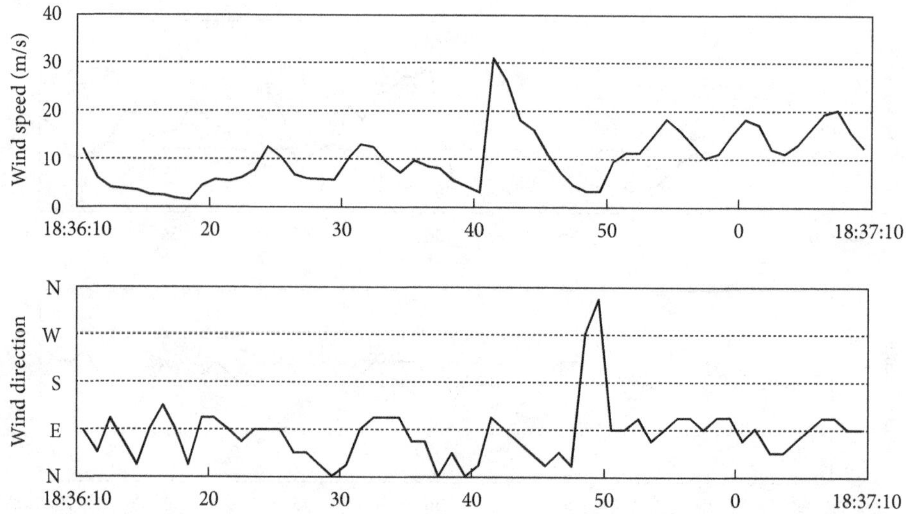

FIGURE 16.35 Example of gust during non-typhoon (April 21, 2014, NeWMeK No. 68, anemometer height 40.3 m).

3.5. Gust Factors for 10 min Including Selected Gusts Using 1-s Averaged Wind Speed

A height correction for 10-min averaged wind speeds based on NeWMeK was performed with a height of 10 m using a power index of 1/7 and for 3-s averaged peak gusts using a power index of 1/11. Gust factors were then calculated based on the mean wind speeds and peak gusts. Figure 16.36 shows the relationship between 10-min averaged wind speeds and gust factors, including selected gusts shown in Table 16.1 using 1-s value during typhoon and non-typhoon situations. As shown in the figure, even when mean wind speeds were large, the gust factors during typhoons were larger than during non-typhoon situations, while gust factors during non-typhoon situations seemed to be larger than during typhoons when mean wind speeds were less than 10 m/s. During average wind speeds of about 10 m/s to about 20 m/s, the lower limit of gust factors was not significantly different regardless of situation.

Figure 16.37 shows gust factors for 10 min using 3-s averaged value, regardless of the inclusion of the selected gust, and it indicates the variation of gust factors against mean wind speeds in the box plot diagram. The top and bottom of the box show the 90th percentiles and the 10th percentiles, respectively. The upper end and the lower end of the bar show the maximum value and the minimum value, respectively, while the black line in the box shows the median. Please note that gust factors in Figure 16.37 include all records at 123 observation points in only 1 year (2009) because of processing limitations. The smaller the mean wind speeds, the larger fluctuations of gust factors occurred, such that the average and variation of gust factors became smaller as wind speeds increased. Let us compare the gust factors at the mean wind speed of 10 m/s in Figures 16.36

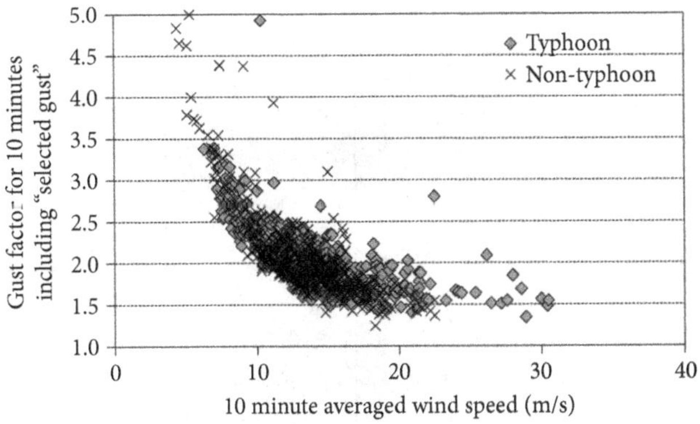

FIGURE 16.36 Relationship between gust factors including selected gusts and 10-min averaged wind speed (2009–2013).

and 16.37. The 10th percentile was 1.3 in Figure 16.37. Meanwhile, the 10th percentile in Figure 16.36 was 2.0. Gust factors using data including selected gusts are significantly larger than those in Figure 16.36. This suggests that sudden rising gusts may occur in a situation where large gust factors occur for 10 min.

3.6. Determination of Criteria that Short-Time-Rising Gusts Occur

From these results, the authors selected a "10-min record" based on 10-min averaged wind speed and 3-s averaged peak gust. The record has an aptitude for causing short-time-rise gusts. In this section, criteria for having an aptitude for causing selected gusts, mentioned in Section 3.2, were obtained for each mean wind speed, focusing on the 10th percentiles of the gust factor for 10 min as shown in Figure 16.36. Figure 16.38 shows the 10th percentile of the gust factor for each mean wind speed and for each source of strong winds. As illustrated in Figure 16.38, variation of the 10th percentile of the gust factor is relatively small within the range of mean wind speed between 10 and 20 m/s. Thus, let us focus on gust factors for 10 min, including selected gusts, regardless of the sources of strong winds. The authors focused on the range of mean wind speed from 8 to 18 m/s because it has enough records to obtain the 10th percentiles of gust factors. The approximate curve in the figure is represented by the following equation:

$$G_{f10} = 14.2\bar{U}^{-1.1} + 1 \; (8 \leq \bar{U} \leq 18) \quad (9)$$

where G_{f10} is the 10th percentiles of gust factors, and U is 10-min averaged wind speed. In the following, 10 min in which gust factors are larger than the value of Equation 9 are taken as periods wherein gusts are expected to occur. Thus, the period for 10 min is defined as the "gust expected period."

Table 16.1 Mode of Parameters of Selected Gusts

		Maximum of Wind Speed (m/s)	Minimum of Wind Speed (m/s)	Gap of Wind Speed (m/s)	Rise Time of Wind Speed (s)	Gust Effect Scale (m)	Wind Speed Ratio	Acceleration of Wind Speed (m/s^2)	Gap of Wind Direction (degree)
Typhoon		22.4	3.5	19.2	7	138.8	0.15	2.8	22.5
Non-typhoon		21.8	3.3	18.9	7	132.7	0.15	2.8	22.5
Non-typhoon	Bomb cyclone	21.5	3.5	18.4	7	134.5	0.16	2.6	22.5
	Others	22.1	3.3	19.2	7	132.4	0.15	2.9	45.0

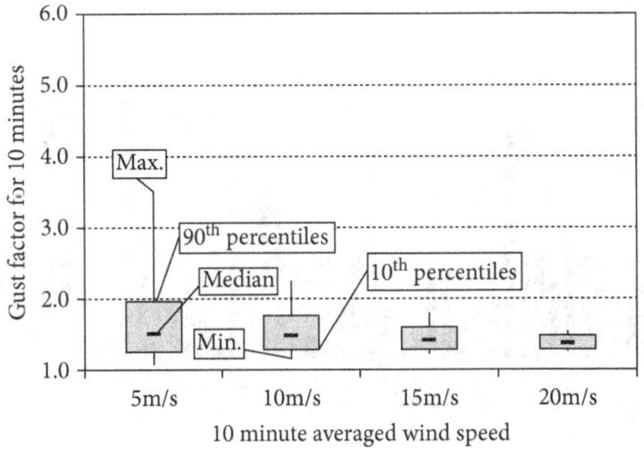

FIGURE 16.37 Variation of gust factors for mean wind speed (2009).

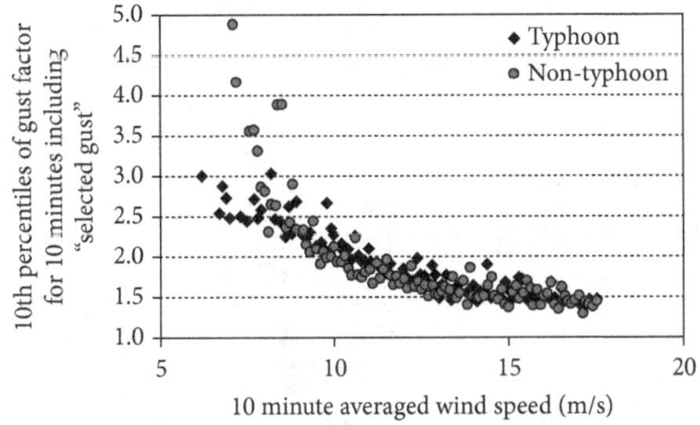

FIGURE 16.38 Relationship between 10th percentiles of gust factor in 10 min, including selected gust and mean wind speed and for each cause of strong winds (2009–2013).

3.7. Selection of "Gust Expected Period" Based on 3-s Averaged Peak Gusts and Relationship of Them and Selected Gusts Based on 1-s Averaged Value

"Gust expected periods" were selected from 10-min records using the gust factors based on Equation 9, and the total of the gust expected periods selected at each observation was calculated for each month. In this study, 262,800 records were analyzed per observation; thus, there were more than 30 million records at all 123 points over 5 years from 2009 to 2013. As shown in Figure 16.39, the number of selected gust expected periods in 5 years at all 123 observation points was 124,120. Although the number of gust expected periods were varied depending on seasons, it was found that there were many gust expected periods during strong winds due to typhoons, bomb cyclones, and seasonal winds.

Figure 16.40 shows the relationship between the number of gust expected periods selected based on Equation 9 and selected gusts, including short-rise-time gusts, based

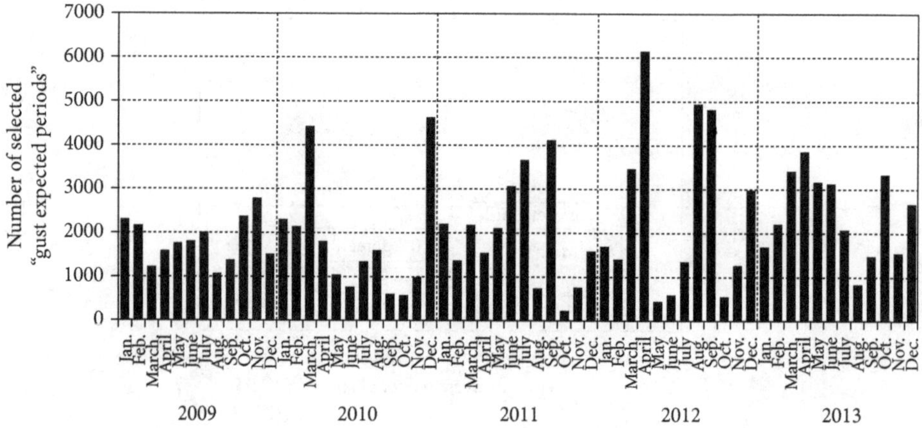

FIGURE 16.39 Number of selected gust expected periods for each month in 5 years.

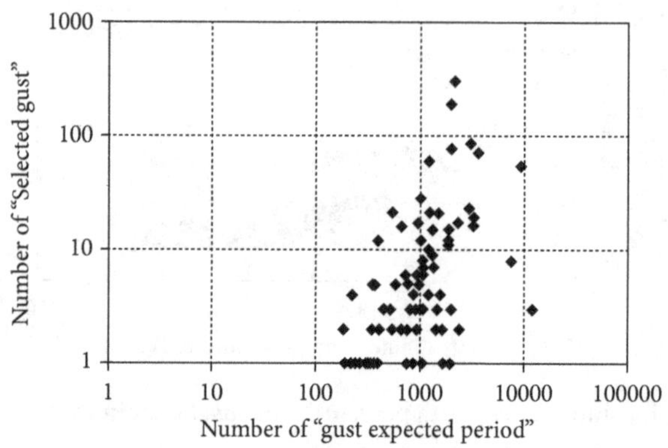

FIGURE 16.40 Relationship between number of selected gusts and gust expected periods.

on the criteria in Section 3.2. Although some relationships were found between them, the number of selected gusts to the number of gust expected periods was quite small, about 0.25 percent.

As a result, even if the gust factors for 10 min are large, the aptitude for occurrence of short-rise-time gusts, which causes overshoot wind force, is not large at all. Thus, it is difficult to expect the occurrence of gusts where the wind speed rises sharply in a few seconds using 10-min averaged wind speed and 3-s averaged values.

4. Remarkable Notes

A gusty wind is said to be a suddenly increasing wind over a short period of time, and the rapid change in wind speed or its direction due to a tornado or a downburst is a representative example of a gust, but the quantitative definitions are not given.

On the other hand, the specificity of a fluid force in a sudden change of a steady flow has received attention since the 1950s. There, "an overshoot phenomenon" in which a rapid change of flow speed in a water tank exceeded fluid force in the steady flow was observed, and an overshoot wind force at the sudden change of wind speed was reported during wind tunnel testing.

This chapter focused on the change of wind force due to a gust and examined the quantitative definition of gust from a viewpoint of aerodynamic forces. First, a wind tunnel experiment was performed to generate modeled step-function-type gusts using a unique device. While this step-function-type gust is different from an actual gust, such as that caused by a tornado or downburst, in several respects, it is possible to examine the specificity of the change process while wind speed increases suddenly and reaches the target speed. As a result, it was found that the occurrence of an overshoot wind force on a structure subjected to a step-function-type gusty wind can be predicted by the non-dimensional rise time calculated using the gusty wind speed, the rise time, and the size of the structure. The current procedure of CFD can reproduce the experimental results for a modeled gusty wind and can predict the occurrence of the overshoot wind force on a real-scale structure. It was found that the overshoot phenomenon was affected by unsteady vortices generated on a body subjected to a short-rise-time gust.

Finally, the occurrence of an overshoot wind force on a specified structure was discussed using a high-resolution wind observation record taken from the viewpoint of an overshoot wind force.

As an example, it was investigated whether a gusty wind that induced an overshoot wind force on a railroad vehicle was found in actual wind observations from the viewpoint of the non-dimensional rise time. As a result, it was possible to confirm a gusty wind that might have induced the overshoot wind force on the vehicle in the 1-s evaluation records of the actual wind observation, but it was difficult with 3-s evaluation records that were often used for official observations. For larger structures, it is expected that the overshoot wind force will be related to a gusty wind even when using the 3-s evaluation records.

Acknowledgments

This study was supported by Grant-in-Aid for Scientific Research (B), No. 26282112, 22360230 and Grant-in-Aid for Scientific Research (C), No. 17K06643, the Japan Society for the Promotion of Science. We would like to thank Kyushu Electric Power Company, Incorporated, and Kyushu Electric Engineering Consultants, Incorporated, for collecting data and for maintaining the NeWMeK.

References

Committee of Wind Disaster Study. (2007). *Investigation report on strong wind disasters caused by Typhoon Shanshan (No. 16, 2006) and Tornado in Saroma, Hokkaido on November 7, 2006* [in Japanese]. Tokyo: Japan Association for Wind Engineering.

Davenport, A. G. (1967). Gust loading factors. *Journal of the Structural Division, 93*(3), 11–34.

Ishida, N., Maeda, J., Kusumoto, T., Honda, M., Tsuru, N., & Noguchi, H. (2000, December). Some effects of topographical features on strong wind characteristics of the western area of Fukuoka City [in Japanese]. In N. Fujisawa (Ed.), *Proceedings of the 16th National Symposium on Wind Engineering* (pp. 7–12). Tokyo: Japan Association for Wind Engineering.

Japan Meteorology Agency. (2007). *Report on change of observation method of instantaneous wind speed* [Press release; in Japanese]. Retrieved from Japan Meteorology Agency website: https://www.jma.go.jp/jma/press/0710/26a/syunkan1026.html

Matsumoto, M., Shimamura, M., Maeda, T., Shirato, H., Yagi, T., Hori, K., Kawashima, Y., & Hashimoto, M. (2007, July). Drag forces on 2-D cylinders due to sudden increase of wind velocity. In Australasian Wind Engineering Society (Ed.), *Proceedings of 12th International Conference on Wind Engineering* (pp. 1727–1734). Cairns, Australia: Australasian Wind Engineering Society.

Morison, J. R., O'Brien, M. P., Johnson, J. W., & Schaaf, S. A. (1950). The force exerted by surface waves on piles. *Journal of Petroleum Technology* 2(5), 149–154.

Noda, M., Hisanobu, S., Waki, T., & Nagao, F. (2012, December). Overshoot coefficient of transient drag forces on square prism in quick change of wind speed [in Japanese]. In H. Shirato (Ed.), *Proceedings of the 22nd National Symposium on Wind Engineering* (pp. 133–138). Tokyo: Japan Association for Wind Engineering.

Nomura, T. (2009). The overshoot coefficients of unsteady drag exerted on the square cylinder by gust wind [in Japanese]. *Wind Engineers, JAWE, 34*(2) (no. 119), 139–140.

Nomura, T. (2010). Numerical simulation and modeling [in Japanese]. *Wind Engineers, JAWE,* 35(2) (no. 123), 139–146.

Nomura, T., Kitamura, N., & Kitagawa, T. (2000, September). *Characteristics of unsteady drag on a square cylinder under sudden change of wind speed.* Paper presented at 3rd International Symposium on Computational Wind Engineering, Birmingham, United Kingdom.

Okauchi, I., Ito, M., & Miyata, T. (1977). *Wind-resistant structure.* Tokyo, Japan: Maruzen [in Japanese].

Sarpkaya, T. (1966). Separated flow about lifting bodies and impulsive flow about cylinders. *AIAA Journal, 4*, 414–420.

Shiraishi, N., Matsumoto, M., & Kitagawa, M. (1982, December). On unsteady aerostatic lift forces induced by the step-like varying wind flow[in Japanese]. In S. Hori (Ed.), *Proceedings of National Symposium on Wind Engineering* (pp. 107–114). Tokyo: Japan Association for Wind Engineering.

Solari, G., Burlando, M., De Gaetano, P., & Repetto, M. P. (2015). Characteristics of thunderstorms relevant to the wind loading of structures. *Wind and Structure, 20*, 763–791.

Takeuchi, T., & Maeda, J. (2010, May). *Effects of inertia force proportional to flow acceleration on unsteady wind forces acting on an elliptic cylinder under short-rise-time gusts.* Paper presented at 5th International Symposium on Computational Wind Engineering. Chapel Hill, NC.

Takeuchi, T., & Maeda, J. (2012). Overshoot of wind forces on a body under step-function-like gusty winds [in Japanese]. *Journal of Structural and Construction Engineering, Transactions of AIJ, 681*, 1629–1635.

Takeuchi, T., & Maeda, J. (2013). Unsteady wind force on an elliptic cylinder subjected to a short-rise-time gust from steady flow. *Journal of Wind Engineering and Industrial Aerodynamics, 122*, 138–145.

Takeuchi, T., Maeda, J., Hayata, T., & Kawashita, H. (2009, November). *Effects of section size on aerodynamic forces on an elliptic cylinder under short-rise-time gusts.* Paper presented at 7th Asia-Pacific Conference on Wind Engineering. Taipei, Taiwan.

Takeuchi, T., Maeda, J., & Kawashita, H. (2008, July). *The overshoot of aerodynamic forces on a railcar-like body under step-function-like gusty winds.* Paper presented at Sixth International Colloquium on Bluff Body Aerodynamics & Applications, Milano, Italy.

Takeuchi, T., Maeda, J., & Otsubo, K. (2014, June). *Numerical investigation of a flow development around an elliptic cylinder subjected to short-rise-time gusts.* Paper presented at 6th International Symposium on Computational Wind Engineering, Hamburg, Germany.

Taneda, S. (1972). The development of the lift of an impulsively started elliptic cylinder at incidence. *Journal of the Physical Society of Japan, 33*, 1706–1711.

Tomokiyo, E., Maeda, J., Ishida, N., & Imamura, Y. (2004). Typhoon damage analysis of transmission towers in mountainous regions of Kyushu, Japan. *Wind & Structures: An International Journal, 7*(5), 345–357.

Tomokiyo, E., & Maeda, J. (2016, December). Selection of short-time-rising gust from wind records at AMeDAS and JMA observations using characteristics of gusts selected from NeWMeK [in Japanese]. In T. Tamura (Ed.), *Proceedings of the 24th National Symposium on Wind Engineering* (pp. 1–6). Tokyo: Japan Association for Wind Engineering.

Tomokiyo, E., Maeda, J., Nogata, K., & Takeuchi, T. (2012, December). Statistical estimation of observed gusty wind and estimate of the associated wind force [in Japanese]. In H. Shirato (Ed.), *Proceedings of the 22nd National Symposium on Wind Engineering* (pp. 13–18). Tokyo, Japan: Japan Association for Wind Engineering.

Wen, Y.-K. (1975). Dynamic tornadic wind loads on tall buildings. *Journal of the Structural Division, 101*, 169–185.

Zhang, S., Solari, G., De Gaetano, P., Burlando, M., & Repetto, M. P. (2018). A refined analysis of thunderstorm outflow characteristics relevant to the wind loading of structures. *Probabilistic Engineering Mechanics, 54*, 9–24.

CHAPTER 17

STRUCTURAL RESPONSE TO NON-STATIONARY THUNDERSTORM OUTFLOWS

DAE KUN KWON, GIOVANNI SOLARI, AND AHSAN KAREEM

1. Introduction

The study of thunderstorm outflows and their loading on structures is a topic of significant interest in wind engineering (Letchford et al. 2002; Solari 2014). Thunderstorms are mesoscale atmospheric phenomena that consist of a set of storm cells that evolve in about 30 m over a few kilometers of space (Fujita 1990). These give rise to intense transient downdrafts that impact the earth's surface, followed by radial outflows with a typical "nose" profile (Goff 1976; Hjelmfelt 1988) and horizontal ring vortices. The ensemble of these air movements is called "downburst" (Figure 17.1) and is divided into macroburst and microburst depending on whether its size is greater or smaller than 4 kilometers (Fujita 1990). The design wind velocity is often related to intense microbursts.

The literature is rich in contributions aiming to determine the dynamic response of reference systems and real structures to thunderstorm outflows. It exhibits an extensive panorama of methods whose complexity matches the complexity of these phenomena. Many other papers did not reach the evaluation of the wind-induced response but provide propaedeutic methods to represent transient wind fields in the time-domain.

The study of *single degree of freedom* (SDOF), *multi-degree of freedom* (MDOF), and *slender beam models* are aimed at depicting the conceptual aspects of the structural response, formulating general procedures to evaluate it and identify the role of the parameters on which the response depends.

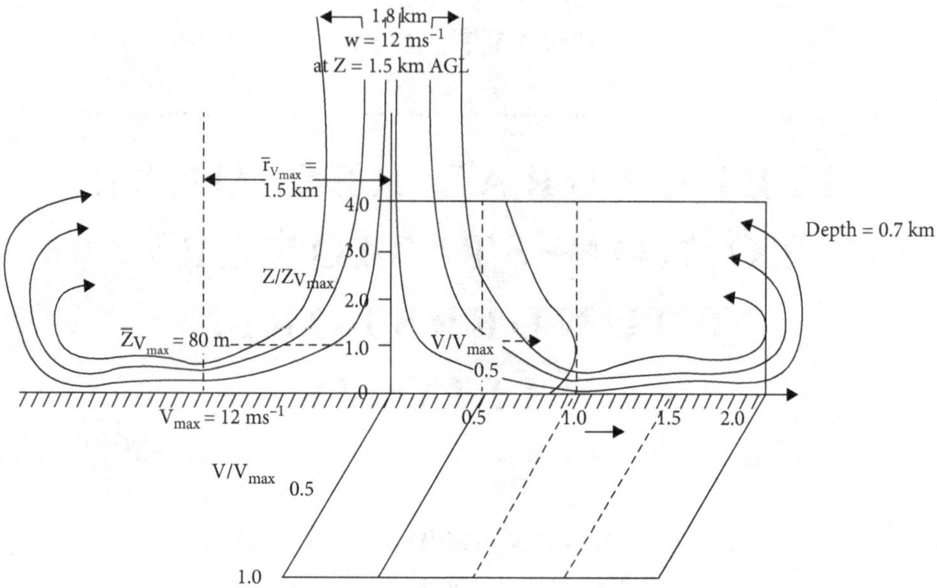

FIGURE 17.1 Thunderstorm downburst and nose velocity profile in the radial outflow (Hjelmfelt 1988).

In 2002, Choi and Hidayat (2002) studied the first dynamic alongwind response of a SDOF system, aiming to extend the gust factor technique (Davenport 1961) from synoptic to transient winds. This method was refined in (Chen & Letchford 2004a, 2004b), where a so-called *maximum dynamic magnification factor* was defined as the ratio between the maximum dynamic response and the static response to the peak load, and in (Chay et al. 2006), where autogressive moving average (ARMA) simulations were used to determine a "dynamic response factor" given by the ratio between the maximum response to a downburst and a stationary wind. All these papers solved the linear equations of motion in the time-domain, schematizing the turbulence as identically coherent on the structural surface exposed to the wind loading.

In 2008, Chen (2008) performed a frequency domain analysis of the wind-excited response of buildings to non-stationary winds, expressing the wind loading as the sum of a deterministic mean function slowly varying with time and a random fluctuating component, rapidly varying with time. The response was determined in turn as the sum of a static response to the mean wind loading and a fluctuating response expressed by an evolutionary power spectrum (EPS).

Kwon and Kareem (2009) applied the EPS to extend the gust factor method from synoptic winds to gust fronts. They determined the equivalent static force on a building as the product of the equivalent static force associated with an atmospheric boundary layer (ABL) profile, the gust front factor (GFF), and a factor that transforms the ABL shape of the force into the gust front shape. The GFF was given in turn by the product of the kinematic effects factor, the pulse dynamic factor, the structural dynamics factor, and the load magnification factor. They take into account, respectively, the different profiles of a gust front and an ABL wind, the variation of the mean wind speed, the non-stationary role of turbulence, and transient aerodynamic effects; the discussion on the latter quantity was purely conceptual.

Also, Huang et al. (2013) expressed the non-stationary wind loading by its equivalent power spectral density (EPSD) and simulated non-stationary loading time-histories for integrating the equations of motion in time-domain; results were compared with those obtained in Chen (2008). More recently, Huang et al. (2020) advanced a data-driven simulation of multivariate non-stationary winds using a hybrid multivariate empirical mode decomposition and spectral representation method. In this context, Zhao et al. (2019) presented a fast convolution integration-based scheme for the non-stationary analysis of linear and nonlinear structures with nonproportional damping. Chen (2015) developed a frequency-domain framework to predict the multimodal coupled buffeting response of long-span bridges to non-stationary winds. The time-varying mean and self-excited and buffeting forces were modeled respectively by static force coefficients, flutter derivatives, and aerodynamic admittances. Kareem et al. (2019) generalized Davenport's chain from stationary to non-stationary winds, based on both wavelet transforms and EPS. This formulation reduces seamlessly to conventional Davenport's chain for stationary cases.

Diversely from EPS and wavelet techniques, the study of the thunderstorm response of structures carried out at the University of Genova was inspired by the consideration that thunderstorm outflows are transient phenomena with short duration, and the structural response to these phenomena, most notably to earthquakes, is usually evaluated by the response spectrum technique (Housner et al. 1953). Based on this remark and some previous studies on synoptic winds (Solari 1989), a "new" method was formulated that generalized the "old" response spectrum technique from earthquakes to thunderstorm outflows.

Initially, this problem was formulated for an SDOF system (Solari et al. 2015) subjected to a perfectly coherent wind field. This led to express the equivalent static force as the product of the peak wind force by a non-dimensional quantity, the thunderstorm response spectrum, depending on the fundamental frequency and the damping ratio.

Later, this formulation was generalized to MDOF systems (Solari 2016) subjected to a partially coherent wind field; the structure was modeled as a slender vertical cantilever beam whose dynamic response was dependent on the sole fundamental mode. Analyses were carried out making recourse of the equivalent wind spectrum technique (EWST; Piccardo & Solari 1998), which makes the application of the response spectrum straightforward. The equivalent static force was expressed as the product of the peak wind force by a non-dimensional quantity, the equivalent response spectrum, depending on the first natural frequency, the damping ratio, and a reference structural size that synthesizes the role of aerodynamic admittance.

In parallel with the thunderstorm response spectrum technique (TRST), time-domain analyses were carried out based on a so-called hybrid simulation strategy (Solari et al. 2017). This study showed that the probability density function of the maximum value of the structural response due to thunderstorm outflows is more spread than that related to synoptic extratropical cyclones. Thus, diversely from synoptic winds (Davenport 1964), it is not appropriate to identify the maximum response with its mean value. On the other hand, many aspects of the response of structures to thunderstorm outflows are qualitatively similar to synoptic winds: the structural displacement is almost unaffected by the contribution of higher vibration modes, the aerodynamic admittance gives rise to analogous effects for thunderstorms and cyclones, and the resonant part of the response to thunderstorms is apparent despite their short duration.

Solari and De Gaetano (2018) carried out a joint calibration and advancement of the two methods just described, proving that the results provided by the response spectrum technique and the time-domain solutions closely agree. This confirms the potential of the response spectrum to be a suitable tool for evaluating the thunderstorm loading and response of structures and the efficiency of hybrid simulation and time-domain integrations to study, with a limited computational burden, advanced structural issues. In particular, relevant conceptual and numerical simplifications may be obtained by embedding the EWST (Piccardo & Solari 1998) within the preceding non-stationary formulation. This generates a multivariate non-stationary wind field through the simulation of a monovariate stationary process, without any relevant loss of precision.

Within this framework, it is worth mentioning the first research (Brusco et al. 2019) aiming to study the dynamic response taking into account the directional shift of the outflow due to the translation of the thunderstorm cell. It proves that such a shift often involves a relevant increase in the response.

Several other contributions were addressed to the dynamic response of structures to thunderstorm outflows making recourse to different approaches. Le and Caracoglia (2015a, 2015b) used the *wavelet-Galerkin method* to determine the non-linear and/or non-stationary response of SDOF and MDOF systems. Su et al. (2015) analyzed the dynamic response of a tall building by a quasi-static approach for the slowly varying mean and the pseudo-excitation method for the fluctuating component. Wang et al. (2017) found a closed-form solution for the buffeting response of a non-linear double-hinged overhead transmission conductor. Le and Caracoglia (2018a) implemented a numerical model of the transient response of a tall building under a digitally simulated downburst consistent with an EPS schematization. Peng et al. (2018) developed an EPS approach including a time-varying coherence function.

Contrary to complex formulations, Miguel et al. (2018) provided a simple engineering model and several critical remarks on the wind loading of buildings due to thunderstorm outflows. Their study confirmed observations previously reported that high-rise buildings are sensitive to extratropical cyclones, whereas low-rise buildings are sensitive to downbursts.

Finally, it is worth mentioning two papers addressing structural safety and performance-based design that used wavelet-Galerkin method to evaluate structural fragility due to downbursts (Le & Caracoglia 2018b) and combined different features to construct a model that takes into account the non-stationary response to downbursts, wind-induced damage, and costs related to structural maintenance (Caracoglia 2018).

The contributions just described highlight a wide range of formulations aiming to extend classical stationary structural analyses to non-stationary phenomena. A dominant aspect of many of these contributions is the striking contrast between elegant methods and the lack of data with which to calibrate and validate these models. The European Projects, Wind and Ports (Solari et al. 2012) and Wind, Ports, and Sea (Repetto et al. 2018), offer a huge amount of data that provide a unique chance to formulate methods robustly coherent with real measurements. In this framework, an extensive wind monitoring network was created in the High Tyrrhenian area, from which many records of thunderstorm outflows were gathered (De Gaetano et al. 2014). Such records were analyzed first aiming to evaluate their statistical properties (Solari et al. 2015; Zhang et al. 2018a), then to develop criteria to determine the loading and response of structures, in particular the TRST (Solari et al. 2015; Solari 2016; Solari & De Gaetano 2018) and the hybrid simulation strategy (Solari et al. 2017; Solari & De Gaetano 2018).

After this brief introduction, Section 2 provides the traditional wind velocity model of thunderstorm outflows. Sections 3 and 4 illustrate, respectively, evolutionary power spectrum-based analysis and wavelet-based analysis. Section 5 describes the TRST. Section 6 examines the hybrid simulation strategy. Section 7 provides some preliminary comparisons between the results obtained using these different methods. Section 8 summarizes the main conclusions.

2. Wind Velocity Model

The horizontal component of wind velocity in a thunderstorm outflow (Figure 17.1) along a vertical axis is expressed here by the classical decomposition rule (Chen & Letchford 2004a, 2004b; Holmes et al. 2008):

$$v(z,t) = \overline{v}(z,t) + v'(z,t) \tag{1}$$

$$v'(z,t) = \sigma_v(z,t)\,\tilde{v}'z(z,t) \tag{2}$$

where z is the height above ground, $t \in [0, \Delta T]$ is the time, ΔT = 10 min, $\overline{v}$ is the slowly varying mean wind velocity, v' is the residual non-stationary fluctuation, σ_v is the slowly varying standard deviation of v', and v' is the reduced turbulent fluctuation dealt with as a stationary Gaussian random field with zero mean and unit standard deviation. The extraction of $\overline{v}$ from v and of σ_v from v' is often carried out here by a moving average filter with an average period T = 30 s (Solari et al. 2015); many other methods are available in the literature to perform the same operation in a more refined way (Chen and Letchford 2007: 42). Replacing Equation 2 into Equation 1:

$$v(z,t) = \overline{v}(z,t)[1 + I_v(z,t)\,\tilde{v}'(z,t)] \tag{3}$$

where I_v is the slowly varying turbulence intensity:

$$I_v(z,t) = \frac{\sigma_v(z,t)}{\overline{v}(z,t)} \tag{4}$$

The decoupling of space and time in $\overline{v}$ and I_v allows us to express these two quantities as:

$$\overline{v}(z,t) = \overline{v}_{\max}(h)\alpha(z)\gamma(t) \tag{5}$$

$$I_v(z,t) = \overline{I}_v(z)\beta(z)\mu(t) \tag{6}$$

where $\overline{v}_{\max}$ is the maximum value of $\overline{v}$ in ΔT; h is a reference height; α is a non-dimensional function of z that defines the shape of the vertical profile of $\overline{v}$ (Oseguera & Bowles 1988; Vicroy 1992; Wood & Kwok 1998), being $\alpha(h) = 1$; and γ is a non-dimensional function of t that expresses the time variation of $\overline{v}$, being $\gamma_{max} = 1$. Similarly, $\overline{I}_v$ is the average value of I_v in ΔT; β is a non-dimensional function of z that defines the shape of the vertical profile of $\overline{I}_v$, being $\beta(h) = 1$; and μ is a non-dimensional function of t that expresses the time variation of I_v, being $\overline{\mu} = 1$.

The reduced turbulent fluctuation v is identified by its cross-power spectral density (CPSD), frequently expressed by the classical models usually adopted for synoptic winds (e.g., Solari & Piccardo 2001):

$$S_{\tilde{v}}(z,n) = \frac{6.868L_v(z)/\overline{v}_{\max}(z)}{\left[1+10.302nL_v(z)/\overline{v}_{\max}(z)\right]^{5/3}} \tag{7}$$

$$Coh_{\tilde{v}\tilde{v}'}(z,z',n) = \exp\left\{-\frac{2nc_z|z-z'|}{\overline{v}_{\max}(z)+\overline{v}_{\max}(z')}\right\} \tag{8}$$

where L_v is the integral length scale and c_z is the exponential decay coefficient of $\tilde{v}'$. Replacing Equations 5 and 6 into Equation 3:

$$v(z,t) = \overline{v}_{\max}(h)\alpha(z)\gamma(t)[1+\overline{I}_v(h)\beta(z)\mu(t)\tilde{v}'(z,t)] \tag{9}$$

A novel directional decomposition rule has been recently proposed in Zhang et al. (2019).

3. Evolutionary Power Spectrum-Based Analysis

In a non-stationary wind velocity model (Equations 1, 2), one popular representation concerning non-stationary turbulent fluctuation is the introduction of the EPS (e.g., Priestley 1965, 1967; Chen & Letchford 2004a, 2004b; Chay et al. 2006; Kareem 2008;

Hu et al. 2013, 2017; Kwon & Kareem 2013, 2019; Wang et al. 2013; McCullough et al. 2014; Huang et al. 2015; Zhou et al. 2015; Solari 2016; Wang et al. 2016, 2018, 2019; Peng et al. 2017; Solari et al. 2017; Solari & De Gaetano 2018; Kareem et al. 2019). In this approach, Equation 2 is recast in the time-frequency domain in terms of the EPS:

$$G_{\tilde{v}}(z,n,t) = |A(z,n,t)|^2 S_{\tilde{v}}(z,n,t) \tag{10}$$

where $G_{\tilde{v}}(z,n,t)$ is the longitudinal EPS, $A(z,n,t)$ is the modulating function, and $S_{\tilde{v}}(z,n,t)$ is the longitudinal zero-mean Gaussian process. In addition, an amplitude modulating function, which is independent of frequency, and a time-independent spectral form (e.g., Equation 10) are often introduced for simplicity:

$$G_{\tilde{v}}(z,n,t) = |A(z,t)|^2 S_{\tilde{v}}(z,n) \tag{11}$$

With the format in Equation 11, cross-EPS is expressed in terms of time-independent coherence function (e.g., Equation 8), akin to the case of boundary layer winds:

$$G_{\tilde{v}}(z,z',n,t) = |A(z,t)|^2 |A(z',t)|^2 S_{\tilde{v}}(z,n,t) S_{\tilde{v}}(z',n,t) Coh_{\tilde{v}\tilde{v}}(z,z',n) \tag{12}$$

Note that the modulating function is generally assumed to use the decoupling of space (z) and time (t) as does Equation 5; thus it may be expressed as a product of $A(z)$ and $A(t)$.

The structural response in the format of the EPS may be computed from: (1) a simulation-based approach in the time domain (e.g., invoking Monte Carlo simulation to generate a series of wind load samples and integrate the equations of motion); (2) a fully non-stationary solution by the direct integration method (e.g., Howell & Lin 1971; Li & Chen 2009; Kareem et al. 2019); (3) a computationally efficient solutions, such as modal superposition and stochastic decomposition (e.g., Sun & Kareem 1989; Li& Kareem 1995), a pseudo excitation method (e.g., Lin 2004; Kareem et al. 2019, etc.; or (4) an approximate closed-form solutions, such as the GFF framework (Kwon & Kareem 2009, 2013, 2019) or TRST (Solari et al. 2015; Solari 2016; Solari & De Gaetano 2018). In particular, the use of the EPS offers an advantage for possibly deriving approximate and simplified closed-form solutions (e.g., GFF and TRST frameworks, which may pave a way toward a codification concerning thunderstorm outflows).

4. Wavelet-Based Analysis

Wavelet transform is one of the popular time-frequency tools that can be used as an alternative formulation to deal with non-stationary fluctuating components. Unlike the EPS, which compromises the resolutions due to the fixed analysis window size, the wavelet transform adopts flexible windows and thus can achieve the finest time-frequency resolution allowed by Heisenberg's uncertainty principle for all different frequency bands (e.g., Kijewski-Correa & Kareem 2003). In addition, the wavelet scale can be directly related to the Fourier frequency.

A formulation using the periodic generalized harmonic wavelet (PGHW) is introduced here, in which the wavelet function is defined as (e.g., Kong & Li 2015; Spanos et al. 2016; Kareem et al. 2019):

$$\Psi^{G}_{(m_j,n_j),k}(t) = \frac{1}{(n-m)} \sum_{q=m_j}^{n_j-1} e^{-iq2\pi\Delta f\left(t-\frac{kT_0}{n-m}\right)} \tag{13}$$

so that wavelet coefficients can be obtained by the wavelet transform:

$$W^{G}_{(m_j,n_j),k}(x) = \frac{n-m}{T_0} \int_0^{T_0} x(t) \Psi^{G}_{(m_j,n_j),k}(t) dt \tag{14}$$

where $x(t)$ is the displacement; $T_o = N_o \Delta t$ is the time period of concern; $j = 1, \ldots, N_f$; $N_f = N_o/(n-m)$ is the j-th wavelet scale and the number of scales; $k, l = 0, \ldots, N_T - 1$; and $N_T = (n-m) = n_j - m_j$ is the wavelet time shift and the number of shifts. The equations of motion may be solved in terms of the PGHW, resulting in the wavelet scalogram of response

$$E\left[\mathbf{W}^{j}_{\mathbf{X}}\left(\mathbf{W}^{j}_{\mathbf{X}}\right)^{T}\right] = \left(\frac{T_0}{n-m}\right)^2 \left[\left(\mathbf{A}^{j}\right)^{-1}\right] \left\{E\left[\mathbf{W}^{j}_{\mathbf{F}}\left(\mathbf{W}^{j}_{\mathbf{F}}\right)^{T}\right]\right\} \left[\left(\mathbf{A}^{j}\right)^{-1}\right]^{T} \tag{15}$$

where $\mathbf{W}^{j}_{\mathbf{X}}$ and $\mathbf{W}^{j}_{\mathbf{F}}$ are vectors of wavelet coefficients of non-stationary fluctuating response and external force respectively; $\mathbf{A}^j$ is a matrix comprised of block matrix $\mathbf{A}^{j}_{k,l} = C^{(2)}_{j;k,l}\mathbf{M} + C^{(1)}_{j;k,l}\mathbf{C} + C^{(0)}_{j;k,l}\mathbf{K}$; **M**, **C**, and **K** are mass, damping, and stiffness matrices, respectively; and $C^{(o)}_{j;k,l}$ are wavelet connection coefficients. Detailed information can be found in Kong & Li (2015) and Spanos et al. (2016). Equation 15 implies a wavelet-based representation of a second-order structural dynamics system; therefore, by analogy, it may be extended to higher order linear systems. Accordingly, the aerodynamics transfer relationship between non-stationary fluctuating winds and attendant force may also be represented in terms of wavelet coefficients when this relationship can be approximated by a high-order linear system with sufficient accuracy.

The wavelet scalogram of buffeting response may be transformed into the EPS by:

$$E\left[\mathbf{W}^{j}_{\mathbf{X}}\left(\mathbf{W}^{j}_{\mathbf{X}}\right)^{T}\right] = \frac{(n-m)}{T} \begin{bmatrix} S_{\mathbf{XX}}(n_j,t_1) & & & 0 \\ & S_{\mathbf{XX}}(n_j,t_k) & & \\ & & \ddots & \\ 0 & & & S_{\mathbf{XX}}(n_j,t_{N_T}) \end{bmatrix} \tag{16}$$

A similar relationship also exists for the external force, $\mathbf{W}^{j}_{\mathbf{F}}$.

Note that another wavelet-based analysis, the wavelet-Galerkin method, has recently been introduced for computing the dynamic response of tall buildings subject to thunderstorm/downburst events. It utilizes discrete orthonormal Daubechies wavelets as the basis functions in the Galerkin expansion method (Le & Caracoglia 2015a, 2015b).

5. Thunderstorm Response Spectrum Technique

5.1. SDOF System

Let us consider an SDOF system (Solari et al. 2015) subjected to a perfectly coherent wind field. The equation of motion is given by:

$$m\ddot{x}(t)+c\dot{x}(t)+kx(t)=f(t)=\frac{1}{2}\rho v^2(t)AC_D \tag{17}$$

where x is the alongwind displacement as a function of time t; m, c, and k are the mass, damping, and stiffness; f is the external force; ρ is the density of air; A is the area exposed to the wind loading; and C_D is the drag coefficient. Equation 17 may be rewritten as:

$$\ddot{d}(t)+2\xi(2\pi n_0)\dot{d}(t)+(2\pi n_0)^2 d(t)=(2\pi n_0)^2 u^2(t) \tag{18}$$

where $u(t)=v(t)/\hat{v}$ is the reduced wind velocity, $d(t)=x(t)/\hat{x}$ is the reduced displacement, $\hat{v}$ is the peak wind velocity, $\hat{x}$ is the peak displacement due to the peak force $\hat{f}$ associated with $\hat{v}$, n_0 is the fundamental frequency, and ξ is the damping ratio. Accordingly, let us define the thunderstorm response spectrum as:

$$S_d=d_{max}(n_0,\xi) \tag{19}$$

It is easy to show that the maximum alongwind displacement is given by $x_{max}=\hat{x}\cdot S_d$, whereas the equivalent static force results in $f_{eq}=\hat{f}\cdot S_d$. Figure 17.2 shows the mean value and the coefficient of variation (cov) of the response spectrum as obtained from 93 records of thunderstorm outflows detected by the Wind and Ports project.

5.2. MDOF System

Let us consider an MDOF system (Solari 2016) subjected to a partially coherent wind field; for the sake of simplicity, the structure is modeled as a slender vertical cantilever beam whose dynamic response depends on the sole fundamental mode with natural frequency n_1. The partially coherent nature of the wind field represents a major difficulty in solving this problem. However, it is overcome by using the EWST (Piccardo & Solari 1998) concerning the stationary component, $v(z,t)$, in Equation 9. This leads to a

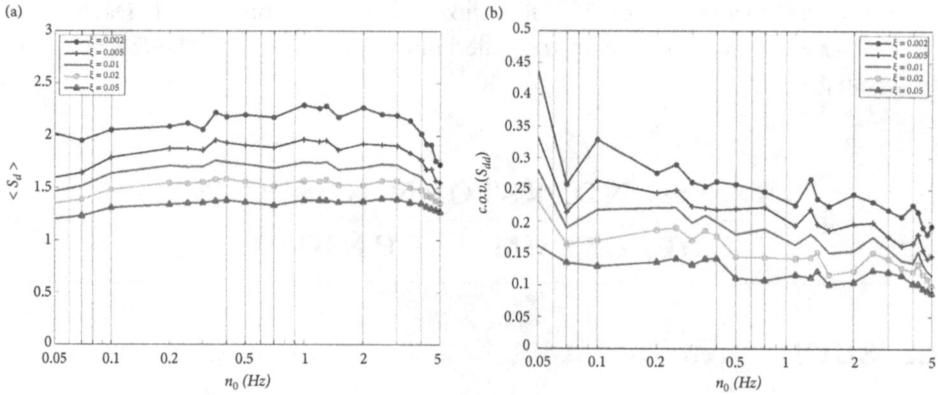

FIGURE 17.2 Mean value (a) and cov. (b) of the thunderstorm response spectrum.

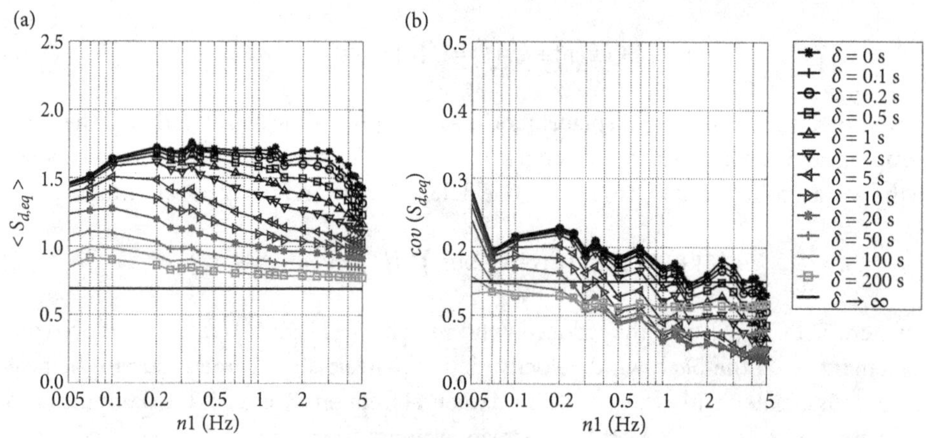

FIGURE 17.3 Mean value (a) and cov. (b) of the equivalent thunderstorm response spectrum for $\xi = 0.01$.

new formulation that represents a generalization of the response spectrum technique described earlier for a SDOF system. Accordingly, Equation 18 may be rewritten as:

$$\ddot{d}_{eq}(t) + 2\xi(2\pi n_1)\dot{d}_{eq}(t) + (2\pi n_1)^2 d_{eq}(t) = (2\pi n_1)^2 u_{eq}^2(\delta, t) \tag{20}$$

where $u_{eq}(\ddot{a}, t) = v_{eq}(h, \ddot{a}, t)/\hat{v}(h)$ is the reduced equivalent wind velocity, $d_{eq}(t) = x(h,t)/\hat{x}(h)$ is the reduced equivalent displacement, and δ is a so-called size factor that takes into account the role of aerodynamic admittance. As a consequence, let us define the equivalent response spectrum as:

$$S_{d,eq} = d_{eq,max}(n_1, \xi, \delta) \tag{21}$$

It is easy to show that the maximum alongwind displacement is given by $x_{max}(z) = \hat{x}(z) \cdot S_{d,eq}$ whereas the equivalent static force results in $f_{eq}(z) = \hat{f}(z) \cdot S_{d,eq}$. Figure 17.3 shows the mean value and the cov of the equivalent response spectrum for

$\xi = 0.01$ as obtained from 93 records of thunderstorm outflows detected by the Wind and Ports project. It is worth noting that the numerical procedure to solve Equation 21 is quite elaborated (Solari 2016; Solari and De Gaetano 2018) while its application to find the equivalent static force is straightforward.

6. Time-Domain Simulation

6.1. Hybrid Simulation

Diversely from classical Monte Carlo simulations of non-stationary vector fields (Li & Kareem 1991; Deodatis 1996; Sakamoto & Ghanem 2002; Huang 2014), the hybrid strategy proposed in Solari et al. (2017) to generate artificial velocity fields of thunderstorm outflows is based on simulating one at a time and then assembling the different ingredients that make up the wind velocity model in Equation 9. A brief description of each step is given here.

Step 1: Velocity scale. The maximum value of $\overline{v}$, $\overline{v}_{max}$, at height h is evaluated by extreme statistics (Zhang et al. 2018a) as a function of the return period.

Step 2: Vertical profiles. The vertical profiles of $\overline{v}$ and I_v are evaluated by assigning $\alpha(z)$ and $\beta(z)$ through expressions coherent with measurements. In this chapter $\alpha(z)$ is provided by the model proposed in Solari et al. (2015) for $J = 4$ values of the height z_m at which $\overline{v}$ is maximum, $z_m = 25, 50, 75, 100$ m; and $\beta(z) = 1$ (Solari et al. 2015).

Step 3: Slowly varying time dependence. The random time-dependence of $\overline{v}$ and I_v is simulated by gathering measured thunderstorm outflow records (Solari et al. 2015) and extracting from them synchronous pairs of $\gamma(t)$ and $\mu(t)$ sample functions. For instance, using the data provided by the Wind and Ports monitoring network, $K = 93$.

Step 4: Turbulence field. The space-time variation of the stationary Gaussian field v is simulated, say L times, by a Monte Carlo algorithm (Carassale & Solari 2006) based on the spectral representation method (Shinozuka & Jan 1972; Li & Kareem 1995). Its computational efficiency is increased by replacing harmonic superimposition by a fast Fourier transform-based implementation (Li & Kareem 1995) and by factorizing the PSD matrix of v by its proper orthogonal decomposition (POD) eigenvalues and eigenvectors (Di Paola 1998, Carassale et al. 2007). Typical expressions of the PSD and the coherence function of v are given by Equations 7 and 8 (Solari & Piccardo 2001).

Step 5: Component assemblage. $M = J \times K \times L$ thunderstorm velocity fields may be simulated for each examined structure. Assuming that each structure is discretized by N nodes, all in all, $M \times N$ artificial velocity histories may be generated. Figure 17.4 shows three sample functions of a wind velocity field with $z_m = 50$ m. The simulations have been carried out for $\overline{v}_{max}(h) = 36.5$ m/s and $L_v(h) = 34.6$ m, with $h = 13$ m and $c_z = 10$. Coherently with measurements, simulations have been carried out with a time step $\Delta t = 0.1$ s in a time interval $\Delta T = 10$ min. The turbulence harmonic content is simulated between 0 and the cutoff frequency $n_c = 5$ Hz, with a frequency resolution $\Delta n = 1/600$ Hz. $L = 1{,}000$ artificial wind fields of v are simulated for each $J = 4$ velocity profiles.

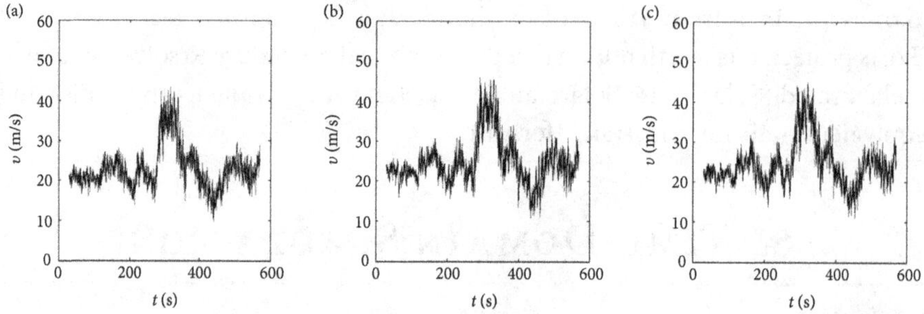

FIGURE 17.4 Sample functions of a wind velocity field $v\,(z_m = 50\text{m})$ at (a) $z = 18$ m; (b) $z = 54$ m; and (c) $z = 80$ m.

Therefore, $M = J \times K \times L = 4 \times 93 \times 1{,}000 = 372{,}000$ thunderstorm velocity fields that have been simulated.

The performance of the whole simulation algorithm has been investigated and discussed in Solari et al. (2017), where it was shown that the hybrid technique is characterized by limited errors and produces wind field features consistent with the measured data.

6.2. Equivalent Hybrid Simulation

The EWST described in Piccardo & Solari (1998) is a method that replaces the actual turbulent field, as a random function of time and space, by an equivalent turbulent fluctuation, as a random function of time, identically coherent in space.

The generalization of this technique to transient thunderstorm outflows was introduced in Solari (2016). Accordingly, the reduced equivalent turbulent fluctuation v_{eq} is defined by its PSD:

$$S_{\tilde{v}',eq}(n,\delta) = S_{\tilde{v}'}(z_{eq},n)C(\delta n) \tag{22}$$

where $S_{\tilde{v}'}$ is the PSD of $\tilde{v}'$, $z_{eq} = 0.6\text{H}$ is the equivalent height, and C is the frequency filter:

$$\begin{aligned} c(\eta) &= \frac{1}{\eta} - \frac{1}{2\eta^2}\left(1 - e^{-2\eta}\right)(\eta > 0) \\ c(0) &= 1 \end{aligned} \tag{23}$$

η is the argument of C; δ is the size factor:

$$\delta = \frac{\kappa c_z H}{\overline{v}_{\max}(z_{eq})} \tag{24}$$

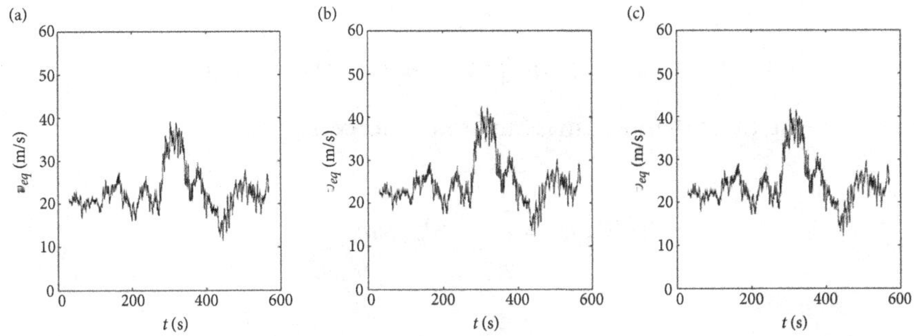

FIGURE 17.5 Sample functions of the equivalent velocity field v_{eq} corresponding to that in Figure 17.4 (z_m = 50 m): (a) z = 18 m; (b) z = 54 m; and (c) z = 80 m.

κ is the modal shape factor. Dealing with slender vertical cantilever structures whose first mode shape may be approximated as $\Psi_1(z) = (z/H)^\zeta$, $\kappa = 0.5/(\zeta + 1)^{0.55}$ (Piccardo & Solari 1998).

Using this method, Equation 9 may be approximated by the relationship:

$$v(z,t) - \overline{v}_{\max}(h)\alpha(z)\gamma(t)[1+\overline{I}_v(h)\beta(z)\mu(t)\,\tilde{v}'_{eq}(t,\delta)] \quad (25)$$

Figure 17.5 shows three sample functions of the equivalent velocity field corresponding to the actual field in Figure 17.4 (z_m = 50 m). The shape and the trend of the diagrams in Figure 17.4 are preserved while the high-frequency harmonic content is filtered out to take into account, in equivalent terms, the coherence of the fluctuations and the aerodynamic admittance.

Equation 25 does not modify the slowly varying mean part of v, whereas it drastically changes its residual fluctuation by replacing the 2-D process $v\,(z,t)$ by the equivalent 1-D process $v_{eq}(t,\delta)$. The step from Equation 9 to Equation 25 is a formidable simplification whose correctness was proved conceptually in Solari & De Gaetano (2018) by stressing that, although used in the non-stationary field, the EWST is strictly applied to the sole stationary part of the turbulent fluctuations.

6.3. Wind-Excited Response

Let us consider a slender vertical structure with linear elastic behavior. Using modal analysis, assuming that natural frequencies are well-separated, and dealing with damping as small and proportional, its wind-excited response is given by:

$$x\,(z,t) = \sum_k \Psi_k(z)\,p_k(t) \quad (26)$$

where Ψ_k is the k-th mode shape and p_k is the k-th principal coordinate:

$$\ddot{p}_k(t) + 2\xi(2\pi n_k)\dot{p}_k(t) + (2\pi n_k)^2 p_k(t) = \frac{1}{m_k} f_k(t) \tag{27}$$

m_k and f_k being the k-th modal mass and modal force, respectively:

$$m_k = \int_0^H m(z)\Psi_k^2(z)dz \tag{28}$$

$$f_k(t) = \int_0^H f(z,t)\Psi_k(z)dz \tag{29}$$

in which m is the structural mass per unit length and f is the aerodynamic wind loading:

$$f(t) = \frac{1}{2}\rho v^2(z,t)b(z)C_D(z) \tag{30}$$

where v is the wind velocity defined by Equation 9 or 25, b is the reference size of the structure cross-section, and C_D is the drag coefficient evaluated by neglecting the transient character of the wind field (Kwon & Kareem 2009).

The integration of the equations of motion is conveniently performed in the state space by introducing a Hamming window at the beginning of the wind loading history to avoid initial transient effects. Each time, step $\Delta t = 0.1$ s is divided into 10 parts and a linear interpolation of the wind loading is applied. Analyses are carried out by neglecting aerodynamic damping and retaining the contribution of the sole first mode.

7. Comparison Between Different Methods

To check the consistency of the different methods just described, this section provides some preliminary comparisons between the results obtained using these procedures.

7.1. Time-Domain Simulation Versus Equivalent Time-Domain Simulation

Three real slender vertical structures are examined here as reference test cases: a steel lighting pole, a steel telecommunication antenna mast, and a reinforced concrete telecommunication tower already studied in Solari et al. (2017) and Solari and De Gaetano (2018). For each of these structures, Figure 17.6 shows a picture, the model scheme, and the first three modal shapes Ψ_1, Ψ_2, Ψ_3. Table 17.1 reports the main properties: H is the

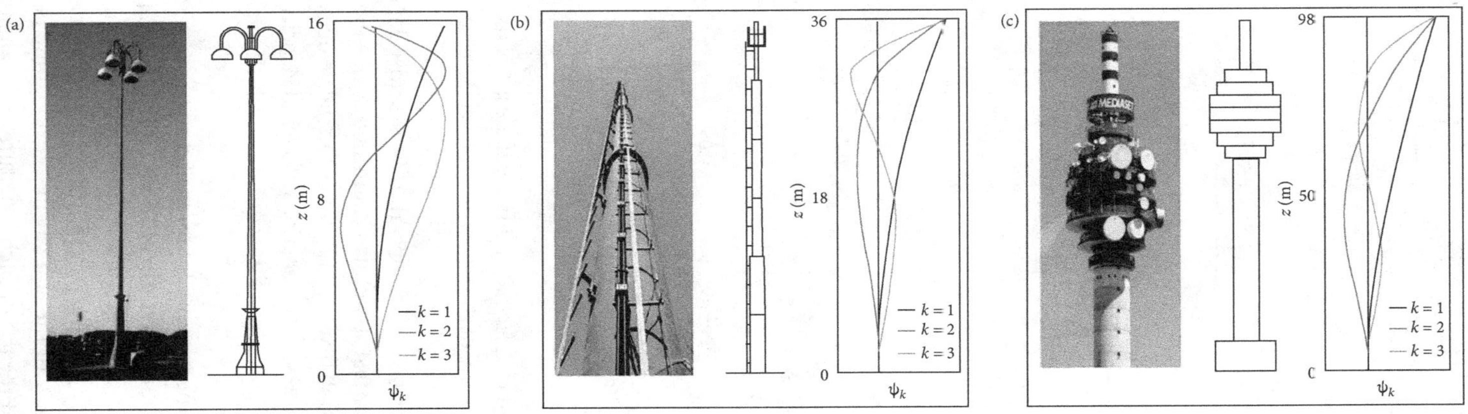

FIGURE 17.6 Structure test cases (Solari & De Gaetano, 2018). (a) Steel lighting pole (S1); (b) steel telecommunication antenna mast (S2); and (c) reinforced concrete telecommunication tower (S3).

Table 17.1 Main properties of the three structure test cases,

Structure	Description	H (m)	n_1 (Hz)	n_2 (Hz)	n_3 (Hz)	ξ	N
1	Steel lighting pole	15.76	0.532	3.186	8.744	0.01	16
S2	Steel telecommunication antenna mast	36.00	0.821	3.106	5.972	0.01	19
S3	R.C. telecommunication tower	98.00	0.494	3.167	6.274	0.02	26

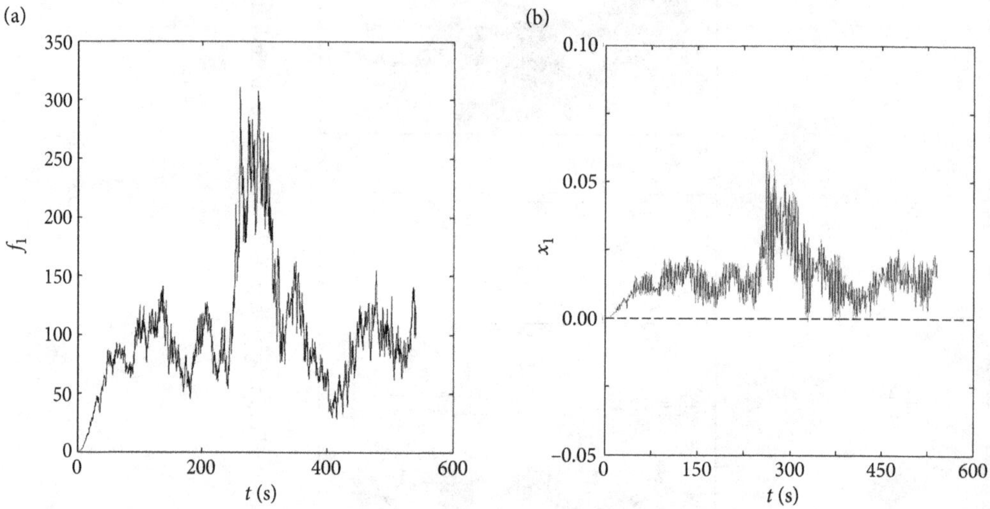

FIGURE 17.7 First modal force f_1 (a) and first modal displacements x_1 (b) of structure S3 for the simulated thunderstorm outflow in Figure 17.4.

total height; n_1, n_2, n_3 are the first three natural frequencies; ξ is the structural damping ratio; and N is the number of nodes of the structural model. More details are provided in Solari and De Gaetano (2018).

$M = J \times K \times L = 4 \times 93 \times 1{,}000 = 372{,}000$ thunderstorm velocity fields have been simulated for each of the three test structures examined. Since they are discretized by $N = 16,19,26$ nodes (Table 17.1), all in all, 22,692,000 artificial velocity histories have been generated. Figure 17.4 shows three sample functions of a wind velocity field with $z_m = 50$ m. Figure 17.5 shows three sample functions of the equivalent velocity field v_{eq} corresponding to that in Figure 17.4.

Figures 17.7 and 17.8 show the first modal force and the first modal displacement of structure S3 for the simulated thunderstorm outflows in Figures 17.4 and 17.5. The first is obtained by Equation 9, whereas the second derives from the simplified Equation 25; nevertheless, the similarity between the corresponding diagrams is apparent.

To check the global efficiency of the equivalent hybrid simulation, 93,000 values of the maximum displacement at the top of each structure are evaluated for the four wind velocity profiles considered here. Thus, all in all, $93{,}000 \times 3 \times 4 = 1{,}116{,}000$ values of

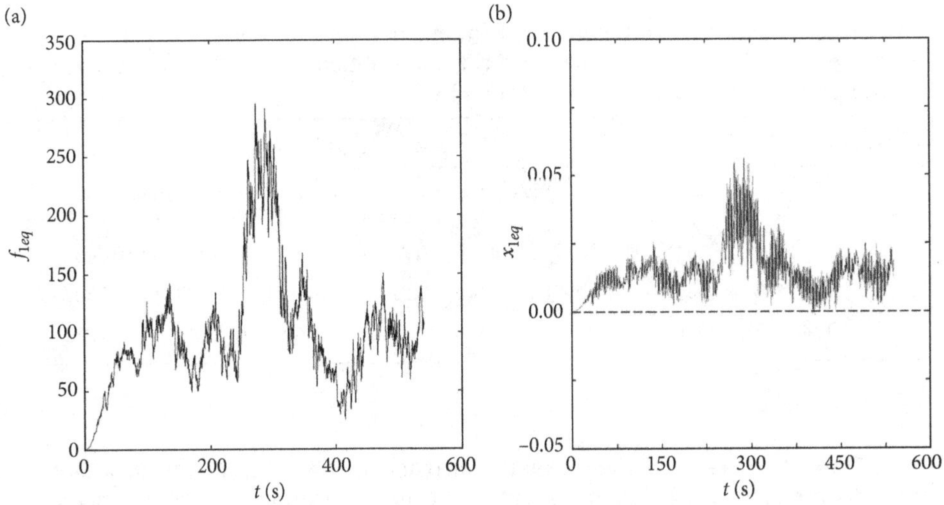

FIGURE 17.8 Equivalent first modal force f_1 (a) and first modal displacements x_1 (b) of structure S3 for a thunderstorm outflow simulated by EWST (Figure 17.5).

Table 17.2 Mean value (m) and cov of the maximum displacement x_{max} at the structure top H through hybrid time-domain simulation

Structure	$\langle x_{max}(H)\rangle$				$\mathrm{cov}(x_{max}(H))$			
	zm = 25 m	zm = 50 m	zm = 75 m	zm = 100 m	zm = 25 m	zm = 50 m	zm = 75 m	zm = 100 m
S1	0.651	0.651	0.643	0.639	0.242	0.246	0.241	0.240
S2	0.326	0.357	0.364	0.370	0.216	0.218	0.218	0.218
S3	0.046	0.072	0.082	0.087	0.184	0.192	0.195	0.193

$x_{max}(H)$ are extracted and analyzed probabilistically. Tables 17.2 and 17.3 summarize the mean values and the cov of $x_{max}(H)$ evaluated, respectively, by the hybrid simulation based on Equation 9 and by the equivalent hybrid simulation based on the approximate Equation 25. The values in parenthesis are the percent errors committed (with reference to Tables 17.2). Despite the complexity of the problem dealt with, the error never exceeds 4%. In particular, the precision with which the EWST replicates not only the mean of $x_{max}(H)$ but also its cov is almost surprising.

7.2. Time-Domain Simulation Versus Equivalent Response Spectrum Technique

Tables 17.4 and 17.5 show, respectively, the mean value and the cov of the maximum displacement at the top of the reference test structures as evaluated by the equivalent response spectrum technique. The values in parenthesis are the percent errors with

Table 17.3 Mean value (m) and cov of the maximum displacement x_{max} at the structure top H through equivalent hybrid time-domain simulation (in parenthesis, the percent error related to Table 17.2)

Structure	$\langle x_{max}(H)\rangle$				$\mathrm{cov}(x_{max}(H))$			
	zm = 25 m	zm = 50 m	zm = 75 m	zm = 100 m	zm = 25 m	zm = 50 m	zm = 75 m	zm = 100 m
S1	0.670 (+3)	0.665 (+2)	0.658 (+2)	0.654 (+2)	0.247(+2)	0.249 (+1)	0.244 (+1)	0.243 (+1)
S2	0.335 (+3)	0.362 (+1)	0.374 (+3)	0.377 (+2)	0.221 (+2)	0.219 (+1)	0.223 (+2)	0.220 (+1)
S3	0.046 (−1)	0.070 (−3)	0.079 (−4)	0.084 (−4)	0.180 (−2)	0.185 (−4)	0.188 (−4)	0.186 (−4)

Table 17.4 Mean value of the maximum displacement x_{max} (m) at the structure top H by the equivalent response spectrum technique (in parenthesis, the percent error with reference to Table 17.3)

Structure	$\xi = \xi a$				$\xi = \xi b$			
	zm = 25 m	zm = 50 m	zm = 75 m	zm = 100 m	zm = 25 m	zm = 50 m	zm = 75 m	zm = 100 m
S1	0.630 (−3)	0.624 (−4)	0.622 (−3)	0.620 (−3)	0.783(−8)	0.775 (−9)	0.773 (−8)	0.771 (−7)
S2	0.323 (−1)	0.351 (−2)	0.359 (−1)	0.363 (−2)	0.395 (−5)	0.429 (−6)	0.439 (−6)	0.444 (−6)
S3	0.046 (=)	0.071 (−1)	0.080 (−2)	0.085 (−2)	0.053 (−4)	0.081 (−7)	0.092 (−8)	0.097 (−8)

Table 17.5 Cov of the maximum displacement x_{max} at the structure top H by the equivalent response spectrum technique (in parenthesis, the percent error with reference to Table 17.4)

Structure	$\xi = \xi a$				$\xi = \xi b$			
	zm = 25 m	zm = 50 m	zm = 75 m	zm = 100 m	zm = 25 m	zm = 50 m	zm = 75 m	zm = 100 m
S1	0.169(−30)	0.169(−31)	0.169(−30)	0.169(−30)	0.240(−18)	0.240(−19)	0.240(−17)	0.240(−17)
S2	0.142(−34)	0.143(−34)	0.143(−34)	0.143(−34)	0.199(−23)	0.200(−24)	0.200(−23)	0.200(−24)
S3	0.095(−48)	0.099(−48)	0.100(−49)	0.100(−48)	0.142(−34)	0.149(−34)	0.151(−36)	0.152(−34)

respect to the hybrid time-domain solution (Tables 17.3 and 17.4), treated here as a reference target.

As far as concerns the errors committed by determining the mean value of $x_{max}(H)$ by $\langle S_{d,eq}\rangle$, they are so small that, in their regard, the response spectrum technique can be judged as highly effective. This remark is strengthened by the tremendous conceptual and computational simplifications involved in this technique, in comparison to the sequential application of hybrid simulation, even if in its equivalent form, and the hybrid time-domain solution. Examining these errors more in detail, for average damping values ($\xi = \xi_a$), they do not exceed 3–4%. For low damping values ($\xi = \xi_b$), errors reach 8–9%. In any case, differently from the hybrid time-domain solution involving the EWST (Table 17.4), the response spectrum technique always underestimates the structural response. In addition, errors are almost independent of the wind speed profile.

Regarding the errors committed by determining the cov of $x_{max}(H)$ by cov($S_{d,eq}$), they are so large (up to 50%) that they cannot be explained other than in relation to conceptual differences between the two methods. An inspection of these conceptual frameworks carried out in Solari and De Gaetano (2018) highlighted two main reasons capable of justifying such differences: the use of different reference values for the wind speed and of different methods to reconstruct the thunderstorm outflow velocity field starting from individual velocity records. A detailed explanation is given in Solari and De Gaetano (2018).

7.3. Evolutionary Power Spectral Density Versus Thunderstorm Response Spectrum Technique

Recently, Kwon and Kareem (2019) compared the evolutionary power spectral density (EPSD)-based GFF/G-GFF framework with the TRST framework. It was found that while underlying principles between the two frameworks to describe the peak wind load were close following the non-stationary model (Equation 1), some differences such as the use of the maximum wind velocity and velocity profile were noted. In addition, the two frameworks used different approaches for the treatment of non-stationary turbulence components. While the TRST employed a data-based equivalent wind velocity approach in the time domain (Equation 20) for estimating the ERS ($S_{d,eq}$) (Equation 21), the GFF/G-GFF utilized a conventional frequency domain approach involving the evolutionary power spectrum model (Equations 10–12) in which the spectrum was described as a product of an amplitude modulating function and a stationary wind component assuming Davenport wind spectrum. To overcome these differences for a comparison, Kwon and Kareem (2019) introduced a non-dimensionalized factor normalized by common parameters used in both frameworks for an indirect comparison. In view of all intrinsic differences between the GFF/G-GFF and TRST approaches, they showed reasonably good agreement with each other (i.e., less than 10% difference). More detailed information can be found in Kwon, Kareem, and Solari in this volume.

8. Conclusion

The literature is rich in contributions aiming to determine the dynamic response of structures to transient thunderstorm outflows whose complexity matches the complexity of the phenomenon itself. Several papers did not offer schemes to evaluate the wind-induced response but only provide propaedeutic methods to represent transient wind fields in the time-domain. A dominant aspect of many of these contributions is the striking contrast between well-designed methods and the lack of data with which to calibrate and validate these models.

This chapter deals with the dynamic response of structures to non-stationary thunderstorm outflows, with particular regard to single degree of freedom, multidegree of freedom, and distributed mass systems. This choice allows us to focus attention on the dominant conceptual aspects of the dynamic response of the structures and identify the main parameters essential to describe the response.

With such a premise, this contribution initially emphasizes the time-frequency domain methods such as the evolutionary power spectrum-based and the wavelet-based approaches, briefly reviewing their respective formulations. In particular, the use of the evolutionary power spectrum offers an advantage for possibly deriving approximate and simplified closed-form solutions (e.g., the GFF framework), which may pave a way toward codification of the response to thunderstorm outflows.

Subsequently, the chapter illustrates the response spectrum technique popular in the seismic field and recently generalized to thunderstorms. Like the GFF method, this procedure also offers ample opportunities for closed-form engineering solutions. The chapter then goes on to discuss a new wind field simulation scheme, referred to as a hybrid strategy, and the use of the equivalent spectrum method to significantly reduce the computational burden without compromising the accuracy of results.

In the final section, the chapter illustrates a series of applications of the various methods outlined here and compares the results. Clearly, preliminary assessment suggests a promising agreement which may promote interesting engineering developments about the design of structures subjected to thunderstorm outflows.

Acknowledgments

The first and third authors gratefully acknowledge the support in part by the NSF Grant CMMI-1562244. The research carried out by the second author is funded by the European Research Council (ERC) under the European Union's Horizon 2020 research and innovation program (grant agreement No. 741273) for the project THUNDERR (Detection, Simulation, Modeling and Loading of Thunderstorm Outflows to Design Wind-Safer and Cost-Efficient Structures), supported by an Advanced Grant 2016. The data used were recorded by the monitoring network realized for the European Projects Winds and Ports (grant No. B87E09000000007) and Wind, Ports, and Sea (grant No. B82F13000100005), funded by the European Territorial Cooperation Objective, Cross-border program Italy France Maritime 2007–2013.

References

Brusco, S., Lerzo, V., & Solari, G. (2019). Directional response of structures to thunderstorm outflows. *Meccanica, 54*, 1281–1306.

Caracoglia, L. (2018). Unified stochastic dynamic and damage cost model for the structural analysis of tall buildings in thunderstorm-like winds. *ASCE-ASME Journal of Risk and Uncertainty in Engineering Systems, 4*, 04018043.

Carassale, L., & Solari, G. (2006). Monte Carlo simulation of wind velocity fields on complex structures. *Journal of Wind Engineering and Industrial Aerodynamics, 94*, 323–339.

Carassale, L., Solari, G., & Tubino, F. (2007). Proper orthogonal decomposition in wind engineering. Part 2: Theoretical aspects and some applications. *Wind Structures, 10*, 177–208.

Chay, M. T., Albermani, F., & Wilson, B. (2006). Numerical and analytical simulation of downburst wind loads. *Engineering Structures, 28*, 240–254.

Chen, L., & Letchford, C. W. (2004a). A deterministic-stochastic hybrid model of downbursts and its impact on a cantilevered structure. *Engineering Structures, 26*, 619–629.

Chen, L., & Letchford, C. W. (2004b). Parametric study on the alongwind response of the CAARC building to downbursts in the time domain. *Journal of Wind Engineering and Industrial Aerodynamics, 92*, 703–724.

Chen, L., & Letchford, C. W. (2007). Numerical simulation of extreme winds from thunderstorm downbursts. *Journal of Wind Engineering and Industrial Aerodynamics, 95*, 977–990.

Chen, X. (2008). Analysis of alongwind tall building response to transient non-stationary winds. *Journal of Structural Engineering ASCE, 134*, 782–791.

Chen, X. (2015). Analysis of multimode coupled buffeting response of long-span bridges to non-stationary winds with force parameters from stationary wind. *Journal of Structural Engineering ASCE*, 141, 04014131.

Choi, E. C. C., & Hidayat, F. A. (2002). Dynamic response of structures to thunderstorm winds. *Progress in Structural Engineering and Materials*, 4408–416.

Davenport, A. G. (1961). The application of statistical concepts to the wind loading of structures. *Proceedings of the Institution of Civil Engineers, 19*, 449–472.

Davenport, A. G. (1964). Note on the distribution of the largest value of a random function with application to gust loading. *Proceedings of the Institute of Civil Engineers (London, UK), 24*, 187–196.

De Gaetano, P., Repetto, M. P., Repetto, T., & Solari, G. (2014). Separation and classification of extreme wind events from anemometric records. *Journal of Wind Engineering and Industrial Aerodynamics, 126*, 132–143.

Deodatis, G. (1996). Non-stationary stochastic vector processes: Seismic ground motion applications. *Probabilistic Engineering Mechanics, 11*, 149–168.

Di Paola, M. (1998). Digital simulation of wind field velocity. *Journal of Wind Engineering and Industrial Aerodynamics, 74–76*, 91–109.

Fujita, T. T. (1990). Downburst: Meteorological features and wind field characteristics. *Journal of Wind Engineering and Industrial Aerodynamics, 36*, 75–86.

Goff, R. G. (1976).Vertical structure of thunderstorm outflows. *Monthly Weather Reviews, 104*, 1429–1440.

Hjelmfelt, M. R. (1988). Structure and life cycle of microburst outflows observed in Colorado. *Journal of Applied Meteorology, 27*, 900–927.

Holmes, J. D., Hangan, H. M., Schroeder, J. L., Letchford, C. W., & Orwig, K. D. (2008). A forensic study of the Lubbock-Reese downdraft of 2002. *Wind Structures, 11*, 19–39.

Housner, G. W., Martel, R. R., & Alford, J. L. (1953). Spectrum analysis of strong-motion earthquakes. *Bulletin of the Seismology Society of America, 43*, 97–119.

Howell, L. J., & Lin, Y. K. (1971). Response of flight vehicles to non-stationary atmospheric turbulence. *AIAA Journal, 9(11)*, 2201–2207.

Hu, L., Xu, Y. L. & Huang. W. F. (2013).Typhoon-induced non-stationary buffeting response of long-span bridges in complex terrain. *Engineering Structures, 57*, 406–415.

Hu, L., Xu, Y. L., Zhu, Q., Guo, A., & Kareem, A. (2017). Tropical storm-induced buffeting response of long-span bridges: Enhanced non-stationary buffeting force model. *Journal of Structural Engineering ASCE 143(6)*, 04017027.

Huang, G. (2014). An efficient simulation approach for multivariate non-stationary process: Hybrid of wavelet and spectral representation method. *Probabilistic Engineering Mechanics, 37*, 74–83.

Huang, G., Chen, X., Liao, H., & Li, M. (2013). Predicting of tall building response to non-stationary winds using multiple wind speed samples. *Wind Structures, 17*, 227–244.

Huang, G., Peng, L., Kareem, A., & Song, C. (2020). Data-driven simulation of multivariate non-stationary winds: A hybrid multivariate empirical mode decomposition and spectral representation method. *Journal of Wind Engineering and Industrial Aerodynamics, 197*, 104073.

Huang, G., Zheng, H., Xu, Y. L., & Li, Y. (2015). Spectrum models for non-stationary extreme winds. *Journal of Structural Engineering ASCE*, 141, 04015010.

Kareem, A. (2008). Numerical simulation of wind effects: A probabilistic perspective. *Journal of Wind Engineering and Industrial Aerodynamics, 96*, 1472–1497.

Kareem, A., Hu, L., Guo, Y., & Kwon, D. K. (2019). Generalized wind loading chain: Time-frequency modeling framework for non-stationary wind effects on structures. *Journal of Structural Engineering ASCE, 145*, 04019092.

Kijewski-Correa, T., & Kareem, A. (2003). Wavelet transforms for system identification in civil engineering. *Computer-Aided Civil and Infrastructure Engineering, 18*, 339–355.

Kong, F., & Li, J. (2015). Wavelet-expansion-based stochastic response of chain-like MDOF structures. *Journal of Sound and Vibration, 359*, 136–153.

Kwon, D., & Kareem, A. (2013). Gust front factor: New framework for wind load effects on structures. *Engineering Structures, 48*, 635–644.

Kwon, D. K., & Kareem, A. (2009). Gust-front factor: New framework for wind load effects on structures. *Journal of Structural Engineering ASCE, 135*, 717–732.

Kwon, D. K., & Kareem, A. (2019). *Towards codification of thunderstorm/downburst using gust front factor: Model-based and data-driven perspectives*. Engineering Structures, *199*, 109608.

Le, T. H., & Caracoglia, L. (2015a). Reduced-order Wavelet-Galerkin solution for the coupled, nonlinear stochastic response of slender buildings in transient winds. *Journal of Sound Vibrations, 344*, 179–208.

Le, T. H., & Caracoglia, L. (2015b). Wavelet-Galerkin analysis to study the coupled dynamic response of a tall building against transient wind loads. *Engineering Structures, 100*, 763–778.

Le, T. H., & Caracoglia, L. (2018a). Computer-based model for the transient dynamics of a tall building during digitally simulated Andrews AFB thunderstorm. *Composite Structures, 193*, 44–72.

Le, V., & Caracoglia, L. (2018b). Computationally efficient stochastic approach for the fragility analysis of vertical structures subjected to thunderstorm downburst winds. *Engineering Structures, 165*, 152–169.

Letchford, C. W., Mans, C., & Chay, M. T. (2002). Thunderstorms: Their importance in wind engineering (a case for the next generation wind tunnel). *Journal of Wind Engineering and Industrial Aerodynamics, 90*, 1415–1433.

Li, J., & Chen, J. (2009). *Stochastic dynamics of structures*. Singapore: Wiley.

Li, Y., & Kareem, A. (1991). Simulation of multivariate non-stationary random processes by FFT. *Journal of Engineering Mechanics, ASCE 117*, 1037–1058.

Li, Y., & Kareem, A. (1995). Stochastic decomposition and application to probabilistic mechanics. *Journal of Engineering Mechanics ASCE*, 121, 162–174.

Lin, J. H. (2004). *Pseudo excitation method in random vibration*, Beijing: Science Press.

McCullough, M., Kwon, D. K., Kareem, A., & Wang, L. (2014). Efficacy of averaging interval for non-stationary winds. *Journal of Engineering Mechanics, ASCE, 140*(1), 1–19.

Miguel, L. F. F., Riera, J. D., & Miguel, L. F. F. (2018). Assessment of downburst wind loading on tall structures. *Journal of Wind Engineering and Industrial Aerodynamics, 174*, 252–259.

Oseguera, R. M., & Bowles, R. L. (1988). *A simple analytic 3-dimensional downburst model based on boundary layer stagnation flow*. NASA Technical Memorandum 100632. Hampton, VA: NASA.

Peng, L., Huang, G., Chen, X., & Kareem, A. (2017). Simulation of multivariate non-stationary random processes: Hybrid stochastic wave and proper orthogonal decomposition approach. *Journal of Engineering Mechanics ASCE*, 143, 04017064.

Peng, L., Huang, G., Chen, X., & Yang, Q. (2018). Evolutionary spectra-based time-varying coherence function and application in structural response analysis to downburst winds. *Journal of Structural Engineering ASCE, 144*, 04018078.

Piccardo, G., & Solari, G. (1998). Generalized equivalent spectrum technique. *Wind Structures, 1*, 161–174.

Priestley, M. B. (1965). Evolutionary spectra and non-stationary processes. *Journal of the Royal Statistical Society. Series B (Methodological), 27*(2), 204–237.

Priestley, M. B. (1967). Power spectral analysis of non-stationary random process. *Journal of Sound Vibrations, 6(1)*, 86–97.

Repetto, M. P., Burlando, M., Solari, G., De Gaetano, P., Pizzo, M., & Tizzi, M. (2018). A GIS-based platform for the risk assessment of structures and infrastructures exposed to wind. *Advances in Engineering Software*, 117, 29–45.

Sakamoto, S., & Ghanem, R. (2002). Simulation of multi-dimensional non-Gaussian non-stationary random fields. *Probabilistic Engineering Mechanics, 17*, 167–176.

Shinozuka, M., & Jan, C. M. (1972). Digital simulation of random processes and its applications. *Journal of Sound and Vibration*, 25, 111–128.

Solari, G. (1989). Wind response spectrum. *Journal of Engineering Mechanics ASCE, 115*, 2057–2073.

Solari, G. (2014). Emerging issues and new scenarios for wind loading on structures in mixed climates. *Wind Structures, 19*, 295–320.

Solari, G. (2016). Thunderstorm response spectrum technique: Theory and applications. *Engineering Structures, 108* 28–46.

Solari, G., Burlando, M., De Gaetano, P., and Repetto, M. P. (2015). Characteristics of thunderstorms relevant to the wind loading of structures. *Wind Structures, 20*, 763–791.

Solari, G., & De Gaetano, P. (2018). Dynamic response of structures to thunderstorm outflows: Response spectrum technique vs time-domain analysis. *Engineering Structures, 176*, 188–207.

Solari, G., De Gaetano, P., & Repetto, M. P. (2015). Thunderstorm response spectrum: Fundamentals and case study. *Journal of Wind Engineering and Industrial Aerodynamics, 143*, 62–77.

Solari, G., & Piccardo, G. (2001). Probabilistic 3-D turbulence modeling for gust buffeting of structures. *Probabilistic Engineering Mechanics, 16*, 73–86.

Solari, G., Rainisio, D., & De Gaetano, P. (2017). Hybrid simulation of thunderstorm outflows and wind-excited response of structures. *Meccanica, 52*, 3197–3220.

Solari, G., Repetto, M. P., Burlando, M., De Gaetano, P., Pizzo, M., Tizzi, M., & Parodi, M. (2012). The wind forecast for safety and management of port areas. *Journal of Wind Engineering and Industrial Aerodynamics, 104–106*, 266–277.

Spanos, P. D., Kong, F., Li, J., & Kougioumtzoglou, I. A. (2016). Harmonic wavelets based excitation-response relationships for linear systems: A critical perspective. *Probabilistic Engineering Mechanics, 44*, 163–173.

Su, Y., Huang, G., & Xu, Y. L. (2015). Derivation of time-varying mean for non-stationary downburst winds. *Journal of Wind Engineering and Industrial Aerodynamics, 141*, 39–48.

Sun, W. J., & Kareem, A. (1989). Response of MDOF systems to non-stationary random excitation. *Engineering Structures, 11(2)*, 83–91.

Vicroy, D. D. (1992). Assessment of microburst models for downdraft estimation. *Journal of Aircraft, 29*, 1043–1048.

Wang, D., Chen, X., & Xu, K. (2017). Analysis of buffeting response of hinged overhead transmission conductor to non-stationary winds. *Engineering Structures, 147*, 567–582.

Wang, H., Wu, T., Tao, T., Li, A., & Kareem, A. (2016). Measurements and analysis of non-stationary wind characteristics at Sutong Bridge in Typhoon Damrey. *Journal of Wind Engineering and Industrial Aerodynamics, 123*, 100–106.

Wang, H., Xu, Z., Feng, D., & Tao, T. (2019). *Non-stationary turbulent wind field simulation of bridge deck using non-negative matrix factorization. Journal of Wind Engineering and Industrial Aerodynamics, 188*, 235–246.

Wang, H., Xu, Z., Wu, T., & Mao, J. (2018). Evolutionary power spectral density of recorded typhoons at Sutong Bridge using harmonic wavelets. *Journal of Wind Engineering and Industrial Aerodynamics, 177*, 197–212.

Wang, L., McCullough, M., & Kareem. A. (2013). A data-driven approach for simulation of full-scale downburst wind speeds. *Journal of Wind Engineering and Industrial Aerodynamics, 123*, 171–190.

Wood, G. S., & Kwok, K. C. S. (1998). An empirically derived estimate for the mean velocity profile of a thunderstorm downburst. *Proceedings of the 7th Australian Wind Engineering Society Workshop*, Auckland, Australia.

Zhang, S., Solari, G., Burlando, M., & Yang, Q. (2019). Directional decomposition and analysis of thunderstorm outflows. *Journal of Wind Engineering and Industrial Aerodynamics, 189*, 71–90.

Zhang, S., Solari, G., De Gaetano, P., Burlando, M., & Repetto, M. P. (2018a). A refined analysis of thunderstorm outflow characteristics relevant to the wind loading of structures. *Probabilistic Engineering Mechanics*, 54, 9–24.

Zhang, S., Solari, G., Yang, Q., & Repetto, M. P. (2018b). Extreme wind speed distribution in a mixed wind climate. *Journal of Wind Engineering and Industrial Aerodynamics, 176*, 239–253.

Zhao, N., Huang, G., Yang, Q., Zhou, X., & Kareem, A. (2019). Fast convolution integration-based non-stationary response analysis of linear and nonlinear structures with nonpropotional damping. *Journal of Engineering Mechanics ASCE, 145(8)*, 04019053.

Zhou, G.-D., Ding, Y.-L. & Li, A.-Q. (2015). Evolutionary spectra estimation of field measurement typhoon processes using wavelets. *Mathematical Problems in Engineering*, 945203.

CHAPTER 18

ASSESSING WIND DAMAGE TO RESIDENTIAL STRUCTURES

TIMOTHY P. MARSHALL AND J. ARN WOMBLE

1. Introduction

THERE is a need in wind science to evaluate building damage in the wake of wind disasters, particularly with tornadoes and hurricanes/typhoons. Part of that need stems from wanting to know how buildings fail to build them better, to help reduce property damage, and to save lives. Building codes have been developed so that buildings can withstand "minimum" wind loads depending on risk level and geographic location. Hardened engineered buildings must be better designed than conventional buildings to avoid catastrophic failures so they stay operational during and after disasters. There is also a need to develop a wind damage database to establish a climatology of wind events. A multibillion-dollar insurance and reinsurance industry depends on such a database to help set equitable insurance rates by developing computer models to estimate certain levels of risk based on geographic location. Modeling and insuring building risks is big business. The insurance premiums you pay to insure your house, car, and apartment all depend on data gathered from prior losses as well as predictions of future losses.

In this chapter, we discuss the complex relationship between wind speed and wind damage to various types of wood-framed residential buildings. We explore some of the problems encountered when trying to estimate the wind speed necessary to cause building damage from tornadoes and hurricanes. Using damage scales to estimate failure wind speeds contains errors, but damage scales become essential as there usually is neither the time nor the personnel required to generate engineering calculations for every wind-damaged building. We also will present certain guidelines when conducting wind damage surveys.

We limit this chapter to the assessment of wind damage to wood-framed residences since they are the most common building type. It would require several book chapters to discuss damage to all other building types. We present some of the key findings made during various damage surveys over the past several decades, and we dissect critical components of various types of residential structures, namely the foundation-to-wall attachments and wall-to-roof attachments.

2. Wind Speed–Damage Correlation

There is a probabilistic relationship between wind load and resistance of buildings. Such load–resistance relationships can be modeled, calculated, and tested in wind tunnels. However, damage evaluators need a quick and easy-to-use method of determining a range of possible wind speeds that cause certain degrees of building damage. Such easy-to-use wind damage scales have been developed by various scientists throughout the world, but they are not without error or controversy. We concentrate on the damage scale that is most widely used in the United States, Canada, China, Japan, and France, which is the EF or Enhanced Fujita Scale. This scale was derived from the original Fujita or F Scale developed by Dr. T. Theodore Fujita during the 1970s.

2.1. The F Scale

Fujita (1971, 1973, 1981) developed the Fujita or F Scale in order to rate various degrees of damage to buildings caused by winds from tornadoes, hurricanes, and straight-line winds. The F Scale (Table 18.1) is a subjective, visual interpretation of the degree of damage (DOD) that simply assigns a numerical value ranging from 0 to 5 based on increasing severity of damage primarily to "well-constructed" or "strong" wood-framed houses. Fujita divided the difference between Beaufort 12 (32.6 m s^{-1}) (a mariner's scale), and

Table 18.1 Fujita Scale

F Scale	Fastest ¼-Mile Wind Speeds, m s^{-1} (mph)
F0	18–32 (40–72)
F1	33–50 (73–112)
F2	51–70 (113–157)
F3	71–93 (158–207)
F4	94–125 (208–280)
F5	126–142 (261–318)

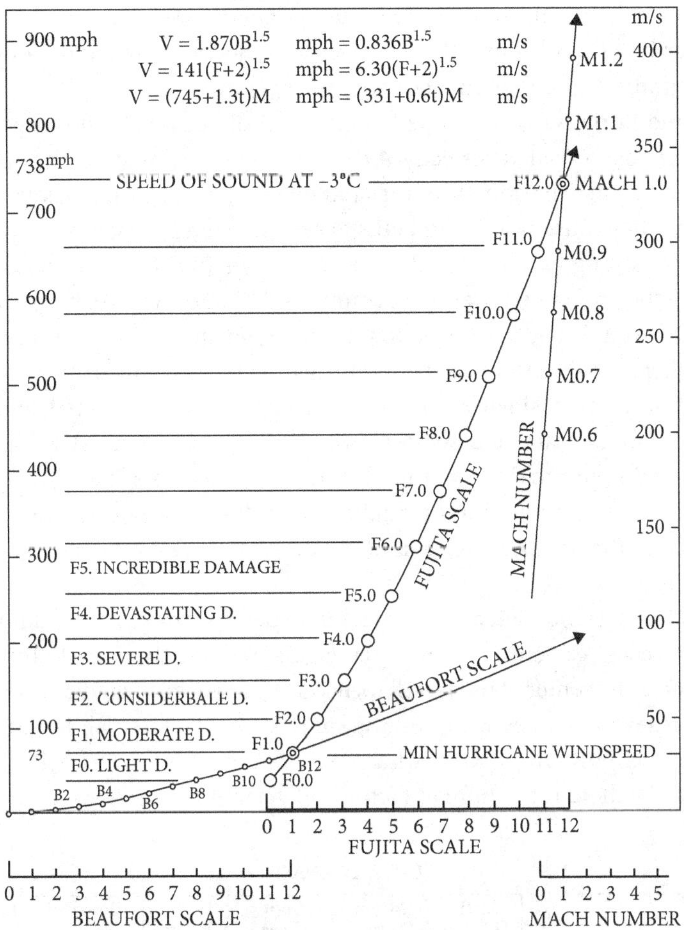

FIGURE 18.1 Graphical depiction of the Fujita scale that connects Beaufort Force, a mariner's scale, with Mach number, an aviator's scale (from Fujita 1971).

Mach 1 (343 m s^{-1}) (an aviator's scale), into 12 nonlinear increments (Figure 18.1). He used the wind speeds in the lowest six divisions, noting that F-scale values of 6 through 12 were not practical for tornadoes and therefore not used. Fujita defined the wind speeds in the F0 to F5 range as being a "fastest ¼-mile speed," usually in the 5- to 10-s range for most tornadoes. Fujita deemed the wind speeds in this scale to be "experimental" and awaited engineering assessments of tornado damage to help "calibrate" the wind speed ranges. The empirical selection of 12 divisions between the Beaufort and Mach scales brought criticisms from many in the scientific community. Engineering assessments of tornado damage by Minor, McDonald, and Mehta (1977) questioned the accuracy of the empirical F-Scale wind speeds, especially when they exceeded 56 m s^{-1} (125 mph).

Grazulis (1993) noted that single-paragraph descriptions of damage given by F Scale are vague and limited in scope and can introduce large errors in assigning F-Scale numbers. For example, homes "swept clean" from their foundations in the LaPlata,

Maryland, tornado initially were assigned an F5 damage rating but were downgraded to F1 when Marshall (2003) found those homes were not anchored and had slid off their split-level foundations as they collapsed in their backyards.

Doswell and Burgess (1988) pointed out that building damage and tornado intensity are related but not correlated perfectly. A destroyed building may have been built poorly, which can lead to an overestimate of tornado intensity. Since tornadoes are rated by the worst damage they cause along their paths, there would be a tendency to overrate tornadoes unless the strengths of the buildings were known. In contrast, tornadoes would be underrated if they remained over open county and did not cause damage. Schaefer and Galway (1982) found that tornadoes that strike populated areas are more likely to be assigned higher F-Scale ratings than those tornadoes that remain in open country. Fujita (1992) recognized that residences are not homogeneously constructed, and he devised corrections to compensate for assigning F-Scale ratings (Figure 18.2). He indicated that a strong-framed house may sustain F2 damage, whereas the same wind speeds might not damage a concrete building or might totally destroy a poorly built outbuilding. Thus, he realized that relative strengths of buildings must be considered when assigning F-Scale ratings.

Phan and Simiu (1998) determined that high-speed winds of longer duration caused greater damage to residences in their damage survey of the Jarrell, Texas, tornado. Residences near the center of the Jarrell tornado experienced tornadic winds for about 3 min. This left several homes with nothing more than concrete slab foundations. Even the flooring and plumbing fixtures had been removed by flying debris during the tornado. Finally, there is the human factor in determining tornado intensity based on

THE FUJITA TORNADO SCALE

Damage f scale		Little Damage	Minor Damage	Roof Gone	Walls Collapse	Blown Down	Blown Away	
		f0	f1	f2	f3	f4	f5	
Windspeed F scale	17 m/s	32	50	70	92	116	142	
		F0	F1	F2	F3	F4	F5	
	40 mph	73	113	158	207	261	319	
To convert f scale into F scale, add the appropriate number								
Weak Outbuilding	−3	f3	f4	f5	f5	f5	f5	
Strong Outbuilding	−2	f2	f3	f4	f5	f5	f5	
Weak Framehouse	−1	f1	f2	f3	f4	f5	f5	
strong Framehouse	0	F0	F1	F2	F3	F4	F5	
Brick Structure	+1	-	f0	f1	f2	f3	f4	
Concrete Building	+2	-	-	f0	f1	f2	f3	

FIGURE 18.2 The original F Scale shown with corrections for building strength (from Fujita 1992).

analyzing damage. An evaluator with knowledge of how buildings are constructed and fails will likely rate buildings differently from an evaluator without such knowledge.

2.2. The Enhanced Fujita Scale

Although the Fujita Scale had been used by the National Weather Service (NWS) and the scientific community for several decades, limitations of the scale were well known to its users. Primary limitations include a lack of damage indicators (DIs), no accounting for construction quality and variability, and no calibration of building damage with respect to wind speed. These limitations have led to inconsistent ratings of tornado damage and, in some cases, over- or underestimates of tornado wind speeds. Recognizing the need to address these limitations, researchers at Texas Tech University (TTU) Wind Science and Engineering Center (WSEC) proposed a project to reexamine the F Scale, revise it where necessary, and attempt to develop a consensus between the meteorological and engineering communities. According to McDonald and Mehta (2001), a steering committee first was organized to initiate the project. Then, a panel of wind damage experts met in 2001 and assigned failure wind speed values to various degrees of damage (DOD) to 28 DIs, which included various building types, power poles, and trees. Through this expert elicitation process, experts in the industry have estimated wind speeds corresponding to the DODs. These estimated wind speeds then determined the EF (Enhanced Fujita) Scale category appropriate for the observed damage. The baseline failure wind speed for the EF Scale was a 3-s gust, at 10 m above ground, in open, unobstructed terrain (for consistency with American Society of Civil Engineers [ASCE] 7 basic wind design). We note that time duration of the wind speed in the EF Scale (3-s gust) is different than the time duration wind speed in the original F Scale (fastest ¼-wind speed). McDonald, Mehta, and Mani (2003) and McDonald, Mehta, Smith, and Womble (2009) summarize steps involved in the development of the EF Scale. They felt it important to preserve as much of the original F Scale as possible. For example, the EF Scale still has the same number range from 0 to 5, but with much lower wind speeds, especially at the higher EF Scale ratings. One concern with adopting a new damage scale was to have a seamless transition to preserve the original F Scale database. After much discussion, the NWS adopted and implemented the new EF Scale in 2007. The EF Scale document can be found at WSEC (2006). Table 18.2 presents the EF Scale numbers and associated wind speed ranges.

A few examples of DIs include one- and two-story residences, single-wide manufactured homes, small professional buildings, elementary schools, large isolated retail buildings, institutional buildings, freestanding towers, and hardwood trees. The number of DODs for each DI depends on the complexity of construction. DODs range from "no visible damage" to "total destruction" of entire buildings. Initial visible damage to a residence might be loss of roofing material, a toppled chimney, or damaged garage doors. With higher wind speeds, damage can occur to gable end walls and the roof deck. Still higher wind speeds can remove the roof and collapse the garage. The progression of

Table 18.2 Enhanced Fujita Scale

EF Scale	Three-Second Gust, m s^{-1} (mph)
EF0	29–38 (65–85)
EF1	39–49 (86–110)
EF2	59–60 (111–135)
EF3	61–74 (136–165)
EF4	75–89 (166–200)
EF5	>90 (>200)

damage continues, with higher wind speeds removing exterior walls, leaving only interior walls standing. Finally, the strongest wind speeds topple all walls and can sweep away debris from the foundations.

In an attempt to account for variability in building resistance, ranges of failure wind speeds were determined for each DOD. These failure wind speeds originally were given the names "expected" (EXP), "upper bound" (UB), and "lower bound" (LB), as shown in Table 18.3. The expected failure wind speed caused damage under "normal" conditions. A weak connection, material deterioration, or a "fatal flaw" of local wind pressure might result in failure at a lower than expected wind speed. On the other hand, stronger-than-normal connections (e.g., using steel "hurricane" clips instead of toenailing) might require higher than expected wind speeds to produce the damage. Thus, a damage evaluator applying the EF Scale has the option to go above or below the expected failure wind speed depending on variations in construction while remaining within the range of the upper and lower bounds.

A major challenge to the EF Scale project was how to obtain correlations between DODs and wind speeds. There were few published scientific studies that calculated failure wind speeds to damaged structures. Minor, McDonald, and Mehta (1977) and Marshall and McDonald (1983) were among the few who found clean structures, with construction plans, with which to calculate a range of failure wind speeds and then compare them to their F Scale ratings. Such calculations are tedious and time consuming and can only be performed on a small percentage of damaged structures. Additionally, calculations would involve assumptions in material strengths of failed members, connections, and load paths that would introduce errors. Therefore, having experts assign a "best guess" to certain DODs for specific DIs ahead of time, before the damage survey, helps expedite the process while, it is hoped, keeping failure wind speed errors relatively low. Expert elicitation has been used successfully to estimate certain unknown parameters related to seismic hazard analyses. The Senior Seismic Hazard Analysis Committee (SSHAC) formalized the process (SSHAC, 1997) while working under the auspice of the US Nuclear Regulatory Commission, the US Department of Energy, and the Electric Power Research Institute. Subsequently, Boissonnade, Hossain, Kimbell, Mensing, and Savy (2000) at Lawrence Livermore National Laboratory applied expert elicitation to estimate parameters for tornado hazard assessment.

Table 18.3 Lower Bound, Expected, and Upper Bound Failure Wind Speeds for One and Two-Story Residences for Each DOD in the EF Scale

DOD[a]	Damage Description	Lower Bound, $m\ s^{-1}$ (mph)	Expected, $m\ s^{-1}$ (mph)	Upper Bound, $m\ s^{-1}$ (mph)
1	Threshold of visible damage	24 (53)	29 (65)	36 (80)
2	Loss of roof covering material (<20%), gutters and/or awning; loss of vinyl or metal siding	28 (63)	35 (79)	43 (97)
3	Broken glass in doors and windows	35 (79)	43 (96)	51 (114)
4	Uplift of roof deck and loss of significant roof covering material (>20%); collapse of chimney; garage doors collapse inward; failure of porch or carport	36 (81)	43 (96)	52 (116)
5	Entire house shifts off foundation	46 (103)	54 (121)	63 (141)
6	Large sections of roof structure removed; most walls remain standing	47 (105)	55 (122)	64 (143)
7	Exterior walls collapsed	51 (113)	59 (132)	68 (153)
8	Most walls collapsed, except small interior rooms	57 (127)	68 (152)	80 (178)
9	All walls down	63 (142)	76 (170)	89 (198)
10	Destruction of engineered and/or well-constructed residence; slab swept clean	74 (165)	89 (200)	98 (220)

Note. From WSEC, 2006.
[a]DOD is degree of damage.

The EF Scale has addressed some of the major limitations of the original F Scale, at the same time preserving the same damage levels as in the original scale. In addition, the EF Scale has expanded to include various building types with failure wind speed ranges depending on the quality of building construction. An important contribution is that the EF Scale has lowered the failure wind speeds (especially at higher EF Scale numbers) from the F Scale when expert elicitation was introduced.

Certainly, the EF Scale has shortcomings. After all, it is a scale that simplifies wind speed–damage correlations. The problem remains for rating the strength of tornadoes in open country when no damage occurs. Additionally, wind duration must be considered. A tornado might last a few seconds over a building, whereas a straight-line wind might last minutes, and a hurricane might last hours. Also, certain buildings respond differently to turbulent gusts over extended periods of time, especially as the wind changes direction. That said, we note that Forbes and Wakimoto (1983) used the F Scale to map downburst damage in central Illinois, while Stiegler and Fujita (1984) and Wakimoto and Fujita (1994) used the F Scale to evaluate damage from Hurricane Andrew. Also, Womble, Smith, Mehta, and McDonald (2009) noted that buildings are likely to have greater wind resistance in hurricane-prone zones due to more stringent building codes and increased awareness of the need for increased resistance in such areas. Therefore, newly constructed buildings should theoretically sustain less damage and consequently might be assigned lower EF ratings (based on the resulting damage). However, individual engineering examinations must be performed to confirm whether code upgrades actually were implemented and whether there were any critical flaws in the structure that contributed to the degree of wind damage.

2.3. The Future EF Scale

In 2014, the ASCE formed the Committee for Tornado Wind Speed Estimation in an effort to upgrade further the EF Scale and to introduce methodologies for wind speed estimation in a formal ASCE standard. The committee was divided into various subcommittees. The EF Scale subcommittee was charged with converting the existing EF Scale into an ASCE standard, improving narratives, and adding commentary and guidance for existing DIs. Additional subcommittees were created to formalize additional methods of wind speed estimation based on tree-fall pattern analysis, remote sensing of damage, forensic engineering analysis, and use of in situ anemometer and radar measurements. A data archival subcommittee also was formed. The standard will be intended for use by the NWS and by those who conduct storm damage evaluations. New knowledge gained from many tornado and hurricane disasters that have occurred since the creation of the EF Scale in 2001 will be incorporated into the new standard. Significant revisions will be made to the EF Scale DIs, DODs, and wind speeds for certain damages; however, the basic six-tier wind damage ratings (EF0 to EF5) will remain unchanged with only minor modifications to the wind speed ranges due to number rounding to the nearest 5 mph. The updated EF Scale by ASCE will be limited to tornado damage, which occurs over relatively short time periods of wind loading, in contrast to longer duration hurricane wind loads. However, many of the referenced works (particularly techniques for damage analysis) in the revised EF Scale include hurricane damage assessments as well as tornado damage assessments. Part of the reason for this is that tornado damage is similar to hurricane damage for most building types.

2.4. International EF Scales

Currently, several countries outside the United States (including Canada, China, Japan, and France) utilize the EF Scale or modified versions of the EF Scale to rate building damage. Sills, McCarthy, and Kopp (2014) indicated that Canada began using the EF Scale in 2011 and with modifications in 2013. Failure wind speeds were lowered on electrical transmission lines based on experience of damage evaluators in Canada. Hardwood and softwood trees were combined into a single DI. Other new DIs were added to include heritage churches, solid masonry houses, farm silos and grain bins, as well as sheds, fences, and outdoor furniture. A publication listing the DIs in Canada is available from Environment Canada (2014).

Meng et al. (2018) presented damage survey results from the deadliest tornado in China over the past 40 years. An EF4 tornado traveled 34.5 km through Funing County in Jiangsu Province on June 23, 2016, resulting in 98 fatalities and 846 injuries. Other damage surveys in China that have utilized the EF Scale have been performed by Meng and Yao (2014), Meng et al. (2016), and Bai, Meng, and Hunag (2017).

According to Niino et al. (2016) and Suzuki and Tanaka (2017), the Japan Meteorological Agency (JMA) began operational use of the Japanese Enhanced Fujita (JEF) scale in April 2016. The JEF is a modified version of the original EF Scale used in the United States to include additional DIs such as small sheds, greenhouses, vending machines, billboards, fences, scaffolding, and gantry cranes.

Mahieu and Wesolek (2011a, 2011b) helped develop an "augmented" EF Scale that has been used in France since 2009. Additional DIs were added to include factories, churches, medieval castles, and tombstones in addition to the EF Scale DIs used in the United States. They proposed expanding use of the EF Scale throughout Europe. Currently, the United Kingdom uses the TORRO or T Scale developed by Meaden (1976), while the rest of Europe uses an International F Scale or IF Scale, which is an "adapted version" of the original F Scale (see Feuerstein et al., 2012, and IF Scale Steering Group, 2018). An important distinction between the EF Scale and IF Scale is that with the IF Scale, wind damage to buildings are not categorized based on their function but rather on their "sturdiness." Also, the IF Scale is applicable to other types of wind events, not just tornadoes.

2.5. The Saffir-Simpson Wind Scale

The most popular method to rate hurricanes intensities in the Atlantic, Gulf of Mexico, and portions of the eastern Pacific oceans is the Saffir-Simpson Hurricane Wind Scale (SSHWS). This scale was developed in 1971 by Dr. Herbert Saffir, renowned engineer, and Dr. Robert Simpson, first director of the National Hurricane Center (NHC; 2019). The scale was divided into five categories, ranging from 1 to 5 in increasing levels of intensity and based on 1-min sustained wind speeds at 10 m over unobstructed terrain. It is important to note that 1-min sustained wind speeds are very different from 3-s gust wind speeds used in the EF Scale or the fastest ¼-mile wind speeds used in the original F Scale, leading to possible confusion when comparing the resulting loads. For a given time history of turbulent (fluctuating) winds, the largest 1-min sustained wind speed will be much lower than the largest 3-s peak wind speed. In 2009, the NHC eliminated barometric pressure and storm surge from the original scale. Further modifications were made in 2012, when the NHC expanded the wind speed range for Category 4 by 1 mph in both directions. In the western North Pacific, the term *supertyphoon* is used for tropical cyclones with sustained 1-min winds exceeding 67 m s^{-1} (150 mph). Interestingly, Dr. Fujita used the F Scale in the 1970s through 1990s to rate building damage intensity to buildings in hurricanes (as well as straight-line winds). Simply put, the SSHWS does not relate the DODs in detail to specific building types as does the EF Scale. Thus, Marshall (2008) utilized the EF Scale to rate degrees of hurricane damage instead of the SSHWS.

3. Wind Damage Surveys

3.1. Logistics/Equipment

There are many challenges to performing high-quality damage surveys. Ongoing rescue operations can restrict ground access and close airspace. Downed power lines and trees can block roads. Sometimes, disaster areas are accessible only by walking. It is important to conduct a damage survey as soon as possible after the disaster in order to preserve evidence before cleanup efforts begins. One of the first tasks is to locate areas of the most severe damage. Local authorities can help provide useful information and even escort the survey team to the damage sites. This helps expedite the survey by avoiding long lines at checkpoints and the need to possibly defend credentials. Having global positioning system tracking is essential as street signs frequently are missing, so street maps usually are not very helpful. Good-quality digital cameras are needed with plenty of batteries and secure digital (SD) memory cards. Wide-angle views can capture overall damage scenes and small-room panoramas. Zoom views capture construction details, such as nailed or bolted connections. Many cameras have geotagging capabilities to record latitude and longitude of each image as well as a date stamp. A digital sound recorder provides a quick way to document image locations and pertinent observations. Tape measures and digital rangefinders help determine distances between objects and dimensions of building components. Hard hats provide safety against striking your head on low objects. Make sure to bring proper identification to enter the disaster areas, including business cards or university/government badges, as well as any documents from officials allowing the survey team to enter the disaster areas. Customized magnetic signs attach quickly to any vehicle to help identify the storm damage survey team.

3.2. Ground Damage Surveys

A ground survey of specific buildings and structures reveals details of damage that otherwise cannot be determined during an aerial survey. Ground surveys can determine critical connections that cannot be observed from the air. Because ground surveys are time consuming, they should be planned carefully to minimize wasted time. When a town or city is devastated, it is not possible to look at every building in detail. Therefore, certain building types should be selected for analysis. Buildings that are destroyed are just as important as buildings that remain as they provided upper and lower bound failure wind speeds. Drones, if used properly, can provide detailed information in a short period of time. It is important to note the (1) building type, (2) building address/location, (3) type of foundation/connections, (4) type of roof/connections, (5) failure locations, (6) maximum DOD, (7) direction of the strongest wind, and (8) any unusual objects or items on site. Similar damage survey methodologies were described by Bunting and Smith (1990), Marshall, Bunting, and Weithorn (2003), and McDonald and Marshall (1984).

The NWS uses a Damage Assessment Toolkit (DAT), an application based on a geographic information system that works on a phone or tablet to collect and deliver storm damage information. The software contains the EF Scale tables and geotag images so it can plot the locations of each selected site on a map, which immediately appears on the Internet. In cases of tornado damage, the NWS needs to know the starting and ending points of the tornado track as well as its widest point and maximum intensity. Sometimes, this means driving back and forth on rural roads in a sawtooth pattern. As Burgess et al. (2014) noted, the DAT worked well during the survey of the El Reno, Oklahoma, tornado in 2013, although heavy use of the DAT quickly drained cell phone batteries. Therefore, it is important that survey teams using the DAT come equipped with car phone chargers.

Eyewitnesses can provide valuable information regarding the wind event. They can furnish a timeline of what happened, describe the event, and indicate where items were before the storm hit. They also might be able to tell you if objects and debris were moved during search and rescue operations. Caution should be used in making conclusions based on locations and positions of debris because items may have been moved or disturbed in rescue cleanup operations before the study team arrived. However, keep in mind there are still popular misconceptions regarding windstorms, as noted by Minor (1982), such as the notion of buildings "exploding" from low barometric pressure used to explain building damage. A twisted tree is more a physical characteristic of the tree rather than a tightly rotating wind.

Documentation of windborne projectiles, especially larger specimens, can be important as they can cause considerable damage to structures. These include, but are not limited to, large-diameter pipes, timber beams, posts and poles, steel beams and trusses, automobiles, storage tanks, and other heavy objects. It can be determined whether these objects flew or not by looking for ground impact marks and noting the distances between them. Other important information includes the dimensions, weight, and travel distances of the projectiles. Aerial or satellite images can be useful for deducing points of origin and travel distances.

During the examination of an engineered structure, it might be possible to acquire or photograph construction drawings of the building for later use in making wind speed estimates. Structural drawings are of most interest and will be marked with the letter "S" in the lower right corner of each page. They will commonly show the layout of the building as well as sizes and dimensions of structural members and details of the connections and anchorages. Floor plans, wall cross sections, and connection details provide important information. In certain instances, samples of critical connections might be obtained for later testing in an engineering laboratory.

High-quality images are essential for examining structures at a later date. First, wide-angle images should be taken of the building in its local environment, showing the terrain and obstacle roughness such as other buildings and trees. Then, images should be acquired from all four sides of the building. These medium-range views will show the type of building and materials used in construction. Adjacent images should overlap prior ones around 20 to 25 percent. Close-up images of critical connections at walls to foundations and roofs to walls should be acquired. A tape measure or ruler can be utilized to measure critical connections and to provide a scale for the images. Each image should have a specific purpose. Photographic documentation of building damage should follow from general to more specific views.

3.3. Aerial Damage Surveys

Fujita and Smith (1993) have demonstrated that near-ground wind fields can be better inferred by viewing damage patterns from above. Typically, a high wing aircraft or helicopter is used to conduct an aerial survey. The best images are obtained using a sliding zoom lens with fast shutter speed, like 1/500th of a second, to freeze-frame the images. Preferable altitudes for aircraft range between 100 m (330 ft) and 900 m (3,000 ft) above the ground, but this depends on flight clearance of the airspace as portions of disaster areas frequently are restricted. Flying in overlapping circles and centering on particular landmarks such as schools or churches works best in later trying to identify exact locations of the images. The use of geotagged images greatly reduces the subsequent work required for image interpretation. The best perspective is obtained from imaging almost directly above the damaged buildings. Clear skies and fast camera shutter speeds provide the best contrast for obtaining sharp images. In most instances, it is not possible to determine how well a structure was built from aerial surveys. This is why it is important to conduct detailed ground surveys. Also, it is best to acquire images through an open window to avoid glare or reflections. The photographer should wear dark colored clothing to minimize window reflection.

The importance of an aerial survey cannot be stressed enough. An aerial survey of the damage can give the ground survey team a quick overview of the damage path and help the team pinpoint certain areas for further study on the ground. So, it is best that the aerial survey be performed at the beginning of the survey. Tree-fall patterns and debris swaths are much easier to identify from the air. Aerial surveys can cover areas where ground surveys cannot. As a result, tornado tracks are found to extend farther when aerial surveys are conducted. It also is much easier to identify the presence of multiple tornado tracks from the air than from the ground. In identifying initial touchdown and lift-up points, one should cover an additional 10 miles to be sure the beginning and end points of the tornado path have been found. Flying several miles in either direction perpendicular to an identified path will also identify any parallel paths that might exist.

Aerial images generally fall into three categories: overall, block views, and building views. Using a sliding zoom lens on one camera works best during an aerial survey as changing lenses is rather difficult while flying. Overall views give an idea of the damage path and show large landmarks and topographic features. Block views show buildings arranged on each particular block. This range is most useful in identifying debris patterns. Areas of interest include ground markings, bounce marks, areas of ground or rooftop gravel scour, diffluent or confluent lines of debris, crops, and trees. Specific building views can show details of roof damage not visible from the ground. Thus, each type of image complements the other. The most useful images are acquired with the sun behind the photographer to avoid lens flare. Ideally, try obtaining multiple images of the same building or block from different perspectives. There is no such thing as taking too many images. After fieldwork is completed, many computer hours are needed to organize the images and store them on external hard drives. The images should be labeled or identified such that other evaluators can readily determine where the images were taken.

Many post disaster studies have been published by various agencies in the wake of hurricanes showing wind and storm surge damage to buildings. Some of the earliest surveys have been conducted by the US Army Corps of Engineers (USACE) after Hurricanes Betsy (1965), Beulah (1968), Camille (1970), Celia (1971), and Frederic (1981). The National Academy of Sciences (NAS) published damage surveys after Hurricanes Iwa (1983), Alicia (1984), Diana (1986), Elena (1991), and Hugo (1994). Also, detailed damage surveys have been published by the Federal Emergency Management Agency (FEMA) after Hurricanes Andrew (1993a), Iniki (1993b), Opal (1996), Fran (1997), Georges (1999a), Charley (2005a), Ivan (2005b), Katrina (2006), Ike (2009), Isaac (2013a), Sandy (2013b), Irma (2018), and Michael (2018). The National Institute of Standards and Technology (NIST; 2006) issued a survey report after Hurricanes Katrina and Rita. Marshall (2006a, 2006b, 2010b) also published damage surveys after Hurricanes Ivan, Katrina, and Ike.

3.4. Additional Remote-Sensing Surveys

While aerial surveys have long served as the workhorse for residential wind damage surveys, several burgeoning technologies have added tremendous power to rapidly collect timely and widespread damage information since the early 2000s. Womble, Wood, and Mohammadi (2018) chronicled the development of additional remote-sensing platforms over the past several decades, including commercial satellite imaging (with increasingly finer spatial resolutions), unmanned aerial systems (UASs) or "drone" imaging, and the ability to create highly detailed electronic three-dimensional (3D) replicas of damaged structures from lidar and digital photogrammetric (or structure-from-motion) techniques. These 3D replicas (known as "point clouds") offer the ability to preserve damage scenes digitally for future detailed analysis as well as the ability to make highly accurate measurements of structural deformations. These imaging platforms can cover large geographic areas quickly after the disaster as long as sky conditions permit scanning. Since wind damages roofing components, the DOD to roofing can be identified and wind direction inferred from analyzing the imagery.

4. Residential Foundation Inspection

There are seven common types of house foundations: (1) concrete slab, (2) pier and beam, (3) poured concrete wall, (4) concrete masonry wall, (5) stacked brick or block, (6) timber piles, and (7) masonry and concrete columns. The damage evaluator may encounter one or more of these foundation types during a damage survey.

4.1. Homes on Concrete Slab Foundations

Building codes require that wall bottom plates be bolted to concrete slab foundations. Steel anchor bolts should be at least 1.3 cm (1/2 in.) diameter and 30 cm (12 in.) long and have a J-shaped profile that will "grab" the concrete and enhance pullout resistance. Current building codes specify that anchor bolts must be inserted to a depth of at least 17.8 cm (7 in.) into the slab while the concrete is poured, and the bolts shall have sufficient height above the slab to pass through the plate and accept a steel nut (at least three turns) and a washer. Anchor bolts are spaced 1 to 2 m (3 to 6 ft) apart and located within 30 cm (12 in.) of the end of the plate and wall corners (Figure 18.3). Anchor bolts shall be

FIGURE 18.3 Properly bolted wall bottom plate in Moore, Oklahoma. Nuts with washers are tightened on the bolts to the wall bottom plates. Wall studs are straight nailed to wall bottom plates, and the wall exteriors are sheathed completely with plywood. Image by Tim Marshall.

positioned along the centerlines of the wall bottom plates, and there shall be a minimum of two bolts per wall bottom plate.

A number of problems can arise when installing anchor bolts in concrete slab foundations, especially since the construction trade that pours the concrete usually does not know where wall bottom plate framing will be installed. Anchor bolts can end up in the wrong location, such as in doorways or where wall studs intersect wall bottom plates. Sometimes, anchor bolts are installed too deep, such that there is insufficient thread to engage the washer and nut over the wall bottom plates. In other instances, anchor bolts appear at the edge or fall outside the wall bottom plates. Then, when wall framing commences, problems arise when bolt stems are too close to the wall studs to permit installation of a nut or washer (Figure 18.4). FEMA (1999b, 2012) found a number of homes with bolted wall bottom plates without nuts or washers during their survey of tornado damage. Therefore, walls may not necessarily be secured to their foundations, compromising their wind resistance.

If wall bottom plates are bolted properly to their concrete slab foundations, the weak link shifts to where the wall studs are straight nailed to the bottom wall plates. In these

FIGURE 18.4 Examples of wall bottom plates bolted to their concrete slab foundations under construction in Moore, Oklahoma. These homes were being rebuilt after a tornado destroyed them in 2013. Examples show improper foundation attachment with (a) bolt stem up against a wall stud, (b) bolt stem outside the wall bottom plate, (c) nut and washer on bolt on the edge of a wall bottom plate, and (d) wall stud bridging bolt stem. Images by Tim Marshall.

instances, anchored wall bottom plates remain as the walls are removed (Figure 18.5). Like links of a chain, the weakest link in the load path is where failure usually initiates. Marshall (2002) and Marshall and Eblen (2008) found numerous houses with bolted wall plates that remained intact but failed where wall studs were straight nailed into wall bottom plates.

Unfortunately, some municipalities have adopted variances in the building codes that allow straps, shot pins, or cut nails in lieu of anchor bolts (Figure 18.6). Metal straps are inserted into the concrete slab like anchor bolts with the anticipation they would be bent and nailed to the tops of the wall bottom plates. The straps typically are made from perforated 18-gauge metal (1.2 mm or 0.0478 in.). Unfortunately, in many instances, the straps are not fastened to the wall bottom plates. Either the straps are broken or cut or hidden in the air cavity between the frame wall and brick masonry (Figure 18.7). Sometimes, fasteners are placed in the perforated holes in the straps.

Steel shot pins are long steel dowels that literally are shot into the foundation with a powder-actuated hammer. They are typically 7.6 cm (3 in.) long and have an accompanying

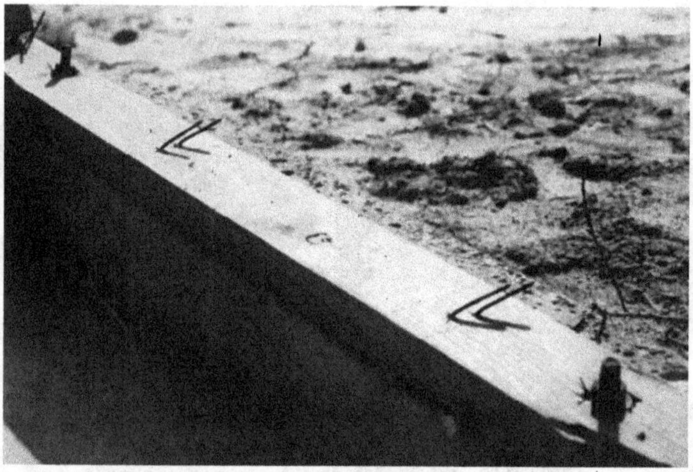

FIGURE 18.5 Wall framing failed where it was straight nailed into the wall bottom plates, leaving the properly bolted wall bottom plates intact in the Moore, Oklahoma, tornado. Image by Tim Marshall.

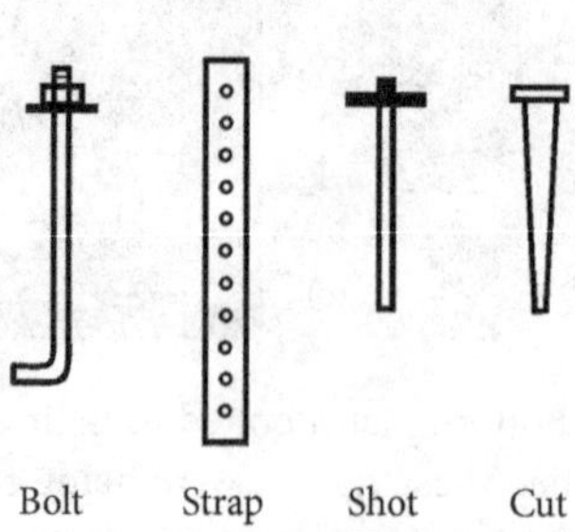

FIGURE 18.6 Various types of attachments of bottom wall plate to foundation.

FIGURE 18.7 Metal straps were bent outside the base of the wall to be hidden in wall cavity behind brick veneer in a newly constructed home in Moore, Oklahoma. Image by Tim Marshall.

washer (Figure 18.8). Unfortunately, the sudden driving of steel dowels can sometimes fracture the concrete, compromising the holding strength of the fasteners. Also, wall bottom plates have split around the shot pins during high-wind events, leading to complete wall failure (Figure 18.8 inset).

Unfortunately, cut nails sometimes are used to secure wall bottom plates to the concrete slab foundations. Cut nails are tapered and flat sided, stamped from steel. These fasteners are not as strong in tension or bending as properly installed anchor bolts. Cut nails must be hand nailed, which is difficult when nailing into concrete. Usually, cut nails do not penetrate very far into the concrete, and, when the nails are removed, they also remove conical areas of concrete clear cover, similar to that when removing carpet tack strip nails. If cut nails actually hold in the concrete, they can pull through the wall bottom plates as the plates are uplifted or rotated, leaving the nail standing alone in the concrete slab. Many building officials do not allow cut nails to be installed in wall bottom plates around the perimeter foundations, but some officials do allow them to be used to secure interior walls. Burgess et al. (2014); Marshall (2002); Marshall, Fox, Cavanaugh, and Huffman (2016); Marshall, Davis, and Runnels (2012); and Marshall, Stefkovich, DeBlock, Ladue, and Karstens (2012) have observed many instances where cut nails were utilized to install perimeter wall bottom plates in violation of local building codes (Figure 18.9).

FIGURE 18.8 Shot pin with attached washer in the bottom plate. Inset image shows aftermath of tornado where bottom plate broke around the fasteners. Images by Tim Marshall.

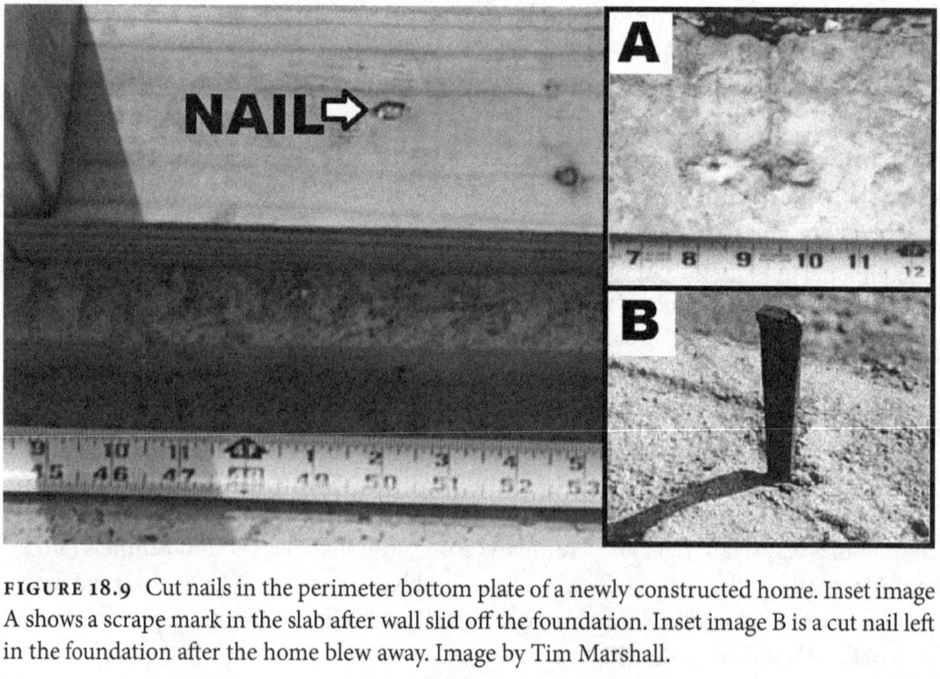

FIGURE 18.9 Cut nails in the perimeter bottom plate of a newly constructed home. Inset image A shows a scrape mark in the slab after wall slid off the foundation. Inset image B is a cut nail left in the foundation after the home blew away. Image by Tim Marshall.

4.2. Homes on Pier-and-Beam Foundations

Pier-and-beam foundations comprise different materials. Piers can be wood, loose block, masonry, or concrete. Perimeter-grade beams can be poured-in-place concrete or masonry. The same types of anchors secure houses to the beams just as with concrete slab foundations. However, floors rarely are anchored to the piers; rather, the floor platform rests on piers, leaving most of the home unanchored. Homes on pier-and-beam foundations tend to have little to no anchorage, making them quite susceptible to sliding or being lifted from their foundations during strong wind events. Marshall (2010a); Marshall, Davis, and Runnels (2012); Marshall, Morrison, Herzog, and Green (2010); and Marshall, Stefkovich, DeBlock, Ladue and Karstens (2012) reported numerous homes on pier-and- beam foundations swept completely clean from their foundations. Usually, the only item remaining is the concrete porch (Figure 18.10).

4.3. Homes on Poured Concrete-Wall Foundations

Poured concrete-wall foundations are found in homes with basements or crawl spaces. A concrete footing supports the foundation wall. Wooden sill plates are either bolted or

FIGURE 18.10 This home shifted off its concrete masonry unit (CMU) pier-and-beam foundation during the Picher, Oklahoma, tornado. Wood sill plates were not attached to the foundation. Image by Tim Marshall.

strapped around the perimeter of the foundation. Floor joists typically are toenailed to the sill plate. Stud walls are straight nailed to the floor indiscriminately, regardless whether they contact a floor joist or not. Most of the time, the nails only penetrate the flooring into the air below. Beams span basements or crawl spaces and are supported at intermediate locations by columns or posts and slotted into perimeter foundation walls. Floor beams and columns are either wood or steel.

Since floor platforms are supported by foundation walls, there are four critical connections necessary to keep the house in place. These connections occur where the sill plates connect to the top of the foundation walls, where the floor joists connect to the sill plates, where wall bottom plates connect to the floors, and where the wall studs connect to the wall bottom plates. Generally, a floor assembly will fail at the weakest link, which can lead to destruction of the house. Failure of connections of the sill plate to the foundation or floor joist to sill plate can result in lateral shifting of the floor platform as wind loads are transferred to the foundation (Figure 18.11). Marshall (2003) termed these homes "sliders" in his survey of tornado damage at LaPlata, Maryland, and found these homes slid off their foundations at relatively low wind speeds, only around 45 m s^{-1} (100 mph). Such wind speeds are well below the Fujita Scale F5 damage level at the time of 117 m s^{-1} (261 mph) denoting a house "swept clean" from its foundation. Many slider homes also were found in Greensburg, Kansas (Marshall, McCarthy, & Ladue, 2008), and

FIGURE 18.11 Home swept clean off a poured concrete foundation during the Parkersburg, Iowa, tornado. The wood sill plate remained bolted to the top of the foundation. However, the floor joists slid away when toenailed connections pulled out of the sill plates. A portion of the roof sits upright in the background. Image by Tim Marshall.

Parkersburg, Iowa (Marshall, Jungbluth, & Baca, 2008), because they were not attached to their foundations.

4.4. Homes on Masonry Wall Foundations

Masonry wall foundations include those constructed with stone, brick, or concrete masonry units (CMUs). Wood sill plates must be anchored to the masonry. Proper anchoring into stone or brick involves drilling into the masonry and securing wood sill plates with wedge anchors. However, a common practice has been not to anchor the sill plates. In the case of CMUs, proper anchoring involves filling cells with grout and embedding anchor bolts or straps in the grout. The entire height of a CMU foundation wall must be grouted with rebar extending downward into the footing. However, it is common practice to fill an occasional top cell of the foundation wall with grout and add an anchor or strap. The result is that most homes on masonry foundations have little-to-no foundation attachment and are susceptible to sliding off their foundations during strong wind events (Figure 18.12).

FIGURE 18.12 This home shifted off its CMU foundation in the Birmingham, Alabama, tornado on April 8, 1998. Only an occasional top block contained an anchor bolt, but none had nuts or washers. There was no wind damage to the vinyl siding and only minor wind damage to the roof covering, suggesting that wind speeds necessary to shift the home off its foundation were rather low. Image by Tim Marshall.

4.5. Homes on Pilings or Column Foundations

Homes elevated on either timber pilings or concrete/masonry piers usually have substantial anchorage such that they rarely fail during expected wind events. Timber piles usually are driven deep into the soil, while concrete and masonry piers usually have significant steel reinforcement. Wood floor platforms usually are well attached to pilings or piers with steel bolts and brackets. Although rare, piers or shallow pilings can fail in high winds when they are not braced adequately, causing the entire house to topple.

Weak points in this type of elevated construction are found where wall bottom plates are nailed to the floors or where the wall studs are nailed to wall bottom plates. Galvanized steel "hurricane" clips or straps can provide much stronger connections provided they are installed properly and are protected from corrosion due to salt exposure. Wind damage to these homes usually involves loss of the walls, leaving the floor platform intact. As Marshall (2006a) noted, wind pushing inward on windward walls can cause them to fail, forming hinges at the bases of the walls. Windward walls are pushed inward at the top and outward on leeward sides (Figure 18.13). Roof failures also are common in high winds, especially if roof–wall attachments are poor or windward portions of the building are compromised.

FIGURE 18.13 Wind damaged this home elevated on concrete masonry piers during Hurricane Ivan. The floor platform remained intact as it was well anchored to the piers. Note the windward wall was pushed inward, forming a hinge at the base of the wall. Image by Tim Marshall.

5. Residential Frame Wall Inspection

Conventional wood-framed walls comprise rows of wall studs connected by single wall bottom plates and double wall top plates; both wall plates are straight nailed to the studs. Common weak links are the straight-nailed connections, which can pull apart rather easily. Marshall (1983) conducted pull tests on pairs of 16d (8.9-cm or 3.5-in.) nails that connected wood plates to studs. For straight-nailed connections, the average pullout strength was 984 N (221 lbf) with a standard deviation of 375 N (84 lbf). For toenailed connections, the average pullout strength was 1,312 N (295 lbf), with a standard deviation of 307 N (69 lbf). Canfield, Niu, and Liu (1991) conducted similar pull tests using three 16d (8.9-cm or 3.5-in.) toenailed connections and found an average pullout strength of 1,846 N (415 lbf) if the wood did not split when nailed. They also conducted pull tests on metal hurricane clips and found an average strength of 5,498 N (1,236 lbf) depending on the type of clip used. Thus, Canfield found that a metal clip can be about three times stronger than a 16d (8.9-cm or 3.5-in.) toenailed connection.

Wood-framed walls must be braced properly to stiffen the frame and resist racking from lateral wind loads. Walls need a double top plate to help support the weight of the roof. Also, one of the top plates must be attached to the intersecting wall. Diagonal braces or "let-ins" stiffen the wall corners and provide lateral support to intersecting walls. If wood is used as a let-in brace, notches in the wall studs are made to receive the board, which extends diagonally from the wall top plate at the wall corner down to the wall bottom plate. Let-in braces are straight nailed to each wall stud. Plywood or oriented strand board (OSB) sometimes is installed at the wall corners, especially when diagonal bracing cannot be installed due to an intervening window or door (and also due to more recent code requirements). It is best to solidly sheath the home to provide greater stiffness and better wind resistance, but this is not often done as it is a more expensive option. Certainly, walls not properly attached to each other cannot transfer loads to intersecting walls and thus are prone to separation by the wind (Figure 18.14).

Typically, critical flaws in building construction are not discovered until the damage has occurred. This tragic possibility makes having a storm shelter or hardened area within a house a reasonable alternative in order to protect life and critical property. FEMA (1999c) published guidelines and even construction plans on how to build substantial in-residence shelters for relatively low financial cost. Marshall, Ladue, Ortega, and Stumpf (2012) found that above-ground shelters performed well even when homes sustained EF4 or greater damage. Many occupants survived without injury due to storm shelters, whereas they would have been seriously injured or killed had they not taken refuge in their shelters.

Windows and doors play critical roles in keeping wind out of buildings, thereby reducing structural damage. Marshall and McDonald (1982) found that 49 percent of homes with attached garages suffered severe structural damage when the garages faced the winds during a tornado in Grand Island, Nebraska. Weak garage doors have routinely

FIGURE 18.14 Second-story wall failure from Hurricane Andrew in Lakes by the Bay residence. Image by Tim Marshall.

been shown to be a critical factor leading to severe structural damage in structures subjected to high winds (Cushman, 2017; Kopp, Oh, & Inculet, 2008; Morrison, Kopp, Gavanski, Miller, & Ashton, (2014). In a study of residences damaged by Hurricane Ike (2008), the Joplin, Missouri, tornado (2011), and the Moore, Oklahoma, tornado (2013), Kovar, Brown-Giammanco, and Lombardo (2018) found that roof structural damage was rare (<5 percent of residences) without a garage door failure, and that about 60 percent of residences experienced structural damage if the garage door was breached.

For example, the garage doors shown in Figure 18.15 failed due to high wind pressures, allowing pressurization of the garage. The combination of the internal pressurization and external suctions led to the removal of the sidewall, roof, or both Therefore, it is important to have wind-resistant garage doors. After suffering damage from four major tornadoes from 1999 to 2013, the city of Moore, Oklahoma, adopted new building code regulations requiring residential garage doors to be rated for 135-mph wind gusts, the high-end range of an EF2 tornado (Wadsworth, 2014). FEMA (2011a) also provides guidance on how to reinforce existing garage doors to make them more wind resistant.

Impact-resistant windows withstand small debris impacts and help prevent wind from breaching a building. Typically, these laminated glazing systems consist of two or more panes and an interlayer of polyvinyl butyral (or equivalent) film laminated into the glazing assembly. Impact standards allow the glass to break during impact, but the pane must remain in place, and even resist cyclic pressure loading after impact. Installation of

FIGURE 18.15 Inward failure of garage door in the Channelview, Texas, tornado led to outward failure of a sidewall and collapse of the garage. Image by Tim Marshall.

shutters or boards over windows helps block the wind. However, shutters must meet impact-resistant standards and be secured properly to the building in order to be effective. They also are not practical to install in advance of a tornado or other short-term high wind events. The importance of preventing wind from getting inside a house cannot be overemphasized enough (Figure 18.16). FEMA (2011a) presents tips on how to protect windows from windborne debris.

Certain wall claddings perform better than others do during high-wind events. Vinyl siding is loose hung and frequently torn away by high winds, usually because the siding is not attached correctly or is not the correct type of siding for high-wind regions. Vinyl siding applied over other siding materials such as wood paneling typically has short fasteners that do not extend into the framing. FEMA (2009) noted that vinyl siding was the cladding most commonly damaged during Hurricane Ike. They also found that gable end-wall claddings frequently were damaged since they were not backed with solid wood sheathing. In many instances, the cladding had to bear the full force of differential wind pressure. Fiber–cement siding is a much better alternative to resist high winds provided it is attached properly to each wall stud.

Marshall (2006a, 2010b) observed considerable damage to EIFSs (exterior insulation and finish systems) on condominium buildings after Hurricanes Ivan and Ike. This composite siding system comprises insulation board attached mechanically or by an adhesive to a substrate. The weathering side of the insulation board is covered with a

FIGURE 18.16 Inward failure of the second-story window during Hurricane Andrew resulted in the outward failure of the second-story wall. Image by Tim Marshall.

polymer reinforcing scrim that holds a cementitious base coat and a textured finish coat. The application is similar to stucco, although stucco is applied over expanded steel mesh on a hard substrate and thus forms a hard surface. FEMA (2010) indicated that EIFS wall failures were commonplace during hurricanes. Such failures were concentrated in high-wind zones near wall corners, in areas where the substrate had deteriorated, or in areas where there was greater internal or differential wind pressure.

Brick veneer walls often topple during high-wind events, especially when steel brick ties have corroded, leaving the walls essentially freestanding. In some instances, there are insufficient numbers of steel ties to secure the wall to the framing. In other instances, steel ties are not installed correctly. Marshall and Foster (2002) documented several brick veneer failures on residences in the Arlington, Texas, tornado. Close examination revealed steel ties were fastened to the wall sheathing but never bent to engage the masonry (Figure 18.17). Such walls were free to move in the wind and were susceptible to toppling, especially on suction sides of the buildings. Freestanding walls that did not fail could be flexed easily by applying hand pressure. Flexing of the walls did not result from steel ties being broken or loosened during the wind, but rather because they did not engage the masonry. In many instances along coasts, steel ties had corroded to the point where they no longer engaged the masonry. Corrosion was most severe at the interface where the steel ties engaged the mortar. Apparently mineral salts in the mortar and salt air as well as periodic dew formation on interior faces of brick eventually accumulated

FIGURE 18.17 Steel ties (circled) on this home were not bent to engage the masonry. As a result, this freestanding wall easily toppled in the wind. Image by Tim Marshall.

on the steel ties, removing the protective zinc galvanizing. Thus, while steel ties may have been installed correctly when the masonry walls were built, they eventually corroded away, leaving freestanding walls.

6. Residential Roof Structure Inspection

Downward-acting gravity forces are the primary loads for which roof structures are constructed to resist. However, wind creates uplift and lateral forces on roof structures. Thus, commonly used connections might be adequate for gravity loads but inadequate to resist wind loads. Building codes require that roof structures resist "minimum" wind uplift loads. This means that roof structures must be fastened adequately to wall top plates to resist wind uplift loads. At a minimum, building codes require rafters or trusses to be toenailed to wall top plates. Toenailing is typically done with two or three nails driven at approximately 45-degree angles on opposite sides of the rafters or trusses where they bear or sit on the wall top plates. A better connector is a galvanized steel

FIGURE 18.18 Improper toenailing of roof trusses where nails were driven too close to the edges of the wall top plates, splintering the wood. Image by Tim Marshall.

hurricane clip, which is an angled plate that connects the rafter/truss to the wall top plates. Fasteners are driven horizontally through holes in the steel plates, placing the connection in shear. Nails are much stronger in shear than in tension. Riley and Sadek (2003) tested 16d toenailed and steel-clipped wall-to-roof connections and found that hurricane clips had substantially stronger uplift resistance than toenailed connections. Marshall (2002) found that toenailing is not always completed correctly. In new construction after the 1999 Moore, Oklahoma, tornado, nails were found protruding through the wall top plates and thereby not securing the roof trusses properly (Figure 18.18). As a result, these roofs were susceptible to being removed by winds lower than code minimums.

Rafters and trusses require proper cross-bracing to resist lateral wind forces. Installation of ridge blocking and metal brackets can prevent the trusses from rotating (Figure 18.19). In the case of residential roof framing, the roof deck diaphragm should not be the sole lateral support. Also, gable end walls should be well secured to the roof structure. In many instances, gable end walls are not well attached since they do not support roof weight. Marshall (2012) documented many gable end walls that had blown inward or outward in Hurricane Andrew, depending on the wind direction (Figure 18.20).

The shape of a roof also determines its wind resistance. Steep gable roofs catch the wind on their windward sides, acting like sails. This results in high uplift loads and

FIGURE 18.19 Roof trusses toppled in Hurricane Andrew due to a lack of lateral bracing. Proper wooden blocking between the trusses would have prevented rotation of the trusses. Image by Tim Marshall.

FIGURE 18.20 Gable end wall blew out due to internal wind pressure in the attic and the lack of attachment to the roof structure. Image by Tim Marshall.

eventual overturning of roofs if connections fail. Flat roofs also experience high uplift loads along their windward edges as sharp corners create high negative pressures. These high uplift loads can remove portions of roof coverings and roof decks if they are not well attached. Shallow-pitched, hip-style roofs are also more streamlined and structurally stronger than other roof geometries as wind loads can be transferred to other framing members on intersecting roof slopes. Also, with a lack of sharp edges, the wind can flow around low-pitched, hip-type roofs more readily, and as a result, there are less uplift and overturning loads. These are reasons why many homes in hurricane prone areas have low-pitched, hip style roofs.

It is important that the roof deck be well attached to rafters or trusses. Sheet decking material is typically plywood or OSB. The decking should be fastened at 15.24-cm (6-in.) intervals along edges and 30.48-cm (12-in.) intervals in the field of the sheet. Pneumatic guns commonly are used to drive fasteners into the decking, and in some instances, the fasteners miss the underlying framing (Figure 18.21). As a result, the roof deck is not attached and is therefore susceptible to removal in the wind. Marshall (2012) found entire sheets of wood decking fully clad with roof shingles in the debris after Hurricane Andrew. The exposure of rafters and trusses (as shown in Figure 18.22) usually indicates inadequate deck attachment.

There are many different types of roofing materials installed on residences. These include asphalt shingles, tile, metal, wood shingles, shakes, and more. Each industry has guidelines available for proper installation of the roofing product. Installation instructions usually are printed on labels accompanying the roofing products indicating the proper size, number, and spacing of fasteners. Roof coverings also must meet the minimum wind uplift requirements as stated in the building codes. The underlayment

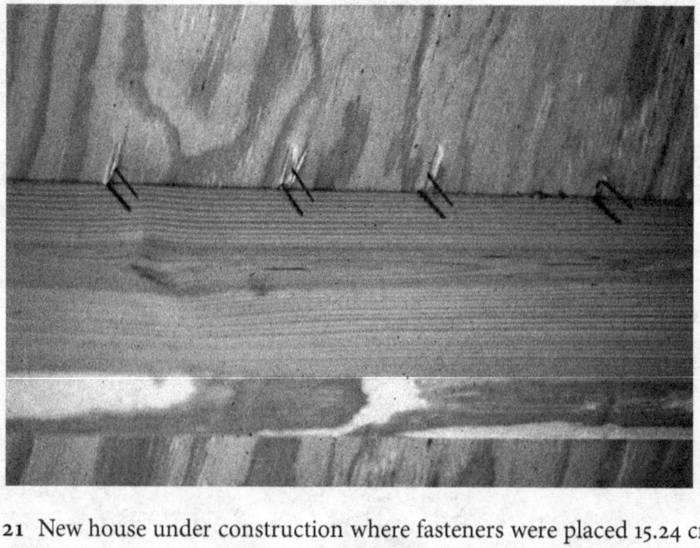

FIGURE 18.21 New house under construction where fasteners were placed 15.24 cm (6 in.) on center as required by code. However, the fasteners missed the truss. Image by Tim Marshall.

FIGURE 18.22 Complete loss of roof deck during the Arlington, Texas, tornado because the decking was not fastened correctly. There was minimal tree damage, and the vinyl spa cover remained. Image by Tim Marshall.

or anchor sheet must be well attached to serve as a redundant water-shedding layer if portions of the roof covering are lost. It is important that the roof covering remain intact as water infiltration can lead to substantial interior damage.

Marshall (2004) and Marshall, Morrison, Herzog, and Green (2010) saw numerous deficiencies with installations of roof coverings that led to their removal during relatively low wind speeds. Common roof installation errors with asphalt shingles include placing fasteners too high on the product, overdriving, underdriving, or orienting the fasteners incorrectly. Marshall (2012) discovered entire roofs stripped of asphalt shingles and underlayment after Hurricane Andrew (Figure 18.23). Although there was little-to-no structural damage to these homes, they sustained substantial interior water damage and had to be gutted to the wall studs due to loss of the roof coverings.

Marshall (2004) found that three-tab asphalt shingles did not perform as well as laminated shingles in his survey of roof coverings after Hurricane Frances. Only 44 percent of roofs with three-tab shingles had little-to-no damage compared to 87 percent of roofs with laminated-type asphalt shingles. Three tabs suffered more damage due in part to the lighter weights of the tabs and their design, which acted like flaps on aircraft wings. Once the sealant bond was broken, the tabs would flip back and forth, forming creases along the tops of the tabs. Eventually, the tabs would tear and blow away. Four primary

FIGURE 18.23 Poor installation of the roof covering on this home led to its complete removal during Hurricane Andrew. Staples were installed crooked and underdriven (inset image). Image by Tim Marshall.

modes of asphalt shingle failures occurred during Hurricane Frances: (1) creasing, (2) flipping, (3) tearing/removal, and (4) abrading from flying or falling debris. It was not uncommon to find multiple modes of failure on the same roof or slope. In some instances, entire shingles tore away when the fasteners pulled through the mats. Noone and Blanchard (1993) reported similar modes of failure for asphalt shingles during wind storms.

Tile roofs are quite wind resistant if the tiles are well fastened to the roof deck. However, tiles not attached well (i.e., with mortar patties) are prone to removal in strong winds, especially along rakes and windward eaves, where wind uplift forces are highest. Ridge tiles also are prone to wind damage as many ridge tiles are not attached to ridge boards and frequently are grouted in place. Once tiles become dislodged, they are flipped or thrown about, breaking additional tiles.

Acknowledgments

We would like to thank the National Oceanic and Atmospheric Administration (NOAA) for being asked to participate in many of the damage surveys mentioned here and to Haag Engineering Company for being allowed to take part in such surveys. Many thanks to C. S. Kirkpatrick and Dr. Christine Alfano for reviewing this work. Remote-sensing funding was obtained in part from the National Science Foundation CAREER Award Number 1751018: Remote Sensing for Enhanced Understanding of Tornado Actions and Broadened STEM Education in Rural West Texas (Womble).

References

Bai, L., Meng, Z., & Hunag, L. (2017). An integrated damage, visual, and radar analysis of the 2015 Foshan, Guangdong, EF3 tornado in China produced by the landfalling Typhoon Mujigae (2015). *Bulletin of the American Meteorological Society, 98*, 2619–2640.

Boissonnade, A., Hossain, Q., Kimbell, J., Mensing, R., & Savy, J. (2000). *Development of a probabilistic wind hazard model for the continental United States* (Report No. UCRL-ID-140922). Livermore, CA: Lawrence Livermore National Laboratory.

Bunting, W. F., & Smith, B. E. (1990). *A guide for conducting damage surveys* (NOAA Tech. Memo. NWS SR-146). Silver Spring, MD: National Oceanic and Atmospheric Administration.

Burgess, D., Ortega, K., Stumpf, G., Garfield, G., Karstens, C., Meyer, T., . . . & Marshall, T. (2014). 20 May 2013 Moore, Oklahoma tornado: Damage survey and analysis. *Weather and Forecasting, 29*, 1229–1237.

Canfield, L. R., Niu, S., & Liu, H. (1991). Uplift resistance of various rafter-wall connections. *Forest Products Journal, 41*, 27–34.

Cushman, T. (2017, May). A Texas tornado: Lessons learned. *Journal of Light Construction*, pp. 39–44.

Doswell, C. A., III, & Burgess, D. (1988). On some issues of United States climatology. *Monthly Weather Review, 116*, 495–501.

Environment Canada. (2014). *Enhanced Fujita Scale: Damage indicators and degrees of damage.* https://www.canada.ca/en/environment-climate-change/services/seasonal-weather-hazards/publications/enhanced-fujita-scale-damage-indicators.html

Federal Emergency Management Agency (FEMA). (1993a). *Hurricane Andrew in Florida: Observations, recommendations, and technical guidance* (Publication 22). https://www.fema.gov/media-library/assets/documents/10827

Federal Emergency Management Agency (FEMA). (1993b). *Hurricane Iniki in Hawaii: Observations, recommendations, and technical guidance* (Publication 23). https://www.fema.gov/media-library/assets/documents/10850

Federal Emergency Management Agency (FEMA). (1996). *Hurricane Opal in Florida: A building performance assessment* (Publication 281). https://www.fema.gov/media-library/assets/documents/10874

Federal Emergency Management Agency (FEMA). (1997). *Hurricane Fran in North Carolina: Observations, recommendations, and technical guidance* (Publication 290). https://www.fema.gov/media-library/assets/documents/10893

Federal Emergency Management Agency (FEMA). (1999a). *Hurricane Georges in the Gulf Coast* (Publication 338). https://www.fema.gov/media-library/assets/documents/2070

Federal Emergency Management Agency (FEMA). (1999b). *Midwest tornadoes of May 3, 1999* (Publication 342). https://www.fema.gov/media-library/assets/documents/647

Federal Emergency Management Agency (FEMA). (1999c). *Taking shelter from the storm: Building a safe room inside your house* (Publication 320). https://www.fema.gov/media-library/assets/documents/2009

Federal Emergency Management Agency (FEMA). (2005a). *Hurricane Charley in Florida: Mitigation assessment team report, observations, recommendations, and technical guidance* (Publication 488). https://www.fema.gov/media-library/assets/documents/905

Federal Emergency Management Agency (FEMA). (2005b). *Hurricane Ivan in Alabama and Florida: Mitigation assessment team report, observations, recommendations, and technical guidance* (Publication 489). https://www.fema.gov/media-library/assets/documents/2338

Federal Emergency Management Agency (FEMA). (2006). *Hurricane Katrina in the Gulf Coast: Mitigation assessment team report, building performance observations, recommendations, and technical guidance* (Publication 549). https://www.fema.gov/media-library/assets/documents/4069

Federal Emergency Management Agency (FEMA). (2009). *Hurricane Ike in Texas and Louisiana: Mitigation assessment team report, building performance observations, recommendations, and technical guidance* (Publication 757). https://www.fema.gov/media-library/assets/documents/15498

Federal Emergency Management Agency (FEMA) (2010). *Home builder's guide to coastal construction* (Publication 499). https://www.fema.gov/media-library/assets/documents/6131

Federal Emergency Management Agency (FEMA). (2011a). *Protect windows and doors with covers.* http://www.safe-wise.com/downloads/WindPreparation-WindowProtection-FEMA.pdf

Federal Emergency Management Agency (FEMA). (2011b). *Reinforce or replace garage doors.* https://www.dgdoors.com/media/general/FEMA-Hurricane-reinforcement.pdf

Federal Emergency Management Agency (FEMA). (2012). *Spring 2011 tornadoes: April 25–28 and May 22. Building performance observations, recommendations, and technical guidance* (Publication 90). https://www.fema.gov/media-library/assets/documents/25810

Federal Emergency Management Agency (FEMA). (2013a). *Hurricane Isaac in Louisiana: Building performance, observations, recommendations, and technical guidance* (Publication 938). https://www.fema.gov/media-library/assets/documents/31386

Federal Emergency Management Agency (FEMA). (2013b). *Hurricane Sandy in New Jersey and New York: Building performance, observations, and recommendations, and technical guidance* (Publication 942). https://www.fema.gov/media-library/assets/documents/85922

Federal Emergency Management Agency (FEMA). (2018). *Hurricane Irma in Florida: Building performance, observations, and recommendations, and technical guidance* (Publication 2023). https://www.fema.gov/media-library/assets/documents/176315

Federal Emergency Management Agency (FEMA). (2018). *Hurricane Michael in Florida: Building performance, observations, and recommendations, and technical guidance* (Publication 2023). https://www.fema.gov/media-library/assets/documents/186057

Feuerstein, B., Groenemeijer, P., Dirksen, E., Hubrig, M., Holzer, A. M., & Dotzek, N. (2012). Towards an improved wind speed scale vs. damage description adapted for Central Europe. *Atmospheric Research, 100*, 547–564.

Forbes, G. S., & Wakimoto, R. M. (1983). A concentrated outbreak of tornadoes, downbursts, and microbursts, and implications regarding vortex classification. *Monthly Weather Review, 111*, 220–235.

Fujita, T. T. (1971). *Proposed characterization of tornadoes and hurricanes by area and intensity* (SMRP Research Rep. 91). Chicago, IL: University of Chicago.

Fujita, T. T. (1973). *Experimental classification of tornadoes in FPP scale* (SMRP Research Rep. 98). Chicago, IL: University of Chicago.

Fujita, T. T. (1981). Tornadoes and downbursts in the context of generalized planetary scales. *Journal of Atmospheric Science, 38*, 1511–1534.

Fujita, T. T. (1992). *Mystery of severe storms.* Chicago, IL: University of Chicago Press.

Fujita, T. T., & Smith, B. E. (1993). Aerial survey and photography of tornado and microburst damage. In C. Church, D. Burgess, C. Doswell, and R. Davies-Jones (Eds.), *The tornado: Its structure, prediction, and hazard* (pp. 479–493). American Geophysical Union, Washington, D.C.

Grazulis, T. P. (1993). *Significant tornadoes.* St. Johnsbury, VT: Environmental Films.

IF Scale Steering Group. (2018). *The International Fujita (IF) Scale: Tornado and wind damage assessment guide*. Wessling, Germany: European Severe Storms Laboratory. Available online at: https://www.essl.org/media/publications/IF-scale_v0.10.pdf

Kopp, G. A., Oh, J. H., & Inculet, D. R. (2008). Wind-induced internal pressures in houses. *Journal of Structural Engineering*, *134*, 1129–1138.

Kovar, R. N., Brown-Giammanco, T. M., & Lombardo, F. T. (2018). Leveraging remote-sensing data to assess garage door damage and associated roof damage. *Frontiers in Built Environment*, *4*, 61. https://doi.org/10.3389/fbuil.2018.00061

Mahieu, P., & Wesolek, E. (2011a, October). *Contribution to a European adaptation of the Enhanced Fujita Scale: Analysis of damage caused by tornadoes in France*. Paper presented at the 6th European Conference on Severe Storms, Palma de Mallorca, Spain. https://www.essl.org/ECSS/2011/programme/presentations/111.pdf

Mahieu, P., & Wesolek, E. (2011b). *Tornado rating in Europe with the EF-scale: Definition of specific damage indicators for accurate tornado ratings in Europe*. Keraunos. http://www.keraunos.org/recherche/publications/tornado-rating-in-europe-enhanced-fujita-scale-ef-scale

Marshall, T. P. (1983). *Utilization of load and resistance statistics in a wind speed assessment* (MS thesis). Texas Tech University, Lubbock, TX. https://ttu-ir.tdl.org/bitstream/handle/2346/20598/31295003396933.pdf?sequence=1

Marshall, T. P. (2002). Tornado damage survey at Moore, Oklahoma. *Weather and Forecasting*, *17*, 582–598. https://doi.org/10.1175/WAF-D-14-00039.1

Marshall, T. P. (2003, February). The La Plata, MD tornado: Issues regarding the F Scale. *Proceedings of the Symposium on the F-Scale and Severe-Weather Damage Assessment*, Long Beach, CA. Boston, MA: American Meteorological Society. https://ams.confex.com/ams/pdfpapers/53280.pdf

Marshall, T. P. (2004, May). *Roof damage issues in hurricanes*. Paper presented at 26th Conference on Hurricanes and Tropical Meteorology, Miami, FL. Boston, MA: American Meteorological Society. https://ams.confex.com/ams/pdfpapers/75153.pdf

Marshall, T. P. (2006a, April). *Hurricane Ivan damage survey*. Paper presented at 27th Conference on Hurricanes and Tropical Meteorology, Monterey, CA. Boston, MA: American Meteorological Society. https://ams.confex.com/ams/pdfpapers/106925.pdf

Marshall, T. P. (2006b, April). *Hurricane Katrina damage survey*. Paper presented at 27th Conference on Hurricanes and Tropical Meteorology, Monterey, CA. Boston, MA: American Meteorological Society. https://ams.confex.com/ams/pdfpapers/106926.pdf

Marshall, T. P. (2008, January). *Wind speed-damage correlation during Hurricane Katrina*. Paper presented at Tropical Meteorology Special Symposium, 88th Meeting of the American Meteorological Society, New Orleans, LA. https://ams.confex.com/ams/pdfpapers/131732.pdf

Marshall, T. P. (2010a, October). *Damage survey of the Picher, Oklahoma tornado*. Paper presented at the 25th Conference on Severe Local Storms, Denver, CO. Boston, MA: American Meteorological Society. https://ams.confex.com/ams/pdfpapers/173923.pdf

Marshall, T. P. (2010b, May). *Hurricane Ike damage survey*, Paper presented at the 29th Conference on Hurricanes and Tropical Meteorology, Tucson, AZ. Boston, MA: American Meteorological Society. https://ams.confex.com/ams/pdfpapers/167534.pdf

Marshall, T. P. (2012, April). *Hurricane Andrew—20 years later: What have we learned?* Paper presented at the 30th Conference on Hurricanes and Tropical Meteorology, Ponte Vedra, FL. Boston, MA: American Meteorological Society. https://ams.confex.com/ams/30Hurricane/webprogram/Manuscript/Paper/206042/Andrew-V4.pdf

Marshall, T. P., Bunting, W. F., & Weithorn, J. D. (2003, February). *Procedure for assessing wind damage to wood-framed residences*. Paper presented at the Symposium on the F-Scale and Severe-Weather Damage Assessment, 83rd Annual Meeting of the American Meteorological Society,Long Beach, CA. https://ams.confex.com/ams/pdfpapers/52226.pdf

Marshall, T. P., Davis, W., & Runnels, S. (2012, November). *Damage survey of the Joplin Tornado: 22 May 2011*. Paper presented at the 26th Conference Severe Local Storms, Nashville, TN. Boston, MA: American Meteorological Society. https://ams.confex.com/ams/26SLS/webprogram/Manuscript/Paper211662/Joplinmerger.pdf

Marshall, T. P., & Eblen, L. (2008, October). *The Rosita Valley, TX tornado: 24 April 2007*. Paper presented at the 24th Conference on Severe Local Storms, Savannah, GA. Boston, MA: American Meteorological Society. https://ams.confex.com/ams/pdfpapers/141532.pdf

Marshall, T. P., & Foster, M. (2002, August). *Damage survey and radar analysis of the Fort Worth and Arlington, TX tornadoes on 28 March 2002. Preprints, 21st Conference on Severe Local Storms*, San Antonio, TX. Boston, MA: American Meteorological Society. https://ams.confex.com/ams/pdfpapers/45859.pdf

Marshall, T. P., Fox, M., Cavanaugh, D., & Huffman, M. (2016, November). *Damage survey of the North Texas tornadoes*. Paper presented at the 28th Conference on Severe Local Storms, Portland, OR. Boston, MA: American Meteorological Society. https://ams.confex.com/ams/28SLS/webprogram/Manuscript/Paper300382/2016-NTXpaperV3.pdf

Marshall, T. P, Jungbluth, K., & Baca, A. (2008, October). *The Parkersburg, IA tornado: 25 May 2008*. Paper presented at the 24th Conference on Severe Local Storms, Savannah, GA. Boston, MA: American Meteorological Society. https://ams.confex.com/ams/pdfpapers/141533.pdf

Marshall, T. P., Ladue, J. G., Ortega, K. L., & Stumpf, G. J. (2012, November). *Performance of residences and shelters in the Oklahoma tornadoes of 24 May 2011*. Paper presented at the 26th Conference of Severe Local Storms, Nashville, TN. Boston, MA: American Meteorological Society. https://ams.confex.com/ams/26SLS/webprogram/Paper211671/OKC_paper_combined.pdf

Marshall, T. P., McCarthy, D., & Ladue, J. (2008, October). *Damage survey of the Greensburg, KS tornado*. Paper presented at the 24th Conference on Severe Local Storms, Savannah, GA. Boston, MA: American Meteorological Society. https://ams.confex.com/ams/pdfpapers/141534.pdf

Marshall, T. P., & McDonald, J. R. (1982, January). *An engineering analysis of the Grand Island tornadoes. Preprints of the 12th Conference on Severe Local Storms, San Antonio, TX*. Boston, MA: American Meteorological Society. pp. 293–296.

Marshall, T. P., & McDonald, J. R. (1983, October). *An engineering analysis of structural damage in the Altus, Oklahoma, tornado: May 11, 1982. Preprints of the 13th Conference on Severe Local Storms, Tulsa, OK*. Boston, MA: American Meteorological Society. pp. 66–69.

Marshall, T. P., Morrison, S., Herzog, R., & Green, J. (2010, May). *Wind effects on asphalt shingles*. Paper presented at the 29th Conference on Hurricanes and Tropical Meteorology, Orlando, FL. Boston, MA: American Meteorological Society. https://ams.confex.com/ams/pdfpapers/167533.pdf

Marshall, T. P., Stefkovich, J., DeBlock, J., Ladue, J., & Karstens, C. (2012, November). *Damage survey of the Tuscaloosa-Birmingham Tornado on 27 April 2011*. Paper presented at the 26th Conference of Severe Local Storms, Nashville, TN. Boston, MA: American Meteorological Society. https://ams.confex.com/ams/26SLS/webprogram/Manuscript/Paper211667/TCLmerged.pdf

McDonald, J. R., & Marshall, T. P. (1984). *Tornado damage documentation.* Lubbock, TX: Institute for Disaster Research, Texas Tech University.

McDonald, J. R., & Mehta, K. C. (2001). *Summary report on Fujita-Scale Forum.* Lubbock, TX: Wind Science and Engineering Research Center, Texas Tech University.

McDonald, J. R., Mehta, K. C., & Mani, S. (2003, February). *F-scale modification process and proposed revisions. Proceedings of the Symposium on the F-Scale and Severe-Weather Damage Assessment, Long Beach, CA.* 83rd Annual Meeting of the American Meteorological Society. Boston, MA. https://ams.confex.com/ams/pdfpapers/53999.pdf

McDonald, J. R., Mehta, K. C., Smith, D. A., & Womble, J. A. (2009, November). *The Enhanced Fujita Scale: Development and implementation. Proceedings, 5th Congress of Forensic Engineering, Washington, DC.* Reston, VA: American Society of Civil Engineers. pp. 719–728.

Meaden, G. T. (1976). Tornadoes in Britain: Their intensities and distribution in space and time. *The International Journal of Meteorology, 1*, 242–251.

Meng, Z., & Yao, D. (2014). Damage survey, radar, and environment analyses on the first-ever documented tornado in Beijing during the heavy rainfall event of 21 July 2012. *Weather Forecasting, 29*, 702–724.

Meng, Z., Yao, D., Bai, L., Zheng, Y., Xue, M., Zhang, X., . . . Wang, M. (2016). Wind estimation around the shipwreck of Oriental Star based on field damage surveys and radar observations. *Science Bulletin, 61*, 330–337.

Meng, Z., Bai, L., Zhang, M., Wu, Z., Li, Z., Pu, M., . . . Li, L. (2018). The deadliest tornado (EF4) in the past 40 years in China. *Weather Forecasting, 33*, 693–713.

Minor, J. E. (1982). Tornado technology and professional practice. *Journal of the Structural Division, 108*, 2411–2422.

Minor, J. E., McDonald, J. R., & Mehta, K. C. (1977). *The tornado: An engineered-oriented perspective* (NOAA Tech. Memo. ERL-NSSL-82). Silver Spring, MD: National Oceanic and Atmospheric Administration.

Morrison, M. J., Kopp, G. A., Gavanski, E., Miller, C., & Ashton, A. (2014). Assessment of damage to residential construction from the tornadoes in Vaughan, Ontario, on 20 August 2009. *Canadian Journal of Civil Engineering. 41*, 550–558. https://doi.org/10.1139/cjce-2013-0570

National Academy of Sciences. (1983). *Hurricane Iwa, Hawaii, November 23, 1982.* Washington, DC: Author. https://www.nap.edu/catalog/19500/hurricane-iwa-hawaii-november-23-1982

National Academy of Sciences. (1984). *Hurricane Alicia: Galveston and Houston August 17–18, 1983.* Washington, DC: Author. https://www.nap.edu/catalog/19411/hurricane-alicia-galveston-and-houston-texas-august-17-18-1983

National Academy of Sciences. (1986). *Hurricane Diana: North Carolina September 10–14, 1984.* Washington, DC: Author. https://www.nap.edu/catalog/19221/hurricane-diana-north-carolina-september-10-14-1984

National Academy of Sciences. (1991). *Hurricane Elena Gulf Coast, August 29–September 2, 1985.* Washington, DC: Author. https://www.nap.edu/catalog/1765/hurricane-elena-gulf-coast-august-29-september-2-1985

National Academy of Sciences. (1994). *Hurricane Hugo: Puerto Rico, the US Virgin Islands, and South Carolina, September 17–22, 1989.* Washington, DC: Author. https://www.nap.edu/catalog/1993/hurricane-hugo-puerto-rico-the-virgin-islands-and-charleston-south-carolina-september-17-22-1989

National Hurricane Center (NHC). (2019). *The Saffir-Simpson Hurricane wind scale.* Miami, FL: Author. https://www.nhc.noaa.gov/aboutsshws.php#.~text=The_Saffir-Simpson_Hurricane_Wind,loss_of_life_and_damage

National Institute of Standards and Technology (NIST). (2006). *Performance of physical structures in Hurricane Katrina and Hurricane Rita: A Reconnaissance Report* (NIST Technical Note 1476). Gaithersburg, MD: Author. https://doi.org/10.6028/nist.tn.1476

Niino, H., Y. Tamura, M. Ito, H. Kikitsu, J. Maeda, Y. Okuda, . . . & Y. Tanaka (2016, November). *Development and implementation of the Japanese Enhanced Fujita Scale.* Paper presented at the 28th Conference on Severe Local Storms, Portland OR. Boston, MA: American Meteorological Society. https://ams.confex.com/ams/28SLS/webprogram/Manuscript/Paper300864/28SLS_Extended_abstract_161209.pdf

Noone, M. J., & Blanchard, W. K. (1993, April). *Asphalt shingles—A century of success and improvement. Proceedings of the 10th Conference on Roofing Technology, Gaithersburg, MD.* Rosemont, IL: National Roofing Contractors Association. pp. 23–33.http://docserver.nrca.net/technical/1817.pdf

Phan, L., & Simiu, E. (1998). *The Fujita tornado intensity scale: A critique based on observations of the Jarrell tornado of May 27, 1997* (NIST Tech. Note 1426). Gaithersburg, MD: National Institute of Standards and Technology. https://tsapps.nist.gov/publication/get_pdf.cfm?pub_id=860192

Riley, M. A., & Sadek, F. (2003). *Experimental testing of roof-to-wall connections in wood-frame houses* (NISTIR 6938). Gaithersburg, MD: National Institute of Standards and Technology. https://tsapps.nist.gov/publication/get_pdf.cfm?pub_id=860470

Schaefer, J., & Galway, J. G. (1982, January). *Population biases in the tornado climatology. Preprints, 12th Conference on Severe Local Storms, San Antonio, TX.* Boston, MA: American Meteorological Society. pp. 51–54.

Senior Seismic Hazard Analysis Committee (SSHAC). (1997). *Recommendations for probabilistic seismic hazard analysis: Guidelines on uncertainty and use of experts* (NUREG/CR6372, UCRL-ID- 122160). Livermore, CA: Lawrence Livermore National Laboratory. https://www.nrc.gov/reading-rm/doc-collections/nuregs/contract/cr6372/vol1/index.html

Sills, D., McCarthy, P., & Kopp, G. (2014, November). *Implementation and application of the EF-scale in Canada.* Paper presented at the 27th Conference on Severe Local Storms, Madison, WI. Boston, MA: American Meteorological Society. https://ams.confex.com/ams/27SLS/webprogram/Manuscript/Paper254999/SLS2014_Sillsetal_EF-scale_ExtendedAbstract_Final2.pdf

Steigler D., & Fujita, T. T. (1984, August). *Mesoscale structure of Hurricane Alicia. Proceedings of Specialty Conference on Hurricane Alicia- Aerospace Division, Galveston, TX.* Reston, VA: American Society of Civil Engineers. pp. 21–39.

Suzuki, S., & Tanaka, Y. (2017, February). *The Japanese Enhanced Fujita Scale: Its development and implementation.* Paper presented at the Special Symposium on Severe Local Storms: Observation needs to advance research, prediction and communication, Seattle, WA. Boston, MA: American Meteorological Society. https://ams.confex.com/ams/97Annual/webprogram/Manuscript/Paper303781/EXTENDED_ABSTRACT_20170228b.pdf

US Army Corps of Engineers (USACE). (1965). *Hurricane Betsy: September 8–11, 1965.* New Orleans District, LA: Author.

US Army Corps of Engineers (USACE). (1968). *Hurricane Beulah September 8–21, 1967.* Galveston District, TX: Author.

US Army Corps of Engineers (USACE). (1970). *Hurricane Camille: 14–22 August 1969.* Mobile District, AL: Author.

US Army Corps of Engineers (USACE). (1971). *Hurricane Celia: 30 July–5 August 1970.* Galveston District, TX; Author.

US Army Corps of Engineers (USACE). (1981). *Hurricane Frederic Post Disaster Report, 30 August–14 September 1979*. Mobile District, AL: Author.

Wadsworth, T. (2014, Summer). First-ever building code requires tornado-resistant garage doors. *Door and Access Systems*, pp. 30–35. https://www.dasma.com/articles/feature/TornadoDoors_Sum2014.pdf

Wakimoto R. M., & Fujita, T. T. (1994). Damage survey of Hurricane Andrew and its relationship to the eyewall. *Bulletin of the American Meteorological Society*, *75*, 189–200.

Wind Science and Engineering Center (WSEC). (2006). *A recommendation for an enhanced Fujita scale*. Lubbock, TX: Texas Tech University.

Womble, J. A., Smith, D. A., Mehta, K. C., & McDonald, J. R. (2009, November). *The Enhanced Fujita Scale: For use beyond tornadoes? Proceedings of the 5th Congress on Forensic Engineering, Washington, DC*. Reston, VA: American Society of Civil Engineers. pp. 699–708.

Womble, J. A., Wood, R. L., & Mohammadi, M. E. (2018). Multi-scale remote sensing of tornado effects. *Frontiers in Built Environment*, *4*, 1–21.

CHAPTER 19

NUMERICAL SIMULATION OF FLIGHT AND IMPACT OF WIND-BORNE DEBRIS

TAKASHI NOMURA

1. Introduction

Damage of wind-induced disasters is caused by not only the large fluid force of the strong wind but also the impact force of flying debris. The damage by flying debris may cause new debris to fly, and this expands the area of disaster. For mitigation of wind-induced disaster, it is important to know how fast and how far debris flies as well as how strong debris hits houses or structures.

Recently, Maruyama (2011) and Noda et al. (2014) have studied the flying behavior of debris. In these studies, swirling flow fields to imitate a tornado are generated by computational fluid dynamics (CFD). In these flow fields, the flying motion of debris is simulated by assuming the debris as particles or rigid bodies. In Maruyama's (2011) study, the flying objects had the drag coefficients of wooden rods or a large truck. Noda et al. (2014) dealt with the flight of flat plates, for which aerodynamics forces were based on the parameters obtained from a wind tunnel experiment by Tachikawa (1983).

Thus, current studies on flight simulation of windborne debris are using wind tunnel experimental data for the aerodynamic constants although such experimental data are limited to fundamental configurations, such as pebbles, rods, or plates as Holmes classified (2001). However, debris has various shapes and various weight balances. Also, the aerodynamic forces change dynamically during the flight of debris.

In addition to this, these aerodynamic constants are used in simulation based on the quasi-static assumption; that is, the aerodynamic forces are determined by the instantaneous flying velocity of the debris as well as the instantaneous wind velocity as if the debris is placed in a steady flow. However, recent investigation by Takeuchi, Maeda, and Kawashita (2008) on wind forces generated by gusty wind suggests that a short time increase of wind speed leads to a stronger wind force than the steady wind. This implies that unsteadiness in debris velocity and wind velocity as well as debris attitude during its flight should be taken into account. Wind tunnel experiments have difficulty achieving these requirements for an arbitrary shape of windborne debris. These considerations motivate the application of CFD to simulate flight of windborne debris.

Flights of debris have three features as (1) debris objects have various shapes and various weight balances; (2) the aerodynamic forces should be evaluated for such objects by changing their attitude dynamically during the flight; (3) flight distance is fairly large in comparison to the size of objects. Based on the features (1) and (2), the flight of windborne debris should be simulated as a fluid–structure interaction problem based on the finite element method so that dynamical effects between debris motion and flow around it can be taken into account for the arbitrary shape of objects.

Taking into consideration feature (3), it may not be practical to cover all flight paths by a single large computational domain as shown in Figure 1a. This consideration leads to a methodology where a computational domain moves with the debris as in Figure 1b. It is assumed that this moving computational domain should be wide and fine enough to evaluate the instantaneous aerodynamic forces.

To analyze flow in a moving computational domain, the arbitrary Lagrangian–Eulerian (ALE) method is employed. The ALE method is one that is widely used to deal with moving boundary problems, among other methods, including the space–time approach of Tezduyar et al. (1992) (Bazilevs, Takizawa, & Tezduyar, 2013). An early work on the ALE approach was conducted by Hirt, Amsden, and Cook (1974), followed by Hughes, Liu, and Zimmerman (1981); Belytscho, Flanagan, and Kennedy (1982); and Donea, Giuliani, and Halleux (1982). Kawahara (2016) recently wrote a comprehensive chapter on the ALE method in his latest book on finite element flow simulation. I have employed the ALE method in some classes of fluid–structure interaction problems (Nomura & Hughes, 1992; Nomura, 1994; Nomura & Shinmyo, 1999).

In the present article, the following two subjects are described: One is flight simulation of a rigid sphere using a moving mesh attached to the sphere; the other is a computational method for an impact procedure when a sphere hits the ground or solid walls. As for the numerical procedure of impact, the Lagrangian multiplier method or the penalty method is usually employed in computational contact mechanics (Wriggers, 2006). The method investigated in the present study is different from them (Nomura, 2018).

In the remainder of the present article, two key methods are described. In Section 2, the numerical method to simulate flights of a projected sphere in wind using moving computational mesh is described. After briefly introducing the basic equations and the fluid–structure interaction algorithm used in the present study, several numerical examples are presented in order to show that aerodynamic forces around the flying sphere are appropriately introduced into the simulation. In Section 3, the computational procedure to deal with the impact process when a sphere hits the ground or solid walls is

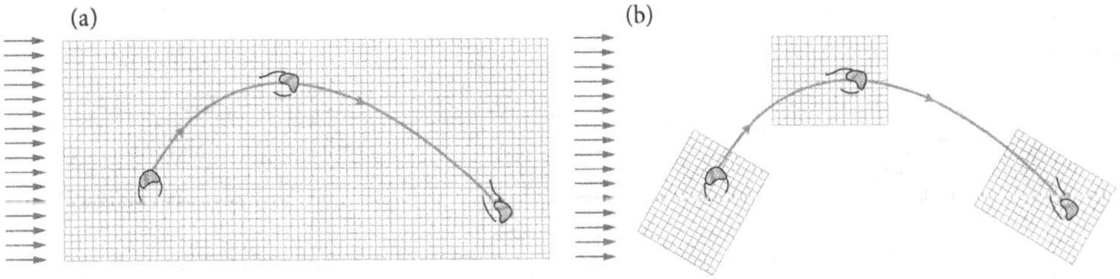

FIGURE 19.1 Schematic view of computational meshes to simulate flight of flying debris: (a) a mesh to cover the whole flight path; (b) a mesh moving with the flying debris.

described. In order to suppress unphysical step-by-step pressure oscillation, the time integration step at the impact is subdivided into multiple short time intervals, and the impact force is expressed by a smooth function of time over this period. Finally, concluding remarks are presented in Section 4.

2. Numerical Simulation of a Projected Sphere in Wind Using Moving Mesh

In this section, the strategy to set up the computational domain moving with flying debris is described.

2.1. The Computational Domain Moving with a Projected Sphere

As an example of flying debris, a spherical object projected into air has been chosen. Figure 2a shows an instant when the velocity of the sphere is $\dot{x}$, where U is the horizontal wind speed, and mg is the gravitational force. As shown in Figure 19.2b, the drag force F_D is exerted in the direction of the relative velocity v_R. In other words, the situation of Figure 19.2a has been regarded as the situation of Figure 19.2b where the sphere is moving in still air with speed v_R. Then, the computational domain covers the area around the sphere so that its two sides are parallel to the direction of the relative velocity v_R. As the direction of the relative velocity v_R changes in space during the sphere's flight, the direction of the computational domain changes in the physical space as well.

Figure 19.2c shows the computational domain fixed in the "computational space." This looks like a familiar computational model in wind engineering that is employed to

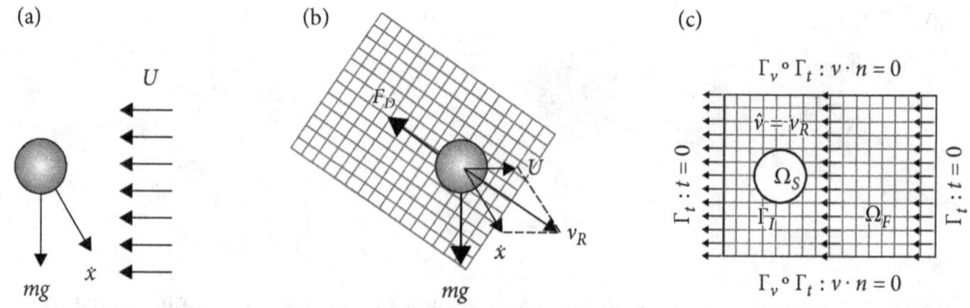

FIGURE 19.2 The strategy to set up the computational domain moving with the flying debris: (a) a projected sphere in horizontal wind (an example problem); (b) a moving computational domain dynamically aligned to the instantaneous relative velocity direction; (c) the computational domain and the boundary conditions.

analyze flows around bluff bodies like bridge girders, buildings, and others. However, there are two special features in the present computational domain:

1. Every node has the same mesh velocity $\hat{v}$ as

$$\hat{v} = \langle v_R \quad 0 \quad 0 \rangle^T. \tag{1}$$

where v_R is the magnitude of $\boldsymbol{v}_R$. The mesh velocity $\hat{v}$ in the ALE method is in Section 2.2.1.

2. The upstream inflow boundary is the traction-free boundary and is the same as the downstream outflow boundary:

$$\boldsymbol{t} = \boldsymbol{0} \text{ on } \Gamma_t. \tag{2}$$

where $\boldsymbol{t}$ is the boundary traction, and Γ_t denotes the boundaries on which the traction condition is specified.

On the other two boundaries, the slip condition $v \cdot \boldsymbol{n} = 0$ is specified, where $\boldsymbol{n}$ is the unit's outward normal vector to the boundary. In Figure 19.2c, the symbols Ω_F, Ω_S, and Γ_I denote the domain of air, the domain of the solid, and the interface between Ω_F and Ω_S, respectively.

2.2. The Basic Equations

In this section, the basic equations of the present problem are described. They are the ALE form of the Navier–Stokes equation of fluid flow around the sphere, the equation of motion of the sphere, and the interface condition between the sphere and the fluid.

2.2.1. *The ALE Form of the Navier–Stokes Equation*

The basic equations of the fluid motion in the present study are the following ALE form of the Navier–Stokes equation and the continuity equation for the viscous incompressible fluid:

$$\frac{\partial v}{\partial t} + v^{\#} \cdot \nabla v + \nabla p = \nu \Delta v + f \quad \text{in } \Omega_F \tag{3}$$

$$\nabla \cdot v = 0 \quad \text{in } \Omega_F \tag{4}$$

where $v = \langle v_x \quad v_y \quad v_z \rangle^T$ is the velocity vector; p is the pressure; $f = \langle f_x \quad f_y \quad f_z \rangle^T$ is the body force vector; and ν is the dynamic viscosity of the fluid. The modified convection velocity $v^{\#} = v - \hat{v}$ is the relative flow velocity to the mesh velocity $\hat{v} = \langle \hat{v}_x \quad \hat{v}_y \quad \hat{v}_z \rangle^T$ of the moving referential frame in the ALE formulation. The first term of Equation 3 is the time derivative with respect to the ALE referential frame. The mesh velocity $\hat{v}$ expresses the motion of the referential frame. If $\hat{v} = 0$, the referential frame is fixed in space; and the equation of fluid motion (Equation 3) coincides with the ordinary Navier–Stokes equation based on the Eulerian description; and $\partial v / \partial t$ is the spatial time derivative. If $\hat{v} = v$, the referential frame moves with the fluid particle (i.e., the Lagrangian description), and $\partial v / \partial t$ is the material time derivative. In finite element flow analysis, the referential frame is finite element mesh. The mesh velocity $\hat{v}$ can be given arbitrarily to express the motion and deformation of finite element meshes.

2.2.2. *The Basic Equations of Sphere Motion*

The sphere is assumed to be a rigid body. The origin of the coordinates $x - \langle x \quad y \quad z \rangle^T$ is placed at the initial position of the center of gravity of the sphere. The equation of motion is written as

$$m\ddot{x} = \mathbf{X} \tag{5}$$

where m is the mass of the body, $\ddot{x} = \langle \ddot{x}\ \ddot{y}\ \ddot{z} \rangle^T$ is the acceleration vector of the center of gravity of the sphere, and $\mathbf{X} = \langle X \ Y \ Z \rangle^T$ is the force vector, which is regarded as exerted on the center of gravity of the body. The force **X** consists of the gravitational force and the resultants of fluid forces.

2.2.3. *The Conditions on the Interface Between the Fluid and the Rigid Body*

Along the interface Γ_I, which lies between the sphere Ω_S and the fluid domain Ω_F, the following conditions must be satisfied:

$$\mathbf{u}_I = \mathbf{T}^T \dot{x} \qquad \text{on} \quad \Gamma_I \tag{6}$$

$$\mathbf{X} = \mathbf{T}\mathbf{F}_I \qquad \text{on} \quad \Gamma_I \tag{7}$$

where $\mathbf{u}_I$ and $\mathbf{F}_I$ are the nodal velocity of the fluid on Γ_I and the nodal traction forces of the fluid on Γ_I, respectively, as shown in Figures 19.3a and 19.3b, respectively, and $\dot{x} = \langle \dot{x} \quad \dot{y} \quad \dot{z} \rangle^T$ is the velocity vector of the center of gravity of the sphere. Equation 6 expresses that the nodal velocity vector $\mathbf{u}_I$ is equal to the sphere velocity $\dot{x}$, while Equation 7 expresses that **X** is the resultant of $\mathbf{F}_I$. The matrix T represents the relations between $\mathbf{u}_I$ and $\dot{x}$ as well as $\mathbf{F}_I$ and **X**.

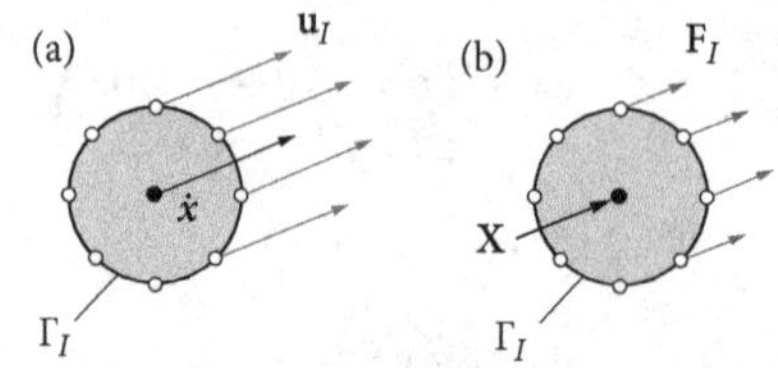

FIGURE 19.3 The conditions on the interface Γ_I: (a) the nodal velocity of the fluid $\mathbf{u}_I$ and the velocity of the rigid body $\dot{x}$; (b) the nodal traction forces of the fluid F_I and the concentrated force exerted on the center of gravity of the rigid body.

2.3. The Finite Element Equations of Fluid and the Fluid–Structure Interaction Procedure

In the present study, the basic equations, Equations 3 and 4, are discretized into finite element equations based on the variational multiscale method of Bazilevs et al. (2007). The eight-node, trilinear, hexahedral element is used in this method.

The time marching algorithm of the fluid–structure interaction analysis used in the present study is a kind of staggered scheme. The features of the algorithm are briefly described. For details, please refer to the work of Nomura (2018).

1. The time integration method for the finite element equations of fluid is the generalized α method of Bazilevs et al. (2007).
2. The time integration method for the sphere motion, Equation 5, is the Newmark β method in the predictor–corrector procedure.
3. The fluid–structure interaction algorithm is a staggered procedure where the fluid equations and the equations of motion of the sphere are solved alternately. Because both time integration methods for fluid and sphere are predictor–corrector algorithms, after the predictor phases, the corrector phases are repeated twice by taking into account the interface conditions (Equation 6) and (Equation 7). Finally, the two solutions are averaged before proceeding to the next time step.

2.4. Computed Examples of a Projected Sphere in Wind Using Moving Mesh

In order to demonstrate the present computational strategy, motions of a sphere projected in wind are computed.

2.4.1. *The Computational Conditions*

Figure 19.4 shows the finite element mesh used in the present computation, where the numbers of nodes and the elements are 4,122 and 3,584, respectively. As the boundary conditions, the slip condition is applied to the four side boundary surfaces, and the

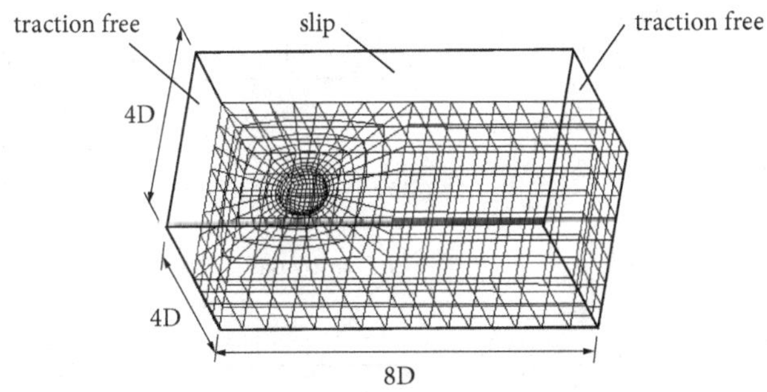

FIGURE 19.4 The finite element mesh for the analysis of a projected sphere.

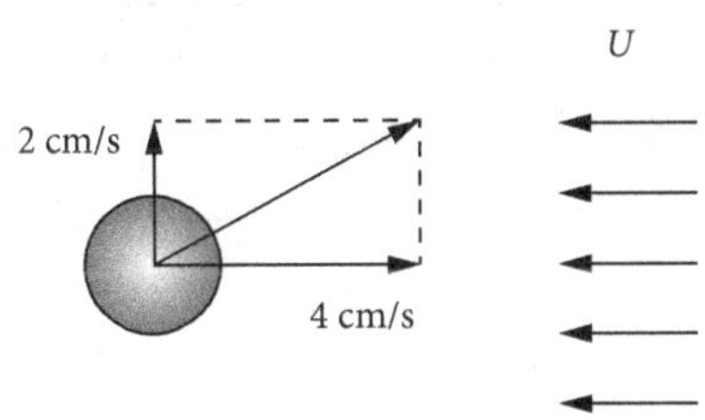

FIGURE 19.5 The initial velocity of a projected sphere.

traction-free condition is applied to both end boundary surfaces. The diameter of the sphere D is 2 cm. As shown in the computations that follow, the density ratio of the sphere material to the air is less than 200; that is, the mass of the sphere is less than 1.026 g. This small mass is employed because the present finite element mesh is relatively small. Figure 19.5 shows the initial velocity given to the sphere.

2.4.2. *Effect of the Density Ratio*

Figure 19.6a shows the computed trajectories of the projected sphere in still air (i.e., no wind) for three cases of density ratio 2, 20, and 200. For the case of the density ratio 200, computed streamlines at three time instants [A], [B], and [C] are presented. These computational domains are parallel to the slopes of the trajectory. The direction of these tilting computational domains is that of the relative velocity v_R in Figure 19.2b.

The trajectory of density ratio 200 is almost parabolic. This indicates that the effect of the drag force is negligible. By reducing the density ratio, the trajectories deflect away from that of density ratio 200 as a consequence of resistance by air.

2.4.3. *Effect of Wind*

Figure 19.6b shows the computational trajectories in horizontal wind. The density ratio of the sphere to the air is 2. The wind speeds U are 8 cm/s (headwind), -8 cm/s (tailwind), and 0 cm/s (no wind). The trajectory for the no wind case is the same as plotted in Figure 19.6a. In the case of the tailwind, the sphere travels farther than the case of no wind. In the case of the headwind, at first the sphere is uplifted and then falls. These differences of trajectories are reasonable as the consequence of horizontal aerodynamic force by wind.

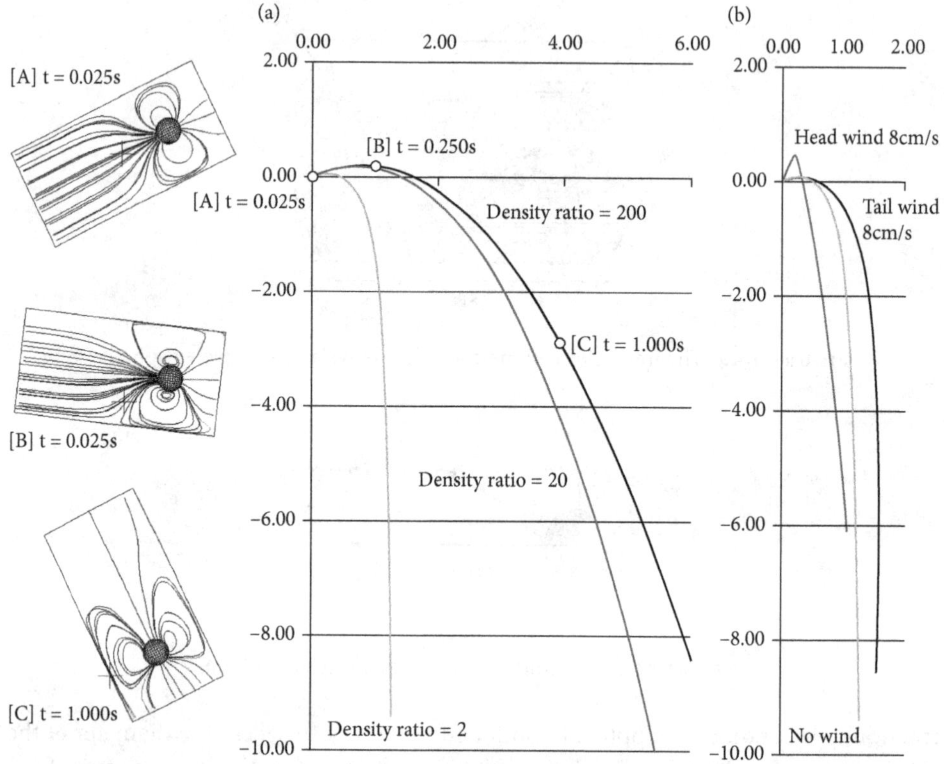

FIGURE 19.6 Computed trajectories of a sphere projected to air: (a) effect of the density ratio (three left figures of computed streamlines correspond to the three time instants along the case of density ratio 200); (b) effect of horizontal wind (the density ratio 2).

3. Impact Forces of a Falling Sphere in Air

In this section, the computational strategy to evaluate impact forces is described.

3.1. The Computational Model and Conditions

In order to develop the computational procedure to simulate the impact process in air, a simple problem as described in Figure 19.7 has been employed: a falling sphere in a rectangular parallelepiped domain. A sphere is released from rest, falls in air, and then hits the rigid, flat ground and bounces back to midair. As the boundary conditions, the slip condition is applied to the four lateral boundaries, the traction free condition is applied to the top horizontal boundary, and the no-slip condition is applied to the bottom horizontal boundary as well as to the sphere's surface. Figure 8a is the finite element mesh,

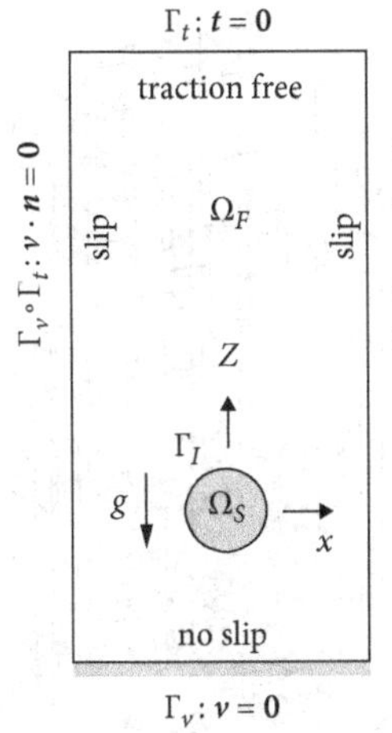

FIGURE 19.7 A falling sphere in a rectangular parallelepiped domain: The origin of the coordinates (x, y, z) is placed at the initial center of the sphere before falling.

which is the same discretization as the mesh of Figure 19.4. The difference from the previous flying sphere analysis is that the computational spaces of Figure 19.7 and Figure 19.8 are fixed in the physical space. Instead of that, as the sphere descends or ascends, the finite element mesh in the lower half portion of the computational domain deforms as shown in Figure 19.8b.

3.2. Computational Procedures at the Impact

In this section, the present computational procedures at the impact is described. A computational strategy to suppress unphysical numerical oscillation is also given.

3.2.1. *A Procedure to Conserve Momentum of the Sphere*

Figure 19.9 shows two consecutive time steps at impact. Step n at $t = t_n$ is the step just prior to the instant when the sphere hits the ground. The symbol δ denotes the gap between the sphere and the ground. Step $n + 1$ at $t = t_{n+1} \left(= t_n + \Delta t\right)$ is the very step when the impact force R is applied to the sphere. The basic procedure to determine the value of the impact force R is to repeat Step $n + 1$ until the computed sphere velocity $\dot{z}_{n+1}$ reaches a prescribed target velocity $\bar{\dot{z}}$. According to the investigation, it has been found that it is appropriate to choose $\dot{z}_{n+1} = \bar{\dot{z}} = 0$. As shown in Figure 19.10, the upward velocity of the sphere at step $n + 2\left[t = t_{n+2}\left(= t_n + 2\Delta t\right)\right]$ after impact at the ground is almost the same as the downward velocity at step n, that is, $\dot{z}_{n+2} = -\dot{z}_n$. On the contrary, if $\dot{z}_{n+1} = \bar{\dot{z}} = -\dot{z}_n$ is chosen, the upward velocity of the sphere at step $n + 2$ is almost double. Figure 19.11 shows the computed time history of the vertical displacement of the sphere. In the case of $\dot{z}_{n+1} = \bar{\dot{z}} = 0$, the sphere bounces up to the level exactly the same as the initial position. This indicates that the conservation of energy is accurately satisfied.

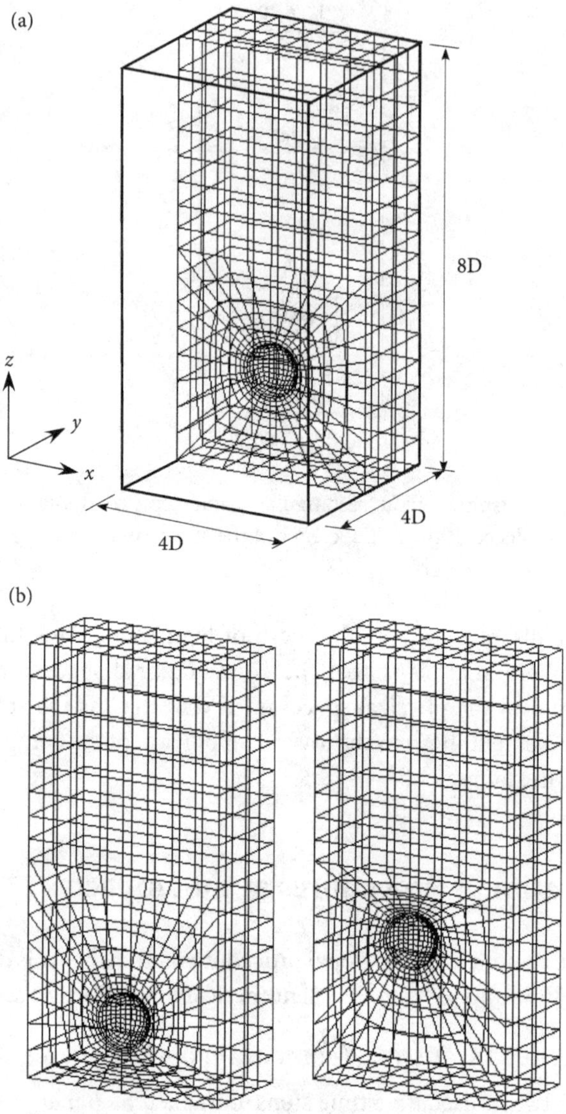

FIGURE 19.8 (a) Finite element mesh; (b) mesh deformation pattern.

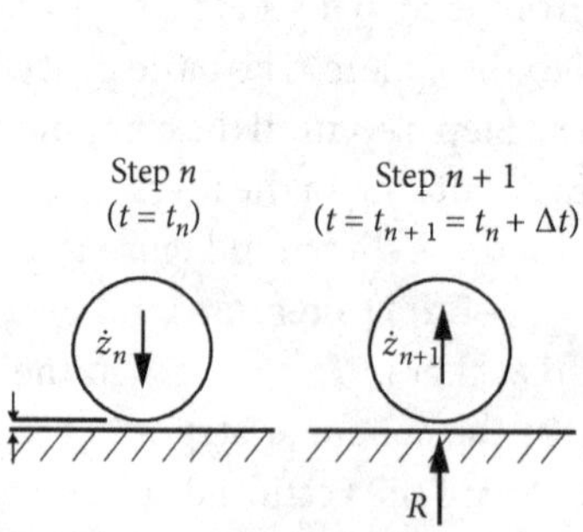

FIGURE 19.9 Time steps of the impact process.

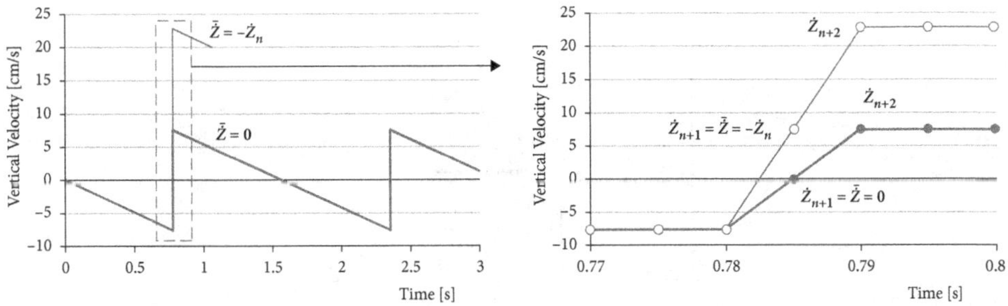

FIGURE 19.10 Two cases of the computed time histories of the vertical velocity of the sphere: thin line $\bar{\dot{z}} = -\dot{z}_n$; thick line $\bar{\dot{z}} = 0$.

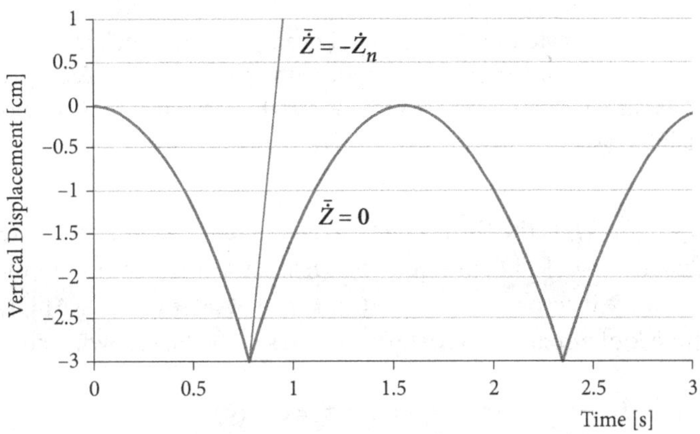

FIGURE 19.11 Two cases of the computed time histories of the vertical displacement of the sphere.

It has also been found that even if the time integration interval Δt is changed, the computed impulse $R\Delta t$ is almost the same. This indicates that the present computational procedure at the impact conserves the momentum of the sphere before and after collision to the ground.

However, the computational procedure at the impact described so far causes an unphysical step-by-step pressure oscillation because the impact force R is applied at a single time instant.

3.2.2. *Subdivision of the Time Step of Impact*

The strategy attempted in the present study to suppress this step-by-step pressure oscillation is to subdivide the single time step of impact into multiple short time steps. The magnitude of impact force during this series of short time steps gradually changes as schematically shown in Figure 19.12. In the present study, the following smooth polynomial function of R_f is used:

$$R_f(t - t_S) = R_f(\tau) = \bar{R}\left(\frac{2}{\Delta t}\right)^6 \{\tau^3(\Delta t^3 - 3\Delta t^2\tau + 3\Delta t\,\tau^2 - \tau^3)\} \tag{8}$$

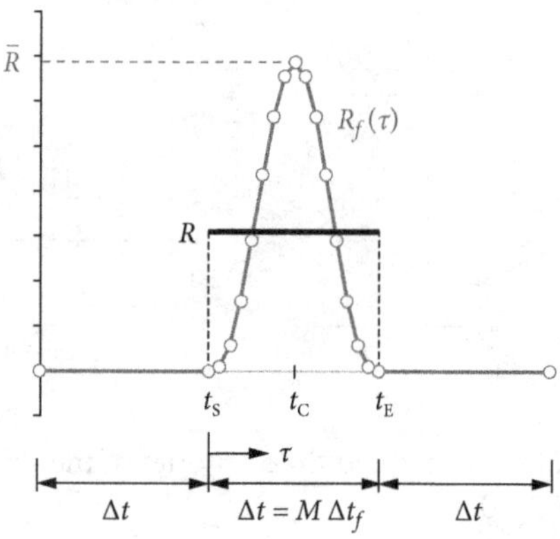

FIGURE 19.12 Schematic view of the function Equation 8 of the applied impact force $R_f(\tau)$ during the multiple short time steps impact procedure (red line). The black line indicates the impact force R of the single-step impact procedure, which can be thought constant during a single time step interval Δt.

where $\tau = t - t_S$, and t_S is the time instant to start the multiple short time step impact procedure. This function $R_f(\tau)$ of Equation 8 is the sixth-order polynomial of τ. Up to the second derivative is zero either at $\tau = 0\left(t = t_S\right)$ or at $= \Delta t\left(t = t_S + \Delta t = t_E\right)$.

The amplitude $\bar{R}$ of Equation 8 is determined to satisfy the following condition:

$$\dot{z}(t=t_C) = \dot{z}_C = \dot{z}_{\bar{C}} = 0 \quad (9)$$

where t_C is the mid time instant $t_c = t_S + \frac{M}{2}\Delta t_f$ of this multiple short time step impact procedure.

Figure 19.13 shows two computed histories of the vertical fluid force. One is the result of the single-step impact procedure. The other is the result of the multiple short time step procedure. The number of subdivision M is 16. The unphysical oscillation of the single-step procedure has been well removed by the multiple short time step procedures.

Figures 19.14a–19.14d show the computed time histories of the vertical velocity and the vertical displacement of the sphere. The magnified velocity history shown in Figure 14b indicates that the vertical velocity of the sphere at the mid time instant t_C during the multiple short time procedures is zero. Then, the sphere ascends to the initial height as shown in Figure 14c.

3.2.3. *Effect of Air*

In the previous computations of Sections 3.2.1 and 3.2.2, after the impact against the ground, the sphere bounces up and ascends to the initial height. These results indicate that the effect of the air surrounding the moving sphere is negligible in these computations in spite of the fluid–structure interaction analysis. In order to examine the effect of the surrounding air on the sphere motion, a computation has been conducted where the density ratio of the sphere to the air is 10.

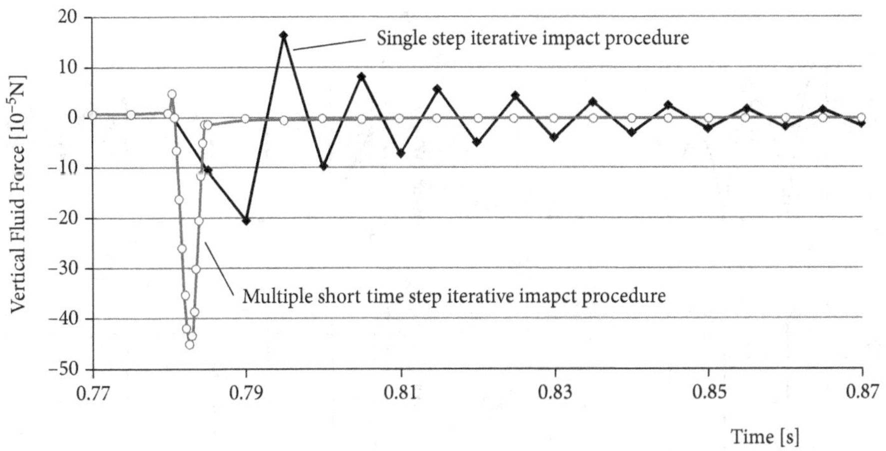

FIGURE 19.13 Computed histories of the vertical fluid force acting on the moving sphere after the first impact at $t = 0.78$s. Black line: the single-step iterative impact procedure. Red line: the multiple short time step iterative impact procedure with the number of subdivisions $M = 16$.

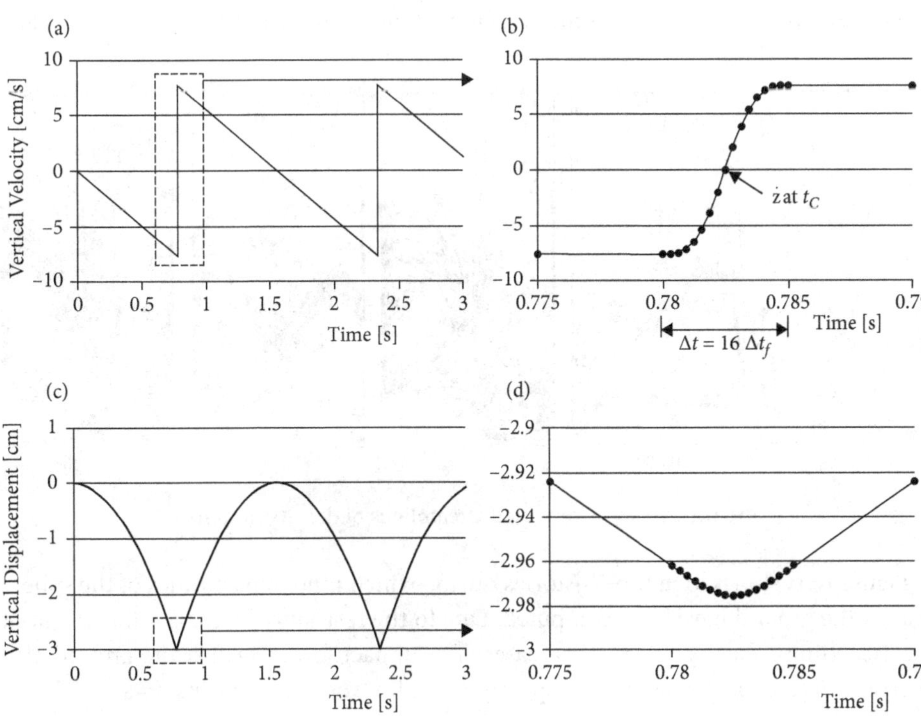

FIGURE 19.14 Computed time histories of the vertical motion of the sphere: (a) the vertical velocity; (b) the vertical velocity during the multiple short time step impact procedure (magnified from the indicated region in Figure 19.14a); (c) the vertical displacement; (d) the vertical displacement during the multiple short time step impact procedure (magnified from the indicated region in Figure 19.14c).

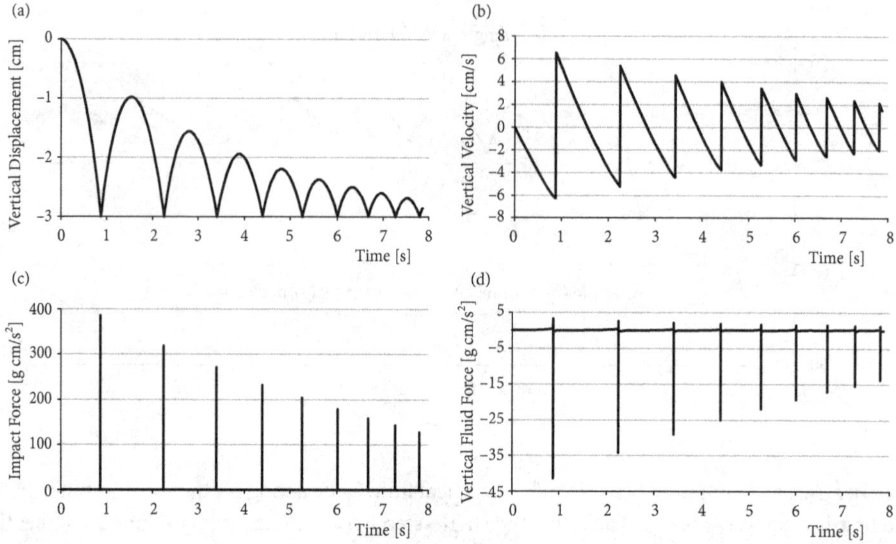

FIGURE 19.15 Computed time histories of density ratio 10: (a) vertical displacement of the sphere; (b) vertical velocity of the sphere; (c) impact force; (d) vertical component of the fluid force.

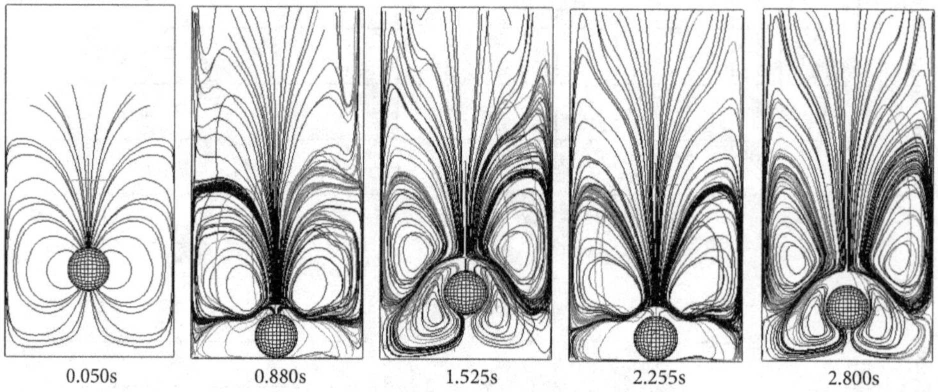

FIGURE 19.16 Computed streamlines of density ratio 10.

Figure 19.15 shows long time histories during which nine impact times of the sphere against the ground have been computed. Due to the resistance by the air, the height of the ascending sphere gradually decreases. The impact forces against the ground also decrease. Figure 19.16 shows the computed streamlines at some time instants.

3.3. Diagonally Projected Sphere in a Box

As shown in Figure 19.17, a sphere is projected diagonally in a rectangular parallelepiped domain. In this case, the four lateral vertical surfaces of this domain are no-slip

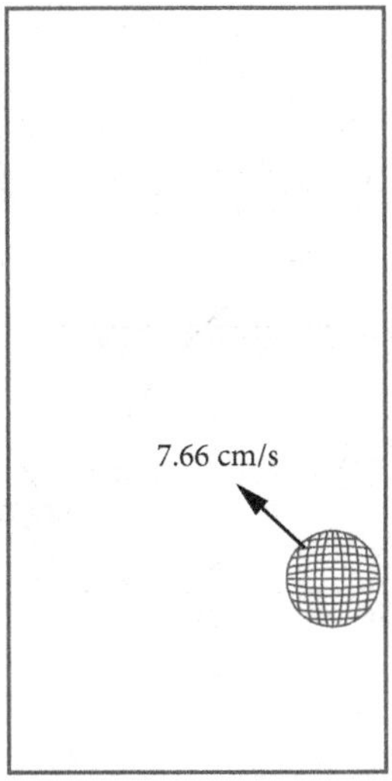

FIGURE 19.17 Diagonally projected sphere in a box.

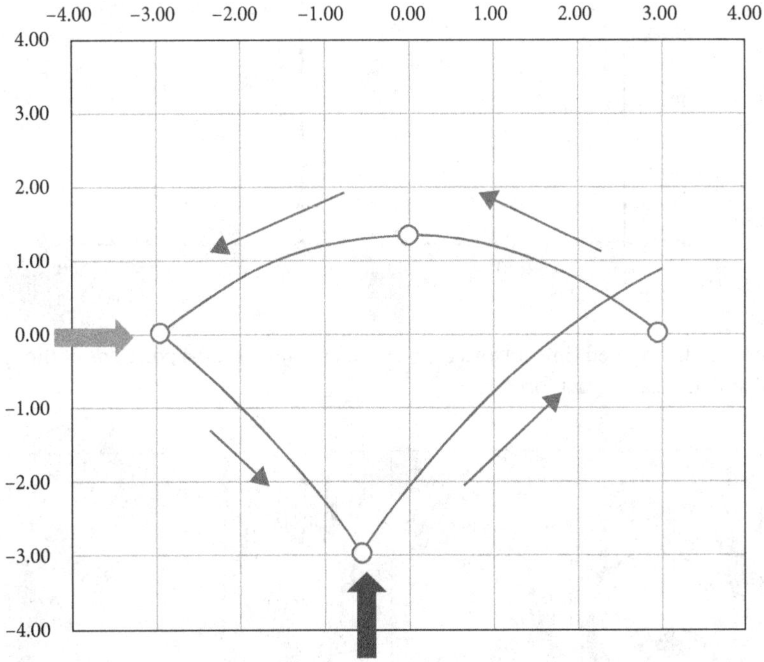

FIGURE 19.18 Computed trajectory of the projected sphere.

boundaries. The density ratio of the sphere to the air is 200. Figure 19.18 shows the computed trajectory of the sphere in the two-dimensional vertical plane. After the sphere collides at the other side of the domain, it falls to the ground and bounces back. Figure 19.19 shows time histories of the sphere's position and the horizontal and the vertical impact forces. Figure 19.20 shows the computed streamlines at some time instants.

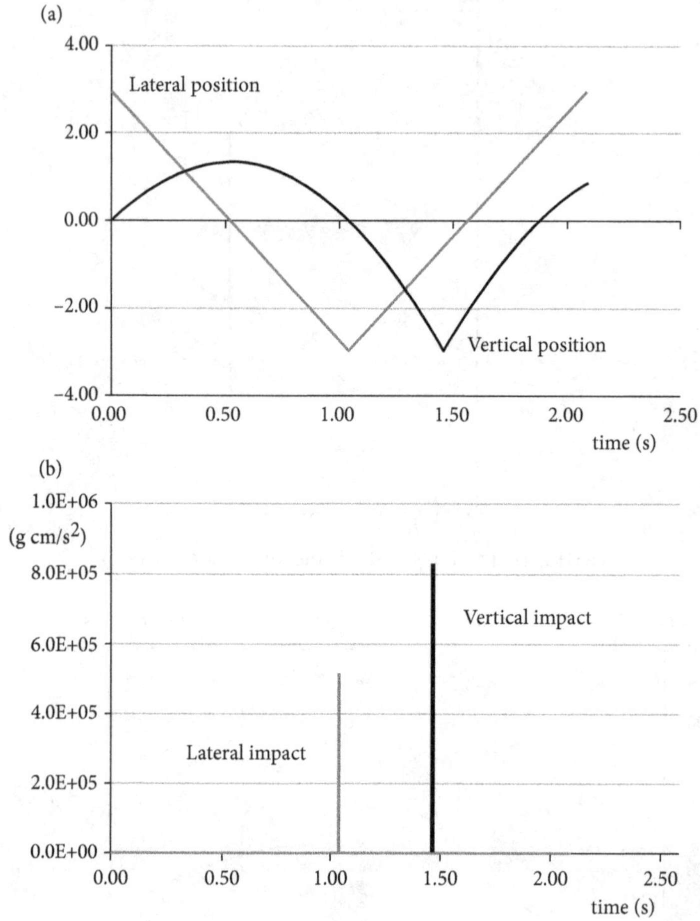

FIGURE 19.19 Computed time histories: (a) the lateral and vertical positions of the sphere; (b) the lateral and vertical impact forces.

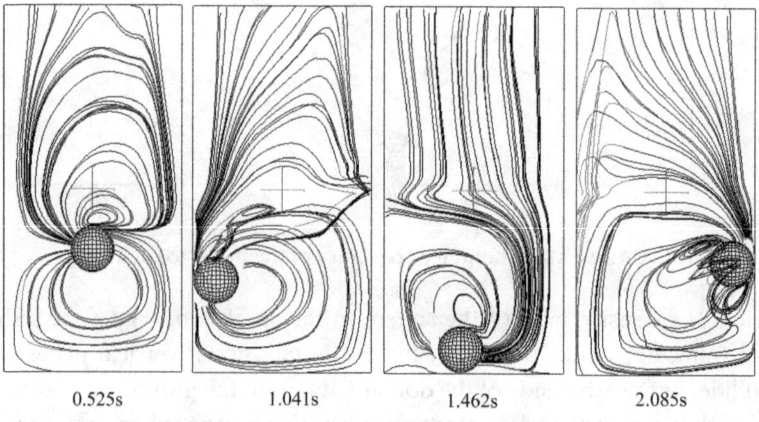

FIGURE 19.20 Computed streamlines.

4. Concluding Remarks

The present chapter has described a numerical method to simulate the flight and impact of windborne debris. The problems are treated as fluid–structure interaction problems and based on the ALE finite element flow analysis.

Two subjects have been described in the present article: (1) flight simulation of a projected rigid sphere using a moving mesh attached to the sphere; and (2) the computational method of impact when a sphere hits the ground or solid walls.

The first subject, the moving mesh attached to the sphere, was investigated because generally flying debris flies a long distance in comparison to its body length. The moving mesh is assumed wide and fine enough to evaluate instantaneous aerodynamic forces exerted on the debris. The direction of the moving mesh is always aligned to that of the relative velocity of the flying sphere to the wind. Several numerical examples have shown that aerodynamic forces are appropriately introduced into the fluid–structure interaction system.

The second subject, the impact procedure, is a unique one because this is different from usual methods used in numerical contact mechanics as the Lagrangian multiplier method or the penalty method. The impact force is determined by iteration so that the sphere velocity after impact reaches a prescribed target velocity. The investigation has shown that the appropriate target velocity is zero. In addition to this, the time integration step at the impact is subdivided into multiple short time intervals, and the impact force is expressed by a smooth function of time in order to suppress unphysical step-by-step pressure oscillation. Several numerical examples have been shown to indicate that the present impact procedure satisfies the energy conservation law as well as the momentum conservation law.

So far, these two numerical methods are separately applied, and the flying object is a sphere only. As future work, these two methods will be combined and applied to an arbitrary shape of windborne flying debris.

Acknowledgments

The research work described here is supported by Grants-in-Aid for Scientific Research, No. 18H01527, from the Japan Society for the Promotion of Science. I am very grateful for this support.

References

Bazilevs, Y., Calo, V. M., Cottrell, J. A., Hughes, T. J. R, Reali, A., & Scovazzi, G. (2007). Variational multiscale residual-based turbulence modeling for large eddy simulation of incompressible flows. *Computer Methods in Applied Mechanics and Engineering*, *197*, 173–201.

Bazilevs, Y., Takizawa, K., & Tezduyar, T. E. (2013). *Computational fluid-structure interaction*. New York, NY: Wiley.

Belytscho, T., Flanagan, D. P., & Kennedy, J. M. (1982). Finite element methods with user-controlled meshes for fluid-structure interaction. *Computer Methods in Applied Mechanics and Engineering, 33*, 669–688.

Donea, J., Giuliani, S., & Halleux, J. P. (1982). An arbitrary Lagrangian-Eulerian finite element method for transient dynamic fluid-structure interactions. *Computer Methods in Applied Mechanics and Engineering, 33*, 689–723.

Hirt, C. W., Amsden, A. A., & Cook, H. K. (1974). An arbitrary Lagrangian Eulerian computing method for all flow speeds. *Journal of Computational Physics, 14*, 227–253.

Holmes, J. D. (2001). *Wind loading of structures*. London, UK: Spon Press.

Hughes, T. J. R., Liu, W. K., & Zimmerman, T. K. (1981). Lagrangian-Eulerian finite element formulation for incompressible viscous flows. *Computer Methods in Applied Mechanics and Engineering, 29*, 329–349.

Kawahara, M. (2016). *Finite element methods in incompressible, adiabatic, and compressible flows—From fundamental concepts to applications*. New York, NY: Springer, Chapter 11.

Maruyama, T. (2011). Simulation of flying debris using a numerically generated tornado-like vortex. *Journal of Wind Engineering and Industrial Aerodynamics, 99*, 249–256.

Noda, M., Kazuhito M., MergumiÂ N., &Â Nagao F. (2014). *Behavior of flying debris in tornado-like flow*. Proceedings of 6th International Symposium on Computational Wind Engineering, 1–8, Hamburg, Germany.

Nomura, T. (1994). ALE finite element computations of fluid-structure interaction problems. *Computer Methods in Applied Mechanics and Engineering, 112*, 291–308.

Nomura, T. (2018). Numerical simulation of impact forces by a falling sphere in air using ALE finite element method. *International Journal of Computational Fluid Dynamics, 32*, 131–145.

Nomura, T., & Hughes, T. J. R. (1992). An arbitrary Lagrangian-Eulerian finite element method for interaction of fluid and a rigid body. *Computer Methods in Applied Mechanics and Engineering, 95*, 115–138.

Nomura, T., & Shinmyo, M. (1999). ALE finite element analysis of flow interaction with a flexible filament. *International Journal of Computational Fluid Dynamics, 13*, 65–82.

Tachikawa, M. (1983). Trajectories of flat plates in uniform flow with application to wind-generated missiles. *Journal of Wind Engineering and Industrial Aerodynamics, 14*(1–3), 443–453.

Takeuchi, T., Maeda, J., & Kawashita, H. (2008, July). *The overshoot of aerodynamic forces on a railcar-like body under step-function-like gusty winds*. Proceedings of 6th International Colloquium on Bluff Bodies Aerodynamics & Applications, Milan, Italy.

Tezduyar, T. E., Behr, M., & Liou, J. (1992). A new strategy for finite element computations involving moving boundaries and interfaces—The deforming-spatial-domain/space-time procedure: I. The concept and the preliminary numerical tests. *Computer Methods in Applied Mechanics and Engineering, 94*, 339–351.

Wriggers, P. (2006). *Computational contact mechanics* (2nd ed.). New York, NY: Springer.

SECTION 5

RISK MODELS RELATED TO NSWS

CHAPTER 20

SIMPLIFIED HAZARD MODELING AND STRUCTURAL RELIABILITY ANALYSIS CONSIDERING NON-SYNOPTIC WIND SYSTEMS (NSWS) IN CANADA

HAN PING HONG AND QIAN HUANG

1. Introduction

High-intensity wind events such as tornadoes and downbursts can be very destructive to structures and infrastructure systems. A tornado is a rapidly rotating column of air extending from the base of a thunderstorm down to the ground. A downburst is defined as "a strong downdraft which induces an outburst of damaging winds on or near the ground" (Fujita, 1985). Statistics indicate that about 100 tornadoes per year occur in Canada. An incomplete tornado catalogue is available in Canada; a historical catalogue for the downburst does not exist. Both tornadoes and downbursts cause significant property damage and fatalities. According to Etkin, Brun, Shabbar, and Joe (2001), the notes compiled by M. J. Newark indicated that the first report of a tornado in Canada, which occurred on July 1, 1792, leveled all houses in its path and uprooted trees along a narrow track. The first systematic Canadian tornado catalogue was established by Newark (1984), indicating that "much of Canada east of the large Manitoba lakes is an extension of the great tornado-prone region centered over the south-central United States" (p. 352). The database was subsequently updated by Sills, Scriver, and King (2004) for Ontario. A tornado catalogue for the Prairie and Northern Regions in Canada from 1826 to 2010 was compiled by Patrick McCarthy (September 17, 2012, personal communication). Additional information on tornado occurrences in Canada is described on the Environment and Climate Change Canada (ECCC) website (https://open.canada.ca/data/en/dataset/f314a39f-009d-430b-97b9-d6e0cae22340) and in Wikipedia (https://en.wikipedia.org/wiki/List_of_Canadian_tornadoes_and_tornado_outbreaks#%20cite_ref-ehfs_1-0; and https://en.wikipedia.org/wiki/List_of_21st-century_Canadian_tornadoes_and_tornado_outbreaks). The path orientation, length, and width for each listed tornado event is not always available.

A qualitative tornado hazard assessment for Ontario was presented by Newark (1984) using the established database, indicating that, on average, there is an F3 tornado every 5 years in southwestern Ontario. Based on very simplistic and practical assumptions, Newark (1991) suggested designing wind velocity pressure by considering the design basis for tornadoes for different regions. More sophisticated modeling and assessment for the tornadic wind hazard for southern Ontario were carried out by Banik, Hong, and Kopp (2007). Their estimate focused on the estimation of the tornado occurrence rate, tornado intensity distribution, probabilistic models of the tornado path length and path width, and distribution of the annual maximum tornado wind velocity at a site within southern Ontario. It was considered that the probabilistic tornado wind field given in the work of Twisdale (1978) and Twisdale, Dunn, Horie, and Davis (1981) can be used adequately to estimate a tornadic wind hazard, and that the spatial inhomogeneity of tornado occurrence near a site of interest can be ignored. They concluded that, at 10-m height, the tornado wind does not dominate a wind hazard for specifying the design wind load for the normal building category defined in the National Building Code of Canada (NBCC) (National Research Council of Canada [NRCC], 2015). However, the tornado winds dominate the wind hazard for the design of very high consequence structures (e.g., nuclear power plants) or transmission lines extended over a few tens of kilometers. Similar conclusions were given by Tan and Hong (2010) by considering the inhomogeneity of tornado occurrence near the site of interest.

In addition, it was observed that the consideration of uncertainty in tornado occurrence rate through the application of the hierarchical Bayesian modeling technique (Wikle & Anderson, 2003) only negligibly affects the estimated tornadic wind hazard for southern Ontario (Tan, 2008). The use of the Bayesian hierarchical model, including the consideration of the impact of local population density on observed tornado occurrences and the predictive relationship between the cloud-to-ground (CG) lightning flash, were considered by Cheng et al. (2013) in mapping the tornado occurrence for Canada. Cheng, Arhonditsis, Sills, Gough, and Auld (2016) further applied such an approach to model the tornado occurrence in North America. However, the probabilistic tornadic wind-velocity hazard map for Canada is unavailable.

One of the possible weaknesses of the tornadic wind hazard assessment presented by Banik et al. (2007) and Tan and Hong (2010) is that the tornadic wind hazard assessment was presented only in terms of wind velocity at a 10-m height. The use of the wind hazard expressed in terms of a 10-m height wind velocity for synoptic winds in structural design codes is a well-accepted practice since the along height wind profile of synoptic winds is relatively consistent and can be represented by a log wind profile or a power law wind profile (NRCC, 2015). However, this is not the case for tornadoes (Hangan & Kim, 2008; Natarajan & Hangan, 2012; Refan & Hangan, 2016; Twisdale et al., 1981; Wurman & Gill, 2000). Therefore, the tornadic wind hazard maps expressed in terms of a 10-m height wind velocity or of the tornado occurrence rate are not very practical for engineers to design new and evaluate existing structures and infrastructure systems. For such a tornadic wind velocity hazard map to be useful for Canada, additional investigation and recommendations on how to codify the tornadic wind profile can be valuable for structural analysis and design.

Unlike the case for tornadoes, a historical downburst catalogue is not available. The lack of catalogue is understandable as a systematic investigation of the downburst was only started in 1985 by Fujita (1985). The downburst wind field models are available in the literature (Holmes & Oliver, 2000; Kim & Hangan, 2007; Oseguera & Bowles, 1988; Sengupta, Haan, Sarkar, & Balaramudu, 2008). Perhaps one may assume a ratio of the tornado occurrence rate to downburst occurrence rate and carry out probabilistic downburst wind velocity hazard mapping to gain some perspectives on the downburst hazard.

In the context of structural reliability analysis and structural design code calibration by considering the natural loads, one effective method is to characterize the structural capacity probabilistically and the load separately (Ellingwood, Galambos, MacGregor, & Cornell, 1980; Madsen, Krenk, & Lind, 2006). The probabilistic characterization of the structural capacity curve can be used to evaluate the exceedance probability of a given performance level conditioned on a specified load measure, which is known as the fragility curve. For example, analyses of transmission tower line systems subjected to downburst wind load can be carried out by nonlinear static or nonlinear increment dynamic load evaluation (Banik, Hong, & Kopp, 2010; Mara & Hong, 2013; Mara, Hong, Lee, & Ho, 2016; Yang & Hong 2016). A review of the tower line system subjected to wind loading was given by Aboshosha, Elawady, El Ansary, and El Damatty (2016) and Abd-Elaal, Mills, and Ma (2018). The combined use of the capacity curve (or surface) and the spatially varying wind hazard was presented by Yang, Liu, and Hong (2017), but only for synoptic winds. The modeling and analysis of a tower subjected to tornado wind load were shown by Savory, Parke, Zeinoddini, Toy, and Disney (2001) and Hamada and El Damatty (2011).

In general, the tornado wind field is more complicated than the downburst wind field, the development of the capacity surface of the tower line system subjected to three-dimensional (3D) stochastic tornado wind field is challenging. The wind load on buildings due to microburst and tornado winds was investigated by Haan, Balaramudu, and Sarkar (2010) and Sengupta et al. (2008), and tornado risk analysis for the residential wood-frame roof was presented by Standohar-Alfano and van de Lindt (2016). However, such an analysis is currently unavailable for Canadian residential building stock, although valuable tornado damage surveys and damage analyses were presented by Kopp, Hong, Gavanski, Stedman, and Sills (2017).

The main objectives of the present study, which focuses on Canadian sites, are to (a) discuss available tornado occurrence information for Canada that can be used as the basis to develop a tornado occurrence rate model; (b) present an overall framework for developing a probabilistic tornadic wind velocity hazard map for Canada; (c) explore possible equivalent along height wind profile that could be used to evaluate tornadic wind loading; and (d) highlight a simple and practical procedure to evaluate the failure probability of structures subjected to high-intensity wind events.

2. Modeling of the Spatial Occurrence of High-Intensity Wind Events

2.1. Tornado Catalogue for Canada

As mentioned in the Introduction, Newark (1984) established the first systematic Canadian tornado catalogue for events that occurred from 1950 to 1979. The data were gathered from a variety of sources, including newspapers, newspaper clipping services, local community histories, field surveys of tornado damage, Atmospheric Environment Service (AES) (which became the ECCC) archive material; AES weather observations; data collected by provincial organizations; public archive material; and human memory. A threshold population density equal to 1 person/km^2 was employed to limit the study area within Canada for the assessment of tornado occurrence rate since the lack of observations is likely to reduce the confidence on the obtained statistics of the tornado occurrence rate. The gathered information was used not only to assess the tornado occurrence rat, but also to assess the tornado length, width, and area. It was indicated that the peak activity of the tornadoes occurred within 15:00 to 17:00 local standard time and that southwestern Ontario and southeastern Manitoba had the maximum tornado activity within Canada.

The shortcomings of Newark's (1984) tornado catalogue include the population bias, rare events bias, and inadequate treatment of the tornado path length. An update to the tornado catalogue for the province of Ontario was presented in Sills et al. (2004). A new tornado catalogue for Ontario was given, and in all cases, an event was classified in categories as confirmed, probable, or possible. The catalogue covers tornado events from 1918 to 1992, each with its characteristics, such as the tornado classification, Fujita Scale (i.e., F Scale), touchdown latitude–longitude, length, and motion direction when available.

An effort was made, within our research group, to update the tornado catalogue from different sources. It was noted that the tornado catalogue for the Prairie and Northern

Regions in Canada from 1826 to 2010 was compiled by Patrick McCarthy (September 17, 2012, personal communication) and that sites within Wikipedia (see Introduction to this chapter) provide a very detailed account of tornadoes that occurred in Canada in chronological order. Moreover, it was noted that further updating of the tornado catalogue from 1980 to 2009 was released by the ECCC (see https://open.canada.ca/data/en/dataset/fd3355a7-ae34-4df7-b477-07306182db69). In this catalogue, there are 1,839 tornado events that occurred in Canada from 1980 to 2009. By combining the four sources of tornado catalogue mentioned (e.g., from P. McCarthy, September 17, 2012, personal communication; Sills et al., 2004; Wikipedia websites mentioned earlier), a tornado catalogue was developed for the present study. It should be noted that the reporting of the tornado intensity in Canada was changed from F Scale to Enhanced Fujita Scale (EF Scale) (Marshall, McDonald, & Forbes, 2004) since 2013. Although the catalogue contains events as recent as 2019, only events up to 2014 were considered since our scrutiny of the data, at present, was carried out for such a date. Moreover, the tornado intensities reported in 2013 and 2014 were categorized as tornadoes using the F Scale.

Events within the newly developed catalogue from 1826 to 2014 are depicted in Figure 1a. In presenting this plot, no distinction was made in the reported tornado intensities in terms of F Scale and EF Scale, noting that tornadoes in 2013 and 2014 were reported in EF Scale. By considering events from 1980 to 2014, the occurrence rate obtained by applying Gaussian kernel smoothing, which was used by Tan and Hong (2010), is shown in Figure 20.1b. For the smoothing and plotting, no correction was made for possible bias in reporting. The figure indicates that the occurrence rate for southwestern Ontario and southeastern Manitoba is the highest. The highest occurrence rate is about 6.0×10^{-5} (per [year × km^2]). For the plotted occurrence rate, no reporting bias correction is considered.

A more detailed plot of the spatial distribution of historical tornado events and the estimated tornado occurrence rate for southern Ontario are shown in Figure 20.2 by using the catalogue from 1980 to 2014. The figure indicates that the maximum occurrence rate was equal to 3.6×10^{-4} (per (year × km^2), and that the occurrence rate for the area between London and Toronto was higher than for other areas. It is noteworthy that the tornado occurrence rate was not equal to the tornado striking rate for a site of interest; the evaluation of the tornado striking rate for a site requires the consideration of tornado path direction, width, and length (Banik et al., 2007; Tan & Hong, 2010).

2.2. Stochastic Occurrence Modeling

The tornado occurrence in time and space is uncertain. It could be modeled by using well-known point processes such as the Poisson process, the zero-inflated Poisson process, and the negative binomial model (Daley & Vere-Jones, 2007; Lambert, 1992). A stochastic process $\{N(t), t \geq 0\}$ follows the Poisson model, if

$$p\big(N(t)=n\big)=\frac{(\lambda t)^n e^{-\lambda t}}{n!} \tag{1}$$

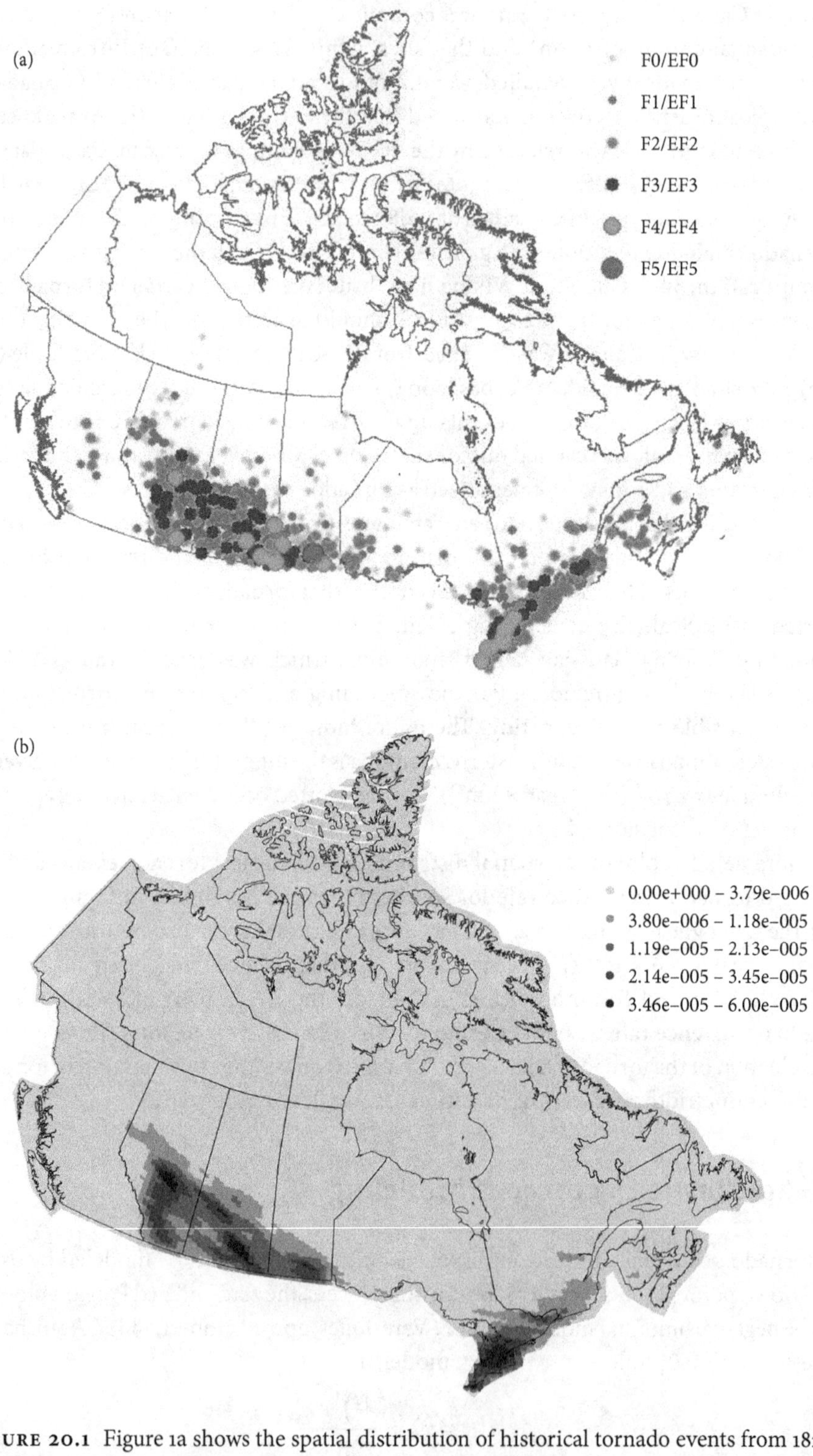

FIGURE 20.1 Figure 1a shows the spatial distribution of historical tornado events from 1826 to 2014, and Figure 1b represents the tornado occurrence rate per (year × km^2) considering the catalogue from 1980 to 2014.

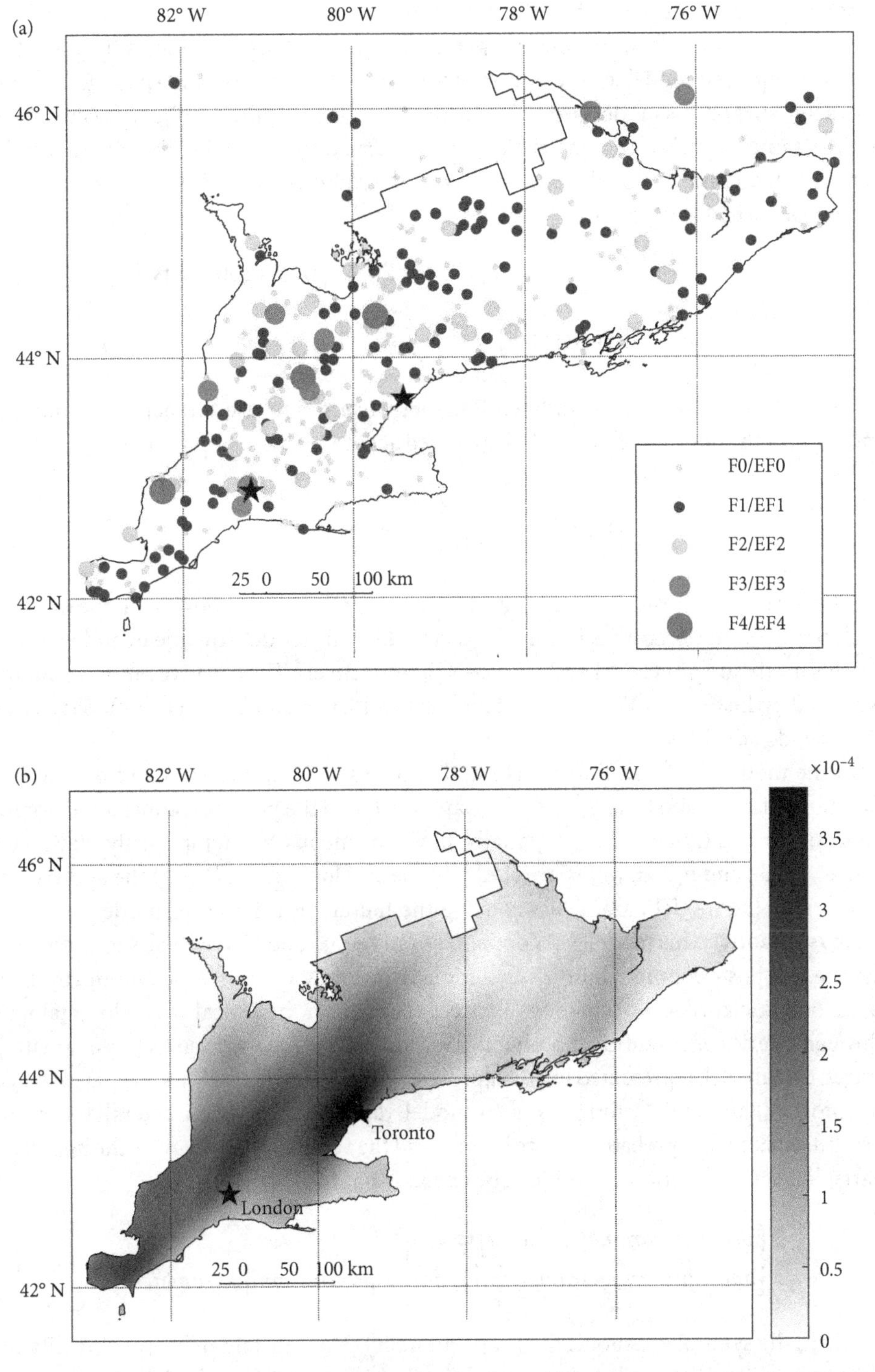

FIGURE 20.2 Figure 2a shows the spatial distribution of historical tornado events from 1980 to 2014. Figure 2b represents the tornado occurrence rate per (year × km²) considering the catalogue from Figure 2a.

where $p(\,)$ denotes the probability mass function, n is the number of events, and λ is the occurrence rate per unit time and per unit area. The Poisson model is the most frequently employed model for the occurrence of rare events in time and space. One of the characteristics of this model is that the mean and variance of the number of occurrences are the same. If such a characteristic is not observed from the data, one may use the zero-inflated Poisson model, which was considered by Wikle and Anderson (2003) to model the tornado occurrence,

$$\{N(t),t\geq 0\}\sim\begin{cases} 0, & \text{with probability } p_0 \\ p\left(N(t)=n\right)=\dfrac{(\lambda t)^n e^{-\lambda t}}{n!}, & \text{with probability } (1\text{-}p_0) \end{cases}, \tag{2}$$

An alternative to the zero-inflated Poisson process is to use the negative binomial model with the number of events N distributed according to

$$p\left(N=n\right)=\frac{\Gamma(n+r)}{n!\Gamma(r)}p_{NB}^{n}\left(1-p_{NB}\right)^{r}, \tag{3}$$

where r is a real positive number, and p_{NB} is the probability of observing the event in each trial. The stationarity is implicitly assumed in this model. The mean and variance for this model are $pr/(1-p)$ and $pr/(1-p)^2$, respectively. The negative binomial model was used by Elsner and Widen (2014) and Jagger, Elsner, and Widen (2015) to describe the tornado activities.

In the model shown in Equations 1 and 2, the model parameters λ and p_o could be functions of space and time. λ, p_o, and p_{NB} are constant if the process is considered homogeneous and stationary. For a spatially inhomogeneous but temporarily stationary process, λ, p_o, and p_{NB} vary in space (i.e., latitude and longitude). One of the approaches to estimate the model parameters is using the hierarchical Bayesian model (Wikle & Anderson, 2003). This model was adopted by Tan (2008) and Cheng et al. (2013, 2016). It assesses the posterior probability distribution of the tornado occurrence rate and updates the stochastic tornado occurrence process based on a historical tornado catalogue through a series of conditional probabilistic models. It can be written as three updating stages: data model, represented by [data|process, parameters]; process model, represented by [process|parameters]; and parameter model, represented by [parameters], where the brackets refer to the "probability distribution" and the vertical line | separates the uncertain variable(s) and condition(s). This can be expressed as

$$\begin{aligned}&\left[process;\ parameters\,|\,data\right]\propto\\&\left[data\,|\,process;\ parameters\right]\left[process\,|\ parameters\right]\left[parameters\right]\end{aligned} \tag{4}$$

where the symbol $\propto$ denotes the proportionality. The constant of proportionality for the posterior probability distribution is difficult to obtain analytically, if not impossible. The use of numerical integration methods for its evaluation can be computationally time consuming, which can be overcome using the Markov chain Monte Carlo (MCMC) sampling approaches (Gilks, Richardson, & Spiegelhalter, 1996).

The formulation given in Equation 4 can include covariates. For example, one may include the population density and the CG lightning flash as covariates, as was done by Cheng et al. (2016). Example plots of the population density and the CG lightning flash are shown in Figure 20.3. A comparison of the plots shown in Figure 20.1 and those shown in Figure 20.3 indicates there are relations between the observed tornado occurrence and population density and between the observed tornado occurrence and the CG lightning flash density. Our preliminary analysis indicated that the use of the maximum likelihood method or the hierarchical Bayesian method provides a consistent estimated spatially varying occurrence rate, provided the model used to represent the spatially varying tornado occurrence model is adequate. Further analysis and detailed discussion of the application of the hierarchical Bayesian model is beyond the scope of the present study, and the reader is referred to the work of Wikle and Anderson (2003).

3. Simulation-Based Tornadic Wind-Velocity Hazard Mapping and Equivalent Wind Profile

3.1. Wind Field Model and Wind Hazard Estimation

A tornado wind field model needs to be considered for mapping the tornado wind hazard in terms of the wind velocity. A review of earlier studies on the tornado wind field was given by Twisdale and others (Twisdale, 1978; Twisdale et al., 1981); they also proposed a probabilistic wind field model for tornadoes. Other wind field models discussed by Wurman and Gill (2000), Natarajan and Hangan (2012), and Refan and Hangan (2016) may also be used for probabilistic tornado wind hazard modeling if the uncertainty characterizations of the model parameters for these wind field models become available.

By following the tornado hazard modeling procedure used by Tan and Hong (2010), which adopted the model suggested by Twisdale et al. (1981), and the tornado path length, width, direction, and intensity variation used by Banik et al. (2007), the assessment of the tornadic wind-velocity hazard mapping is carried out in this section for southern Ontario. For the assessment, the updated tornado occurrence rate shown in Figure 20.2 is considered.

The wind field model given by Twisdale et al. (1981) is a steady-state model and is illustrated schematically in Figure 20.4. The model depends on the ratio of the radial velocity to tangential velocity, the radius to the maximum tangential velocity at a reference height, the linear variation of the reference radius with height, the reference boundary

(a)

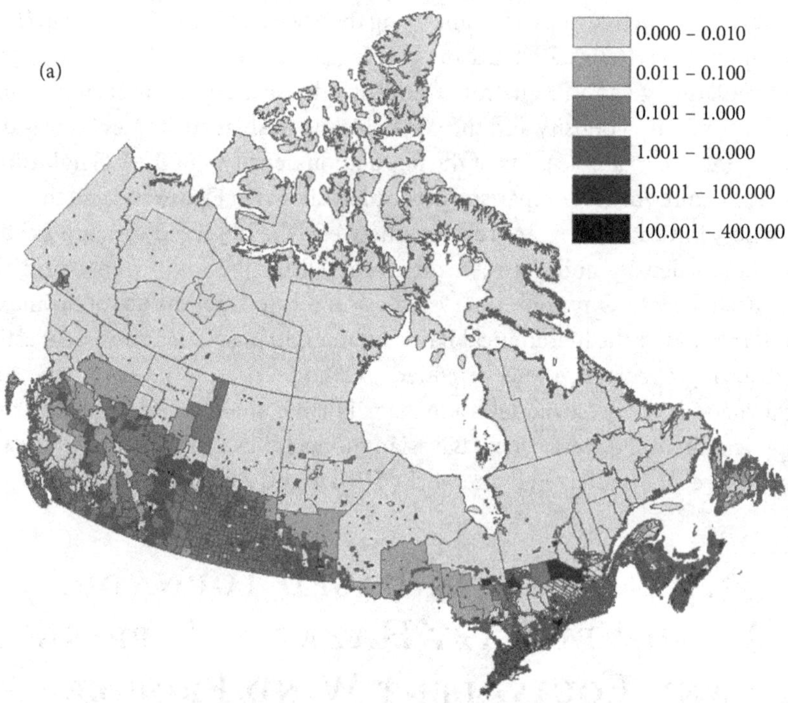

(b)

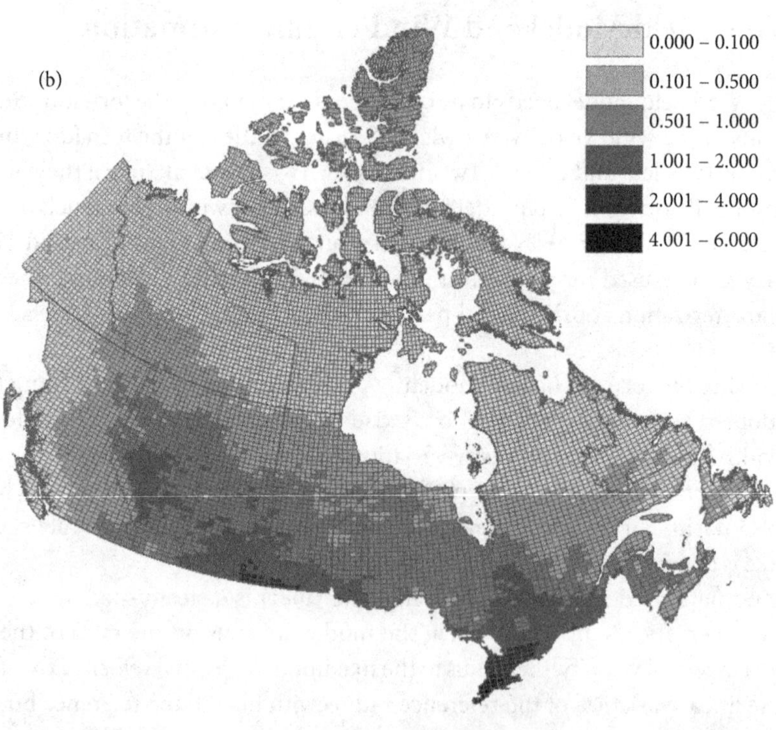

FIGURE 20.3 Figure 3a shows the spatial variability of the population density (number of people per km²). Figure 3b represents the cloud-to-ground (CG) lightning flash (per [year per km²]).

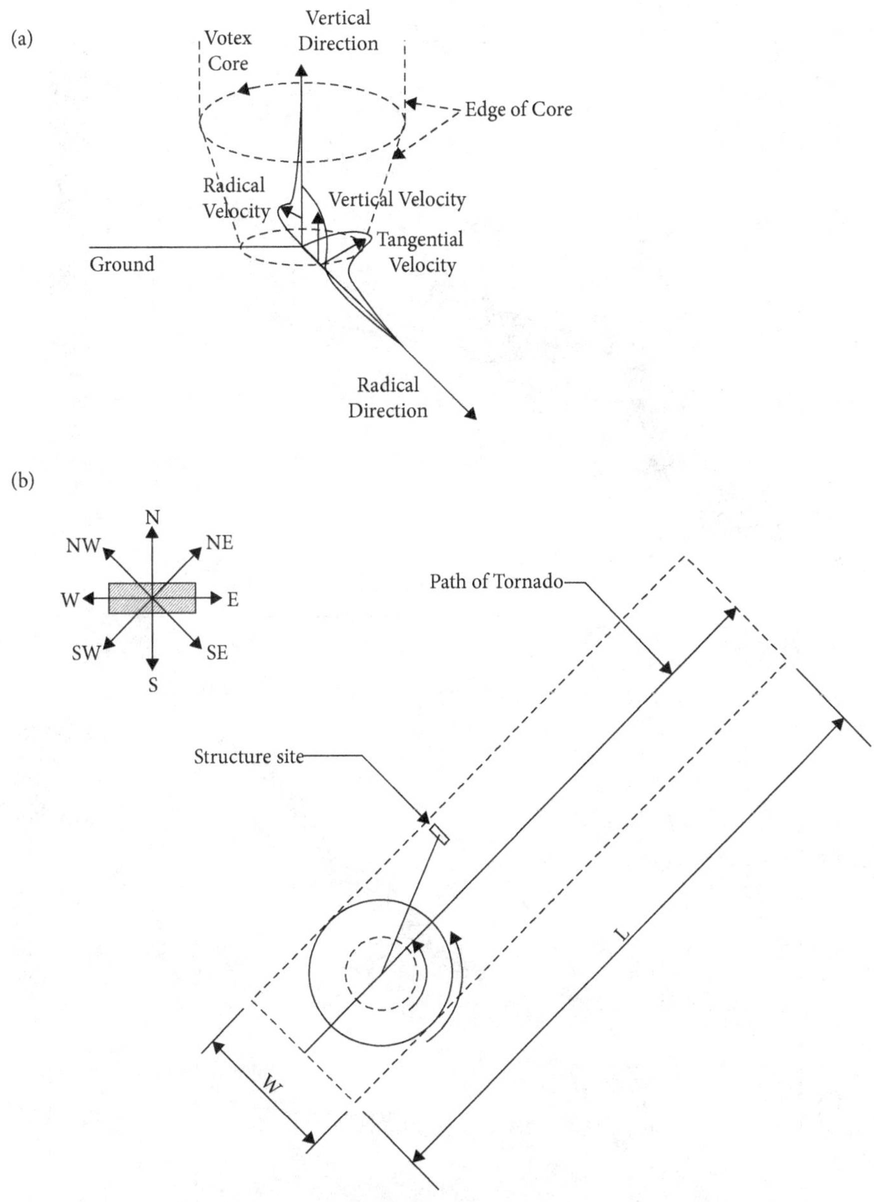

FIGURE 20.4 Figure 4a illustrates schematically the tornado wind field, and Figure 4b shows the tornado path information (W and L represents the path length and path width, respectively).

layer thickness, the reference rotational velocity, the translational speed, and the tornado width. The probabilistic model parameters account for the natural variability observed among tornadoes, including the tornado intensity, path length and width, translational speed, core radius, ratio of radial to tangential wind velocity components, core radius, and boundary layer thickness. For a horizontal plane, the tangential velocity $U_{è}$ and the

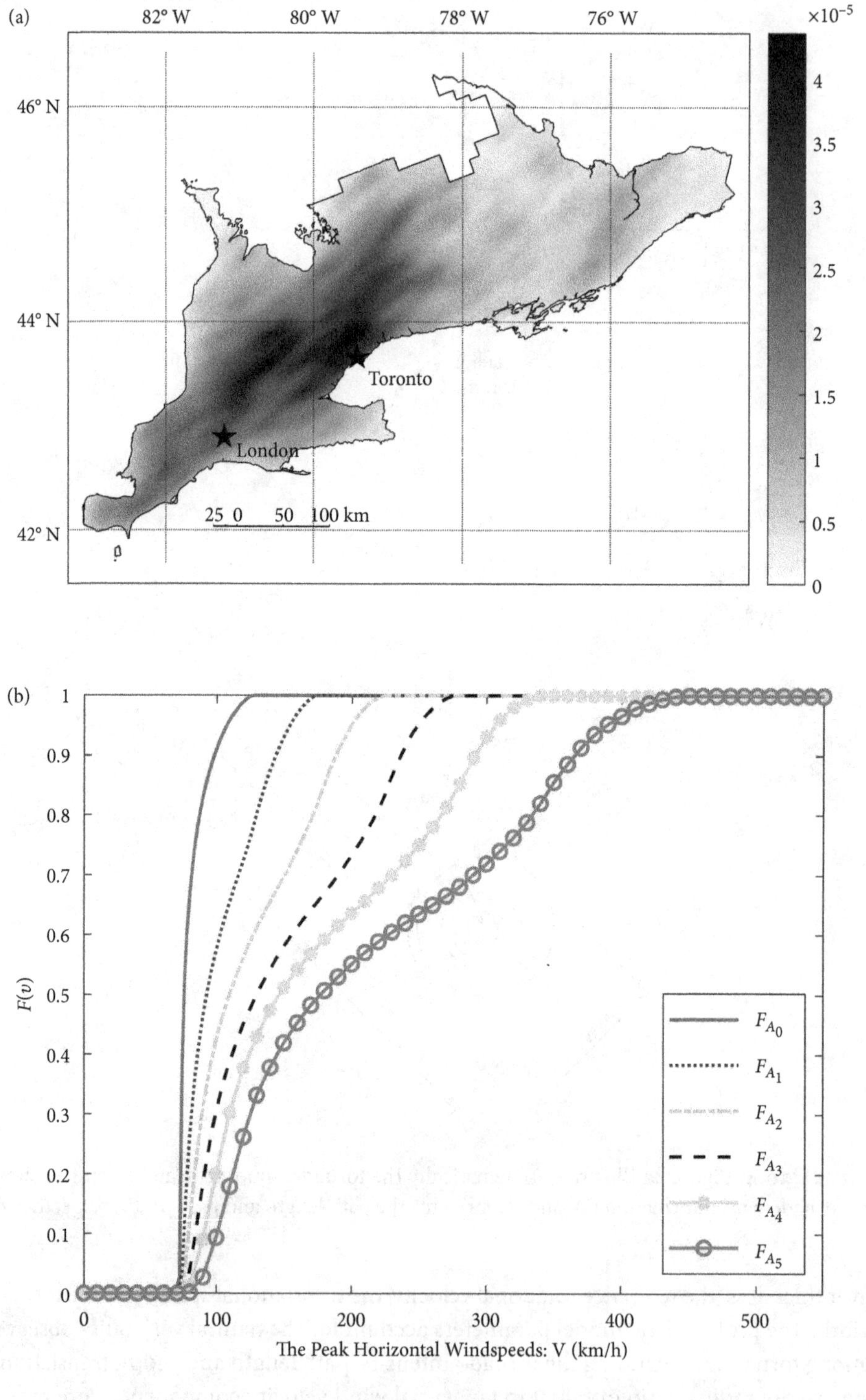

FIGURE 20.5 Figure 5a shows the estimated striking rate γ_{Aj} for southern Ontario (per year × km²), and Figure 5b depicts the estimated $F\left(V < v \mid F_{Ai}\right)$.

radial velocity U_r can be calculated as a function of the radial distance n and height z. The calculated wind field includes the effect of the translation velocity U_T.

It is well known that the intensity of a tornado varies along the tornado path length. Only a portion of the recorded tornado length for a reported tornado with F Scale F_i, $i = 0, \ldots, 5$, is actually equal to the given intensity classification. Let F_{Aj} denote the actual striking intensity. The probability distribution of the percentage of path length where the intensity equals F_{Aj}, $j = 0, \ldots, i$, given that the reported intensity equals F_i, was given by Twisdale et al. (1981). Therefore, a convenient approach to assess the tornadic wind-velocity hazard is to

1. Estimate the striking rate at a site for F_{Aj}, γ_{Aj}, based on the occurrence rate, such as those shown in Figures 20.1 and 20.2;
2. Assess the probability distribution function of the maximum horizontal wind velocity at a selected height given the site is stricken by a tornado with F_{Aj}, $F\left(V < v \mid F_{Ai}\right)$; and
3. Combine γ_{Aj} and $F\left(V < v \mid F_{Aj}\right)$ to assess the probability distribution of the annual maximum tornadic wind velocity, $P\left(V < v\right)$.

Although the γ_{Aj} for southern Ontario were all calculated based on the simulation procedure for $j = 0, \ldots, 5$, only γ_{A3} is shown in Figure 20.5a since γ_{Aj} for different j values exhibits similar spatial patterns. The obtained pattern is consistent with that obtained by Tan and Hong (2010), which is based on a shorter historical tornado catalogue (from 1950 to 1992). The estimated $F\left(V < v \mid F_{Aj}\right)$ based on the simulation technique is shown in Figure 5b as well, where the wind velocity represents a 3-s gust wind velocity at a 10-m height. Finally, the combined use of γ_{Aj} and $F\left(V < v \mid F_{Aj}\right)$ results in the tornadic wind-velocity hazard map for southern Ontario that is shown in Figure 20.6 for a return period value of 10^5 years.

3.2. Equivalent Wind Profile for Tornado Wind Loading

The tornadic wind-hazard map for a specified return period, such as the one shown in Figure 20.6, provides an overall picture of the extreme tornadic wind velocity that a structural element at a 10-m height is likely to experience at a site. However, such a map alone does not facilitate the design checking or evaluation of existing structures since along height wind profile or the tornadic wind load effects are unknown.

To facilitate the use of a tornadic wind-velocity hazard map, consider Toronto as the site of interest. The evaluation of the return period value of the annual maximum tornadic wind velocity at a height ranging from 0.1 to 60 m is carried out by following the same procedure used in the previous section. The obtained T-year return period values of the annual maximum tornadic wind velocity at a height z m $V_{Tornado,T}(z)$ normalized with respect to $V_{Tornado,T}(10)$ are shown in Figure 20.7 for a few selected return periods. Also shown is the wind profile for synoptic wind velocity defined by the power law with

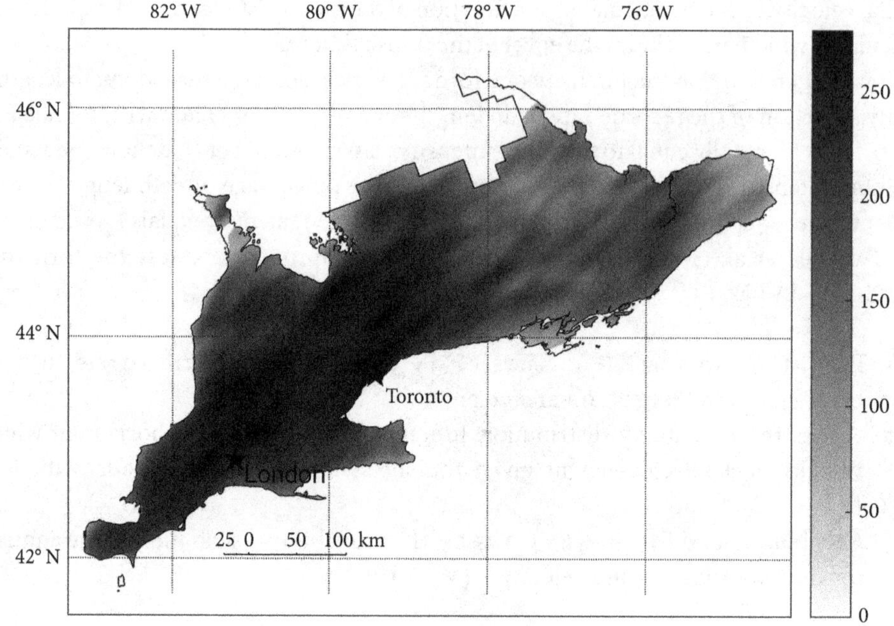

FIGURE 20.6 Tornado wind-velocity hazard map for a return period of 10^5 years (the wind velocity [km/h] is given in terms of a 3-s gust wind velocity).

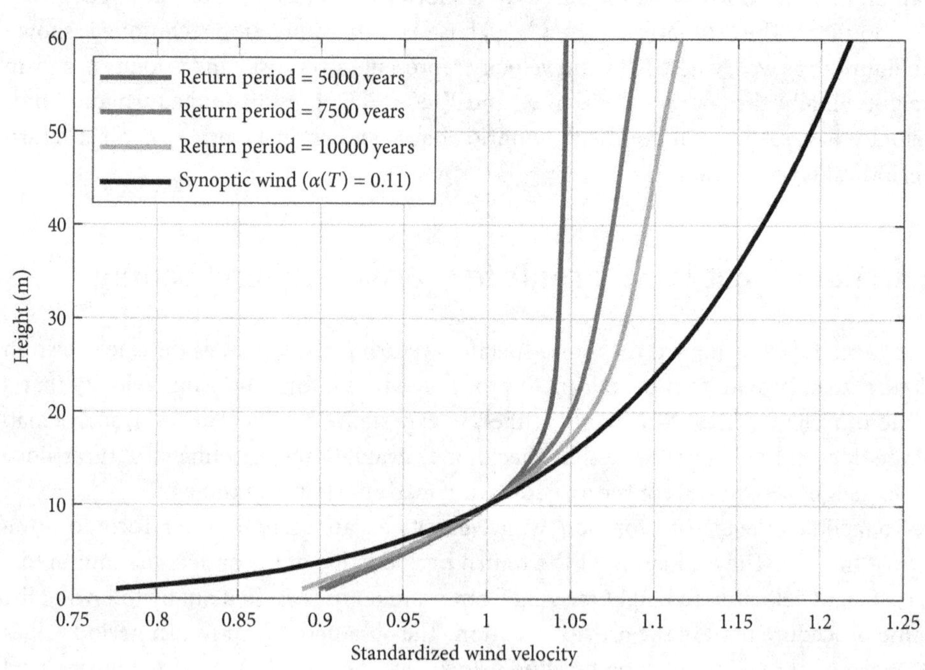

FIGURE 20.7 Wind profile based on the normalized annual maximum tornadic wind velocity.

an exponent of 0.11. In addition, Figure 20.7 shows that the wind profile for the annual maximum tornado wind velocity (where the wind may come from different tornadoes) has a much steeper slope, at least for heights greater than 10 m. The figure also suggests that $V_{Tornado,T}(z)$ can be approximated with an equivalent power law wind profile,

$$V_{Tornado,T}(z) = V_{Tornado,T}(10) \times \left(\frac{z}{10}\right)^{\alpha(T)} \tag{5}$$

A simple regression analysis indicates that $\alpha(T)$ equal to 0.013, 0.026, 0.042 and 0.074 could be used for T equal to 500, 5,000, 20,000 and 60,000 years, respectively. Further analysis suggests that $\alpha(T)$ equal to 0.072 may be considered for T greater than 60,000 years, at least for the considered site. The increased $\alpha(T)$ implies that as the return period increases, the wind profile is more likely dominated by tornadoes of higher intensity.

To see the adequacy of using such an equivalent wind profile in evaluating the wind loading, a circular prismatic structure with a constant width of B and height of H m that is located in Toronto was considered. The bending moment and shear force at a height Z due to a tornado can be calculated using

$$M(z) = \frac{1}{2} C_D \rho_{air} B \int_z^H V^2(x)(x-z)dx \tag{6}$$

and

$$S(z) = \frac{1}{2} C_D \rho_{air} B \int_z^H V^2(x)dx \tag{7}$$

where $M(z)$ and $S(z)$ are the bending moment and shear at height z, respectively; C_D denotes the drag coefficient; and ρ_{air} is the air density.

By carrying out simulation analysis similar to the case of estimating $V_{Tornado,T}(z)$, the return period value of $M(z)$ and $S(z)$, denoted as $M_T(z)$ and $S_T(z)$, respectively, are estimated using the simulated samples for a selected return period and at a given height. The obtained results are shown in Figure 20.8. Note that, again, $M_T(z)$ and $S_T(z)$ at different heights could be associated with different tornadoes.

Rather than carrying out the simulation, if $V_{Tornado,T}(z)$ is used to calculated $M(z)$ and $S(z)$, the obtained values are compared to $M_T(z)$ and $S_T(z)$ in Figure 20.8. The comparison indicates that the calculated bending moment and shear by using $V_{Tornado,T}(z)$ are conservative but agree well with $M_T(z)$ and $S_T(z)$. Therefore, the use of the equivalent wind profile and $V_{Tornado,T}(10)$ could facilitate the structural design code making for tornadic wind load. It is noteworthy that the results presented in Figures 20.7 and 20.8 are for circular prismatic structures and considering all possible wind directions. For definite recommendations, the effects of structural orientation for noncircular prismatic structures, the wind direction, and spatial variation of the equivalent wind profile need to be further investigated. This is beyond the scope of the present study.

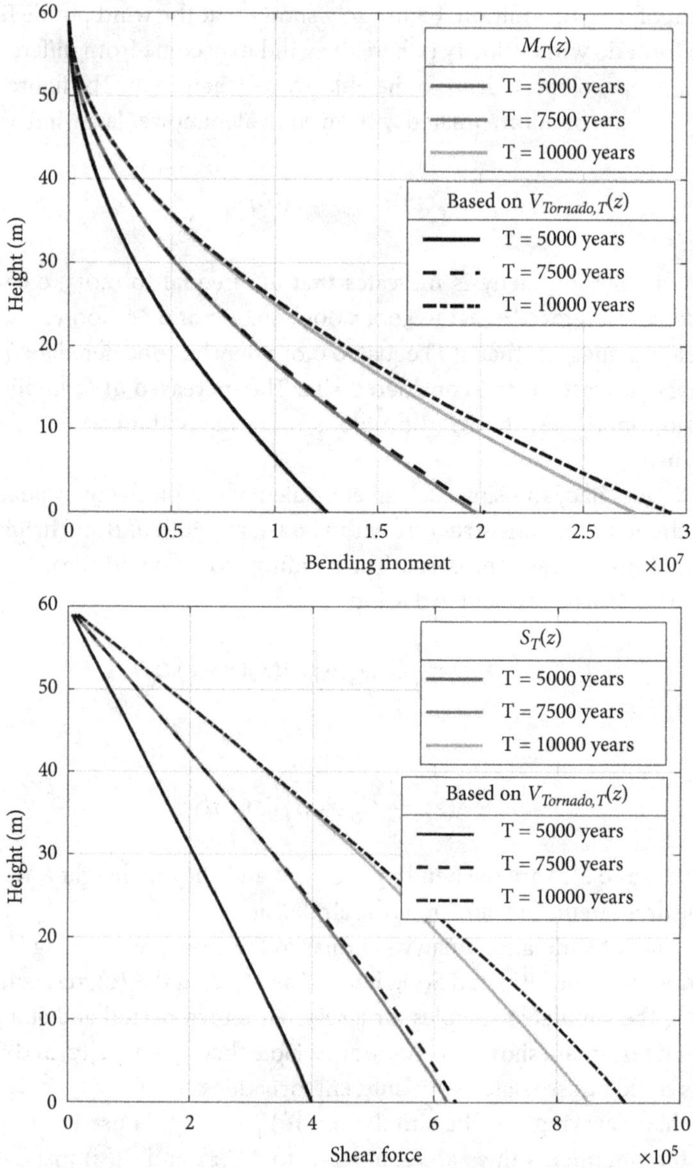

FIGURE 20.8 Comparison of the estimated $M_T(z)$ and $S_T(z)$ (standardized with respect to $C_D \rho_{air} B$) based on simulation to those obtained based on an equivalent wind profile.

4. Discussion of a Simple Procedure to Assess Structural Capacity and Reliability of Towers Subjected to High-Intensity Wind Events

This section focuses on the analysis of transmission towers, although the outlined procedure could be adopted for other structural types, such as prismatic buildings. The advances in the assessment and definition of the capacity curve (or surface) for structures, such as the transmission towers, subjected to tornadic winds are limited because of the complexity of the time histories of the wind patterns. However, the capacity curve and surface of towers subjected to downburst wind loading can be characterized with relative ease, as shown by Mara et al. (2016) and Yang and Hong (2016). This is partly due to the simplicity of the downburst wind field as compared to that of the tornado wind field considered in engineering applications. Simple downburst wind field models were presented by Oseguera and Bowles (1988) and Holmes and Oliver (2000). A transient downburst wind field modeled using a stationary axisymmetric impinging jet was presented by Hangan, Roberts, Xu, and Kim (2003). The spatiotemporally varying mean wind field obtained for their model is depicted in Figure 20.9 for a site located at the distance r from the center of the downburst, where the horizontal mean wind velocity (i.e., outflow velocity) at a height z (m) and time t (s), $V_{hor}(z, r, t)$ (m/s), is a function of the jet diameter D_{jet} (m), jet velocity V_{jet} (m/s), and r. Also illustrated in the figure is the along height wind profile of the downburst wind field as a function of $V_{hor}(z, r, t)/V_{jet}$ for the scenario downburst.

To take into account the fluctuating components of the wind, several researchers (Chay, Albermani, & Wilson, 2006; Chen & Letchford, 2007; Kwon & Kareem, 2009) considered that the fluctuating winds can be modeled as an amplitude-modulated stochastic process. It was argued by Hong (2016) that such a representation ignores the frequency modulation—It can be inferred from the well-known wind spectra, such as the von Karman spectrum, the Davenport spectrum, and the Kaimal spectrum, for synoptic winds that the frequency content changes with mean wind velocity. By applying the model in Hong (2016), which includes both amplitude and frequency modulation functions, the simulated wind time history at Points 1–3 are illustrated in Figure 20.10. For the simulation, the same coherence model given by Hong (2016) was considered. The consideration of the frequency modulation could be of importance for some nonlinear inelastic structures.

By carrying out a series of nonlinear static and nonlinear dynamic analyses, it was observed (Mara et al., 2016; Yang & Hong, 2016) that the capacity curve of a typical transmission tower subjected to downburst wind loading, defined by the tip displacement (or tower drift ratio) of the tower versus the base shear, can be adequately obtained by using the nonlinear static pushover analysis. This significantly simplifies the task of

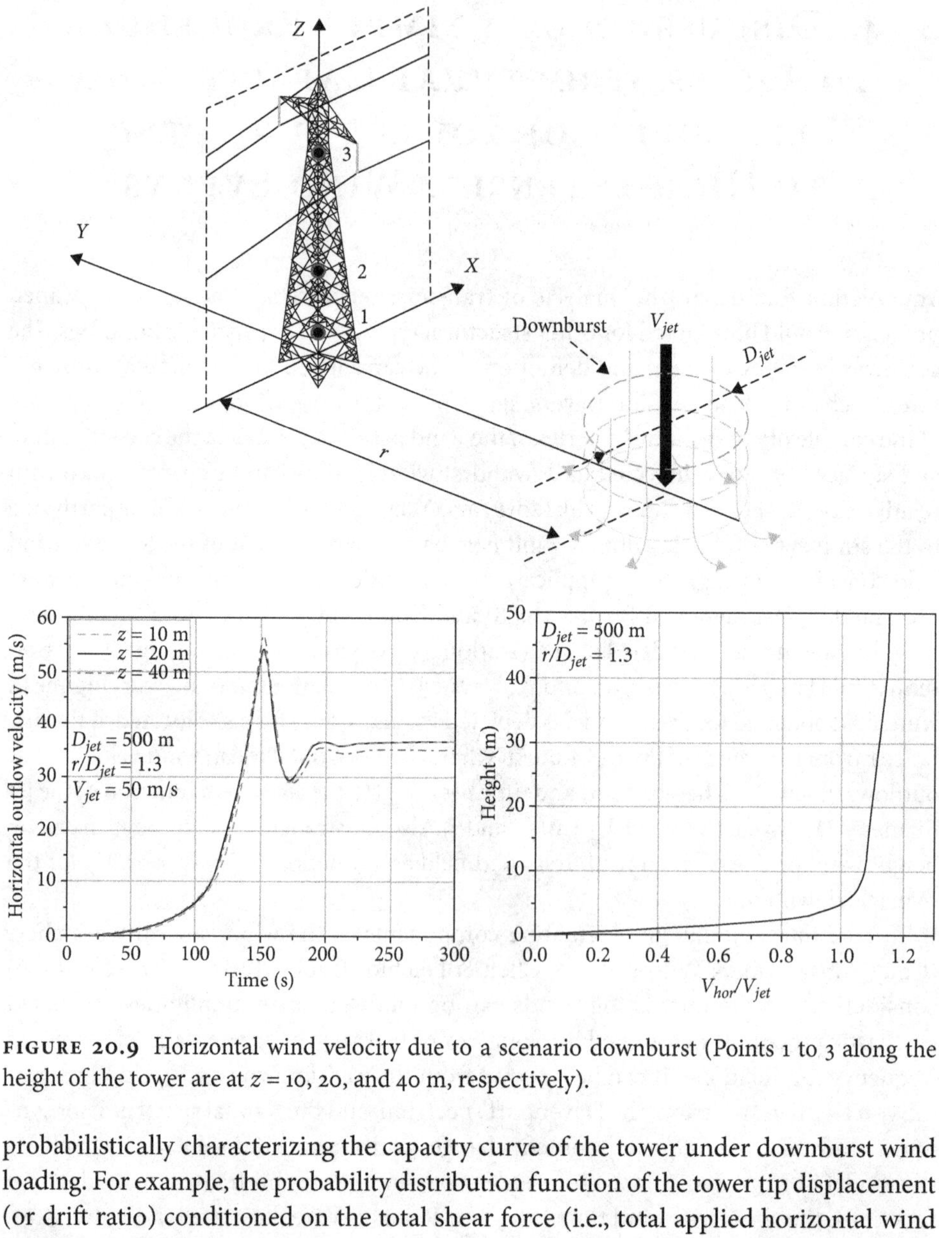

FIGURE 20.9 Horizontal wind velocity due to a scenario downburst (Points 1 to 3 along the height of the tower are at $z = 10$, 20, and 40 m, respectively).

probabilistically characterizing the capacity curve of the tower under downburst wind loading. For example, the probability distribution function of the tower tip displacement (or drift ratio) conditioned on the total shear force (i.e., total applied horizontal wind load) can be assessed using the simple technique by including uncertainty in material properties and geometric variables. The probabilistic model of the capacity curve is essential for reliability analysis. In particular, the probability at the incipient yield and at the incipient collapse of the tower for a scenario downburst defined by D_{jet}, V_{jet} (m/s), and r can then be obtained from the conditional probability density function.

If a reasonable assumption on the occurrence rate of the downburst (e.g., based on the ratio of the tornado occurrence rate to downburst occurrence rate) can be made, one could carry out the downburst wind-velocity hazard analysis by following the same

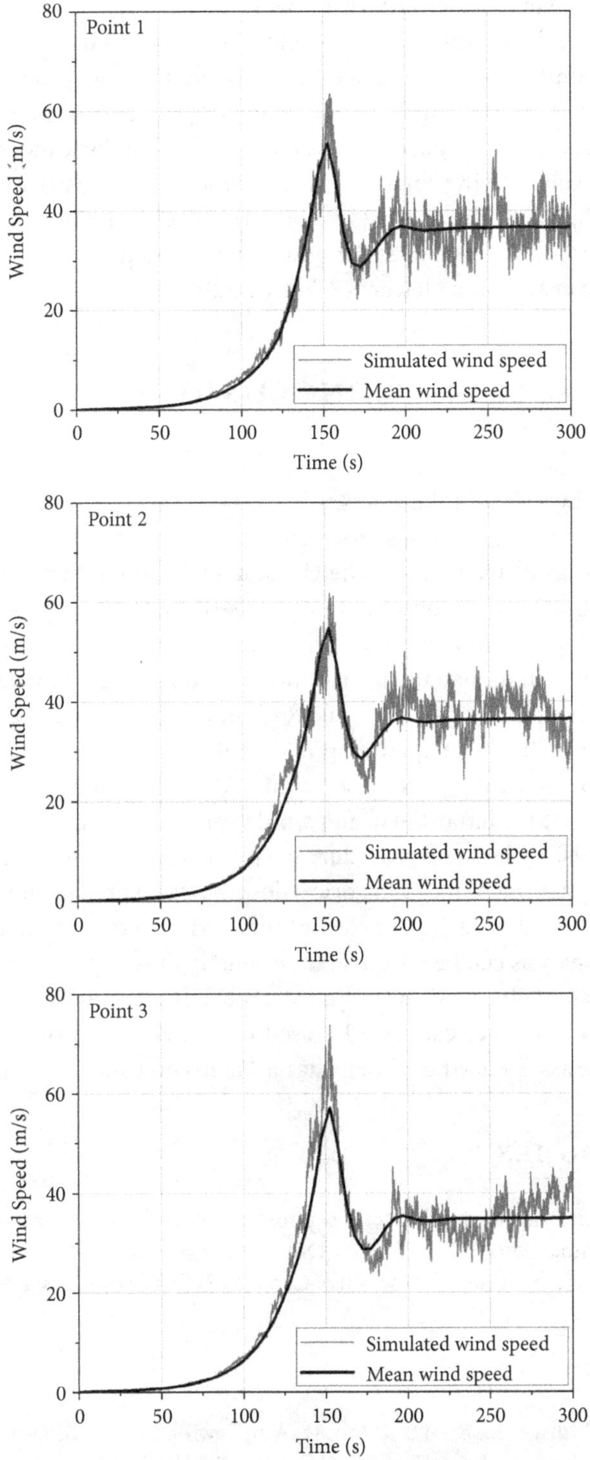

FIGURE 20.10 Simulated wind–time history considering both amplitude and frequency modulations by Mara et al. (2016) and Yang and Hong (2016) considering the vertical coherence.

procedure employed to evaluate the tornadic wind-velocity hazard. Using the probabilistic characterization of the downburst wind velocity at 10-m height $V_D(10)$ and an equivalent downburst wind profile, one could evaluate the resultant wind force that acts on the considered tower. The probabilities at the incipient yield and at the incipient collapse can then be evaluated. The evaluation is based on the probabilistic model of the capacity curve and the resultant wind force, which is a function of $V_D(10)$ and the equivalent downburst wind profile. The analysis can be carried out using one of several available reliability analysis methods, such as the simulation technique and first-order reliability method (Madsen et al., 2006; Melchers & Beck, 2018).

5. Conclusions

In this study, an overview of the assessment of the wind hazard due to tornadoes and downbursts for Canadian sites was provided. It was pointed out that the historical tornado catalogue is available to assess the Canadian tornado occurrence rate. For wind engineering applications, a procedure to evaluate the tornadic wind-velocity (at 10-m height) hazard map and to assess an equivalent wind profile was recommended. The procedure was applied to map the tornado wind-velocity hazard for southern Ontario and to develop an equivalent wind profile for Toronto by considering specified return periods. The use of the developed map and profile facilitates the engineer's task in designing new or evaluating existing structures subjected to tornado wind hazards. It was pointed out that the hazard map and wind profile development can be carried out for a downburst using the same procedure if a justifiable occurrence rate of downburst can be reasonably assumed since a historical downburst catalogue is unavailable.

It was indicated that for a certain class of prismatic structures, the use of nonlinear static pushover analysis can be adequate for evaluating the capacity curve of the structure subjected to downburst wind loading. A probabilistic model of the capacity curve obtained in such a manner can then be used to evaluate the structural reliability by incorporating the assessed wind-velocity hazard map and equivalent wind profile.

Acknowledgments

We gratefully acknowledge the financial support received from several agencies: Western University, the China Scholarship Council (No. 201607980004, for Q. H.) and the National Science and Engineering Research Council of Canada (RGPIN-2016-04814, for H. P. H.).

References

Abd-Elaal, E. S., Mills, J. E., & Ma, X. (2018). A review of transmission line systems under downburst wind loads. *Journal of Wind Engineering and Industrial Aerodynamics*, 179, 503–513.

Aboshosha, H., Elawady, A., El Ansary, A., & El Damatty, A. (2016). Review on dynamic and quasi-static buffeting response of transmission lines under synoptic and non-synoptic winds. *Engineering Structures, 112*, 23–46.

Banik, S. S., Hong, H. P., & Kopp, G. A. (2007). Tornado hazard assessment for southern Ontario. *Canadian Journal of Civil Engineering, 34*(7), 830–842.

Banik, S. S., Hong, H. P., & Kopp, G. A. (2010). Assessment of capacity curves for transmission line towers under wind loading. *Wind and Structures, 13*(1), 1–20.

Chay, M. T., Albermani, F., & Wilson, R. (2006). Numerical and analytical simulation of downburst wind loads. *Engineering Structures, 28*(2), 240–254.

Chen, L., & Letchford, C. W. (2007). Numerical simulation of extreme winds from thunderstorm downbursts. *Journal of Wind Engineering and Industrial Aerodynamics, 95*(9–11), 977–990.

Cheng, V. Y., Arhonditsis, G. B., Sills, D. M., Auld, H., Shephard, M. W., Gough, W. A., & Klaassen, J. (2013). Probability of tornado occurrence across Canada. *Journal of Climate, 26*(23), 9415–9428.

Cheng, V. Y., Arhonditsis, G. B., Sills, D. M., Gough, W. A., & Auld, H. (2016). Predicting the climatology of tornado occurrences in North America with a Bayesian hierarchical modeling framework. *Journal of Climate, 29*(5), 1899–1917.

Daley, D. J., & Vere-Jones, D. (2007). *An introduction to the theory of point processes: Volume II: General theory and structure*. New York, NY: Springer Science & Business Media.

Ellingwood, B. R., Galambos, T. V., MacGregor, J. G., & Cornell, C. A. (1980). *Development of a probability based load criterion for American National Standard A58: Building Code requirements for minimum design loads in buildings and other structures*. Washington, DC: National Bureau of Standards.

Elsner, J. B., & Widen, H. M. (2014). Predicting spring tornado activity in the central Great Plains by 1 March. *Monthly Weather Review, 142*(1), 259–267.

Etkin, D., Brun, S. E., Shabbar, A., & Joe, P. (2001). Tornado climatology of Canada revisited: Tornado activity during different phases of ENSO. *International Journal of Climatology: A Journal of the Royal Meteorological Society, 21*(8), 915–938.

Fujita, T. T. (1985). *The downburst* (SMRP Research Paper No. 210; NTIS PB-148880). Chicago, IL: University of Chicago.

Gilks, W. R., Richardson, S., & Spiegelhalter, D. (1996). *Markov chain Monte Carlo in practice*. Boca Raton, FL: Chapman and Hall/CRC.

Haan, F. L., Jr., Balaramudu, V. K., & Sarkar, P. P. (2010). Tornado-induced wind loads on a low-rise building. *Journal of Structural Engineering, 136*(1), 106–116.

Hamada, A., & El Damatty, A. A. (2011). Behaviour of guyed transmission line structures under tornado wind loading. *Computers & Structures, 89*(11–12), 986–1003.

Hangan, H., & Kim, J. D. (2008). Swirl ratio effects on tornado vortices in relation to the Fujita scale. *Wind and Structures, 11*(4), 291–302.

Hangan, H., Roberts, D., Xu, Z., & Kim, J. (2003, June). Downburst simulation: Experimental and numerical challenges. In *Proceedings of the 11th International Conference on Wind Engineering*, Lubbock, TX.

Holmes, J. D., & Oliver, S. E. (2000). An empirical model of a downburst. *Engineering Structures, 22*(9), 1167–1172.

Hong, H. P. (2016). Modeling of nonstationary winds and its applications. *Journal of Engineering Mechanics, 142*(4), 04016004.

Jagger, T. H., Elsner, J. B., & Widen, H. M. (2015). A statistical model for regional tornado climate studies. *PloS One, 10*(8), e0131876.

Kim, J., & Hangan, H. (2007). Numerical simulations of impinging jets with application to downbursts. *Journal of Wind Engineering and Industrial Aerodynamics, 95*(4), 279–298.

Kopp, G. A., Hong, E., Gavanski, E., Stedman, D., & Sills, D. M. (2017). Assessment of wind speeds based on damage observations from the Angus (Ontario) tornado of 17 June 2014. *Canadian Journal of Civil Engineering, 44*(1), 37–47.

Kwon, D. K., & Kareem, A. (2009). Gust-front factor: New framework for wind load effects on structures. *Journal of Structural Engineering, 135*(6), 717–732.

Lambert, D. (1992). Zero-inflated Poisson regression, with an application to defects in manufacturing. *Technometrics, 34*(1), 1–14.

Madsen, H. O., Krenk, S., & Lind, N. C. (2006). *Methods of structural safety*. North Chelmsford, MA: Courier Corporation.

Mara, T. G., & Hong, H. P. (2013). Effect of wind direction on the response and capacity surface of a transmission tower. *Engineering Structures, 57*, 493–501.

Mara, T. G., Hong, H. P., Lee, C. S., & Ho, T. C. E. (2016). Capacity of a transmission tower under downburst wind loading. *Wind and Structures, 22*(1), 65–87.

Marshall, T. P., McDonald, J. R., & Forbes, G. S. (2004, October). The enhanced Fujita (EF) scale. In *Preprints, 22nd Conf. on Severe Local Storms, Hyannis, MA, Amer. Meteor. Soc. B* (Vol. 3).

Melchers, R. E., & Beck, A. T. (2018). *Structural reliability analysis and prediction*. New York, NY: Wiley.

Natarajan, D., & Hangan, H. (2012). Large eddy simulations of translation and surface roughness effects on tornado-like vortices. *Journal of Wind Engineering and Industrial Aerodynamics, 104*, 577–584.

National Research Council of Canada (NRCC). (2015). *National building code of Canada*. Ottawa, Canada: Institute for Research in Construction, National Research Council of Canada.

Newark, M. J. (1984). Canadian tornados, 1950–1979. *Atmosphere-Ocean, 22*, 343–353.

Newark, M. J. (1991). A design basis tornado. *Canadian Journal of Civil Engineering, 18*(3), 521–524.

Oseguera, R. M., & Bowles, R. L. (1988). *A simple, analytic 3-dimensional downburst model based on boundary layer stagnation flow* (NASATM-100632). Hampton, VA: NASA Langley Research Center.

Refan, M., & Hangan, H. (2016). Characterization of tornado-like flow fields in a new model scale wind testing chamber. *Journal of Wind Engineering and Industrial Aerodynamics, 151*, 107–121.

Savory, E., Parke, G. A., Zeinoddini, M., Toy, N., & Disney, P. (2001). Modelling of tornado and microburst-induced wind loading and failure of a lattice transmission tower. *Engineering Structures, 23*(4), 365–375.

Sengupta, A., Haan, F. L., Sarkar, P. P., & Balaramudu, V. (2008). Transient loads on buildings in microburst and tornado winds. *Journal of Wind Engineering and Industrial Aerodynamics, 96*(10–11), 2173–2187.

Sills, D. M. L., Scriver, S. J., & King, P. W. S. (2004). The tornados in Ontario project (TOP). In *Preprints of the 22nd AMS Conference on Severe Local Storms (Hyannis, MA)* (CD-ROM 7B.5). Boston, MA: American Meteorological Society.

Standohar-Alfano, C. D., & van de Lindt, J. W. (2016). Tornado risk analysis for residential wood-frame roof damage across the United States. *Journal of Structural Engineering, 142*(1), 04015099.

Tan, L. (2008). *Assessment of tornado hazard maps for southern Ontario* (MESc thesis). University of Western Ontario, London, Ontario, Canada.

Tan, L., & Hong, H. P. (2010). Influence of spatial inhomogeneity of tornado occurrence on estimated tornado hazard. *Canadian Journal of Civil Engineering, 37*(2), 279–289.

Twisdale, L. A. (1978). Tornado data characterization and wind speed risk. *ASCE Journal of Structural Engineering, 104*, 1611–1630.

Twisdale, L. A., Dunn, W. L., Horie, Y., and Davis, T. L. (1981). *Tornado missile simulation and design methodology: Model verification and database updates* (Vol. 2). Report by Research Triangle Park, NC: Research Triangle Institute.

Wikle, C. K., & Anderson, C. J. (2003). Climatological analysis of tornado report counts using a hierarchical Bayesian spatiotemporal model. *Journal of Geophysical Research: Atmospheres, 108*(D24), 1–15.

Wurman, J., & Gill, S. (2000). Finescale radar observations of the Dimmitt, Texas (2 June 1995), tornado. *Monthly Weather Review, 128*(7), 2135–2164.

Yang, S. C., & Hong, H. P. (2016). Nonlinear inelastic responses of transmission tower-line system under downburst wind. *Engineering Structures, 123*, 490–500.

Yang, S. C., Liu, T. J., & Hong, H. P. (2017). Reliability of tower and tower-line systems under spatiotemporally varying wind or earthquake loads. *Journal of Structural Engineering, 143*(10), 04017137.

SECTION 6

REDUCTION OF EXPOSURE TO NSWS

CHAPTER 21

PREDICTION OF NON-SYNOPTIC WIND SPEEDS

FRANKLIN T. LOMBARDO AND
ALEXANDER S. ZICKAR

1. Introduction

Losses by non-synoptic wind storms (i.e., thunderstorms, tornadoes) are a major issue. Reinsurer data on natural disasters that occurred in 2019 reveal that non-synoptic wind storm disasters occurred globally and were responsible for two of the five highest insured losses in 2019 (Munich RE, 2019). The United States experienced 14 disaster events in 2019 that each exceeded $1 billion in total losses. Of those 14 events, 8 were caused by hazards produced from non-synoptic wind storms (National Oceanic and Atmospheric Administration [NOAA], 2020).

In the last 20 years, and more recently spurred by some rather significant events, non-synoptic wind storms have gained additional traction in wind engineering research. Numerous studies since 2000 have been carried out and published on non-synoptic wind hazards, including specialized field experiments (e.g., Orwig & Schroeder, 2007); wind speed prediction from observed data (e.g., Zhang, Solari, Yang, & Repetto, 2018)—the focus of this chapter; numerical (e.g., Vermeire, Orf, & Savory, 2011) and wind tunnel (e.g., Mason, Letchford, & James, 2005) experiments. Following on from non-synoptic wind hazards, countless other works have been carried out on non-synoptic wind loading (e.g., Kim, Hangan, & Eric Ho, 2007; Sarkar, Haan, Balaramudu, & Sengupta, 2006); structural response (e.g., Shehata, El Damatty, & Savory, 2005; Solari, 2016) and structural damage (e.g., Loredo-Souza et al., 2019). Additional details on studies focusing on something other than non-synoptic wind speed prediction can be found elsewhere in this handbook. These studies suggest the importance of non-synoptic wind events in the design of structures, and the reader need not look further than this handbook for evidence of their importance.

Despite the significant losses and explosion of recent studies, non-synoptic events have struggled to find their own place in so-called traditional wind engineering practice (e.g., having their own wind tunnel testing procedure, etc.). This struggle is largely due to lack of field data and the interrelated small temporal and spatial scales of these events. These challenges make using traditional analysis techniques and comparisons to the well-studied atmospheric boundary layer (ABL), or *synoptic*, events difficult. It seems logical that the ABL analogue for non-synoptic winds when it exists in practice would have some component or components of the wind loading chain (Davenport, 1961), which is used exclusively in wind engineering practice and in codes and standards. In fact, a number of recent articles have highlighted frameworks that reflect components of the chain for non-synoptic winds (Hangan, Romanic, & Jubayer, 2019; Kwon & Kareem, 2009; Solari, 2016). Details about those frameworks can be found elsewhere in this handbook.

In the wind loading chain, the first, and arguably the most important, link in the chain owing to the squaring of wind speed is the prediction of the extreme wind climate. It follows then that the accurate prediction of the wind climate is most critical. Due to some of the difficulties mentioned, prediction of non-synoptic winds can present more challenges than synoptic winds, as is described in this chapter. However, the wind climate link for non-synoptic winds has been studied longer than the other links in the chain. Early works recognized the importance of non-synoptic wind events in wind engineering and specifically in the extreme wind climate. Gomes and Vickery (1978), through initially understanding that tropical cyclone winds needed to be separated out in so-called mixed wind climates, separated out thunderstorm winds for analysis. They then predicted extreme non-synoptic wind speeds and combined these predictions with other storm types, specifically synoptic and tropical cyclone wind speeds. Further discussion of these techniques occurs throughout this chapter.

Subsequent analysis of extreme wind speeds generated from thunderstorms has shown that for annual probabilities p of the occurrence (or exceedance) of interest for wind load design (e.g., $p \leq .1$), thunderstorm events are likely to produce the highest wind speeds (e.g., De Gaetano, Repetto, Repetto, & Solari, 2014; Holmes, 2002; Kasperski, 2002; Kruger, Goliger, Retief, & Sekele, 2010; Lombardo, Main, & Simiu, 2009; Mohr, Kunz, Richter, & Ruck, 2017) for a significant number of global locations. The role non-synoptic winds play in the wind climate in numerous regions has prompted some work on the explicit inclusion of thunderstorms in a number of wind load codes and standards worldwide (e.g., Holmes, 2002; Kruger et al., 2017; Pintar, Simiu, Lombardo, & Levitan, 2015; Solari et al., 2012; Vallis, Loredo-Souza, Ferreira, & de Lima Nascimento, 2019).

This chapter focuses only on non-synoptic wind speeds generated from a *thunderstorm*. Detailed descriptions of definitions and subclassifications of those wind speeds (e.g., microburst, macroburst) and the different types of thunderstorms that generated them (e.g., bow echo, supercell) are not presented in detail in this chapter. The acknowledgment of these differences and suggestions for how to handle differences for wind speed prediction are discussed. Readers can turn to the work of

Wakimoto (2001) for a more detailed discussion of the differences within the relatively broad category of thunderstorm.

Due to the severe lack of observational data, methods for predicting tornado wind speeds have generally been difficult and have revolved around using areal damage swaths to estimate p. Please see, for example, Twisdale and Dunn (1983), Widen et al. (2013), and Kuligowski, Lombardo, Phan, Levitan, & Jorgensen (2014) for prediction methods of tornado-generated wind speeds. The methods discussed in this chapter, including newer methods, to account for the relative lack of data for thunderstorms could work much the same for tornadoes. By the same token, most of the methods discussed here could also work for synoptic wind events (and have been to a significant extent). Synoptic wind data will be of importance in the chapter for mixed wind climates, which in theory include all storm types.

Analysis of events in this section mostly are limited to the frequency, intensity, and geographic location and character of extreme wind speeds (i.e., *extreme wind analysis*) at a single point. Since the direction of the chapter has wider application in designing structures for wind loading, *prediction* does not refer to "real-time" prediction and instead focuses on prediction of non-synoptic wind speeds for some relatively low p. This chapter focuses on standardized values of "peak" non-synoptic wind speeds (0.2- to 3-s averaging time, roughly 10-m height, "open" terrain). Definitions of a standard wind speed vary slightly depending on the character of the data, which usually depends on the region/country.

Although the majority of this chapter assumes that non-synoptic wind speed data have already been standardized, it should be noted that methods to arrive at a standard value may be different from that of the ABL due to characteristics such as non–log law wind profiles and short duration of the events. Resorting to methods like the Durst curve, for example, in converting between averaging times could be problematic because it assumes a single deterministic factor independent of storm type. Also, it should be stated again that non-synoptic wind speeds are generally measured at a single point, and a complete four-dimensional picture of the event, and hence the complete prediction of the event, is unknown at this point. Discussions on improvement in the prediction process are presented in Section 3 of this chapter.

2. Non-Synoptic Wind Speed Prediction

A step-by-step discussion of the entire prediction process follows in this section, with illustrative examples at each step. General analysis results from other publications using the processes laid out are also discussed throughout. Techniques discussed throughout this chapter are illustrated with the same example data set. The data set is from ISD 3505 and recorded at the Champaign, Illinois, airport (KCMI).

2.1. Data Recording, Archiving, and Extraction

Wind speed measurments that have been used to predict non-synoptic wind speeds have almost exclusively recorded from "single-point" (i.e., anemometer at only one height), surface-based networks. A large majority of these measurements were taken at airport locations, with the duration of the record being approximately 5–50 years. Over the length of data recording, measurement characteristics (e.g., anemometer height, averaging time, sampling frequency) have typically varied frequently due to the needs of multiple communities such as aviation and meteorology (McKee, Doesken, & Kleist, 1996). As technology has advanced, the type of anemometers has changed from cup to sonic anemometers. Changes in these characteristics can be dealt with in part using standardization techniques discussed in the previous section. Presently, at least for the Automated Surface Observing System (ASOS) in the United States, recorded data are standardized based on World Meteorological Organization requirements (e.g., 3-s moving average, 10-m height) and are sampled at 1 Hz. A number of additional networks with a denser station resolution have popped up in recent years in the United States, including state-based mesonets (e.g., Schroeder et al., 2005). The denser networks can help with improved identification, classification, and prediction of non-synoptic wind speeds. Further discussion of station network density takes place further in this chapter.

Due to size and space requirements, not all the data that are measured become archived. For example, although the ASOS network records the data every second, only certain statistics (e.g., the peak 3-s gust over a 1-min period) are archived for some data sets, and for others, wind gusts over some arbitrarily high threshold are archived. Limited data archival is not a new problem. Riera and Nanni (1989) noted that mean wind speed data were not available from Brazilian stations when performing an extreme wind prediction within that country. The NOAA, especially in the United States (although data from other countries exist within NOAA databases), has archived a significant number of wind data sets important to non-synoptic wind speed prediction. Examples of these data sets are Integrated Surface Databases (ISDs) 3505 and ASOS 1-min data (6401) and can be found online (https://www.noaa.gov).

Portions of these data sets have been made available in tabular form to make extraction and subsequent analysis relatively easy, while others remain in a mostly "raw" form. Some of these raw data sets have been postprocessed and curated for the purpose of extreme wind speed prediction. A large number of these data sets for the United States are currently available on the National Institute of Standards and Technology (NIST) website (https://www.itl.nist.gov/div898/winds/overview.htm). These data sets include older ones that were used for the earliest versions of the US wind maps for wind load design, which only contain annual maximum wind speeds. It should be noted it would be extremely difficult to do an analysis of non-synoptic wind speeds only from these annual maximum data sets alone as only one speed exists per year regardless of storm

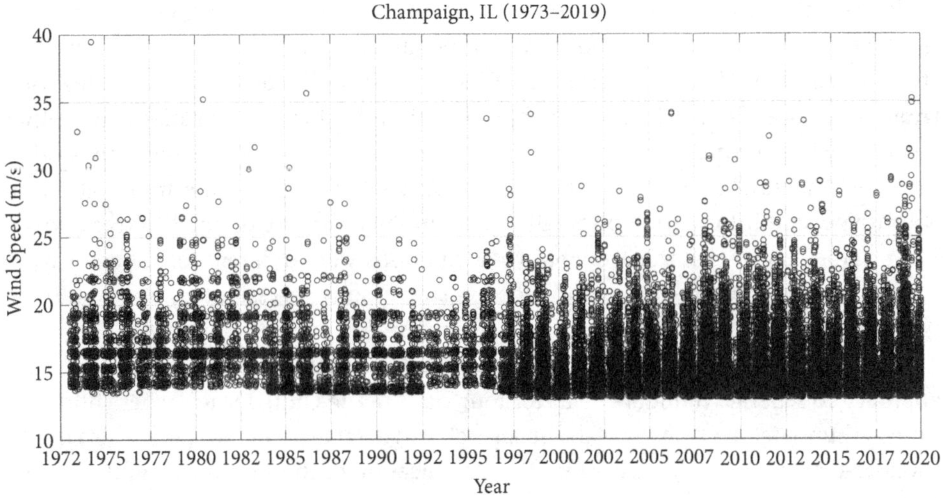

FIGURE 21.1 Time history of standardized gust wind speeds at Champaign, Illinois, airport (KCMI) from 1973 to 2019.

type. The NIST website also hosts more rigorously quality-controlled data, which include wind speeds over some threshold and wind speeds classified as non-synoptic. Wind data for these datasets have been extracted out of raw meteorological data in the United States by means of software programs (Lombardo et al., 2009), and other more specialized network data extraction has worked in a similar manner (Lombardo, 2012).

The data were extracted using the methodologies presented in Lombardo et al. (2009) and Lombardo (2012). The data span from 1973 to 2019, the length of "gust" data available. This location harbors a fairly representative data set for what is available in the United States at larger commercial airports. As mentioned, these gust data have been corrected to standard conditions using methods described by Lombardo (2012). Figure 21.1 shows a time history of the standardized wind speeds over the course of the record. In Figure 21.1, it is important to note potential issues, such as slightly different wind speed thresholds over time, "layering" (i.e., disproportionate amount of wind speeds at certain values) of data, and volume of reports. These are common features of most long-term data sets and stem from mostly reporting requirements and changes made over the course of the data, as mentioned previously.

2.2. Identification, Classification, and Separation of Non-Synoptic Winds

Once the archived data are extracted, the next logical step is to then identify wind speeds that are associated with non-synoptic events. Gomes and Vickery (1978) and Riera et al.

(1977) were the first to recognize a need to separate extreme wind storm mechanisms in the context of designing for wind extremes for Australia and Brazil, respectively. Most studies, including the early ones, generally separated wind storms into five categories: tropical cyclones, synoptic, *non-synoptic* (i.e., thunderstorms), tornadoes, and other localized wind storms (e.g., topographic-induced effects). It appears the early studies performed identification manually as older data sets provided little information on storm type, and information on weather conditions was more difficult to track down. Both studies noted that the probability distributions were different for different wind storms, which for both physical and probabilistic reasons necessitates the separation of non-synoptic wind storms from their synoptic counterparts.

Data quality was improving when Twisdale and Vickery (1992) used a "thunder-day" approach to separate thunderstorms for four cities in the United States where information on thunderstorms was available in an archived database. Holmes (2002) also used a thunder-day approach for Australia. The classification of thunderstorms in the United States was derived using thunderstorm "flags" from 1,000+ meteorological stations located mostly at airports throughout the United States (Lombardo et al., 2009) using an updated database that included data used by Twisdale and Vickery (1992).

For networks that do not have explicit thunderstorm flags, classification of winds generated from thunderstorms has been deduced from other methods. Kasperski (2002) used what amounted to a thresholding and gust factor approach to identifying non-synoptic winds in Germany. De Gaetano et al. (2014) and Solari, De Gaetano, and Repetto (2015) applied similar methods to a high-resolution network in Italy. Mohr et al. (2017) rigorously inspected surface measurements in conjunction with radar and satellite images and lightning data to classify non-synoptic events. As non-synoptic winds also are accompanied by changes in temperature (usually decreases) and pressure (generally increases), metrics describing these changes have been used as further classification techniques (e.g., Vallis et al., 2019).

Manual analyses always should be a preferred method at least for a portion of the classification step, but the size and scope of some datasets make it near impossible. Automated techniques using lightning or radar data, for example, can be beneficial. Other challenges can arise using statistical criteria on data with multiple height measurements when criteria are met at one height and not the others (Zhang, Solari, Burlando, & Yang, 2019). Simple thresholding methods applied to wind speed or seasonality can also be used to separate out storm type but will likely only yield the highest extremes of a non-synoptic phenomena. Newer techniques, such as using machine learning to classify storm events, are discussed in Section 3.

Figure 21.2 shows the classification of thunderstorm winds from KCMI. The classification was done using the methods described by Lombardo et al. (2009) and Lombardo (2012). Approximately 8 percent of *all* wind speeds in this dataset are thought to have originated from non-synoptic wind events. Note a number of the most extreme recorded wind speeds were from thunderstorms (non-synoptic), including all of those greater than 35 m/s.

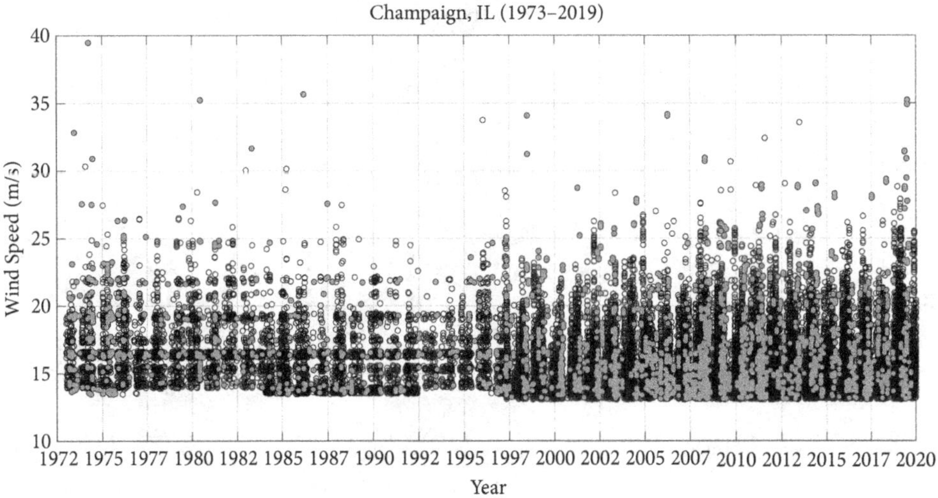

FIGURE 21.2 Time history of standardized gust wind speeds at Champaign, Illinois, from 1973 to 2019 with thunderstorm-generated wind speeds identified as *gray* circles.

2.3. Analysis and Prediction Techniques

Now that the non-synoptic wind data have been classified, the non-synoptic wind speed data can go through the numerous analysis steps and techniques required to arrive at a prediction.

2.3.1. *Data Independence*

As a lot of analysis techniques discussed here are probabilistic in nature, a requirement of probability distributions is that the data used are identically distributed (roughly satisfied by storm-type separation) and independent. To ensure reduced dependence, a time interval between "events" has generally been used, with the maximum wind speed value in this time interval utilized for the analysis. These methods range from simply setting a fixed time interval to more detailed setups, such as analyzing drops in pressure data to assess storm passage, otherwise known as the method of independent storms (MIS) (Harris, 1996, 1999). For the purposes of this chapter, all time interval–type methods are classified as MIS. Studies of synoptic winds have generally used a time interval of 4 days between events, suggested by analyzing wind power spectra by van der Hoven (1957) and used in multiple studies after that. Other time periods such as 1 day have been used with similar success. For non-synoptic winds, given their much shorter duration, shorter time intervals for reduced dependence have usually been selected (e.g., Lombardo et al., 2009). Time periods in the numerous studies referenced here have ranged from 6 to 24 hr.

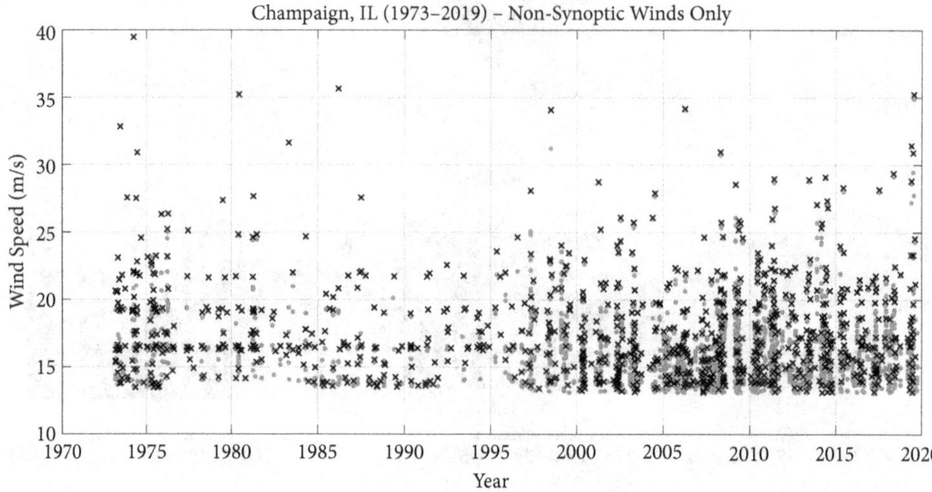

FIGURE 21.3 Time history of non-synoptic wind speeds only at Champaign, Illinois, from 1973 to 2019, with independent wind speeds identified with a black *x*.

2.3.2. *Rate of Occurrence*

Figure 21.3 shows the KCMI data set and a subset with reduced dependence. A value of 6 hr was used for the interval. As suggested, occurrence frequencies are usually much lower than synoptic wind speeds for most locations, which can enable alternative techniques, to be discussed further in this chapter. In the KCMI data set, the occurrence rate λ is approximately 21.2 events/year ($\lambda = 21.2$). The λ value is important and really the first step in the prediction of extreme wind speeds. Obviously, this value will change depending on the technique that is used. For example, if annual maximum non-synoptic wind speeds are used, $\lambda = 1$ with a total number of data points equal to 47 ($N = 47$). If methods such as peaks over threshold (POT) are used (Simiu & Heckert, 1996), λ is a function of the threshold. The KCMI data set has a built-in wind speed threshold due to the recording and archival characteristics of the data (Figures 21.1 and 21.2), so λ can be thought of as the number of non-synoptic wind events that produce a wind speed over the recording threshold. Also, note that the frequency of thunderstorm reports changes over time (apart from interannual variability)—certainly due to reporting techniques—and it is significantly lower than synoptic counterparts for KCMI. Handling frequency changes or general nonstationarities in the data are discussed briefly in Section 3. Methods to deal with very low values of λ (i.e., infrequent thunderstorms) are discussed throughout the chapter.

2.3.3. *Probability Distributions*

Modern-day computer programs allow easy and rapid probability distribution fitting to non-synoptic wind speeds with reduced statistical dependence without much, if any, description of the process. This description is given here. Initially, an empirical probability is assigned to the data sorted in ascending order. Works in the literature exist debating how to assign this probability (e.g., Makkonen, 2006), but in this chapter, the following empirical probability P_e is assigned in Equation 1:

$$P_e = \frac{i}{N+1} \quad (1)$$

where i is the rank of data ($i = 1$ being the lowest value), and N is the number of data points. When dealing with a data set that is considered "extreme," the next step is to "fit" the P_e values to an existing probability distribution through methods such as maximum likelihood estimation (MLE) or method of moments (MOM) to estimate the parameters of the probability distribution. The choice of probability distribution has varied, but for prediction of non-synoptic wind speeds and wind speeds in general, the choice has been largely part of the generalized extreme value distribution (GEVD). The cumulative distribution function of the GEVD is shown in Equation 2, where U is the wind speed, u is the location parameter (measure of central tendency), a is the scale parameter (measure of dispersion), and c is the shape parameter (measure of tail behavior).

$$F(U) = \exp\left[-\left(1 + c\left(-\frac{U-u}{a}\right)\right)^{-1/c}\right] \quad (2)$$

The GEVD can be "split" into three separate distributions based on the value of the shape parameter c. For $c > 0$, a Type II or Frechet distribution is obtained, and for $c < 0$ a Type III or reverse Weibull distribution is obtained. As c à 0, Equation 2 tends to the Gumbel or Type I distribution. The equation for the Gumbel distribution is shown in Equation 3.

For non-synoptic winds generally, the Type III and Type I distributions have been used in a number of publications up to the present, some of which include use of both distributions. In some of the first publications on prediction of extreme winds, a Type II was suggested on various occasions (e.g., Thom, 1960). It was soon realized that Type II distributions were found to be a good fit for occasions when various storm types were commingled together and used for prediction of extremes and hence violated the identically distributed requirement. This finding set the stage for considering non-synoptic wind speeds in a separate manner. Based on extreme value theory, a number of exceedances over some large threshold converge to the generalized Pareto distribution (GPD). The GPD has been used on a number of occasions (e.g., Pintar et al., 2015) to predict extreme non-synoptic wind speeds. Table 1 shows a non-exhaustive list of references and the probability distributions used to predict non-synoptic winds.

$$F(U) = \exp\left[-\exp\left(-\frac{U-u}{a}\right)\right] \quad (3)$$

As a large number of studies focus on annual maximum wind speeds ($\lambda = 1$), then the corresponding probability distribution would correspond to an annual probability of exceedance (or non-exceedance). The annual probability is a reasonable framework to start from, although any time interval for prediction can be used. Ideally, the analyzer

Table 21.1 Sample of Studies That Contain Specific Details on Prediction of Non-Synoptic Wind Speeds and Their Methods and Probability Distributions Used for Analysis

Study	Method	Distribution(s)
Gomes and Vickery (1978)	Annual maxima	Type I (Gumbel)
Riera et al. (1977)	Annual maxima	Type II
Riera and Nanni (1989)	Annual maxima	Log normal, Type I, Type II
Twisdale and Vickery (1992)	Annual maxima	Log normal, Type I
Holmes (2002)	POT	GPD
Kasperski (2002)	MIS	Type III
Lombardo et al. (2009)	Annual maxima, POT	GPD, Type I
Mohr et al. (2017)	POT	GPD
Zhang et al. (2018)	Annual Max	Type I
Lombardo and Zickar (2019)	Annual Max, MIS	Type I (Gumbel), Type III

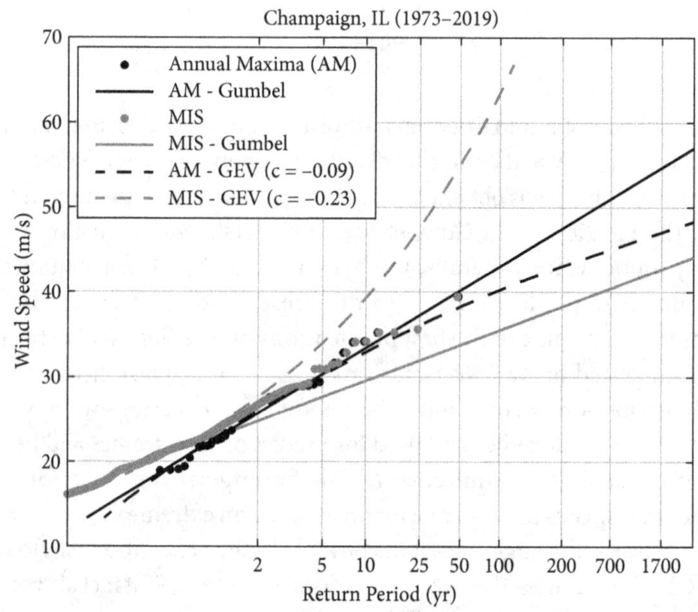

FIGURE 21.4 Prediction estimates of non-synoptic wind speeds at Champaign, Illinois, from 1973 to 2019 using annual maxima (black dots) and all independent non-synoptic events (gray).

would want to transition a probability distribution built with data at a given rate to an annual probability. To do this, λ is assumed to represent the number of independent trials, and the rate value is used as an exponent to the corresponding probability distribution (Cook, 2014). For example, Equations 1–3 would simply be taken to the exponent λ to derive p or, correspondingly, non-exceedance $(1 - p)$. From the annual probability of non-exceedance the return period R in years can easily be derived as $R = (1 - (1 - p))^{-1}$.

To continue the illustrative example, the non-synoptic data with reduced dependence (Figure 21.3) from KCMI were analyzed using a few separate methods highlighted previously. The results of this analysis are illustrated in Figure 21.4 using Gumbel coordinates on the x-axis (i.e., $-ln(-ln(P))$). Two sets of the non-synoptic wind speeds were used (1) annual maxima and (2) all independent non-synoptic events over recording threshold (i.e., MIS). The Gumbel distribution and GEVD were used for both sets of data. Parameters of the distributions were estimated using MLE. It should first be noted in Figure 21.4 that the different methods give different predictions of wind speed at various R. As R increases, the differences between wind speed predictions become greater, as expected. For this particular case, especially with the full dataset (MIS), the differences are significant. The differences are due to the artificial threshold imposed by the data collection and the majority of wind speeds residing at lower values, suggesting additional weight being placed on them in the parameter estimation procedure. The character of the data results in a highly positive shape parameter of $c = 0.23$. POT methods (Simiu & Heckert, 1996) or similar (Harris, 1999) may work better in this case. Fixing a shape parameter other than $c = 0$ (Gumbel) may also be recommended if warranted by the data (Holmes, 2002).

It is important to assess non-synoptic wind speeds in a holistic manner with other storm types for mixed wind climates (e.g., significant contribution from multiple storm types in the extremes). The mixed probability distribution is used for mixed wind climates and considers the product of the assumed independent probability distributions of different wind storm types (Lombardo, 2014; Twisdale & Vickery, 1992), as shown in a simplified form in Equation 4, where V is wind speed, and the subscript NS is non-synoptic and the subscript S is synoptic.

$$P(V \leq v) = P(V_{NS} \leq v)P(V_S \leq v) \tag{4}$$

Previous analysis using the mixed distribution approach has shown the true significance of non-synoptic wind speeds for prediction of extreme wind speeds. This significance includes locations that receive infrequent, but extreme, non-synoptic wind speeds (e.g., maximum overall recorded wind speed), locations that had similar non-thunderstorm and thunderstorm extreme wind climates, and obviously non-synoptic–dominated wind climates (e.g., nearly all annual maxima are non-synoptic).

The original works on non-synoptic extreme wind prediction included analysis of the mixed wind climate. Gomes and Vickery (1978) analyzed the mixed climate for various regions in Australia. They found thunderstorms (non-synoptic winds) did not contribute much to the extreme wind climate in tropical cyclone-prone regions, but predictions of non-synoptic winds tended toward higher wind speeds at return periods greater than 10 years in areas where synoptic winds were prevalent. Holmes (2002) in Australia found similar results, with the exception of Perth, where synoptic winds dominated. In Brazil, Riera and Nanni (1989) found similar results in addition to a higher modal wind speed and higher variability in thunderstorm winds. Twisdale and Vickery (1992) showed a non-synoptic–dominant wind climate in the United States for Dallas, Texas, and Minneapolis, Minnesota, and a "true" mixed wind climate for Charlotte, North Carolina, and a dominant synoptic climate for LaGuardia Airport in New York City. Kasperski (2002) found for Germany that non-synoptic winds were of limited importance at higher wind speeds (low annual probability of occurrence). Kruger et al. (2010) found specific regions in South Africa that were dominated by non-synoptic winds, and Kruger, Goliger, Retief, and Sekele (2012) found higher variability for non-synoptic wind speeds as compared to synoptic.

The work in Lombardo (2014), which considered the entire contiguous United States, demonstrated that not considering the non-synoptic wind speed probability distributions was unconservative for low-probability events of interest to wind load design for most areas of the US Central Plains and Great Lakes. Lombardo (2014) also found the statistics of annual maximum thunderstorm wind speeds on average produced higher coefficients of variation and scale parameters than synoptic wind storms, as well as all wind speeds grouped, or comingled, together regardless of storm type. Cook, Harris, and Whiting (2003) reanalyzed stations used by Gomes and Vickery (1978) using additional data and found the mixed distribution was appropriate for areas with strong non-synoptic wind speeds, and they explored other options for considering multiple storm types probabilistically.

Figure 21.5 shows an example of mixed wind climate analysis using the KCMI data set. The data used for this illustration are both the annual maxima non-synoptic and

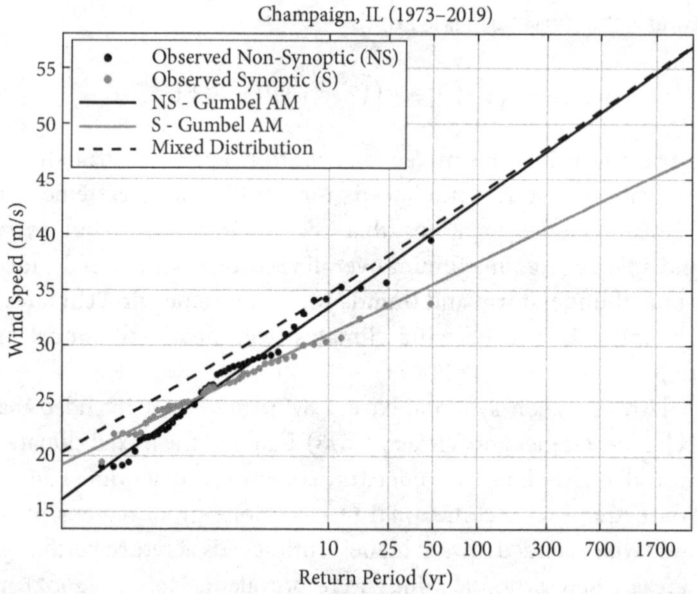

FIGURE 21.5 Prediction estimates of non-synoptic and synoptic wind speeds at Champaign, Illinois, from 1973 to 2019 using annual maxima with the derived mixed distribution (black dotted line).

synoptic wind speeds over the same time period. The tropical climate does not have significant influence at KCMI. Perhaps surprisingly, KCMI shows similar behavior to other worldwide locations in regard to the mixed distribution in that, at around a return period of 10–25 years, the mixed distribution converges to the non-synoptic probability distribution. Although the modal extreme wind speed (location parameter) for non-synoptic and synoptic winds are practically equivalent for KCMI, the dispersion (scale parameter) for non-synoptic winds is roughly 1.5 times higher than synoptic, giving the non-synoptic probability distribution a steeper slope when plotted in Gumbel coordinates.

It has been described here and in numerous other publications referenced in this chapter that there are clear statistical and hence probabilistic differences for non-synoptic winds when compared to their synoptic counterparts. It is important then to make the distinction, where possible, of reasons for the differences and whether they are more physical or data driven in nature. The next paragraphs discuss methodologies for better assessment of the differences.

2.3.4. *Simulation Techniques*

To this point, the chapter has addressed a point estimate or prediction of non-synoptic wind speeds. In other words, a single non-synoptic wind speed magnitude is given for a single frequency (i.e., return period) at a single location (height, averaging time, latitude, longitude). The prediction method has been the main method of non-synoptic wind speed analysis in the literature. A natural progression of this type of point estimate is to assess the uncertainties (in both space and time) surrounding the estimate through stochastic simulation techniques. A number of simulation techniques are described here, most of which can be used for synoptic winds. However, due to the smaller spatial and temporal scales of non-synoptic winds, a handful of techniques will be more useful for non-synoptic winds due to the relative lack of data.

Once a set of parameters for a chosen probability distribution has been estimated, a stochastic simulation can be carried out. If $\lambda > 1$ from the recorded data, then the interannual variability in λ can be simulated with a Poisson distribution. For each simulated "event" within a given year over the total length of the simulation, the magnitude can then be estimated by randomly sampling from the chosen probability distribution using Monte Carlo methods. The ability to randomly sample from a probability distribution allows for some uncertainty quantification in the prediction of extreme non-synoptic wind speeds. Simulation of non-synoptic wind speeds can also assist with detecting the presence of nonrandom trends in the observed data, which can be valuable for climate change and standardization assessment, discussed further in this section. A few references in the literature on non-synoptic wind speed prediction have discussed simulation procedures, which are focused largely on time history generation (e.g., Twisdale & Vickery, 1992). However, as stated previously, simulation methods are less interested in long-term prediction and have not been carried out to a significant extent. Given this, an example using the KCMI data set is described in the following paragraphs.

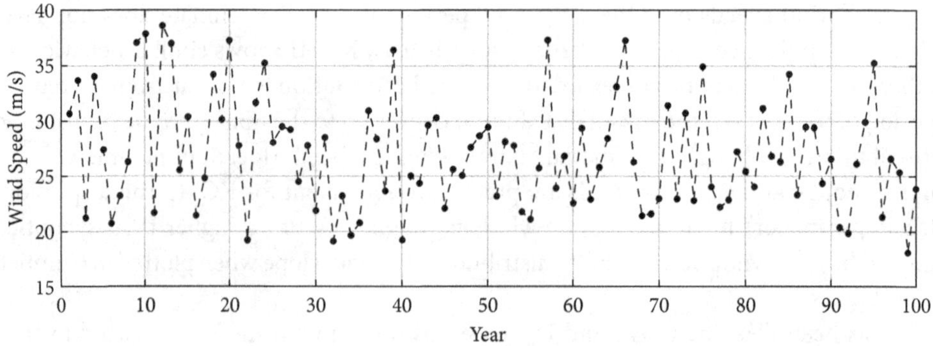

FIGURE 21.6 Stochastic simulation of non-synoptic wind speeds at Champaign, Illinois, of annual maxima for a 100-year period.

Figure 21.6 shows a single simulation of non-synoptic wind speeds using the annual maximum parameters from KCMI. In this example, 100 data points were simulated using Monte Carlo methods, which represent 100 years of data given that annual maximum parameters were used.

One potential technique to quantify or address uncertainty is to run a large number of simulations similar to the one shown in Figure 21.6. As an example, a case of parametric bootstrapping (Cook, 2013) is highlighted. As the actual KCMI data set is 47 years long or $N = 47$ and MLE-based Gumbel parameters have been estimated, a large set of 47-year time histories can be simulated using these MLE Gumbel parameters. For each set of 47-year time histories, Gumbel distributions can be fit to estimate the possible range of parameters (and hence wind speeds) for this given length of data. In this example, 10,000 of these sets were generated (i.e., 10,000 × 47 wind speed matrix), which yields 10,000 sets of Gumbel parameters. From this Gumbel parameter set, confidence intervals can be estimated. For this simulated set, the mean Gumbel parameters were $a = 4.1$ and $u = 24.2$ (in m/s). The 5th and 95th confidence intervals on the parameters were for $a = [3.3, 4.8]$ and $u = [23.2, 25.3]$. Distributions of these parameters, although not shown, appear to come from a Gaussian distribution.

Figure 21.7 shows the uncertainty in these parameters through the predicted non-synoptic wind speeds based on the 5th and 95th confidence intervals. Note these parameters represent the MLE parameters (best fit) for each simulation. As shown in Figure 21.7, there is not much uncertainty at lower return periods, but the differences become significant at higher return periods. These differences are largely due the uncertainty in the scale parameter a, as can be surmised from the confidence intervals for the parameters given. The challenges that the scale parameter present are further highlighted by doing the following: First, assume that the parameters of a given distribution (e.g., a, u, λ) are already known. Once this assumption is made, it is possible to ascertain how much of an issue sample size (N)—or in this case number of years—will be on estimating those parameters. Numerous studies (e.g., Diniz, Sadek, & Simiu, 2004) have shown uncertainties can be large when sample sizes are low, and it may cause "noise" in

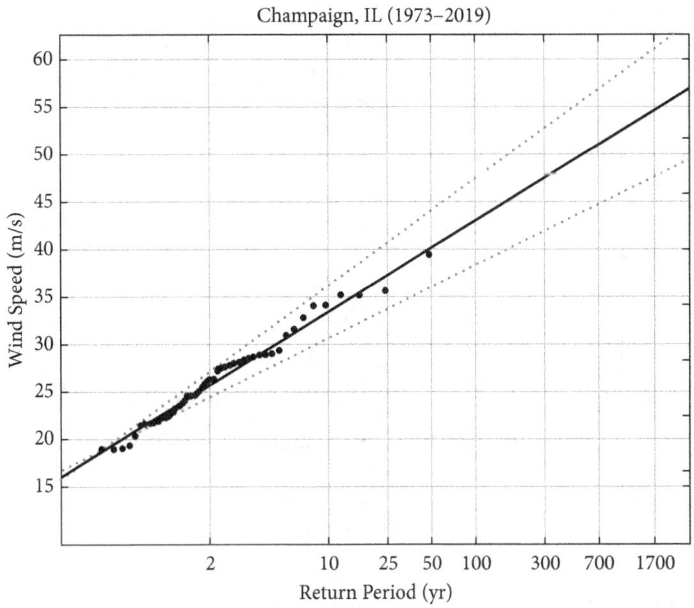

FIGURE 21.7 Confidence intervals on Gumbel parameters and associated *expected* non-synoptic wind speed prediction.

the extreme wind data to be mistaken for a "signal." For this example, Gumbel parameters based on the MIS wind data from KCMI were used.

The process illustrated here consists of simulating 20,000 years of data and "sampling" after each year of records. For example, a probability distribution is fit after Year 1, then data from Year 2 are added and a probability distribution is fit again and so on. Approaching the problem in this manner allows an analyst to determine how close the parameters of the fitted probability distribution compare to the true parameters and estimate the amount of time it takes for the parameters to converge to their true values. This procedure is illustrated for a single case in Figure 21.8. It is clear in Figure 21.8 that the location parameter converges around 10 years, but the scale and rate parameters take quite a bit longer and in some cases fail to converge after 100 years (which is longer than all available data sets used to predict non-synoptic wind speeds). Convergence for the scale and rate parameters is very sensitive to outliers—the presence of which is difficult to determine from an observed data set.

This behavior is the case for simulations with different parameters and ratios than those shown in Figure 21.8. For example, non-synoptic wind climates that have high average annual rates have on average less uncertainty in the estimation of parameters as compared to those with lower rates. This finding from the simulations seems to highlight the importance of further understanding non-synoptic wind climates in locations with infrequent occurrence but one or two significantly high wind speed values. For example, in the work of Lombardo (2014), it was noted that Boston, Massachusetts, a location that receives less than 20 percent of its annual maxima from non-synoptic

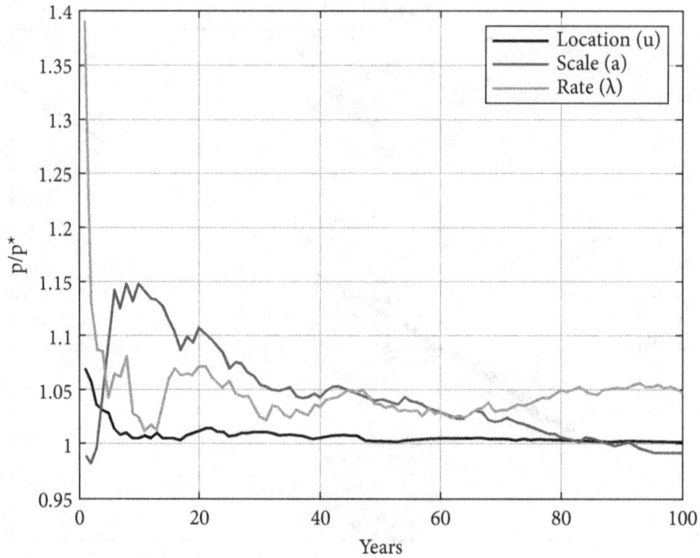

FIGURE 21.8 One realization of parameter (p) estimates (u, a, λ) for the Gumbel distribution given as a ratio with the true parameter estimates (p^*) as additional years of data up to 100 years are added. True parameters were $u = 16.1$ m/s, $a = 2.7$ m/s, and $\lambda = 23.0$ events/yr.

winds, had its highest recorded wind speed due to a non-synoptic event, and this event had a significant effect on the prediction of the mixed wind climate, especially at long return periods. This type of phenomena is discussed further in this chapter. The magnitude of the variability (i.e., scale parameter) seems to have less of an effect on average.

It should be noted that variability in the annual rate of non-synoptic winds λ was simulated using a Poisson distribution. Different distributions of λ could influence the results shown here. Although errors may be perceived as small, errors in the parameters can be magnified when simulating over significant time periods. Similar qualitative results were achieved when using annual maxima, but the accuracy was worse. It is also possible to fit the data to a GEVD and see how long the simulation needs be run before it reasonably converges to $c = 0$ (if starting with Gumbel as the "true" distribution).

As mentioned, other tests and methods can be used to assess uncertainty in prediction (Grigoriu, 2009; Minciarelli, Gioffrè, M., Grigoriu, M., & Simiu, 2001; Pandey, Van Gelder, & Vrijling, 2003). This uncertainty is especially important considering that about 50 years of data (at a maximum) exist when trying to predict non-synoptic wind speeds for return periods that are orders of magnitude higher.

As has been shown, prediction of non-synoptic extreme winds is very sensitive to parameter estimation, specifically parameters other than the location parameter. Based on these findings, Lombardo and Zickar (2019) advocated for a multimethod approach to prediction of non-synoptic wind speeds. This approach suggests using a site-specific location parameter (which is less sensitive to the extremes) u and a more regional a (or scale) parameter to dampen out some of the variation that is inherent for non-synoptic wind speeds. This sort of concept can be important for emerging philosophies, such as performance-based design (PBD) (American Society of Civil Engineers [ASCE], 2019) that will allow for more rigorous analyses of a specific site or location.

Regions of homogeneous non-synoptic climates (based on station data) can and have been assessed through methods such as *k*-means cluster analysis (Hong & Ye, 2014; Kruger et al., 2012; Lombardo & Zickar, 2019) and also can be achieved through more traditional or statistical methods (Hosking & Wallis, 1993). Data from these regions can either be "pooled" together in methods such as the superstation approach or sort of regional simulation approach (Hosking & Wallis, 1993) to predict non-synoptic wind speeds. Although regional maps have been provided in some of these references, overall discussion of the regional character of the non-synoptic wind climate has been limited.

When discussing spatial properties of non-synoptic wind events (and the wind speeds they produce), it should be noted that, unlike synoptic events, these events have relatively small spatial dimensions (e.g., Fujita, 1990) that may affect prediction of not only non-synoptic speeds but also constructs of how wind loading is predicted, especially on larger buildings. Up to this point, spatial dimensions have been ignored for non-synoptic wind speed prediction as defined in this chapter. Of particular note are the dimensions of these events relative to the density of most observing station networks, which are exclusively used to predict non-synoptic wind speeds. Figure 21.9 shows an example of a non-synoptic event (downburst) footprint generated from the work of Holmes and Oliver (2000) overlaid with a typical station network density near KCMI. Figure 21.9 shows that the footprint is barely even visible when showing scales of typically spaced airport stations that would be used in the prediction of non-synoptic wind speeds. Footprint models like these are based on very limited field data and their resemblance to actual events has not been rigorously evaluated, so it is difficult to predict true non-synoptic wind speeds and the risks at return periods typically needed. Section 3 discusses potential avenues for future research that can improve models used in prediction of non-synoptic wind speeds.

In recent years, as mentioned previously, denser station networks have been cropping up in the United States and across the world (e.g., Schroeder et al., 2005; Solari et al., 2012). Earlier works have shown using denser station networks can lead to different descriptions of the extreme wind climate. For example, denser networks tend to capture more extreme, small-scale thunderstorm events (Lombardo, 2012), and station density or some similar metric should be accounted for when performing an extreme value analysis for non-synoptic events. For example, values like the annual rate of non-synoptic wind speeds when considering a threshold or MIS-type method will certainly be a function of the station density. For denser, non–airport-based station networks, however, thunderstorm flags are usually not available. Automation of the non-synoptic identification process can be done using techniques discussed previously in this chapter.

Figure 21.10 and Figure 21.11 further highlight an example of potentially how many non-synoptic events are not even being captured using traditional assessment methods (i.e., from surface-based airport recording stations). For this particular example, two

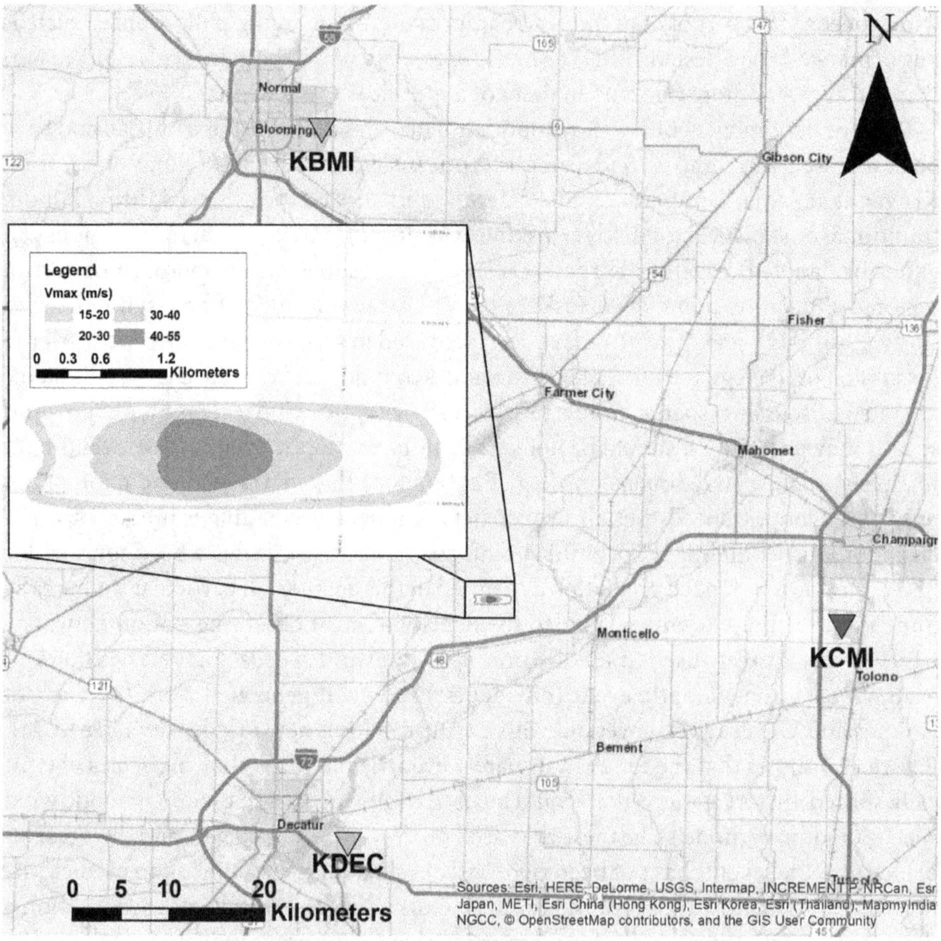

FIGURE 21.9 Simulated wind field footprint of non-synoptic event (downburst) from Holmes and Oliver (2000) overlaid on a map of Illinois, United States, showing Champaign, Illinois, station and others that would reasonably be used for non-synoptic wind speed prediction.

networks in the state of Oklahoma were used. One network is the traditional aviation-based surface network (i.e., ASOS). The second network is known as the Oklahoma Mesonet (OKM), a dense surface-based network of stations that was developed for a number of applications, including agriculture (Brock et al., 1995). Although not specifically intended for wind speed prediction, the data produced from the OKM can be used for such a purpose.

Figure 21.10 highlights the locations of both networks. In this example, the OKM has 108 stations, while the ASOS network has 39 total that were used for the analysis.

For approximately 25 years of overlapping data collection between OKM and ASOS, all OKM events were identified that had over 20 m/s peak gusts and were used as the "control." If an event was recorded (not recorded) by an ASOS station a "hit" ("miss")

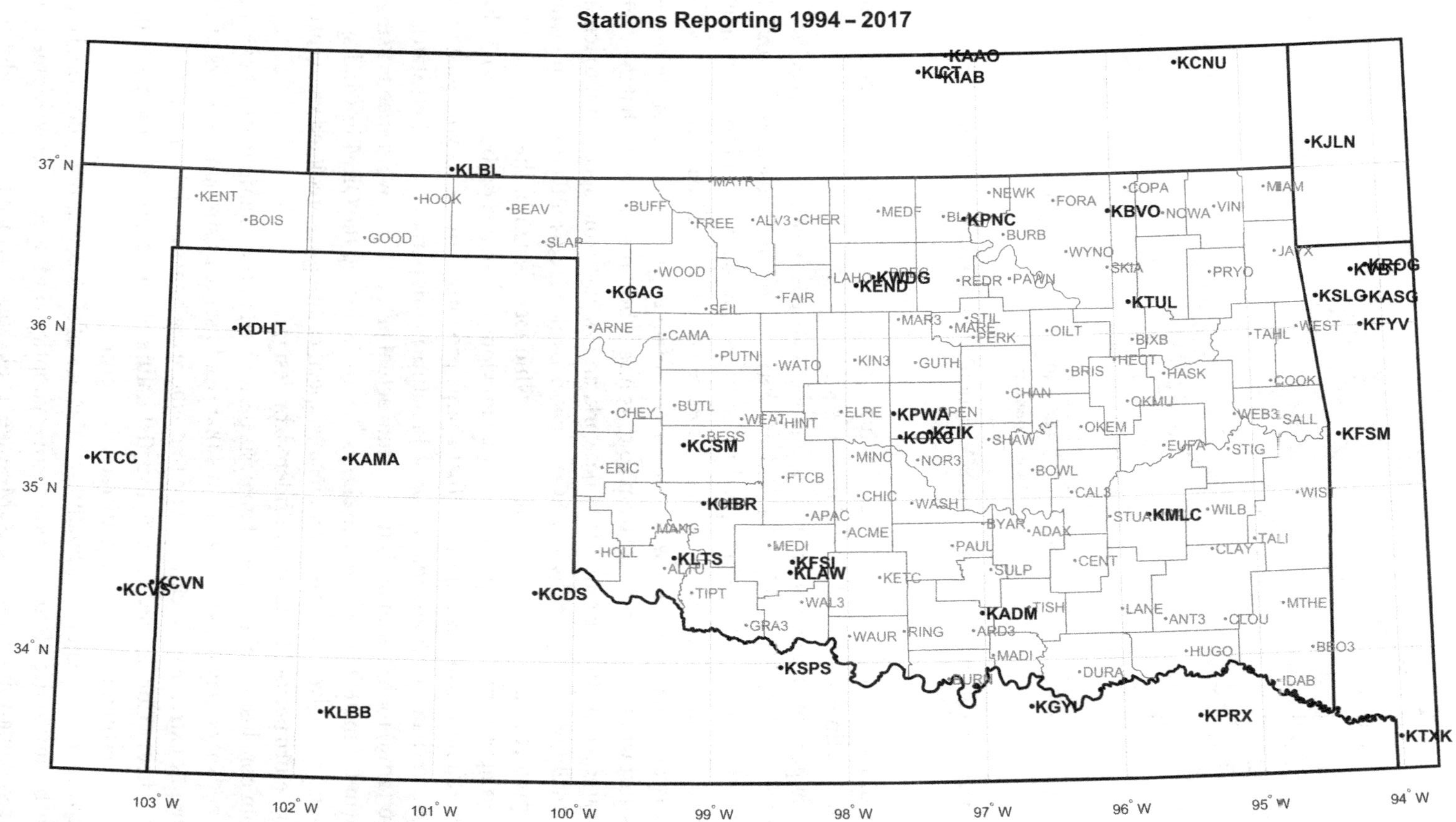

FIGURE 21.10 Location of ASOS stations (black dots) and Oklahoma Mesonet stations (gray dots).

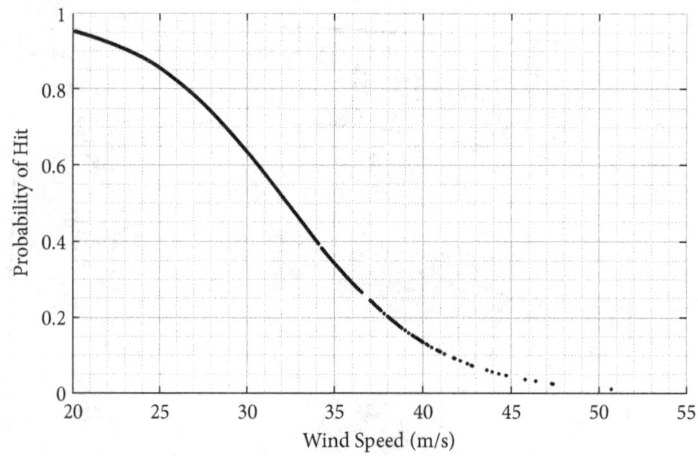

FIGURE 21.11 Likelihood of "hit" at *any* ASOS station given an event of a certain wind speed magnitude from OKM.

was noted. Hit or misses were defined as functions of wind speed magnitude, wind direction, and time of event (Zickar, 2019).

For this particular example, non-synoptic wind speeds have not been specifically classified for OKM. However, a convenient metric that can be used as a proxy to non-synoptic wind speeds—wind speed magnitude—can be used to infer how likely non-synoptic events, especially those of an extreme nature, will be recorded by the ASOS network. Figure 21.11 shows the likelihood of an event hit using logistic regression. As gust wind speeds increase from 30 to 35 m/s, it becomes unlikely that the event will be captured by the current ASOS network, and as gust wind speeds approach 40 m/s, the likelihood of capture by an ASOS network (which is currently the only network used to predict non-synoptic wind speeds for wind load design in the United States) reduces to approximately 10 percent or less. Previous research (Lombardo, 2012) has shown that a fairly safe assumption is that a large proportion of events with gust wind speed greater than 30 m/s are non-synoptic in nature in this part of the United States.

Zickar (2019) took this analysis further into actual prediction of wind speeds using a mixed wind climate. His Figure 6.26 shows that by including the OKM in wind speed prediction, the analysis yields different results, specifically higher predicted wind speeds for given return periods typically considered for design.

A main conclusion of the type of analysis performed in this example is that existing surface networks may miss crucial details that would serve as strong predictors of the non-synoptic extreme wind climate, and that a more regional description may serve as a better predictor. Conversely, the small spatial nature of non-synoptic wind gusts may cause an overestimation of risk as illustrated in Figure 21.9.

To further the concept of regionality and the lack of surface stations, additional data sets can be utilized to better predict non-synoptic wind speeds. Development of grid-based reanalysis data sets (e.g., Mesinger et al., 2006), which have typically been assimilated over relatively long time periods with near global coverage, has also given researchers the opportunity to look at the large-scale environments responsible for extreme non-synoptic winds. The overall spatial coverage (usually global or continental) is unmatched by observational techniques; however, grid boxes usually span multiple kilometers, which eliminates the ability to ascertain localized parameters exactly at the site of interest, for example. Therefore, the reanalysis data, although large scale, can be especially useful when coupled with flagged or otherwise identified non-synoptic surface wind speed data for both stochastic simulation and prediction of non-synoptic wind speeds (e.g., Lombardo & Ayyub, 2015).

2.3.5. *Directionality*

Work on predictions of non-synoptic wind speeds as discussed have largely focused on nondirectional wind climates (i.e., all directions together). The framework of these predictions can easily be extended to the concept of analyzing non-synoptic wind speeds by direction. Although there has been some research on wind directionality of non-synoptic winds, it has garnered much less attention. Wind directionality in this chapter describes the relative frequency of wind from all directions. This lack of work in directionality is most likely due to the lower frequency of non-synoptic wind speeds, which becomes even less as the wind speeds are partitioned further into directional sectors (most 10°, 22.5°, 45°). A potential drawback of this method is that the larger the sector, the more data are needed for the prediction but there is also potential for loss in accuracy or, in other words, a washing out of the directional wind climate.

Studies mentioned in the next paragraph show non-synoptic winds have different directional distributions from their synoptic counterparts for the same location or in studies that are more meteorological in nature, significant directionality (e.g., Geerts, 2001), which influences the applicability of nondirectional predictions, especially for structures whose wind-induced loadings are sensitive to direction.

Riera and Nanni (1989) appear to be the first to tackle directionality specifically in regard to prediction of non-synoptic wind speeds for a small set of stations in Brazil. They showed some relatively minor differences in non-synoptic wind direction compared to synoptic, which suggests a specific non-synoptic generation mechanism aligned with synoptic flow (more on that further in the chapter). Twisdale and Vickery (1992) discussed a directional prediction model for thunderstorm winds using daily extremes and showed different directions for the peaks of synoptic versus non-synoptic winds. Kruger et al. (2012) noted that non-synoptic wind speeds in areas of South Africa generally came from the west. Lombardo and Zickar (2019) also noted most non-synoptic wind speeds in the United States, especially those considered "severe" (>30 m/s), came mostly from a direction with a westerly component when considering four directional sectors. When pooling the data in regions, they also found a preference for a westerly direction and large differences in non-synoptic wind prediction when considering direction. Zhang et al. (2019) noted a preference for a specific directional sector and directional shifts for non-synoptic wind speeds in Italy.

Although not specifically predictions of non-synoptic wind speeds, Yeo (2014) used the directional data from Lombardo et al. (2009) to generate long synthetic (stochastic) time histories of directional non-synoptic winds from which predictions could be made, although it was mentioned that the relatively low frequency of non-synoptic wind speeds could affect the simulations. Habte, Chowdhury, Yeo, and Simiu (2015) used the same data to generate directionality factors similar to those used in codes and standards for non-synoptic winds.

To continue the example illustration, directional polar histograms of both non-synoptic and synoptic wind speeds (those greater than 13 m/s) for KCMI are shown in Figure 21.12. Directional sectors in this example are shown at every 45°. Figure 21.12 shows that the two different storm types have different directional distributions, with a lesser directional preference (westerly in this case) in non-synoptic winds. However, the distributions have some correlation suggesting a portion of the non-synoptic wind climate may be driven by the synoptic wind climate (Lombardo & Zickar, 2019). For example, thunderstorms producing non-synoptic wind speeds along a synoptic frontal passage (e.g., Dowdy & Catto, 2017; Kasperski, 2002). The implications of this correlation between synoptic and non-synoptic is addressed in the final section of this chapter.

For KCMI, there are approximately 1,000 independent thunderstorm wind speeds over the 47-year history of the station. When partitioned into eight sectors in a clearly

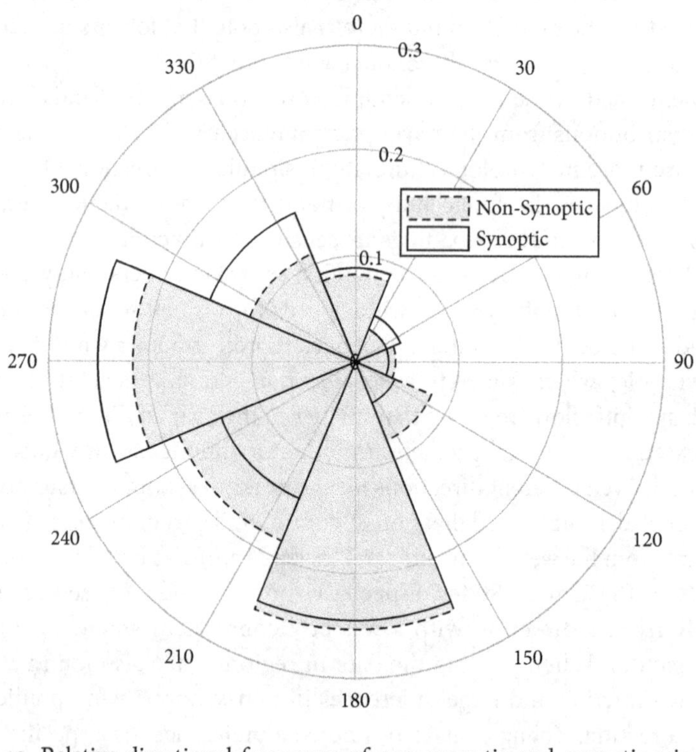

FIGURE 21.12 Relative directional frequency of non-synoptic and synoptic wind speeds for KCMI. A wind direction of 0° is a wind coming from the north.

nonuniform directional wind climate, both the sample sizes and the wind speeds can become precariously low for a large number of sectors. For the eight 45° sectors, only six had a non-synoptic λ greater than 1. These lower rates by direction should open the door to alternative simulation and prediction measures using data that is not from a surface-observing station. If a threshold of 20 m/s is used, only two sectors have λ greater than 1. As stated previously, the differences in choice of sector width will influence predictions of the non-synoptic wind climate. Additional work and research avenues considering wind directionality of non-synoptic wind speeds are discussed in the final section of the chapter.

2.3.6. *Climate Change*

Even less discussed than directionality is the idea of climate change becoming part of the prediction process of non-synoptic wind speeds. Assessing climate change effects in modifying the frequency and intensity of extreme wind events or at the very least the conditions conducive for these events has been stressed as a priority in the last decade (Intergovernmental Panel on Climate Change [IPCC] 2012). However, the general consensus is that both the quality of data and the understanding of the phenomena when talking about non-synoptic winds are the least understood when trying to attribute changes to climate (Lubchenco & Karl, 2012). Understanding possible changes in thunderstorm (non synoptic) parameters in a probabilistic sense is a research priority for wind engineering. Most of the studies of this topic have been pretty far removed from wind engineering contexts and have focused on the possibility of the conditions (e.g., instability) that are conducive to thunderstorm development being modified by changes in climate. The feasibility of larger scale climatic features influencing thunderstorm frequency and location has been well documented (e.g., Allen & Allen, 2016; Diffenbaugh, Scherer, & Trapp, 2013; Trapp et al., 2007) using climate models, and the large majority of research suggests that these specific thunderstorm conditions are likely to change in the future.

Since the discussion in this chapter revolves around non-synoptic gust wind speeds measured very close to the ground over terrain, there are a number of factors other than climate change that could significantly influence the magnitude over time, and hence prediction, of non-synoptic wind events. These factors, which were discussed briefly in the introduction can be averaging time, type of recording system, height above ground, and surface roughness, among other factors. A trend, for example, in the annual maximum non-synoptic wind speed over time due to different recording and standardization techniques will significantly complicate such analyses (Lombardo & Ayyub, 2015). Methods to deal with possible trends either through wind speed correction or extreme value parameterization can be found in the work of Wang, Wang, and Khoo (2013) and Hundecha, St.-Hilaire, Ouarda, El Adlouni, and Gachon (2008). However, the conditions or large-scale parameters that are used in studies of a more meteorological nature are typically included in reanalysis data sets. It is possible therefore to assess how the large-scale parameters change over time and infer how near-surface non-synoptic wind speeds may change. A similar methodology was employed by Lombardo and Ayyub

(2017). For example, reanalysis data could be used to assess changes in *thunderstorm frequency*. Frequency changes could then be incorporated into stochastic simulations and prediction models as was done by Cechet et al. (2010).

3. Future Work/Research Needs

The relatively limited breadth and depth of prediction of non-synoptic wind speeds opens the door for novel research on the topic, especially in an engineering context. Part of the difficulty in developing a wide body of research on the topic stems from the difficulty in recording events due to the numerous factors discussed in this chapter.

Speaking to the issue of recording events, high-resolution measurements of non-synoptic wind events in both space and time need to be routinely collected worldwide. A handful of studies, both long- (Solari et al., 2012) and short-term projects (Gunter & Schroeder, 2015; Orwig & Schroeder, 2007), have been up to the significant task of data collection and to some extent archival; however, it is not enough to adequately characterize the true details and variability of non-synoptic wind events. These studies, to date, have not lasted long enough to rigorously predict non-synoptic wind speeds. Current archived data for high-resolution measurements at most fixed existing surface networks only report a peak gust over 1- to 10-min which allows for the potential of "losing" important details, especially those for classification of thunderstorm events (which may not last longer than the intervals mentioned) at the initial stages of the prediction process. Continuous updating and reanalysis of non-synoptic wind speed prediction should take place as more data in both time and space become available.

Following collection of data, there is a need to improve classification techniques for non-synoptic events, including for networks that have thunderstorm flags. These include using archived radar imagery to better pinpoint times and locations of non-synoptic wind occurrence. Other emerging techniques, such as machine learning, can be used to initially classify (and possibly subclassify; see next paragraph) thunderstorm events (e.g., Spassiani, Mason, Krupar, & Cheng, 2018). However, harking back to data collection and archiving, a relatively coarse spatial and temporal network makes some of the improved classification techniques harder to implement.

In many recent works, as non-synoptic data has been more routinely collected, analysis has hinted at the possibility of a subdivision within the general category of "non-synoptic" and is apparent and discussed to some extent even in earlier works (e.g., Kasperski, 2002). Not including subdivisions could potentially be analogous to including tropical cyclone wind speeds together with synoptic wind speeds and then fitting a single probability distribution. Physically, the generating mechanisms are different depending on the type of non-synoptic wind event (e.g., microburst vs. derecho) and are dependent on the environmental conditions where the storm occurs (Wakimoto, 2001).

For wind engineering, and more specifically prediction, purposes, it has been noted that the duration of high winds and "sharpness" of the peak in wind speed time histories

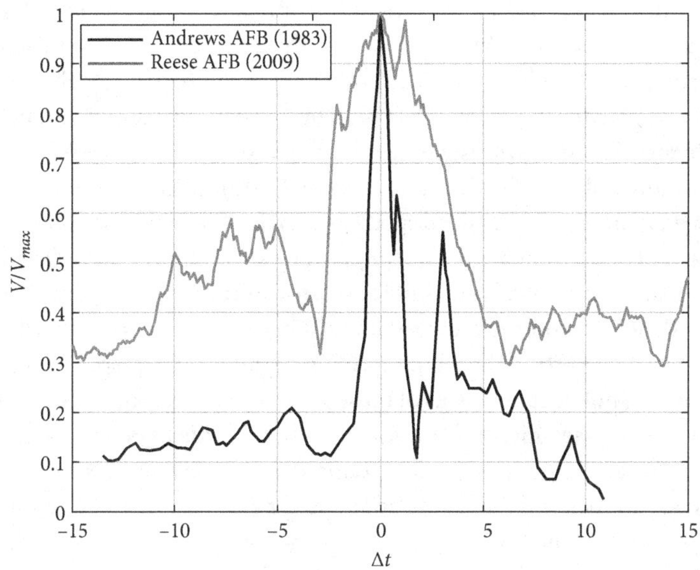

FIGURE 21.13 Wind speed time history (x-axis in minutes) characteristics from two different non-synoptic wind storm types. The Andrews Air Force Base (AFB) event was attributable to a microburst and the Reese AFB event attributable to a bow echo.

can be largely dependent on the type of thunderstorm (Lombardo, Smith, Schroeder, & Mehta, 2014). Indeed, other research (Solari, 2016; Zhang et al., 2018) has shown a wide variation in the temporal character of thunderstorm events. More research on this topic is needed. An example of two wind speed time histories generated from different non-synoptic events is shown in Figure 21.13. Figure 21.13 shows the significant time and velocity differences associated with these two events, one of which was a bow echo, or more linear non-synoptic wind storm similar to a derecho (Lombardo et al., 2014), and the other is the Andrews Air Force Base downburst, or microburst (Fujita, 1985). The difference in the sharpness of the peak is evident. Time histories were smoothed and scaled similar to the work of Lombardo et al. (2014) and Solari (2016), respectively. Other research (Lombardo et al., 2014) has shown that the near-surface turbulence characteristics differ as well. Detailed radar analyses of storm type or classifying time histories of other meteorological variables (e.g., pressure rise/pressure drop) could also be used.

Another issue to be considered with classification is the idea of "joint" or intermediate events, as suggested by Kasperski (2002). For example, Lombardo (2014) noted that the strongest ever recorded wind speed in Boston, Massachusetts (KBOS) was due to a non-synoptic event (i.e., had a thunderstorm flag). A further investigation into this event revealed that although a thunderstorm was reported at the station at the time of the peak gust, it was clear that the gust occurred in significant background synoptic flow due to a historically strong synoptic system. A similar event occurred in Brazil based on the analysis of Vallis et al. (2019), with a peak gust of over 48 m/s. More detailed discussion and research of this "hybrid" phenomenon should happen for cases like these (Dowdy & Catto, 2017). Even an event such as the leading edge of a cold front, a synoptic feature, can produce localized, transient gusts (e.g., Colquhoun, Shepherd, Coulman, Smith, & McInnes, 1985) and could be considered a non-synoptic event. Any of the events mentioned previously may modify wind loading to be different from that of the ABL. Improved classification and subclassification will likely reduce some of the variability seen in non-synoptic wind speed properties discussed in this chapter.

As the focus of prediction has been a single height and standard conditions, it should be noted that there are no current studies on, for example, non-synoptic wind speed prediction with height. Some tall tower data exist in the literature (Lombardo et al., 2014; Zhang et al., 2019) for vertical distributions of non-synoptic wind speeds, but prediction of wind speeds has not been the focus of these studies. Overall four-dimensional characterization of spatial assessment of non-synoptic wind fields to better assess overall risk to infrastructure and comparison or updating of existing non-synoptic "footprint" models such as Holmes and Oliver (2000) should be carried out. Spatial assessment of non-synoptic wind events, including hazard-mapping research for both individual events and longer term predictions, has been classified as a priority research and development item by wind engineering researchers and practitioners. Overall, spatial assessment is important, especially as non-synoptic wind speeds begin to control the design in wind load codes and standards at 10 m height.

Proper standardization of non-synoptic wind speed values for use in codes and standards requires knowledge of the physical characteristics that generate the flow fields and the modifying effects of terrain associated with these events. Simplified regionally based mapping as is currently employed requires conversion into standardized micrometeorological conditions as discussed in the previous section and more thought should be given on best ways to standardize non-synoptic data, even if it is simply using conventional ABL methods.

In general, a greater push, when appropriate, to include more information about non-synoptic wind speeds and their prediction in codes and standards should be a priority. For example, a lingering question is how these short-duration events impact the response of high-rise buildings (e.g., crosswind response), which are historically based on hourly mean values and ABL wind profiles. From a broader design perspective, the trend, especially in the United States, is toward a PBD framework (ASCE, 2019). Using PBD methods will allow for analysis methods specific to non-synoptic winds where warranted. For example, instead of using a single directionality factor and wind speed from a mixed distribution, specific directional distributions of non-synoptic wind speeds could be used and coupled with non-synoptic building aerodynamics.

A framework to assess overall risk from non-synoptic wind events is long overdue. There seems to be a clear need for a comprehensive (at all scales) convective storm model published in the archived literature, similar to those generated for tropical cyclones (Vickery, Wadhera, Powell, & Chen, 2009). Assessment of the occurrence probabilities should also incorporate the joint independent probabilities of different non-synoptic wind storm types (if warranted) as well as the joint concurrent probabilities of multiple hazards (e.g., wind and storm surge, wind and hail). Numerical modeling solutions on various temporal and spatial scales. including large-eddy simulation (Aboshosha, Bitsuamlak, & El Damatty, 2015; Sengupta et al., 2008), numerical weather prediction models such as Weather Research and Forecasting (WRF) (e.g., Vermeire et al., 2011), and reanalysis data can and should be used in conjunction with traditional methods of non-synoptic wind speed prediction.

In summary, the importance of considering non-synoptic wind speeds when discussing the extreme wind climate (i.e., design wind speed) is clear. Of course, the prediction of extreme non-synoptic wind speeds is only one piece of the puzzle. A single peak wind speed is part of a larger four-dimensional wind field, which is largely complex for non-synoptic wind storms. Discussion of these related phenomena takes place elsewhere in this handbook.

References

Aboshosha, H., Bitsuamlak, G., & El Damatty, A. (2015). Turbulence characterization of downbursts using LES. *Journal of Wind Engineering and Industrial Aerodynamics, 136*, 44–61.

Allen, J. T., & Allen, E. R. (2016). A review of severe thunderstorms in Australia. *Atmospheric Research, 178*, 347–366.

American Society of Civil Engineers (ASCE). (2019). *Prestandard for performance-based wind design*. https://doi.org/10.1061/9780784482186

Brock, F. V., Crawford, K. C., Elliott, R. L., Cuperus, G. W., Stadler, S. J., Johnson, H. L., & Eilts, M. D. (1995). The Oklahoma Mesonet: A technical overview. *Journal of Atmospheric and Oceanic Technology, 12*(1), 5–19.

Cechet, R. P., Arthur, W. C., Wehner, M., Sanabria, L. A., Thomas, C., Nadimpalli, K., & Yang, T. (2010, June). *National wind risk assessment (NWRA) for the Australian region*. The 5th International Symposium on Computational Wind Engineering (CWE 2010), Chapel Hill, NC.

Colquhoun, J. R., Shepherd, D. J., Coulman, C. E., Smith, R. K., & McInnes, K. (1985). The southerly burster of south eastern Australia: An orographically forced cold front. *Monthly Weather Review, 113*(12), 2090–2107.

Cook, N. J. (2013). Calibration of EVA methods for peak-over-threshold wind data using bootstrapping. *Journal of Wind Engineering and Industrial Aerodynamics, 120*, 91–95.

Cook, N. J. (2014). Consolidation of analysis methods for sub-annual extreme wind speeds. *Meteorological Applications, 21*(2), 403–414.

Cook, N. J., Harris, R. I., & Whiting, R. (2003). Extreme wind speeds in mixed climates revisited. *Journal of Wind Engineering and Industrial Aerodynamics, 91*(3), 403–422.

Davenport, A. G. (1961). The application of statistical concepts to the wind loading of structures. *Proceedings of the Institution of Civil Engineers, 19*(4), 449–472.

De Gaetano, P., Repetto, M. P., Repetto, T., & Solari, G. (2014). Separation and classification of extreme wind events from anemometric records. *Journal of Wind Engineering and Industrial Aerodynamics, 126*, 132–143.

Diffenbaugh, N. S., Scherer, M., & Trapp, R. J. (2013). Robust increases in severe thunderstorm environments in response to greenhouse forcing. *Proceedings of the National Academy of Sciences of the United States of America, 110*(41), 16361–16366.

Diniz, S. M., Sadek, F., & Simiu, E. (2004). Wind speed estimation uncertainties: Effects of climatological and micrometeorological parameters. *Probabilistic Engineering Mechanics, 19*(4), 361–371.

Dowdy, A. J., & Catto, J. L. (2017). Extreme weather caused by concurrent cyclone, front and thunderstorm occurrences. *Scientific Reports, 7*, 40359.

Fujita, T. T. (1985). The downburst. *Satellite and Mesometeorology Research Project 210*, 112.

Fujita, T. T. (1990). Downbursts: Meteorological features and wind field characteristics. *Journal of Wind Engineering and Industrial Aerodynamics, 36*, 75–86.

Geerts, B. (2001). Estimating downburst-related maximum surface wind speeds by means of proximity soundings in New South Wales, Australia. *Weather and Forecasting, 16*(2), 261–269.

Gomes, L., & Vickery, B. J. (1978). Extreme wind speeds in mixed wind climates. *Journal of Wind Engineering and Industrial Aerodynamics, 2*(4), 331–344.

Grigoriu, M. (2009). *Algorithms for generating large sets of synthetic directional wind speed data for hurricane, thunderstorm, and synoptic winds*. Gaithersburg, MD: US Department of Commerce, National Institute of Standards and Technology.

Gunter, W. S., & Schroeder, J. L. (2015). High-resolution full-scale measurements of thunderstorm outflow winds. *Journal of Wind Engineering and Industrial Aerodynamics, 138*, 13–26.

Habte, F., Chowdhury, A. G., Yeo, D., & Simiu, E. (2015). Wind directionality factors for non-hurricane and hurricane-prone regions. *Journal of Structural Engineering, 141*(8), 04014208.

Hangan, H., Romanic, D., & Jubayer, C. (2019). Three-dimensional, non-stationary and non-Gaussian (3D-NS-NG) wind fields and their implications to wind–structure interaction problems. *Journal of Fluids and Structures, 910*(8), 102583.

Harris, R. I. (1996). Gumbel re-visited-a new look at extreme value statistics applied to wind speeds. *Journal of Wind Engineering and Industrial Aerodynamics, 59*(1), 1–22.

Harris, R. I. (1999). Improvements to the method of independent storms. *Journal of Wind Engineering and Industrial Aerodynamics, 80*(1–2), 1–30.

Holmes, J. D. (2002). A re-analysis of recorded extreme wind speeds in region A. *Australian Journal of Structural Engineering, 4*(1), 29–40.

Holmes, J. D., & Oliver, S. E. (2000). An empirical model of a downburst. *Engineering Structures, 22*(9), 1167–1172.

Hong, H. P., & Ye, W. (2014). Estimating extreme wind speed based on regional frequency analysis. *Structural Safety, 47*, 67–77.

Hosking, J. R. M., & Wallis, J. R. (1993). Some statistics useful in regional frequency analysis. *Water Resources Research, 29*(2), 271–281.

Hundecha, Y., St.-Hilaire, A., Ouarda, T. B. M. J., El Adlouni, S., & Gachon, P. (2008). A non-stationary extreme value analysis for the assessment of changes in extreme annual wind speed over the Gulf of St. Lawrence, Canada. *Journal of Applied Meteorology and Climatology, 47*(11), 2745–2759.

Intergovernmental Panel on Climate Change (IPCC). (2012). *Special report on managing the risks of extreme events and disasters to advance climate change adaptation*. Cambridge, UK: Cambridge University Press.

Kasperski, M. (2002). A new wind zone map of Germany. *Journal of Wind Engineering and Industrial Aerodynamics, 90*(11), 1271–1287.

Kim, J., Hangan, H., & Eric Ho, T. C. (2007). Downburst versus boundary layer induced wind loads for tall buildings. *Wind and Structures, 10*(5), 481–494.

Kruger, A. C., Goliger, A. M., Retief, J. V., & Sekele, S. (2010). Strong wind climatic zones in South Africa. *Wind and Structures, 13*(1), 37–55.

Kruger, A. C., Goliger, A. M., Retief, J. V., & Sekele, S. S. (2012). Clustering of extreme winds in the mixed climate of South Africa. *Wind and Structures, 15*(2), 87–109.

Kruger, A. C., Retief, J. V., & Goliger, A. M. (2017). Development of an updated fundamental basic wind speed map for SANS 10160-3. *Journal of the South African Institution of Civil Engineering, 59*(4), 12–25.

Kuligowski, E. D., Lombardo, F. T., Phan, L. T., Levitan, M. L., & Jorgensen, D. P. (2014). *Technical investigation of the May 22, 2011, tornado in Joplin, Missouri* (Rep. NIST NCSTAR, 3). Gaithersburg, MD: US Department of Commerce, National Institute of Standards and Technology.

Kwon, D. K., & Kareem, A. (2009). Gust-front factor: New framework for wind load effects on structures. *Journal of Structural Engineering, 135*(6), 717–732.

Lombardo, F. T. (2012). Improved extreme wind speed estimation for wind engineering applications. *Journal of Wind Engineering and Industrial Aerodynamics, 104*, 278–284.

Lombardo, F. T. (2014). Extreme wind speeds from multiple wind hazards excluding tropical cyclones. *Wind and Structures, 19*(5), 467–480.

Lombardo, F. T., & Ayyub, B. M. (2015). Analysis of Washington, DC, wind and temperature extremes with examination of climate change for engineering applications. *ASCE-ASME Journal of Risk and Uncertainty in Engineering Systems, Part A: Civil Engineering, 1*(1), 04014005.

Lombardo, F. T., & Ayyub, B. (2017). Approach to estimating near-surface extreme wind speeds with climate change considerations. *ASCE–ASME Journal of Risk and Uncertainty in Engineering Systems, Part A: Civil Engineering, 3*(3), A4017001.

Lombardo, F. T., Main, J. A., & Simiu, E. (2009). Automated extraction and classification of thunderstorm and non-thunderstorm wind data for extreme-value analysis. *Journal of Wind Engineering and Industrial Aerodynamics, 97*(3–4), 120–131.

Lombardo, F. T., Smith, D. A., Schroeder, J. L., & Mehta, K. C. (2014). Thunderstorm characteristics of importance to wind engineering. *Journal of Wind Engineering and Industrial Aerodynamics, 125*, 121–132.

Lombardo, F. T., & Zickar, A. S. (2019). Characteristics of measured extreme thunderstorm near-surface wind gusts in the United States. *Journal of Wind Engineering and Industrial Aerodynamics, 193*, 103961.

Loredo-Souza, A. M., Lima, E. G., Vallis, M. B., Rocha, M. M., Wittwer, A. R., & Oliveira, M. G. (2019). Downburst related damages in Brazilian buildings: Are they avoidable? *Journal of Wind Engineering and Industrial Aerodynamics, 185*, 33–40.

Lubchenco, J., & Karl, T. R. (2012). Extreme weather events. *Physics Today, 65*(3), 31.

Makkonen, L. (2006). Plotting positions in extreme value analysis. *Journal of Applied Meteorology and Climatology, 45*(2), 334–340.

Mason, M. S., Letchford, C. W., & James, D. L. (2005). Pulsed wall jet simulation of a stationary thunderstorm downburst, Part A: Physical structure and flow field characterization. *Journal of Wind Engineering and Industrial Aerodynamics, 93*(7), 557–580.

McKee, T. B., Doesken, N. J., & Kleist, J. (1996). *Climate data continuity with ASOS: Report for the period September 1994–March 1996*. Fort Collins, CO: Colorado Climate Center, Department of Atmospheric Science, Colorado State University.

Mesinger, F., DiMego, G., Kalnay, E., Mitchell, K., Shafran, P. C., Ebisuzaki, W., . . . Ek, M. B. (2006). North American regional reanalysis. *Bulletin of the American Meteorological Society, 87*(3):343–360.

Minciarelli, F., Gioffrè, M., Grigoriu, M., & Simiu, E. (2001). Estimates of extreme wind effects and wind load factors: Influence of knowledge uncertainties. *Probabilistic Engineering Mechanics, 16*(4), 331–340.

Mohr, S., Kunz, M., Richter, A., & Ruck, B. (2017). Statistical characteristics of convective wind gusts in Germany. *Natural Hazards & Earth System Sciences*, *17*(6), 957–969.

Munich, RE. (2019). *Relevant natural catastrophe loss events worldwide 2019*. Retrieved from https://www.munichre.com/content/dam/munichre/global/content-pieces/documents/media-relations/2019-nat-cat-world-map.pdf/_jcr_content/renditions/original./2019-nat-cat-world-map.pdf

National Oceanic and Atmospheric Administration (NOAA). (2020). *Billion-dollar climate and weather disasters: An overview*. Retrieved from https://www.ncdc.noaa.gov/billions/

Orwig, K. D., & Schroeder, J. L. (2007). Near-surface wind characteristics of extreme thunderstorm outflows. *Journal of Wind Engineering and Industrial Aerodynamics*, *95*(7), 565–584.

Pandey, M. D., Van Gelder, P. H. A. J. M., & Vrijling, J. K. (2003). Bootstrap simulations for evaluating the uncertainty associated with peaks-over-threshold estimates of extreme wind velocity. *Environmetrics*, *14*(1), 27–43.

Pintar, A. L., Simiu, E., Lombardo, F. T., & Levitan, M. L. (2015). *Maps of non-hurricane non-tornadic wind speeds with specified mean recurrence intervals for the contiguous United States using a two-dimensional Poisson process extreme value model and local regression* (NIST SP-500-301). Gaithersburg, MD: US Department of Commerce, National Institute of Standards and Technology.

Riera, J. D., & Nanni, L. F. (1989). Pilot study of extreme wind velocities in a mixed climate considering wind orientation. *Journal of Wind Engineering and Industrial Aerodynamics*, *32*(1–2), 11–20.

Riera, J. D., Viollaz, A. J., & Reimundin, J. C. (1977). Some recent results on probabilistic models of extreme wind speeds. *Journal of Wind Engineering and Industrial Aerodynamics*, *2*, 271–287.

Sarkar, P. P., Haan, F. L., Jr., Balaramudu, V., & Sengupta, A. (2006, May). Laboratory simulation of tornado and microburst to assess wind loads on buildings. In E. Cross & J. Finke (Eds.), *Structures Congress 2006: Structural Engineering and Public Safety* (pp. 1–10). Reston, VA: ASCE.

Schroeder, J. L., Burgett, W. S., Haynie, K. B., Sonmez, I., Skwira, G. D., Doggett, A. L., & Lipe, J. W. (2005). The West Texas mesonet: A technical overview. *Journal of Atmospheric and Oceanic Technology*, *22*(2), 211–222.

Sengupta, A., Haan, F. L., Sarkar, P. P., & Balaramudu, V. (2008). Transient loads on buildings in microburst and tornado winds. *Journal of Wind Engineering and Industrial Aerodynamics*, *96*(10–11), 2173–2187.

Shehata, A. Y., El Damatty, A. A., & Savory, E. (2005). Finite element modeling of transmission line under downburst wind loading. *Finite Elements in Analysis and Design*, *42*(1), 71–89.

Simiu, E., & Heckert, N. A. (1996). Extreme wind distribution tails: A "peaks over threshold" approach. *Journal of Structural Engineering*, *122*(5), 539–547.

Solari, G. (2016). Thunderstorm response spectrum technique: Theory and applications. *Engineering Structures*, *108*, 28–46.

Solari, G., De Gaetano, P., & Repetto, M. P. (2015). Thunderstorm response spectrum: Fundamentals and case study. *Journal of Wind Engineering and Industrial Aerodynamics*, *143*, 62–77.

Solari, G., Repetto, M. P., Burlando, M., De Gaetano, P., Pizzo, M., Tizzi, M., & Parodi, M. (2012). The wind forecast for safety management of port areas. *Journal of Wind Engineering and Industrial Aerodynamics*, *104*, 266–277.

Spassiani, A. C., Mason, M. S., Krupar, R. J., & Cheng, V. (2018, October). *An Australian climatology of severe convective wind storms using Bayesian hierarchical modelling.* Presented at the 29th Conference on Severe Local Storms. American Meteorology Society, Stowe, VT.

Thom, H. C. S. (1960). Distributions of extreme winds in the United States. *Transactions of the American Society of Civil Engineers, 126*(2), 450–462.

Trapp, R. J., Diffenbaugh, N. S., Brooks, H. E., Baldwin, M. E., Robinson, E. D., & Pal, J. S. (2007). Changes in severe thunderstorm environment frequency during the 21st century caused by anthropogenically enhanced global radiative forcing. *Proceedings of the National Academy of Sciences of the United States of America, 104*(50), 19719–19723.

Twisdale, L. A., & Dunn, W. L. (1983). Probabilistic analysis of tornado wind risks. *Journal of Structural Engineering, 109*(2), 468–488.

Twisdale, L. A., & Vickery, P. J. (1992). Research on thunderstorm wind design parameters. *Journal of Wind Engineering and Industrial Aerodynamics, 41*(1–3), 545–556.

Vallis, M. B., Loredo-Souza, A. M., Ferreira, V., & de Lima Nascimento, E. (2019). Classification and identification of synoptic and non-synoptic extreme wind events from surface observations in South America. *Journal of Wind Engineering and Industrial Aerodynamics, 193*, 103963.

Van der Hoven, I. (1957). Power spectrum of horizontal wind speed in the frequency range from 0.0007 to 900 cycles per hour. *Journal of Meteorology, 14*(2), 160–164.

Vermeire, B. C., Orf, L. G., & Savory, E. (2011). Improved modelling of downburst outflows for wind engineering applications using a cooling source approach. *Journal of Wind Engineering and Industrial Aerodynamics, 99*(8), 801–814.

Vickery, P. J., Wadhera, D., Powell, M. D., & Chen, Y. (2009). A hurricane boundary layer and wind field model for use in engineering applications. *Journal of Applied Meteorology and Climatology, 48*(2), 381–405.

Wakimoto, R. M. (2001). Convectively driven high wind events. In C. A. Doswell III (Ed.), *Severe convective storms* (pp. 255–298). Boston, MA: American Meteorological Society.

Wang, C. H., Wang, X., & Khoo, Y. B. (2013). Extreme wind gust hazard in Australia and its sensitivity to climate change. *Natural Hazards, 67*(2), 549–567.

Widen, H. M., Elsner, J. B., Cruz, R. B., Xing, G., Fraza, E., Migliorelli, L., & Michaels, L. (2013). Adjusted tornado probabilities. *E-Journal of Severe Storms Meteorology, 8*(7), 1–12.

Yeo, D. (2014). Generation of large directional wind speed data sets for estimation of wind effects with long return periods. *Journal of Structural Engineering, 140*(10), 04014073.

Zhang, S., Solari, G., Burlando, M., & Yang, Q. (2019). Directional decomposition and properties of thunderstorm outflows. *Journal of Wind Engineering and Industrial Aerodynamics, 189*, 71–90.

Zhang, S., Solari, G., Yang, Q., & Repetto, M. P. (2018). Extreme wind speed distribution in a mixed wind climate. *Journal of Wind Engineering and Industrial Aerodynamics, 176*, 239–253.

Zickar, A. S. (2019). *Enhancement of the United States extreme wind database and implications for extreme wind climatology* (MS thesis, University of Illinois at Urbana-Champaign).

CHAPTER 22

TOWARD THE CODIFICATION OF THUNDERSTORM/ DOWNBURST WINDS

DAE KUN KWON, AHSAN KAREEM, AND GIOVANNI SOLARI

1. Introduction

Non-synoptic wind storms such as hurricanes, typhoons, and cyclones, thunderstorms and downbursts, tornadoes, etc., have been of great interest in structural and wind engineering communities due to their massive impacts on structures. From the codification viewpoint, the hurricane or typhoon has generally been treated as increased wind speeds (as compared to synoptic winds) in specific areas where the probability of occurrence of the storm is much higher (e.g., American Society of Civil Engineers [ASCE] 7–16 wind map; ASCE 2016). It has been thought that they are a large synoptic-scale event based on experimental observations (US dropsondes, etc.). However, recent studies reported the existence of non-stationarity, and its vertical profile showed a distinct difference from synoptic boundary layer winds (e.g., Kareem 2008; Giammanco et al. 2013; Hu et al. 2013, 2017; Zhou et al. 2015; Huang et al. 2015; Wang et al. 2016, 2018, 2019). The recent ASCE 7–16 Commentary includes more information concerning tornadoes, although tornadoes are not considered in the wind load provision because of their very low probability of occurrence (ASCE 2016). The Commentary included Enhanced Fujita (EF) records and design wind speed maps for tornado safe room design in the United States. In addition, it suggested the calculation of equivalent static wind pressure/load procedure using the gust loading/effect factor. A simplified method was also introduced in terms of the Tornado Factor (TF).

Thunderstorms are almost ubiquitous around the world (e.g., as many as 40,000 thunderstorm occur each day worldwide, and an estimated 100,000 thunderstorms occur annually in the United States as reported by NOAA/NWS [2014]). Thunderstorms produce two extreme wind events—tornadoes and downbursts—and the latter is known to be much more frequent than tornadoes. Winds associated with thunderstorm/downburst cause a variety of damage to residential houses, buildings, transmission towers, wind turbines, and temporary structures, threatening human lives and property loss. Munich Re (2017) shows an escalation of losses due to thunderstorms. Yet any guidelines for design practice in regards to thunderstorm/downburst have thus far not evolved. Thus, design loads in gust front winds obtained from conventional analysis frameworks included in codes and standards, such as the gust loading/effect factor approach (e.g., Kwon & Kareem 2013b), would not be appropriate. This fact calls for a careful examination of traditional design procedures.

In an effort to establish a new design procedure for this type of wind load effect on structures, two frameworks were reported in the past literature. One is a gust front factor (GFF)-based framework first proposed by Kwon and Kareem (2009), which accounts for a description of the genesis of the overall wind load effects on structures under both gust front and boundary layer winds. Rather than introducing a different framework for such non-synoptic winds, the proposed methodology based on the GFF consisted of a multiplier, embodying unique kinematic and dynamic features of non-synoptic winds to ASCE 7 gust loading/effect factor formulation for boundary layer winds. In the absence of a definitive procedure to quantify the effect of gust front winds on structures, this approach has offered an effective framework by which to capture, in a rational manner, the influence of various distinctive features of gust fronts that distinguishes them from synoptic winds. A modified framework for assessing gust front wind loading effects on wind turbine towers has also been presented in Kwon et al. (2012). A generalized version of the GFF (G-GFF) framework was later introduced in Kwon and Kareem (2013a); this was intended not only to analytically encapsulate dynamic load effects associated with gust front winds independent of any reference design standard (e.g., ASCE 7), but also to highlight other general features like those found in conventional gust loading factor scheme. In view of the complexity on the treatment of non-stationarity, a preliminary web-based portal for the evaluation of GFF and associated loads in an e-design format (Kwon et al. 2008) was made available at https://vortex-winds.org to facilitate expeditious utilization of this framework in design practice. More recently, Kwon and Kareem (2019) revisited the GFF approach in a model-based and data-driven perspective. The other framework is a thunderstorm response spectrum technique (TRST) reported in Solari (2016). It has been adopted from the concept of the response spectrum being widely used in the field of earthquake engineering as well as blast design. For this reason, the TRST also demands several representative datasets to establish a reliable graphic response spectrum; thus, this is a data-based approach.

This chapter reviews the GFF and the TRST frameworks as a step toward the codification of gust front winds. Gleaning from the success of Vickery's study for the original gust loading factor (Vickery 1970) which focused on its reliability, it is apparent that codification based on a design framework would benefit from a study focusing on its development and assessment of its efficacy. A preliminary uncertainty analysis using the GFF/G-GFF framework is also carried out to assess the uncertainties associated with the load effects of gust front winds. In addition, a comparison is made between the two frameworks to assess the performance of the two approaches. Finally, a living codification concept through learning and updating, invoking the emerging "design thinking" approach, is discussed.

2. Gust Front Factor and Generalized GFF

2.1. Concept of the Gust Front Factor

With the exemplary success of the gust loading factor (Davenport 1967) in capturing the dynamic wind effects introduced by the buffeting action of winds and its popularity in design codes and standards worldwide, Kwon and Kareem (2009) proposed formulation of a framework along the lines of the gust loading/effect factor formulation existing in ASCE 7 (Solari & Kareem 1998) that encapsulates critical features of gust front winds to capture their load effects. To accomplish this objective, the GFF was therefore introduced for use in conjunction with the existing design codes and standards. The design wind loading in a gust front (F_{Design}), which is represented by an equivalent static wind load (ESWL), is then expressed by

$$F_{Design} = F_{ASCE7} \cdot K_{z,G-F} \cdot G_{G-F} \tag{1}$$

where $F_{ASCE\,7}$ represents the ESWL in ASCE 7, the G_{G-F} is the GFF that relates F_{Design} in a gust front to the $F_{ASCE\,7}$ recommendations in conventional boundary layer winds, and $K_{z,G-F}$ accounts for the velocity/pressure profile in a gust front as opposed to boundary layer winds in ASCE 7. The G_{G-F} can be best captured in terms of four underlying factors (Kwon & Kareem 2009):

$$G_{G-F} = I_1 \cdot I_2 \cdot I_3 \cdot I_4 \tag{2}$$

In this format, the G_{G-F} explicitly takes into account the following features: variation in the vertical profile of wind speed (kinematic effects factor [mean load effects]), I_1; dynamic effects introduced by the sudden rise in wind speed (pulse dynamics factor [rise-time effects]), I_2; non-stationarity of turbulence in gust front winds (structural dynamics factor [non-stationary turbulence effects]), I_3; and transient aerodynamics (potential load modification factor [transient aerodynamics effects]), I_4. It is worth noting that for conventional boundary layer winds, both the GFF (G_{G-F}) and the velocity

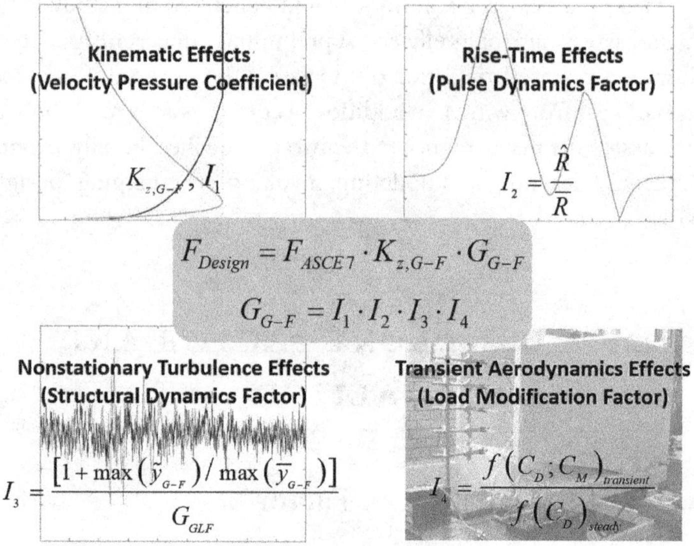

FIGURE 22.1 Schematic diagram of gust front factor framework.

From Kwon and Kareem (2009).

pressure coefficient ($K_{z,G-F}$) in gust front winds reduce to unity. Therefore, the conventional gust loading/effect factor approach in ASCE 7 becomes a special case of the proposed GFF approach. A schematic diagram that portrays the genesis of the design wind loads in gust fronts is given in Figure 22.1.

2.2. Formulation of GFF and Its Generalized Version

Assuming that $x_{B-L}(z,t)$ is the structural displacement in boundary layer winds, the expected maximum response, $\max[x_{B-L}(z,t)]$, can be expressed following the gust loading factor format given in Davenport (1967):

$$\max[x_{B-L}(z,t)] = \max[\bar{x}_{B-L}(z) + \tilde{x}_{B-L}(z,t)] = \bar{x}_{B-L}(z)\left[1 + \frac{g_{B-L} \cdot \sigma_{B-L}(z)}{\bar{x}_{B-L}(z)}\right] = \bar{x}_{B-L}(z) \cdot G_{GLF} \tag{3}$$

where, subscript B-L represents boundary layer; superscripts – and ~ represent mean and fluctuating components of displacement, respectively; g_{B-L} is the peak factor; $\sigma_{B-L}(z)$ is the root-mean-square (RMS) displacement by fluctuating wind components; and G_{GLF} is the gust loading factor. As such, the gust loading factor accounts for the dynamics of wind fluctuations and any load amplification introduced by the dynamics of the structure, and this factor is utilized to estimate design loads of structure (i.e., ESWL).

In the case of gust front winds, structural displacement by gust front winds [$x_{G\text{-}F}(z,t)$] may be described in terms of a non-stationary wind model that includes storm-translation effect (Kwon & Kareem 2009),

$$x_{G-F}(z,t) = \bar{x}_{G-F}(z,t) + \tilde{x}_{G-F}(z,t) + x_{s-m} \tag{4}$$

where, subscript *G-F* represents gust front; superscripts – and ~ represent time-varying mean and non-stationary fluctuating components of displacement, respectively; and $x_{s\text{-}m}$ is the displacement introduced by the storm-translation speed. The same model does not take into account, in this stage, the background atmospheric boundary layer (ABL) wind flow into which the downburst is embedded. Accordingly, the corresponding maximum response may be approximately expressed as

$$\begin{aligned} \max\left[x_{G-F}(z,t)\right] &\approx \max\left[\bar{x}_{G-F}(z,t)\right] + \max\left[\tilde{x}_{G-F}(z,t)\right] + \bar{x}_{s-m} \\ &\approx \max\left[\bar{x}_{G-F}(z,t)\right]\left[1 + \frac{g_{G-F}\cdot\max\left[\sigma_{G-F}(t)\right]}{\max\left[\bar{x}_{G-F}(z,t)\right]} + \frac{x_{s-m}}{\max\left[\bar{x}_{G-F}(z,t)\right]}\right] \end{aligned} \tag{5}$$

where $g_{G\text{-}F}$ and $\sigma_{G\text{-}F}(t)$ represent mean peak factor and time-dependent RMS displacement by non-stationary fluctuating components of gust front winds.

In terms of the factorized analytical/empirical models of gust front winds (e.g., Kwon & Kareem 2009, 2013a), the maximum displacement of the time-varying mean component may be divided into static and dynamic parts for GFF and G-GFF, respectively:

$$\begin{aligned} \max\left[\bar{x}_{G-F}(z,t)\right] &= \bar{x}_{st,G-F}(z)\cdot\frac{\max\left[\bar{x}_{G-F}(t)\right]}{\bar{x}_{st,G-F}(z)}(GFF) \\ \max\left[\bar{x}_{G-F}(z,t)\right] &= \bar{x}_{st,G-F}(z)\cdot\max\left[\bar{x}_{G-F}(t)\right](G-GFF) \end{aligned} \tag{6}$$

The relationship between GFF ($G_{G\text{-}F}$) with associated underlying sub-factors (Equation 2) and generalized GFF ($G_{G,G\text{-}F}$) can be described as

$$\begin{aligned} G_{G-F} &= \frac{\bar{y}_{st,G-F}(z)}{\bar{y}_{B-L}(z)}\cdot\frac{\max\left[\bar{y}_{G-F}(z,t)\right]}{\bar{y}_{st,G-F}(z)}\cdot\frac{\left[1 + \frac{g_{G-F}\cdot\max\left[\sigma_{G-F}(t)\right]}{\max\left[\bar{y}_{G-F}(z,t)\right]} + \frac{y_{s-m}}{\max\left[\bar{y}_{G-F}(z,t)\right]}\right]}{G_{GLF}}\cdot\frac{C_{D,G-F}}{C_D} \\ &= I_1\cdot I_2\cdot I_3\cdot I_4 \end{aligned} \tag{7a}$$

$$G_{G,G-F} = \frac{\max\left[\bar{x}_{G-F}(t)\right]}{\bar{x}_{st,G-F}(z)}\left[1 + \frac{g_{G-F}\cdot\max\left[\sigma_{G-F}(t)\right]}{\max\left[\bar{x}_{G-F}(z,t)\right]} + \frac{x_{s-m}}{\max\left[\bar{x}_{G-F}(z,t)\right]}\right] = I_2\cdot\left(I_3\cdot G_{GLF}\right) \tag{7b}$$

In these formats, the $G_{G\text{-}F}$ embodies an intuitive picture of the underlying mechanisms that represent the fundamental building blocks of both the kinematic (I_1) and dynamic (I_2, I_3, I_4) load effects associated with gust front winds on structures, distinguishing them from those experienced in conventional boundary layer flows. On the other hand, the $G_{G,G\text{-}F}$ can best capture the dynamic effects of gust front winds akin to the concept of conventional gust loading factor. Note that, strictly speaking, $x_{s\text{-}m}$ may be categorized as a static component, but since its contribution becomes relatively small due to a large value in the denominator (i.e., the maximum displacement of time-varying mean component), this term may be treated as small in the consideration of the overall static effect. However, this hypothesis deserves to be investigated more carefully.

The ($I_3 \cdot G_{GLF}$) in $G_{G,G\text{-}F}$ indeed relates to the non-stationary fluctuating component. Please note that in order to delineate the influence of $C_{D,G\text{-}F}$ on the $G_{G\text{-}F}$, the drag coefficient in the gust front and boundary layer wind formulations is separated by redefining response x as y in Equation 7a, where y response is evaluated by setting the drag force coefficient equal to unity. Regardless of this format, the relationship in Equation 7b remains valid because, in the formulation of $G_{G,G\text{-}F}$, the effect of drag coefficient ($C_{D,G\text{-}F}$) cancels out in each term. It is worth noting that, in the case of boundary layer winds, the sub-factors I_2 and I_3 become unity in the absence of gust front wind effects. Accordingly, the resulting generalized GFF ($G_{G,G\text{-}F}$) also reduces to conventional gust loading factor (G_{GLF}) due to the relationship between the two factors shown in Equation 7b. Accordingly, the design loads (ESWL) associated with the G-GFF are expressed in the conventional format as used in the case of boundary layer winds:

$$F_{G,G-F}(z) = \overline{F}(z) \cdot G_{G,G-F} = \frac{1}{2}\rho \cdot A \cdot V_{G-F}^2(z) \cdot G_{G,G-F} \cdot C_{D,G-F} \tag{8}$$

where $V_{G\text{-}F}(z)$ is the gust front wind profile. In this format, the G-GFF can analytically encapsulate dynamic load effects associated with gust front winds independent of any reference design standard, unlike the GFF that is specifically associated with ASCE 7 (Equation 1).

2.3. Toward a Simplified Codification in the Frequency Domain

To implement and simplify the frequency domain approach to non-stationary turbulence of gust front winds in which a closed-form solution for a codification purpose may possibly be derived, a number of assumptions are introduced for the GFF/G-GFF framework. First, the non-stationary fluctuating wind is described as a non-stationary model comprised of a time-varying standard deviation wind component and a normalized fluctuating wind component, and the latter is assumed to be a zero-mean stationary Gaussian process (e.g., Howell & Lin 1971; Chen & Letchford 2004; Chay et al. 2006; Kareem 2008; Chen 2008; Kwon & Kareem 2009, 2013a, 2019; Kareem et al. 2019).

In addition, the non-stationary fluctuating wind is described by the evolutionary power spectrum (Priestley 1965, 1967). An intrinsic roadblock in this approach is the presence of time-varying features and spatiotemporal dependency. As a way to overcome this complexity, a decomposition in space (z) and time (t) is introduced. This separable type description may be an adequate representation of the velocity field for codification. However, recent studies reported that, in some cases, spatiotemporal dependence has been noted for the cases of thunderstorms or downbursts (Lombardo et al. 2014; Scott Gunter & Schroeder 2015). Certainly, we present it here with the provision that this treatment may be modified should additional data suggest an alternative description.

Second, the normalized fluctuating wind component is represented by the Davenport wind spectrum and conventional coherence functions in the horizontal and vertical directions (Davenport 1967) are used. Initial studies suggest that the fluctuating wind component in their format is similar to that in boundary layer winds (e.g., Choi & Hidayat 2002; Wang & Kareem 2005). Owing to the height-independent characteristics of the Davenport wind spectrum, the normalized spectrum can be taken out of the double integration, which alleviates a more extensive calculation.

Third, for simplification purpose, analytical forms of the time function [$V_{G\text{-}F}(t)$] (half-sine function) and time-independent vertical wind profile [$V_{G\text{-}F}(z)$] (Vicroy model, Vicroy 1991, 1992) are introduced for deriving a closed-form (Kwon & Kareem 2009, 2013a, 2019).

$$V_{G-F}(t) = \sin\left(\frac{\pi}{t_d} t\right) \tag{9}$$

$$V_{G-F}(z) = A \cdot V_{\max}\left[e^{b_1(z/z_{\max})} - e^{b_2(z/z_{\max})}\right]$$

where t_d is the pulse duration of the excitation; V_{max} is the maximum horizontal wind speed; z_{max} is the a height where V_{max} occurs; A is the a constant that can be determined from model constants b_1 and b_2; and $b_1 = -0.22$, $b_2 = -2.75$, and $A = 1.354$ following Vicroy (1991). The simplicity of the half-sine function is an attractive feature as it requires only a single parameter, pulse duration, to define the time function in a gust front wind event while encapsulating the essential features of a sudden rise and drop in wind speed. Solari et al. (2015a) confirmed that the half-sine shape was close to the inner envelope of full-scale data. However, a question in this regard remains regarding the duration for codification purposes. A study by Solari et al. (2015a) reported from the extensive analysis of full-scale data showed great variability in duration among the data. They also found that the mean value of the durations was 248 seconds while the minimum was 57 seconds, thus revealing the possibility of a rapid change in wind velocity. These observations imply that a dynamic effect by the pulse duration may be negligible based on the relationship between pulse duration and the fundamental period of a structure (Kwon & Kareem 2009). Although the temporal characteristics of gust front winds have been well reported in the literature, spatial features like the vertical wind profile have rarely been clearly found yet. Recently, Scott Gunter and Schroeder (2015) attempted high-resolution full-scale measurements using two mobile Doppler radars. Repetto et al. (2018) implemented ground-based pulsed coherent LiDAR system to measure wind profiles up to about 250 meters in height. Junayed et al. (2019) and Hangan et al. (2019) performed large-scale experiments using WindEEE Dome to identify the characteristics of downburst, and they have successfully simulated temporal-spatial variations of wind profiles. These are promising avenues by which to better identify the characteristics of the gust front wind profile. Note that, strictly speaking, the vertical profile of the gust front evolves with time due to the translation of the storm. A rigorous approach may include the effects of profile evolutions in a probabilistic manner; however, a single-profile model with an adjustment in a gust front wind map may be a better approach at this early stage of codification.

Finally, another parameter to be considered in a codification of gust front winds is the potential impact of terrain roughness. For synoptic winds, most codes and standards classify the exposure in three to five categories for design purposes (Kwon & Kareem 2013b). Nonetheless, most studies concerning gust front winds have usually been made under open terrain. For example, vertical profile models by, for example, Oseguera and Bowles (1988) and Vicroy (1991, 1992) were derived based on limited full-scale data gathered from airport areas (Joint Airport Weather Studies, JAWS), and such areas generally represent open terrain exposure. Although gust front winds are different from synoptic winds in many aspects, studies using wind tunnel experiments have shown that a terrain roughness effect exists, although the results have been limited in showing trends in different terrain conditions without any quantitative estimates. The general trends exhibited higher z_{max} (e.g., Hangan & Xu 2005) and lower V_{max} (e.g., Wood et al. 2001; Choi 2004) in a built-up terrain. In conjunction with these parameters, terrain exposure would affect other wind parameters (e.g., turbulence intensity or TI). Thus, without a general consensus and lack of definitive evidence, it may be reasonable to assume that a certain terrain roughness effect on both z_{max} and V_{max} in the Vicroy model (Equation 9) is present, whereas the model constants defined in the profile models are assumed to be constant irrespective of terrain roughness. Indeed, Kwon and Kareem (2009) were the first to attempt to quantitatively address the effects of terrain conditions through this approach. They considered two criteria: (1) gust front wind speed at 10-meter height, $[V_{G\text{-}F}(10)]$ is set equal to the boundary layer gust speed at 10 meters [Criterion 1: $V_{G\text{-}F}(10) = V_{B\text{-}L,3\text{-}s}(10)$]; and (2) the maximum gust front wind speed (V_{max}) is equal to gust speed at the gradient height (z_G) in boundary layer winds [Criterion 2: $V_{max} = V_{B\text{-}L,3\text{-}s}(z_G)$]. Then, the gradient height defined for boundary layer winds was used to take into account the effects of terrain conditions. In this definition, Criterion 1 leads to higher V_{max} than Criterion 2, which reflects the distinct feature of each profile. One may conveniently introduce additional criteria that may better reflect the data or may meet other site-specific requirements when it becomes available in the future. With proper assumptions, vertical profiles of gust front winds and TI for an arbitrary terrain exposure condition were established. Details can be found in Kwon and Kareem (2009). Note that the z_{max} has been reported in the range of 50 to 100 meters based on full-scale data; however, a recent study exhibited the possibility of a higher altitude, beyond 100 meters (e.g., Scott Gunter & Schroeder 2015).

With these assumptions, more recently, a more complete closed-form solution concerning non-stationary fluctuations was derived by Kwon and Kareem (2019) to facilitate rapid assessment of non-stationary turbulence effects for the case of a long pulse duration similar to the typical gust front life cycle (e.g., Solari et al. 2015a). The assumption of the long pulse duration leads to simplifying the derivation of the RMS of the non-stationary fluctuation and complex mean peak factor (Michaelov et al. 2001; Kwon & Kareem 2009, 2013a, 2019). Detailed information can be found in Kwon and Kareem (2019): they reported that results from numerical and closed-form solutions showed excellent agreement, which is promising in possibly taking a step toward more simplified codification for the evaluation of gust front wind effects on structures.

2.4. Preliminary Uncertainty Analysis

A preliminary uncertainty analysis was performed to assess the uncertainties associated with the load effects of gust front winds. In view of the GFF/G-GFF formulations, a total of six uncertain parameters were chosen: four wind parameters such as 3-second gust speed ($V_{3\text{-}s}$), Vicroy profile model parameter (z_{max}), pulse duration in the time function (t_d), and TI related parameter (c_1); and two structural parameters such as n_0 (fundamental natural frequency) and ζ (damping ratio). The choice of the $V_{3\text{-}s}$ parameter was made because it was directly related to the determination of V_{max} in the Vicroy profile model (Equation 9) through the two criteria (Criteria 1 and 2) mentioned earlier. The c_1 parameter represents the amplitude modulating coefficient that plays the role of TI (Kwon & Kareem 2009). All parameters were assumed to follow a lognormal distribution, and their coefficients of variation (COV) were all set to 0.2 except the fundamental natural frequency and damping ratio, whose COVs were defined as 0.05 and 0.4, respectively, following the suggestion in Kwon et al. (2015). Note that, for this analysis, the Criterion 2 condition was utilized to determine the V_{max}.

To assess the level of uncertainties associated with the wind-induced load effects, a wind load factor was considered for the base moment. Based on the recent study reported in Kwon et al. (2015), a load factor definition (γ_w), which assumes that the resulting wind response has a lognormal distribution, was expressed as

$$\gamma_w = \left(\frac{\mu_W}{W_n}\right) \frac{\exp\left[\beta \varepsilon_\ell \sqrt{\ln\left(1+V_W^2\right)}\right]}{\sqrt{1+V_W^2}} \tag{10}$$

where, μ_W/W_n is the bias or ratio between the mean and the nominal wind loads/response; the reliability index β is 3.0 based on Ellingwood and Tekie (1999); $\varepsilon_\ell = 0.72$ (e.g., Lind 1971; Haldar & Mahadevan 2000); and V_W is the COV in the wind loads/response. In this format, for the case in which no uncertainties are involved, the load factor (γ_w) reduces to unity (1.0).

The Commonwealth Aeronautical Advisory Research Council (CAARC) standard tall building (e.g., Holmes et al. 1990) was utilized as an example building for this uncertainty analysis (width $B = 46$ m; depth $D = 30$ m; height $H = 183$ m; building bulk density $\rho_B = 157$ kg/m^3). The building was assumed to be located in an open environment within the continental United States, which corresponds to Exposure C defined in ASCE 7. For the load factor defined in Equation 10, the nominal wind loads/response (W_n), which is the base moment in this study, was determined in terms of the GFF/G-GFF approach assuming that there were no uncertainties involved, while μ_W and V_W were determined by Monte Carlo simulations (10,000 samples) taking uncertainties in the parameters into account.

Table 22.1 shows the results of uncertainty analysis, in which the mean values of the uncertain parameters were also listed. It was observed that the load factor (γ_w) associated with all uncertainties was significant, more than 3. For reference, the wind load factor defined in ASCE 7–05 was 1.6 based on an analysis of the performance of rigid buildings, while Kwon et al. (2015) suggested 1.9 in light of the uncertainties associated with the dynamic characteristics of flexible buildings, such as natural frequency and damping ratio. Among uncertain parameters considered in this study, it was identified that the effect of the basic wind speed ($V_{3\text{-}s}$) was dominant as its COV of the base moment was much higher than other individual parameters. This observation can be interpreted

Table 22.1 Results of a preliminary uncertainty analysis

	Inputs		Load factor (γw)	Base moment		GG,G-F	
	Mean	COV		Mean[b]	COV	Mean	COV
Exposure C							
All[a]	–	–	3.270	2.296	0.629	1.448	0.073
V3-s	40 m/s	0.20	2.315	2.151	0.427	1.448	0.007
zmax	60.35 m	0.20	1.269	2.027	0.124	1.449	0.005
td	200 sec	0.20	1.022	2.065	0.011	1.446	0.011
c1	0.134	0.20	1.141	2.067	0.062	1.447	0.062
n0	0.17 Hz	0.05	1.002	2.068	0.001	1.448	0.001
ζ	0.01	0.40	1.005	2.069	0.002	1.448	0.002
Exposure B							
All[a]	–	–	3.115	2.499	0.600	1.661	0.094
V3-s	40 m/s	0.20	2.292	2.377	0.419	1.660	0.006
zmax	80.47 m	0.20	1.186	2.227	0.092	1.661	0.005
td	200 sec	0.20	1.027	2.277	0.013	1.658	0.013
c1	0.201	0.20	1.185	2.278	0.008	1.659	0.080
n0	0.17 Hz	0.05	1.0047	2.279	0.002	1.660	0.002
ζ	0.01	0.40	1.003	2.280	0.002	1.660	0.002

[a] All six uncertain parameters were considered.
[b] Unit in [10^9N·m].
COV, coefficient of variation.

as twofold: (1) wind loads are proportional to the square of the wind velocity (e.g., Equation 10), thus its contribution on the load effects is considerable; and (2) dynamic effects by gust front winds are less significant than static/quasi-static and kinematic effects. Indeed, it turned out that the z_{max} parameter, which explains the kinematic effect (profile effect), was another influential parameter, second only to $V_{3\text{-}s}$. Although the TI (c_1) showed some importance (the third most influential parameter), however, its COV for the base moment was far less than the first two parameters, such as $V_{3\text{-}s}$ and z_{max}, which corroborated less dynamic effects in this example. It is worth noting that the effects of the fundamental frequency (n_0) and damping ratio (ζ) were negligibly small, suggesting less dynamic effects as well. The effect of pulse duration (t_d) was rather marginal. From the viewpoint of the generalized GFF ($G_{G,G\text{-}F}$) as shown in the last column of Table 22.1, which encapsulates dynamic load effects associated with gust front winds (Equation 7b), the variation of $G_{G,G\text{-}F}$ was governed by the c_1 parameter, implying that the TI is a key parameter among other parameters for non-stationary turbulence. Nonetheless, overall uncertainty effects on the $G_{G,G\text{-}F}$ were much less than those observed for the base moment. Uncertainty analysis for the example building under the city/suburban environment (Exposure B in ASCE 7) was also carried out, and almost the same trends with the case under Exposure C were observed (Table 22.1).

The preceding should be taken with caution in that such a high load factor associated with gust front winds was obtained through the assumption of probability distribution and COV as a result of a preliminary assessment, mainly focusing on random uncertainty. In addition to this, epistemic uncertainty, such as modeling of gust front winds, etc., would certainly exist. A more accurate evaluation would need to be done through a detailed uncertainty quantification with more supportive data and better knowledge.

2.5. A Data-Driven Approach

A data-driven approach has gained its popularity in many fields of research, mainly owing to the rapid advance of measurement and monitoring systems that enable us to collect long-term data. In the context of the treatment of time-dependent non-stationary gust front winds, two approaches concerning the term "data-driven" may be feasible: *modeling* and *simulation*. Indeed, any empirical models can be regarded as data-driven approaches because those models are in general derived from data. From this viewpoint, a data-driven model differs from mechanics-based models which are generally formulated given the basic mechanics involved. An example of data-driven modeling in the GFF/G-GFF is the use of a half-sine function for describing a time-varying mean feature of thunderstorm/downburst events, given in Equation 4. Data-driven simulation approaches that utilize the instantaneous features of available data can offer a straightforward and statistically meaningful approach to non-stationary simulation (e.g., Wang et al. 2013).

A major drawback of such data-driven approaches inevitably involves the paucity of data. To date, only limited datasets are present due to the inherent nature of thunderstorms or downbursts (i.e., a short time frame as well as a local spatiotemporal event), which often hinders the capture of such events in full-scale. More attempts in full-scale measurements may lead to better identifying time dependency and wind characteristics along with height; these include wind profile, wind spectrum/coherence, and TI. Another potential issue is the reliability of measured data and potential biases. For example, Solari et al. (2015a) reported the characteristics of measured full-scale datasets that showed significant variability among extreme events as well as measurement locations.

Utilizing the simulation-based approach, one could design structures for a prescribed event and its statistical emulations to form a suite of time histories. A time-domain approach may then be pursued to estimate load effects, and statistical averages may be derived based on these simulations. Akin to the earthquake case, one may utilize scaling of gust front intensity for design purposes to select a most probable event or one associated with a given return period.

Assuming that a large set of datasets is available, the very first choice for codification is to extract salient features from datasets then fit them to analytical model(s) to produce a conventional code and standard format. However, in light of uncertainties, a data-driven design involving uncertainty quantification may offer a better option to probabilistically evaluate gust front wind effects on structures. For a better description of such effects, it is desirable to identify probabilistic and statistical information of wind parameters from datasets such as statistical moments (mean, standard deviation, etc.) and probability distributions. Then, this information involving uncertainties associated with parameters is taken into account in a design procedure, similar to the treatment used in codes and standards. With the success of data-enabled designs introduced in recent ASCE 7 (ASCE 7–05, 7–10, and 7–16), a web-enabled data-driven design in this regard may be a better approach. An example of this approach regarding the treatment of uncertainties in a web format is a performance-based data-enabled design for occupant comfort in a high-rise building (PBDED-OC) (Bernardini et al. 2015), which not only allows users to input structural/wind parameters and their associated uncertainties but also offers analysis/design on-the-fly through Monte Carlo simulation to establish the probability of the exceedance curve of response owing to the embedded computational engine in the server.

It is worth noting that the analysis/design framework and its workflow were preliminarily introduced within a web-based portal to facilitate expeditious utilization of the GFF framework in design practice and potential inclusion in codes and standards (Kwon & Kareem 2009). This eliminates the need for an in-depth understanding of the background within the framework as well as associated computational effort. The portal has a user-friendly interface and is available at https://vortex-winds.org (Kareem & Kwon 2017), permitting convenient analysis of several design scenarios with a host of potential loading conditions, including the ASCE 7–05 procedure in the boundary layer winds for immediate comparison in an e-design format (e.g., Kwon et al. 2008). Particularly, the current GFF portal has in part embedded the concept of a data-driven approach by allowing user-defined profile parameters to be inputted, such as V_{max} and z_{max} (Equation 9). Once the input is completed, the following quantities are then displayed in the results interface: input parameters selected by the user to confirm user inputs; GFF ($G_{G\text{-}F}$) and its sub-factors (I_1 to I_4); and a figure that displays the ESWL given in ASCE 7 and in gust front winds. More generic and flexible implementation would enable a better data-driven approach for a host of models associated with wind parameters.

3. Thunderstorm Response Spectrum Technique

Recently, Solari et al. (2015b) and Solari (2016) have introduced an alternative approach for the design of thunderstorms, one focusing on downburst events. The TRST has been adopted from the concept of the response spectrum being widely used in the field of earthquake engineering as well as blast design. For this reason, the TRST also demands several representative datasets to establish a reliable response spectrum, and such datasets have been obtained through a wide range of full-scale wind monitoring networks, such as Wind, Ports, and Seas (Solari et al. 2012). This is also a data-based approach and is similar to the simulation-based approach described in the previous section.

In this section, the underlying principles of the GFF and TRST approaches are examined and discussed. For convenience, symbols used in Solari (2016) are changed to follow those employed in the GFF/G-GFF. Akin to the GFF/G-GFF, the TRST adopted the non-stationary model, and the decomposition of space (z) and time (t) was also assumed. The time-varying mean component in the TRST was modeled as

$$V_{G-F,Sol}(z,t) = V_{\max}(h)V_{G-F,Sol}(z)V_{G-F,Sol}(t) \tag{11}$$

where, $V_{max}(h)$ is the maximum value of time-varying mean at a height h; h is the height of an anemometer that measures the wind velocity or a reference height (e.g., 10-m height); $V_{G\text{-}F,Sol}(z)$ is a normalized wind vertical profile used in Solari (2016); and $V_{G\text{-}F,Sol}(z)$ and $V_{G\text{-}F,Sol}(t)$ are the wind profile and time function used in Solari (2016), respectively.

The ESWL in the TRST has a format similar to Equation 8 used in the G-GFF framework, which is expressed as (Solari 2016)

$$\begin{aligned} f_{eq}(z) &= \hat{f}(z) \cdot S_{d,eq} \\ \hat{f}(z) &= \frac{1}{2}\rho V_{\max}^2(h)\hat{G}^2(h)V_{G-F,sol}^2(z)A(z)C_D(z) \end{aligned} \tag{12}$$

where, $\hat{f}(z)$ is the peak static wind load by thunderstorm outflows; $S_{d,eq}$ is the equivalent response spectrum (ERS); and $\hat{G}(h)$ is the gust velocity factor at h. Note that Solari (2016) used $\hat{G}(h)$ to relate between gust peak wind speed (1-sec moving-average) and

$V_{max}(h)$ (30-sec moving-average) as detailed in Solari et al. (2015a). In this format, the peak wind load defined in Equation 12 is essentially similar to the one used in the G-GFF framework given in Equation 8, provided that the two approaches employ the same normalized vertical profile $[V(z)/V_{max}]$. Nonetheless, the difference for the peak static wind load lies in the definition of the maximum wind speed, V_{max}. In the GFF/G-GFF, two criteria compared to ASCE 7 gust profile were established to estimate V_{max} (Kwon & Kareem 2009), while Solari (2016) described it as a product of $V_{max}(h)$ and a gust peak factor $[\hat{G}(h)]$ (Equation 12).

Despite the similarity in the description of the peak static wind load, the two frameworks used different approaches for the treatment of the non-stationary turbulence component. While the TRST has employed a data-based equivalent wind velocity approach in the time domain for estimating the ERS ($S_{d,eq}$), the GFF/G-GFF utilized a conventional frequency domain approach involving an evolutionary power spectrum model (Priestley 1965, 1967) in which the spectrum was described as a product of an amplitude modulating function and a stationary wind component, assuming Davenport wind spectrum and coherence functions. In the TRST, a reduced equivalent wind velocity (u_{eq}) was introduced as

$$u_{eq}(\tilde{t},\tilde{\delta}) = \frac{v_{eq}(h,\tilde{t},\tilde{\delta})}{V_{\max}(h)\hat{G}(h)} \tag{13}$$

where, $\tilde{\delta}$ and $\tilde{t}$ are the reduced size factor and time defined in Solari (2016), respectively; and v_{eq} is the equivalent wind velocity invoking the equivalent wind spectrum (EWS) technique (Piccardo & Solari 1998), which is a method to replace the actual turbulent fluctuation as a function of time (t) and space (z) by an equivalent turbulent fluctuation as a function of time identically coherent in space (Solari 2016). The calculation of the equivalent wind velocity (v_{eq}) needs a rather complex procedure using datasets, which is a key part of the assessment of the ERS ($S_{d,eq}$). Solari (2016) suggested three different methods depending on the use of dimensional/non-dimensional parameters, though the underlying procedure is similar in each. For completeness, a summary of the method to use non-dimensional parameters is provided here. Given the N-representative time series of gust front winds, $(N \times J)$ sets of v_{eq} need to be computed individually following the EWS technique, where J is the total number of the reduced size factors ($\tilde{\delta}$) considered. Then, in light of the combination of the three key parameters such as the reduced size factor (size of J), the reduced first natural frequency ($\tilde{n}_1$, size of L), and the damping ratio (ζ, size of M), a large set of the reduced equivalent displacements (d_{eq}) are computed by solving an equation of motion. Finally, the ERS ($S_{d,eq}$) is determined using the relationship, $S_{d,eq} = max(d_{eq})$ at each case. Similar to the response spectra in earthquake engineering, the ERS is indeed a graphical representation portraying the mean ERS and its COV for the preceding three parameters ($\tilde{n}_1$, $\tilde{\delta}$, ζ) where M-sets of plots (for different damping ratios) are provided.

It is worth noting that the dependence of the response spectrum on the reduced frequency n_1 and the damping ratio ζ is a classical and general property of this quantity. The dependence of $S_{d,eq}$ on the reduced size factors δ derives from the partial correlation of the wind field on the structural surface exposed to the wind, traduces the concept of aerodynamic admittance, and represents the main difficulty and peculiarity of the generalization of the response spectrum technique from earthquakes to thunderstorms.

Aside from the applicability of the EWS technique for thunderstorm/downburst events, an inherent shortcoming in the TRST framework is currently a lack of representative datasets. Although Solari et al. (2015a) and Solari (2016) reported the availability of 93 datasets, their monitoring system has measured wind velocity at limited heights (mostly around a typical reference height, 10 m), thus implying little information about vertical wind characteristics such as vertical profiles of wind velocity and TI, wind spectrum and coherence with heights. Accordingly, the current TRST reported in Solari (2016) assumed wind characteristics at the reference height to be used at each elevation, constant TI, V_{max} at the reference height, etc., although a non-dimensional wind spectrum was introduced. The wind profile and the coherence function were also assumed, which will be discussed in the next section.

It is worth noting that full-scale spatial information regarding thunderstorms and downbursts such as vertical profile, TI, etc. is also being measured through a LiDAR system in the range of 40–250 meter heights and sampling rate of 1 Hz in the Wind, Ports, and Sea project (Repetto et al. 2018). Therefore, it is anticipated that additional data will become available in upcoming years and offer a better knowledge of the spatial information that has been mostly limited in past studies concerning thunderstorm/downburst events.

4. Comparison of the GFF/G-GFF and the TRST Frameworks

As discussed in the previous sections, the two frameworks have adopted different approaches with different wind characteristics even though some underlying models are quite similar. For this reason, a direct comparison of design loads (ESWL) induced by both approaches is not feasible, especially because the TRST heavily depends on the availability of data while the GFF/G-GFF is a model-based approach in which the adopted models may be improved given additional data in the future. Solari (2016) has demonstrated the TRST with an application in which data-based parametric values such as $V_{max}(h)$, $\hat{G}(h)$, etc. were given in the study. Therefore, a plausible way to make an indirect comparison is to employ those values in the GFF/G-GFF and compare a non-dimensionalized factor normalized by common parameters used in both frameworks.

Indeed, the generalized GFF (G-GFF) in Equation 8 and the TRST in Equation 12 show some similarities with each other. Not only are the static wind forces defined in both frameworks fundamentally the same, but also $G_{G,G\text{-}F}$ in the G-GFF is a similar concept with $S_{d,eq}$ in the TRST as both are intended to capture the dynamic effects of gust front winds on structures, akin to the conventional gust loading/effect factor approach for boundary layer winds. In light of this aspect, a non-dimensional factor (λ) is introduced here as

$$\lambda = \frac{f_{ESWL}(z)}{1/2\rho V_{\max}^2 V_{G-F}^2(z)A(z)C_D(z)} \tag{14}$$

Note that $V_{\max}^2$ in this equation is $V_{\max}^2$ for the GFF/G-GFF (Equations 8 and 9) and $V_{\max}^2(h)$ for the TRST (Equation 12). In an application reported in Solari (2016), a modified format of the Wood model (Wood et al. 2001) was employed to describe wind profile while the GFF/G-GFF utilized the Vicroy model (Equation 9). The two profile models are distinctly different, thus such a non-dimensionalization by the static load term in Equation 14 may reduce the related discrepancy between the two models. To further minimize the difference in profiles, for a comparison purpose, $V_{max}(z_{max})$ in the GFF/G-GFF was determined from the Wood model, though it is expected that some discrepancies due to inherently different profile models between the two approaches still remain to some degree. Using this treatment, the factor λ relating to the two approaches may be best expressed in terms of Equations 8, 12, and 14.

$$\lambda \approx G_{G,G-F} \approx \hat{G}^2(h)S_{d,eq} \tag{15}$$

where, $\hat{G}(h)$ is equal to 1.2 in Solari et al. (2015a) and Solari (2016).

Three example structures with square cross-sections were considered to compare the two approaches. The following features were common for these examples: linear mode shape; open terrain condition; air density ρ_a = 1.225 kg/m^3; reference height h = 13 m; $V_{max}(h)$ = 36.5 m/s; z_{max} = 50 m; and $V_{max}(z_{max})$ = 41.1 m/s (obtained using Wood model for the GFF/G-GFF framework). Structural dimensions and other parameters are given for (1) a mid-rise structure with width and depth $B = D = 1$ m, height $H = 30$ m, the first natural frequency $n_1 = 0.9$ Hz, damping ratio $\zeta_1 = 0.005$, and a structural density $\rho_B = 67.1$ kg/m^3 (this example simplified the application used in Solari 2016); (2) a high-rise structure, width and depth $B = D = 25$ m, height $H = 100$ m, the first natural frequency $n_1 = 0.5$ Hz, damping ratio $\zeta_1 = 0.01$, and structural density $\rho_B = 120$ kg/m^3; and (3) a tall structure, $B = D = 40$ m, $H = 200$ m, $n_1 = 0.2$ Hz, $\zeta_1 = 0.01$, and structural density $\rho_B = 180$ kg/m^3. Note that since GFF/G-GFF have adopted the half-sine pulse as a time function, the input of pulse duration (Equation 9) is additionally required. For the sake of a reasonable comparison, the mean value of the duration reported in Solari et al (2015a), $t_d = 248$ sec, was used for the GFF/G-GFF in these examples.

Table 22.2 shows the results of the normalized factor (λ) and COV of $S_{d,eq}$ for the TRST, which was also obtained through the graphs shown in Solari (2016) by computing non-dimension parameters (n_1, δ) with given example properties. Note that the numerical solution of the $G_{G,G\text{-}F}$ was computed using the web-enabled portal (https://vortex-winds.org) developed by Kwon and Kareem (2009) owing to the feature that allows user-defined inputs for V_{max} and z_{max}.

Despite the intrinsic difference between two approaches, an interesting trend was observed in that the GFF/G-GFF resulted in a somewhat larger value than the TRST for Examples 1 and 2, while the two approaches exhibited very close results for Example 3. This is intriguing, especially since the treatment of the non-stationary fluctuation component between the two differs significantly. As mentioned earlier, the GFF/G-GFF was formulated based on models and Davenport wind spectrum/coherence function in the frequency domain using the evolutionary wind spectrum while the TRST employed a data-based EWS technique in terms of wind velocity data at the measurement heights and non-dimensional parameters in the time domain. As mentioned earlier, the mean duration of 248 seconds (t_d) reported in Solari et al. (2015a) was determined based on 93 measured records in which its average COV was 0.52 and the minimum duration was 57 seconds, while no maximum was reported. Given this large variation on the duration, the effect of the duration was examined for the GFF/G-GFF first. Note that the minimum duration reported in Solari et al. (2015a) was 1.48 times the standard deviation; thus the maximum duration was assumed to be the summation of the mean value and 1.48 times the standard deviation, which resulted in 439 seconds. With these minimum and maximum bounds, three truncated probability distributions such as uniform, normal, and lognormal distributions were invoked to generate random samples of durations (10,000 samples), and the non-dimensional factors (λ) were computed in the GFF/G-GFF. It was observed that the variation effects of durations were marginal for all three examples and their COVs were negligibly small (Table 22.3) (i.e., mean of samples were almost identical to those using a single duration; see Table 22.2). It is probably because even the minimum duration (57 sec) used in the examples is large enough when compared to the natural periods of example buildings, which makes the pulse dynamic effects (I_2) become almost unity for all sampled durations as reported in Kwon and Kareem (2009). In addition, the results indicated that the variations of the non-stationary turbulence effects (I_3) were relatively small and converged to the results using the single duration (t_d = 248 sec) for all three examples. A series of small samples (100 realizations) were also tested to examine the duration effects given the number of data points used in the TRST (93 datasets in Solari 2016); however, because no discernible difference was found, the results were omitted here for the sake of brevity.

Among many other influencing factors on the observed trend in Table 22.2, the effect of TI may be more prominent as observed in the preliminary uncertainty analysis performed earlier (Table 22.2). The alongwind dynamic effects are, in general, proportional to the TI (e.g., the gust loading factor). Note that the GFF/G-GFF modeled and employed structural height-dependent TI for different exposure conditions (Kwon & Kareem 2009). On the other hand, the TRST in Solari (2016) assumed constant TI based on the value estimated at the measurement heights: In most cases, the height was about 10 meters and the mean TI ranged from 0.08 to 0.17 (Solari et al. 2015a). Accordingly, a mid-rise structure (H = 30 m) in Example 1 would be exposed to much higher TI in the GFF/G-GFF (about 0.18) than in the TRST, which might lead to higher GFF/G-GFF results. On the other hand, a tall structure (H = 200 m) in Example 3 experiences less significant effects of TI in the GFF/G-GFF (TI = 0.13, which lies in the middle range of mean TI reported in Solari et al. [2015a]). In view of this perspective, additional

Table 22.2 Comparison of the gust front factor/generalized gust front factor (GFF/G-GFF) and the thunderstorm response spectrum technique (TRST)

	GFF/G-GFF	TRST	
	$\lambda \approx GG, G\text{-}F$	$\lambda \approx \hat{G}^2(h)S_{d,eq}$	$COV(S_{d,eq})$
Example 1 ($H = 30$ m)	2.11 (1.81)[a]	1.61	0.185
Example 2 ($H = 100$ m)	1.65 (1.58)[a]	1.41 (1.79, 1.73)[b]	0.145 (0.115, 0.110)[b]
Example 3 ($H = 200$ m)	1.45	1.47 (1.80, 1.65)[b]	0.140 (0.165, 0.155)[b]

[a] Values in the parenthesis were obtained using Example 3's turbulence intensity.
[b] Using Solari and De Gaetano (2018): results obtained through Wood profile (first value) and Vicroy profile (second value), respectively.

Table 22.3 Non-dimensional factor λ of the gust front factor/generalized gust front factor (GFF/G-GFF) using randomly generated t_d with different truncated probability distributions (10,000 samples for t_d = 57 ~ 439 sec)

	Uniform		Normal		Lognormal	
	Mean	COV	Mean	COV	Mean	COV
Example 1	2.09	0.031	2.10	0.026	2.07	0.025
Example 2	1.65	0.022	1.65	0.019	1.64	0.018
Example 3	1.44	0.023	1.45	0.019	1.44	0.018

COV, coefficient of variation.

normalized factors (λ) for Examples 1 and 2 were also computed with the TI value of Example 3 (0.13) and shown in parentheses in Table 22.2; these turned out to be closer to the TRST results in Solari (2016). The other major factor may be the treatment of a vertical wind profile and wind spectrum for the calculation of non-stationary fluctuation response, which the GFF/G-GFF considers a Vicroy profile (Vicroy 1991) and Davenport spectrum model inside the formulation, while the TRST utilizes a data-based EWS approach and Wood profile (Wood et al. 2001).

Recently, Solari and De Gaetano (2018) reported improved graphs of the TRST after recognizing an incorrect computational step in Solari (2016) and comparing the results by time-domain analysis. They recommended using the dimensional form of the spectrum (n_1, δ, ζ) instead of the non-dimensional form (n_1, δ, ζ) employed in the preceding calculations. The size factor, δ, is expressed as

$$\delta = \frac{kc_z H}{\overline{v}_{\max}(z_{eq})} \tag{16}$$

where k is the mode shape factor; c_z is the exponential decay coefficient; and z_{eq} is equivalent height (0.6H). Detailed information can be found in Solari and De Gaetano (2018).

For a comparison purpose, new results of the normalized factor (λ) and COV of $S_{d,eq}$ obtained through the improved graphs in Solari and De Gaetano (2018) are added in parentheses in Table 22.2: Note that there was no graph for the case of ζ = 0.005 in the literature, thus Example 1 was excluded in this computation. Overall, it is observed that normalized factors (λ) in the new TRST are slightly increased while their COVs are decreased as compared to those in Solari (2016). The normalized factor of Example 2 becomes somewhat closer to the GFF/G-GFF and new TRST, while example 3 reveals some discrepancy despite the previous TRST (Solari 2016) showing good agreement with the GFF/G-GFF. It is worth noting that, with the use of the dimensional size factor (Equation 16), certain degrees of profile effects become predominant due to the presence of the denominator term, $\overline{v}_{\max}(z_{eq})$, in which the profile term was not defined in the non-dimensional size factor (δ): Its denominator was simply expressed as z_{eq}. That is, inherent profile discrepancies between the two approaches, Vicroy (GFF/G-GFF) and Wood profiles (TRST), become more relevant in the new TRST because the size factor (δ) is one of the key parameters to select $S_{d,eq}$ and its COV in the TRST graphs. To examine such profile effects, two normalized factors using Solari and De Gaetano (2018) are computed and shown in parentheses in Table 22.3, where the first value was obtained using a Wood profile and the second utilized a Vicroy profile for the calculation of the dimensional size factor (δ): more exactly, the denominator term, $\overline{v}_{\max}(z_{eq})$ in Equation 16. As expected, Wood profile-based results for both Examples 2 and 3 show higher values than Vicroy profile-based ones, and Example 3, with its higher building height than Example 2, reveals a larger difference. These trends agree with the profile difference, where the Wood profile model has higher wind velocity than the Vicroy and the height and wind velocity above z_{max} shows more discrepancy between the two profiles with the increase of height. Certainly, Vicroy profile-based results in the TRST show better agreement with GFF/G-GFF than those based on the Wood profile, which explains the inherent profile effects.

Note that both the GFF/G-GFF and TRST approaches have been respectively validated with Monte Carlo simulations in the time domain (Kwon & Kareem 2013a; Solari & De Gaetano 2018), thus such discrepancies between the two approaches observed in Table 22.2 are rarely caused by respective theoretical/computational deficiency.

Rather, it is more likely due to inherent distinctions between the two approaches, employing different wind characteristics such as wind spectrum, profile, TI, etc. Given all intrinsic differences, they showed reasonably good agreement with each other: less than 10 percent discrepancies when the GFF/G-GFF results were compared to Vicroy profile-based ones in TRST. Nevertheless, this calls for further in-depth comparison to examine the results more systematically.

5. Living Codification Through Learning and Updating Using a Design Thinking Approach

It is vital to critically assess the efficacy of a design framework in light of the existing knowledge base if we are to embark on the next step toward the codification of gust front winds and rationally and holistically quantify design loads such as ESWL in the same way as those provided in the current codes and standards for boundary layer winds. This includes quantification of uncertainty associated with the data (meteorological, aerodynamic, and structural data) and the reliability of frameworks; this will lead to users, code/standard writers, and researchers having confidence in the codification of design frameworks. Once a framework is established and implemented, it is suggested that a working group of potential users be assembled from academia and practice to review, learn, and, in turn, introduce additional changes if needed. To facilitate this procedure, it is suggested that community engagement in implementing these revisions follow the emerging "design thinking" approach that entails identifying options, assessing their impact, and evaluating tradeoffs (e.g., Plattner et al. 2010). Accordingly, implementation and analysis take place concurrently to establish a codification framework through scoping exercises combined with advances made through learning and updating (Figure 22.2). As such, a design framework would benefit from further refinements over

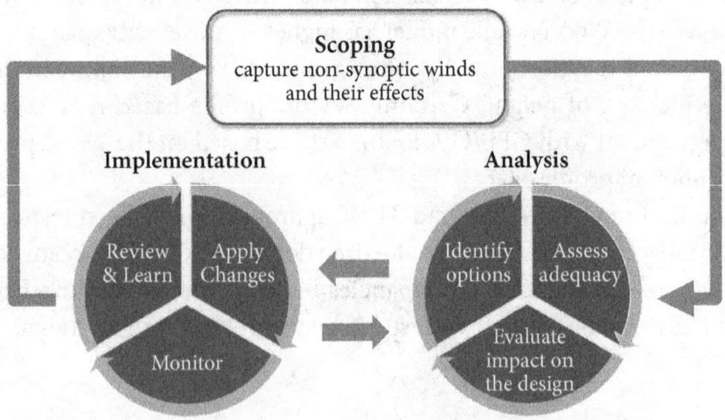

FIGURE 22.2 Codification through learning and updating.

time as additional knowledge about the gust front wind field and associated interactions with structures is accumulated. It is anticipated that new datasets and data-based information resulting from recent research on thunderstorms and downbursts, such as the full-scale monitoring of thunderstorms by the European Research Council's project THUNDERR (Solari et al. 2020) and large-scale experiments at the WINDEEE Dome (Junayed et al. 2019; Hangan et al. 2019), will play important roles in offering a better understanding and modeling of thunderstorm/downburst events.

It is worth noting that many subsequent developments to date have been introduced in the gust loading factor approach (e.g., Vickery 1970; Solari 1993; Solari & Kareem 1998; Zhou & Kareem 2001) since its introduction in 1967 (Davenport 1967); these have enhanced its accuracy and versatility through learning and updating. It should be noted that the proposed learning and updating concept (Figure 22.2) within a web-enabled design framework (e.g., https://vortex-winds.org; Kareem & Kwon 2017) can be a natural fit for encompassing such a reliability analysis, which will also offer expeditious utilization of the GFF and TRST frameworks in design practice and their potential inclusion in codes and standards.

6. Conclusion

Design wind loads based on conventional approaches in codes and standards may not accurately describe fundamentally different non-synoptic non-stationary wind events that depart from typical boundary layer winds both in their kinematics and dynamics. To begin working toward establishing a new design procedure for this type of wind load effect on structures, this study reviews two design frameworks reported in the past literature—the GFF (GFF/G-GFF) and TRST frameworks—concerning gust front winds originating from thunderstorms and downbursts.

The GFF and its generalized frameworks that realistically capture characteristics of gust front winds and their attendant load effects were reviewed as a step toward codification by promoting wind design practice that enhances the disaster resilience of the built environment. In light of limited full-scale data to derive a consensus on quantifying key parameters to date, a model-based approach utilizing analytical models appears to be more appealing at this early juncture of codification, subject to further refinements of underlying models over time, similar to the many subsequent developments in the original procedure of conventional gust loading factor for boundary layer winds. A data-driven design approach involving simulation-based schemes and those capturing uncertainties can also become an attractive alternative for those cases where a suite of datasets will become available in the future.

A GFF/G-GFF-based closed-form solution concerning non-stationary fluctuations can facilitate rapid assessment of non-stationary turbulence effects for a case of a long pulse duration, similar to a typical gust front life cycle. This promises a possible step toward more simplified codification for the evaluation of gust front wind effects on structures. In addition, the results of a preliminary uncertainty analysis suggested that parameters concerning the wind profile, such as wind velocity ($V_{3\text{-}s}$, V_{max}) and z_{max}, were the two most influential factors for the load effects associated with gust front winds and that TI (c_1) was a key parameter for non-stationary turbulence effects. Also, it turned out that dynamic effects in gust front winds were less significant than static/quasi-static and kinematic effects for the example studied because gust front winds are relatively short time events as compared to conventional synoptic winds. A more accurate evaluation will continue through detailed uncertainty quantification with more data and a better support knowledge base.

Data-based TRST is another promising design framework given the fact that the amount of full-scale data is increasing over time. A comparative example between the GFF/G-GFF and TRST frameworks demonstrated that both approaches exhibited a reasonably good agreement despite their underlying differences, thus establishing the efficacy of both frameworks.

Given the lessons learned from the history of the gust loading factor on codes and standards, it is vital not only to critically assess the efficacy of the two design frameworks in light of the existing knowledge base to embark on the next step toward the codification of non-synoptic gust front winds, but also to review, learn, and, in turn, introduce additional changes if needed. It is envisaged that these validations and revisions can be promptly and efficiently made using a proposed learning and updating concept that invokes the emerging design thinking approach that entails identifying options, assessing their impact, and evaluating tradeoff. It is anticipated that new datasets and data-based information resulting from recent researches on thunderstorms/downbursts, such as full-scale monitoring (THUNDERR) and large-scale experiments (WINDEEE Dome), will offer a better understanding and modeling of thunderstorm/downburst events in this context. Accordingly, implementation and analysis can take place concurrently to establish a codification framework through scoping exercises combined with advances made through the learning and updating concept. A web-enabled design framework (e.g., https://vortex-winds.org) embedding the proposed concept can be a natural fit for encompassing such a reliability analysis, which will also offer expeditious utilization of this framework in design practice and potential inclusion in codes and standards. The two frameworks may serve as a building block for an exercise that would benefit from continued advances and input from the design community and work toward a possible codification in the years to come.

Acknowledgments

The first and second authors gratefully acknowledge the support in part by the National Science Foundation (NSF) Grant CMMI-1562244 and 1612843. The research carried out by the third author is funded by the European Research Council (ERC) under the European Union's Horizon 2020 research and innovation program (Grant Agreement No. 741273) for project THUNDERR—Detection, Simulation, Modeling and Loading of Thunderstorm Outflows to Design Wind-Safer and Cost-Efficient Structures—supported by an Advanced Grant 2016. The data used were recorded by the monitoring network realized for the European Projects "Winds and Ports" (Grant No. B87E09000000007) and "Wind, Ports, and Sea" (grant No. B82F13000100005), funded by the European Territorial Cooperation Objective, Cross-border Program Italy-France Maritime 2007–2013.

References

American Society of Civil Engineers (ASCE). (2016). *Minimum design loads for buildings and other structures.* ASCE 7-16. Reston, VA: ASCE.

Bernardini, E., Spence, S. M. J., Kwon, D. K., and Kareem, A. (2015). Performance-based design of high-rise buildings for occupant comfort. *Journal of Structural Engineering,* ASCE, 141(10), 04014244.

Chay, M., Albermani, F., and Wilson, R. (2006). Numerical and analytical simulation of downburst wind loads. *Engineering Structures, 28*(2), 240–254.

Chen, L., and Letchford, C. W. (2004). A deterministic–stochastic hybrid model of downbursts and its impact on a cantilevered structure. *Engineering Structures, 26*(5), 619–629.

Chen, X. (2008). Analysis of alongwind tall building response to transient 821 nonstationary winds. *Journal of Structural Engineering, ASCE, 134*(5), 782–791.

Choi, E. C. C., and Hidayat, F. A. (2002). Dynamic response of structures to thunderstorm winds. *Progress in Structural Engineering and Materials,* 4(4), 408–416.

Choi, E. C. C. (2004). Field measurement and experimental study of wind speed profile during thunderstorms. *Journal of Wind Engineering and Structural Aerodynamics, 92*(3–4), 275–290.

Davenport, A. G. (1967). Gust loading factors. *Journal of Structural Division, ASCE, 93*(3), 11–34.

Ellingwood, B., and Tekie, P. (1999). Wind load statistics for probability-based structural design. *Journal of Structural Engineering, ASCE, 125*(4), 453–463.

Giammanco, I. M., Schroeder, J. L., and Powell, M. D. (2013). GPS dropwindsonde and WSR-88D observations of tropical cyclone vertical wind profiles and their characteristics. *Weather & Forecasting, 28*(1):77.

Haldar, A., and Mahadevan, S. (2000). *Probability, reliability and statistical methods in engineering design.* New York: John Wiley & Sons.

Hangan, H., Romanic, D., and Jubayer, C. (2019). Three-dimensional, non-stationary and non-Gaussian (3D-NS-NG) wind fields and their implications to wind–structure interaction problems. *Journal of Fluids and Structures, 91,* 102583.

Hangan, H., and Xu, Z. (2005). Scale and roughness effects in impinging jets with application to downburst simulations. Paper No. 068, AAWE. CD-ROM. *Proceedings of the 10th Americas Conference on Wind Engineering.* Baton Rouge, LA.

Holmes, J. D., Melbourne, W. H., and Walker, G. R. (1990). *A commentary on the Australian standard for wind loads: AS 1170 Part 2.* Australian Wind Engineering Society.

Howell, L. J., and Lin, Y. K. (1971). Response of flight vehicles to nonstationary atmospheric turbulence. *AIAA Journal,* 9(11), 2201–2207.

Hu, L., Xu, Y.-L., and Huang, W.-F. (2013). Typhoon-induced non-stationary buffeting response of long-span bridges in complex terrain. *Engineering Structures,* 57, 406–415.

Hu, L., Xu, Y. L., Zhu, Q., Guo, A., and A. Kareem. (2017). Tropical storm-induced buffeting response of long-span bridges: Enhanced non-stationary buffeting force model. *Journal of Structural Engineering, ASCE, 143*(6), 04017027.

Huang, G., Zheng, H., Xu, Y.-L., and Li, Y. (2015). Spectrum models for non-stationary extreme winds. *Journal of Structural Engineering, ASCE, 141*(10), 04015010.

Junayed, C., Jubayer, C., Parvu, D., Romanic, D., and Hangan, H. (2019). Flow field dynamics of large-scale experimentally produced downburst flows. *Journal of Wind Engineering and Industrial Aerodynamics, 188*, 61–79.

Kareem, A. (2008). Numerical simulation of wind effects: A probabilistic perspective. *Journal of Wind Engineering and Industrial Aerodynamics, 96*, 1472–1497.

Kareem, A., Hu, L., Guo, Y., and Kwon, D. K. (2019). Generalized wind loading chain: Time-frequency modeling framework for non-stationary wind effects on structures. *Journal of Structural Engineering, ASCE, 145*(10), 04019092.

Kareem, A., and Kwon, D. K. (2017). A cyber-based data-enabled virtual organization for wind load effects on civil infrastructures: VORTEX-Winds. *Frontiers in Built Environment, 3*, 1–32.

Kwon, D., and Kareem, A. (2009). Gust front factor: New framework for wind load effects on structures. *Journal of Structural Engineering, ASCE, 135*(6), 717–732.

Kwon, D., Kareem, A., and Butler, K. (2012). Gust front loading effects on wind turbine tower systems. *Journal of Wind Engineering and Industrial Aerodynamics*, 104–106, 109–115.

Kwon, D., and Kareem, A. (2013a). Gust front factor: New framework for wind load effects on structures. *Engineering Structures, 48*, 635–644.

Kwon, D. K., and Kareem, A. (2013b). Comparative study of major international wind codes and standards for wind effects on tall buildings. *Engineering Structures, 51*, 23–35.

Kwon, D. K., and Kareem, A. (2019). Towards codification of thunderstorm/downburst using gust front factor: Model-based and data-driven perspectives. *Engineering Structures, 199*, 109608.

Kwon, D. K., Kareem, A., Stansel, R., and Ellingwood, B. R. (2015). Wind load factors for dynamically sensitive structures with uncertainties. *Engineering Structures, 103*, 53–62.

Kwon, D., Kijewski-Correa, T., and Kareem, A. (2008). e-Analysis of high-rise buildings subjected to wind loads. *Journal of Structural* Engineering, *ASCE, 134*(7), 1139–1153.

Lind, N. C. (1971). Consistent partial safety factors. *Journal of the Structural Division, ASCE, 97*(ST6) 1651–1670.

Lombardo, F. T., Smith, D. A., Schroeder, J. L., and Mehta, K. C. (2014). Thunderstorm characteristics of importance to wind engineering. *Journal of Wind Engineering and Industrial Aerodynamics, 125*, 121–132.

Michaelov, G., Lutes, L. D., and Sarkani, S. (2001). Extreme value of response to non-stationary excitation. *Journal of Engineering Mechanics, ASCE, 127*(4), 352–363.

Munich Re. (2017). *TOPICS Geo: Natural Catastrophes 2017*. Munich: Munich Re.

NOAA/NWS (2014). *Thunderstorms, tornadoes, lightning: Nature's most violent storms: A preparedness guide* (NOAA/PA 201051). Washington, DC: US Department of Commerce.

Oseguera, R. M., and Bowles, R. L. (1988). A simple, analytical 3-dimensional downburst model based on boundary layer stagnation flow (NASA Technical Memorandum–100632). Hampton, VA: National Aeronautics and Space Administration (NASA).

Piccardo, G., and Solari, G. (1998). Generalized equivalent spectrum technique. *Wind and Structures, 1*(2), 161–174.

Plattner, H, Meinel, C., and Leifer, L. (2010). *Design Thinking: Understand-Improve-Apply*. Berlin: Springer.

Priestley, M. B. (1965). Evolutionary spectra and non-stationary processes. *Journal of the Royal Statistical Society. Series B (Methodological), 27*(2), 204–237.

Priestley, M. B. (1967). Power spectral analysis of non-stationary random process. *Journal of Sound and Vibration, 6*(1), 86–97.

Repetto, M. P., Burlando, M., Solari, G., De Gaetano, P., Pizzo, M., and Tizzi, M. (2018). A web-based GIS platform for the safe management and risk assessment of complex structural and infrastructural systems exposed to wind. *Advances in Engineering Software, 117*, 29–45.

Scott Gunter, W., and Schroeder, J. L. (2015). High-resolution full-scale measurements of thunderstorm outflow winds. *Journal of Wind Engineering and Industrial Aerodynamics, 138*, 13–26.

Solari, G. (1993). Gust buffeting. II: Dynamic along-wind response. *Journal of Structural Engineering, ASCE, 119*(2), 383–397.

Solari, G. (2016). Thunderstorm response spectrum technique: Theory and applications. *Engineering Structures, 108*, 28–46.

Solari, G., Burlando, M., De Gaetano, P., and Repetto, M. P. (2015a). Characteristics of thunderstorms relevant to the wind loading of structures. *Wind and Structures, 20*(6), 763–791.

Solari, G., Burlando, M., and Repetto, M. P. (2020). Detection, simulation, modelling and loading of thunderstorm outflows to design wind-safer and cost-efficient structures. *Journal of Wind Engineering and Industrial Aerodynamics, 200*, 104142.

Solari, G., and De Gaetano, P. (2018). Dynamic response of structures to thunderstorm outflows: Response spectrum technique vs time domain analysis. *Engineering Structures, 176*, 188–207.

Solari, G., De Gaetano, P., and Repetto, M. P. (2015b). Thunderstorm response spectrum: Fundamentals and case study. *Journal of Wind Engineering and Industrial Aerodynamics, 143*, 62–77.

Solari, G., and Kareem, A. (1998). On the formulation of ASCE 7–95 gust effect factor. *Journal of Wind Engineering and Industrial Aerodynamics*, 77–78, 673–684.

Solari, G., Repetto, M. P., Burlando, M., De Gaetano, P., Pizzo, M., Tizzi, M., and Parodi, M. (2012). The wind forecast for safety management of port areas. *Journal of Wind Engineering and Industrial Aerodynamics*, 104–106, 266–277.

Vickery, B. J. (1970). On the reliability of gust loading factors. In R. D. Marshall and H. C. S. Thom (Ed.), *Proceedings of a Technical Meeting concerning Wind Loads on Buildings and Structures* (pp. 296–312). Building Science Series 30. Washington, DC: National Bureau of Standards (NBS).

Vicroy, D. D. (1991). A simple, analytical, axisymmetric microburst model for downdraft estimation (NASA Technical Memorandum–104053). Hampton, VA: National Aeronautics and Space Administration (NASA).

Vicroy, D. D. (1992). Assessment of microburst models for downdraft estimation. *Journal of Aircraft, AIAA, 29*(6), 1043–1048.

Wang, L., and Kareem, A. (2005). Modeling and simulation of transient winds: Downbursts/Hurricane. *Proceedings of 10th Americas Conference on Wind Engineering* (CD-ROM), Baton Rouge, LA, Paper No. 097, AAWE.

Wang, L., McCullough, M., and Kareem, A. (2013). A data-driven approach for simulation of full-scale downburst wind speeds. *Journal of Wind Engineering and Industrial Aerodynamics, 123*, 171–190.

Wang, H., Wu, T., Tao, T., Li, A., and Kareem, A. (2016). Measurements and analysis of non-stationary wind characteristics at Sutong Bridge in Typhoon Damrey. *Journal of Wind Engineering and Industrial Aerodynamics, 123*, 100–106.

Wang, H., Xu, Z., Feng, D., and Tao, T. (2019). Non-stationary turbulent wind field simulation of bridge deck using non-negative matrix factorization. *Journal of Wind Engineering and Industrial Aerodynamics, 188*, 235–246.

Wang, H., Xu, Z., Wu, T., and Mao, J. (2018). Evolutionary power spectral density of recorded typhoons at Sutong Bridge using harmonic wavelets. *Journal of Wind Engineering and Industrial Aerodynamics, 177*, 197–212.

Wood, G. S., Kwok, K. C. S., Motteram, N. A., and Fletcher, D. F. (2001). Physical and numerical modelling of thunderstorm downbursts. *Journal of Wind Engineering and Industrial Aerodynamics, 89*(6), 535–552.

Zhou, G.-D., Ding, Y.-L., and Li, A.-Q. (2015). Evolutionary spectra estimation of field measurement typhoon processes using wavelets. *Mathematical Problems in Engineering*, 945203.

Zhou, Y., and Kareem, A. (2001). Gust loading factor: New model. *Journal of Wind Engineering and Industrial Aerodynamics, ASCE, 127*(2), 168–175.

CHAPTER 23

ECONOMIC AND POLICY ISSUES FOR WIND-RESISTANT CONSTRUCTION

KEVIN M. SIMMONS

1. Introduction

In 1992, Hurricane Andrew caused a record-setting $16 billion (unadjusted) in insured losses, driving 11 insurance companies bankrupt as a result (Fronstin & Holtmann, 1994). But, that number was dwarfed during the high hurricane years of 2004 and 2005, which included Hurricanes Charley, Katrina, and Wilma. Losses from those storms were overtaken in 2017 by those incurred by Hurricanes Harvey, Irma, and Maria, causing most of the over $300 billion nationally in natural hazard losses for the year (National Centers for Environmental Information [NCEI], 2018). A clear trend, shown in Figure 23.1, suggests that damage from wind storms is increasing. While it is possible that climate change may play a role in the uptick in damage, it is primarily due to increased population in regions vulnerable to wind storms. Over the last 40 years, cumulative damage from wind storms is approaching $1 trillion, with 85 percent of the amount incurred since 2000 based on data from the NCEI (2018).

In addition to increasing monetary damage, wind storms can be lethal. This was dramatically illustrated in 2011, with the highest daily death toll from wind storms since records have been kept on April 27, with over 300 fatalities. And, in May of that year, the largest death toll from a single tornado occurred in Joplin, Missouri, with 158 fatalities. Two years later, an Enhanced Fujita (EF) Scale Level 5 (EF5) tornado struck Moore, Oklahoma, killing 24, including 7 children at Plaza Towers Elementary School (Storm Prediction Center [SPC] Tornado Archive; SPC, 2018).

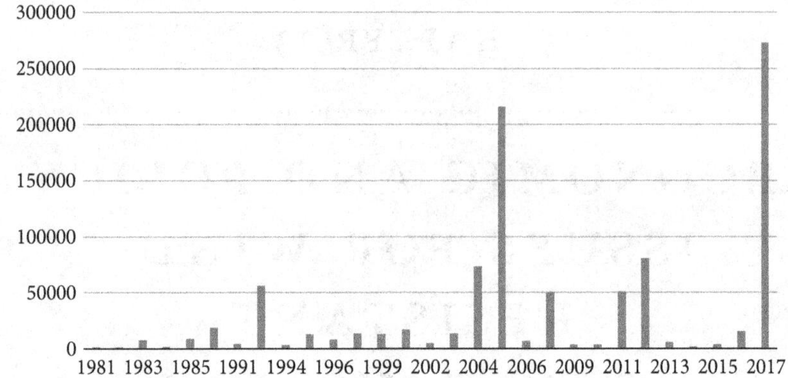

FIGURE 23.1 Annualized damage from US wind storms.

While most fatalities from hurricanes come from drownings (Czajkowski, Simmons, & Sutter, 2011), the numbers are chilling nonetheless. Hurricane Katrina claimed over 1,500 lives.[1] And though the official death toll in Puerto Rico from Hurricane Maria is less than 100, estimates are that more than 1,000 deaths can be attributed to the storm (Santos-Lozada, 2018).[2] A recent paper published in the *New England Journal of Medicine* estimated that almost 5,000 perished from the hurricane and its effects (Kishore et al., 2018).

In 2015, the National Windstorm Impact Reduction Act was reauthorized by Congress. In the strategic plan of the National Windstorm Impact Reduction Program (NWIRP), the mission of the NWIRP (NWIRP, 2018) is defined as follows:

> To achieve major measurable reductions in the losses of life and property from windstorms through a coordinated Federal effort, in cooperation with other levels of government, academia, and the private sector, aimed at improving the understanding of windstorms and their impacts and developing and encouraging the implementation of cost-effective mitigation measures to reduce those impacts.

To achieve the goals of the NWIRP, a combination of regulatory and market-based approaches should be considered. Regulation needs to be cost effective, meaning that the costs to implement a new regulation are outweighed by the benefits in lower damage and morbidity from future wind storms. But, in areas where the explicit benefits of additional regulations prove not to exceed explicit costs, market solutions may provide channels to increased resilience. This is possible because for some consumers implicit benefits of resilient construction in combination with the explicit benefits do exceed cost.

A purpose of this chapter is, first, to discuss the cost-effectiveness of existing building codes designed to mitigate wind. Next, this knowledge is used to search for regions where new regulation in the form of enhanced codes would be cost effective. Ending this section is insight into public support for statewide building codes. Following this discussion of the regulatory approach, I examine the market impact of voluntary wind-enhanced construction. This leads to results from a recent study on consumers' willingness to pay (WTP) for enhanced construction.

2. Cost-Effectiveness of Wind-Enhanced Construction

Economic theory suggests that society is better off when consumers have the freedom to choose goods they prefer and sufficient competition exists to provide a wide selection of suppliers. This is the essence of what Adam Smith called the "invisible hand of the marketplace" in his seminal work *Wealth of Nations* (Smith, 1776/2000). However, when it comes to the purchase of significant assets, such as a home, much of the information needed to make an informed decision is unknown. Classical economic theory relies on knowledgeable consumers. And, understanding the risk wind storms could bring should influence the choice of what features a home should include. But, there is evidence that people underestimate or ignore risks that are considered low-probability high-consequence (LPHC) (Kunreuther, 1976, 1979). When a significant risk is underestimated or ignored, it may be necessary to assist the "invisible hand of the marketplace" with the "visible hand of the policymaker" in the form of regulation (Camerer & Kunreuther, 1989).

While regulation may be called for, public policy is best when it can be shown to be cost effective. For this, the benefits should exceed cost. In this section, I first examine how effective building codes to mitigate wind hazard have proven to be, followed by a benefit/cost analysis of those codes. Next I search for states where the risk of wind storms is high enough to provide justification for the implementation of a statewide building code to mitigate wind. The section finishes by considering public support for enhanced codes in a state where violent tornadoes are common, Oklahoma.

2.1. Florida Statewide Building Code

When Hurricane Andrew struck Florida in August 1992, it became apparent that construction standards and practices had deteriorated over time (Sparks, Schiff, & Reinhold, 1994). Inspections of damaged homes in the aftermath of Andrew revealed inferior practices had unnecessarily inflated the loss from Andrew (Fronstin & Holtmann, 1994; Keith & Rose, 1994). The counties of Dade, Broward, and Monroe all participated in the South Florida Building Code (SFBC) and upgraded the SFBC in 1994. Beyond the counties of the SFBC, several communities voluntarily adopted stronger local codes. In 1996, the Florida Building Code Commission began a review of the state's building codes in consultation with wind engineers. They provided the state legislature with recommendations in 1998 that authorized the changes and created the

2001 statewide Florida Building Code (FBC). Legal action delayed the full implementation of the FBC until March 2002, when the code was fully enacted statewide. The FBC is based on model codes of the International Code Council. It is one of the strictest in the nation, earning the highest-in-nation score of 95 in the 2018 Rating the States report by the Insurance Institute for Business and Home Safety (IBHS, 2018). Included in the FBC is a heavy emphasis on wind engineering principles, including that for hurricane protection (Dixon, 2009).

2.2. Moore, Oklahoma, Enhanced Building Code

On May 20, 2013, an EF5 tornado in Moore, Oklahoma, caused 24 fatalities and over $3 billion in damage (SPC, 2014; Swiss, 2014). This tornado was particularly tragic in that seven children perished at Plaza Towers Elementary School. The raw tragedy of this event combined with the fact that Moore had suffered three violent tornadoes in less than 15 years prompted the city of Moore to adopt a new residential building code based on wind engineering principles. Design wind speed for the new code increased from 90 mph to 135 mph, and the code was implemented beginning on April 1, 2014 (City of Moore, 2014; Ramseyer, Floyd, Holliday, & Roswurm, 2014). A rare feature of this code change was that it was supported by the Moore Association of Homebuilders, which worked with a consulting engineer to design the updated requirements. Changes to the code included narrower spacing of roof joists, the use of hurricane straps to secure the roof to the exterior wall, wind-rated garage door, and the use of oriented strand board (OSB) sheathing to add rigidity to the walls. The expected increase in cost was $1.00 per square foot (Cannon, 2014; Hampton, 2014).

2.3. Effectiveness of Wind-Enhanced Building Codes

Florida did not have to wait long to test the effectiveness of the FBC and local code enhancements. During 2004 and 2005, the state was struck by seven major landfalling hurricanes. The IBHS compared claims after the 2004 Category 4 Hurricane Charley between homes built before and after 1996 (IBHS, 2004). They found that homes built after 1996 suffered lower claim frequency (60 percent less) and lower damage severity (42 percent less) than homes built prior to 1996. Another study conducted by Applied Research Associates (ARA, 2008) using 2002 as the benchmark date found similar results. They used policy-level claims from eight insurance companies for claims filed in 2004 and 2005; these showed significant damage reductions for post-2002 homes. Further, their findings indicated that losses were reduced beginning with homes built in the mid-1990s. While indirectly associated with actual incurred damages, stronger building codes reduced federal disaster spending in 795 unique Florida ZIP codes impacted at least once by the 2004 hurricanes Charley, Frances, Ivan, and Jeanne, as well as Tropical Storm Bonnie (Deryugina, 2013).

And, Moore did not have to wait long to test their new building code. On March 25, 2015, almost a year after enactment of the upgraded code, an EF2 tornado struck Moore. The tornado damaged parts of Moore and southwestern Oklahoma City.[3] The associated system caused four fatalities.[4] But, poststorm assessments found that homes built to the new code performed as expected. "The robust structural system led to only minor damage, which could be quickly repaired" (Ramseyer, Floyd, & Holliday, 2017).

2.4. Benefit/Cost Analysis of Wind-Enhanced Construction

The National Institute for Building Sciences (NIBS) found that mitigation for wind storms returned a benefit-to-cost ratio (BCR) of 5 to 1 in its 2017 report (NIBS, 2017). This report is an overall estimate of the benefits to society compared to the costs and does not speak to specific state or local building codes. But, the estimate illustrates how effective mitigation for wind storms can be. After the SFBC enhanced its standards following Hurricane Andrew, Englehardt and Peng (1996) found that the upgraded code provided a net present value benefit of $7 billion, with a BCR greater than 1. In their 2002 study of the expected effect of the FBC, ARA found that expected average loss reductions due to the FBC were 54 percent, with a further expectation that the resulting losses would exceed cost particularly in coastal areas (ARA, 2002). Finally, Torkian, Pinelli, Gurley, and Hamid (2014) conducted a benefit/cost analysis for retrofit wind mitigation on a synthetic portfolio of existing housing types. They estimated BCRs ranging from 0.41 to 1.83 for the retrofit mitigation activities to existing housing.

As mentioned, the state of Florida and Moore, Oklahoma, have adopted explicit residential building codes to mitigate wind storms. The FBC is administered statewide, while the code for Moore only pertains to new homes built there. Two important studies, those of Simmons, Kovacs, and Kopp et al., 2015, and Simmons, Czajkowski, and Done, 2018, performed benefit/cost analysis on first the Moore (tornadic region) upgraded code and next the FBC (hurricane region). Hurricanes affect large areas of a state, while tornadoes are more local. As a result, the cost-effectiveness for hurricane mitigation may appear more obvious. However, tornadoes are more common, and when the sum of loss from these events is tallied, cost-effectiveness can be found.

2.4.1. *Moore*

Moore, Oklahoma, suffered three violent tornadoes in less than 15 years: 1999, 2003, and 2013. The 2013 tornado was particularly violent and struck two elementary schools, Briarwood and Plaza Towers, killing seven children at Plaza Towers. The quick succession of violent tornadoes combined with the raw tragedy of losing children sheltered in their school motivated the city council to consider stronger codes.

To determine if the cost to implement the code changes exceeded the expected benefits of lower future losses, Simmons et al. (2015) conducted a benefit/cost analysis assuming the changes in Moore would be implemented statewide. Based on historic loss data, the state of Oklahoma averaged more than $1 billion in paid insured tornado-related losses annually, not including deductibles, with about two thirds of that amount affecting residential losses. The study used an estimated life for a residential home of 50 years, so the present value of future loss from tornadoes would equal $35 billion. Engineering estimates were that the new code would reduce 30 percent of direct damage compared to a home built prior to the updated code. If 30 percent of that damage could be mitigated, the present value of reduction in loss amounted to just over $10 billion. This estimate was for direct losses only and did not consider the effect of reduced claims due to the new code.[5] Using the estimated increase in cost of $1 per square foot and a reduction in direct loss of 30 percent across the life of the structure, homes built to the new code should experience more than $3 in reduced loss for every $1 in increased cost.

2.4.2. *Florida*

Florida was the first state to institute a statewide building code incorporating wind engineering principles. Motivation for this move was Hurricane Andrew, which devastated southern Florida, causing record losses from wind and creating a crisis for property insurance in the state. Andrew caused the insolvency of several insurers, leading to the creation of a state-supported insurer, Citizens, as remaining insurers stopped issuing policies in the most vulnerable regions of the state (Lecomte & Gahagan, 1998; McChristian, 2012).

Andrew also motivated the state to examine its building codes. During the 1970s and 1980s, Florida experienced a large increase in population. Construction standards fell, creating the conditions for high losses from a large hurricane (Keith & Rose, 1994). The code was fully implemented in March 2002. As mentioned, the high hurricane years of 2004 and 2005 provided evidence that the homes built to the new code performed better than older homes. But, do the new codes provide protection that is cost effective? In other words, do the benefits of the statewide codes exceed the added cost of construction? This question was put to the test by Simmons, Czajkowski, and Done (2018). Using realized insured loss data from 2001 to 2010 from over 1 million insured customers, they found that the FBC was indeed cost effective.

The new code reduces loss in two ways. First, it lowers the direct loss when two damaged homes are compared, one built to the FBC and the other not. Here, they found that the reduction in direct loss was 53 percent on average. This differs from the estimated direct loss reduction of 30 percent for the Moore study (Simmons et al., 2015). The Moore study used an estimate based on changes in the degree of damage (DOD) within the categories of the EF Scale. That may reflect an underestimation of the impact of the Moore code, which was similar in many ways to the FBC. Or, it could reflect that loss reduction for hurricanes differs from that for tornadoes. Additional structure-level research, using paid losses from insurers, is necessary to examine the realized effect of the Moore code on loss.

In addition to the direct reduction in loss, stronger construction also affects the probability of suffering a loss at all, thus lowering claims for homes built to the FBC compared to those built prior to its enaction. When this indirect effect is included, the overall reduction in estimated loss for homes built to the FBC was 72 percent.

Table 23.1 Increased Cost of the FBC

	Nominal (2002)			CPI[6] Adjusted (2010)		
	Low	Average	High	Low	Average	High
N-WBDR	0.23	0.76	1.28	0.28	0.92	1.55
WBDR-Cover	1.04	1.77	2.49	1.26	2.15	3.02
WBDR-Glazed	3.25	5.35	7.45	3.94	6.48	9.03
Statewide	0.95	1.37	1.78	1.15	1.66	2.16

Assuming the same relationship holds for the Moore code, the full effect would be a 46 percent reduction in loss. But, this estimate may be low because no analysis has been done using realized paid losses on homes built before and after enactment of the Moore code.

According to Hamid, Pinelli, Chen, and Gurley (2011), the state of Florida can expect $3 billion in wind-related insured losses annually in 2007 dollars. If deductibles are included, that number increases to $5 billion. Averaged across the almost 9 million homes in Florida according to the 2010 census, that is over $27,000 in loss across the 50-year expected life of the home. Reducing that loss by 72 percent provides a benefit of the FBC of almost $20,000.

To determine if the reduction in loss is cost effective, an estimate of the increase in construction cost to comply with the added regulations of the FBC is needed. In 2002, ARA conducted a comprehensive estimate of the increase in costs to comply with the FBC (ARA, 2002) shown in Table 23.1. The state was divided into two regions: the Wind Borne Debris Region (WBDR), which was mostly in southern Florida and along the coasts; and the Non–Wind Borne Debris Region (N-WBDR), which is mostly in the interior of the state. Homes in the WBDR required impact protection on all windows and doors, a feature not required for homes in the N-WBDR. Based on the ARA 2002 report, the range of increased cost for homes in the N-WBDR is a low of $.28 per square foot (2010 dollars) and a high of $1.55 per square foot. This is an average cost of $0.92 per square foot or very close to the estimate in Moore of $1.00 per square foot. For homes in the WBDR, there are two options for impact protection. The less expensive option is to have exterior coverings on all windows and doors. A more expensive option is to have impact-resistant glass for all windows and doors. The range for the covered option was a low estimate of $1.26 and a high estimate $3.02, giving an average of $2.15. Impact-resistant glass has a low estimate of $3.94, a high estimate of $9.03, and an average of $6.48.

This first analysis of the FBC is statewide, so I calculated the percentage of homes in each region, then found a weighted average increase in cost of $1.66 in 2010 dollars. The average square footage of a home in Florida is 1,960 square feet, giving an average increase in construction cost of $3,254. Dividing the reduction in loss by the cost of the FBC gives a BCR of 6.06, meaning that for every $1 in increased cost, losses from wind storms in Florida are reduced by $6.

Table 23.2 Loss Reduction by Wind Borne Debris Region, Florida

Zone	PV[7] of Loss, 50-Year Total–Per Unit	Direct Reduced Loss Percentage	Direct Reduced Loss Per Unit	Full Reduced Loss Percentage	Full Reduced Loss Per Unit
N-WBDR	$15,923	57%	$9,076	81%	$12,897
WBDR	$43,691	32%	$13,981	64%	$27,962
Statewide	$27,396	53%	$14,520	72%	$19,725

Having the two regions within Florida means that the cost-effectiveness of the FBC may vary from the N-WBDR and WBDR. First, the effectiveness of the mitigation may be different along the coast, where wind fields are higher than in the interior of the state. But, it also varies within the WBDR depending on which type of impact protection is chosen because impact-resistant glass is more expensive. Simmons, Czajkowski, and Done (2019) conducted a follow-up to the study of the FBC published in *Land Economics*. This study performed the same analysis, but conducted the examination on the effect of the code by region as well as the cost-effectiveness by region.

To begin, the annual average loss cannot be calculated at the state level but must be allocated to each region. The N-WBDR accounts for 23 percent of the statewide loss from wind storms, leaving the remaining 77 percent of the loss in the WBDR. As shown in Table 23.2, an average home in the N-WBDR can expect almost $16,000 in loss (with deductibles) across its 50-year life. The same home in the WBDR can expect almost $44,000 in loss with deductibles. This is expected because hurricanes lose strength as they proceed inland. Exposure to higher winds also has an impact on the amount of loss that can be mitigated. Not surprisingly, the effectiveness of the code does vary, with the best results found in the N-WBDR. Here, the FBC reduced direct loss by 57 percent. When avoided claims are considered, the reduction in loss is 81 percent. But in the WBDR, the reduction in direct loss is 32 percent, suggesting that the code is not as effective when facing the full brunt of an approaching wind storm. When avoided claims are added, the reduction in loss is 64 percent, again lower than what is found in the N-WBDR. Homes in the N-WBDR have an average reduction in loss of almost $13,000, while homes in the WBDR have an average reduction in loss of $28,000.

Using the average estimated increase in cost for the N-WBDR and both impact resistance options with the WBDR, the increase in construction cost is $1,803 for the N-WBDR, $4,214 for the lower cost option in the WBDR, and $12,701 for the higher cost option in the WBDR. BCRs range from 2.20 for the higher cost option in the WBDR to 6.64 for the lower cost option in the WBDR and 7.15 for the N-WBDR (Table 23.3).

The cost-effectiveness of the FBC is higher in the N-WBDR, but results in the WBDR are still robust, especially for the lower cost option of using covers for the windows and doors instead of impact-resistant glass. A reasonable question arises regarding how consumers can justify purchasing a home with the impact-resistant glass instead of exterior covers. The answer may be as simple as aesthetics. Exterior coverings are

Table 23.3 Benefit/Cost Ratio by Wind Borne Debris Region, Florida

Zone	Full Reduced Loss Per Unit	Per Unit Added Construction Cost (1960 Average Square Feet)	BCR
N-WBDR	$12,897	$1,803	7.15
WBDR-Covered[8]	$27,962	$4,214	6.64
WBDR-Glazed[9]	$27,962	$12,701	2.20
Statewide	$19,725	$3,254	6.06

not as attractive, and consumers who value aesthetics over cost will choose the higher cost option for that reason.

2.4.3. *Other States*

The results from both Moore and Florida are encouraging. They illustrate that building codes designed with wind engineering principles can be cost effective. The next question is, Are there other states beyond Florida that could find similar results to justify a statewide code for wind hazard? To address this question, Simmons, Kovacs, and Smith (2020) identified the top 20 states in terms of wind damage from both tornadoes and hurricanes. The states came from the expected regions, Plains States in the middle of the country (Tornado Alley) and the Southeast. Using data and results from the Moore study for tornado-prone states and from the Florida study for states that suffer hurricanes, eight states were identified that have estimated benefits that exceed cost. So, it is not unreasonable to suggest that other states would benefit from a statewide code for wind beyond Florida. The top state is Mississippi, followed by Oklahoma, Alabama, and Missouri. Mississippi has an estimated BCR of 4.13, Oklahoma 4.10, Alabama 3.18, and Missouri 2.13. The remaining states showing cost-effectiveness were Kansas 1.68, Iowa 1.31, Arkansas 1.27, and Tennessee 1.15.

Mississippi and Alabama suffer both tornadoes and hurricanes. Isolating damage from tornados alone shows that tornado damage in Mississippi is somewhat less than damage from hurricanes. But, Alabama shows tornado damage greatly exceeding damage from hurricanes. If the BCRs for these two states are calculated using only damage from tornadoes, Mississippi would be 1.69 and Alabama 2.64. Both are cost effective based on tornadoes alone, but the high level of tornado damage in Alabama provides a stronger case. In the first 4 months of 2019, Alabama had experienced 53 tornadoes, killing 23.[10]

For a state to adopt a similar code to the FBC will require the political will of policymakers. States that may benefit from a wind-enhanced code but require many years for a payoff may find too much resistance to make the effort worthwhile. So, to identify the states most likely to adopt a statewide code for wind, an additional test was performed. One way to choose the best states is to calculate a payback for the investment in stronger codes. All of these studies used an expected structure life of 50 years. To estimate how long it takes to pay back the investment in stronger codes, divide the expected structure

Table 23.4 Benefit/Cost Ratios and Payback for States Most Likely to Benefit from Statewide Wind-Enhanced Building Codes

State	BCA Average (Tornado Damage Only)	BCA Average	Payback in Years
Mississippi	1.69	4.13	12.11
Oklahoma	4.10	4.10	12.20
Alabama	2.64	3.18	15.75
Missouri	2.13	2.13	23.47
Kansas	1.68	1.68	29.76
Iowa	1.31	1.31	38.17
Arkansas	1.27	1.27	39.37
Tennessee	1.15	1.15	43.48

life of 50 years by the BCR. A state with a ratio of 2 would have a payback of 25 years. Table 23.4 shows the BCR and payback time frames for the eight states identified as cost effective.

The typical mortgage is 30 years, and most families move before the mortgage on their home is paid. For that reason, it may be doubtful that residents would support a policy that would not find benefits within the time they plan to live in the home. Mississippi, Oklahoma, and Alabama all have payback periods that are less than 20 years. Missouri is 23, which comes close to the typical mortgage. For this reason, the top three states appear to have payback periods short enough to have a good chance of public approval.

Mississippi, Oklahoma, and Alabama are not a surprise. They are at high risk of violent wind storms and are not known for having strong statewide codes. In fact, Mississippi and Alabama were recently identified by the IBHS as having some of the weakest statewide building codes from the hurricane-prone states evaluated by their Rating the States report in 2018 (IBHS, 2018). Only Delaware scored lower on their scale than either Mississippi or Alabama. In contrast, Florida stands at the top of their rankings, with a score of 95 on a 100-point scale. The score for Mississippi is 28 and 27 for Alabama.

2.4.4. *Public Support for Enhanced Building Codes*

Ultimately, for increased regulation to have a chance to be adopted, the proposed change needs public support. Many of the states that suffer significant damage from wind storms are conservative politically, and conservative ideology is suspicious of regulation generally. In these states, it may be necessary for the public to lead politicians toward reasonable standards by indicating their support for more resilient construction.

Oklahoma is a good example. It is clearly in the heart of Tornado Alley, suffering an average of more than 65 tornadoes per year (SPC, 2018). And politically, the state is conservative. But, there is evidence that residents of Oklahoma do indicate at least weak support for stronger construction standards. In a new study, Ripberger et al. (2018) conducted a survey of Oklahoma residents to test their support for enhanced standards for building codes costing about $2,000. They found that push dynamics that include objective data such as how they perceive risk and personal experience with wind storms increased support, but pull dynamics, which include skepticism of climate change or conservative views, lessened support. An estimate of the average support for stronger codes suggests that most residents supported more regulation, with an average probability of support of .64. When push dynamics prevailed, support increased to 0.76. But, when pull dynamics prevailed, probability of support dropped to .40. This result showed that pull dynamics provide a stronger effect than push dynamics.

The article offers insight into how support for stronger codes is formed through push or pull dynamics. To increase support for enhanced codes, programs that influence push dynamics should be attempted. For instance, ongoing permanent education is needed to provide information about wind storms to assist residents in forming an accurate perception of risk. And, an opportunity presents itself each time an event occurs. At that moment, the attention of the public is heightened, and proponents of stronger codes should be ready to educate the public on the actual risk they face. This effort is more successful if it is part of a regular education effort rather than appearing to be an opportunistic attempt to use a tragedy for furthering an agenda.

Affecting the pull dynamics is more challenging as it deals with an individual's belief system. Conservative beliefs place more value on a market-driven result than a policy-driven one. However, markets can create externalities or unintended consequences which are outcomes that do not coincide with an efficient market solution. Full reliance on a market to provide resilient construction allows for a well-built home to be next to one poorly built. A tornado that damages the poorly built home may also damage the well-built home from the poorly built home debris field. To lessen the impacts of an externality requires all homes to be constructed to a safe standard. Bringing attention to this effect might reduce political opposition and could be an effective communication strategy of the benefits of resilient construction.

3. Market Effects of Wind-Enhanced Construction

Economists look to the behavior of markets to understand which products are demanded by the public. But, we can also examine how markets respond to public policy decisions. When consumers decide how to spend their money, they are implicitly revealing their preferences and attitudes about the options available to them. In this section, I first examine how the market for new homes responded to a decision on the part of policymakers in Moore, Oklahoma, to upgrade their building codes when neighboring communities did not. Also, I examine how mitigation from wind storms is valued when it is only available to consumers voluntarily rather than by code. Finally, I use results from a study on consumer WTP for wind mitigation for insights into the variables that affect that decision.

3.1. Market Response to Code Change

Municipalities are often reluctant to increase their building code standards for fear that the increased cost drives development to nearby cities with less expensive requirements. This reluctance is understandable. New residents mean higher tax revenue and more sales for local merchants; new residents make the community more attractive to other businesses looking for places to expand. Developers and builders want to find business opportunities that allow them to make a profit. Economic theory suggests that a city with higher costs for building homes will not be as attractive as a neighboring city with lower costs.

The action of Moore, Oklahoma, provides an ideal opportunity to test that hypothesis. If economic theory is correct, Moore would have experienced a negative impact on the market for new homes with fewer sales and permits after the code change. Simmons and Kovacs (2018) collected sales and permit data from Moore and Norman, Oklahoma, for years preceding the code change and after. Sales data came from Multiple Listing Service data collected by realtors. Permit data came from both the city of Moore and the city of Norman.

The procedure to test the effect of the code change is called a difference in differences (D-I-D) test. It has been used in studies to examine the effect of training programs on earnings (Ashenfelter & Card, 1985) and on the effect in employment of an increase in the state minimum wage in one state next to a state with no change in the minimum wage (Card & Krueger, 1994). To conduct a D-I-D test, Norman is considered the control group compared against the target group, Moore. Changes in Norman, before the code change (treatment) and after is compared to what occurred in Moore. A statistical test is conducted on what occurred in Moore (actual change) and what would have occurred in Moore if it experienced the same change as Norman (expected change). If there is no statistical difference between the expected and actual change, then the conclusion is that the code change had no effect. In Simmons and Kovacs (2018) three possible market effects were studied: price per square foot, weekly sales, and monthly permits. Using a null hypothesis of no difference between expected and actual change, all three market effects failed to reject the null hypothesis. As an example, consider weekly sales. The average weekly sales in Norman prior to the code change was 14.45. After the code change, that declined slightly to 14.34. In Moore, before the code change, average weekly sales were 8.73. If they changed the same percentage as Norman, weekly sales would decline slightly to 8.66. But the actual change was a slight increase to 8.85. The difference between the expected change and the actual change is 0.19 and is not statistically significant. Similar tests on price per square foot and monthly permits yield the same result, suggesting that despite economic theory, there was no negative impact on the real estate market in Moore after enacting a building code using wind engineering principles.

Finding a result that contradicts theory requires a search for an explanation. There are several possible explanations, but the most likely is that consumers in central Oklahoma prefer the additional resilience that homes built to the Moore code offer. This increase in demand vis-à-vis homes without those features keeps builders from leaving Moore for neighboring communities.

3.2. Market Valuation of Perceived Hazard and Mitigation

The possible explanation for what occurred in Moore offers a way forward for increasing resilience in states where the benefits to cost of stronger construction fail to make public efforts viable. If consumers value resilient construction such that the price they are willing to pay is higher compared to homes without those features, then increased resilience can occur simply from market pressure on firms to provide that product.

For consumers to increase the price they are willing to pay for hazard mitigation, they must be aware of the threat and use that information to form a price response. An early paper on the impact of earthquake and volcano notifications (Bernknopf, Brookshire, & Thayer, 1990) found that hazard notices in the Mammouth Lake area of California caused home prices to decline between $9,400 and $13,000. Three studies (MacDonald, Murdock, & White, 1987; Shilling, Benjamin, & Sirmans, 1985; Speyer & Ragas, 1991) examined the effect of location in a flood plain on resale prices of residential structures. All studies found that location in a flood plain caused homes to sell for a discount relative to homes outside the flood plain.

While perceiving the risk of a hazard can decrease home value, do homes that mitigate the effects of the hazard increase value? Three studies that evaluated the effect of wind mitigation on home prices suggested that they do. Simmons, Kruse, and Smith (2002) looked at the effect of hurricane mitigation on resale prices for homes on Galveston Island and found that homes with hurricane shutters sold for almost 6 percent higher than similar homes without shutters on the island. Awondo, Hollans, Powell, and Wade (2016) found that homes built to the IBHS FORTIFIED™ standard commanded a 7 percent premium in Alabama. And Dumm, Sirmans, and Smersh (2011) found that homes built in Miami–Dade county after the SFBC was upgraded in 1994 sold for a premium in the years between 2000 and 2007. They further observed that the premium declined after 2004 as most of the hurricanes affecting Florida in that year did not impact Miami–Dade County. But, the premium reappeared after 2004, when Hurricane Wilma devastated southern Florida.

Resale premiums are a powerful incentive for homeowners who are considering a retrofit project that may allow for wind mitigation, but other incentives are available in some areas. For instance, homes in Florida covered by the state insurer Citizens may qualify for discounts on their property insurance if they install wind mitigation.[11] And in Alabama, the Strengthen Alabama Homes program of the Alabama Department of

Insurance offers grants to homeowners in the coastal counties of Baldwin and Mobile counties. The grants are managed by the Alabama Department of Insurance but funded by insurance companies.[12] Efforts such as these provide financial incentives to build resilient homes and match the priorities set by the Obama administration through their programs on disaster mitigation.[13]

3.3. Willingness to Pay for Private Wind Mitigation

Studies cited previously used hedonic pricing models to estimate the price premium homes with wind mitigation may command. They are useful to understand how mitigation is valued in the aggregate, but they don't provide insight into which variables influence individual consumers or which incentives may increase demand. To do that, direct questioning of residents in an area that suffers from wind storms is needed. Carson, McCullough, and Pooser (2013) found that homes in Florida that have high wind insurance premiums are more likely to mitigate. But, even states without specific wind insurance, like Oklahoma, also provide a positive response to mitigation. Czajkowski et al. (2018) performed a survey of Oklahoma residents on their WTP for home features that mitigate damage from wind storms. Overall, about 45 percent of surveyed residents stated that they were willing to pay for wind mitigation. The first finding was that WTP decreased with the price to mitigate. That result is no surprise. It simply confirms a basic tenet of economics that the quantity demanded of most products is inversely related to price. The article found other effects that were also expected: Higher family income and higher education increased WTP as well as the residents' perception of the threat.

One interesting result from Czajkowski et al. (2018) is that residents appeared not to substitute insurance for mitigation. As their perception of the threat increased, they tended to increase insurance coverage and their investment in mitigation. That is interesting because mitigation can be thought of as a form of self-insurance (Ehrlich & Becker, 1972). This result minimizes the potential for moral hazard.

A final contribution of Czajkowski et al. (2018) is the role that incentives may play in influencing WTP. Two incentives are considered: discounts on property insurance and subsidized loans to install mitigation. For low-income respondents, the subsidized loans were more attractive, but for moderate- to high-income respondents, discounts on their property insurance proved to be more appealing. Knowledge about which incentives work with different consumers is important as incentive programs can be better targeted.

4. Conclusion

Florida led the way in designing statewide building codes to mitigate wind hazard after Hurricane Andrew in 1992. Strong evidence exists that this decision has reduced losses from hurricanes both in the high hurricane years of 2004 and 2005 and for the Hurricane Irma in 2017. Further evidence showed that the Florida decision not only reduced losses but also was cost effective, providing a BCR of $6 in reduced loss for every $1 in added cost. Other states that experience high activity for tornadoes and hurricanes can also show cost-effectiveness. At least eight states beyond Florida show positive benefit/cost comparisons, with three of those states having BCRs robust enough to pay back the investment in less than 20 years. However, the decision to enhance building codes can be political, and pressure is mounting on officials in Florida to weaken their statewide code (Federal Alliance for Safe Homes, 2017). And, the states with the most promising cost-effectiveness are conservative politically and may be suspicious of additional regulations. This presents a formidable challenge for advocates of resilient housing. Strategies to broaden acceptance of wind-enhanced building codes must consider the reluctance to adopt more regulation and find ways to address and overcome those objections.

In states where cost-effectiveness of statewide codes cannot be shown, increased resilience is still possible. Real estate markets value mitigation in a similar fashion to other home attributes. Studies indicated that consumers have a WTP for wind mitigation. Education of homebuyers regarding the risk wind storms bring will increase the number of consumers willing to pay more for wind mitigation, particularly in the immediate aftermath of a wind storm event. Other incentives, such as discounts on property insurance and subsidized loans to low-income families who wish add mitigation features, show promise. These results suggest that in regions where regulation is either not cost effective or not politically feasible, increased resilience is within reach.

Notes

1. https://www.nhc.noaa.gov/data/tcr/AL122005_Katrina.pdf (page 11)
2. http://www.newsweek.com/hurricane-maria-killed-more-puerto-ricans-we-know-877518
3. http://tornatrix.net/?p=2131
4. https://newsok.com/article/5406130/death-toll-from-march-25s-Tstorms-in-oklahoma-increases-to-four
5. Based on a regression model in Simmons, Czajkowski and Done (2018), the overall reduction would be 46% if lower claims were taken into account.
6. https://www.weather.gov/bmx/tornadodb_2019
7. https://www.citizensfla.com/discounts
8. https://strengthenalabamahomes.com/
9. https://obamawhitehouse.archives.gov/blog/2016/08/03/getting-smart-about-how-we-finance-disaster-resilience
10. Consumer Price Index
11. Present Value
12. Window and Door coverings
13. Impact resistant window glazing

References

Applied Research Associates (ARA). (2002). *Florida Building Code cost and loss reduction benefit comparison study*. Tallahassee: Florida Department of Community Affairs.

Applied Research Associates (ARA). (2008). *2008 Florida residential wind loss mitigation study*. Tallahassee: Florida Office of Insurance Regulation. Retrieved from https://www.floir.com/siteDocuments/ARALossMitigationStudy.pdf

Ashenfelter, O., & Card, D. (1985). Using the longitudinal structure of earnings to estimate the effect of training programs. *The Review of Economics and Statistics*, *67*(4), 648–660.

Awondo, S., Hollans, H., Powell, L., & Wade, C. (2016). *Estimating the effect of FORTIFIED Home construction on home resale value* (working paper). Retrieved from https://www.smarthomeamerica.org/assets/uploads/UniversityofAL_Value-Study_FORTIFIEDReport_V2__2.pdf

Bernknopf, R. L., Brookshire, D. S., & Thayer, M. A. (1990). Earthquake and volcano hazard notices: An economic evaluation of changes in risk perceptions. *Journal of Environmental Economics and Management*, *18*(1), 35–49.

Camerer, C. F., & Kunreuther, H. (1989). Decision processes for low probability events: Policy implications, *Journal of Policy Analysis and Management*, *8*(4), 565–592.

Cannon, J. G. (2014, March 17). Moore city council takes historic step in bolstering homes. The Oklahoman. Retrieved from https://oklahoman.com/article/3944338/moore-city-council-takes-historic-step-in-bolstering-homes

Card, D., & Krueger, A. (1994). Minimum wages and employment: A case study of the fast-food industry in New Jersey and Pennsylvania. *American Economic Review*, *84*(4), 772–793.

Carson, J. M., McCullough, K. A., & Pooser, D. M. (2013). Deciding whether to invest in mitigation measures: Evidence from Florida. *Journal of Risk and Insurance*, *80*(2), 309–327.

City of Moore. (2014). Ordinance No. 768 (14). Retrieved from https://www.cityofmoore.com/uploads/City-Council-Meetings-and-Agendas/2014-Minutes/3-17-14.citycouncil.min.pdf

Czajkowski, J., Ripberge, J., Jenkins-Smith, H., Silva, C., Kunreuther, H., Michel-Kerjan, E., & Simmons, K. M. (2018). Homeowner willingness to pay for tornado risk mitigation and the role of economic incentives (working paper).

Czajkowski, J., Simmons, K. M., & Sutter, D. (2011). Hurricanes and moving water, a deadly combination: An analysis of coastal and inland hurricane fatalities. *Natural Hazards*, *59*(3).

Deryugina, T. (2013, July 18). Reducing the cost of ex post bailouts with ex ante regulation: Evidence from building codes. SSRN 2314665. Retrieved from https://ssrn.com/abstract=2314665

Dixon, R. (2009). *The Florida Building Code: Florida's response to hurricane risk*. Florida Building Commission presentation. Tallahassee: Department of Business and Professional Regulation. Retrieved from https://www.sbafla.com/method/portals/methodology/WindstormMitigationCommittee/2009/20090917_DixonFLBldgCode.pdf

Dumm, R. E., Sirmans, G. S., & Smersh, G. (2011). The capitalization of building codes in house prices. *Journal of Real Estate Finance and Economics*, *42*, 30–50.

Ehrlich, I., & Becker, G. S. (1972). Market insurance, self-insurance, and self-protection. *Journal of Political Economy*, *80*(4), 623–648.

Englehardt, J. D., & Peng, C. (1996). A Bayesian benefit-risk model applied to the South Florida Building Code. *Risk Analysis*, *16*(1), 81–91.

Federal Alliance for Safe Homes. (2017). The case for preserving Florida's building code system. *FLASH Newsroom*. Retrieved from http://newsroom.flash.org/commentary/the-case-for-preserving-floridas-building-code-systems.htm

Fronstin, P., & Holtmann, A. G. (1994). The determinants of residential property damage caused by Hurricane Andrew. *Southern Economic Journal, 61*(2), 387–397.

Hamid, S. S., Pinelli, J.-P., Chen, S.-C., & Gurley, K. (2011). Catastrophe model based assessment of hurricane risk and estimates of potential insured losses for the state of Florida. *Natural Hazards Review, 12*(4), 171–176.

Hampton, J. (2014, March 18). Moore city council approves tornado building code amendments. Norman transcript. Retrieved from https://www.normantranscript.com/news/local_news/moore-city-council-approves-tornado-building-code-amendments/article_8de68cad-952a-533e-a6ca-8c4f2840acc1.html

Insurance Institute for Business and Home Safety (IBHS). (2004). Hurricane Charley executive summary. Retrieved April 18, 2018, from http://disastersafety.org/wp-content/uploads/hurricane_charley.pdf

Insurance Institute for Business and Home Safety (IBHS). (2018). Rating the states 2018. Retrieved March 20, 2018, from https://disastersafety.org/ibhs-public-policy/rating-the-states/

Keith, E. L., & Rose, J. D. 1994. Hurricane Andrew: Structural performance of buildings in south Florida. *Journal of Performance of Constructed Facilities, 8*(3), 178–191.

Kishore, N., Marques, D., Mahmud, A., Kiang, M. V., Rodriguez, I., Fuller, A., . . . Buckee, C. (2018). Mortality in Puerto Rico after Hurricane Maria. *New England Journal of Medicine*, Retrieved from https://www.nejm.org/doi/full/10.1056/NEJMsa1803972?query=featured_home+

Kunreuther, H. (1976). Limited knowledge and insurance protection. *Public Policy, 29*(2), 227–261.

Kunreuther, H. (1979). The changing societal consequences of risks from natural hazards. *The Annals of the American Academy of Political and Social Science, 443*(1). https://doi.org/10.1177%2F000271627944300111

Lecomte, E., & Gahagan, K. (1998). Insurance protection in Florida. In H. Kunreuther & R. Roth (Eds.), *Paying the price: The status and role of insurance against natural disasters in the United States*. Washington, DC: Joseph Henry Press.

MacDonald, D. N., Murdock, J. C., & White, H. L. (1987). Uncertain hazards, insurance, and consumer choice: Evidence from housing markets. *Land Economics, 63*(4), 361–371.

McChristian, L. (2012, August). Hurricane Andrew and insurance: The enduring impact of an historic storm. Insurance Information Institute. Retrieved from https://media.local10.com/document_dev/2017/08/24/paper_HurricaneAndrew_final_10338020_ver1.0.pdf

National Centers for Environmental Information (NCEI). (2018). Billion-dollar weather and climate disasters: Overview. Retrieved from https://www.ncdc.noaa.gov/billions/

National Institute for Building Sciences. (2017, December). Natural hazard mitigation saves: 2017 interim report. Prepared by the National Institute of Building Sciences Multihazard Mitigation Council. Retrieved from https://www.fema.gov/media-library-data/1516812817859-9f86633obd6a1a93f54cdc61088f310a/MS2_2017InterimReport.pdf

National Windstorm Impact Reduction Program (NWIRP). (2018). *Strategic plan*. Washington, DC: National Institute for Standards and Technology, U.S. Department of Commerce.

Ramseyer, C., Floyd, R., & Holliday, L. (2017). Performance of enhanced residential building code requirements during the March 25, 2015, Moore tornado. *Journal of Performance of Constructed Facilities, 31*(5).

Ramseyer, C., Floyd, R., Holliday, L., & Roswurm, S. (2014). Influence of lateral load bracing systems on damage and survivability of residential structures impacted by the Moore, Oklahoma, tornado of May 20, 2013. *Proceedings of the Structures Congress 2014*, Boston, MA, ASCE, 1484–1507. doi:10.1061/9780784413357.131

Ripberger, J., Jenkins-Smith, H., Silva, C., Czajkowski, J., Kunreuther, H., Michel-Kerjan, E. & Simmons, K. M. (2018). Tornado damage mitigation: Homeowner support for enhanced building codes in Oklahoma. *Risk Analysis, 38*(11), 2300–2317.

Santos-Lozada, A. R. (2018). In Puerto Rico, counting deaths and making deaths count. *Health Affairs, 37*(4). doi:10.1377/hlthaff.2018.0219

Shilling, J. D., Benjamin, J. D., & Sirmans, C. F. (1985). Adjusting comparable sales for floodplain location. *The Appraisal Journal, 53*(3), 429–436.

Simmons, K. M., Czajkowski, J., & Done, J. (2018). Economic effectiveness of implementing a statewide building code: The case of Florida. *Land Economics, 94*(2), 155–174.

Simmons, K. M., Czajkowski, J., & Done, J. (2019). Building code economic performance under variable wind risk. *Mitigation and Adaptation Strategies for Global Change, 24*(2), 235–258.

Simmons, K. M., & Kovacs, P. (2018). Real estate market response to enhanced building codes in Moore, OK. *International Journal of Disaster Risk Reduction, 27*, 85–93.

Simmons, K. M., Kovacs, P., Kopp, G. A. (2015). Tornado damage mitigation: Benefit–cost analysis of enhanced building codes in Oklahoma. *Weather Climate and Society, 7*, 169–178.

Simmons, K. M., Kovacs, P., & Smith, A. (2020). State by state analysis of the benefits to cost from wind enhanced building codes, *Natural Hazards Review*, forthcoming in 2020.

Simmons, K. M., Kruse, J. B., & Smith, D. A. (2002). Valuing mitigation real estate market response to hurricane loss reduction measures. *Southern Economic Journal, 68*(3), 660–671.

Smith, A. (2000). *The wealth of nations/Adam Smith*. Introduction by R. Reich; edited, with notes, marginal summary, and enlarged index by E. Cannan. New York, NY: Modern Library. (Original work published 1776)

Sparks, P. R., Schiff, S. D., & Reinhold, T. A. (1994). Wind damage to envelopes of houses and consequent insurance losses. *Journal of Wind Engineering and Industrial Aerodynamics, 53*(1–2), 145–155.

Speyrer, J. F., & Ragas, W. R. (1991). Housing prices and flood risk: An examination using spline regression. *Journal of Real Estate Finance and Economics, 4*(4), 395–407.

Storm Prediction Center (SPC). (2014). Storm Prediction Center WCM page. NOAA. Retrieved from https://www.spc.noaa.gov/wcm.

Storm Prediction Center (SPC). (2018). Storm Prediction Center WCM Page. NOAA. Retrieved April 18, 2018, from https://www.spc.noaa.gov/wcm/

Swiss, R. E. (2014). Natural catastrophes and man-made disasters in 2013: Large losses from floods and hail; Haiyan hits the Philippines. *Sigma*, No. 1/2014, 49 pp.

Torkian, B., Pinelli, J.-P., Gurley, K., & Hamid, S. (2014). Cost-and-benefit evaluation of windstorm damage mitigation techniques in Florida. *Natural Hazards Review, 15*(2), 150–157.

CHAPTER 24

CONVECTIVE STORMS AND THE INSURANCE INDUSTRY: CHALLENGES AND OPPORTUNITIES

KIRSTEN D. ORWIG

1. Introduction

It is worth taking a moment to discuss the lexicon used in the insurance industry. The insurance industry generally utilizes the term *convective storm* or *severe convective storm* when referring to damage or losses arising from thunderstorms. From a risk assessment point of view, convective storms consist of the subperils hail, straight-line winds, and tornadoes, the last two of which are non-synoptic wind storms. Straight-line winds in this context refer to thunderstorm winds (aside from tornadoes), such as those generated by downdrafts, and include microbursts, macrobursts, and the associated gust fronts and outflow boundaries. Lightning, while originating from convective storms, is generally viewed from a fire risk perspective, and flash flooding is generally considered as flood. Furthermore, it is common for the risk assessment and losses from hail, tornadoes, and straight-line winds to be aggregated and viewed collectively. Therefore, the focus here is on convective storms as a whole, emphasizing non-synoptic wind storms (i.e., tornadoes and straight-line winds) where possible.

There are two risk assessment perspectives that the insurance industry generally uses: occurrence basis or aggregate basis. The former utilizes the occurrence exceedance probability (OEP) distribution, which is the likelihood that an event loss will exceed a specified amount, while the latter is determined via an aggregate exceedance probability (AEP) distribution, which is the likelihood that the sum of all losses in a given period will exceed a specified amount. Figure 24.1 and Figure 24.2 show the conceptual process

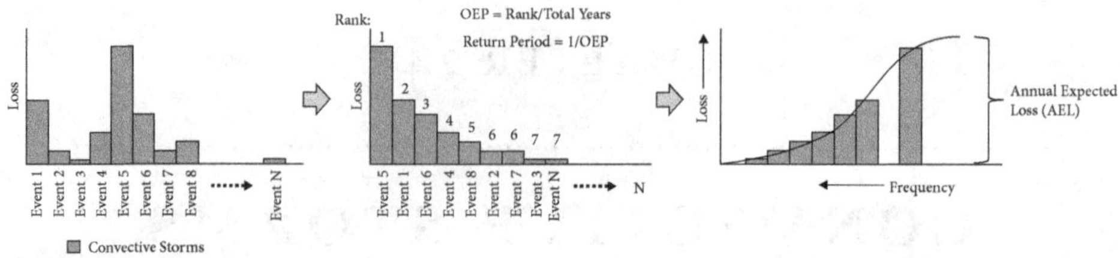

FIGURE 24.1 The conceptual process for calculating the occurrence exceedance probability and associated annual expected loss. N is the total number of events.

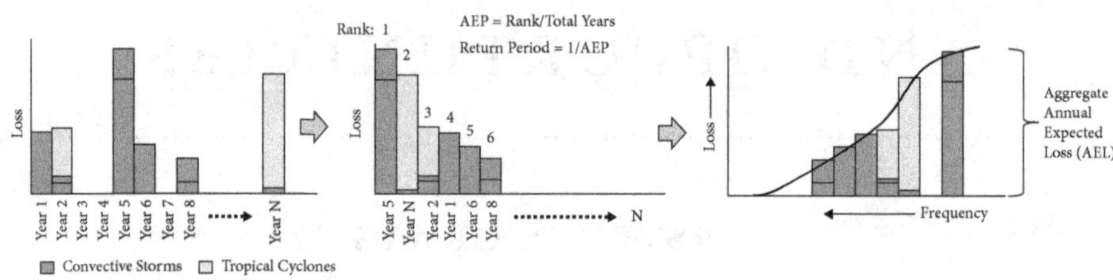

FIGURE 24.2 The conceptual process for calculating the aggregate exceedance probability and associated annual expected loss. The period for aggregation is 1 year, and N is the total number of years.

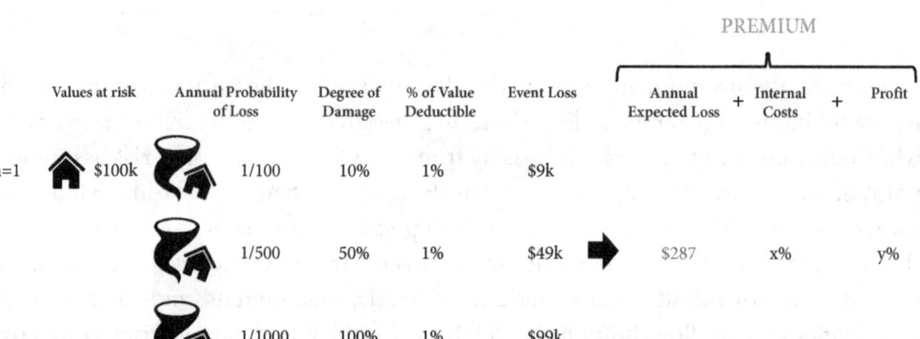

FIGURE 24.3 Conceptual view of the risk premium calculation, assuming a single location (n = 1), where x and y are arbitrary percentages of the AEL to reflect internal costs and profit margin, respectively.

for how these distributions are developed. The annual expected loss (AEL) is then calculated by integrating the respective exceedance probability curves. The AEL (or pure premium) is the basis behind how risk premiums are generally determined, as shown conceptually in Figure 24.3. This process provides an outline to see how changes in the frequency or severity of losses can affect the resulting AEL and thereby the risk premium. In the example, the AEL is calculated by

$$AEL = \sum_{N}^{i=1} L(e_i)P(e_i)$$

such that

$$L(e_i) = (TD(e_i) - d)$$

where T is the total value at risk; $D(e_i)$ is the degree of damage for each event; e_i is expressed as the ratio of values damaged to T; d is the deductible (in this case, expressed in percentage of T); $P(e_i)$ is the annual exceedance probability of the event; N is the number of events; and $L(e_i)$ is the insured loss for each event.

2. Loss Trends

Convective storms pose a unique challenge to the insurance industry due to their inherent nature and recent loss trends. They cause billions of dollars in total economic losses and dozens of casualties worldwide every year. According to the Swiss Re Institute (2015), the global total economic losses attributed to convective storms grew at an annual rate of 7.7 percent per year since 1990, which contrasted with 6.4 percent for all natural catastrophes combined. The rates are higher for insured losses at 9.0 percent and 6.6 percent, respectively. The proportion of weather-related losses (i.e., including from tropical cyclones, winter storms, convective storms, droughts, floods, etc.) caused by convective storms has also grown. In the 1990's, the proportion was 29 percent, while in the 2000's it was 38 percent despite the hurricane seasons of 2004 and 2005. Early indications are that the 2010's will reflect a similar or even higher proportion of the losses.

While Europe, Asia, Africa, Australia, and South America see their fair share of convective storms, with hundreds of millions to billions in economic losses every year, losses are predominantly driven by activity in the United States. As shown in Figure 24.4, convective storms are now the primary source of insured loss from natural catastrophes in the United States and in recent years have surpassed the average losses from tropical cyclones (i.e., hurricanes and tropical storms). For comparison, the 10-year moving averages of flood (i.e., losses reported by the Federal Emergency Management Agency [FEMA] National Flood Insurance Program [FEMA, 2018]) and earthquakes are also shown. The annual insured losses from convective storms since 2008 have consistently exceeded $10 billion and are increasing substantially year over year. From 1998 to 2007, the average annual increase was approximately $70 million, while over the last 10 years from 2008 to 2017 this escalated more than tenfold to over $900 million per year. Risk Management Services has observed a +7.5 percent per year trend in claim severity coupled with a +3.3 percent rise in claim frequency per year from 2001 to 2017, resulting in a total claims increase of +11 percent per year (Cusack, 2018).

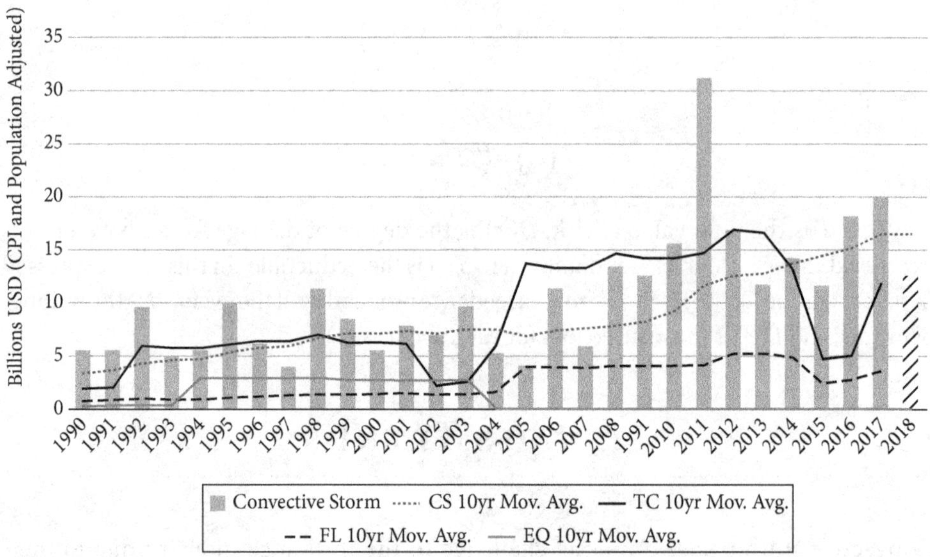

FIGURE 24.4 Historical insured losses for convective storms (bars), along with 10-year moving averages of convective storms (red line), tropical cyclones (black line), floods (blue line), and earthquakes (green line). Note that losses are trended by the Consumer Price Index (CPI) and US population. (Bureau of Labor Statistics 2018; FEMA, 2018; Property Claims Services, 2018; US Census Bureau 2000, 2018c.)

3. Key Loss Drivers

There are many drivers for the increasing loss trends; population growth, increasing numbers of properties exposed, and escalating property values have all been contributing factors. Table 24.1 details the changes since 1990 for the top 15 states impacted by convective storms. Additionally, according to the US Department of Housing and Urban Development and US Department of Commerce Economics and Statistics Administration (HUD and DOC, respectively) (2017), the median square footage of homes since 1990 has grown from 1,905 ft^2 to 2,426 ft^2 in 2017 or +27.3 percent. This statistic suggests that more area and materials are exposed to potential damage.

Another key contributing factor is the evolution of building materials and construction practices. Building codes and standards generally consider seismic, wind, fire, and snow loads when establishing design criteria. However, design criteria for wind fall short of reflecting the wind characteristics and peak loads of tornadoes (Bienkiewicz & Dudhia, 1993; Jischke & Light, 1983; Kuligowski, Lombardo, Phan, Levitan, & Jorgensen, 2013; Yang, Sarkar, & Hu, 2011) and thunderstorm-generated straight-line winds (Choi & Hidayat, 2002; Holmes, Hangan, Schroeder, Orwig, & Letchford, 2008; Letchford, Mans, & Chay, 2002; Orwig, 2010; Orwig & Schroeder, 2007). For example, as demonstrated by Orwig and Schroder (2007) and Orwig (2010), thunderstorm outflows tend to have significantly higher gust factors, spatial coherence, as well as first- and second-order nonstationarities when compared to hurricane or synoptic wind environments.

There are a few exceptions with respect to design criteria for tornadoes: (1) nuclear power plants; (2) tornado shelters and safe rooms; and (3) the community of Moore, Oklahoma. In 2014, Moore adopted provisions requiring that new construction must be able to withstand Enhanced Fujita Scale 2 (EF2) level winds (an increase from 90 to 135 mph), after suffering damage from the fourth EF4–EF5 tornado in less than 15 years. There was some justification for doing this as Kuligowski et al. (2013) found that 90 percent of the damage from the Joplin EF5 tornado was associated with wind speeds of EF3 or lower and that 96 percent of all tornadoes in the United States were EF0–EF3 intensity. Therefore, Moore's criteria would in theory result in a building stock that would be able to withstand the vast majority of tornadoes and substantially reduce the damage and losses to the community in the future.

Building code adoption and enforcement vary from country to country. In the United States, adoption and enforcement is left to the states and even down to the community level to decide which provisions to adopt. Similarly, Australia, Germany, and Canada, for example, adopt a national model code but leave it to the provinces and territories to adopt, adapt, and regulate. In other countries, adoption and enforcement are centralized at a federal level or there are no building regulations whatsoever. In the United States, the Building Code Effectiveness Grading Schedule (BCEGS) grades the performance of

Table 24.1 Percentage Change From 1990 to 2017 for the Top 15 States Affected by Convective Storms

Rank	State	Population	Housing Units	Median Home Value
1	Texas	+66.6%	+25.6%	+68.2%
2	Oklahoma	+25.0%	+18.9%	+59.8%
3	Missouri	+19.5%	+21.3%	+39.9%
4	Illinois	+12.0%	+15.9%	+30.4%
5	Colorado	+70.2%	+38.1%	+118.4%
6	Minnesota	+27.4%	+24.2%	+58.8%
7	Tennessee	+37.7%	+31.5%	+54.3%
8	Mississippi	+15.9%	+23.7%	+58.8%
9	Kansas	+17.6%	+18.0%	+59.9%
10	Indiana	+20.2%	+22.2%	+38.7%
11	Ohio	+7.5%	+16.0%	+22.1%
12	Georgia	+61.0%	+38.4%	+35.4%
13	Alabama	+20.7%	+26.1%	+48.6%
14	Kentucky	+20.8%	+24.1%	+55.6%
15	Nebraska	+21.6%	+21.1%	+66.1%
	United States	**+30.9%**	**+25.6%**	**+53.1%**

Note. The 1990 Median Home Values were adjusted to 2017 dollars by the Consumer Price Index. Data sources: Bureau of Labor Statistics (2018); US Census Bureau (1996, 2000, 2011, 2012, 2018a–2018c).

building regulations, and community participation is voluntary. The BCEGS shows an inverse relationship between the community's score and, for example, hail damage or number of fires; the higher the score is, the lower the incidence will be (BCEGS, 2015), further reinforcing the benefits of strong building regulations.

4. Challenges for Insurers

The insurance industry faces many challenges in managing convective storm losses, such as risk assessment, data quality, and affordability.

4.1. Risk Assessment and Data Quality

Risk assessment and data quality go hand in hand; without good quality historical data, risk assessment has much higher uncertainty. The fundamental principle behind risk transfer is the law of large numbers, which is simply put as follows: As the portfolio of individual exposures becomes larger, the greater the probability is that the actual loss converges to the expected loss. In other words, the greater diversification of risk, the greater the predictability of that risk will be. However, this premise assumes that all risks and losses are independent from one another. Therein lies the difficulty for assessing natural catastrophe risk; the losses for individual exposures are not independent. Hence, the necessity of risk assessment models, which address the issue of correlated loss and strive to reflect all possible scenarios of loss.

In general, risk assessment models consist of four modules: (1) hazard; (2) exposure; (3) vulnerability; and (4) financial. Briefly, these modules are described as follows:

Hazard—reflects the event frequency, intensity, and spatial footprint of the affected area

Exposure—reflects the location or portfolio of locations exposed to a particular hazard, how those locations are represented (i.e., as point locations, areas, etc.), and characteristics of those locations (e.g., building occupancy type, year built, number of stories, etc.)

Vulnerability—reflects the relationship between event intensity, the exposure characteristics, and loss sustained (i.e., damage functions)

Financial—reflects the insurance policy conditions (e.g., deductibles, limits, etc.)

For convective storms, there are high levels of uncertainty for both the hazard and vulnerability components.

4.1.1. *Hazard*

Risk assessment models generally utilize historical data as the foundation for developing the underlying hazard. For example, tropical cyclone models take historical tracks, intensities, and footprints, and these historical storms are perturbed in a variety of ways to generate tens to hundreds of thousands of simulated tracks. The outcome is a historical and simulated event set, as well as a historical and simulated hazard set. The simulated sets are necessary to achieve statistically significant event sample sizes at any specific location such that probability exceedance distributions can be derived. The event set is generally a reflection of event frequency and track characteristics, while the hazard set reflects the intensity footprints, intensity frequency, and spatial correlation. For convective storms, millions of simulated events are necessary to achieve statistical convergence at any given location due to smaller spatial areas of tornado paths or hail swaths. Therefore, convective storm modeling is still somewhat in its infancy relative to other perils due to the computational resources needed.

The historical data quality of convective storms is also of considerably poorer quality or not regularly collected. Tornadoes, for example, are underreported worldwide, and researchers piece together historical reports (Holzer, 2001; Kahraman & Markowski, 2014; Rauhala, 2012; Taszarek & Brooks, 2015) or rely on proxies to estimate climatology. These proxies include reanalysis data (Brooks, Lee, & Craven, 2003; Romero, Gayà, & Doswell, 2007; Sugimoto, Nohara, & Hirakuchi, 2016) and combinations of lightning and population data (Cheng et al., 2013; Sills et al., 2012). Arguably, the United States has the most complete database of tornadoes, hail, and straight-line winds, and yet it has many issues that render it difficult to use for risk assessment modeling purposes. These issues include the following with some examples noted:

Changes in reporting
- Tornado mean path width was recorded prior to 1994, while today the maximum path width is recorded (McCarthy, 2003)
- Change from Fujita Scale to EF Scale in 2007

Binning of data (Figure 24.5)
- Reports of hail size or tornado path width or length are rounded to the nearest 0.25 inch, 100 yards, or 1 mile, respectively

Biases in reporting (Figure 24.6)
- Higher incidence of hail reports in urban areas or along roadways
- Higher incidence of tornado reports in urban areas or near Doppler radars (Doswell, Moller, & Brooks, 1999; Frisbie, 2006; Schaefer & Galway, 1982).
- Increased reporting of F/EF0 and F/EF1 tornadoes (McCarthy, 2003)

Incomplete information
- Start and end points of tornado path provided but no details on path curvature or change in width or intensity
- No information on hail swaths or outflow footprints

Subjectivity of reports
- Hail sizes are rarely precisely measured and instead are relative to common objects, such as golf balls or baseballs
- Tornado intensities are based on damage surveys (Doswell & Burgess, 1988; Edwards, 2003; Marshall, 2003; McCarthy, 2003)

Duplicate or erroneous entries
- When a tornado crosses administrative boundaries (Storm Prediction Center [SPC], 2010)
- Long tornado paths may instead be a series of shorter track tornadoes (Doswell & Burgess, 1988)

The hazard also depends on the risk perspective: single risk or a portfolio of risks. Depending on the perspective, a different approach to modeling the hazard is necessary. For single risk, the concern is with single tornadoes or single hail swaths affecting a particular location. The frequency of occurrence will be very low, but the severity of impact may be high. However, if the perspective is that of a portfolio of risks or locations, then the event definition changes. It is no longer a single tornado or convective storm that is of concern, it is the combination of convective storm occurrences over a given period. Typically, in insurance and reinsurance policies, there are hours clauses that define the loss event, and this generally ranges from 72 to 120 hours or more. Therefore, all convective storm–related losses will be lumped together, and the maximum loss during a running window of 72 or 120 hours will be the loss paid. As a result, the concern is then with outbreaks of convective storms. With more locations, the probability of a loss or a series of losses occurring increases, and the loss exceedance frequency is determined by the event definition (i.e., the hours clause). Therefore, the risk assessment models need to take into account the temporal correlation as well as the spatial correlation of individual convective storms.

4.1.2. *Vulnerability*

The vulnerability of structures to tornadoes and straight-line winds is somewhat better understood (for some occupancies, that is, e.g., residential). The EF Scale, for example, characterizes the damage to buildings of different types, construction characteristics, and construction quality relative to the wind speed (Texas Tech University Wind Science and Engineering Center, 2004). The EF Scale is an improvement on the original Fujita

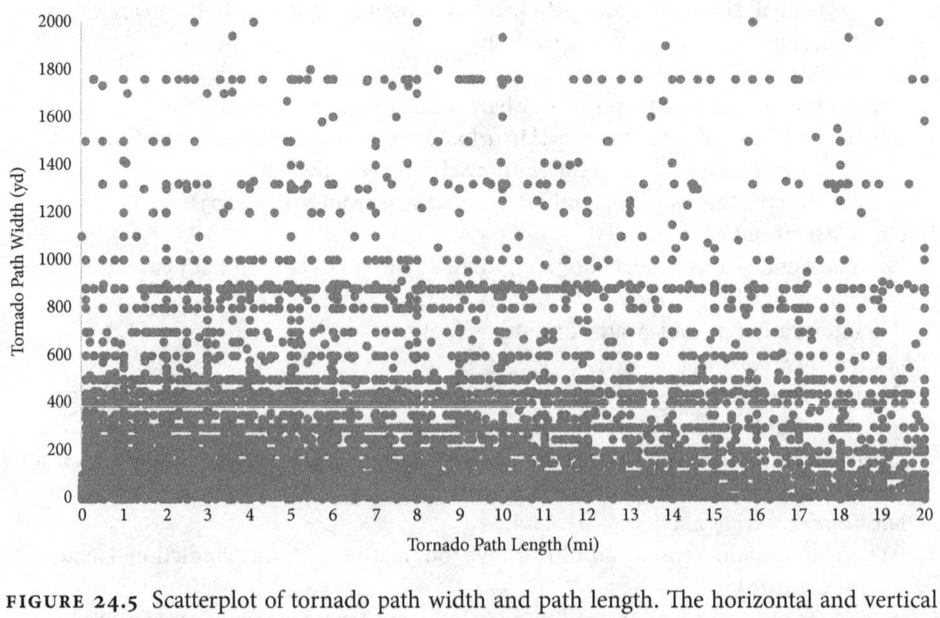

FIGURE 24.5 Scatterplot of tornado path width and path length. The horizontal and vertical lines that appear in the data are due to the binning of reports, illustrating the imprecise reporting.

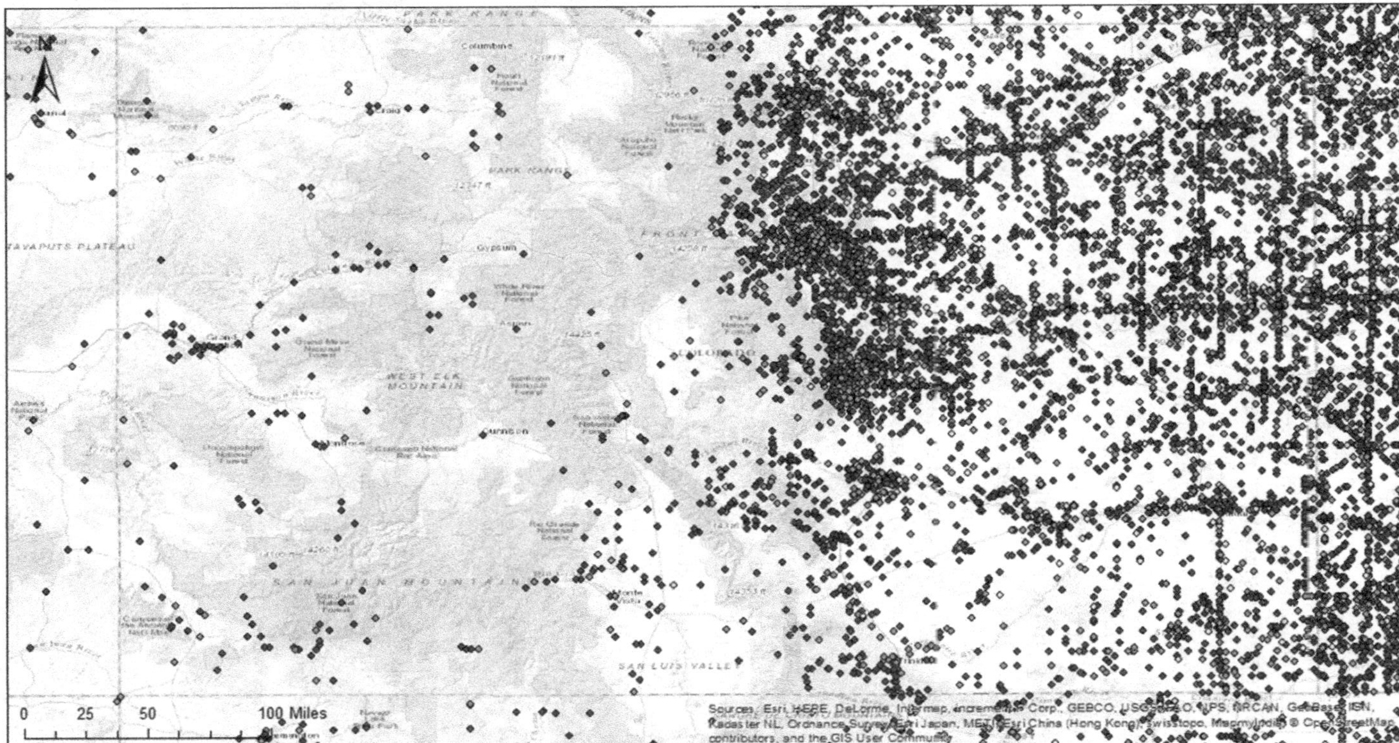

FIGURE 24.6 Hail reports from the Storm Prediction Center Storm Report Database in the state of Colorado. Lines tend to follow roadways and highways.

Scale and strives to remove some of the subjectivity involved in damage surveys. However, a tornado is classified by the maximum observed rating, and information on the individual structural ratings is not regularly or consistently documented. Additionally, it is common for damage assessments to be performed by individuals without engineering expertise (e.g., in the United States, surveys are performed by meteorologists in the National Weather Service). Detailed and accurate information on site-specific surveys is essential for risk assessment model developers to verify and benchmark the vulnerability assumptions. It is also necessary to couple these damage surveys with the insurance loss adjustments to translate the physical damage to an insured loss estimate. This translation is often the key piece missing, not only for convective storm models, but also for other peril models as well.

There has been a lot of research done on fragility curves or damage curves that relate the physical damage to the intensity of the event. However, physical damage is only one aspect of the loss. Other loss factors include time element losses (e.g., lost revenue during recovery), expenses to mitigate an imminent threat or remove debris, loss adjustment expenses, post-loss amplification due to demand for materials and labor, the deductible or limit that applies, and whether the insurance reimbursement is in terms of the replacement cost or actual cash value. The insured losses compiled by the Property Claims Services, for example, are one of the few market-level insured loss databases; the data are very high level, broken down by commercial, residential, and auto. The losses are net of deductibles and limits and must meet a threshold of $25 million or more per event to be reported. Additionally, losses are not separated by subperil and are only available on an aggregate level for convective storms. Ideally, for verifying the vulnerability curves, model developers would need both the total economic loss for each location and the insured loss, as well as broken down by subperil because the modeling framework is structured on a subperil basis. Detailed loss information simply is not available on a market level or federal level, and there are discrepancies between data that are available (Cutter, 2001; Gall, Borden, & Cutter, 2009).

4.1.3. *Loss Data Trending*

It is worth taking a moment to discuss loss data trending. In the absence of detailed loss data, researchers utilize what is available and make assumptions regarding how to relate these losses in today's dollars. As mentioned, insured losses are net of deductibles and coverage limits. These insurance policy conditions can vary year over year depending on changes in the market or changes in the view of risk. Ideally, one would need to know how these conditions changed year over year to generate as if losses in today's terms. Additionally, it is essential to incorporate changes in exposure and price inflation (Collins & Lowe, 2001; Pielke et al., 2008; Pielke & Landsea, 1998). There is still uncertainty introduced when trending these losses due to probabilities of new properties being impacted, construction market dynamics, community resiliency changes, changes in insurance take-up rates, infusion of government emergency funds for recovery, and a variety of other reasons.

4.1.4. *Model Consensus (or Lack Thereof)*

As a result of the data quality issues and technical difficulties in convective storm risk assessment, there is little consensus between models. Every model utilizes its own approach, assumptions, and foundational science and engineering. Every year, the Reinsurance Association of America holds a conference during which a study is presented comparing vendor models for various perils and regions. In 2014, a comparison between US convective storm models was performed and presented. The models were evaluated for the Ohio Valley, Tornado Alley, and the Southeast. The exceedance probability curves varied substantially for all three regions, even for high frequencies where there should be more consensus (due to the available historical data). Each model utilized different probability distributions with varying spatial granularities, had challenges smoothing small footprints, and used different event set sizes and event definitions as well as different assumptions on subperil contributions to the total expected loss (Keeton, 2014).

5. Conclusions and Recommendations

A number of issues and challenges have thus far been raised, and there are a number of solutions that can help alleviate them, a few of which are discussed here, namely:

- More stringent building codes and codes that reflect expected loads from tornadoes, straight-line winds, and hail.
- More investment in mitigation and resilience.
- More granular and consistent data collection for tornadoes, straight-line winds, and hail.
- More granular exposure and loss databases.
- More research in the translation of building damage to economic and insured loss, particularly for commercial, industrial, and public facilities.

5.1. Improved Building Resilience

As indicated, building codes, construction practices, and construction materials all have an impact on the resilience of structures to convective storms. One clear recommendation is therefore to have more stringent building codes and broader utilization of resilient materials. For example, the Insurance Institute of Building and Home Safety (IBHS) FORTIFIED™ program enables residential and commercial construction to be more resilient to straight-line winds, hail, and hurricanes. Engineering design and full-scale simulations have been utilized to demonstrate that improved standards can result in significant reductions in damage (IBHS, 2009; Reinhold, 2009, 2011).

Several other studies have demonstrated how improved building codes and code enforcement have resulted in less damage (BCEGS, 2015; Czajkowski & Simmons, 2013; FEMA, 1992; IBHS 2004, 2009, 2018; Simmons, Czajkowski, & Done, 2019). More resilient materials have also demonstrated better performance and reduced damage in high wind and hail impact (IBHS, 2014, n.d.; Liu, Pogrzelski, Masters, Tezak, & Reinhold, 2010). There are a number of certification programs and standards, such as Underwriters Laboratories (UL) 2218A (UL, 2019); UL 580 (UL, 2006); UL 1897 (UL, 2015); FM 4471 (FM Approvals, 2010); ANSI/FM 4474 (FM Approvals, 2011); ASTM D7158 (ASTM International, 2019b); ASTM D3161 (ASTM International, 2019a); and ASTM C1570 (ASTM International, 2016), to improve performance in high wind or moderate-sized hail (up to 2 inches).

There is a cost associated with more resilient design; however, several studies (FEMA, 1997; National Institute of Building Sciences, 2017; Simmons, Kovacs, & Kopp, 2015) have shown that the benefits substantially outweigh these costs. For example, Simmons et al. (2015) found that if the entire state of Oklahoma implemented the same provisions as those in Moore, the benefits of mitigation outweighed the costs 3:1. Additionally, the study expected that the increase in home cost would be roughly $1 per square foot, suggesting that the total costs would only be a few thousand dollars for the median home size. A follow-up article by Simmons and Kovacs (2018) indicated that the realized costs were closer to $2 per square foot, but still reasonably affordable. According to the authors, economic theory suggests that more regulation would result in less construction, leading to lower availability of homes, and increased demand would push up home prices. However, they found that the codes implemented in Moore had no statistically significant effect on home sales or new building permits.

Hazard mitigation grants are another way to achieve greater resilience. Rose et al. (2007) demonstrated on average a 4:1 benefit-to-cost ratio for FEMA hazard mitigation grants.

5.2. Improved Historical and Exposure Database

The current historical record for convective storms is highly biased and incomplete. It does not provide the level of detail or quality required for developing sophisticated and reliable risk assessment modeling. One suggested improvement includes creating and compiling detailed footprints from tornado paths, hail swaths, and straight-line wind swaths. Following the Joplin, Missouri, tornado of 2011, the US Army Corps of Engineers (2011) generated a detailed damage footprint, which serves as an example of what could be done. Ideally, the footprints would reflect the geospatial damage as well as the intensity of the event. For the database to be effective, it would need to include not only the most severe events, but also events of l severity levels. This capability is already available through satellite or aerial imagery mapping of damage and geospatial analyses (Camp, 2008; Gokon & Koshimura, 2012; Matsuoka & Yamazaki, 2005; Womble, 2005; Womble et al., 2010). These techniques, however, still have challenges in rural areas, such as the prairies and grasslands commonly found in Tornado Alley.

Ideally, the database would also contain address-level details on the damage sustained and estimated intensity. This historical database could be complemented with a building stock database that captures details of the building characteristics, historical damage, total economic losses, and insured losses. The two databases together would substantially reduce the primary and secondary uncertainties (i.e., event occurrence uncertainty and loss estimation uncertainty, respectively) in convective storm risk assessment models.

6. Concluding Remarks

Finally, a word on insurance affordability. As convective storm claim frequency and severity continue to climb, premiums will inevitably follow suit. To manage this, insurers will be faced with difficult decisions of increasing deductibles, reducing coverage, or pulling out of the market altogether, while consumers will be paying more for less and governments subsidize the costs; this benefits no one. Mitigation and resilience are essential to effectively manage and transfer risk. Researchers, insurers, risk managers, economists, and others have been recommending solutions for decades. What is lacking is the impetus of governing bodies and consumers to voluntarily implement cost-effective mitigation measures. Until this behavior changes, losses will continue to grow. However, there are opportunities to create incentives for mitigation, whether through cost-sharing programs, discounts on insurance premiums, lending programs or requirements, or other innovative financial instruments. Making these programs and tools more readily available will certainly help move the needle in the right direction. Meanwhile, governing bodies should follow the example of Moore, Oklahoma, recognizing that stronger building codes and enforcement provide substantial benefits to citizens (as taxpayers), as well as to government budgets.

Author's Note

The views and opinions expressed here are those of the author and do not necessarily reflect the views of Swiss Re or its subsidiaries.

References

ASTM International. (2016). *Standard test method for wind resistance of concrete and clay roof tiles (air permeability method) (ASTM C1570-03(2016))*. Retrieved from: http://www.astm.org/cgi-bin/resolver.cgi?C1570

ASTM International. (2019a). *Standard test method for wind-resistance of steep slope roofing products (fan-induced method) (ASTM D3161/D3161M-19a)*. Retrieved from: http://www.astm.org/cgi-bin/resolver.cgi?D3161D3161M

ASTM International. (2019b). *Standard test method for wind resistance of asphalt shingles (uplift force/uplift resistance method) (ASTM D7158/D7158M-19a)*. Retrieved from: http://www.astm.org/cgi-bin/resolver.cgi?D7158D7158M

Bienkiewicz, B., & Dudhia, P. (1993, June). *Physical modeling of tornado-like flow and tornado effects on building loading*. Paper presented at the Seventh US National Conference on Wind Engineering, Los Angeles, CA, pp. 95–106.

Brooks, H. E., Lee, J. W., & Craven, J. P. (2003). The spatial distribution of severe thunderstorm and tornado environments from global reanalysis data. *Atmospheric Research, 67–68*, 73–94.

Building Code Effectiveness Grading Schedule (BCEGS). (2015). *National building code assessment report*. Retrieved from https://www.isomitigation.com/bcegs/

Bureau of Labor Statistics. (2018). *Consumer price index*. United States Department of Labor. Retrieved from https://www.bls.gov/cpi/

Camp, P. J. (2008, January 20–24). *Integrating a geographical information system into storm assessment: The southeast Alabama tornado outbreak of 1 March 2007*. Paper presented at the 24th Conference on Interactive Information Processing Systems, New Orleans. American Meteorological Society. Retrieved from https://ams.confex.com/ams/88Annual/techprogram/MEETING.HTM

Cheng, V. Y. S., Arthonditsis, G. B., Sills, D. M., Auld, H., Shephard, M. W., Gough, W. A., & Klaassen, J. (2013). Probability of tornado occurrence across Canada. *Journal on Climate, 26*. doi:10.1175/JCLI-D-13-00093.1

Choi, E. C. C., & Hidayat, F. A. (2002). Gust factors for thunderstorm and non-thunderstorm winds. *Journal of Wind Engineering and Industrial Aerodynamics, 90*, 1683–1696.

Collins, D. J., & Lowe, S. P. (2001, Winter). *A macro validation dataset for US hurricane models*. Paper presented at the Casualty Actuarial Society Forum, Arlington, VA. Casualty Actuarial Society. Retrieved from https://www.casact.org/pubs/forum/01wforum/

Cusack, S. (2018, August 3). *US severe convective storm claims going through the roof* [Blog post]. Risk Management Solutions. Retrieved September 24, 2018, from http://www.rms.com/blog/2018/08/03/u-s-severe-convective-storm-claims-going-through-the-roof/

Cutter, S. L. (Ed.). (2001). *American hazardscapes: The regionalization of hazards and disasters*. Washington, DC: Joseph Henry Press. doi:10.17226/10132

Czajkowski, J., & Simmons, K. M. (2013). Convective storm vulnerability: Quantifying the role of effective and well-enforced building codes in minimizing Missouri hail property damage. *Land Economics, 90*(3), 482–508. Retrieved from http://le.uwpress.org/content/90/3/482.short

Doswell, C. A., III, & Burgess, D. W. (1988). On some issues of United States tornado climatology. *Monthly Weather Review, 116*, 495–501.

Doswell, C. A., III, Moller, A. R., & Brooks, H. E. (1999). Storm spotting and public awareness since the first tornado forecasts of 1948. *Weather Forecasting, 14*, 544–557.

Edwards, R. (2003, February). *Rating tornado damage: An exercise in subjectivity*. Paper presented at the Symposium on the F-Scale and Severe-Weather Damage Assessment, Long Beach. American Meteorological Society. Retrieved from https://ams.confex.com/ams/annual2003/techprogram/programexpanded_149.htm

Federal Emergency Management Agency. (1992). *Building performance: Hurricane Andrew in Florida*. Retrieved from https://www.fema.gov/media-library/assets/documents/10827

Federal Emergency Management Agency. (1997). *Costs and benefits of natural hazard mitigation*. Retrieved from https://www.fema.gov/media-library/assets/documents/3459

Federal Emergency Management Agency. (2018). *NFIP statistics by calendar year*. Retrieved from https://www.fema.gov/statistics-calendar-year

FM Approvals. (2011). *ANSI/FM 4474 (R2010) American national standard for evaluating the simulated wind uplift resistance of roof assemblies using static positive and/or negative differential pressures.* Retrieved from https://www.fmapprovals.com/-/media/Standards/American-National-Standard-for-Evaluation-of.pdf?isGated=true&itemId={E0A18467-3BC9-43AD-AECD-4BA9E831DD98}

FM Approvals. (2010). *FM 4471 approval standard class 1 panel roofs.* Retrieved from https://www.fmapprovals.com/-/media/Standards/Class-1-Panel-Roofs.pdf?isGated=true&itemId={65543CE2-90AD-4188-AEF7-484857537D8C}

Frisbie, P. (2006). The population bias of severe weather reports west of the Continental Divide. *National Weather Digest, 30.* Retrieved from https://nwas.org/publications/digest/papers/2006/Vol30No1/Pg11-Frisbie.pdf

Gall, M., Borden, K. A., & Cutter, S. L. (2009, June). When do losses count? Six fallacies of natural hazards loss data. *Bulletin of the American Meteorological Society.* Retrieved from https://www.spc.noaa.gov/wcm/data/2008bams.pdf

Gokon, H., & Koshimura, S. (2012). Mapping of building damage of the 2011 Tohoku earthquake tsunami in Miyagi Prefecture. *Coastal Engineering Journal, 54*(1), 1250006-1–1250006-12.

Holmes, J. D., Hangan, H. M., Schroeder, J. L., Orwig, K. D., & Letchford, C. W. (2008). A forensic study of the Lubbock-Reese downdraft of 2002. *Wind and Structures, 11*, 137–152.

Holzer, A. M. (2001). Tornado climatology of Austria. *Atmospheric Research, 56*(1–4), 203–211.

Insurance Institute of Building & Home Safety. (2004). *Hurricane Charley: Nature's Force vs. Structural Strength Executive Summary.* Retrieved from https://21r01f2g715d2qfj633cfomm-wpengine.netdna-ssl.com/wp-content/uploads/wpmembers/files/Hurricane-Charley-Natures-Force-vs-Structural-Strength-Executive-Summary_IBHS.pdf

Insurance Institute of Building & Home Safety. (2009). *Hurricane Ike: Nature's force versus structural strength.* Retrieved from https://21r01f2g715d2qfj633cfomm-wpengine.netdna-ssl.com/wp-content/uploads/wpmembers/files/Hurricane-Ike-Natures-Force-vs-Structural-Strength_IBHS.pdf

Insurance Institute of Building & Home Safety. (2014). *Relative impact resistance of asphalt shingles, summary of UL 2218 impact tests.* Retrieved from https://21r01f2g715d2qfj633cfomm-wpengine.netdna-ssl.com/wp-content/uploads/wpmembers/files/Relative-Impact-Resistance-of-Asphalt-Shingles_IBHS.pdf

Insurance Institute of Building & Home Safety. (2018). *Hurricane Harvey wind damage investigation.* Retrieved from https://21r01f2g715d2qfj633cfomm-wpengine.netdna-ssl.com/wp-content/uploads/wpmembers/files/Hurricane-Harvey-Wind-Damage-Investigation_IBHS.pdf

Insurance Institute of Building & Home Safety. (n.d.). *Wind loads on components of multi-layer wall systems with air-permeable exterior cladding.* Retrieved from https://21r01f2g715d2qfj633cfomm-wpengine.netdna-ssl.com/wp-content/uploads/wpmembers/files/Wind-Loads-Multi-Layer-Wall-Systems-Air-Permeable-Exterior-Cladding_IBHS.pdf

Jischke, M. C., & Light, B. D. (1983). Laboratory simulation of tornadic wind loads on a rectangular model structure. *Journal of Wind Engineering and Industrial Aerodynamics, 13*(1–3), 371–382.

Kahraman, A., & Markowski, P. M. (2014). Tornado climatology for Turkey. *Monthly Weather Review, 142*, 545–552. doi:10.1175/MWR-D-13-00364.1

Keeton, D. (2014, February). *2014 model comparison—Severe convective storms.* Presented at Cat Modeling 2014: Adapting for Today—Preparing for Tomorrow, Orlando, FL. Reinsurance Association of America.

Kuligowski, E. D., Lombardo, F. T., Phan, L. T., Levitan, M. L., & Jorgensen, D. P. (2013). *Technical investigation of the May 22, 2011, tornado in Joplin, Missouri*. US Department of Commerce, National Institute of Standards and Technology. Retrieved from https://www.nist.gov/publications/final-report-national-institute-standards-and-technology-nist-technical-investigation

Letchford, C. W., Mans, C., & Chay, M. T. (2002). Thunderstorms—Their importance in wind engineering (a case study for the next generation wind tunnel). *Journal of Wind Engineering and Industrial Aerodynamics, 90*, 1415–1433.

Liu, Z., Pogrzelski, H., Masters, F., Tezak, S., & Reinhold, T. (2010). Surviving nature's fury: Performance of asphalt shingle roofs in the real world. *Disaster Safety Review, 2*. Retrieved from https://ibhs.org/wp-content/uploads/wpmembers/files/Surviving-Natures-Fury-Performance-of-Asphalt-Shingle-Roofs-in-the-Real-World_IBHS.pdf

Marshall, T. (2003, February). *The La Plata, MD tornado: Issues regarding the F-scale*. Paper presented at the Symposium on the F-Scale and Severe Weather Damage Assessment, Long Beach. American Meteorological Society. Retrieved from https://ams.confex.com/ams/annual2003/techprogram/programexpanded_149.htm

Matsuoka, M., & Yamazaki, F. (2005). Building damage mapping of the 2003 Bam, Iran, earthquake using Envisat/ASAR intensity imagery. *Earthquake Spectra, 21*(S1), 285–294.

McCarthy, D. (2003, February). *NWS tornado surveys and the impact on the National Tornado Database*. Paper presented at the Symposium on the F-Scale and Severe-Weather Damage Assessment, Long Beach. Retrieved from https://ams.confex.com/ams/annual2003/techprogram/programexpanded_149.htm

National Institute of Building Sciences. (2017). *Natural hazard mitigation saves: 2017 interim report*. Retrieved from https://www.fema.gov/natural-hazard-mitigation-saves-2017-interim-report

Orwig, K. D. (2010). Examining strong winds from a time-varying perspective (Doctoral dissertation). Texas Tech University, Lubbock. Retrieved from https://ttu-ir.tdl.org/handle/2346/66786

Orwig, K. D., & Schroeder, J. L. (2007). Near-surface wind characteristics of extreme thunderstorm outflows. *Journal of Wind Engineering and industrial Aerodynamics, 95*(7), 565–584.

Pielke, R. A., Gratz, J., Landsea, C. W., Collins, D., Saunders, M. A., & Musulin, R. (2008). Normalized hurricane damage in the United States: 1900–2005. *Natural Hazards Review, 9*(1), 29–42. Retrieved from https://www.nhc.noaa.gov/pdf/NormalizedHurricane2008.pdf

Pielke, R. A., & Landsea, C. W. (1998). Normalized hurricane damages in the United States: 1925–95. *Weather Forecasting, 13*(3), 621–631.

Property Claims Services. (2018). *Catastrophe history reporter*. Retrieved from https://pcs.iso.com/pcs/app/rpt/reportchoose.do

Rauhala, J. (2012). Tornado climatology of Finland. *Monthly Weather Review, 140*(5). doi:10.1175/MWR-D-11-00196.1

Reinhold, T. A. (2009). Steps taken in building and insurance industries for extreme wind related disasters. *Global Environmental Research, 13*(2), 189–196. Retrieved from http://www.airies.or.jp/attach.php/6a6f75726e616c5f31332d32656e67/save/0/0/13_2-12.pdf

Reinhold, T. A. (2011). Much more than a grand opening. *Disaster Safety Review, 1*, 4–9. Retrieved from https://ibhs.org/?a=download&key=a9d5b189f07d663d336b4476c2d99a39

Romero, R., Gayà, M., & Doswell, C. A., III. (2007). European climatology of severe convective storm environmental parameters: A test for significant tornado events. *Atmospheric Research, 83*(2–4), 389–404.

Rose, A., Porter, K., Dash, N., Bouabid, J., Huyck, C., Whitehead, J., . . . Taylor West, C. (2007). Benefit-cost analysis of FEMA hazard mitigation grants. *Natural Hazards Review, 8*(4), 1–15.

Schaefer, J. T., & Galway, J. G. (1982, January). *Population biases in tornado climatology*. Paper presented at the 12th Conference on Severe Local Storms, San Antonio. American Meteorological Society.

Sills, D., Cheng, V., McCarthy, P., Rousseau, B., Waller, J., Elliott, L., . . . Auld, H. (2012, November). *Using tornado, lightning and population data to identify tornado prone areas in Canada*. Paper presented at the 26th Conference on Severe Local Storms, Nashville. American Meteorological Society. Retrieved from https://ams.confex.com/ams/26SLS/webprogram/Paper211359.html

Simmons, K. M., Czajkowski, J., & Done, J. M. (2019). Building code economic performance under variable wind risk. *Journal on Mitigation and Adaptation Strategies for Global Change, 24*(2), 235–258. doi:10.1007/s11027-018-9808-x

Simmons, K. M., & Kovacs, P. (2018). Real estate market response to enhanced building codes in Moore, OK. *International Journal of Disaster Risk Reduction, 27*, 85–93.

Simmons, K. M., Kovacs, P., & Kopp, G. A. (2015). Tornado damage mitigation: Benefit-cost analysis of enhanced building codes in Oklahoma. *Weather, Climate, and Society, 7*, 169–178. doi:10.1175/WCAS-D-14-00032.1

Storm Prediction Center (SPC). (2010). *SPC tornado, hail, and wind database format specification*. Retrieved September 23, 2018, from https://www.spc.noaa.gov/wcm/

Sugimoto, S., Nohara, D., & Hirakuchi, H. (2016). Climatological and regional characteristics of supercell tornado occurrence based on a long-term reanalysis of tornado parameters. *Denryoku Chuo Kenkyusho Hokoku*, 48(3), 1–22.

Swiss Re Institute. (2015). *Natural catastrophes and man-made disasters in 2014: Convective and winter storms generate most losses* (Swiss Re sigma No 2/2015). Retrieved from https://www.swissre.com/institute/research/sigma-research/sigma-2015-02.html

Taszarek, M., & Brooks, H. E. (2015). Tornado climatology of Poland. *Monthly Weather Review, 143*(3). doi:10.1175/MWR-D-14-00185.1

Texas Tech University Wind Science and Engineering Center. (2004). *A recommendation for an Enhanced Fujita Scale (EF-Scale)*. Retrieved from http://www.depts.ttu.edu/nwi/Pubs/FScale/EFScale.pdf

Underwriters Laboratories. (2006). *UL 580 standard for tests for uplift resistance of roof assemblies*. Retrieved from https://standardscatalog.ul.com/standards/en/standard_580_5

Underwriters Laboratories. (2019). *UL 2218A standard for impact resistance of roofing systems*. Retrieved from https://standardscatalog.ul.com/standards/en/standard_2218a_1

Underwriters Laboratories. (2015). *UL 1897 standard for uplift tests for roof covering systems*. Retrieved from https://standardscatalog.ul.com/standards/en/standard_1897_7

US Army Corps of Engineers. (2011). *USACE Joplin area tornado recovery mission, commander's final report*, p. 4. Retrieved from https://silverjackets.nfrmp.us/State-Teams/Missouri/Joplin_RFO_Presentation_to_SilverJackets_MO.pdf

US Census Bureau. (1996). *Population of states and counties of the United States: 1790 to 1990*. Retrieved from https://www.census.gov/population/www/censusdata/pop1790-1990.html

US Census Bureau. (2000). *Historical national population estimates: July 1, 1900 to July 1, 1999*. Retrieved from https://www.census.gov/population/estimates/nation/popclockest.txt

US Census Bureau. (2011). *Historical census of housing tables, crowding*. Retrieved from https://www.census.gov/hhes/www/housing/census/historic/crowding.html

US Census Bureau. (2012). *Historical census of housing tables, home values*. Retrieved from https://www.census.gov/hhes/www/housing/census/historic/values.html

US Census Bureau. (2018a). *2017 American community survey 1-year estimates, median value (dollars), owner-occupied housing units*. Retrieved from https://factfinder.census.gov/faces/tableservices/jsf/pages/productview.xhtml?pid=ACS_17_1YR_B25077&prodType=table

US Census Bureau. (2018b). *2017 population estimates, annual estimates of housing units for the United States, regions, divisions, states, and counties: April 1, 2010 to July 1, 2017*. Retrieved from https://factfinder.census.gov/faces/tableservices/jsf/pages/productview.xhtml?pid=PEP_2017_PEPANNHU&prodType=table

US Census Bureau. (2018c). *2017 population estimates, annual estimates of resident population for the United States, regions, divisions, states, and counties: April 1, 2010 to July 1, 2017*. Retrieved from https://factfinder.census.gov/faces/tableservices/jsf/pages/productview.xhtml?pid=PEP_2017_PEPANNRES&src=pt

US Department of Housing and Urban Development & US Department of Commerce Economics and Statistics Administration. (2017). *2017 characteristics of new housing*. Retrieved from https://www.census.gov/construction/chars/

Womble, J. A. (2005). *Remote-sensing applications to windstorm damage assessment* (Doctoral dissertation). Texas Tech University, Lubbock. Retrieved from https://ttu-ir.tdl.org/bitstream/handle/2346/17733/Womble_James_diss.pdf

Womble, J. A., Smith, D. A., Liang, D., Schroeder, J. L., Brown, T. M., & Mehta, K. C. (2010, May). *Imagery-based wind damage functions*. Paper presented at Structures Congress 2010, Orlando. American Society of Civil Engineers.

Yang, Z., Sarkar, P., & Hu, H. (2011). An experimental study of a high-rise building model in tornado-like winds. *Journal on Fluids and Structures, 27*, 471–486.

SECTION 7

CONCLUSION

CHAPTER 25

CONCLUDING SUMMARY

HORIA HANGAN AND AHSAN KAREEM

THE *Oxford Handbook of Non-Synoptic Wind Storms* (NSWSs) aimed to cover a large palette of aspects related to these type of winds, which prove to be on one side damaging and on the other side way less understood compared to synoptic wind systems. The handbook characterized non-synoptic winds in terms of their climatology and surface layer flow field and presented the current knowledge related to the aerodynamic and structural impacts of this type of winds on buildings, structures, and the natural habitat. It followed by reviewing the implications related to risk analysis, engineering guidelines and codes, socioeconomic aspects, and insurance policies.

The Introduction (Section 1) started with the definition and meteorological characterization of NSWSs. These wind storms cover a rather large range of subsynoptic-scale systems associated with convective storms, including tornadoes, downbursts, or microbursts. It also included systems not associated with convective storms, including downslope winds, bores, katabatic winds, sting jets, dryline bulges, and low-level jets. In a sense, this part referred to the *incidence* of NSWSs, which invokes aspects related to storm characterization, intensity, and occurrence.

It then continued with tornado warnings and subsequent response by people, tornado hazard at a location, tornado–structure interaction, and design for tornadoes for life safety. In a sense, this part referred to the *exposure* to NSWSs as related to preparedness.

It followed with aspects related to full-scale thunderstorm monitoring and detection and the identification of the physical processes responsible for downburst and downslope wind formation, including new aspects related to their kinematics, dynamics, and thermodynamics.

The difficulties related to the forecasting for tornadoes and downbursts have also been covered, and challenges were identified for the physical understanding and mathematical modeling for these localized, spatially limited, and time-dependent phenomena.

The NSWSs relate to their larger parent environment, and therefore they are implicitly influenced by climatic changes that might result in changes in their intensity, frequency, occurrence, and possibly other characteristics. Long-term trends of these NSWS characteristics were analyzed, and global and regional climate model simulations and downscaling aspects were reviewed.

Section 2 was dedicated to the full-scale data acquisition for NSWSs. This includes observational data acquired based on met station observational networks and remote sensing techniques and model-based data basis, ranging from global analyses to high-resolution, large-eddy simulations. Data deficiencies and suitability aspects were discussed.

This atmospheric science data acquisition is complemented by closer to the surface, engineering-oriented data gathering based on LIDARs, Doppler radars, Doppler on wheels, and more conventional met tower data. The advantages and limitations of these techniques were underlined.

Mesoscale, microscale, numerical, and physical simulations of NSWSs were covered in Section 3. The limitations of the mesoscale and microscale simulations from an NSWS perspective and the way these two types of simulations can be combined were reviewed.

The evolution of a new generation of physical simulators to reproduce mainly tornadoes and downburst winds and the problem of laboratory scaling was also examined. Several of these new simulators were presented, and their limitations and advantages were examined.

Finally, data-driven approaches such as detrending, wavelet analysis, empirical mode decomposition, or blind source separation were also presented as a modern way to involve machine learning for the characterization of these complex and nondeterministic phenomena.

Section 4 was dedicated to structural and other impacts of NSWSs. An overview of the aerodynamic loading on structures, such as buildings, bridges, and transmission lines, due to NSWS events, including a brief description of the provisions in the building codes and standards for NSWS loads, have been presented. Aerodynamic forces associated with a rapid increase of gusty wind in short rise time, which is often observed in the NSWSs, were investigated through a series of wind tunnel tests, and it was found that the overshoot wind force is indeed greater than under the corresponding steady synoptic-type winds if the non-dimensional rise time is less than a certain value.

From the structural response viewpoint, the response caused by NSWSs differs from that by synoptic winds due to the inherent nature of the difference in wind characteristics between the two types of winds. This raises serious concerns regarding the applicability of conventional stationary process-based theories, thus calling for a paradigm shift. A review was conducted for popular approaches concerning the structural analysis involving nonstationary thunderstorm outflows, such as evolutionary power spectrum–based analysis, wavelet-based analysis, thunderstorm response spectrum technique involving the equivalent wind spectrum, and a hybrid simulation-based analysis in the time domain. As a result, some preliminary comparisons between the results obtained using these different methods were presented.

An attempt was made to identify similar failure modes in residential structures based on information from more than four decades of storm damage surveys. The information presented highlighted some of the lessons learned in evaluating storm damage to wood-framed residential structures. In this context, windborne debris, which is one of the wind damage sources, was also examined through numerical methods from the fluid–structure interaction problem perspective. It was based on the ALE (arbitrary Lagrangian–Eulerian) finite element flow field analysis to deal with the arbitrary shape of flying objects as well as the dynamical change of aerodynamic forces. Several numerical examples of a rigid sphere showed satisfactory features of those methods.

Section 5 addressed probabilistic aspects related to the prediction of NSWS wind speeds and hazards. Simple procedures for predicting tornado- and downburst-generated wind speeds were put forward for both the United States and Canada.

For the United States, the procedure involves data recording/archival, storm-type classification, reducing data dependence, estimating a rate of occurrence, and fitting a probability distribution, including consideration of multiple storm types. Based on illustrative examples and the existing literature, physical and probabilistic differences from synoptic prediction methods and analysis were also elucidated. These differences include higher variability of thunderstorm-generated winds, regionality of the wind climate, and topics such as wind directionality and changes in frequency and intensity. Future work would include additional parsing of non-synoptic storms into subtypes and framing a comprehensive non-synoptic model similar to what has been done for tropical cyclones.

For Canada, a historical tornado catalogue is available to assess the Canadian tornado occurrence rate. A procedure to evaluate tornadic wind-velocity (at 10 m height) hazard map and to assess an equivalent wind profile was put forward and applied to map tornado wind-velocity hazard for southern Ontario. For downbursts, a similar hazard map and wind profile development can be carried out but in the absence of a historical downburst catalogue.

A probabilistic model of the capacity curve obtained by a nonlinear static pushover analysis can be used to evaluate structural reliability by incorporating the assessed wind-velocity hazard map and an equivalent wind profile.

Section 6 discussed aspects related to reducing the exposure to NSWSs. One of the most important aspects is codification. It was pointed out that the nonstationarity of NSWSs poses questions on the application of the gust loading factor approach as applied for synoptic winds. Two major frameworks, the gust front factor and the thunderstorm response spectrum technique, were examined to step toward the codification of gust front winds. Comparison between the two frameworks' performance was provided, and a codification concept through learning and updating invoking the emerging "design thinking" approach was discussed.

Economic and policy issues related to NSWSs were discussed next. As non-synoptic systems, such as tornadoes, cause high levels of damage to communities, some states adopted new codes to address this. An analysis of the cost effectiveness of these codes was conducted. Research was reviewed on how real estate markets value voluntary mitigation. Using markets for above-code construction provides opportunities to increase resilience in states where stronger building codes are not adopted.

Finally, the challenges and opportunities presented by NSWSs to the insurance industry were analyzed. It was shown that NSWSs constitute the key insured loss driver in the

United States, even after considering the losses sustained by tropical cyclones in 2017. Since 2008, total insured losses from these convective storms have exceeded $10 billion per year. Additionally, these losses continue to increase not only based on drivers such as population, buildings, vehicles, and property value increases but also due to construction materials and practices, as well as building code adoption and enforcement. These increasing loss trends pose challenges for the insurance industry and broader society. These challenges were discussed, and some recommendations were presented.

The *Oxford Handbook of Non-Synoptic Wind Storms* brought together for the first time the vast majority of aspects related to these wind systems for which science has still yet not matured. The handbook attempted to cover aspects related the general *vulnerability* to NSWSs in terms of (1) *incidence*, including the flow field and intensity and the frequency and occurrence of these storms, and (2) *exposure*, including preparedness for NSWSs. In doing so, it presented the state of the art regarding full-scale data acquisition and analysis, mesoscale and microscale numerical modeling, physical simulations, structural analysis, risk modeling, building code implementation, and insurance analysis. For each of these aspects, the presentation aimed at being informative, reviewing a large palette of approaches and presenting their advantages and limitations. It also, when possible, tried to stress the need for future research.

Clearly, there are many research aspects that need further investigation and some that have not been addressed at all. The handbook is an attempt to present the state of knowledge of NSWSs and provide, for the first time, a cohesive frame for future research.

Index

Note: Tables and figures are indicated by an italic "*t*" and "*f*", respectively, following the page number.

For the benefit of digital users, indexed terms that span two pages (e.g., 52–53) may, on occasion, appear on only one of those pages.